XIV International Conference on Calorimetry in High Energy Physics 2010

(CALOR 2010)

Journal of Physics: Conference Series Volume 293

Beijing, China
10-14 May 2010

ISBN: 978-1-61782-912-3
ISSN: 1742-6588

Printed from e-media with permission by:

Curran Associates, Inc.
57 Morehouse Lane
Red Hook, NY 12571

proceedings.com

Some format issues inherent in the e-media version may also appear in this print version.

Printed by Curran Associates, Inc. (2011)

For permission requests, please contact the Institute of Physics
at the address below.

Institute of Physics
Dirac House, Temple Back
Bristol BS1 6BE UK

Phone: 44 1 17 929 7481
Fax: 44 1 17 920 0979

techtracking@iop.org

Additional copies of this publication are available from:

Curran Associates, Inc.
57 Morehouse Lane
Red Hook, NY 12571 USA
Phone: 845-758-0400
Fax: 845-758-2634
Email: curran@proceedings.com
Web: www.proceedings.com

Table of contents

Journal of Physics: Conference Series

Volume 293

2011

XIV International Conference on Calorimetry in High Energy Physics (CALOR 2010)
10–14 May 2010, Beijing, China

Preface

Opening Session

Materials and detectors

Readout techniques

Algorithms and simulations

Astrophysics and neutrino calorimetry

Operating calorimeters and calibration

Future calorimetry

PREFACE

The International Conferences on Calorimetry in High Energy Physics (also known as the Calor Conference series, started in October 1990 at Fermilab) address all aspects of calorimetric particle detection and measurement, with an emphasis on high energy physics experiments. The XIV International Conference on Calorimetry in High Energy Physics (Calor 2010) was held at the campus of the Institute of High Energy Physics, Beijing, China, from May 10–14, 2010.

This conference brought together more than 110 participants from 20 countries, including senior scientists and young physicists. During the five days of the conference, 98 presentations were given in seven plenary sessions. The attendees had in-depth discussions on the latest developments and innovations in calorimetry, including the exciting new LHC results. From the presentations, 83 papers were published in this proceedings.

The success of the conference was due to the participants' enthusiasm and the excellent talks given by the speakers, and to the conveners for organizing the individual sessions. We would like to thank the International Advisory Committee for giving us the opportunity to host this Conference in Beijing.

Finally we would like to thank all the people involved in the organization of the Conference, who have provided valuable local support.

Yifang Wang
Chair of Local Organizing Committee

International Advisory Committee

M Danilov	*ITEP Moscow*
M Diemoz	*INFN Roma I*
A Ereditato	*Bern*
F L Fabbri	*INFN Frascati*
T Kobayashi	*ICEPP Tokyo*
M Livan	*Pavia University & INFN*
P Lubrano	*INFN Perugia*
S Magill	*ANL Argonne*
A Maio	*LIPP Lisbon*
H Oberlack	*MPI Munich*
A Para	*Fermilab*

R Wigmans	*TTU Lubbock*
R Yoshida	*ANL Argonne*
R Zhu	*Caltech*

Local Organizing Committee

Y Wang	*IHEP* (Chair)
Y Gao	*Tshinghua University*
T Hu	*IHEP* (Scientific secretary)
C Li	*USTC*
W Li	*IHEP*
J Lu	*IHEP*
P Wang	*IHEP*
T Xu	*IHEP*
L Zhou	*IHEP*

Session Conveners

1) Materials and detectors – Junguang Lu (IHEP), Francesca Nessi (CERN)

2) Algorithm and simulation – Nural Akchurin (Texas Tech University), Weidong Li (IHEP)

3) Readout techniques – Gerald Eigen (University of Bergen), Zheng Wang (IHEP)

4) Operating calorimeters and calibration – Marat Gataullin (CERN), Francesco Lanni (BNL)

5) Future calorimetry – Tohru Takeshita (Shinshu University), Lei Xia (Argonne National Laboratory)

6) Astrophysics and neutrino calorimetry – Giuliana Fiorillo (INFN), Hiro Tajima (SLAC)

List of Participants

AKCHURIN, Nural	*Texas Tech University*
AN, Zhenghua	*IHEP*

AUFFRAY, Etiennette	*CERN*
BANFI, Danilo	*Università degli Studi di Milano, INFN*
BASHARINA-FRESHVILLE, Anastasia	*University College London*
BEAUCHEMIN, Pierre-Hugues	*University of Oxford*
BENAGLIA, Andrea Davide	*University of Milano – Bicocca and INFN*
BIAN, Jianmin	*IHEP*
BIINO, Cristina	*INFN*
BILKI, Burak	*University of Iowa*
BLAHA, Jan	*LAPP*
BOUDRY, Vincent	*LLR / CNRS-IN2P3*
CAI, Xiao	*IHEP*
CAPONE, Antonio	*Physics Department University "La Sapienza" and INFN*
CAVALLARI, Francesca	*CERN and INFN Rome*
CECCHI, Claudia	*University di Perugia e INFN*
CHANG, Jinfan	*IHEP*
CHEN, Hucheng	*Brookhaven National Laboratory*
CHILDERS, Taylor	*Universität Heidelberg – Kirchhoff-Institut für Physik*
DAO, Valerio	*Geneva University – DPNC*
DE LA TAILLE, Christophe	*IN2P3/OMEGA-LAL*
DIEMOZ, Marcella	*INFN Roma*
DOTTI, Andrea	*CERN*
EIGEN, Gerald	*University of Bergen*
EPIFANOV, Denis	*Budker Institute of Nuclear Physics*
FAIVRE, Julien	*LPSC Grenoble France*
FANG, Jian	*IHEP*
FANG, Shuangshi	*IHEP*
FANTONI, Alessandra	*INFN – LNF*
FERRI, Federico	*CEA/Saclay Irfu/SPP*
FERRONI, Fernando	*Sapienza University & INFN Roma*
FISK, Henry Eugene	*Fermilab*
GABALDON, Carolina	*CERN*
GARUTTI, Erika	*DESY*
GAUDIO, Gabriella	*Istituto Nazionale di Fisica Nucleare – Sezione di Pavia*

GILLBERG, Dag	*Carleton University*
GIOVANNINI, Paola	*Max-Planck-Institut für Physik*
GLAZOV, Alexander	*DESY*
GRACHOV, Oleg	*University of Kansas*
HAPPACHER, Fabio	*INFN*
HE, Miao	*IHEP*
HORI, Yasuto	*University of Tokyo, CNS*
HU, Tao	*IHEP*
HULTH, Per-Olof	*Stockholm University*
JUN, Soon Yung	*Carnegie Mellon University*
JURK, Stefan	*ISEG Spezialelektronik gmbH*
KAVATSYUK, Myroslav	*KVI, University of Groningen*
KHRAMOV, Evgeny	*Joint Institute for Nuclear Research*
KISTENEV, Edouard	*Brookhaven National Laboratory*
KUO, Chia Ming	*NCU*
KWON, Youngil	*Yonsei University*
LAKTINEH, Imad	*IPNL*
LECOQ, Paul	*CERN*
LI, Cheng	*University of Science and Technology of China*
LI, Ningbo	*NanJing Normal University & IHEP*
LI, Weidong	*IHEP*
LI, Weiguo	*IHEP*
LIU, Chunxiu	*IHEP*
LIU, Lijiao	*University of Bergen, Norway*
LIVAN, Michele	*Dipartimento di Fisica Nucleare e Teorica, University di Pavia , Italy*
LU, Junguang	*IHEP*
LUBRANO, Pasquale	*INFN Sezione di Perugia*
MACHIKHILIYAN, Irina	*LAPP, Annecy, France*
MAVROMANOLAKIS, Georgios	*CERN*
MILLER, David W	*SLAC and Stanford University*
NECESAL, Petr	*Institute of Physics AS CR, v.v.i.*
NEMECEK, Stanislav	*FZU AVCR Praha*
NIESS, Valentin	*LPC, Clermont (CNRS/IN2P3)*
NOVOTNY, Rainer W	*2nd Physics Institute, University Giessen, Germany*

OBERLACK, Horst	*MPI für Physik, Munich*
PARA, Adam	*Fermilab*
PARAMATTI, Riccardo	*INFN Rome & CERN*
PEPE, Monica	*INFN Perugia*
POSPELOV, Gennady	*Max-Planck-Institut für Physik*
REPOND, Jose	*Argonne National Laboratory*
ROSSETTI, Valerio	*IFAE - Barcelona*
SCHACHT, Peter	*MPI/Munich*
SEFKOW, Felix	*DESY*
SFYRLA, Anna	*CERN*
SGUAZZONI, Giacomo	*INFN Section of Florence*
SIMON, Frank	*Max-Planck-Institute for Physics*
SIMONYAN, Margar	*Niels Bohr Institute, Copenhagen University*
SPADARO, Tommaso	*LNF INFN*
SUN, Xilei	*IHEP*
TAKESHITA, Tohru	*Shinshu University*
TANG, Zebo	*University of Science and Technology of China*
TAPAN, Ilhan	*Uludag University*
THEOFILATOS, Konstantinos	*ETH Zurich*
THOMSON, Mark	*University of Cambridge*
TOKUNO, Hisao	*Department of Physics, Graduate School of Science and Engineering*
TRONCON, Clara	*INFN*
UOZUMI Satoru	*Kyungpook National University*
USAI, Giulio	*University of Texas at Arlington*
VAZQUEZ GOMEZ, Ricardo	*Universitat de Barcelona*
VIDEAU, Henri	*LLR – Ecole polytechnique*
VOLOBOUEV, Igor	*Texas Tech University*
WAN, Renzhuo	*IOPP-CCNU, Wuhan, China & IPHC-CNRS, Strasbourg, France*
WANG, Zheng	*IHEP*
WANG, Zhigang	*IHEP*
WENZEL, Hans	*Fermilab*
WIGMANS, Richard	*Texas Tech University*
WILKENS, Henric	*CERN*

XIA, Lei	*Argonne National Laboratory*
XIE, Yuguang	*IHEP*
YE, Jingbo	*SMU*
YETKIN, Taylan	*University of Iowa*
YIN, Zhongbao	*Huazhong Normal University*
YU, Boxiang	*IHEP*
ZHANG, Aiwu	*IHEP*
ZHANG, Huaqiao	*Centre de Physique des Particules de Marseille (CPPM)*
ZHANG, Liyuan	*Caltech*
ZHANG, Qingmin	*IHEP*
ZHANG, Yunlong	*University of Science and Technology of China*
ZHENG, Yangheng	*GUCAS*
ZHU, Renyuan	*Caltech*

XIV International Conference on Calorimetry in High Energy Physics (CALOR 2010) IOP Publishing
Journal of Physics: Conference Series **293** (2011) 012001 doi:10.1088/1742-6596/293/1/012001

Performance of calorimeters at the LHC

Francesca Cavallari
INFN Sezione di Roma Italy and CERN
E-mail: Francesca.Cavallari@roma1.infn.it

Abstract. The LHC is the new CERN hadron accelerator that has recently started to collide proton beams at 7 TeV center of mass energy. It features two general purpose experiments ATLAS and CMS and two specific experiments: LHCb to study b-quark physics and Alice to study heavy ions collisions. Electromagnetic and hadron calorimeters play a crucial role in the ATLAS and CMS experiments. High energy electrons and photons and missing transverse energy are used in most new particle searches. Calorimeters are also used in the experiment trigger.

This paper reviews the design of the various electromagnetic and hadronic calorimeters employed in the LHC experiments, their commissioning and the first results on the performance with beam.

1. Introduction

The LHC is the new CERN hadron accelerator that has recently started to collide proton beams at 7 TeV center of mass energy and may in the future reach energies up to 14 TeV. The design luminosity is 10^{34} cm^{-2}s^{-1} and the bunch crossing rate is 40 MHz.

It features two general purpose experiments: ATLAS and CMS and two dedicated experiments: LHCb that will study b-quark physics in the forward direction and Alice for heavy ion collisions.

The LHC is primarily a discovery machine to explore an unprecedented range of energies. Its main goal is to search for the Higgs boson and to study the physics beyond the Standard Model. In addition standard particles like top and beauty quarks can be produced and studied with large statistics.

The Higgs boson search will be performed exploiting the H$\rightarrow \gamma\gamma$ decay channel in the low Higgs mass range and the H$\rightarrow ZZ \rightarrow 4l$ decay to cover up to 1 TeV Higgs mass. These channels have electrons or photons in the final state and require an excellent energy and position resolutions for electrons and photons. The ATLAS and CMS electromagnetic calorimeters were optimized with these channels as bench-marks.

Supersymmetric particle searches will study channels with leptons, jets and neutralinos in the final state. Good jet energy and position resolution and good missing transverse energy reconstruction are important for these studies. The hadronic calorimeters of ATLAS and CMS have been optimized to have the best possible jet energy resolution and missing transverse energy (E_T^{miss}) resolution compatible with the space constraint of the experiment.

At the LHC the inelastic cross-section is about 70 mb while the new physics channels have a small cross-section. Only a few 100 Hz event rate can be recorded on tape so the trigger system is crucial. The calorimeter signals are integrated in the trigger logic.

Journal of Physics: Conference Series **293** (2011) 012001 doi:10.1088/1742-6596/293/1/012001

The detectors were installed in the experimental cavern few years ago and at the beginning they were commissioned using cosmic rays. Then the LHC beams started to circulate in the accelerator and several *beam splash* events were recorded where single bunches of 10^9 protons were sent on purpose on the closed collimators 150 metres upstream of the experiments. These events have proved to be important for timing and calibration studies. At the end of 2009 proton collisions were recorded at the injection energy (450 GeV per beam) and in the course of 2010 an energy of 3.5 TeV per beam was reached.

This paper describes the ATLAS and CMS electromagnetic and hadronic calorimeters design, commissioning and performance with first beams. Section 2 reviews the design principles for the LHC electromagnetic and hadronic calorimeters while section 3 and 4 describe the general experiment layout. The following sections describe in detail the ATLAS and CMS electromagnetic and hadronic calorimeters and review the status and performance of the detectors as it has been studied with cosmic rays, beam splashes and collision events.

Section 7 and 8 briefly outline the LHCb and Alice calorimeters.

2. Electromagnetic and hadronic calorimeters requirements

The ATLAS and CMS electromagnetic calorimeters were optimized for the Higgs boson search. In the H$\rightarrow \gamma\gamma$ channel, the typical photon energy is in the range above 50 GeV. This channel requires a very good energy and position resolution in order to reconstruct the Higgs invariant mass with a resolution of 1% and detect the Higgs mass peak over the large continuous background. The H$\rightarrow ZZ \rightarrow$ 4 electrons decay requires a full $\eta - \phi$ coverage up to $|\eta| = 3$ in order to detect the four electrons. The calorimeter energy range must be able to reach up to 1.5 TeV in order to detect electrons from possible Z$'$ bosons. The detectors must be fast to cope with the LHC bunch crossing rate of 40 MHz and give an input to the trigger. Radiation hardness has also been taken into account in the calorimeter project because of the unprecedented LHC design luminosity.

The energy resolution of a calorimeter can be parametrized as:

$$\frac{\sigma(E)}{E} = \frac{a}{\sqrt{E}} \oplus \frac{b}{E} \oplus c \qquad (1)$$

where a is the stochastic term, b is the electronic noise term and c is a constant term that includes effects of detector instabilities and mis-calibration. The stochastic term takes into account the statistical fluctuations in the shower detection and it is small for homogeneous calorimeters and larger for sampling calorimeters, but its effect decreases with growing energy.

Physics channels with jets require good energy and position resolution for high energy jets up to about 1 TeV. Simulations show that such energetic jets require at least 11 interaction lengths to be contained. New physics channels require good energy measurement for jets, hermeticity and a good resolution on missing transverse energy. Jets are made of an electromagnetic component as well, so good combined hadronic and electromagnetic energy resolution is required.

The space available for the ATLAS hadronic calorimeter is larger than for CMS due to the limited volume available inside the CMS solenoid. So the CMS hadronic calorimeter design was limited by space constraint.

Both ATLAS and CMS have installed forward calorimeters to tag forward jets and cover the solid angle up to $\eta = 5$. The radiation levels in the forward region are so high that the detector design was limited to very radiation hard materials. Nevertheless forward calorimeters have a reasonable forward energy measurement for missing E_T resolution and are also employed for luminosity measurement.

Journal of Physics: Conference Series **293** (2011) 012001 doi:10.1088/1742-6596/293/1/012001

3. The ATLAS experiment

ATLAS [1] is the largest of the four LHC experiments. It is 40 metres long and 22 metres high. It has a muon spectrometer embedded in a system of three toroidal magnets, a highly segmented lead-liquid argon electromagnetic calorimeter and a tile calorimeter for hadronic activity. A smaller 2 T solenoid surrounds the tracker, which has tracking and particle identification capability. The volume inside the muon spectrometer that is available for the calorimeters has a diameter of 8.5 metres.

4. The CMS experiment

The CMS experiment [2] is smaller in size than ATLAS. It is made of a superconducting solenoid that contains a full silicon tracker and the calorimeters. Muon chambers are embedded in the magnet iron return yoke. The electromagnetic calorimeter is a homogeneous crystal calorimeter. The hadron calorimeter is a compact tile calorimeter. The volume inside the solenoid that is available for the calorimeters has a diameter of 6 metres.

5. Electromagnetic calorimeters

5.1. The ATLAS electro-magnetic calorimeter

The ATLAS electromagnetic calorimeter is a sampling calorimeter made of accordion-shaped lead absorbers and liquid argon operated at a temperature of 87 K. The signal is given by the particle ionization in the liquid argon gap that is collected on the electrodes. The signal level is high, stable and radiation hard.

The calorimeter has a barrel and two endcap parts housed in three cryostats. The barrel covers $|\eta| < 1.475$ while the endcaps extend the coverage up to $|\eta| = 3.2$. The barrel readout is segmented in depth in three sections: a front, a middle and a back section. The central section cells are subdivided in $\Delta\eta \text{x} \Delta\phi = 0.025\text{x}0.025$. The front section has a more granular η segmentation.

The design energy resolution is parametrized by:

$$\frac{\sigma(E)}{E} = \frac{10\%}{\sqrt{E}} \oplus \frac{170MeV}{E} \oplus 0.7\%. \tag{2}$$

During the construction one of the challenges for this calorimeter was the high mechanical precision required in the thickness of the lead absorber. It had to be machined to 2 mm with a precision of 10 μm. Indeed a variation of 1% of the lead thickness would cause a 0.6% signal drop.

Another point of concern is the temperature stability in the cryostat due to a signal dependence on the temperature of -2 %/K. The calorimeter is equipped with 500 temperature probes. During the 2009 run it was shown that a temperature stability of 1.6 mK (maximum 5 mK) of each probe over a period of 10 days was achieved and the uniformity of the temperature across the calorimeter was between 50 and 70 mK for the barrel and endcaps, well below the required level of 100 mK.

The liquid argon signal is shaped, amplified and sampled by the front-end electronics. The energy calculation relies on the accurate knowledge of the electronics response. Regular calibration runs measure the electronics stability: the pedestal and noise levels, the electronic gains and the pulse shape with a test-pulse. During the 2009-2010 data-taking the pedestals, noise levels, gains and pulse shapes have been measured and proven to be stable. Figure 1 shows the pulse shape for the barrel middle layer of the calorimeter and the comparison with the predicted values.

The total number of channels is about 180 thousands and 98.5% of the channels are operational. Most of the faulty channels are due to non-functioning optical transmitters on the front-end electronics. This failure is expected to be repaired during the next LHC shutdown.

Journal of Physics: Conference Series **293** (2011) 012001 doi:10.1088/1742-6596/293/1/012001

For what concerns the timing, the calorimeter has been preliminarily timed in using splash events.

The LAr calorimeter is used in the trigger. Figure 2 shows the trigger efficiency turn-on curve, produced using the collisions data at 7 TeV. This curve is given by the probability that an event with an offline reconstructed cluster of a given transverse energy has fired the Level1 trigger selecting electromagnetic clusters above 5 counts (approximately 5 GeV). As the Figure shows, this trigger is fully efficient above 7-8 GeV.

Figure 3 shows the comparison between the LAr signal in Monte Carlo simulation and in 900 GeV collisions data. The excellent level of understanding of the calorimeter gives a superb agreement between data and simulation over several orders of magnitude.

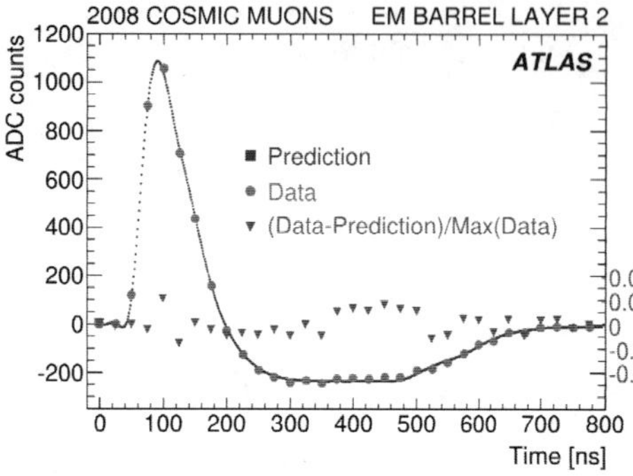

Figure 1. Pulse shape of the ATLAS LAr Barrel calorimeter acquired with cosmic muons. The detector signal is shown superimposed with the expected signal. The acquired and predicted pulse shape are shown to agree at the percent level.

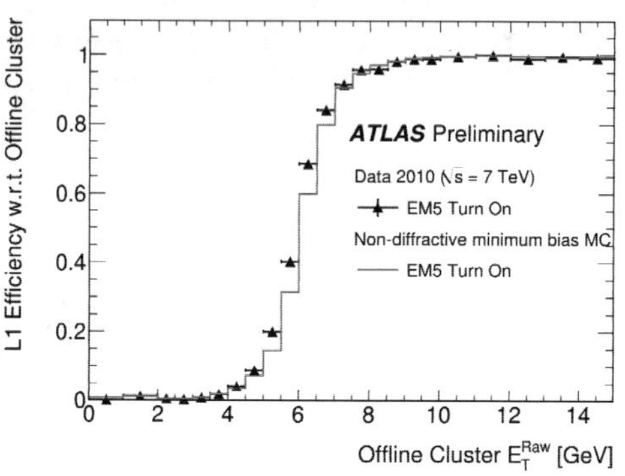

Figure 2. ATLAS LAr calorimeter trigger efficiency turn-on curve: it represents the probability that an event with an offline reconstructed cluster of a given transverse energy has fired the 5 GeV Level1 trigger.

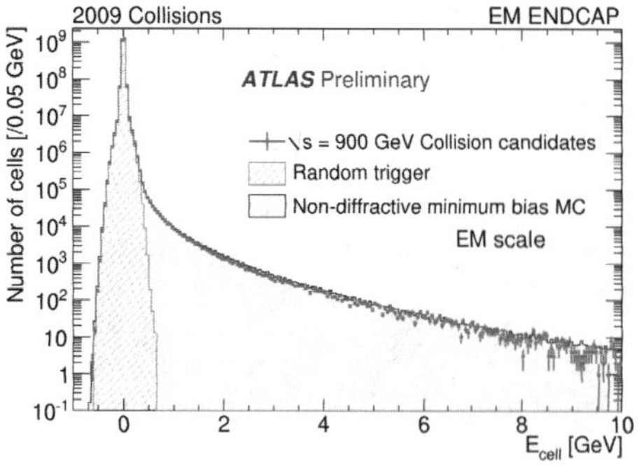

Figure 3. The cell energy distribution for the ATLAS electromagnetic endcap calorimeter acquired during the 2009 data-taking. The data and Monte Carlo simulation are superimposed.

5.2. The CMS electro-magnetic calorimeter

The CMS electro-magnetic calorimeter (ECAL) is a homogeneous crystal calorimeter made of 75648 lead tungstate crystals ($PbWO_4$). The crystals are arranged in a barrel and two endcap sections. The barrel covers $|\eta|$ up to 1.47 while the endcaps extend the coverage up to $|\eta| = 3$. The barrel crystals cover $\Delta\eta x \Delta\phi = 0.0174x0.0174$.

The lead tungstate crystals were specially developed through a 10 year R&D project by SIC (China) and BTCP (Russia). $PbWO_4$ is a very dense material, its radiation length is 0.89 cm and it has a Molière radius of 2.2 cm. This property has allowed to build a very compact calorimeter of 25 X_0 in only 23 cm. The scintillation light is low compared to other scintillating materials, and it is peaked around 420 nm. So photo-detectors with internal amplification were chosen: Avalanche Photo-Diodes (APD) in the barrel and Vacuum Photo-Triodes (VPT) in the endcaps.

A lead-silicon strip pre-shower is installed in front of the endcap crystals.

The ECAL target energy resolution is below 0.5% for high energy unconverted photons. The resolution measured in test-beams with electrons is:

$$\frac{\sigma(E)}{E} = \frac{2.8\%}{\sqrt{E}} \oplus \frac{120MeV}{E} \oplus 0.3\%. \tag{3}$$

Since it is a homogeneous calorimeter, the effect of the very small stochastic and noise term become negligible at 50 GeV. The real challenge for the ECAL energy resolution at high energy is the constant term.

During the construction special care was devoted to making sure that the crystal light collection was uniform along the crystal depth. In fact there can be large fluctuations in the position of the first interaction of electrons and photons. Thus the shower maximum position can vary considerably from event to event. If the crystal light yield is not uniform along its depth, the energy resolution is sensitive to these fluctuations. Thanks to the crystal quality control, the contribution to the constant term due to this effect is limited to 0.3%.

Both the crystal light yield and APD gain have a strong temperature dependence, which give a signal variation of -4%/^{o}C. Measurements taken during the 2009 data-taking show that the calorimeter cooling apparatus stabilizes the temperature at the level of 0.03^{o}C.

Another point of concern is the crystal radiation hardness. Under irradiation lead tungstate crystals tend to loose transparency and consequently the signal deteriorates. A very sophisticated monitoring system based on blue laser light monitors the crystal transparency constantly. Figure 4 shows a typical crystal laser signal acquired for a period of 350 hours during 2010 collisions. A stability of 0.1% is observed. This shows both that the calorimeter environment variables (temperature, high voltage, low voltage,...) are stabilized at the permill level and that the signal from the monitoring system is under control at the same level.

The total number of channels is 75648 and more than 99% are operational.

The calorimeter has been preliminarily timed in using splash events and collisions at the 0.3 ns level.

The calorimeter is used in the Level1 trigger and preliminary efficiency turn-on curves have been produced with cosmics in 2008 and with collisions in 2009-2010.

During the 2009-2010 data-taking, anomalous crystal signals were observed in the barrel. These signals appear as isolated single crystal high energy deposits. This effect is under investigation and it seems to be due to low energy neutron interactions in the APDs. Algorithms have been studied to reject these signals based on timing and shower shape variables. After this selection, the agreement between data and Monte Carlo is superb (see Figure 5).

5.3. Detector commissioning and performance

In this section few common points are discussed and compared for ATLAS and CMS. In particular the calibration and the material budget in front of the calorimeters.

5.3.1. Calibration The ATLAS LAr calorimeter is by construction mechanically very uniform. The electronics calibration system measures the gain and response of each signal with an electronics path that is almost identical to the ionization signal one. Ten percent of the

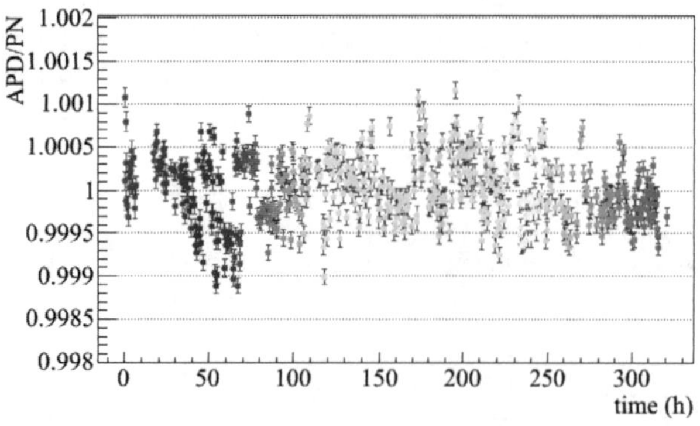

Figure 4. A typical crystal laser signal in the CMS ECAL acquired for a period of 350 hours during 2010 collisions.

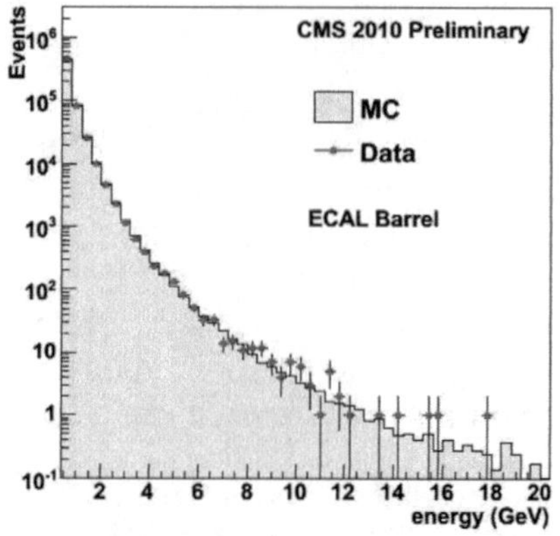

Figure 5. Signal from the Barrel ECAL crystals acquired during the 2010 data-taking at 7 TeV. Data and Monte Carlo simulation are superimposed.

calorimeter modules were exposed to test-beam. The uniformity of response was verified in the test-beams and measured at the level of 0.45% within one module. The long range non uniformity will be measured using physics events (like Z→ e^+e^- or W→ $e\nu$) and is expected to limit this effect at the level of 0.7%.

The CMS ECAL crystals and photon sensors were measured during construction and provide a starting calibration at the 4% level. All barrel modules were calibrated with cosmic rays at the 1-2% level. Twenty-five percent of the ECAL barrel modules were calibrated with high energy electron beams at the 0.4%. This calibration allows to set a preliminary absolute scale. Various physics channels are exploited to improve the inter-calibrations and to check the energy scale, in particular $\pi^o \to \gamma\gamma$, Z→ e^+e^- or W→ $e\nu$ and ϕ simmetry in minimum bias events.

5.3.2. Material budget A common problem that the ATLAS and CMS calorimeters have faced, compared with previous collider calorimeters, is the material budget in front of them. The ATLAS and CMS trackers constitute 0.5-1.5 X_0 of material depending on η. In front of the ATLAS LAr calorimeter there is also the solenoid coil that brings the material to 2 X_0 at $\eta = 0$.

This amount of material is such that a large fraction of photons convert in electron-positron pairs before reaching the calorimeters. Figure 6 and Figure 7 shows the conversion vertex for photons measured in 2009-2010 collision data by ATLAS and CMS. The position of these vertices can be attributed to the various pixel and tracker layers.

For what concerns the electrons, the combined effect of a heavy tracker and a strong magnetic field is such that the radiated Bremsstrahlung photons may hit the calorimeter at a considerable distance from the impact position of the electron shower. In particular in CMS, where the magnetic field is 3.8 T in the tracker volume, 50% of the electrons give a non negligible radiation. A special electron tracking algorithm has been studied to reconstruct tracks with big kinks, and a special clustering algorithm has been developed to recover radiated photons in the direction of the track curvature. Figure 8 shows the fraction of radiated energy for electrons in 2009 collisions data. This variable is calculated as the difference between the track momentum measured by the innermost track hits P_{in} and the outer ones P_{out} divided by P_{in}. Data and Monte Carlo events show a good agreement on this complex effect.

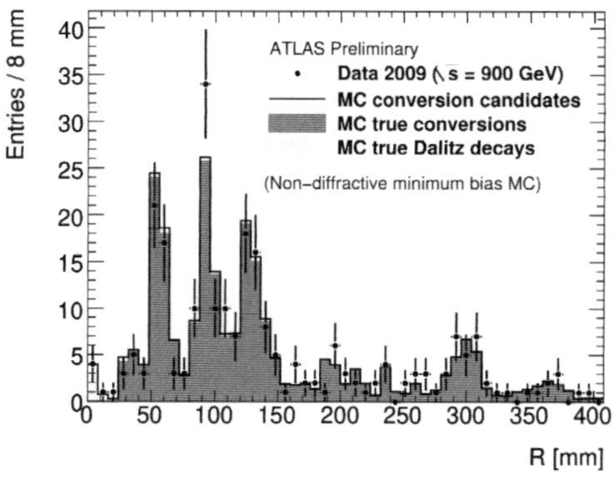

Figure 6. Location of the conversion vertex for converted photon events in ATLAS (2009 collision data).

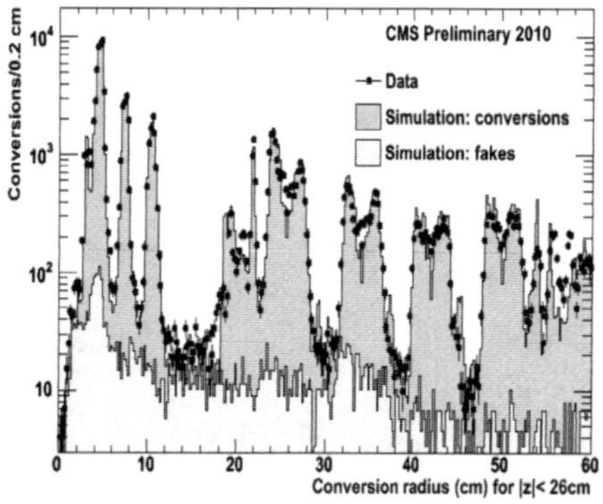

Figure 7. Location of the conversion vertex for converted photon events in CMS (2010 collision data).

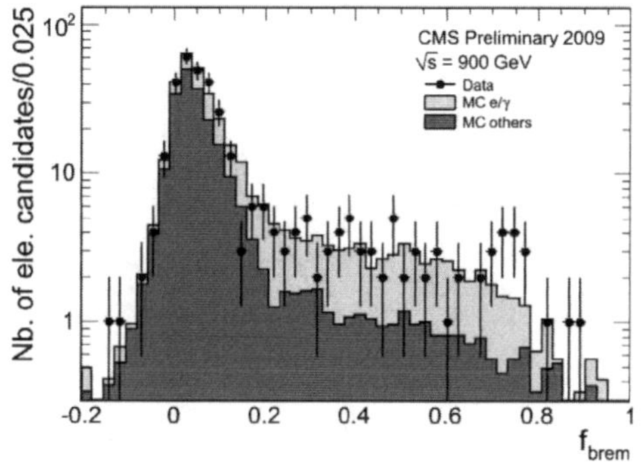

Figure 8. Fraction of radiated energy for electrons in 2009 collisions data. This variable is defined as the difference between the track momentum measured by the innermost track hits P_{in} and the outer ones P_{out} divided by P_{in}.

5.4. Performance

During the 2009-2010 LHC run both the ATLAS and CMS electromagnetic calorimeters have proven to be ready for data-taking: the detectors are well timed in, they are used in the trigger, the design stability on the environment parameters is reached and there is a superb level of accuracy in the Monte Carlo description of the detector.

Both experiments have started to analize beam collision data and have produced preliminary photon pairs invariant mass plots in minimum bias events that show a clear sign of the π^0 peak. Photons are selected with a transverse energy above 300 MeV for ATLAS and 400 MeV for CMS. The calorimeters are not designed for optimal resolution at these low energies. In fact electronics noise levels and channel readout in zero-suppression mode have a non negligible effect not allowing an optimal shower reconstruction. Nevertheless the peak position is well centered on the real π^0 mass. The resolution for the π^0 mass is 19 MeV for ATLAS and 14 MeV for CMS and it is well described by Monte Carlo simulation.

6. Hadronic calorimeters

6.1. The ATLAS hadronic calorimeter

The ATLAS hadronic calorimeter is made of several parts:

- a sampling calorimeter made of steel and plastic scintillator tiles covers the barrel part up to $|\eta| = 1.7$. The tiles are arranged perpendicular to the beam axis. Wavelength shifting fibers carry the light to photo-multiplier tubes located at the back of the calorimeter modules.

- an endcap part made of copper and liquid argon is embedded in the endcap e.m. calorimeter cryostat.
- a forward calorimeter (FCal) installed in the cryostat extends the coverage up to $|\eta| = 5$. It is made of copper and liquid argon (for the electromagnetic section) and tungsten and liquid argon (for the hadronic section).

The standalone tile calorimeter energy resolution obtained in test-beams for single pions is:

$$\frac{\sigma(E)}{E} = \frac{52.9\%}{\sqrt{E}} \oplus 5.7\%. \tag{4}$$

The electronic noise is negligible and so it is not used in the fit.

The detector is installed and taking data since long. It amounts to 9800 channels, out of which 97.3% are operational. The detector signal is used in the trigger to detect events with high energy jets.

The endcap liquid argon hadronic calorimeter and forward calorimeter constitute respectively 5600 and 3500 channels and are fully efficient.

6.2. The CMS hadronic calorimeter

The CMS hadronic calorimeter (HCAL) is a sampling calorimeter made of brass and plastic scintillator tiles. The tiles are arranged parallel to the beam axis in the barrel. The scintillation light is shifted in the visible region via wavelength shifting fibers and detected with HPD (hybrid photodiodes). One HPD can read multiple channels. The Barrel section covers $|\eta| < 1.2$, the endcaps extend the coverage up to $|\eta| < 3$. A forward calorimeter (HF) made of steel and quartz fiber covers $|\eta| < 5.2$.

The barrel and endcap parts have about 5000 channels in total and are fully operational. The HF has 1700 channels and is fully operational.

The detector was well timed using splash events and is used in the jet trigger.

The CMS HCAL is taking data since the beginning and is already starting to see some very high energy jet events, as the one represented in Figure 9.

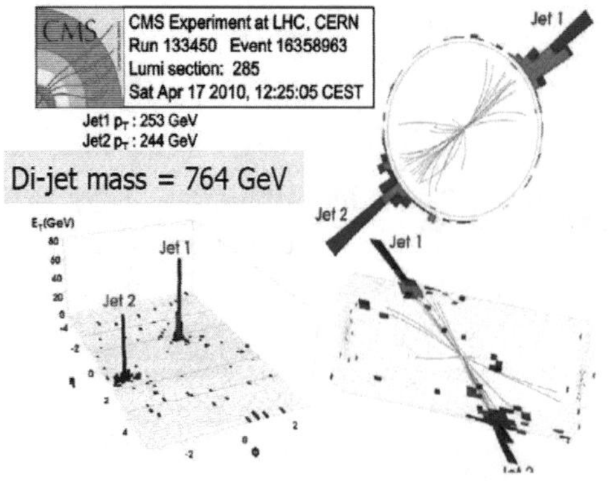

Figure 9. A very high energetic di-jet event recorded by the CMS experiment during 2010 collisions at 7 TeV.

6.3. Detector performance

Two effects limit the hadron calorimeter resolution: non containment and non compensation.

The ATLAS calorimeter reaches 10 λ_I at $\eta = 0$ and more than 12 λ_I in the endcaps.

Due to the limited space available in the experiment design inside the magnet coil, the CMS ECAL and HCAL reach 7 λ_I at $\eta = 0$. Data collected during beam tests have shown that 5% of the energy deposited by a 300 GeV pion escape detection. The installation of an additional layer of scintillator outside the magnet coil (called HO) improves the energy resolution for high energy pions and recovers linearity.

Non compensation is a common problem for the calorimeter system of both experiments. Indeed both hadronic calorimeters have an electron over hadron ratio of approximately 1.4 and the electromagnetic part in front is non compensating, so the combined energy measurement of the hadronic and electromagnetic section is highly non compensating. This causes intrinsic sensitivity to fluctuations in the electromagnetic component of the shower and of the jets, and non linearity. In general the non linearity can be corrected, but the sensitivity to the fluctuation remains and causes a poor resolution. Indeed test-beam analysis shows that before and after non-linearity correction the combined energy resolution of the CMS HCAL+ECAL for pions is:

$$\frac{\sigma(E)}{E} = \frac{110.7\%}{\sqrt{E}} \oplus 7.3\% \text{ (before correction) and } \frac{\sigma(E)}{E} = \frac{84.7\%}{\sqrt{E}} \oplus 7.4\% \text{ (after correction)}. \quad (5)$$

In conclusion the choices made for the hadronic central section by ATLAS and CMS are similar: sampling calorimeters with scintillator as active material. In both cases the dominant factor on resolution and linearity is the non compensation. CMS HCAL space limitation have conditioned the design and performance ATLAS higher segmentation and containment gives a better total resolution.

6.4. Particle flow analysis technique

The particle flow technique principle is to follow the particle signals through the experiment and measure their energy with the most performant detector for each particle type and energy. This technique identifies each energy deposit in the calorimeters and matches it to tracks. The energy of matched objects is evaluated using track momentum. Unmatched energy deposits are attributed to neutral hadronic or electromagnetic objects, depending where they occur and measured by the calorimeters.

Normally this technique requires highly segmented calorimeters (to disentangle overlapping showers) and powerful tracker detectors. However an attempt has been done to use it also in CMS, which has a very powerful tracker, a fine grain electromagnetic calorimeter, but a hadron calorimeter with poor granularity. The results of this analysis are encouraging. The power of the particle flow technique is best seen in the calculation of the missing transverse energy and of the total transverse energy and it will be shown in Section 6.5

6.5. Missing transverse energy and total energy measurements

The measurement of the missing transverse energy is a key ingredient of many important analyses: W bosons, top quark events and supersymmetric particle searches.

Normally the missing transverse energy and the total energy measurements are calculated using the calorimeter cells only. These quantities are particularly sensitive to noise levels in the detectors. Using higher level objects like clusters, rather than single calorimetric energy deposits, helps to reduce the sensitivity to noise.

At colliders the missing transverse energy is defined as minus the vector sum of the transverse energy deposits

$$\vec{E}_T^{miss} = -\Sigma_i \vec{E}_T^i \quad (6)$$

Also its components along the x and y axes are often used $(E_T^{miss})_{x,y}$. Another important quantity that is often referred to is the total transverse energy, which is the scalar sum of the transverse energy deposits:

$$\Sigma E_T = \Sigma_i E_T^i. \quad (7)$$

This last quantity is very challenging because the calorimeter noise adds up in the sum.

The missing transverse energy components along the x and y axes are very useful because their distribution allows to obtain the missing transverse energy resolution from data directly. This helps considerably in the understanding of subtle detector effects that may enter the computation. However it should be stressed that the resolution in $(E_T^{miss})_{x,y}$ should always be quoted as a function of the total transverse energy defined above. Indeed a strong cut on the minimum cluster energy considered in the calculation may falsely improve the resolution on $(E_T^{miss})_{x,y}$, while reducing the total transverse energy.

Among the subtle effects that were cleared during this analysis, it is worth mentioning for CMS the anomalous ECAL signals and HF noise, and for ATLAS the misalignement of the forward calorimeter.

CMS particle flow studies have shown that the resolution in the total transverse energy is more gaussian than for the standard calorimetric cells computation.

Figure 10 and 11 compares the $(E_T^{miss})_{x,y}$ distribution for data and simulated events at $\sqrt{s}$ =7 TeV in ATLAS and CMS. The ATLAS analysis is based on topological clusters where a basic noise suppression is used in the clustering algorithm. The CMS data are based on the particle flow technique discussed in Section 6.4. Figure 12 and 13 shows the $(E_T^{miss})_{x,y}$ resolution as a function of the total transverse energy. A fit to the points gives:

$$\sigma((E_T^{miss})_{x,y}) = 37\%\sqrt{\Sigma E_T} \text{ for ATLAS and } \sigma((E_T^{miss})_{x,y}) = 45\%\sqrt{\Sigma E_T} \text{ for CMS.} \quad (8)$$

However the ATLAS hadron calorimeter is calibrated at the electro-magnetic energy scale, and, since the calorimeter is non compensating, the ΣE_T horizontal scale cannot be directly compared with the CMS plot.

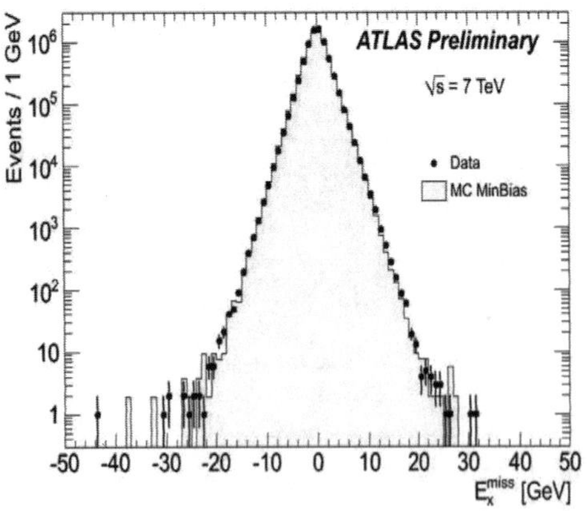

Figure 10. Distribution of the missing transverse energy in the x direction in ATLAS calculated with topological clusters.

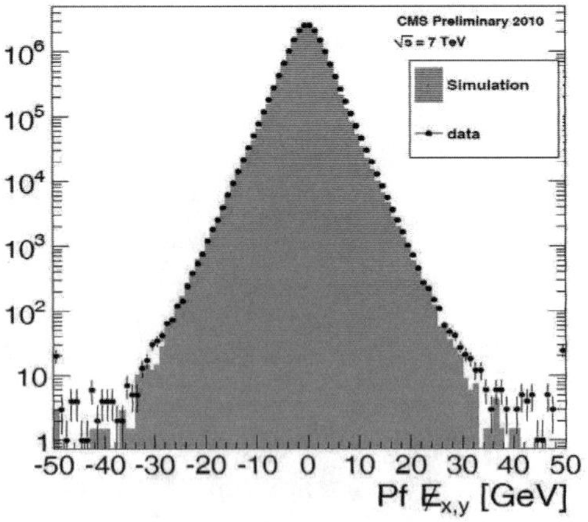

Figure 11. Distribution of the missing transverse energy in the x,y direction in CMS calculated using the particle flow technique.

7. The LHCb experiment

The LHCb experiment [3] uses high production rate of b-quarks at small angle at the LHC to perform a thorough study of the CP-simmetry violation in the neutral B meson systems. It is a single arm spectrometer that covers an angle from the beam of 10-300 mradians. The

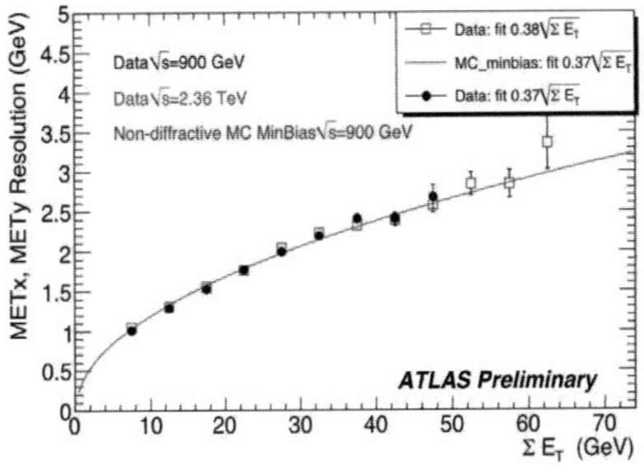

Figure 12. Resolution on the missing transverse energy in the x,y direction as a function of the total transverse energy in ATLAS. The calorimeter energy is calibrated at the e.m. scale.

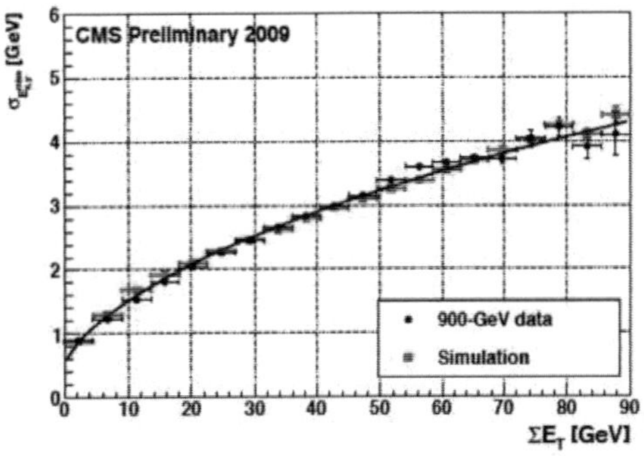

Figure 13. Resolution on the missing transverse energy in the x,y direction as a function of the total transverse energy in CMS using the particle flow technique.

experiment is 20 metres long. It has powerful tracking capability with a silicon vertex locator and particle identification detectors installed near a dipole magnet.

An electromagnetic calorimeter made of lead and scintillators with wavelength shifting fibers in shashlik configuration is installed at 12 metres from the interaction vertex. This large lever arm allows to obtain a very performant π^0 two-photon separation. The calorimeter is employed in the Level0 trigger and its electron/hadron separation capability is already used at the trigger level.

The energy resolution of the calorimeter measured in test-beams is:

$$\frac{\sigma(E)}{E} = \frac{10\%}{\sqrt{E}} \oplus 1\%. \tag{9}$$

The detector is well timed in (at the level of 1 ns) and 99.8% of the channels are operational.

A hadron calorimeter made of iron and scintillator tiles completes the calorimeter system of the experiment. It constitutes 5.6 λ_I.

There are high radiation levels at small angle. A monitoring system based on LED light is installed to monitor the calorimeter stability. The detector was not calibrated with cosmics underground due to the small statistics of cosmic rays that match the horizontal geometry of the LHCb experiment. The calibration is being performed using collision events and in particular with the energy flow method (equalizing the detector signals in symmetric regions) and with $\pi^0 \rightarrow \gamma\gamma$ decay. Simulation shows that the detector can be calibrated at the 1% level with 50 million events. Figure 14 shows the π^0 peak reconstructed by LHCb. The reconstructed mass is in good agreement with expectations and the resolution is of 10 MeV.

8. The Alice experiment

Alice [4] is an experiment dedicated to heavy ion collisions. Its purpose is to study strongly interacting matter at high densities where the formation of quark-gluon plasma is expected. In order to do this, Alice needs to study and measure the spectra of hadrons, electrons, photons and muons produced in the collisions.

The Alice experiment is installed inside a large magnet that provides an axial field of 0.5 T. In order to measure the spectra of photons, π^0 and η in the range of 0.5-10 GeV it has installed a precise electromagnetic calorimeter that covers a limited region in η and ϕ at 4.6 metres from

Journal of Physics: Conference Series **293** (2011) 012001 doi:10.1088/1742-6596/293/1/012001

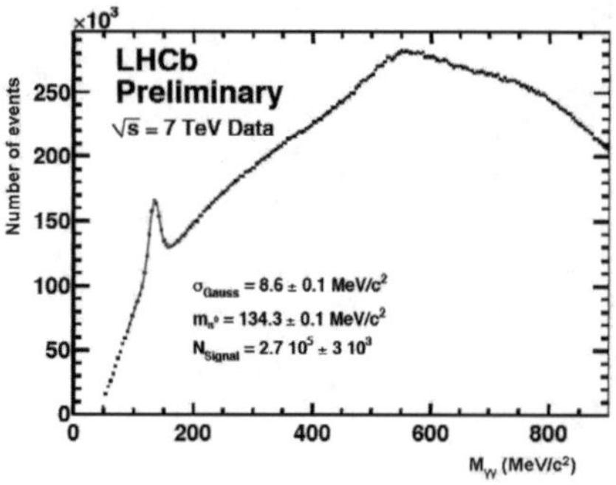

Figure 14. Invariant mass of the two-photon pairs reconstructed with the LHCb e.m. calorimeter during 2009 collisions.

the interaction point. It is made of 18000 lead tungstate crystals read by avalanche photodiodes. Three of the five modules were installed for the 2009-2010 data-taking. The crystals measure 18 cm in length (20 X_0) and they are operated at -25 ^oC in order to increase the light yield and improve the signal to noise ratio. The crystals were not calibrated prior to installation.

The energy resolution measured in test-beams is:

$$\frac{\sigma(E)}{E} = \frac{3.3\%}{\sqrt{E}} \oplus \frac{18\ MeV}{E} \oplus 1.1\%. \qquad (10)$$

9. Conclusions

It was a challenging adventure to build calorimeters with this unprecedented number of channels and the required level of stability and reliability for 10 years or more in the LHC environment. It took almost 20 years from design to construction and operation. All calorimeters are in good shape, they are well timed in and are used in the first level trigger algorithms.

They are well understood detectors with spectacular agreement between data and Monte Carlo simulation on basic and complex observables.

They are precious instruments for a new era in high energy physics going to explore a completely new range of energies.

9.1. Acknowledgments

The author wishes to thank Isabelle Wingerter and Mauro Donega of the ATLAS collaboration, Marie-Noelle Mirard of LHCb and Dave Barney, Francesca Nessi, Roger Rusack, Marcella Diemoz and Egidio Longo of the CMS collaboration for providing the material on which this paper is based and for help in the preparation of the presentation at the conference.

References

[1] ATLAS Collaboration, *The ATLAS technical proposal*, CERN/LHCC/94-43.
[2] CMS Collaboration, *The CMS Technical proposal*, CERN/LHCC/94-38.
[3] LHCb Collaboration, *The LHCb Technical proposal*, CERN/LHCC/98-4.
[4] Alice Collaboration, *The Alice Technical proposal*, CERN/LHCC/95-71.

EM Calorimeter in BESIII Experiment

Weiguo Li, representing BESIII Collaboration
19 Yuquan Road, Beijing, 100049, China
E-mail: liwg@ihep.ac.cn

Abstract. BESIII EM Calorimeter (EMC) is based on CsI crystals, working at upgraded Beijing Electron Positron Collider (BEPC), called BEPCII. The overall BEPCII and BESIII design and current status are briefly described. The design and construction of EMC and its performance are presented. EMC works very well and its performances reach the design expectations, it plays an important roll in detector operation and in the BESIII trigger. There are some special features in the EMC construction and reconstruction: the crystals are assembled without support wall to reduce dead area; the time information of the EM clusters can help to suppress the machine background; the detector noise is smaller than similar detectors; the energy reconstruction with TOF signals helps to improve the energy resolution, especially for low energy showers. Several physics papers mainly based on EMC information have been published.

1. Introduction

Beijing Electron Positron Collider (BEPC) has been approved for major upgrade, called BEPCII, at the end of 2003. The original machine has been upgraded to a two rings machine, with up to 93 bunches in each ring and mini-β achieved by super-conducting quadruples, to increase its luminosity by a factor of about 100, to $10^{33}cm^{-2}s^{-1}$. The electrons and positrons collider with a cross-angle of 2 $\times$ 11 mill radians. Beijing Spectroscopy (BESIII)[1] is the detector to do experiment at BEPCII, it is completely built as a new detector. The project started at the beginning of 2004, and the machine and detector had been completed in the summer of 2008, and first collision event at center of mass system of 3.68 GeV was observed on July 20, 2008. After almost one year of commissioning run, the project passed the Chinese Government review at July 2009.

Until now the machine reached the maximum luminosity of 3 $\times 10^{32}cm^{-2}s^{-1}$, 30% of the designed luminosity to be reached after a few years from commissioning. Most of BESIII systems work well. So far, BESIII has collected 100 M hadronic events at $\psi(2S)$, 220 M at J/ψ and about 610 pb^{-1} so far at $\psi(3770)$.

BESIII consists of the Main Drift Chamber (MDC), Time of Flight (TOF), CsI EM calorimeter, Muon chambers made of Resistive Plate Chamber. The center magnetic field is 1.0 tesla provided by a super-conducting magnet.

MDC has 43 sense wire layers, using He based working gas. The achieved momentum resolution is 11 MeV/c for Bhabha electrons at J/ψ energy of 3.1 GeV, equivalent to a momentum resolution better than 0.6% for charged tracks of 1 GeV. The dE/dx resolution of Bhabha electrons reached 4.9%, better than designed.

For Bhabha events, the time resolution of TOF is about 80 ps for barrel counters combined hits in both layers and about 136 ps for endcap counters. The endcap TOF suffers from material at the endcap region. For di-muons events, the time resolution reaches 95 ps.

BESIII, as an international collaboration, consists of about 300 members, from 43 institutes from China, Japan, Korea, Pakistan, United States, German, Russia, Italy and Netherland.

2. Construction of EMC

BESIII EMC is based on CsI(Tl) crystals, with a designed energy resolution of 2.5% and a position resolution of 6 mm for 1 GeV showers. The EMC consists barrel and endcap parts as shown in Fig. 1. In the barrel, there are 120 sectors in phi direction, each consists of 44 crystals in z direction. The number of barrel crystals is 5280 in total. The crystals are 28 cm long, and typical size of the crystal is 5.2 cm by 5.2 cm in the front surface. The crystals are nearly projective but arranged that they are tilted by 1.5^o in ϕ direction and $1-3^o$ in θ direction to avoid photons go through gaps between crystals. In the endcap, the crystals with similar dimensions as in those in the barrel are arranged in 6 rings in radial direction, there are total of 960 crystals in two endcaps.

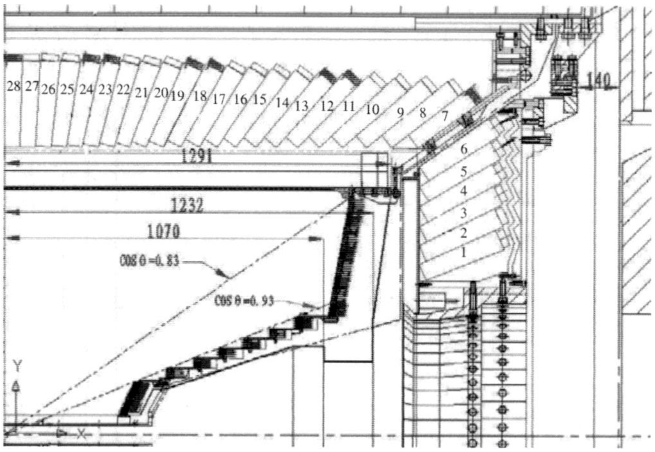

Figure 1. one quarter of cross-section of BESIII EM Calorimeter.

Two photodiodes are attached to the back face of every crystal, each is readout by an pre-amplifier. The readout units is housed in an Al box. The crystals are held by 4 screws drilled into the crystal, then mounted to the support structure. there are fibers emitting light from LED to each crystal for monitoring purpose. During the assembling, one unit, consists two ϕ sectors with 88 crystals, firstly assembled in a jig structure, after carefully adjusting the crystal positions and testing the readout using LED, these two sectors are mounted on a fixture, so the crystals are fixed in relative stable positions and can be moved together. The 60 such units are assembled into a barrel structure. The total weight of the barrel crystal structure is about 54 tons, inserted in the stainless steel structure of detector magnet and mounted. The endcaps crystals used similar structure to hold the crystals from the back. Fig.2 shows how the assembled crystal unit looks, Fig.3 shows the detailed assembling parts and how they were assembled together.

The crystals were produced by 3 companies, Sanit-Gobain in France, Shanghai Institute of Ceramics and Beijing Hamamatsu in China, the crystals delivered to IHEP were tested to check their dimensions, the light yield and the radiation resistivity. The light yield of crystals is required to be more than 35% of the light from a standard small crystal, the average light

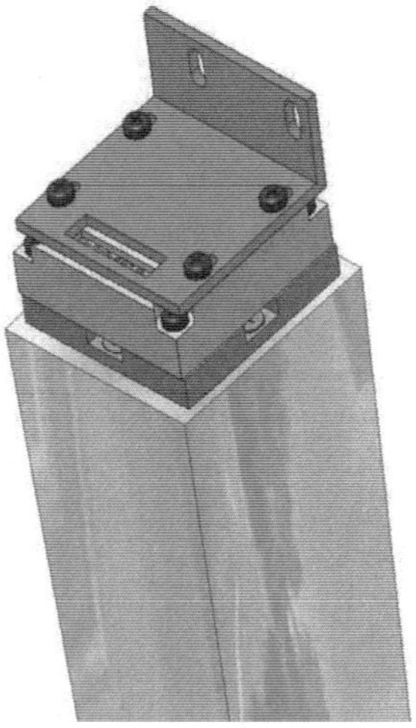

Figure 2. The assembling of a crystal unit.

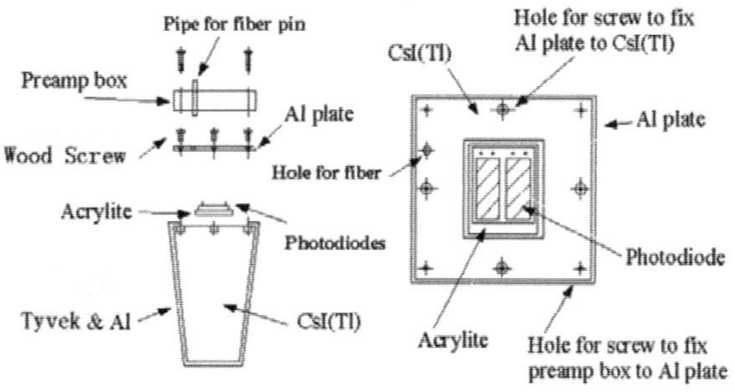

Figure 3. The detailed assembling of crystal unit showing various parts and how they are assembled together.

of all accepted crystals is about 58% of the standard crystal, more than 35% in the design. There is a demand of the light uniformity along the length of the crystal to be less than 7%, the actual average number of this quantity for all the crystals is about 4%. The crystals are tested for their radiation resistivity, by measuring its accompanied small sample in each crystal production batch for its radiation property. The samples were subject to radiation dose up to 1000 rads in a gamma source. The requirement is that the light reduction after 100 rads should be less than 9% and less than 20% after 1000 rads. If a sample did not meet those requirements, all the crystals produced with this sample in a batch will be shipped back to the company to be replaced. 482 crystals are replaced for the reasons of bad radiation hardness, unqualified dimensions etc.

The photon sensors are Hamamatsu S2744-08 (1cm × 2cm) photodiodes, there are two diodes attached to the back surface of one crystal, each diode is readout by one preamplifier. The diodes and preamplifiers are housed in an Al cover mounted on an Al plate frame, which is screwed onto the crystals, as shown in Fig. 2 and Fig. 3.

Each photodiode is tested for its dark current, capacitance and quantum efficiency. All the preamplifiers were tested for their gains. The two preamplifiers on the same crystal are selected to have the gain difference less than 3%, to improve the overall energy measurement accuracy.

The signals from the preamplifier was sent to the main amplifier 18 meters away, the shaping time of the amplifier is 1 micro second. Then the signals are digitized in a Q module. The digitization of the signals is realized by three 20.8 Mhz 10 bit ADCs to cover a dynamic range of 15 bits. And the signal peaking time is also read out with 6 bits accuracy. The electronics are carefully designed, including careful shielding and grounding. The noise of the electronics is required to be less than 1100 electrons, equivalent to shower energy of 220 keV, the average noise of 384 channels tested in electronics lab is 973 electrons.

The detailed information about BESIII EMC electronics can be referred to the talk by Mr. Jinfan Chang at this conference.

3. Operation experience

In order to ensure the safety of the crystals from the radiation dose from the machine. The radiation dose is carefully monitored by dose probes mounted inside the detector. After the BESIII moved into the beam line, the machine operation was carefully planned. At BEPCII/BESIII, the detector dose from machine comes from beams circulated in the ring as well as from the beam injection from Linac to the ring, the later caused more dose if the injection was not well tuned. To avoid sudden dose accumulation to damage the crystals, at the very beginning of BEPCII/BESIII commissioning, the beam injection from Linac used 1 Hz rate compared to the normal 50 Hz Linac operation. After the machine dose is under control, then the machine injection shifted to 5 Hz operation, then to 25 or 50 Hz operation.

There are two kinds of radiation monitor probes, one is pin diode and another is RadFET, the formal can withstand more dose and used to monitor the machine radiation dose online, and the later is more stable and used to estimate the real integrated dose in the crystals. There are 12 pin diodes probes mounted on the beam pipe structure, 6 at east side and 6 at west side, 60 degree apart in phi direction. The RadFET probes are distributed at a radius similar to where the crystals are, in the barrel 10 probes are positioned along the Z direction, at one phi angle. There are 4 rows of probes at phi of 0, 90, 180 and 270 degrees, so there are 40 probes in barrel. at the endcap, each side placed 4 probes along the radius direction at one phi, and probes are placed in phi direction of 0, 90, 180 and 270 degrees, so there are 32 probes in total in the endcaps. The relative doses at pin diodes and at the RadFET were calibrated before the detector was moved in the beam line by IHEP radiation office. When machine is in operation, if some dose from pin diodes exceeds certain level, there will be a warning and the machine people have to take action to reduce the doses on the detector.

There is a LED system to monitor the light output from LED, to check the system during the EMC assembling and in data taking, by reading the LED light from each crystals. Actually this system helped to find out some problems, such as loose gluing of photodiode to the crystal, broken cables, noisy channels because of bad grounding, etc during the assembling. During data taking, the system serves to find out bad channels and monitor the performance of the EMC, including the gain reduction by radiation in each crystal. The details are in the talk given by Jian Fang at this conference.

Up till now, the average drop of the light of all the crystals are less than 3%, but for a small fraction of the crystals, mostly at the endcaps, the light reduction reaches about 10% for some crystals. These light reduction is consistent to the gain reduction calculated from the EMC calibration using Bhabha events. For the worst case, the gains reduction for a small number of crystals reachs about 15%, whether it is due to radiation dose, or some other factors such as reduced light coupling between crystal and the photodiode, should be further investigated.

4. EMC in the trigger

EMC plays important roll in the BESIII trigger. In the barrel, 4×4 crystals form an trigger cell, in the endcaps, neighboring 15 crystals form a trigger cell. The threshold for trigger cell is set to about 70-80 MeV, a cluster finding scheme is carried out to form a cluster using nearby crystals above the threshold. By analyzing the data, it is confirmed that the trigger is fully efficient for clusters with energy lager than 200 MeV. There are several trigger conditions from EMC, they are the number of clusters, the total energy of the EMC energy and the energy balance in EMC. For the condition of total energy, there are two thresholds, one is Etotl which is set to about 200 MEV, and is 100% efficient at 400 MeV from the trigger study. Another is Etotm which is set to about 700 MeV, and is 100% effieicnt at 1000 MeV, which is mainly used for triggering neutral events. At the global level, there is a neutral trigger, which requires Etotm and number of clusters to be large than 1. There are several triggers for events containing charge tracks, besides conditions from MDC and TOF, for EMC, there is requirement of one or more clusters, or requirement of Etotl. At earlier runs at J/ψ and $\psi(2S)$, there was an trigger condition that demand 2 or more tracks and 2 or more TOF hits. But at $\psi(3770)$ runs, the total trigger rate became too higher because of higher beam currents and higher energy compared with the conditions at two earlier beam energies at J/ψ and $\psi(2S)$. Careful studies shown that dropping this global trigger will not result in the loss of good events with an addition trigger for back-back events, so this global trigger condition is no longer used at $\psi(3770)$ data taking. Now all the global triggers require some signals from EMC, either having one cluster at barrel or endcaps or satisfying Etotl condition, except the barrel back to back trigger, which requires back to back tracks and back to back TOF hits, the rate of this trigger is relatively small, so no information from EMC is needed. From this change in the trigger, it shows how powerful the EMC is in its ability to reduce the background in the trigger.

Special data were taken to study the trigger efficiency, the results show that for most of the good events, including Bhabha, hadronic events, barrel di-muons, BESIII trigger is 100% efficient, only for endcap di-muons events, the trigger efficiency is about 95%.

5. EMC calibration and monitoring

After data were taken, major effects were spent to calibrate the detector and monitor the performance of various detector components. In BESIII, Bhabha events are used to calibrate the gain of each crystal, radiative Bhabha, di-photons events and π^0 are used for energy scale and understand the energy response and the resolution for showers at lower energies. To ensure the consistent of Monte-carlo (MC) and data, the detector material were carefully accounted for and put into the MC. At the beginning, the agreement at the endcaps was not good, it then found out that the material from the cables at the endcaps were not correctly put into the data base, after the correct material was added, the agreement between MC and data became quite good. Talks given by Liu Chunxiu and Bian Jianming at this conference provide details about the EMC calibration and improvement of reconstruction.

Fig. 4 shows the average energy deposited in EMC of Bhahha events at c.m. energy of 3.68 GeV, as a function of shower Z position in EMC in the data as well as in the MC. Fig. 5 shows the energy resolution in EMC of same events as a function of shower Z position in EMC in the data as well as in the MC.

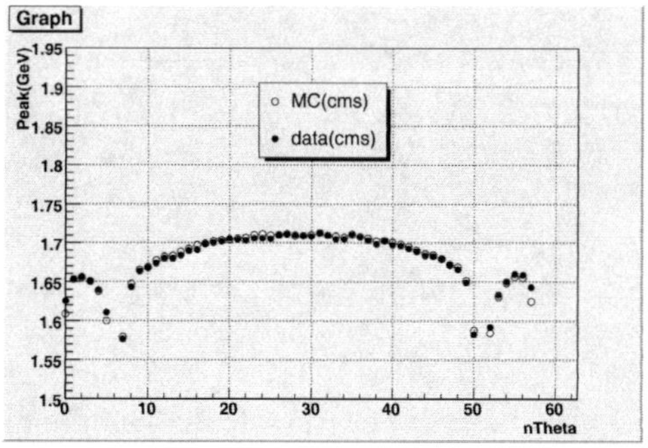

Figure 4. The Comparison of average energy deposited in EMC crystals along Z direction as a function of shower position between data and MC for Bhabha events.

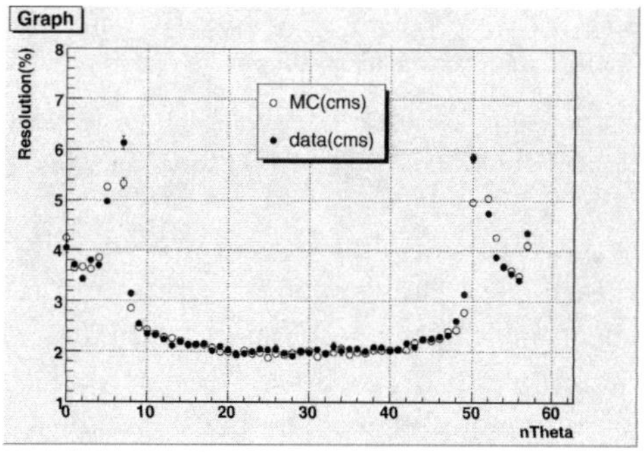

Figure 5. The Comparison of energy resolution in EMC crystals along Z direction as a function of shower position between data and MC for Bhabha events.

Fig. 6 shows the position resolution of Bhabha events at c.m. energy of 3.68 GeV, the corresponding shower position resolution is 4.4 mm. Fig. 7 shows the energy resolution as a function of shower energy, it can be seen that the energy resolution at 1 GeV is about 2.5%.

To prevent causing problem in EMC electronics, the EMC electronics is powered all the time when the machine is in operation, even at the dedicated synchrotron runs. So it is handy to

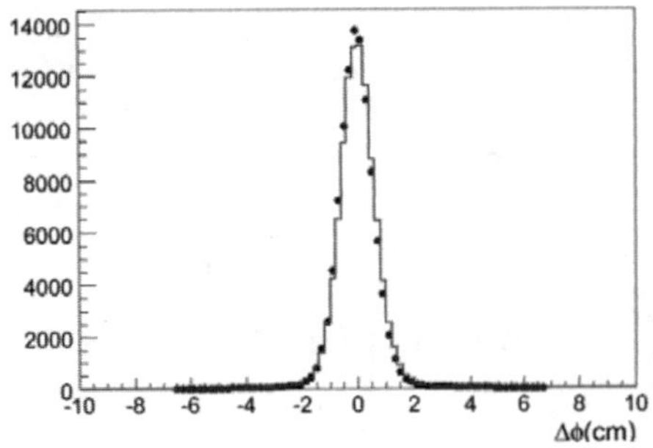

Figure 6. The position resolution of data and MC for Bhabha events at c.m. energy of 3.68 GeV.

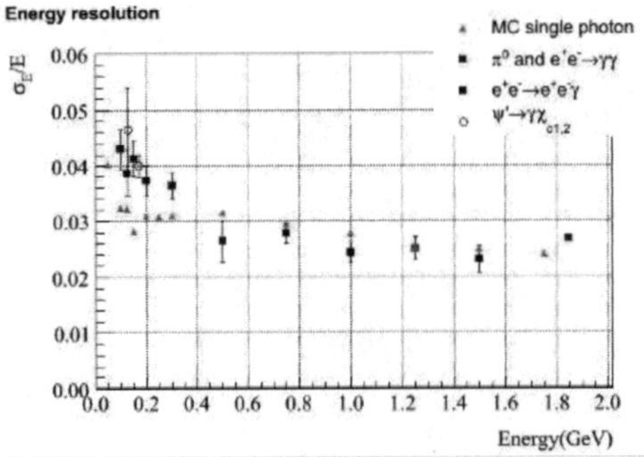

Figure 7. The energy resolution as a function of shower energy as calibrated using radiative Bhahbha, π^0s and physics events.

monitor the EMC conditions and make luminosity measurement easier.

6. EMC Performance

The EMC works fine now after about two years in the beam line. The noise level at the collision point in data taking is similar to that tested at the lab. From the data analyzed, the average noise for all the channels is at a level of 200 MeV as shown in Fig. 8, better than most of other detectors in the world with similar complexity based on CsI(Tl) crystals. Because mechanically it is very difficult to make repair for EMC, and if one crystal is lost, the energy at the area this crystal covers will be degraded, it is critical to keep all the channels working at all times. There is no single crystal lost so far. The number of crystal which has only one photodiode plus preamplifier working was two at the beginning of the data taking in July 2008, now the number of such crystal has increased to about 10, but the effect to the performance is minimum.

After the calibration and correctly input of the detector material, the agreement of Bhabha energy distribution along phi and Z between data and MC is very good. And the energy resolution as a function of shower energy is as expected, and at 1 GeV the energy resolution is about 2.5% as designed.

It worths pointed out that the shower time information is very useful in the rejection of the background in EMC. The time is distributed in a relative narrow window for physics events and the distribution is flat for background in the data taking window, which is much larger than the time window in which the real events occurs. For the physics data, after selecting real showers by a chosen time window cut, the background is reduced by a large factor. Also, the combined reconstruction of shower energies by adding associated TOF energy deposits can improve the shower reconstruction efficiency and the energy resolution is also improved, especially for low energy showers, as shown in Fig.9. In the figure, the solid dots are for shower reconstructed without TOF, and the circles are with the TOF in the reconstruction.

There are large sample of χ_{cJ} decays, the radiative photons in the $\psi(2S)$ to χ_{cJ} decays can be used to study the performance of EMC. The energy distributions of radiative gammas in EMC and TOF agree very well between data and MC. The difference between data and MC in reconstructed energy scale is less than 0.5 % and the difference in energy resolution is less than 5%, by study of the real data. By using TOF in the energy reconstruction, the low energy gamma reconstruction efficiency is improved by more than 7%, the improvement in low energy π^0 reconstruction efficiency is about 12%. These are significant in the physics analyzes.

On the whole, BESIII EMC performs very well from data analyzed, and it plays crucial roll

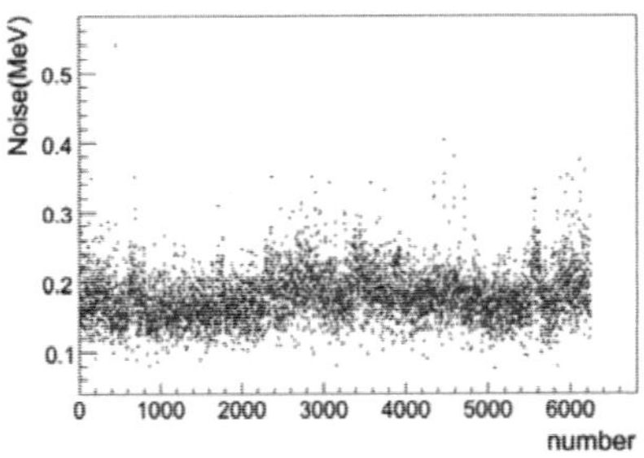

Figure 8. The equivalent noise of each channel as a fuction of channel number for all channels.

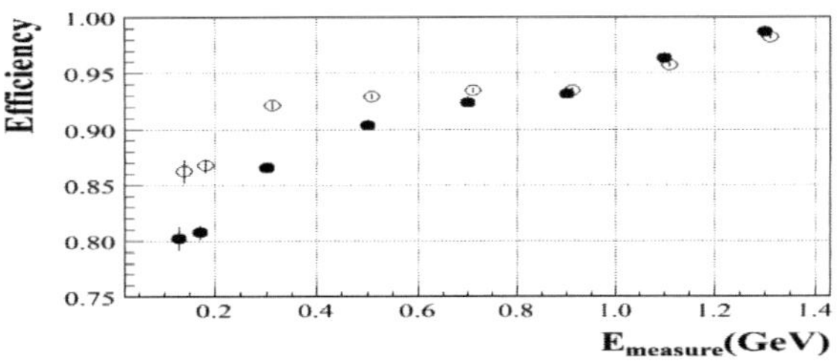

Figure 9. The Comparison of shower reconstruction efficiency with or without using TOF hits, the dots are for the case without TOF in the reconstruction, the circles are with TOF.

in physics analyzes.

7. EMC in BESIII Physics Analyzes

At BESIII, EMC is well understood, so several papers are published mainly using EMC information, which are the BESIII first publications.

One paper published [2] is the BF measurement of $\psi(2S)$ to $\gamma\pi^0\pi^0$ and $\gamma\eta\eta$ through χ_{cJ} states. the χ_0 and χ_2 peak are very clear and almost background free, and they can be nicely fitted. Fig. 10 is the recoiled gamma spectrum for $\psi(2S) \to \gamma\pi^0\pi^0$, Fig. 11 is the recoiled gamma spectrum for $\psi(2S) \to \gamma\eta\eta$. The fit used the MC generated gamma spectrum for χ_0 and χ_2. The

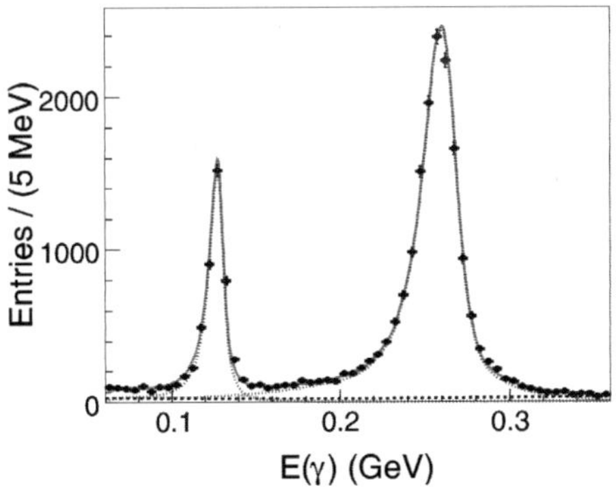

Figure 10. The radiative photon energy spectrum of selected $\chi_c \to \pi^0\pi^0$ events. Dots with error bars are data. The solid curve is the fit. The dotted curve is the χ_{cJ} signals. The dashed curve is the background polynomial.

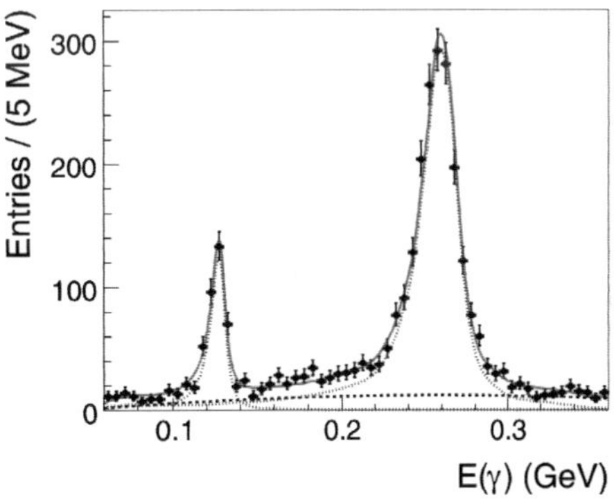

Figure 11. The radiative photon energy spectrum of selected $\chi_c \to \eta\eta$ events. Dots with error bars are data. The solid curve is the fit. The dotted curve is the χ_{cJ} signals. The dashed curve is the background polynomial.

results are in the Table 1 and 2. Compared with PDG and CLEOc, the results are consistent with each other and the BESIII results have smaller errors.

Table 1. Results of χ_{cJ} to $\pi^0\pi^0$

BF(10^{-3})	χ_{c0} to $\pi^0\pi^0$	χ_{c2} to $\pi^0\pi^0$
BESIII	$3.23 \pm 0.03 \pm 0.23 \pm 0.14$	$0.88 \pm 0.02 \pm 0.06 \pm 0.04$
PDG	2.43 ± 0.20	0.71 ± 0.08
CLEOc	$2.94 \pm 0.07 \pm 0.32 \pm 0.15$	$0.68 \pm 0.03 \pm 0.07 \pm 0.05$

Table 2. Results of χ_{cJ} to $\eta\eta$

BF(10^{-3})	χ_{c0} to $\eta\eta$	χ_{c2} to $\eta\eta$
BESIII	$3.44 \pm 0.10 \pm 0.24 \pm 0.15$	$0.65 \pm 0.04 \pm 0.05 \pm 0.03$
PDG	2.4 ± 0.4	< 0.5
CLEOc	$3.18 \pm 0.13 \pm 0.31 \pm 0.16$	$0.51 \pm 0.05 \pm 0.05 \pm 0.03$

Table 3. BESIII results of hc compared to CLEOc

	BESIII	CLEOc (E1-tagged)
M(hc)	$3525.40 \pm 0.13 \pm 0.18$ MeV	$3525.35 \pm 0.23 \pm 0.15$ MeV
$\Gamma(hc)$	$0.73 \pm 0.45 \pm 0.28$ MeV < 1.44 MeV at 90% C.L.)	no measurement
$BF(\psi(2S) \to \pi^0 hc)\times$ $BF(hc \to \gamma\eta_c)$	$(4.58 \pm 0.40 \pm 0.50) \times 10^{-4}$ ($\Gamma(hc)$ float)	$(4.22 \pm 0.44 \pm 0.52) \times 10^{-4}$ ($\Gamma(hc)$ fixed to 0.9 MeV)
$BF(\psi(2S) \to \pi^0 hc)$	$(8.4 \pm 1.3 \pm 1.0) \times 10^{-4}$	no measurement
$BF(hc \to \gamma\eta_c)$	$(54.3 \pm 6.7 \pm 5.2)\%$	no measurement

Another paper published[3] is the results of inclusive $\psi(2S)$ decays to hc. Because of the much better EMC properties compared with these of BESII, BESIII has able to see the hc signal clearly with inclusive $\psi(2S)$ decays, with the tag on the radiative photon of $hc \to \gamma\eta_c$ as well as without the tag. Fig. 12 shows the π^0 recoil mass spectrum with the tag, fits are indicated by solid lines, background by dashed lines, the smaller plot is the one with background subtraction. The data is fitted by treating the signal as a Breit-Wigner convoluted with a detector Gaussian resolution plus background, which is represented by the π^0 recoil mass spectrum in the sideband of the E1 photon, and the background normalization is allowed to float. In the fit, the mass and width of hc are allowed to float. The fitted results are: mass of hc is 3525.40 ± 0.13 MeV, width of hc is 0.73 ± 0.45 MeV, with statistical errors only. The signal has a significance of 18.6 σ with number of signal events of 3679 ± 319. Fig. 13 is the inclusive π^0 recoil mass spectrum in $\psi(2S)$ decays, the signal corresponding to hc is obvious. When fitting the spectrum, The mass and width of hc are fixed to the values obtained from E1-tagged analysis. Fits are indicated by solid lines, background by dashed lines. The background is parameterized by a 4th-order Chebychev polynomial, and all of its parameters are allowed to float. The signal has a significance of 9.5 σ, with a fitted number of signal events of 10353 ± 1097. From the above two fits, the $BF(\psi(2S) \to \pi^0 hc) \times BF(hc \to \gamma\eta_c)$ and the two separate BFs can be obtained. The systematic errors are mainly due to the background shape in the fit, the energy scale, the number of π^0 in the events, the total number of $\psi(2S)$ events. The final results are in the Table 3.

The results of separate BFs of $BF(\psi(2S) \to \pi^0 hc)$ and $BF(hc \to \gamma\eta_c)$ are the first measurements, can be compared with predictions.

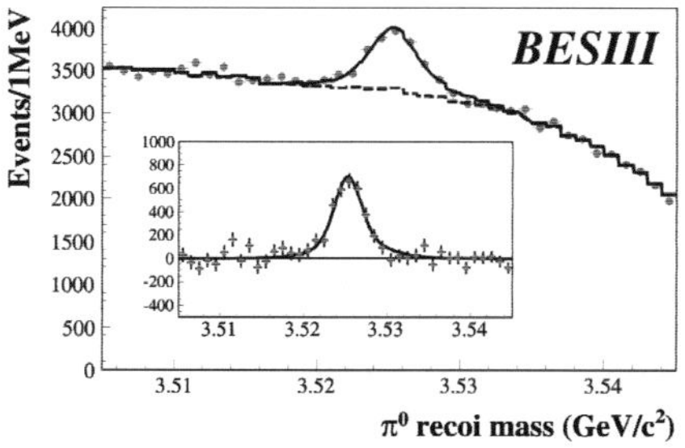

Figure 12. The π^0 recoil mass spectrum and the fit for E1-tagged analysis of $\psi(2S) \to \pi^0 hc$, $hc \to \gamma\eta_c$. The inset is the background subtracted spectrum.

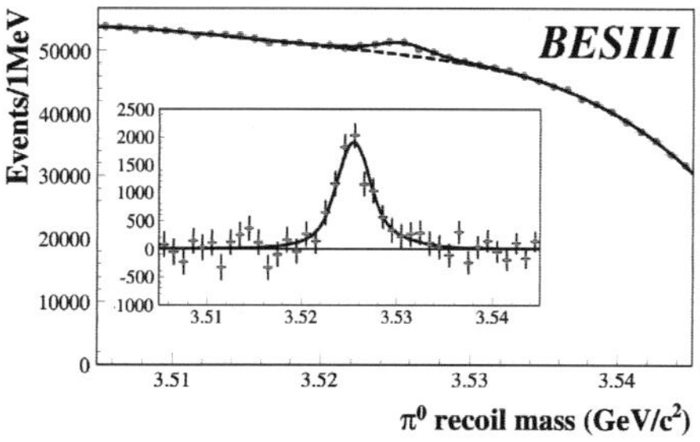

Figure 13. The π^0 recoil mass spectrum and the fit for the inclusive analysis of $\psi(2S) \to \pi^0 hc$. The inset is the background subtracted spectrum.

References

[1] M. Ablikim, *et al.* (BESIII Collaboration), "Design and construction of the BESIII detector", arXiv:0911.4960, accepted by Nucl. Instrum. Meth. A.
[2] M. Ablikim, *et al.* (BESIII Collaboration),*Phys. Rev.* **D81**, 052005 (2010).
[3] M. Ablikim, *et al.* (BESIII Collaboration),*Phys. Rev. Lett.* **104**, 132002 (2010).

Journal of Physics: Conference Series **293** (2011) 012003 doi:10.1088/1742-6596/293/1/012003

High-quality PWO crystals for the PANDA-EMC

R. W. Novotny, D. Bremer, V. Dormenev, P. Drexler, T. Eissner, T. Kuske, M. Moritz and for the PANDA collaboration

2nd Physics Institute, Justus-Liebig-University Giessen, Heinrich-Buff-Ring 16, 35392 Giessen, Germany

rainer.novotny@exp2.physik.uni-giessen.de

Abstract. The paper provides a status report on the crystal production and quality control of a major part of the $PbWO_4$ crystals for the PANDA-EMC. The results confirm the excellent performance of the new generation of PWO-II. The mechanism of stimulated recovery provides an additional tool to recover radiation damage at room and low temperatures by applying an external infrared light source. Even on-line recovery can be considered if the photo sensor is insensitive in that particular wavelength region.

1. The target electromagnetic calorimeter of PANDA

The PANDA detector, one of the approved major experiments at the future FAIR facility at GSI (Darmstadt, Germany) is dedicated to the hadron-physics program exploiting the stored and cooled beam of anti-protons. The research program is focusing on charmonium spectroscopy, gluonic excitations, open and hidden charm in nuclei and gamma-ray spectroscopy of hypernuclei [1]. Figure 1 provides an overview of the target and forward section of the PANDA detector and marks the electromagnetic calorimeter EMC.

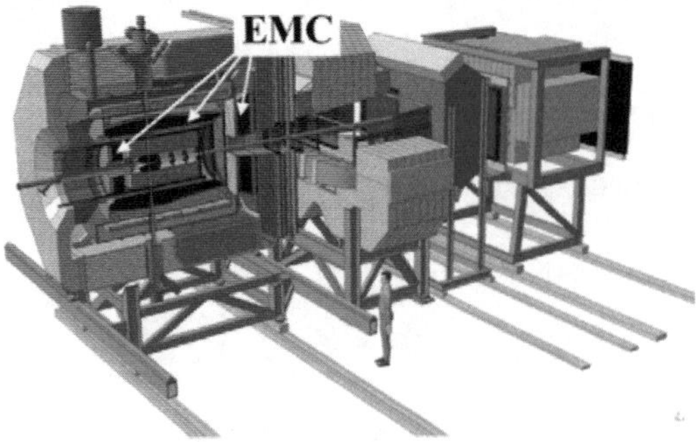

Figure 1. Schematic view of the PANDA detector including the EMC composed of a barrel and two endcap sections.

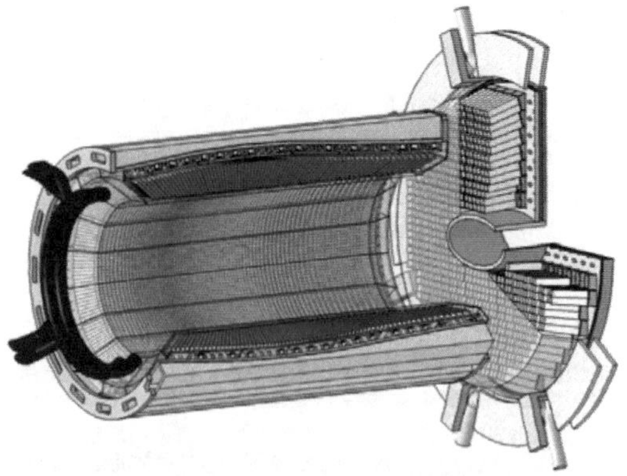

Figure 2. Schematic figure of the barrel and forward endcap as major components of the electromagnetic calorimeter of the PANDA target spectrometer.

The electromagnetic calorimeter (EMC) of the target spectrometer is one of the major detector components to provide high-resolution photon and meson spectroscopy [2,3]. The calorimeter will have to cope with photons in the energy range starting already at a few tens of MeV up to several GeVs. A pair of large area avalanche photo diodes (LAAPD) or a vacuum photo triode (VPT) are foreseen as photo sensor of the individual crystal and both options allow operating in the magnetic field of the superconducting solenoid. Figure 2 illustrates two of the three major components of the calorimeter the barrel and forward endcap.

Irrespectively of the finally collected percentage of the light yield, additional cooling of the scintillator reduces the thermal quenching of the scintillation processes and allows a significant improvement of the energy resolution, when photon statistics dominates. A temperature of T = -25°C was chosen as the operating temperature leading to a light increase of a factor of ~4 compared to room temperature.

1.1. The barrel of the target calorimeter

The layout of the barrel part comprises 16 identical slices. Each one consists of 710 tapered PWO crystals of 200mm length shaped in eleven geometries of two different symmetries. All crystals have a transversal size slightly above the Molière radius of 2.1cm. The 11.360 crystals are pointing slightly off the asymmetrically located target position.

Groups of four crystals are contained in a carbon fiber alveole including the two rectangular LAAPDs (active area 7x14mm^2), the pre-amplification and shaping circuits. Low noise (<2MeV), power consumption and a dynamic range of 10.000 are achieved by a custom designed ASIC (*APFEL*). The concept of the electronics is based on typical count rates of >100kHz per module. All latter components will be integrated into the cold section (T=-25°C) thermally shielded and temperature stabilized on the level <0.1°C. The mechanical support structure will house the complete electronics for control and digitization to minimize the required space for signal transfer out of the detector.

1.2. The endcaps

Both endcaps are designed in a similar manner. However, crystals are of identical and rectangular or slightly tapered shape pointing off the target. In order to cope with count rates well above 500kHz vacuum photo triodes (VPTs) are chosen at least for the most forward region.

2. The quality of PWO-II crystals produced at BTCP

Within the collaboration, the quality of PWO has been significantly improved primarily due to a reduced concentration of defects, achieved by raw material selection and growing technology. As a consequence, the co-doping with La- and Y-ions could be reduced to a level of <40 ppm. Therefore, full size crystals of 200mm length and nearly rectangular shape deliver on average 18 photoelectrons per MeV measured at room temperature (RT) using a phototube with bialkali photocathode. The careful selection of the raw material has strongly reduced those impurities, which lead to slow decay components in particular when the crystals are cooled.

After the approval of the Technical Design Report [3] the mass production at BTCP (Bogoroditsk Techno-Chemical Plant, Bogoroditsk, Tula Region, Russia) was started with very selective quality limits with respect to optical transmittance, homogeneity, light yield, scintillation kinetics and radiation hardness determined at room temperature.

The acceptance tests are performed in close collaboration with the CMS/ECAL collaboration at CERN exploiting the ACCOS [4] machine, which was adapted to the PANDA geometry and specifications. The measurements of radiation hardness and absolute light yield including crosschecks at low temperature as well as the final analysis are done at Giessen. So far, 7.355 crystals have been delivered and completely tested with an overall rejection rate of 7%.

2.1. The optical performance

The optical transmissions are measured longitudinally as well as perpendicularly to the crystal axis in order to control the homogeneity. Figure 3 shows the measured longitudinal transmission values at the relevant wavelengths of 360, 420 and 620nm, respectively. The figure confirms the consistency of the measurements obtained at the three testing locations. All values are well above the specification limits (T≥35% @ 360nm; T≥60% @ 420nm and T≥70% @ 620nm) and none of the crystals had to be rejected due to insufficient optical quality. The maximum variation of the absorption edge along the crystal length, which is a sensitive measure of the optical homogeneity, is determined by the wavelength corresponding to a transmission value of T=50% and remains well below Δλ<3nm.

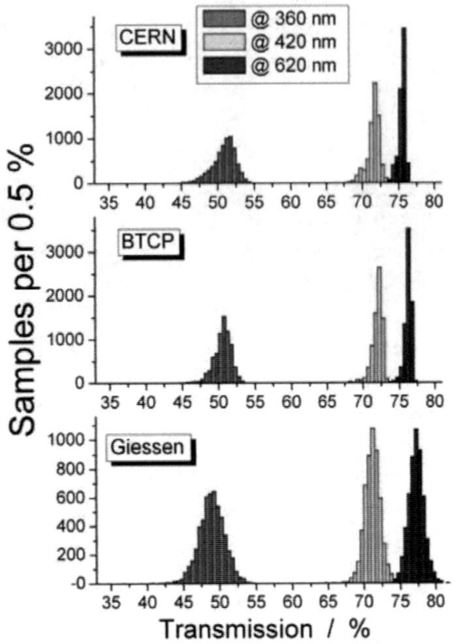

Figure 3. The distributions of the longitudinal transmission at three selected wavelengths of 200mm long PWO-II crystals in PANDA geometry measured at three locations for comparison

2.2. Light yield

The significantly improved light yield is documented in Figure 4. The detected number of photoelectrons per MeV deposited energy is determined at room temperature (T=18°C) with a standard photomultiplier with fused silica window and bialkali photocathode (Hamamatsu R2059, QE(420nm)=21%). The signal amplitude caused by γ-rays of a ^{137}Cs-source (E_γ=662keV) is calibrated with the single-photoelectron peak.

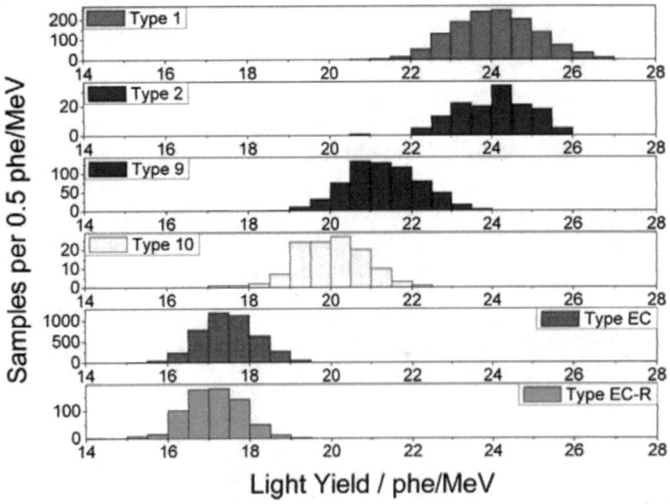

Figure 4. The distribution of the light yield measured at T=18°C with a photomultiplier tube for PWO-II crystals of 6 different PANDA geometries. The strongly tapered shapes (type 1, 2, 9 and 10) indicate the impact of light focusing.

Only a few samples do not pass the limit of 16phe/MeV. The further enhanced light yield of the barrel crystals reflects the light collection due to the more tapered crystal shape compared to the nearly straight geometry of the endcap module. The non-uniformity of light collection will be compensated by an appropriate reflector foil. More than 90% of the light is collected within 100ns, which even holds when crystals are cooled down to T=-25°C.

2.3. Radiation hardness

The most critical parameter is the change of the longitudinal absorption coefficient Δk due to radiation damage. The crystals are irradiated at room temperature at the facility at Giessen with an integral dose of 30Gy using a set of ^{60}Co sources (dose rate 4Gy/minute). The longitudinal transmission is measured up to a wavelength of 900nm delayed by 30 minutes after irradiation in order to exclude fast spontaneous recovery processes. The required value $\Delta k \leq 1.1 m^{-1}$ at 420nm guarantees that light losses due to radiation damage within an experimental period of 6 months remain tolerable even if the detectors are operated at low temperature, which will be discussed in more detail in the next chapter. Figure 5 shows the distribution of the Δk value at 420nm and indicates the significant percentage of extremely radiation hard crystals in the tail of the distribution towards lower values. All crystals above the limit of $1.1 m^{-1}$ are rejected. Radiation hardness appears to be the most selective quality parameter. The measurement at Giessen is compared to the data provided by the manufacturer.

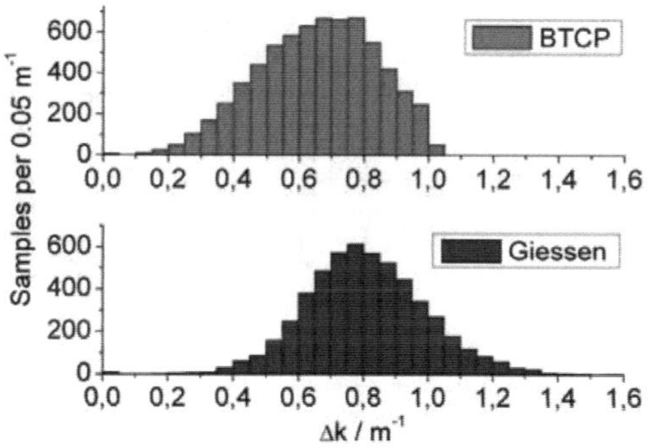

Figure 5. Distribution of the radiation induced absorption coefficient Δk measured at room temperature for PWO-II crystals. The integral dose amounts to 30Gy (^{60}Co). The specification limit amounts to $\Delta k \leq 1.1 m^{-1}$ determined at the radiation facility at Giessen.

3. Recovery of radiation damage by stimulation with infrared light

3.1. Radiation damage of cooled PWO crystals

As investigated within the collaboration [4], [5], the recovery mechanisms become extremely slow at T=-25°C with time constants in the order of τ >400h. Dominated by the damaging process due to irradiation the light yield deteriorates asymptotically towards a final value. This is correlated with the radiation hardness expressed by Δk being a measure of the concentration of defect centers in the crystal.

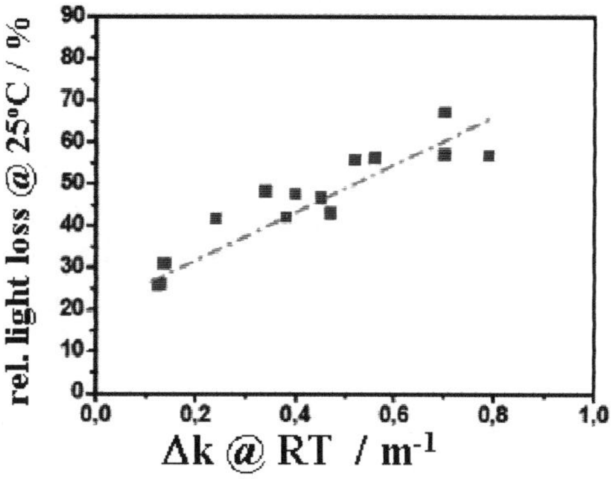

Figure 6. Correlation of the relative loss in light yield of cooled PWO-II detectors versus the induced absorption value at room temperature of investigated crystals. In both cases the integral dose amounts to 30Gy. The dotted line corresponds to a correlation coefficient of 0.95.

Such a correlation for a few accepted crystals is shown in Figure 6, which relates the relative loss of light yield of a cooled detector with the Δk value at room temperature. The detector response is measured either with low energy γ-rays or the energy deposition of cosmic muons, respectively. Both values are determined applying an integral dose of 30Gy. Since only the most inner part of the forward endcap - corresponding to about 10% of the modules - has to cope with a dose rate of typically 20mGy/h for the maximum event rate of $2 \cdot 10^7$ $p\bar{p}$- annihilations, a detector module will not lose more than 50% of the signal response if crystals with $\Delta k < 0.6 m^{-1}$ are selected. In that case the gain due to cooling still provides a net increase of at least a factor of two. In the barrel section, radiation damage will be 1-2 orders of magnitude lower.

3.2. Stimulated recovery of radiation damage at room temperature

It is well known that color centers, which are created during irradiation due to point structure defects or traps for electrons and holes, are leading to absorption bands in a wide spectral region but can be spontaneously relaxed after the end of irradiation via thermo-activation. Therefore, heating of the crystal is the standard procedure to anneal radiation damage, which requires the disassembly of the detector system. Even keeping the crystals in the dark, relaxation processes at room temperature are relatively fast and efficient to recover the crystals nearly completely during a typical beam-off period of ~ 6 months.

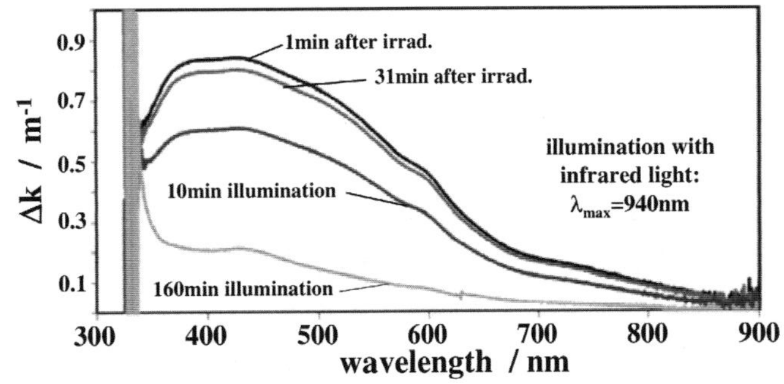

Figure 7. Change of the full spectral distribution of the induced absorption coefficient of a PWO-II crystal after irradiation with an integral dose of 30Gy measured at room temperature. The spectra are measured 1 and 31 minutes after irradiation as well as after additional illumination for 10 and 160 minutes, respectively, with infrared light at 940nm.

During the quality control of the radiation hardness of the PWO-II crystals an unexpected effect was observed at room temperature for the first time. After irradiation one recognizes besides the slow spontaneous recovery a significant additional acceleration of the change of the induced absorption coefficient Δk by switching on an external light source. Figure 7 shows as an example the change over the entire investigated spectral range due to the illumination with infrared light (LED, λ_{max}=940nm) and indicates wavelength-selective effects. At the most relevant wavelength of 420nm the recovery develops with a time constant of τ=270minutes. Similar measurements have been performed successfully even for pure or Mo-doped PWO samples or crystals of CMS/ECAL quality [7]. The stimulated recovery can be observed up to a wavelength of ~1400nm but with increasing recovery times.

3.3. Stimulated recovery at low operating temperatures

More relevant for the PANDA-EMC application is the behavior at low temperature, which imposes a strong impact on all recovery mechanisms. Therefore, we have integrated into each detector module, comprising of a crystal and a photomultiplier as sensor, an additional set of four LEDs as light sources. The in total 10 identical detectors were prepared with crystals of different radiation hardness. To study the wavelength dependence of recovery LEDs with different colors were used. After cooling the modules down to T=-25°C the response to a low energy ^{60}Co γ-source was measured before and after irradiation with 30Gy to determine the integral loss of light output. Illuminating afterwards with light of different wavelength, the recovery of the light loss was studied as a function of the integral illumination time up to two days.

Figure 8 shows the impressive results for the reduction of the relative light loss achieved with visible light sources centered at emission wavelengths of 470nm, 525nm, 640nm and 935nm, respectively. For blue light, nearly 90% of the original signal amplitude was restored already within the first 200 minutes. The used LEDs have imposed a photon flux of approx. 10^{16} photons/s. Presently, detailed studies and correlations with the crystal quality, the most efficient region of wavelengths, the required photon flux and the characterization of the recovered color centers via EPR measurements are under investigation. These results will provide a concept if the radiation damage during PANDA operation can be either compensated on-line without impact on signal readout or periodically after short shutdowns or breaks in an off-line mode. The concept and application of stimulated recovery is part of the patent TM 382_DE.

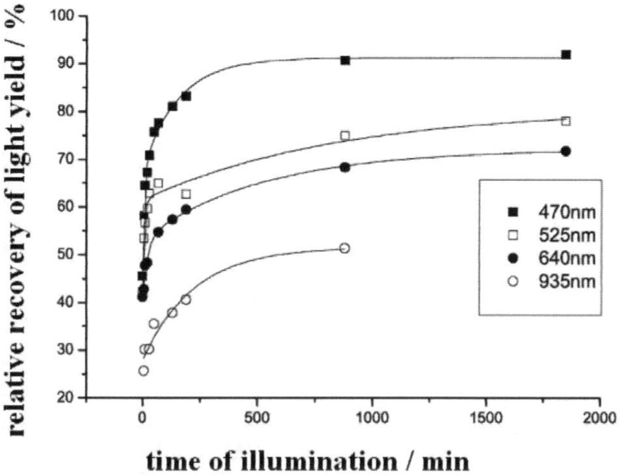

Figure 8. Recovery of the loss of the signal amplitude caused by irradiation as a function of the integral duration of illumination using different light sources. All detectors are operated at a temperature of T=-25°C. The maximum light loss at the beginning reflects the radiation hardness of the individual crystals.

4. Conclusion

The final production of the PANDA-EMC has been started by the mass production of the crystals and confirms the excellent quality of the new generation of PWO-II. The delivered crystals meet the requirements to achieve sufficient resolution down to a few tens of MeV photon energy and to keep the response stabile over the envisaged experimental periods of 6 months ensured in addition by the application of the stimulated recovery. Since the bialkali photocathode of the foreseen vacuum triodes is blind in the infrared region the external light can be even applied during the running experiment.

References

[1] Physics Performance Report (March 2009), arXiv:0903.3905v1.

[2] R. W. Novotny, "Fast and Compact Lead Tungstate-Based Electromagnetic Calorimeter for the PANDA Detector at GSI," *IEEE Trans. Nucl. Sci..*, vol. 51, no. 6, pp.3076-3080, December 2004.

[3] "Technical Design Report for: PANDA Electromagnetic Calorimeter (EMC)", arXiv:0810.1216v1, October 2008.

[4] F. Pauss (Editor-in Cief), CMS. The Electromagnetic Calorimeter. Technical Design Report, CERN, 1997.

[5] P. A. Semenov et al., "First Study of Radiation Hardness of Lead Tungstate Crystals at Low Temperatures," *Nucl. Instr. and Meth.*, A 582 (2007) 575-580.

[6] R. W. Novotny, S. F. Burachas, W. Döring, V. Dormenev et al., "Radiation Hardness and Recovery Processes of PWO crystals at –25°C," *IEEE Trans. Nucl. Sci..*, vol. 55, no. 3, pp.1283-1288, June 2008.

[7] V. Dormenev, private communication, internal PANDA-EMC report.

LSO/LYSO Crystals for Future HEP Experiments

Rihua Mao, Liyuan Zhang and Ren-Yuan Zhu
256-48, HEP, Caltech, Pasadena, CA 91125, USA
E-mail: zhu@hep.caltech.edu

Abstract. Because of their high stopping power (X_0 = 1.14 cm), fast (t = 40 ns) and bright (4 times of BGO) scintillation and good radiation hardness, cerium doped silicate based heavy crystal scintillators (LSO and LYSO) have attracted a broad interest in the high energy physics community pursuing precision electromagnetic calorimeter in severe radiation environment. We present in this paper current status of large size LSO and LYSO crystals adequate for HEP applications. The optical and scintillation properties and their radiation hardness are discussed.

1. Introduction

In the last two decades, cerium doped silicate based heavy crystal scintillators have been developed for the medical industry. As of today, mass production capabilities of lutetium oxyorthosilicate ($Lu_2(SiO_4)O$, LSO) [1] and lutetium-yttrium oxyorthosilicate ($Lu_{2(1-x)}Y_{2x}SiO_5$, LYSO) [2, 3] are established. Because of their high stopping power (> 7 g/cm^3), high light yield (200 times of PWO) and fast decay time (40 ns) this material has also attracted a broad interest in the physics community pursuing precision electromagnetic calorimeter, such as the proposed SuperB forward endcap calorimeter [4], the KLOE experiment [5] and the Mu2e experiment [6]. Our initial investigation on large size (2.5 × 2.5 × 20 cm) LSO/LYSO crystals also shows that this new generation of heavy crystal scintillators has a superb radiation hardness against γ-rays [7], neutrons [8] and charged hadrons [9]. They are thus an excellent material to be used in a severe radiation environment, such as the proposed high luminosity large hadron collider (HL-LHC). The main obstacles of using these crystals in the experimental physics are two fold: the availability of high quality crystals in sufficiently large size and the high cost associated with their high melting point (~2,000°). This report is a part of an on-going R&D program to further develop this material to be of practical use at the HL-LHC.

2. Properties of Scintillating Crystals

Table 1 lists basic properties of heavy crystal scintillators: NaI(Tl), CsI(Tl), BaF_2, CeF_3, bismuth gemanade ($Bi_4Ge_3O_{12}$ or BGO), lead tungstate ($PbWO_4$ or PWO) and LSO [10]. As shown in the table, all crystals, except CeF_3, have either been used in, or actively pursued for, high energy and nuclear physics experiments. Figure 1 is a photo showing twelve crystal samples. In addition to samples listed in Table 1 pure CsI, CsI(Na), LYSO as well as $LaCl_3$ and $LaBr_3$ are also shown in this photo although the last two are not yet in mass production stage. Samples are arranged in an order of their density, or radiation length. All non-hygroscopic samples are wrapped with white Tyvek paper as reflector. Hygroscopic NaI, CsI, $LaBr_3$ and $LaCl_3$ are sealed in packages with two ends made of quartz plates of 3 or 5 mm thick to avoid

Table 1. Properties of Heavy Crystal Scintillators with Mass Production Capability

Crystal	NaI(Tl)	CsI(Tl)	BaF_2	CeF_3	BGO	$PbWO_4$	LSO(Ce)
Density (g/cm^3)	3.67	4.51	4.89	6.16	7.13	8.3	7.40
Melting Point (°C)	651	621	1280	1460	1050	1123	2050
Radiation Length (cm)	2.59	1.86	2.03	1.70	1.12	0.89	1.14
Molière Radius (cm)	4.13	3.57	3.10	2.41	2.23	2.00	2.07
Interaction Length (cm)	42.9	39.3	30.7	23.2	22.7	20.7	20.9
Refractive Index[a]	1.85	1.79	1.50	1.62	2.15	2.20	1.82
Hygroscopicity	Yes	Slight	No	No	No	No	No
Luminescence[b] (nm) (at Peak)	410	560	300 220	340 300	480	425 420	420
Decay Time[b] (ns)	245	1220	650 0.9	30	300	30 10	40
Light Yield[b,c]	100	165	36 4.1	7.3	21	0.30 0.077	85
d(LY)/dT[b,d] (%/°C)	-0.2	0.4	-1.9 0.1	~0	-0.9	-2.5	-0.2
Experiment	Crystal Ball	CLEO *BaBar* BELLE BES III	TAPS	-	L3 BELLE	CMS ALICE PrimEx Panda	SuperB KLOE

a At the wavelength of the emission maximum.
b Top line: slow component, bottom line: fast component.
c Relative light yield of samples of 1.5 X_0 and with the PMT quantum efficiency taken out.
d At room temperature.

surface degradation. To minimize the uncertainties caused by the sample size dependence in the light output measurement all samples have a cubic shape of $1.5 \times 1.5 \times 1.5\ X_0^3$, except NaI(Tl)

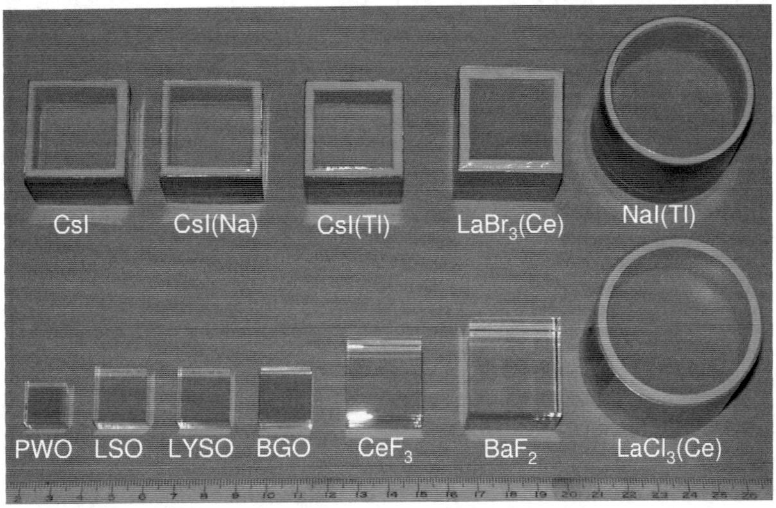

Figure 1. A photo shows twelve crystal scintillators with dimension of 1.5 X_0.

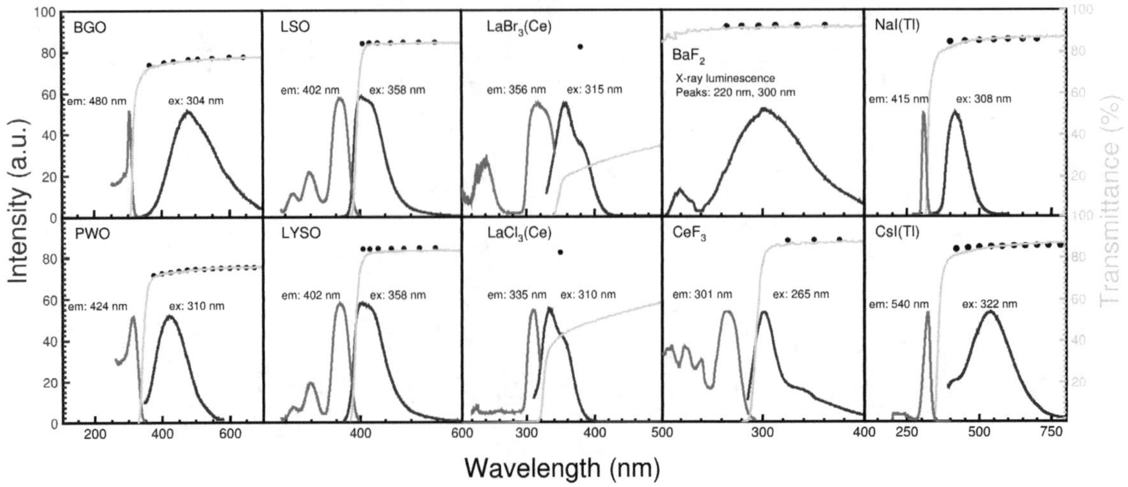

Figure 2. The excitation (red) and emission (blue) spectra (left scale) and the transmittance (green) spectra (right scale) are shown as a function of wavelength for ten crystal scintillators. The solid black dots are the theoretical limit of the transmittance.

and $LaCl_3$ which are a cylinder with a length of 1.5 X_0 and areas at two ends equaling to 1.5×1.5 X_0^2 to match the 2 inch diameter of the PMT cathode.

Figure 2 shows a comparison of the transmittance, emission and excitation spectra as a function of wavelength for ten samples. The solid black dots in these plots are the theoretical limit of the transmittance, which was calculated by using corresponding refractive index as a function of wavelength taking into account multiple bounces between the two parallel end surfaces and assuming no internal absorption [11]. Most samples, except $LaBr_3$ and $LaCl_3$, have their transmittance approaching the theoretical limits, indicating a negligible internal absorption. The poor transmittance measured for $LaBr_3$ and $LaCl_3$ samples is probably due to scattering centers inside these samples.

It is interesting to note that BaF_2, BGO, NaI(Tl), CsI(Tl) and $PbWO_4$ have their emission

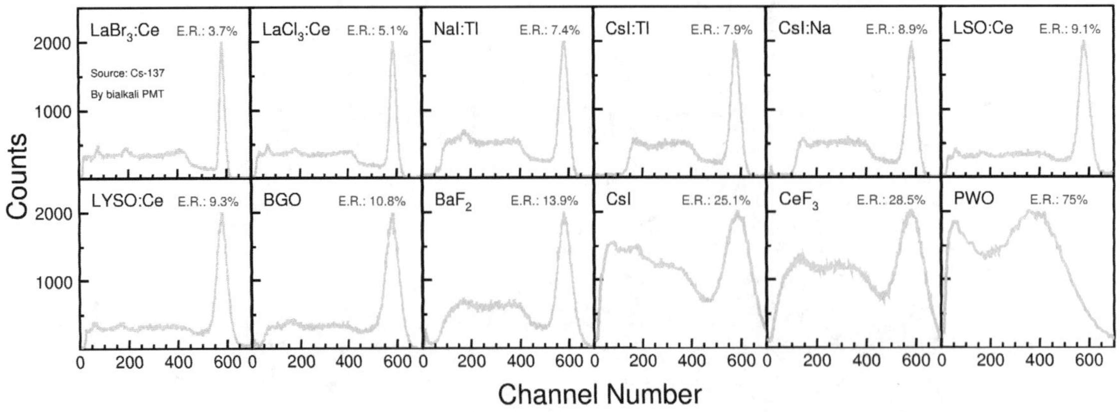

Figure 3. ^{137}Cs γ-ray pulse hight spectra measured by a Hamamatsu R1306 PMT are shown for twelve crystal samples. The numerical values of the FWHM resolution (E.R.) are also shown in the figure.

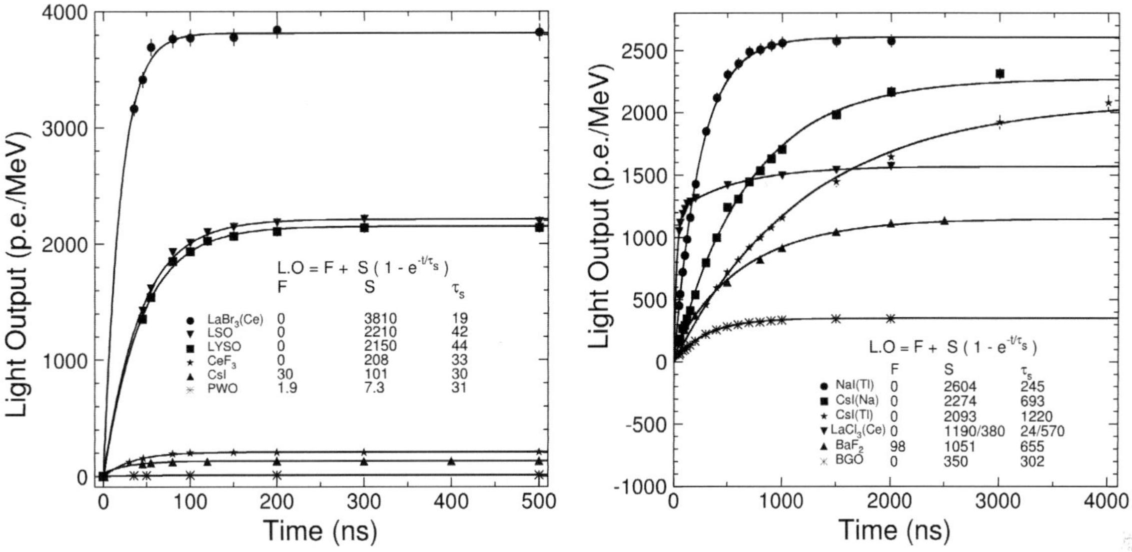

Figure 4. Light output measured by using a XP2254b PMT is shown as a function of integration time for six fast (Left) and six slow (Right) crystal scintillators.

spectra well within the transparent region showing no obvious self-absorption effect. The UV absorption edge in the transmittance spectra of LSO, LYSO, CeF_3, $LaBr_3$ and $LaCl_3$, however, cuts into the emission spectra and thus affects crystal's light output. This self-absorption effect is more serious in long crystal samples used in high energy and nuclear physics experiment as discussed for LSO and LYSO crystals [12]. It is well known that a good light response uniformity is crucial to maintain a small constant term for a crystal calorimeter.

Figure 3 shows the ^{137}Cs γ-ray pulse hight spectra measured by a Hamamatsu R1306 PMT with bi-alkali cathode for twelve crystal samples. Also shown in these figures are the corresponding FWHM energy resolution (E.R.). γ-rays spectroscopy with a few percents resolution is required to identify isotopes for the homeland security application. It is clear that only $LaBr_3$ approaches this requirement. All other crystals do not provide sufficient energy resolution at low energies.

Figure 4 shows light output in photo-electrons per MeV energy deposition as a function of the integration time, measured by using a Photonis XP2254b PMT with multi-alkali photo cathode, for six fast crystal scintillators (Left): $LaBr_3$, LSO, LYSO, CeF_3, undoped CsI and $PbWO_4$ and six slow crystal scintillators (Right): NaI(Tl), CsI(Na), CsI(Tl), $LaCl_3$, BaF_2 and BGO. The corresponding fits to the exponentials and their numerical results are also shown in these

Table 2. Emission Weighted Quantum Efficiencies (%)

Emission	LSO/LYSO	BGO	CsI(Tl)
Hamamatsu R1306 PMT	12.9±0.6	8.0±0.4	5.0±0.3
Hamamatsu R2059 PMT	13.6±0.7	8.0±0.4	5.0±0.3
Photonis XP2254b	7.2±0.4	4.7±0.2	3.5±0.2
Hamamatsu S2744 PD	59±4	75±4	80±4
Hamamatsu S8664 APD	75±4	82±4	84±4

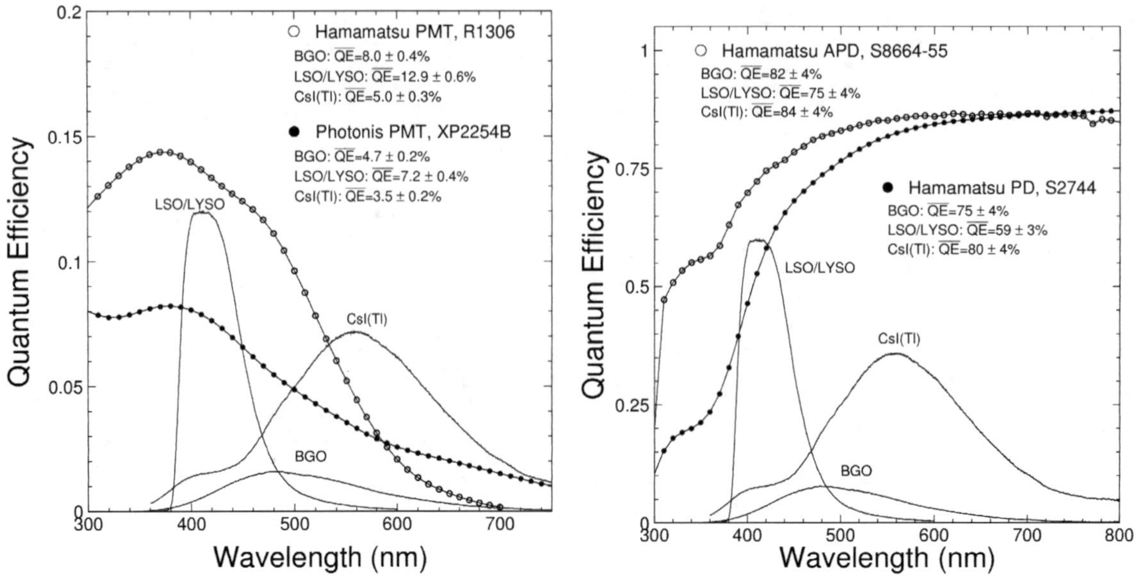

Figure 5. Left: Quantum efficiencies of a Hamamatsu 1306 PMT with bi-alkali cathode (open circles) and a Photonis 2254B PMT with multi-alkali cathode (solid dots) are shown as a function of wavelength together with the emission spectra of the LSO/LYSO, BGO and CsI(Tl) samples, where the area under the emission curves is proportional to their corresponding absolute light output. Right: The same for a Hamamatsu S8664 Si APD (open circles) and a Hamamatsu S2744 Si PIN diode (solid dots).

figures. The undoped CsI, $PbWO_4$, $LaCl_3$ and BaF_2 crystals are observed to have two decay components. Despite its poor transmittance the cerium doped $LaBr_3$ is noticed by its bright fast scintillation, leading to the excellent energy resolution for the γ-ray spectroscopic applications. The LSO and LYSO samples have consistent fast decay time ($\sim$40 ns) and photo-electron yield, which is 6 and 230 times of BGO and $PbWO_4$ respectively.

Since the quantum efficiency of the PMT used for the light output measurement is a

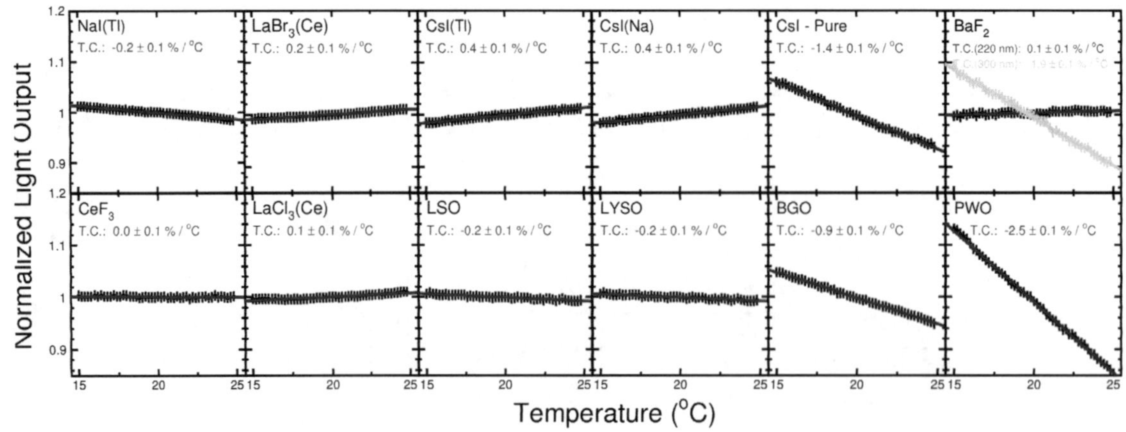

Figure 6. Light output temperature coefficient obtained from linear fits between 15°C and 25°C for twelve crystal scintillators.

function of wavelength, it should be taken out to directly compare crystal's light output. Figure 5 shows typical quantum efficiency as a function of wavelength for a PMT with bi-alkali cathode (Hamamatsu R1306) and a PMT with multi-alkali cathode (Photonis 2254B), a Si APD (Hamamatsu S8664) and a Si PIN PD (Hamamatsu S2744). The emission spectra of LSO/LYSO, BGO and CsI(Tl) crystals are also shown in these figures. Table 2 summarized numerical result of the emission weighted average quantum efficiency for several readout devices. The light output values in Table 1 are listed with the PMT quantum efficiency taken out. The light output of LSO and LYSO crystals is a factor of 4 and 200 of that of BGO and $PbWO_4$ respectively.

Scintillation light yield from crystal scintillators may also depends on the temperature. Fig 6 shows light output variations for twelve crystal samples between 15°C and 25°C. The corresponding temperature coefficients obtained from linear fits are also listed in the figure. The numerical result of these fits is also listed in Table 1.

3. LSO/LYSO Crystal Calorimeter

Large size LSO and LYSO crystals are routinely grown in industry [12]. Figure 7 shows four long crystal samples of $2.5 \times 2.5 \times 20$ cm^3 from CTI Molecular Imaging (CTI), Crystal Photonics, Inc. (CPI), Saint-Gobain Ceramics & Plastics, Inc. (Saint-Gobain) and Sichuan Institute of Piezoelectric and Acousto-optic Technology (SIPAT). Figure 8 shows the spectra of 0.51 MeV γ-rays from a ^{22}Na source observed with coincidence triggers [12]. The readout devices used are a Hamamatsu R1306 PMT (Left) and two Hamamatsu S8664-55 APDs (Right). The FWHM resolution for the 0.51 MeV γ-ray is about 12% for the PMT readout, which can be compared to 15% for the BGO sample. With the APD readout, the γ-ray peaks are clearly visible. The energy equivalent readout noise was less than 40 keV for these long LSO and LYSO samples.

LSO/LYSO crystals is also found to be much more radiation hard than other crystals commonly used in high energy and nuclear physics experiment, such as BGO, CsI(Tl) and $PbWO_4$ [13]. Their scintillation mechanism is not damaged by γ-ray irradiation. Radiation damage in LSO and LYSO crystals recovers very slow under room temperature but can be completely cured by thermal annealing at 300°C for ten hours. The energy equivalent γ-ray induced readout noise was estimated to be about 0.2 MeV and 1 MeV respectively in a

Figure 7. A photo shows four long crystal samples with dimension of $2.5 \times 2.5 \times 20$ cm^3.

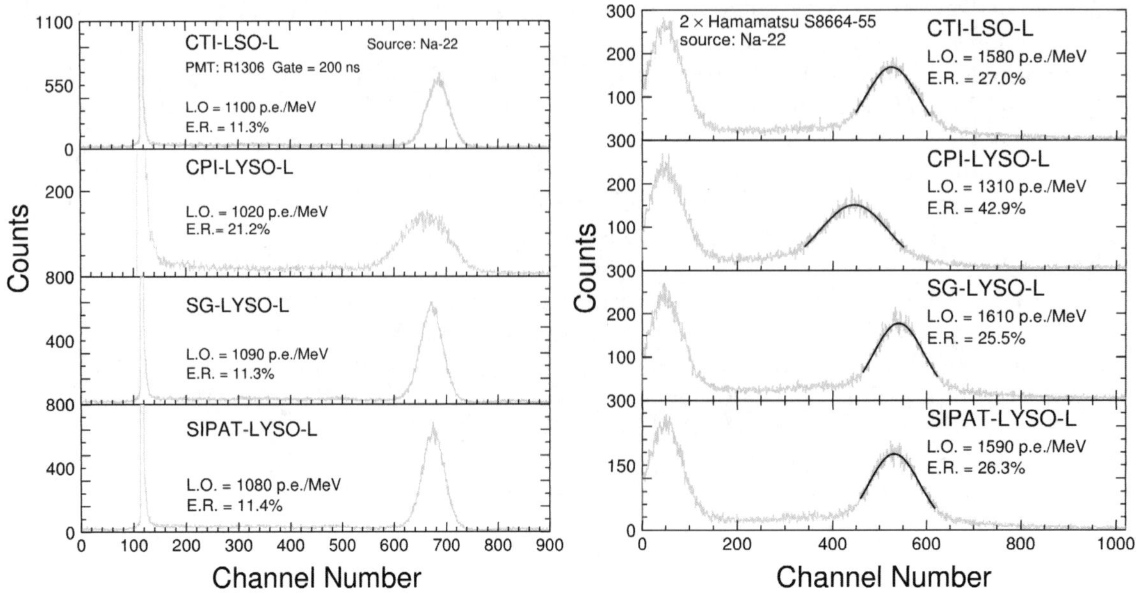

Figure 8. The spectra of 0.511 MeV γ-rays from a ^{22}Na source, measured by a Hamamatsu R1306 PMT (Left) and two Hamamatsu S8664-55 APDs (Right), with a coincidence trigger for four long LSO and LYSO samples from CTI, CPI, Saint-Gobain and SIPAT.

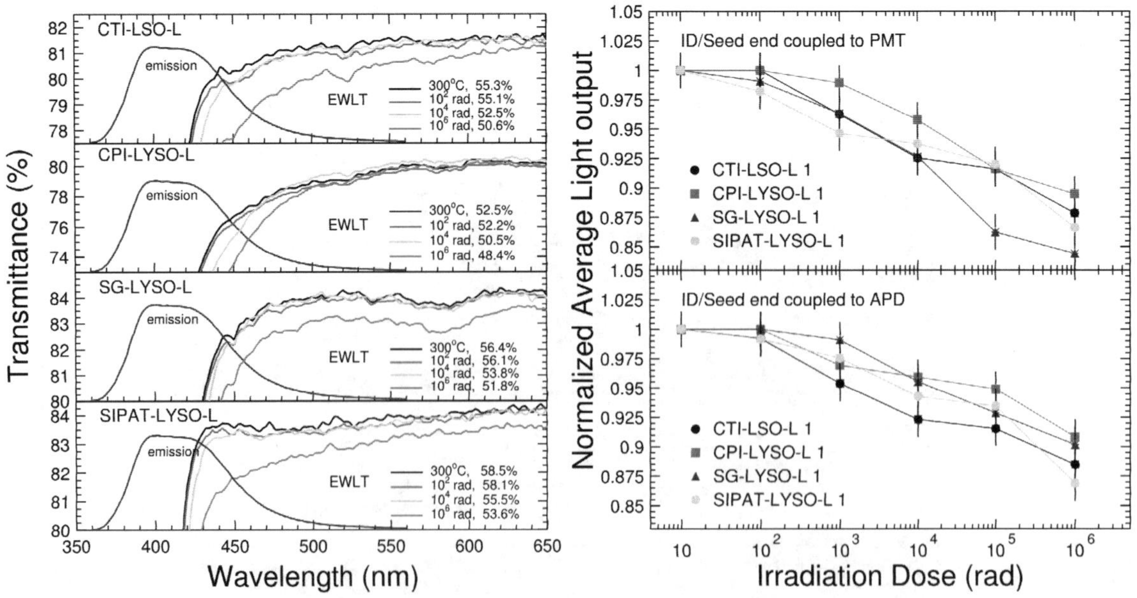

Figure 9. Left: The longitudinal transmittance spectra are shown as a function of wavelength in an expanded scale together with the photo-luminescence spectra for four LSO and LYSO samples before and after the irradiation with integrated doses of 10^2, 10^4 and 10^6 rad. Right: The normalized light output is shown as a function of the integration dose for four long LSO and LYSO samples with PMT (top) and APD (bottom) as the readout devices.

radiation environment of 15 rad/h and 500 rad/h for LSO and LYSO samples of $2.5 \times 2.5 \times 20$ cm^3 [7]. Figure 9 (Left) shows an expanded view of the longitudinal transmittance spectra for three samples before and after several steps of the γ-ray irradiation with integrated dose of 10^2, 10^4 and 10^6 rad. Also shown in the figure is the corresponding numerical values of the photo-luminescence weighted longitudinal transmittance ($EWLT$). Figure 9 (Right) shows the normalized average light output as a function of integrated dose for five long samples from various vendors. It is interesting to note that all samples show consistent radiation resistance with degradations of both the light output and transmittance at 10 to 15% level after γ-ray irradiation with an integrated dose of 1 Mrad.

4. Summary

In a brief summary, LSO/LYSO crystals are an excellent material for a total absorption electromagnetic calorimeter for a future high-energy physics experiment in a severe radiation environment. Assuming the same APD-based readout scheme used for the CMS PWO calorimeter, the expected energy resolution of a LSO/LYSO crystal based electromagnetic calorimeter would be

$$\sigma_E/E = \mathbf{2}\%/\sqrt{E} \oplus \mathbf{0.5}\% \oplus 0.001/E, \qquad (1)$$

where the stochastic term is dominated intrinsically by the fluctuation of shower leakage as indicated by GEANT simulations. This represents a fast calorimeter over a large dynamic range with very low noise. Thanks also to the LSO/LYSO low temperature coefficient, such a calorimeter is less demanding to the experimental environment. Because of the crystals excellent radiation hardness, a LSO/LYSO crystal calorimeter is capable of making precision measurement for electrons, photons and jets and thus provides a great physics discovery potential in a severe radiation environment, like the HL-LHC.

Acknowledgments

This work is partially supported by the U.S. Department of Energy Grant No. DE-FG03-92-ER40701 and the U.S. National Science Foundation Award PHY-0612805 and PHY-0516857.

References

[1] C. Melcher and J. Schweitzer, *Cerium-doped lutetium oxyorthosilicate: a fast, efficient new scintillator, IEEE Trans. Nucl. Sci.* **39** (1992) 502–505.
[2] D.W. Cooke, K.J. McClellan, B.L. Bennett, J.M. Roper, M.T. Whittaker and R.E. Muenchausen, *Crystal growth and optical characterization of cerium-doped* $Lu_{1.8}Y_{0.2}SiO_5$, *J. Appl. Phys.* **88** (2000) 7360–7362.
[3] T. Kimble, M Chou and B.H.T. Chai, *Scintillation properties of LYSO crystals*, in *Proc. IEEE Nuclear Science Symposium Conference* (2002).
[4] C. Cecchi. "A LYSO Calorimeter for a SuperB Factory ," in these Proceedings.
[5] F. Happacher , *CCALT: Crystal Calorimeter at KLOE2*, in these Proceedings.
[6] The Mu2e Experiment, see http://mu2e.fnal.gov/.
[7] R.H. Mao, L.Y. Zhang and R.-Y. Zhu, *Gamma Ray Induced Radiation Damage in PWO and LSO/LYSO Crystals*, Paper N32-5 in NSS 2009 Conference Record (2009).
[8] R.H. Mao, L.Y. Zhang and R.-Y. Zhu, *Effect of Neutron Irradiations in Various Crystal Samples of Large Size for Future Crystal Calorimeter*, Paper N32-4 in NSS 2009 Conference Record (2009).
[9] F. Nessi-Tedaldi, G. Dissertori, P. Lecomte, D. Luckey and F. Pauss, *Studies of Ceriun Fluoride, LYSO and Lead Tungstate Crystals Exposed to High Hadron Fluences*, Paper N32-3 in NSS 2009 Conference Record (2009).
[10] R.H. Mao, L.Y. Zhang and R.-Y. Zhu, *Optical and Scintillation Properties of Inorganic Scintillators in High Energy Physics, IEEE Trans. Nucl. Sci.* **NS-55** (2008) 2425–2431.
[11] D.A. Ma and R.-Y. Zhu, *Nucl. Instr. and Meth.* **A333** (1993) 422.
[12] J.M. Chen, R.H. Mao, L.Y. Zhang and R.-Y. Zhu, *IEEE Trans. Nucl. Sci.* **54** (2007) 718.
[13] J.M. Chen, R.H. Mao, L.Y. Zhang and R.-Y. Zhu, *IEEE Trans. Nucl. Sci.* **54** (2007) 1319.

Journal of Physics: Conference Series **293** (2011) 012005 doi:10.1088/1742-6596/293/1/012005

Calorimetry at 10mK

Fernando Ferroni

Dipartimento di Fisica, Universita' di Roma Sapienza & INFN Roma, Roma I-00185, Italy

E-mail: `fernando.ferroni@roma1.infn.it`

Abstract.
The search for neutrinoless double beta decay requires extreme technologies since the process searched is nearly impossibly rare. Calorimetric technique is one of the best promising line of attack to the problem expecially when the materials are operated at ultra-low temperatures as bolometers. This allows, at least in principle, the use of a large variety of different isotopes and a control of the main parameter dominating the game: the background index. In this paper we will discuss the CUORE calorimeter and the industrial processes with associated QC/QA for the production of radiopure TeO_2 crystals and mention other type of materials that could suit even better the goal of this fundamental yet extremey difficult measurement.

1. Introduction

In the field of fundamental particle physics the neutrino has become more and more important in the last few years, since the discovery of its mass. In particular, the ultimate nature of the neutrino (if it is a Dirac or a Majorana particle) plays a crucial role not only in neutrino physics, but in the overall framework of fundamental particle interactions and in cosmology.

One shall notice that the ambiguity inherent to the measurement of squared mass differences in the oscillation process leaves two possibilities for the hierarchical mass arrangements of neutrinos. There could also be a common baseline. The measured values of the neutrino mass differences are indeed tiny. Many orders of magnitude smaller than the mass of the lightest of charged leptons, the electron. Long ago E. Majorana formulated an elegant and minimal description [1] of the neutrino field. The question is whether Nature makes use of this simplicity. The only way to disentangle its ultimate nature is to search for the so-called Neutrinoless Double Beta Decay ($0\nu2\beta$) [2]. The DBD are extremely rare processes. In the two neutrino decay mode their half- lives range from $T_{1/2} \simeq 10^{18}y$ to $10^{25}y$. The rate for this process will go as

$$1/\tau = G(Q,Z)|M_{nucl}|^2 m_{\beta\beta}^2$$

where G is the easily calculable phase space factor and M is the challenging nuclear matrix element that is known [3] with still large uncertainties. The experimental investigation of these phenomena requires a large amount of DBD emitter, in low-background detectors with the capability for selecting reliably the signal from the background. The sensitivity of an experiment will go as

$$S^{0\nu} \propto a(\frac{MT}{b\Delta E})^{1/2}\epsilon$$

Isotopic abundance (a) and efficiency (ϵ) will end up in a linear gain, while mass (M) and time (T) only as the square root. Also background level (b) and energy resolution (ΔE) behaves as a square root. In the case of the neutrinoless decay searches, the detectors should have a sharp energy resolution, or good tracking of particles, or other discriminating mechanisms. The choice of the emitters should be made also according to its two-neutrino half-life (which could limit the ultimate sensitivity of the neutrinoless decay), according also to its nuclear factor-of-merit and according to the experimental sensitivity that the detector can achieve. One of the best technology for studying this extremely challenging problem is the bolometric one.

2. Bolometric technique

Bolometers are low-temperature-operated particle detectors which provide better energy resolution, lower energy thresholds and broader material choice than conventional devices. They can be thought as perfect calorimeters, able to thermalize fully the energy released by a particle. The best features of bolometric detectors are:

- They can contain the candidate nuclei with a favorable mass ratio and be massive
- They exhibit good energy resolution. This parameter is crucial since the signal is a peak in the energy spectrum of the detector positioned exactly at the Q-value of the reaction. This peak must be discriminated over the background and therefore has to be narrow.
- They can be built in a way to be characterized by low intrinsic background.

Up to now, the choice for bolometers as $0\nu 2\beta$ detectors has fallen on natural TeO_2 that has very good mechanical and thermal properties together with a very large (27% in mass) content of the candidate ^{130}Te. The success of CUORICINO [4] and the excellent prospects for CUORE [5] are based on this approach. Bolometer-based $0\nu 2\beta$ searches require however extremely low levels of background. Even if you reduce drastically the one arising from radioactive contaminants in the bolometers themselves, you still have the problem of the surrounding materials. Surface contamination is of particular concern. In fact, α particles arising from radioactive contaminations located on the surfaces of the detector or of passive elements facing them can lose part of their energy in a few microns and deposit in the detector an energy close to that of the signal, thus mimicking a signal event. A realistic possibility to improve substantially the background rejection capability is to join the bolometric technique proposed for the CUORE experiment with the bolometric light detection technique used in cryogenic dark matter experiments. The bolometric technique allows an extremely good energy resolution while its combination with the scintillation detection offers an ultimate tool for background rejection. Preliminary tests on several double-beta-decay detectors have clearly demonstrated the excellent background rejection capabilities that arise from the simultaneous, independent, double readout (heat + scintillation). One of the great advantages of the bolometric technique is that most of the candidate elements for DBD searches can form chemical compounds suited to be grown as crystals.

3. The TeO_2 show case

In principle, almost any crystal containing DBD candidates could be used in cryogenic experiments. Tellurite crystals have the advantage of very good thermal and mechanical properties. Furthermore, the high natural abundance of ^{130}Te (33.8%) in practice eliminates the need for the extremely costly enrichment processes. TeO_2 crystal is particularly convenient for the use in a cryogenic particle detector because of its thermodynamic char- acteristics. TeO_2 is a dielectric and diamagnetic crystal with a specific heat at low temperature following the Debye law ($\sim (T/T_D)3$). The crystal absorber (1000 modules, cubic, 125cm^3, 790gr each) forms the core of a cryogenic bolometer: the CUORE experiment, preceded by its demonstrator Cuoricino. The energy deposited by a particle is measured through the temperature increase of the crystal. The numerology says that an energy deposit of 1 MeV will produce a temperature

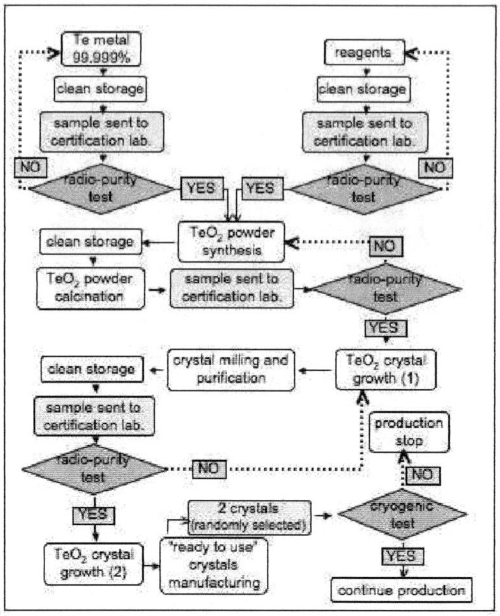

Figure 1. Sketch of the protocol for the production of the TeO_2 crystals of the CUORE experiment at SICCAS.

increase of 100 μK. A high level of crystal perfection is requested for high sensitivity bolometers because the presence of bulk defects like bubbles and metallic inclusions or veil sandcracks may drastically reduce the signal intensity and the energy resolution of such detectors. Mechanical perfection is however only a part of the story. In Double Beta Decay calorimetric experiments events can be identified by their energy spectrum. In practice the sensitivity of the experiment is limited by the presence of environmental radioactivity and radioactive contaminants in the detector. Therefore,it is critical for the detector to be be free of any contaminant that can mimicthe signal. The contamination may come from long-lived, naturally occurring isotopes, such as ^{238}U, ^{232}Th, ^{40}K and their daughters and from cosmogenic activation of the detector after production. To minimize the influence of long-lived nuclei, great care must be devoted to the selection of all material sand ancillaries used forthe preparation of the crystals. A sketch of the production protocol [6] used at SICCAS for the production of the TeO_2 crystals of the CUORE experiment is shown in Fig. 1.

The performances of the final product can be seen looking at the energy resolution (Fig. 2) in the final bolometric test that is used, on a sample basis, to evaluate the quality of the on-going production at SICCAS.

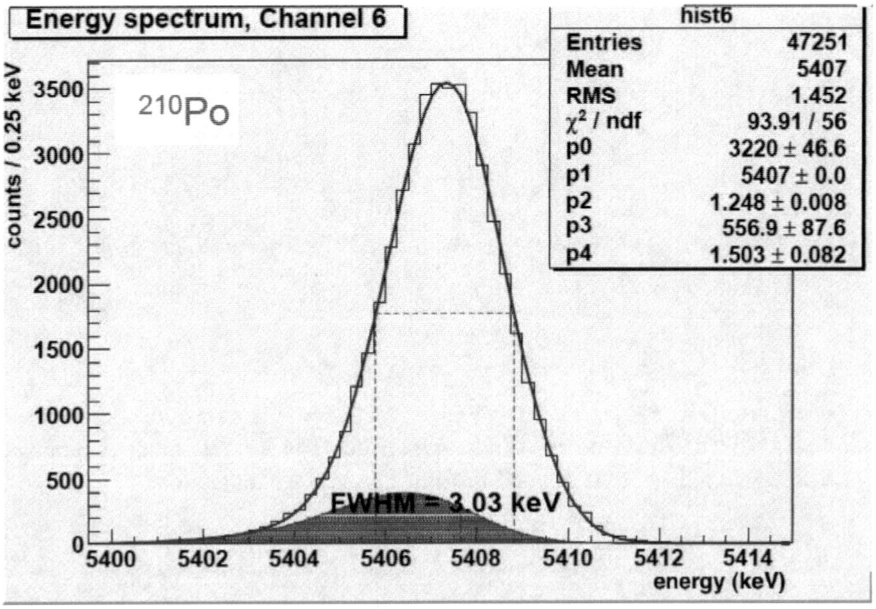

Figure 2. Energy spectrum in a TeO_2 crystal of the CUORE experiment at the ^{210}Po α line.

4. Scintillating bolometers

Although TeO_2 crystals are extremely good bolometers there are two problems that cannot be avoided when making a DBD search with them. The first is that the transition energy, the so-called Q-value is below the last important photon line coming from the U, Th chains, the ^{208}Tl. The second and most important is that the shape of the bolometric response does not allow any discrimination of α particle signals with respect to β/γ ones. The problem is best elucitated in Fig. 3.

Scintillating bolometers to the search for $0\nu2\beta$ bring in an enormous added value, by allowing the use of high Q-value candidates first, and second by providing a substantial α/β discrimination power. When the energy absorber in a bolometer is an efficient scintillator at low temperatures, a small but significant fraction of the deposited energy (up to a few %) is converted into scintillation photons, while the remaining dominant part is detected as usual in the form of heat. The simultaneous detection of the scintillation light is a very powerful tool to identify the nature of the interacting particle. In particular, α particles can be discriminated (see Fig. 3) with respect to beta and gamma interaction because of the different quenching factor (QF).

A scintillating bolometer for $0\nu2\beta$ is no new concept in the field and was proposed more than one decade ago for ^{48}Ca with CaF_2 crystals [7]. Nature has kindly provided us with a few isotope candidates presenting a transition energy higher than 2615 keV and forming chemical compounds suitable for the growth of large scintillating crystals, which proved to work as highly performing bolometers as well. The most suited are based on Cd, Mo and Se with the drawback of a need for an isotopic enrichment that brings their natural abundances (less than 10%) to a much higher

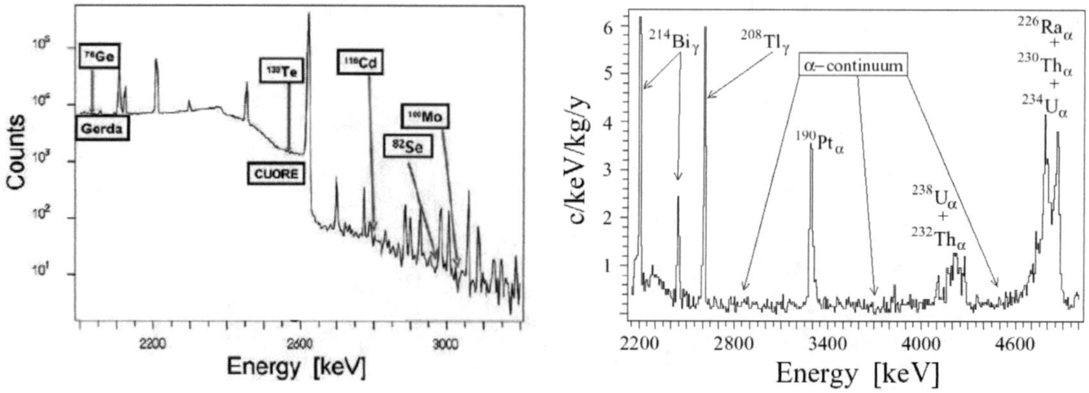

Figure 3. Left: Radiative nuclear transitions. It clearly shows that above the ^{208}Tl this contribution to the DBD background becomes negligible. Right: Cuoricino background in the DBD region and above. It clearly shows the dominance of degraded α's.

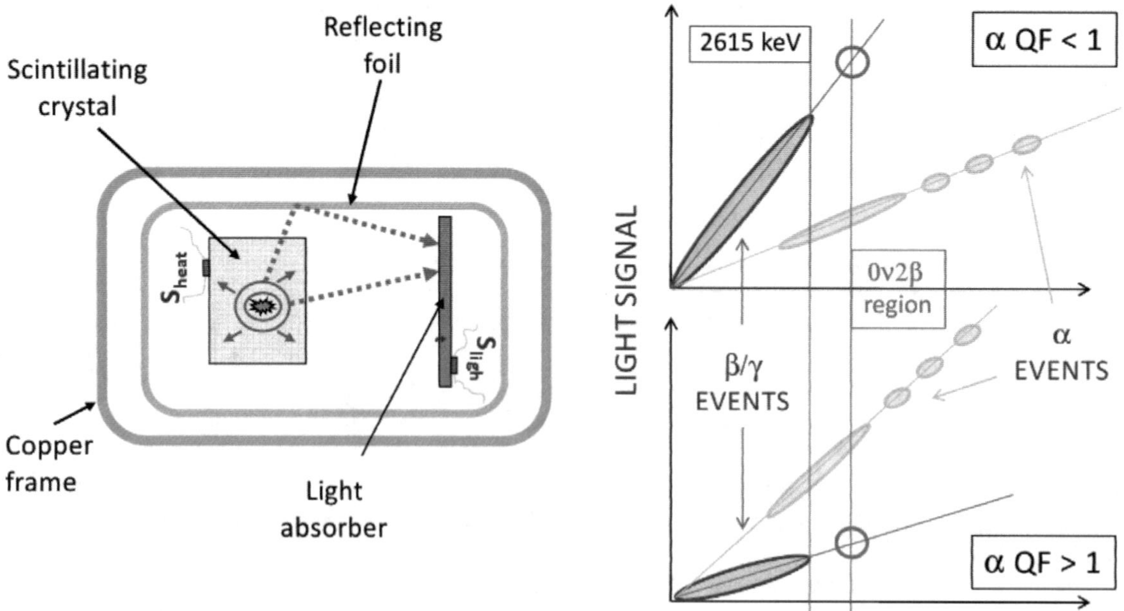

Figure 4. Left: schematic structure of a double read-out scintillating bolometer. All the basic elements of the detector are shown. Right: schematic scatter plots of light signals vs. heat signals corresponding to events occurring in the scintillating bolometer. Cases with QF larger or smaller than 1 are illustrated. In both circumstances α events can be efficiently rejected and the $0\nu2\beta$ signal region is background free.

value. This means in practice that although results [8] obtained by using $CdWO_4$ have basically proven (see Fig. 5) the power of this approach the final choice for a practical experiment cannot make use of this crystal. Cd presents some drawback like the residual, unavoidable presence of ^{109}Cd and ^{113}Cd that could be too much of a nuisance even after enrichment. Mo does not offer at this point any convincing cristalline compound and it is an element heavily contaminated by the presence of U, Th. When applying different materials to this scheme and considering all the relevant elements (scientific, technical, economical), the final balance is in favour of ^{82}Se (ZnSe crystals) although ^{116}Cd still deserves attention.

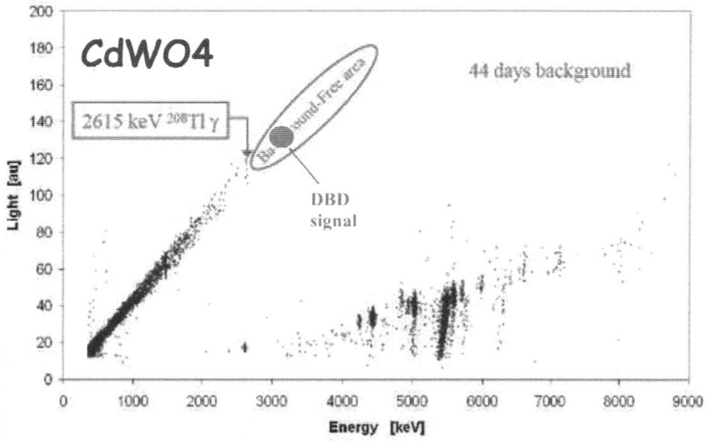

Figure 5. Results from a run on a $CdWO_4$ crystal with double (heat and light) readout

5. LUCIFER: a prototype of a double read out cold calorimeter

One of the most striking features of ZnSe is the abnormal QF, higher than 1 unlike all the other studied compounds. Although not really welcome, this unexpected property does not degrade substantially the discrimination power [9] of this material compared to the others and makes it compatible with the requirement of a high sensitivity experiment. An additional very useful feature is the possibility to perform α/β discrimination on the basis of the temporal structure of the signals, both in the heat and light channel (see Fig. 6).

The detector configuration proposed for LUCIFER resembles closely the one selected and extensively tested for CUORE, with an additional light detector, designed according to the recipes developed during the scintillating-bolometer R&D and consisting of an auxiliary bolometer, opaque to the light emitted by the ZnSe crystals (see Fig. 7). A preliminary version of the LUCIFER structure consists of an array of 48 crystals, divided in 12 elementary modules with 4 crystals each arranged in a tower, which would fit exactly the experimental volume of the Cuoricino cryostat. This structure assumes that a single light detector, quite large in order to monitor four scintillating crystals simultaneously, is sensitive enough to perform efficiently the α/β discrimination. The total detector mass would be 25 kg, with about 14 kg of enriched material assuming an enrichment level of 97% A preliminary evaluation of the LUCIFER sensitivity can be made on the basis of the structure discussed above and on the background expectations after α/β rejection. Assuming 5 year live time, an energy window of 20 keV and a specific background coefficient of 10^{-3} counts/keV/kg/y, less than a few background counts are expected in the region of interest (the transition energy for ^{8}2Se is 2995 keV). The most

Journal of Physics: Conference Series **293** (2011) 012005 doi:10.1088/1742-6596/293/1/012005

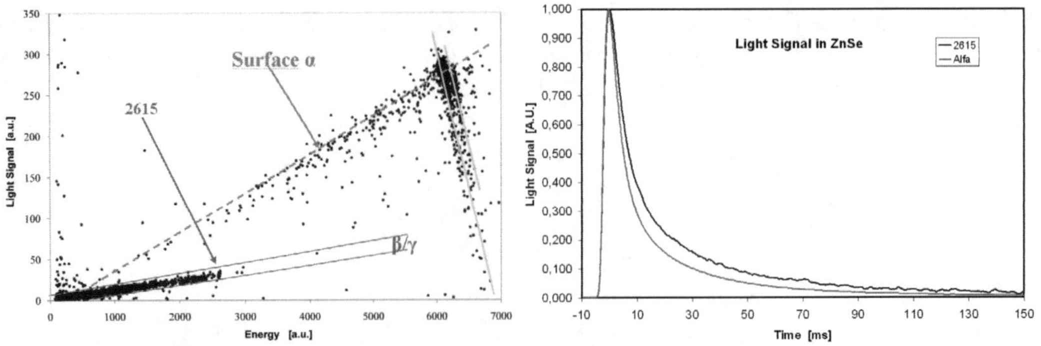

Figure 6. Results from a run on a ZnSe crystal with double (heat and light) readout exposed to radiactive sources. Left: scatter plot Light vs. Heat. Right: Decay time of the scintillation light for α's and ^{208}Tl line.

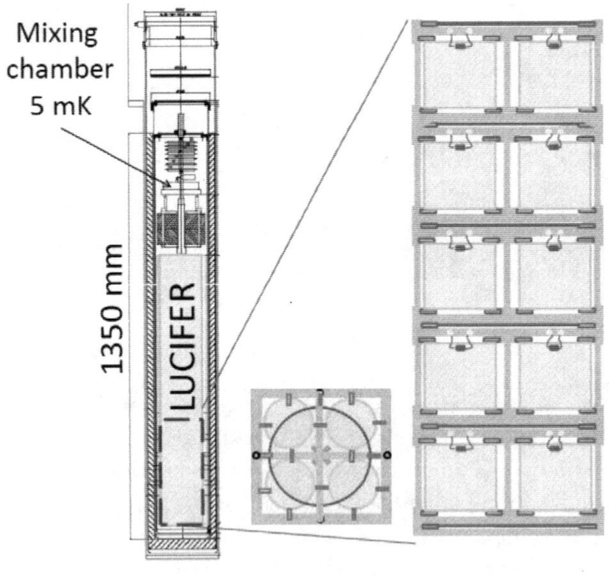

Figure 7. Schematics of Lucifer detector. Left: Cuoricino cryostat with Lucifer inserted. Center: Top view of 2×2 crystal plane with Ge light detector on top. Right: Side view of the detector array

important goal for LUCIFER is to be a demonstrator of the scintillating bolometer technology, with a significant mass and a full test of all the critical elements of this approach:

- large scale enrichment
- efficient chemical purification meeting radioactive requirements
- large size crystals grown with high efficiency in using the precious (100$/gr) material
- background rejection invesigated in many modules simultaneously operated.

It has the ambition to indicate the way to the experiment for the search of $0\nu2\beta$ able to span over the whole inverted hierarchy region.

6. Acknowledgments

The project LUCIFER has received funding from the European Research Council under the European Unions Seventh Framework Programme (FP7/2007-2013/) ERC grant agreement n. 247115.

References

[1] Majorana E , *Il Nuovo Cimento* 1937 **14** 171
[2] Avignone F T, Elliot R S and Engel J 2000 *Rev. Mod. Phys.* **80** 481
[3] Simkovic F, Faessler A, Muther H, Rodin V and Stauf M 2009 *Phys. Rev.* C **79** 055501
[4] Arnaboldi C et al. 2008 *Phys. Rew.* C **78** 035502
[5] Ardito R et al. 2005 *Preprint* hep-ex/0501010
[6] Arnaboldi C et al. 2010 *Preprint* cond-mat.mtrl-sci/0501010
[7] Alessandrell A et al. 1998 *Phys. Lett.* B **420** 109
[8] Arnaboldi C et al. 2010 *Preprint* nucl-ex/1005.1239
[9] Arnaboldi C et al. 2010 *Preprint* nucl-ex/1006.2721

Journal of Physics: Conference Series **293** (2011) 012006 doi:10.1088/1742-6596/293/1/012006

Optimization of Crystals for Applications in Dual-readout Calorimetry

Gabriella Gaudio
on behalf of the Dream Collaboration
E-mail: gabriella.gaudio@pv.infn.it

Abstract. Dual-Readout Calorimetry is a promising new technique for high resolution hadron and jet calorimetry. It is based on simultaneous measurements of the scintillation and Cherenkov light generated in the shower development process. Due to the fact that the Cherenkov light is only produced by the electromagnetic shower component, the relative contribution of this component to the signals can be measured on the event by event basis, resulting in reduction of fluctuations. This leads to an important improvement in the hadronic calorimeter performance. Further improvement on both the electromagnetic and hadronic resolution can be achieved by using homogeneous, dense crystals. This reduces both the sampling fluctuations and the quantum fluctuations. We present a systematic study of lead tungstate crystals doped with a small fraction of molybdenum, varying between 0.1% and 5% and exploring different readout configurations.

1. Introduction

We have performed systematic studies of lead tungstate crystals ($PbWO_4$) doped with a small fraction of Molybdenum (Mo). These crystals were exposed to a beam of 50 GeV electrons and the signals were unraveled into scintillation and Cherenkov contributions, using the time structure of the signals and/or different types of transmission filters. These studies were carried out in view of the possible application of such crystals in dual-readout calorimetry. Key elements for application of the dual readout technique are a clear Cherenkov/Scintillation separation, a good response uniformity and a high light yield, in order to reduce contributions to the resolution due to p.e. statistics. Differences among scintillation and Cherenkov lights which can be exploited in the signal separation are listed in Table 1.

2. The Molybdenum doped Lead Tungstate Crystals

For this systematic study we developed 5 crystals with different Mo concentrations, namely $0.1\%, 0.2\%, 0.3\%, 1\%, 5\%$, among which the last two were already tested in a previous test beam [1]. The effect of Mo on $PbW0_4$ crystals is to shift the scintillation spectrum to longer wavelengths and to achieve a longer decay time. The first effect is illustrated in Figure 1. Unfortunately, Molybdenum has a detrimental effect on crystal transparency at short wavelength, as shown in Figure 2 where UV self-absorption edge clearly depend on doping fraction. The scintillation light is selected using a yellow filter (GG495) which transmits only wavelengths larger than 495 nm, while on Cherenkov side we used three short-pass filters: UG11

Table 1. Properties of Cherenkov and Scintillation lights exploited in signal separation for dual-readout technique.

Properties	Cherenkov	Scintillation
Angular distribution	Light emitted at a characteristic angle by the shower particles that generate it	Light emission is isotropic: excited molecules have no memory of the direction of the incoming particle
Time structure	Instantaneous, short signal duration (few ns)	Characterized by one or several time constants. Long tails are not unusual (slow component)
Optical spectra	λ^{-2} spectrum	Strongly dependent on the crystal type, usually concentrated in a (narrow) wavelength range

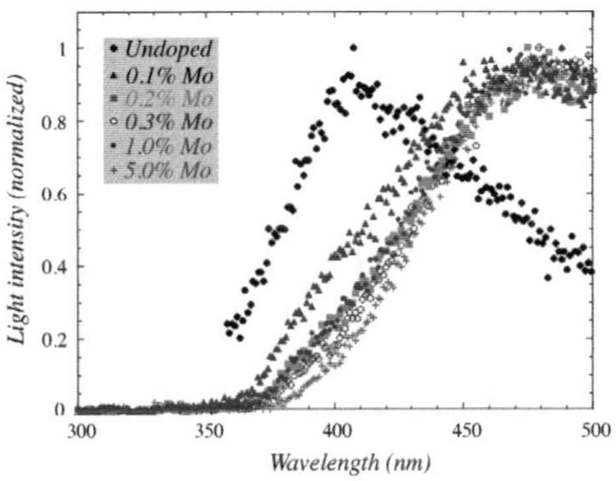

Figure 1. Normalized emission spectra for $PbWO_4$ crystals doped with different fractions of Molybdenum, measured with radioluminescence.

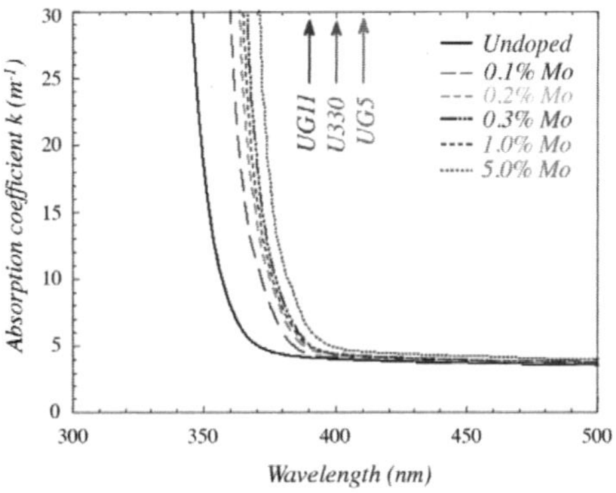

Figure 2. Absorption coefficient as a function of wavelength for $PbWO_4$ crystals doped with different fractions of Molybdenum.

(cut-off $\sim 390\ nm$), U330 ($\sim 400\ nm$) and UG5 ($\sim 420\ nm$)[1] More details on this measurements can be found in [2]

3. Figures of merit

Figures of merit have been identified in order to compare different configuration of optical filters and Mo dopant concentration. Purity of Cherenkov and scintillation lights can be obtained by measuring C/S ratio over an angular scan performed on the crystals, while response uniformity as function of the position is derived by a measurement of the signal attenuation. Light yield is also measured.

3.1. C/S ratio

In order to evaluate the achievable separation power the ratio of C/S at the Cherenkov angle (30° in the data taking configuration) and C/S at the anti-Cherenkov angle (-30°) are measured. In Figure 3 the dependence of these ratios as function of the Mo concentration is shown: the

[1] The UG11, UG5, and GG495 were produced by Schott, the U330 filter by Hoya.

lower the Mo concentration, the shorter the wavelength at which the scintillation emission starts (see Figure 1), causing more scintillation contamination in the Cherenkov signal and therefore giving a lower C/S ratio. The filter cut-off effect is shown in Figure 4: for lower filter cut-off, the scintillation contamination is smaller and therefore one obtains a larger C/S ratio.
A word of caution should be said about these plots. Each of these crystals was the result of a completely separate production process. Even if the Mo concentration was the only parameter that was deliberately varied, other factors may have deviated as well. Looking at Figure 1 it can be noticed that the red shift of the emission spectrum in the 1% crystal was about the same as for the 0.2% crystal. That means that the scintillation contamination was similar for both crystals, as can be seen in the C/S results and in the behavior the the other figures of merit.

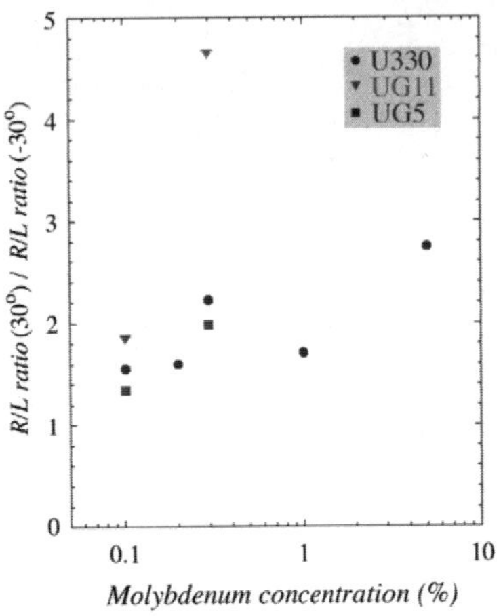

Figure 3. The ratio of the C/S signal ratios measured at 30° and -30° as function of the Mo concentration.

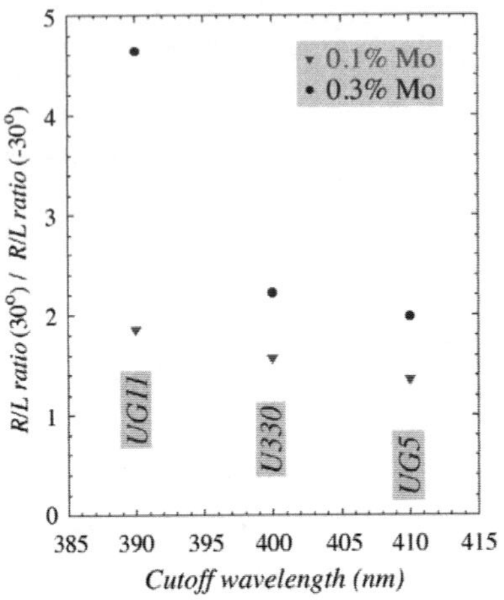

Figure 4. The ratio of the C/S signal ratios measured at 30° and -30° for the different filters used and two Mo concentration.

3.2. Light attenuation

As a consequence of the small effective bandwidth between the self-absorption edge and the filter cut-off (Figure 2), the Cherenkov signal is strongly attenuated. We measured the relative change in the Cherenkov signal over a distance of 10 cm of crystal by means of a longitudinal scan. We found that for smaller Mo concentration, the self-absorption edge is more blue-shifted (see Figure 2), and therefore the effective bandwidth is larger reducing in turn the effect on light attenuation (Figure 5). In Figure 6 the effect of the filter cut-off effect is shown: for cut-off at shorter wavelength, as seen, the Cherenkov signal effective bandwidth is narrower, and consequently the light attenuation is larger.

3.3. Cherenkov light yield

The calibration for the light yield measurement has been carried on using the scintillation signal as a measurement of the deposited energy and assuming from Monte Carlo that the average deposited energy is 0.578 GeV in the experimental configuration. The number of Cherenkov photoelectrons is determined by the fractional width of the corresponding Cherenkov signal distribution σ_{rms}/C_{mean}. Results for light yield are reported in Table 2 for different filter and Mo concentration configurations. It can be noticed that the smaller the Mo concentration, the lower the self-absorption edge (see Figure 2). This allows for a larger effective bandwidth for

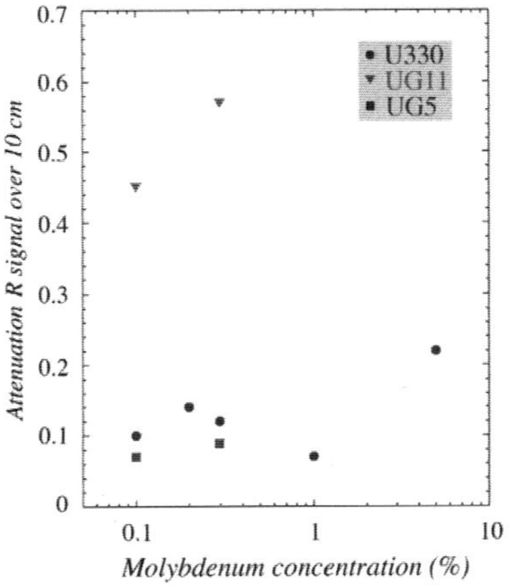

Figure 5. Relative change in the Cherenkov signal over a distance of 10 cm as a function of the Mo concentration in the $PbWO_4$ crystals.

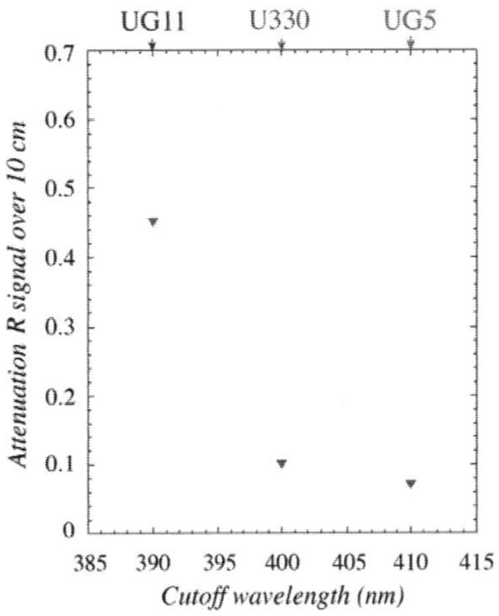

Figure 6. Relative change in the Cherenkov signal over a distance of 10 cm as a function of the optical filter cutoff in the $PbWO_4$ crystals.

the Cherenkov signal and therefore a larger light yield. From the point of view of the filter cut-off, the shorter it is, the smaller the Cherenkov signal effective bandwidth is and, in turn, the smaller the light yield.

Table 2. Cherenkov Light Yield measured for $PbWO_4$ crystals doped with different fractions of Mo. The results are given in p.e./GeV and measured with 50 GeV electrons traversing the crystal at an angle of 30°.

Mo concentration	UG11	U330	UG5
0.1	5	62	
0.2		57	
0.3	6	55	65
1		58	
5		38	

4. Conclusions

We performed a systematic study of $PbWO_4$ crystals doped with small amounts of Molybdenum and using different optical filters. This study was carried out to investigate if this type of crystal is suitable for dual readout calorimetry. An optimal configuration with a dopant concentration of 0.3% and the use of U330 filter shows a sufficiently high light yield (55 p.e./GeV), a reasonable low attenuation ($< 10\%/10\ cm$) while keeping a good Cherenkov to scintillation separation (2.4). A matrix of 7 crystals in this configuration will be tested in 2010 in a test beam.

References

[1] N. Akchurin et al, NIM A604 (2009)
[2] N. Akchurin et al, NIM A621 (2010)

Journal of Physics: Conference Series **293** (2011) 012007 doi:10.1088/1742-6596/293/1/012007

New development in high resolving power W-Si calorimeters

E.Kistenev

PHENIX Collaboration, Brookhaven National Laboratory, NY, Upton 1973

E-mail: kistenev@bnl.gov

Abstract.
The Relativistic Heavy Ion Collider (BNL) collides heavy nuclei and creates a strongly coupled medium at unprecedented density and temperature. Characteristic event structures may be efficiently selected with calorimeters, which can provide triggers on high-pT particles, "jets" of particles, or large transverse energy, along with precision measurements of the structures. The importance of calorimeters in studies of ultrarelativistic heavy-ion collisions was first recognized by W.Willis [1]. The key requirements are photon identification and measurements, and high resolving power to handle extreme occupancies common to this kind of interactions. We present a fully developed and beam tested concept of the W-Si sampling calorimeter built to this specifications. Novel features of this design are concepts of silicon micromodules, use of microconnectors for the silicon alignment purposes and passive signal summation to form readout towers. A prototype calorimeter was built in collaboration between BNL and a number of University groups from USA, Russia, Korea, Finland and Czech Republic and exposed to particle beams at CERN PS and SPS.

1. Introduction

Calorimetry is one area of the instrumentation in physics which is well developed and reasonably well understood. Library of performant simulation and analysis tools describing particle propagation in matter is enormous and constantly growing providing the experimentalist with nearly unlimited possibilities to verify his basic ideas in detector construction well before the detector is even considered for implementation.

New developments in tracking and particle identification resulted in RHIC and LHC experiments being is scale comparable to that of D0/CDF but much more performant. To continue, this trend must be supported by a similar progress in calorimetry which takes over energy and momentum measurements for particles in the momentum range of $\tilde{5}0$ GeV/c and up. Probably the biggest challenge for such a takeover are measurements of π^0's and photons which tend to produce nearly identical showers in traditionally structured hodoscopic calorimeters.

The FOCAL[2] (W-Si tracking electromagnetic calorimeter) detector concept was developed for the PHENIX [3] experiment at RHIC to study electromagnetic probes emanating from quark gluon plasma created in heavy ion collisions. A system prototype which has all main features of proposed detector was built in spring of 2009. The detector was bench tested and preassembled at BNL and delivered to CERN for beam testing in June 2009. The prototype was exposed in the electron and π-meson beams at the CERN PS (1-6 GeV/c momentum range) and SPS

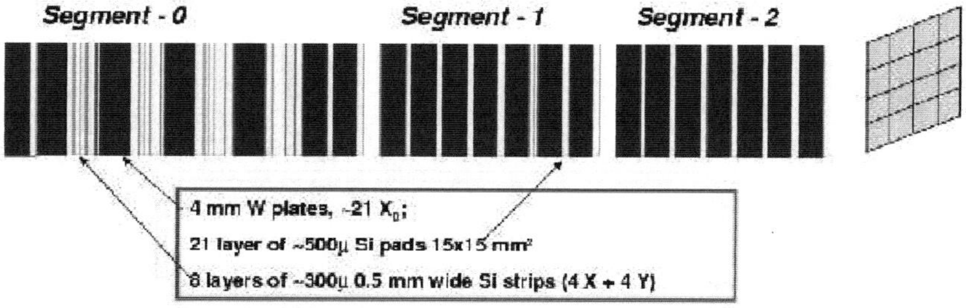

Figure 1. Longitudinal structure of a single calorimeter supertower (one sensor wide) showing the locations of the three calorimetric segments, S0, S1, and S2, and the high-resolution position sensitive layers.

(10-100 GeV/c momentum range). Results of the test beam experiments are summarized in Dr. Y.Kwon contribution to these Proceedings.

2. The FOCAL Design: A Tracking Calorimeter

The FOCAL is a highly segmented tracking calorimeter designed to reconstruct and identify electromagnetic signals at intermediate rapidities in close proximity to the production vertex. FOCAL is composed of an active preshower segment (S0) and two identically structured energy sampling segments (S1 and ES). Each of the three segments is $\sim 8\ L_{rad}$ deep and have seven energy sampling cells: a layer of tungsten followed by pad-structured ($15.5 \times 15.5\,\mathrm{mm}^2$) silicon sensors. Sampling cells 2-5 of the preshower segment also include two strip structured (pitch 500μm) position sensing sensors each (at a depths of $\sim 2X_0$ (L0), $\sim 3X_0$ (L1), $\sim 4X_0$ (L2) and $\sim 5X_0$ (L3)). Strip layers serve to count hits, measure hit-to-hit separation, and estimate the energy sharing between possible contributors to the high energy tracks constructed from clusters seen in the pad-structured calorimeter segments. The longitudinal structure of the calorimeter tower is sketched in Fig. 1.

The FOCAL is proposed for positioning around the beam pipe at 44 cm from the nominal collision point, on the pole of the PHENIX central magnet, and has a geometrical depth of 16 cm. The Molière radius of FOCAL is $\sim 14\,mm$.

An extensive set of performance measures based upon a standalone simulation chain implementing a free standing FOCAL in GEANT3 was developed to select its design parameters. The FOCAL design was then integrated into the complete PHENIX GEANT simulation, which includes all material in PHENIX, allowing for the study of all aspects of operating the FOCAL as part of the PHENIX experiment. The pattern recognition algorithms were further refined to optimize energy and position resolution, along with efficiency and sensitivity to shower shape measurements.

By design, the FOCAL segments are structured into mechanically non-projective towers. The sub-towers in the FOCAL (readout towers in individual segments) have an aspect ratio of ~ 2 (ratio of the sub-tower depth to its lateral size measured in the diagonal direction) which helps to keep occupancy in the individual segments low – even on the periphery of the detector. Figure 2 shows, for each of the three calorimeter segments, how the number of towers that contribute to a cluster depends on the impact angle. The tower cluster multiplicity per track rises by less than 50 % between minimum and maximum impact angles.

Digital projectivity due to longitudinal segmentation have a dramatic effect on the FOCAL

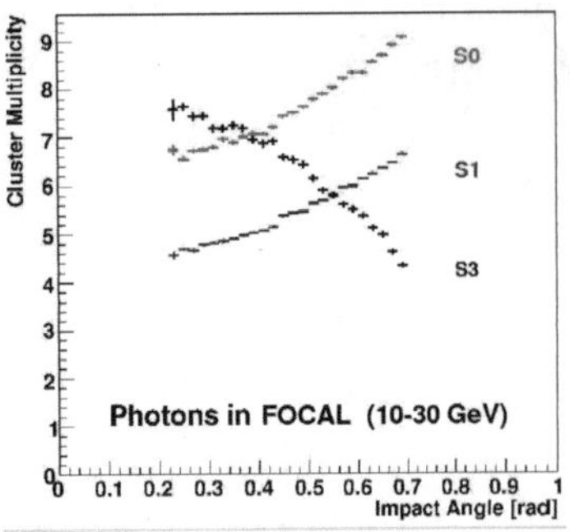

Figure 2. Angular dependence of the tower cluster multiplicity in three FOCAL segments.

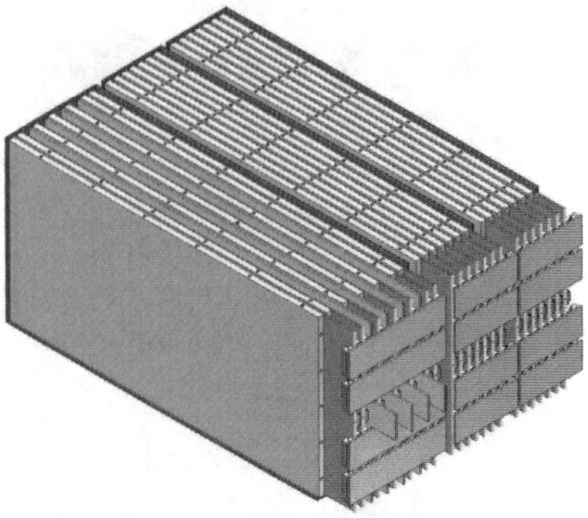

Figure 3. FOCAL building brick. Top skins removed.

performance as a particle identification detector, allowing it to effectively reject hadronic showers by comparing the measured longitudinal shower development to the parametrized electromagnetic shower shape.

2.1. Mechanical design

A modular design was developed for the FOCAL to allow for a relatively easy industrialization of the construction project. The calorimeter is built of "bricks", three bricks (preshower brick and two energy sampler bricks) stacked together form a "superbrick" (see Fig. 3). All bricks are a double $6.2 \times 6.2\,\text{cm}^2$ silicon sensor wide. Each brick has a tungsten plate facing upstream and a copper skin enclosing it on all other sides, save one for connections to external electronics.

2.2. Silicon sensors and concept of micromodules

The basic element of the *FOCAL* readout chain is micromodule which consists of a silicon sensor (pad or strip structured), an interconnect board hosting passive and/or active readout components, connections to a grounding plane, and ceramic backing tile(s) which protect sensors and insure micromodule rigidity. Components of the micromodule are glued together using silicon-based adhesive.

Two kinds of silicon sensors (and two different micromodules) are used: pad structured and strips. All silicon detectors in the FOCAL are $6.2 \times 6.2\,\text{cm}^2$ diced either from 4" or from 6" wafers 300 μm (strips) or 525 μm (pads) thick. Critical to the success of the experiment will be the reliable operation of these sensors in a high radiation environment. The estimates based upon ionization losses of collision related particles produced in pp interactions at $\sqrt{s}$=500 GeV at a luminosity of $10^{32} cm^{-2} s^{-1}$ result in radiation dose of ~10 Krad/year close to the beam pipe. Albedo neutrons from the calorimeter absorber will constitute 3 to 10% of the value of the MIP fluence but will induce ×10 higher damage effectively doubling the radiation dose to ~ 20 Krad/year of operation [4, 5]. In 10 years of running at RHIC luminosity the sensors in the central region of FOCAL will begin showing signs of radiation damage. To alleviate the possible consequences of exposure, all sensors in *FOCAL* are used in AC coupled mode (external RC decoupling network) thus allowing us to handle at least a ×10 increase in the leakage current. In addition, the mechanical design of the readout units allows for replacement of problematic sensors without collateral losses (although this cannot be done without moving calorimeter bricks to a "gray" repair area).

The silicon sensors for FOCAL are built using single-sided single-metal p+ on n− bulk silicon devices. A short summary of sensor parameters is given in Table 1.

Both pad and strip sensors have first been prototyped at ELMA (Russia), ON Semiconductor's (Czech Republic) then at ETRI (South Korea) and BARC (India).

2.3. Pad-Structured Readout Layers

An assembled and exploded (for clarity) micromodule is shown in Fig. 4.

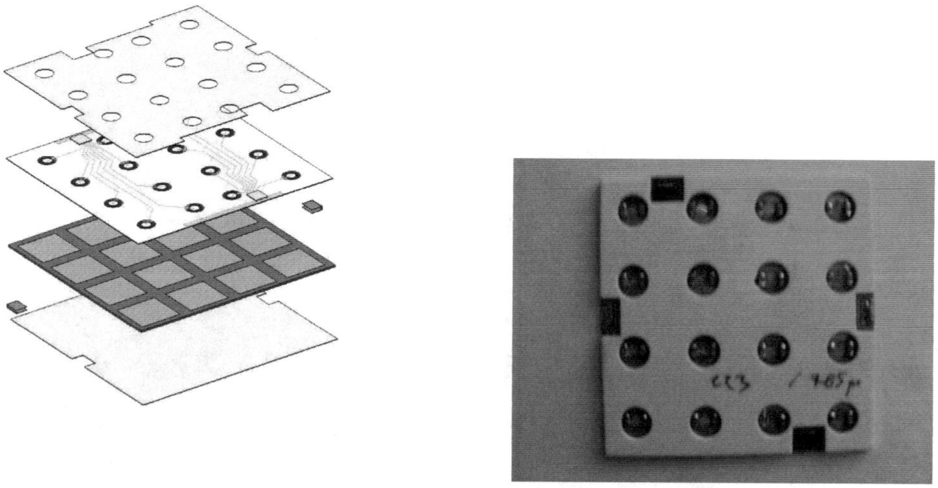

Figure 4. Pad-sensor micromodule sandwich components. From bottom to top: backing ceramic tile, silicon sensor, interconnect board, and ceramic spacer.

The bonding pads on the interconnect board are wire bonded to the aluminized pad centers

Table 1. FOCAL Silicon Sensor parameters

Specification	Pad-structured sensors	Strip sensors
Wafer thickness	525 μm	300 μm
Depletion voltage	100-120 V	80-100 V
Diode capacitance	25 pF	5 pF
Bias voltage	Full dep. V + 20 V	Full dep. V + 20 V
Leakage current	< 300 nA total, < 20nA/pad	< 1 nA per strip
Junction breakdown	> 300 V	> 300 V
Implant area	15×15 mm^2	
Al area	15.02×15.02 mm^2	0.45×61 mm^2.
Interpad (strip) capacitance	< 2 pF (pad-pad) or <8 pF (pad-all neighbors)	< 1 pF (strip-strip)
Maximum heat dissipation from the bulk material	<~ 50 mW / sensor	<~ 50 mW / sensor
Heat dissipation from on-the-sensor electronics	no major heat sources	0.5 W / sensor

through 3 mm diameter vias. Low-profile RC-network components are placed on the interconnect board close to the bonding point locations – hidden inside large diameter vias in the ceramic spacer separating the interconnect and carrier boards. Gold plated copper foil is bonded and glued to the aluminum plating on the sensor common side and soldered to the ground pad on the interconnect for the ground connection. Two 20 pin low profile (0.9 mm total height) connectors are installed close to the edges of the interconnect board to connect the sensor stack to traces on the carrier board which is conductively glued (grounded) to the tungsten plate.

The laminate of the 4 mm tungsten plate and carrier board with installed sensor micromodules form a readout unit (sampling cell) which is used throughout the whole calorimeter depth (see Fig. 5).

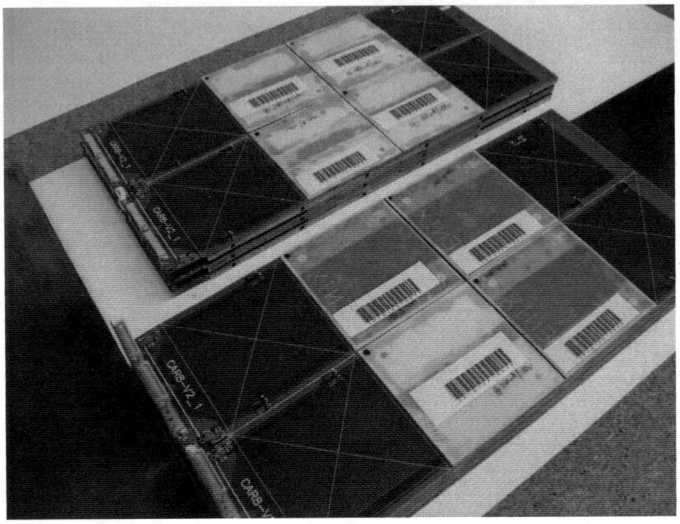

Figure 5. FOCAL pad-structured readout unit.

The interplate gap reserved for the carrier board with installed micromodules is 2.5 mm wide. 0.5 mm of this space is silicon, ~0.2 mm is FR4 (carrier board), and 0.2 mm is the FR4 interconnect board. 0.9 mm, partially filled with ceramic spacer, is used to accommodate connectors and RC-network distributing bias voltage to individual diodes. 0.4 mm is used by ceramic backing which protects the ground side of the silicon sensor. The remaining ~0.3 mm gap is to ensure that no pressure is applied to the surface of the sensors.

2.4. Strip Readout Layers
The readout system for the strip-structured silicon detectors is based on the SVX4 [6] chip developed at FNAL for the D0-detector upgrade. Data from SVX4 chips is collected by custom developed FEM's which are optically connected to Data Collection Modules (DCM's) which are part of the PHENIX DAQ system. Similarly to the pad sensors, strip sensors are used in AC-mode. They are connected to SVX4 inputs through a discreet component based RC network, which is located on the Strip Readout Control (SRC) boards glued to sensors (see Fig. 6).

3. Resolving close showers in PreShower segment
In order to understand the performance of the proposed design, we conducted extensive simulation studies, intended to closely mimic what we would expect to observe with the *FOCAL* in real collisions. To that end, a simulation chain was put in place, which included the detector response, reconstruction, and several stages of analysis. The reconstruction identifies

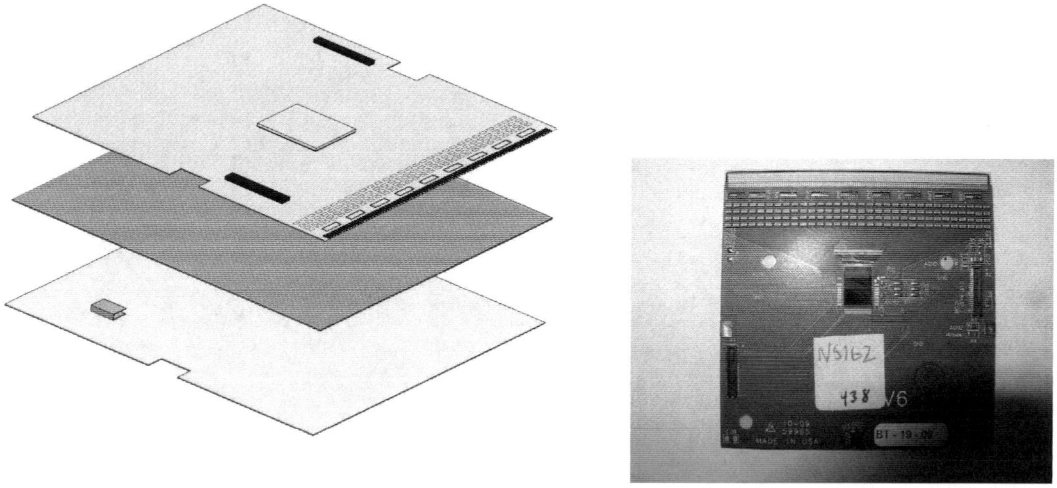

Figure 6. Strip-sensor micromodule sandwich components. From bottom to top: backing ceramic tile, silicon sensor, and interconnect board.

the measured energy clusters in the *FOCAL* subtowers and strip layers, associates clusters into tracks and assigns to each track a χ^2 value describing the quality of its fit to an electromagnetic hypothesis. It is this input data that is then turned over to the strip pattern recognition routines.

The strip tracking uses a Hough-type [7] methodology to determine the angular direction of photons within the *FOCAL*. Hough tracking transforms all the hit positions into Hough parameters — in the case of straight line tracking, these are simply the slope and the intercept. As the number of hit positions is small, the vertex position is used to form the intercept so the slope is the only free parameter available for tracking.

Strips are analyzed independently for x-type and y-type strips. The methodology used in the tracking was developed with emphasis on the difference between direct-γ and π^0's in the *FOCAL*.

All strips inside the regions-of-interest defined by track energies are scanned for distinct maxima in the hit distributions weighted directly by energies in the strips. The two tracks may merge in the x and/or y directions, one of the photons may be lost due to conversion probabilities or, in the case of very asymmetric decays, land outside region of interest, which leads to some π^0 contamination of the γ signal. In PHENIX geometry, π^0's below $E <\sim 5$ GeV are reconstructed as two single γ. In the energy range ($5 < E < 10$ GeV — Fig. 7), a prominent π^0 peak is clearly visible when two distinct maxima are found in strip distributions.

In the range $10 < E < 25$ GeV (two photons from π^0 decay are separated by $> 5mm$ in the first strip layer), the reconstruction is at its best. The two photon decay tracks are generally close enough together to be considered (by the strips, but not the pads) as two tracks, but far enough apart that finite resolution effects are not yet prevalent. In this region, a clear, dominant, π^0 peak is seen.

The finite strip resolution effects begin to come into play above $E >\sim 25GeV$ As the decay tracks from the π^0 become closer together, and the energy increases, the strip width determines a minimum separation at which the π^0 can be reconstructed correctly. The effects of this do not actually limit the reconstruction until $E >\sim 60$ GeV ($< 3mm$ two shower separation).

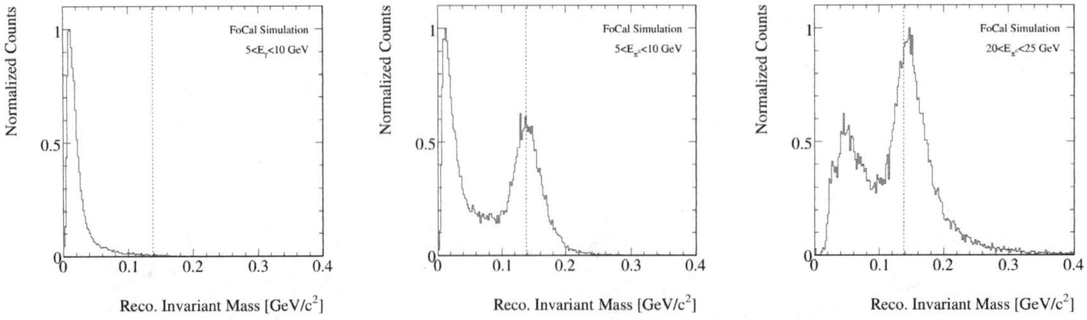

Figure 7. Reconstruction of single particles in *FOCAL*. $5 < E < 10$ GeV γ are on the left, $5 < E < 10$ GeV π^0 are in the center, $20 < E < 25$ GeV π^0 are on the right. Data are integrated over all input η and ϕ.

4. Summary

A conceptual design for the novel W-Si electromagnetic tracking calorimeter with fine lateral and longitudinal segmentation is presented. Simulated data are used to demonstrate its ability to resolve electromagnetic showers from π^0 decays down to 3 mm separation (from 60 GeV π^0's at 40 cm from production vertex). The prototype calorimeter implementing major ideas of new design was built and tested in the particle beams at CERN (see the talk of Dr. Y.Kwon from Younsei University, Korea) in 2009.

5. Acknowledgments

We thank the PHENIX Collaboration for continuous support of the FOCAL Project. We thank Physics Departments at BNL for their vital contributions. We thank ALICE Collaboration at CERN for providing material support to our group during beam tests and viable discussions.

We acknowledge support from the Office of Nuclear Physics in DOE Office of Science and NSF (U.S.A.), MEXT and JSPS (Japan), MSMT (Czech Republic), DAE and DST (India), NRF (Korea), MES, RAS, and FAAE (Russia), HIFP (Finland).

References

[1] Willis, W.J. 1981 *Preprint CERN-EP/81-21*; and *Proc. Workshop Future Relativistic Heavy Ion Experiments, Darmstadt*

[2] Bonvicini, V. and others 2005 *Silicon-tungsten calorimeter for the forward direction in the PHENIX experiment at RHIC, IEEE Trans. Nucl. Sci.* **52** 874

[3] PHENIX Collaboration 1993 *PHENIX Conceptual Design Report(PX20, BNL48922, internal report)*

[4] Spieler, H. 2005 *Radiation Detectors and Signal Processing* 85

[5] Spieler, H. 2005 *Semiconductor detector systems, Oxford, UK: Oxford Univ. Pr.* 489

[6] L. Christofek and others 2005 *SVX4 User's Manual* http://www-d0.fnal.gov/d0pub/d0_private/4252/m_svx4_user_guide.pdf"

[7] Hough,P.V.C. 1959 *Machine Analysis of Bubble Chamber Pictures, Proc. Int. Conf. High Energy Accelerators and Instrumentation*

XIV International Conference on Calorimetry in High Energy Physics (CALOR 2010) IOP Publishing
Journal of Physics: Conference Series **293** (2011) 012008 doi:10.1088/1742-6596/293/1/012008

Test beam study of the Si/W calorimeter for the PHENIX forward upgrade

Y. Kwon

PHENIX FOCAL collaboration,

Department of Physics, Yonsei Univ., 262 Seongsanno, Seodaemun-gu, Seoul, 120-749, Korea

E-mail:ykwon@yonsei.ac.kr

Abstract. The Si/W calorimeter prototype for the PHENIX forward upgrade has been tested using the electron beam of various energies available from PS and SPS at CERN. The measured energy shows a good linearity and the obtained energy resolution σ_E/E can be parametrized as $3.8\% \oplus 22\% / \sqrt{E(GeV)}$. The energy deposition profile measured for 6 GeV electron beam features a characteristic electromagnetic shower and agrees with the Monte Carlo simulation based on GEANT4. Position and angle of the incident electron can be reconstructed from the measured energy deposition profile. The resolution of the obtained position is 2 (mm).

1. Introduction

The PHENIX experiment at the relativistic heavy ion collider is considering an electromagnetic calorimeter FOCAL as the forward upgrade. The calorimeter is expected to replace the 20 cm thick copper absorber located at 40 cm forward from the nominal collision location. A compact calorimeter is a necessity for the upgrade due to the large particle multiplicities at forward and close proximity to the collision location. A Si/W sandwich sampling calorimeter is the natural choice to meet the needs [1].

The lateral and longitudinal segmentation of the calorimeter enables reconstruction of the electromagnetic shower initiated by the incident electron as well as its energy measurement. We can determine the incident position and angle of the beam electron up to a high precision to promote the calorimeter to a tracking device.

Possible enhancement in the PHENIX measurement capabilities achieved by FOCAL is under investigation along with the prototype performance study. General aspects of the detector FOCAL will be described in another write-up of the current proceedings [2]. This write-up focuses on the beam test of the prototype.

2. Experimental Setup

We describe relevant information and define nomenclatures for the later discussions.

2.1. Test beam

Test was performed at CERN in the July of 2009. Data collection for electron beams of lower energies (1, 2, 3, 4, 5, 6 GeV) was made at PS T10 site. Data collection for electron beams of higher energies (20, 30, 40, 50, 60, 75 GeV) was made at SPS H2 site. Total duration of data collection was 13 days.

Two homemade scintillators were positioned in front of the calorimeter and valid beam events required coincident signals from them in addition to the valid beam logic by the beam line detectors. Electron identification was also obtained from the beam line detectors. Calorimeter was rotated by 5°, 10°, 15°, 20°, and 25° to get electron beam with the corresponding incidence angle. 10k – 50k valid beam events were recorded for a given collection setting. Offline data analysis tagged and rejected about 20% of the recorded SPS events as the hadron event.

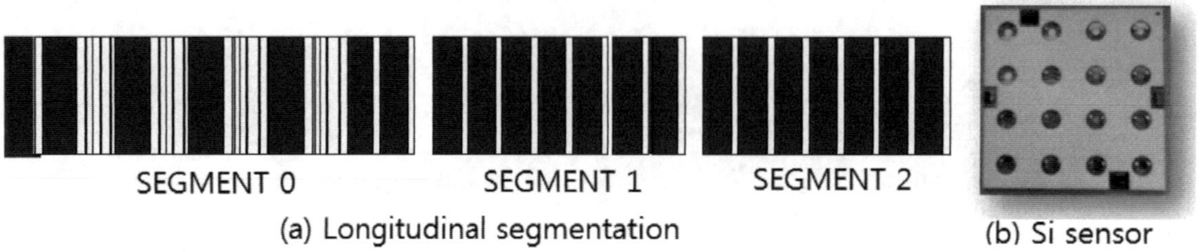

Figure 1 Longitudinal detector structure (a), and Si sensor unit (b).

2.2. Detector and readout electronics

The prototype detector is a Si/W sandwich sampling calorimeter and has the longitudinal and lateral structure relative to the beam direction. The longitudinal direction refers to *z*(beam) direction, and the lateral direction refers to *x*(horizontal) or *y*(vertical) direction perpendicular to beam direction.

Longitudinal structure of the detector is depicted in Fig. 1 (a). The detector contains 21 sandwich layers where each sandwich layer includes a 4 mm thick tungsten plate (black) and a 525 micron thick silicon pad sensor layer (yellow). Additional silicon strip sensor layers were inserted as depicted in figure, but not included in the current analysis. Fig. 1 (b) shows an actual Si sensor. Each Si sensor layer is patched and covered by this Si sensor unit with the areal size of 6 cm x 6 cm. Each Si sensor unit is segmented into 16 pads with the areal size of 1.5 cm x 1.5 cm each.

Signals from pads at the same lateral location in the first, the middle, and the last 7 sensor layers are electrically summed and labeled as segment 0, 1, and 2. The minimum detector granule *pad tower* is associated to this signal. The signal produced by a given *pad tower* is processed by the preamp a trans-impedance amplifier circuit, digitized by the 64 MHz flash ADC, and recorded with the valid beam trigger. ADC values recorded for 20 sample time bins around beam trigger are displayed in Fig. 2. The signal shape matches the preamp circuit simulation by the SPICE[5] program, and we attribute the signal shape to the preamp electronics.

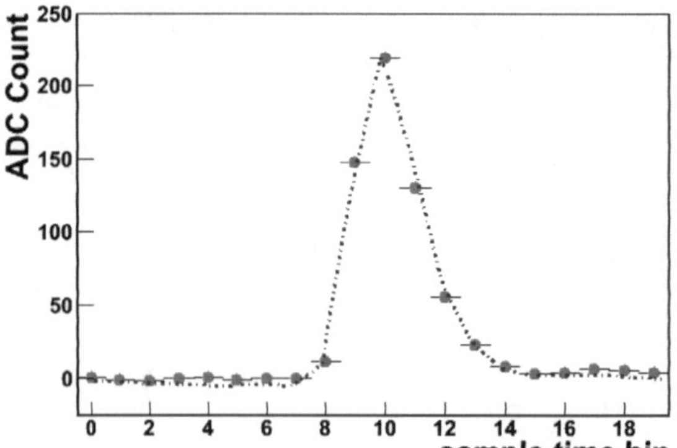

Figure 2 ADC signal distribution around beam trigger.

2.3. Calibration and Monte Carlo simulation

ADC values recorded for each *pad tower* is fit with the predetermined signal shape after pedestal subtraction (Fig. 2). The fit parameter is the amplitude, and this signal amplitude is the energy response of the *pad tower*. The energy response is considered to be proportional to the energy deposited in the *pad tower*. The proportionality constant converts the detector signal into the energy.

We ignore variation of the proportionality constants over *pad towers* after exploring two equalization methods. The approach to utilize the penetrating muon signals can see only crude variations due to the low preamp gain and the finite ADC dynamic range. The other approach was to seek the optimized proportionality constant set to minimize electron energy resolution. However, no significant improvement in electron energy resolution was observed from the minimization process.

We ignore variation of the proportionality constant over time. Sensor and readout electronics are expected to be stable over time, and the method to utilize penetrating muon signals is not accurate to determine the difference.

GEANT4 [3] program is utilized to simulate the energy deposition into *pad towers* by the electron beam. Simulation includes the detailed description of detector and all processes involved with energy deposition. Beam transport in the beam line is not included in the current simulation.

3. Calorimeter performance

We discuss key characteristics of the calorimeter, the energy measurement and the trajectory determination.

3.1. Energy measurement

We sum signal amplitudes obtained for all *pad towers* to get the *energy sum*.

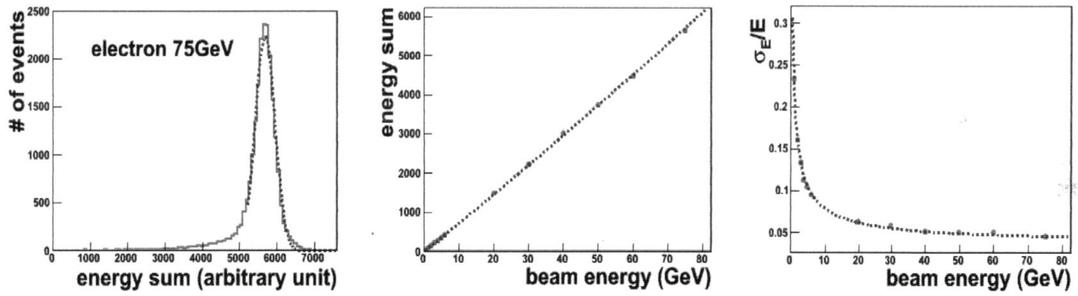

(a) (b) (c)

Fig. 3 Energy sum, the distribution (a), the relation to beam energy (b), and the resolution (c). Data (Red solid line), Fit (Blue dotted line)

3.1.1. Line shape

The distribution of *energy sum* for 75 GeV electron beam is depicted in Fig. 3 (a). The shape is characterized by a clear peak and the long lower energy tail. According to the simulation study, energy leakage such as detector inefficiency causes the tail. We determine the mean energy response E and the absolute resolution σ_E by fitting the peak with Gaussian distribution as shown. The Centroid (width) of the Gaussian distribution corresponds to E (σ_E).

Conversion constant to absolute energy is determined from the 75 GeV electron beam run by adjusting E to 75 GeV the nominal beam energy.

3.1.2. Linearity & resolution

The mean energy response E is shown for various test beam energies in Fig. 3 (b). E increases linearly with the beam energy and deviation from the line fit $y = ax$ is by less than 2.5% for all test setting.

Another fit by the function $y = ax(1 + bx)$ with the nonlinearity parameter b was also performed. Non-linearity parameter b was determined to be 0.085%/GeV.

The (fractional) energy resolution σ_E/E is shown for various test beam energies in Fig. 3 (c). Solid lines are fit by the function $\frac{\sigma_E}{E} = \frac{a}{\sqrt{E}} \oplus b$, where E is in GeV. The coefficient *a* (*b*) is 22% (3.8%). The contribution from stochastic fluctuation term dominates the resolution at low energies and is comparable to the one from constant term at the highest energy.

3.2. Trajectory determination

The measured energy deposition profile for electron beam features the electromagnetic shower in the longitudinal and the lateral direction. The calorimeter can determine the incident position and angle of the shower initiating particle accurately.

3.2.1. Lateral and longitudinal energy deposition profile

Fig. 4 shows the measured longitudinal and lateral energy response profile for 6 GeV electron beam. Measured energy sum are displayed for data (red) while mean energy depositions are displayed for simulation (blue).

The energies distributed in segment 0, 1, and 2 add up to 6 GeV the beam energy. Segment 0 contains the largest fraction of energy and segment 2 contains 1% of the total energy. Simulation suggests the electromagnetic shower initiated by the incident electron reaches maximum quickly within segment 0 and decays slowly along the longitudinal direction.

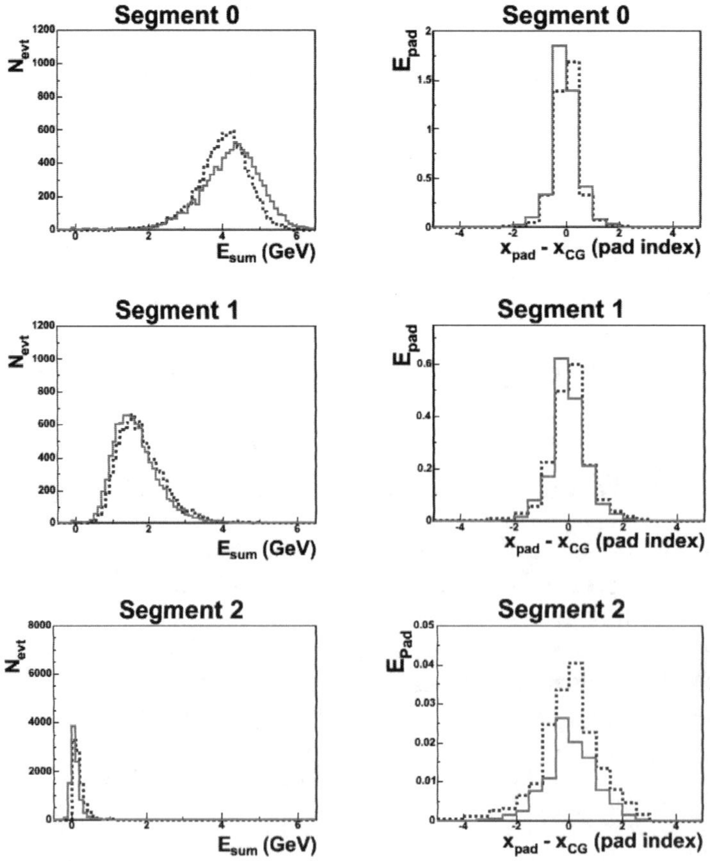

Figure 4 Longitudinal and lateral energy response (data in red) and deposition (simulation in blue)

The lateral energy response shows strong peak within ±1 pad unit from maximum. Pad size 1.5 (cm) was chosen to match the Molière radius of the calorimeter and the observed feature is consistent to the characteristics of the typical electromagnetic shower [4]. Simulated lateral energy deposition profile also agrees with data.

An interesting fact is noted. Agreement between the data and the simulation matches best when we add material of 0.3 radiation length thickness (1 mm thick tungsten plate) in front of the calorimeter. Detectors used to define valid beam adds up to 0.1 radiation length and was not included in the simulation.

3.2.2. Position and angle measurement

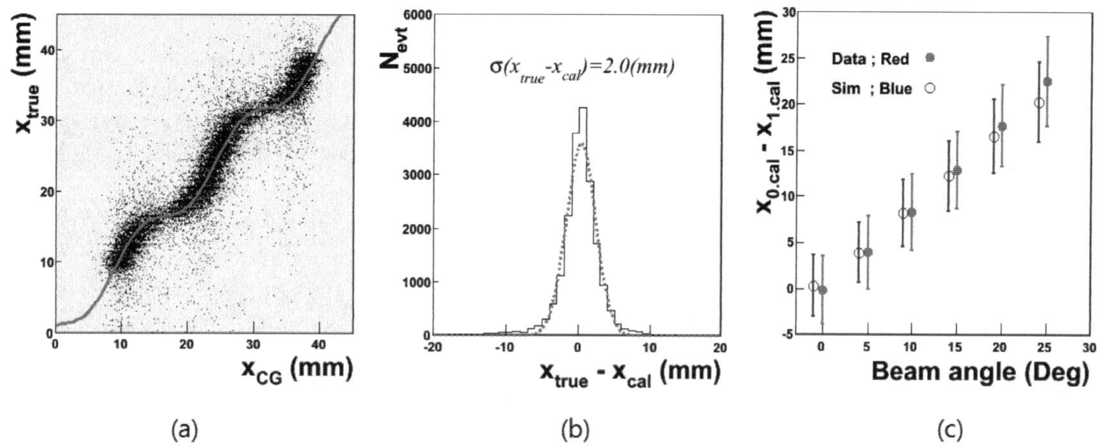

Figure 5 Relation between x_{CG} and x_{true} (a), x_{cal} resolution plot (b), beam angle and Δx (c). Simulation points for (C) were shifted left to avoid overlap with data points.

Each pad tower has a well defined lateral coordinate (x_i, y_i). We construct effective lateral position (x_{CG}, y_{CG}) of the electron beam by weighting pad tower positions with their energies and summing

$$x_{CG} \equiv \frac{\sum_i^{ALL} E_i x_i}{\sum_i E_i}, x_{0,CG} \equiv \frac{\sum_i^{SEG0} E_i x_i}{\sum_i E_i}, \text{ and } x_{1,CG} \equiv \frac{\sum_i^{SEG1} E_i x_i}{\sum_i E_i}.$$

$x_{0,CG}$ ($x_{1,CG}$) denotes weighted sums over the segment 0(1) *pad towers*. Similar definition is given to y_{CG}, $y_{1,CG}$, and $y_{2,CG}$.

x_{CG} has relation to the true x position x_{true}. An example is shown in Fig. 5 (a). The plot is generated from the simulation for normal beam incidence case and the correlation is between x_{CG} and the incident beam x position (= x_{true}). Inherent nonlinearity in x_{CG} can be corrected by empirically parameterized function $y = x + a\sin\left\{2\pi \cdot \frac{(x - x_i)}{1.5(cm)}\right\}$ and subsequent fit (red line). We get x_{cal} the estimator of x_{true} after the nonlinearity correction to x_{CG}. Fig. 5 (b) shows distribution of the difference between x_{true} and x_{cal} for the simulated events. Judging from this plot, we estimate 2.0 mm as the x_{calc} resolution.

Estimation of the incident angle is also possible using Δx the difference between $x_{0,cal}$ and $x_{1,cal}$. Fig. 5 (c) shows the relation between the incident angle and Δx for data (red) and for simulation (blue).

Simulation explains the relation clearly. $x_{0,cal}$ ($x_{1,cal}$) represents mean of the lateral shower profile at some effective longitudinal position $z_{0,eff}$ ($z_{1,eff}$) positioned inside segment 0(1). Simulation estimates Δz, the difference between $z_{0,eff}$ and $z_{1,eff}$, to be 42 mm for 6 GeV electrons. The Δx increases as the incident beam angle and hence the shower slope $\Delta x/\Delta z$ increases.

Some broadening occurs in Δx due to the finite resolution in $x_{0,cal}$ and $x_{1,cal}$. The amount of broadening is depicted as error bars in Fig. 5 (c) for data (red) and for simulation (blue). This agreement asserts the $x_{0,\,cal}$ ($x_{1,\,cal}$) variable resolution is properly simulated supporting the previously estimated x_{cal} resolution of 2 (mm).

4. Conclusions

We performed beam test of the Si/W calorimeter prototype for the PHENIX forward upgrade. Detailed study on energy measurement and tracking by the prototype calorimeter was made using electron beam.

Measured energy shows a good linearity and the achieved resolution meets the design goal. We measured the longitudinal and lateral energy response profile, which features a typical electromagnetic shower and agrees with the GEANT4 simulation. Incident electron position and angle can be calculated from the calorimeter measurements. Resolution for the incident x position is estimated to be 2.0 mm.

To achieve the results, we utilized simultaneous longitudinal and lateral segmentation of the prototype allowed for the Si/W calorimeter while adopting simple calibration process. We recognize these characteristics as the strength of current prototype.

References

[1] G.Barbiellini et al. Energy resolution and longitudinal shower development in a Si/W electromagnetic calorimeter, Nucl.Instr. and Meth.in Phys. Res. A 235(1985), 55
G.Ferri et al. The structure of the lateral shower development in Si/W and Si/U calorimeters, Nucl.Instr. and Meth.in Phys. Res. A 273(1988), 123

[2] See contribution by E. Kistenev

[3] S. Agostinelli et al. GEANT4-a simulation toolkit, Nucl.Instr. and Meth.in Phys. Res. A 506(2003), 250
J.Allison et al. Geant4 developments and applications, IEEE Transactions on Nuclear Science 53(2006), 270

[4] C. Amsler et al., The Review of Particle Physics, Physics Letters B667(2008), 1

[5] http://en.wikipedia.org/wiki/SPICE

Journal of Physics: Conference Series **293** (2011) 012009 doi:10.1088/1742-6596/293/1/012009

Electromagnetic calorimeters of the CMD-3 detector

Denis Epifanov [1]

Budker Institute of Nuclear Physics
Academician Lavrentiev Avenue 11, Novosibirsk, 630090, Russia

E-mail: D.A.Epifanov@inp.nsk.su

Abstract.
Recently the CMD-3 detector has started to record the experimental information produced at the VEPP-2000 e^+e^- collider at BINP (Novosibirsk). CMD-3 is a general purpose detector for a study of e^+e^- annihilation to hadrons in the wide center-of-mass energy range, $\sqrt{s} = 0.3\div2$ GeV. The calorimetry in the detector is based on three subsystems: closest to the beam pipe barrel liquid xenon calorimeter, outer barrel calorimeter based on CsI scintillation crystals and endcap calorimeter made of BGO scintillation crystals. Construction and tests of the calorimeters are described, first results from the experimental runs are presented.

1. Introduction

The new electron-positron collider VEPP-2000 [1, 2] is operating now at Budker Institute of Nuclear Physics. Using round beam technique the collider will provide the luminosity up to 10^{32} cm^{-2}s^{-1} at the maximum center-of-mass (c.m.) energy $\sqrt{s} = 2$ GeV. The physical program includes a measurement of the $e^+e^- \to$ hadrons cross section in the wide c.m. energy range up to 2 GeV [3], a study of known and search for new vector mesons, study of $n\bar{n}$ and $p\bar{p}$ production cross sections in the vicinity of the threshold and search for exotic hadrons. It requires a detector with high efficiency for multiparticle events and good energy and angular resolutions for charged particles as well as for photons.

At present, two detectors, SND (Spherical Neutral Detector) [4] and CMD-3 (Cryogenic Magnetic Detector) [5, 6], are installed in two interaction regions of the collider. CMD-3 is a general purpose detector, its layout is shown in Fig. 1. The tracking system consists of a drift chamber [7] and two layers of a proportional Z-chamber inside a thin superconducting solenoid with 1.35 T magnetic field. The calorimetry in the detector is based on three subsystems: closest to the beam pipe barrel liquid xenon calorimeter (LXe) [8], outer barrel calorimeter based on the CsI scintillation crystals [9] and endcap calorimeter made of BGO scintillation crystals [10]. The detector is also equipped with a barrel-like arrangement of time-of-flight scintillation counters [11], located between LXe and CsI calorimeters, and muon-range system.

The main tasks of the calorimeter are: precise measurement of energy and coordinates of photons in the wide energy range $E_\gamma = 10$ MeV -1 GeV, the electron-hadron separation and generation of the fast signals for the neutral trigger of the CMD-3 detector.

[1] on behalf of the CMD-3 collaboration

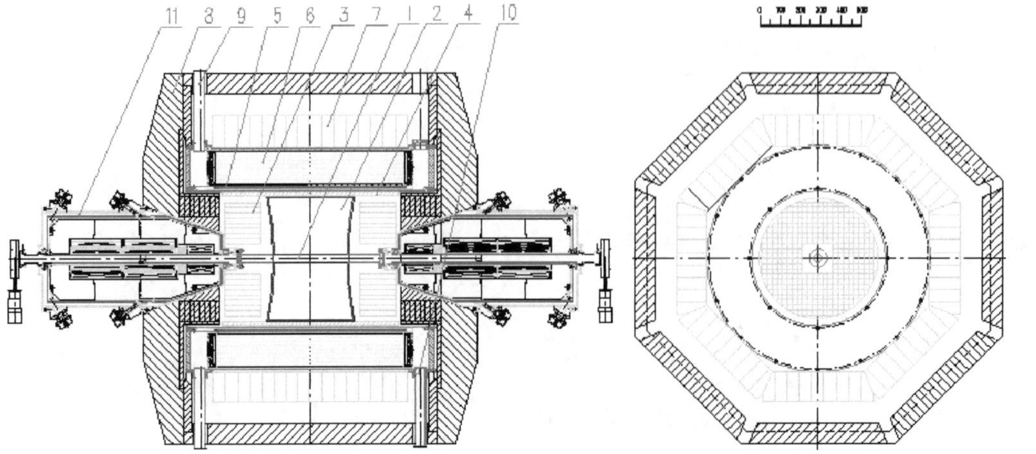

Figure 1. CMD-3 detector layout: 1 - vacuum chamber, 2 - drift chamber (DC), 3 - BGO endcap calorimeter (BGO), 4 - Z-chamber (ZC), 5 - superconducting solenoid, 6 - liquid Xe calorimeter(LXe), 7 - CsI barrel calorimeter (CsI), 8 - iron yoke, 9 - liquid He supply, 10 - vacuum pumpdown, 11 - VEPP-2000 superconducting magnetic lenses.

2. LXe calorimeter

The LXe barrel calorimeter contains 400 l of LXe (1200 kg) and has a depth of 15 cm that corresponds to $5X_0$, where X_0 is the radiation length. The total radiation thickness of the barrel part ($5X_0$(LXe)+$8X_0$(CsI)) is equal to $13X_0$. The LXe calorimeter covers polar angles from 38° to 142° with a solid angle of about $0.79\times 4\pi$. To reduce the thickness of passive materials in front of the calorimeter down to the $0.38X_0$, both the LXe calorimeter and a superconducting solenoid are placed inside the common vacuum vessel. The working temperature of the calorimeter is 170 K with a xenon pressure of 1.4 atm.

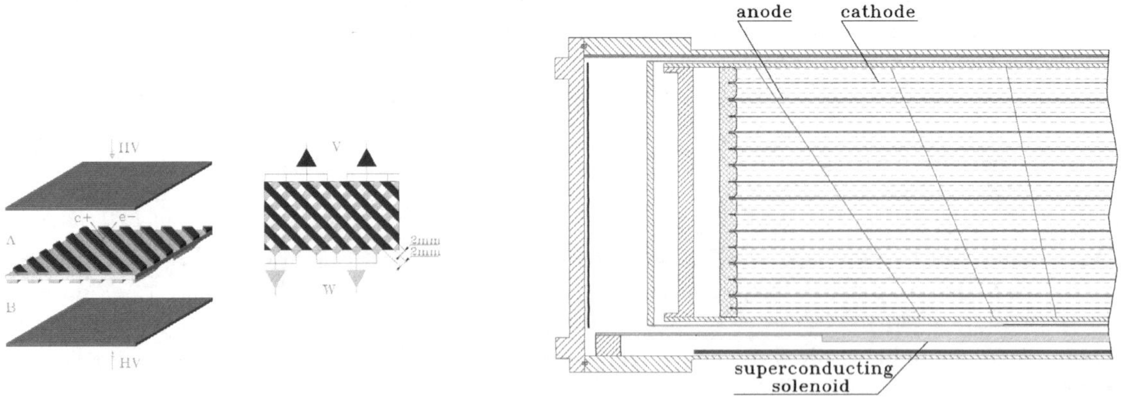

Figure 2. Design of electrodes of the LXe calorimeter

The LXe calorimeter consists of a set of ionization chambers with 7 cylindrical cathodes and 8 anodes, see Fig. 2. The electrodes were made of copper-plated G-10. Both conductive surfaces of the cathode electrodes are divided into strips. A set of four strips is connected to one superstrip read by one channel of electronics, as shown in Fig. 2. The width of the superstrips varies from 10.8 to 15.2 mm depending on the layer number. An induced charge is integrated for

4.5 μs (that is the time required for collection of electrons in an electric field of 1.5 kV/cm) and is used for the measurements of the photon conversion point coordinates and energy losses of the charged particles. Conductive surfaces of anode electrodes are divided into rectangular pads forming towers oriented to the interaction point. The induced current is integrated for 0.4 μs and then used to measure the energy deposition as well as provides signals for the neutral trigger of the detector. The total numbers of superstrips and towers are 2124 and 264, respectively. There are 33 rows of the towers along the beam direction with 8 towers in each row.

Each channel of the electronics includes a charge-sensitive preamplifier (CSP), a shaping amplifier and ADC. The shaping amplifier and ADC of the cathode channel are mounted on one Amplifier-Shaper-Digitizer board, which will be described in the next section. The circuit designs of the anode and cathode channels are similar. They differ only in terms of some active parts and time constants of the shaping RC-CR circuits. The equivalent input noise charge (ENC) of the anode preamplifier is about 3000 el. with an input capacitance of 500 pF. The ENC of the cathode CSP is about 2200 el. with an input capacitance of 1000 pF. The sensitivity of the anode channel is about 13500 el/MeV, while the typical amplitude from the cathode channel, measured with the cosmic ray particles, is about 36000 el.

The LXe calorimeter was installed in the detector and its parameters were studied with help of the cosmic-ray particles. The performed tests showed good stability of the electronics and expected level of the electronic noise. The spatial resolution varies from 0.7 to 1.4 mm depending on the layer number, in particular, because the strip capacitance increases from 400 pF to 700 pF.

3. CsI calorimeter

The CsI calorimeter consists of 8 octants, located around the LXe calorimeter, and contains 1152 counters with the total mass of crystals of about 2.8 tons. Each octant contains 9 linear modules oriented along the beam axis. Seven central modules of each octant consist of rectangular crystals, while two side modules are assembled from the crystals of special shape to avoid gaps between octants. Each module is assembled of 16 CsI crystals fixed on the steel framework to organize a linear array of 16 counters. The layout of the counter is shown in Fig. 3. It is based

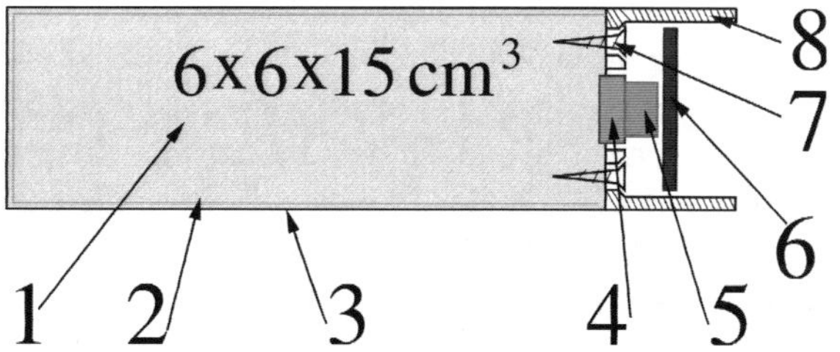

Figure 3. The layout of the CsI counter: 1 - crystal, 2 - porous teflon, 3 - aluminized lavsan, 4 - photodiode, 5 - rubber packing, 6 - preamplifier board, 7 - screws, 8 - steel framework.

on a CsI(Tl) or CsI(Na) crystal (1) of $6 \times 6 \times 15$ cm^3 size. To increase the scintillation light collection, it is wrapped with a layer of 200 μm thick porous teflon (2) and covered by the 20 μm thick aluminised mylar (3) to isolate a crystal from the external light. The scintillation light readout is performed by a Hamamatsu S2744-08 silicon PIN photodiode (4) with a sensitive area of 1×2 cm^2. Charge from the photodiode is converted to the voltage signal by the low noise charge sensitive preamplifier (6), attached to the photodiode. The electronic channel scheme is shown in Fig. 4. Signals from the preamplifiers come to the 32-channel Amplifier-Shaper-Digitizer board (ASD-32) via the twisted pairs. These boards are used to process signals in all calorimeters of the CMD-3. In the spectrometric part of the ASD-32 the amplification, shaping and amplitude digitization of the signal are performed. Digitized amplitudes are stored in the

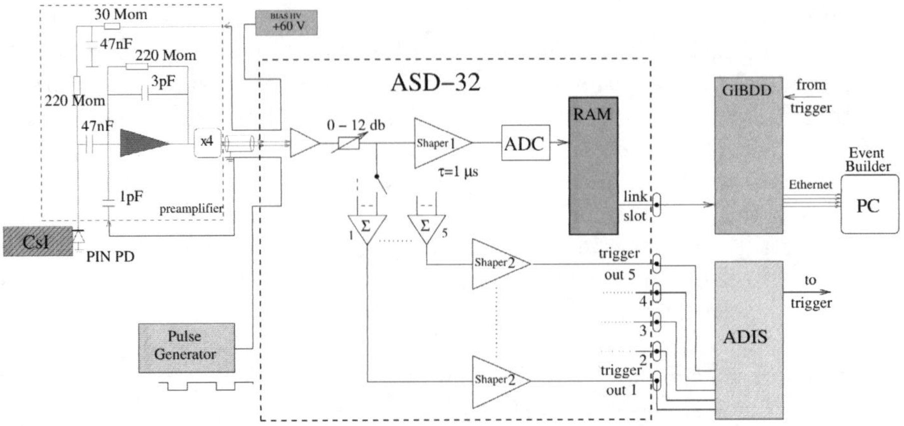

Figure 4. Scheme of the electronic channel.

on-line storage of the board, from which they are transferred to the online PC-farm through the special interface board [12]. The ASD-32 also forms fast analog sum signals to use them in the neutral trigger of the detector. The average noise over all channels is about 700 electrons and the sensitivity varies from about 2500 el./MeV for the CsI(Na)-counters to 5000 el./MeV for the CsI(Tl)-counters.

The calibration coefficients converting the ADC counts to energy are determined with help of cosmic-ray particles.

4. BGO calorimeter

The endcap calorimeter consists of two identical endcaps, each comprising 340 crystals. It covers the polar-angle regions from 16° to 49° and from 131° to 164°, the corresponding solid angle being $0.3{\times}4\pi$ sr. The length of each crystal is 150 mm, which corresponds to $13.4X_0$. The crystal transverse dimensions of 25 × 25 mm^2 were chosen on the basis of optimizing the calorimeter spatial resolution and the number of electronics channels. All crystal faces were polished, and light readout was accomplished on the basis of total internal reflection.

To simplify the mechanical structure of the endcap calorimeter and facilitate assembling, crystals were combined into blocks. The calorimeter was assembled from blocks of two types: 116 blocks of four crystals and 36 blocks of six crystals. Each block is equipped with PIN photodiodes with a sensitive area of 1 cm^2 and a board of charge-sensitive preamplifiers. The average noise over all channels is about 450 electrons and the sensitivity is about 500 el./MeV. The structure of a four-crystal block is shown in Fig. 5. In each endcap of the calorimeter, the blocks are mounted vertically on special supports into a copper ring of inner diameter 576 mm. The ring is fixed at the endface iron of the magnet yoke by means of four stainless-steel brackets.

The light output of BGO crystals decreases by about 1.5% as the temperature increases by 1°C. Since the crystal light output increases and the electronic noise decreases with temperature decrease, it is advantageous to work at low temperatures. A thermal-stabilization system allows us to maintain the temperature of the endcap calorimeter at the level of 10-12°C with long term stability better than 1°C.

The tests and calibration of the channels were carried out via accumulating spectra of the energy deposition caused by the propagation of cosmic-ray muons. For each crystal the spectrum of energy deposition is approximated by a Landau distribution function in the peak region. The value obtained in this way corresponds to the most probable energy release of 22.7 MeV, determined from a simulation. In this procedure the conversion coefficients from ADC channel

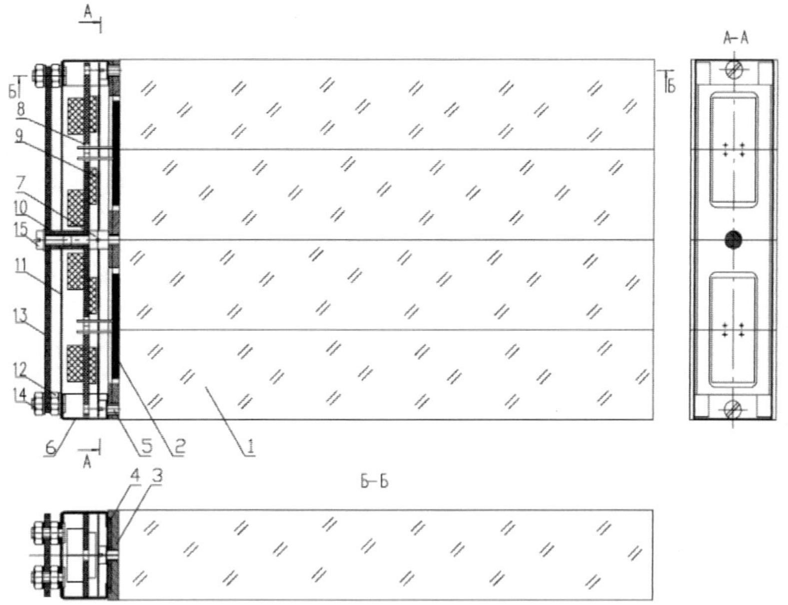

Figure 5. Layout of the four-crystal block: 1 - crystals, 2 - photodiodes, 3 - brass base plate, 4 - brass insertion plate, 5 - 3 screws, 6 - end caps, 7 - pin, 8 - preamplifier, 9 - insulator, 10 - bushing nut, 11 - side lits, 12 - nut, 13 - rear cap, 14 - nut, 15 - 5 screws.

to the deposited energy are calculated.

5. Neutral trigger system

Signals from all calorimeters are used in the neutral trigger system of the CMD-3 [13, 14]. To this end, channels of each calorimeter are combined in a certain number of trigger cells (TC): 80 TC in the LXe, 80 TC in the CsI and 48 TC in the BGO calorimeter. Analog signals from the TC are processed by 14 boards of the amplitude discriminators and summators (ADIS). ADIS boards digitize fast analog signals and generate logic signals for a cluster-finding module, which makes a decision and generates a start signal for data acquisition (DAQ) [12].

6. First results

Data taking with the completely assembled CMD-3 detector started in the first half of 2010. All calorimeters showed good performance, only a few dead channels were found. High stability of the DAQ system allowed us to record a data sample with the integrated luminosity of about 1.5 pb^{-1}. Figure 6(a) shows experimental events with two collinear tracks at the c.m. energy $\sqrt{s} = 1514$ MeV on the two-dimensional plot: total energy deposition in the LXe+CsI calorimeter versus momentum of the charged particle in DC. While signals from the cosmic-ray particles are suppressed by selection criteria, clearly seen are events of e^+e^- elastic scattering (Bhabha), $e^+e^- \to \mu^+\mu^-$ and events of beam interactions with the residual gas. Figure 6(b) demonstrates a distribution of the energy deposition in one endcap of the BGO calorimeter for the experimental events at the c.m. energy $\sqrt{s} = 1900$ MeV. One can see a clear peak coming from the Bhabha and $e^+e^- \to \gamma\gamma$ events producing electromagnetic showers in the BGO.

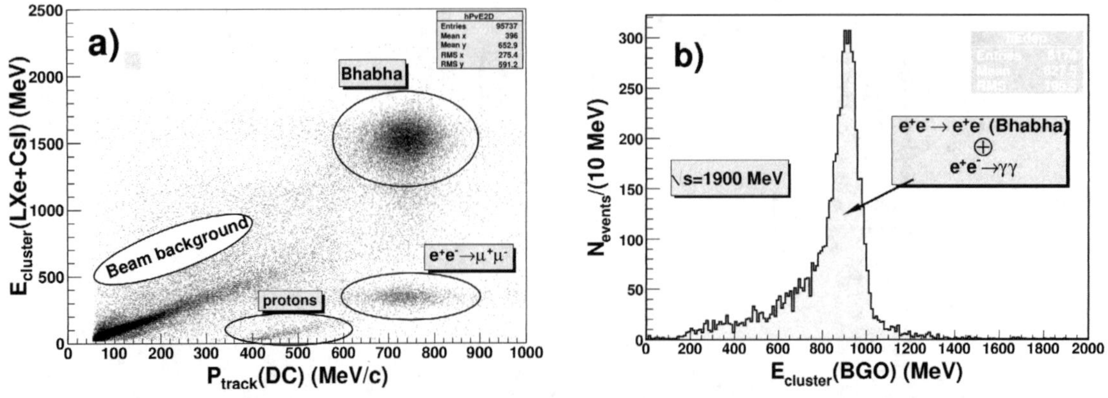

Figure 6. (a) Total energy deposition in the LXe+CsI calorimeter versus momentum of the charged particle in DC for the experimetal data collected at the c.m. energy $\sqrt{s} = 1514$ MeV. (b) Energy deposition in one endcap of the BGO calorimeter for the experimental events at the c.m. energy $\sqrt{s} = 1900$ MeV.

7. Conclusion

All calorimeters are installed in the CMD-3 detector. Their electronics is integrated in the DAQ system. The first data taking session showed good performance of the calorimeters. Monitoring of the calorimeters channels with the use of the generator calibration signal showed long term stability of the DAQ electronics. Energy calibration is done with help of the cosmic-ray muons. Work on the absolute energy calibration of the entire calorimeter with Bhabha events is in progress. A neutral trigger system is tested and will be included in DAQ soon.

8. Acknowledgments

This work was supported in part by Russian Federal Program "Scientific and scientific-pedagogical personnel of innovative Russia in 2009-2013".

References

[1] Yu. M. Shatunov *et al.*, *Prepared for 7th European Particle Accelerator Conference (EPAC 2000), Vienna, Austria, 26-30 Jun 2000*
[2] Yu. M. Shatunov, Phys. Part. Nucl. Lett. **5** (2008) 566.
[3] S. Eidelman, Nucl. Phys. Proc. Suppl. **162** (2006) 323.
[4] M. N. Achasov *et al.*, Nucl. Instrum. Meth. A **598** (2009) 31.
[5] V. M. Aulchenko *et al.*, "CMD-2M detector project. (In Russian)," BUDKER-INP-2001-45.
[6] G. V. Fedotovich [CMD-3 Collaboration], Nucl. Phys. Proc. Suppl. **162** (2006) 332.
[7] F. Grancagnolo *et al.*, Nucl. Instrum. Meth. A **598** (2009) 105.
[8] A. V. Anisyonkov *et al.*, Nucl. Instrum. Meth. A **598** (2009) 266.
[9] V. M. Aulchenko *et al.*, "Electromagnetic calorimeter based on the CsI scintillation crystals for the CMD-3 detector. (In Russian)," BUDKER-INP-2008-39.
[10] R. R. Akhmetshin, D. N. Grigoriev, V. F. Kazanin, S. M. Tsaregorodtsev and Yu. V. Yudin, Phys. Atom. Nucl. **72** (2009) 477 [Yad. Fiz. **72** (2009) 512].
[11] D. A. Drozhzhin *et al.*, Nucl. Instrum. Meth. A **598** (2009) 203.
[12] A. Ruban, A. Aulchenko, K. Kakhuta, A. Kozyrev, A. Selivanov, V. Titiov and Yu. Yudin, Nucl. Instrum. Meth. A **598** (2009) 317.
[13] K. I. Kakhuta, A. N. Kozyrev, A. A. Ruban and Yu. V. Yudin, Phys. Atom. Nucl. **72** (2009) 647 [Yad. Fiz. **72** (2009) 687].
[14] A. N. Kozyrev, A. A. Ruban and Yu. V. Yudin, Nucl. Instrum. Meth. A **598** (2009) 345.

CCALT: A crystal calorimeter for the KLOE-2 experiment

M. Cordelli, F. Happacher*, M. Martini, S. Miscetti, I. Sarra
Laboratori Nazionali di Frascati dell'INFN,
Frascati, Italy
*E-mail: fabio.happacher@lnf.infn.it

M. Schioppa, S. Stucci
INFN and department of physics, University of Calabria,
Cosenza,Italy

Abstract. The angular coverage extension of the KLOE-2 electromagnetic calorimeter, from a polar angle of 20° down to 10°, will increase the multiphoton detection capability of the experiment enhancing the search reach for rare kaon, η and η' prompt decay channels. The basic layout of the calorimeter extension consists of two small barrels of LYSO crystals readout with APD photosensors aiming to achieve a timing resolution between 300 and 500 ps for 20 MeV photons. The first test of a (5.5×6×13) cm^3 prototype for such a detector was carried out at the Beam Test Facility of Laboratori Nazionali di Frascati of INFN. We present here the results of this test.

1. Introduction

In the last years, a new machine layout based on the Crab-waist and a large Piwinsky angle has been proposed and tested [1] to improve the reachable luminosity at the Frascati ϕ-factory, DAΦNE. The success of this test motivated the startup of a new experiment, named KLOE-2 [2]. The first running phase will start at the end of 2010 with the goal of collecting ~ 5 fb^{-1} in one year. A second phase, for a longer data taking, will require a set of detector upgrades all concentrated around the beam-pipe. We are studying the insertion of a calorimeter between the interaction point, IP, and the first inner quadrupole (See Fig. 1) to extend the angular coverage of the main electromagnetic calorimeter. This will enhance the multiphoton detection capability of the detector for the search of rare decays of kaons, η and η' mesons.

2. CCALT: a Crystal Calorimeter with Time

The discussion of the previous section indicates that this calorimeter has to be very dense, with a small value of radiation length, X_0, and Moliere radius, R_M, not hygroscopic and with a large light output to improve photon detection efficiency at low energy (from 20 to 500 MeV). Moreover, the calorimeter has to be extremely fast in order to allow for prompt photon reconstruction Preliminary simulation studies indicates the need to reach a time resolution of $300 \div 500$ ps for 20 MeV photons.

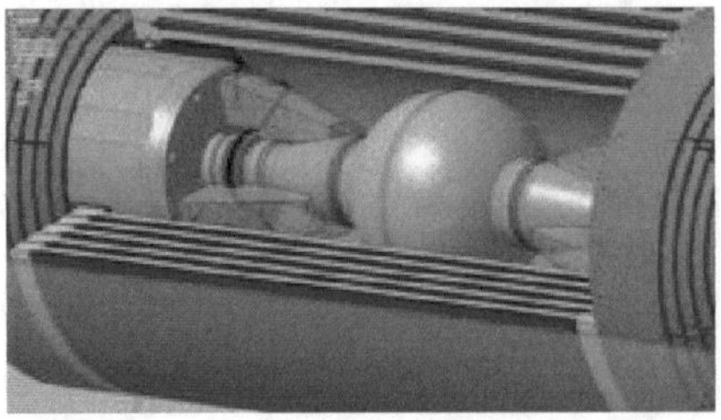

Figure 1. Zoomed-view of the IP region. The area available for the new calorimeter lies between the inner sphere and the closest quadrupoles.

A first detector layout consists of two concentrical barrels of 24 crystals each, with transversal dimensions of 2×2 cm^2 and longitudinal length between 13 and 15 cm. The best crystal choice matching the requirements is provided by LYSO, which has X_0 and R_M values (1.1 and 2 cm) and a scintillation emission time τ_{LYSO} = 40 ns.

In the final location of the CCALT inside KLOE-2, the presence of an axial magnetic field of 0.52 kGauss forces the usage of silicon based photodetectors. Due to the high photon yield, the readout with APDs is a valid solution. In the following, we specifically considered only the Hamamatsu S8664-55, which has an active area of 0.5×0.5 cm^2, fast timing characteristics and a quantum efficiency between 65 and 85% in the wavelenght range of interest (390 $\sim$ 500 nm) for the LYSO emission spectra.

In march 2009, we have built a medium size crystal matrix prototype with transversal radius larger than 2 R_m, longitudinal dimensions being constrained by budget limits to be between 13 and 15 cm (corresponding to 11 $\div$ 12 X_0 of longitudinal containment). The prototype consists of an inner matrix of 10 small(15×15 mm^2 and 20×20 mm^2) LYSO crystals readout by APD and an outer matrix, for leakage recovery, composed by 8 $PbWO_4$ (20×20 mm^2 and (30×30 mm^2)) cristals readout by standard Hamamatsu Bialcali photomultipliers of 1,1/8" diameter. To test the quality of the crystals offered by different vendors, the inner matrix has been assembled in three rows composed by: LFS crystals from Zecotek, LYSO crystals from Saint Gobain and LYSO crystals from Scionix. The LFS from Zecotek is a Luthetium Fine Silicate crystal, with very similar properties to LYSO. Each crystal is wrapped with 100 μm of tyvek on the lateral faces, leaving free both the front and end faces, thus allowing one to bring calibration light pulses through an external LED and a fast change of the photosensors readout. The optical connection of the photosensors with the crystals is done with optical grease. The amplifiers are based on the MAR8A+ chip from Minicircuits, with a gain factor of 25 and a bandwidth of 1 GHz.

3. Test results with electron beams.

We have taken data at the Beam Test Facility, BTF, of LNF for two weeks in april 2009. The BTF plant provides electron and positron beams with a momentum ranging from 50 to 500 MeV/c and can reach a single particle multiplicity. The matrix was positioned at the center of the beam axis with an area delimited by a cross of two finger BC408 scintillators of $1 \times 0.5 \times 5$ cm^3 dimensions, $f1, f2$. The scintillators were used to discriminate single electron events from higher multiplicity ones looking at the pulse height distributions as shown in Fig. 2. In most of the tests, the fingers were aligned in such a way to define a beam spot of 1×1 cm^2. In front of the fingers it was also present a beam position monitor, BPM, of the BTF group, consisting of sixteen horizontal and vertical scintillator strips readout by two Multi Anode PMs.

We have triggered by using a replica of the spill signal from the Linac adjustable from remote

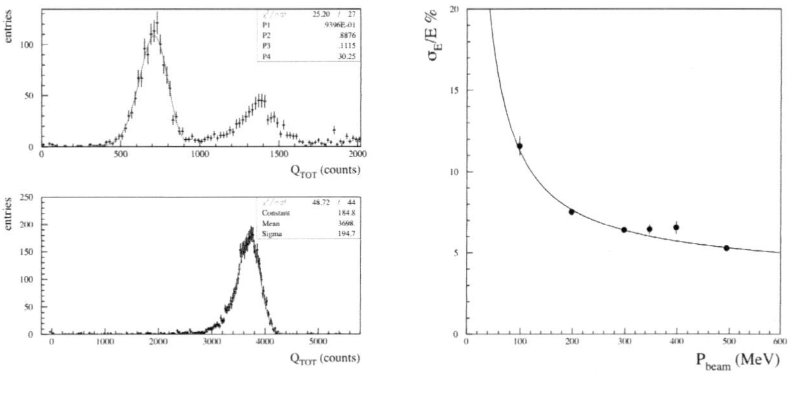

Figure 2. (Left) Distribution of Q_{TOT} for single electron events at 100 MeV (top) with a logG fit superimposed and at 500 MeV (bottom) with a gaussian fit. (Right) Dependence of the energy resolution on beam momentum.

in order to correctly put the signals in time. We acquire data with the KLOE-2 daq system, VME based, reading out KLOE ADC and TDC boards with a sensitivity of 100 fC/count and 50 ps/count respectively.

Observing the response of the prototype to single electrons, we realized that the outer matrix was not properly working due to an unexpected optical cross-talk between crystals. We observe large cross-talk only on the $PbWO_4$ of the outer matrix. We believe this to be due by a cooperation between a light leak through the tyvek and the different amplification gains between PMs and APDs. In the following, we therefore report only results related to the inner matrix wwhich has an overall dimension of 60×55 mm^2 ($\leq$1.5 R_M).

By using the UV LED, we have first equalized each channel at 10% level by proper HV adjusting. We have then calibrated the calorimeter response of each channel with minimum ionizing particles, m.i.p., crossing the calorimeter hortogonally to the crystal axis. We get σ_{ped} of 5 counts and a m.i.p. peak, M_i, of around 100 counts for the smaller size crystals. The total response of the detector is then defined as: $Q_{TOT} = \sum(Q_i - P_i) \times M_0/M_i$, where Q_i and P_i are the collected charge and the pedestal of the i-th channel, M_0 represents an average calibration of all channels in counts and the calibration for the larger crystals is corrected for the different size. In Fig. 2 left, we show the distribution of Q_{tot} for a beam of 100 and 500 MeV respectively after having selected single electron events with a cut on the finger scintillators. We have fit the distribution corresponding to one electron either with a simple gaussian, centered around the peak, or with a logarithmic gaussian, logG, function as follows: $\mathrm{N} \cdot \exp(-\frac{1}{2\sigma_0^2} ln(1 - \frac{\eta}{\sigma_E}(E - E_{peak}))^2 - \frac{\sigma_0^2}{2})$ where N is a normalization factore, η represents the asymmetry, E_{peak} the most probable value of the distribution, $\sigma_0 = \frac{2}{2.36} sinh^{-1}(2.36\eta/2)$ and $\sigma_E = \frac{\mathrm{FWHM}}{2.36}$ is the resolution.

By performing a linear fit to the distribution of E_{peak} vs P_{beam}, we get a slope of 7.5± 0.1 counts/MeV which sets the M_i value to be $\sim$ 16 MeV for a small crystal consistent with an expected energy loss of $\sim$ 10 MeV/cm. At the running voltages of 410 V, the expected APD gain varies between 300-500 from which we estimate the light yield to be between 500 and 800 p.e./MeV.

To understand the different terms of the energy dependence of the energy resolution, we are carring on a full simulation of the prototype based on Geant-4. Studies on photoelectron

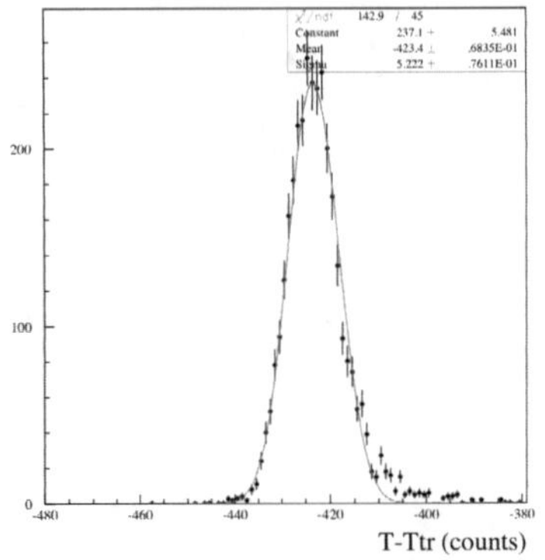

Figure 3. Distribution of the average timing reconstructed with the inner matrix for 500 MeV electron beam.

collection efficiency are still underways. This simulation indicates that there is a large leakage term between 5 and 4 % from 100 to 500 MeV.

In Fig. 2 right, we show the energy dependence of the energy resolution measured on data which has been fit with the following equation: $\sigma_E/E = a \oplus \mathrm{b}/(\mathrm{E/GeV}) \oplus \mathrm{c}/\sqrt{\mathrm{E/GeV}}$, where, accordingly to MC, we have fixed the constant term to be 5 %. We found $b = 1.1\%$ and $c = 1.4\%$ when using the gaussian fits to the spectra. If we repeat this procedure, for the fits with the logG function, we get $b = 0.8\%$ and $c = 2.4\%$.

We have investigated the large b/E term by measuring the total detector noise with a gaussian fit to Q_{TOT} in events without any electron beam impinging. We find $\sigma_Q = 4.2 \div 4.8$ MeV which is slightly larger than the incoherent sum of σ_{ped} resulting to be 3.6÷3.8 MeV showing that a not negligible coherent noise is present. However, the noise does not fully explain the large b term found. We are unable to understand the origin of this contribution. A measurements of the LYSO longitudinal response has been carried out and we found a difference below 5% scanning along the crystals.

In the attempt of understanding the single terms to the energy resolution we developed a full GEANT4 simulation of the calorimeter to check the leakage contributions and the fluctuation of the shower shape, that, convoluted with the longitudinal disuniformity of the crystal response, could give rise to a sizeable contribution to the resolution. We find an effect at th level of 1-2 %, whereas the leakage contribution is at the level of 1%.

We are now planning a new test at the MAMI tagged photon beam facility at Mainz to esclude that the observed resolution is due to the BTF beam intrinsic spread.

We have also determined the position resolution by comparing the reconstructed centroid done with the crystals with the position provided by the Beam Position Monitor of BTF; The centroids were defined as $X_{pos} = \sum Q_i X_i / Q_{tot}$. We observe a position resolution of $2.8 \div 3$ mm at 500 MeV.

We have finally reconstructed the calorimeter timing after correcting it, event by event, for the arrival time of the electrons in the LINAC spill. This was done by measuring the timing with the scintillators f1,f2. The weighted energy average over all calorimeter, T_{clu}, was done after subtracting the average T_0 of each cell. As shown in Fig. 3, a clean gaussian response is observed with a time resolution, σ_T, of $\sim$ 49 ps ($\sim$ 120 ps) at 100 MeV (500 MeV) after correcting for trigger jitter.

References

[1] P.Raimondi, in: "Crab Waist Collisions in DAΦNE AND SUPER-B DESIGN", Proceedings of EPAC08, Genoa, Italy (2008).
[2] F. Bossi et al, for the KLOE-2 collaboration,"A proposal for the Roll-In of the KLOE-2 detector", LNF-Internal Note, 07/19 (2007).
[3] F. Bossi et al, for the KLOE collaboration, "Precision Kaon and Hadron Physics with KLOE", Riv. Nuovo Cimento 031:531-623 (2009)

Study of Various Photomultiplier Tubes for Window Events: Upgrade R&D for CMS Hadron Forward Calorimeters

Burak Bilki
on behalf of CMS HCAL Collaboration
University of Iowa, Iowa City, IA 52242-1479, U.S.A.
E-mail: Burak.Bilki@cern.ch

Abstract. The PMTs of the CMS Hadron Forward calorimeters were found to generate a large amount of signal when their windows were traversed by energetic charged particles. This signal, which is due to Čerenkov light production at the PMT window, could interfere with the calorimeter signal and mislead the measurements. In order to find a viable solution to this problem, the response of different types of PMTs to muons traversing their windows at different orientations is measured at the H2 beam-line at CERN. Certain kinds of PMTs with thinner windows show significantly lower response to direct muon incidence. For one specific type - the four anode PMT- a simple and powerful algorithm to identify such events and recover the PMT signal using the signals of the quadrants without window hits is also presented. For the measurement of PMT responses to Čerenkov light, the Hadron Forward calorimeter signal was mimicked by two different setups in electron beams and the PMT performances were compared with each other. Superior performance of particular PMTs was observed.

1. Introduction

The Compact Muon Solenoid (CMS) [1] is a general-purpose detector designed to run at the highest luminosity provided by the CERN Large Hadron Collider (LHC). Coverage between pseudorapidities of 3.0 and 5.0 is provided by the steel/quartz fiber Hadron Forward (HF) calorimeter. The front face is located at 11.2 m from the interaction point and the depth of the absorber is 1.65 m. The signal originates from Čerenkov light emitted in the quartz fibers, which is then channeled by the fibers to photomultipliers. The absorber structure is created by machining 1 mm square grooves into steel plates, which are then diffusion welded. The diameter of the quartz fibers is 0.6 mm and they are placed 5 mm apart in a square grid. The quartz fibers, which run parallel to the beamline, have two different lengths (1.43 m and 1.65 m) which are inserted into grooves, creating two effective longitudinal samplings. There are 13 towers in η, all with a size given by $\Delta\eta \approx 0.175$, except for the lowest-η tower with $\Delta\eta \approx 0.1$ and the highest-η tower with $\Delta\eta \approx 0.3$. The ϕ segmentation of all towers is $10°$, except for the highest-η one which has $\Delta\phi = 20°$. This leads to 900 towers and 1800 channels in the two HF modules [2]. Details of the HF design, together with test beam results and calibration methods, can be found in [3].

In the framework of the Super LHC (SLHC) upgrade plans, one of the problems to be solved is the large signal generated by the photomultiplier tubes (PMTs) of CMS HF calorimeters (hereon called HFPMT) when the PMT window is traversed by relativistic charged particles. The primary reason for this signal is Čerenkov light production at the PMT window, followed by the liberation of photoelectrons at the photocathode of the PMT. The expected rate is around 0.1 % of the events at low luminosity conditions of LHC. This would have impact on the hadronic energy measurements of HF and missing transverse energy calculations, and could even result in fake triggers and mismeasured online luminosities.

2. Experimental Setup and Data Acquisition

Table 1 summarizes typical properties of the PMTs used in this study. Additionally, HFPMT window is 2 mm thick at the center and gets thicker towards the rim, single and four anode PMT windows are slightly less than 1 mm thick and miniPMT window is around 0.5 mm thick.

Table 1. Typical properties of PMTs tested in this study. Information is obtained from Hamamatsu [4].

PMT Type	PMT Type Number	Photocathode	Quantum Efficiency (max. %)	Typical Gain	Window Area (mm^2) app.
Four Anode PMT	R7600U-100-M4	Super Bialkali	35	1.3×10^6	324 (square)
Four Anode PMT	R7600U-200-M4	Ultra Bialkali	43	1.3×10^6	324 (square)
Four Anode PMT	R8900U-100-M4	Super Bialkali	35	1.0×10^6	324 (square)
Single Anode PMT	R7600U-100	Super Bialkali	35	1.0×10^6	324 (square)
Single Anode PMT	R7600U-200	Ultra Bialkali	43	1.0×10^6	324 (square)
miniPMT	R9880U-110	Super Bialkali	40	2.0×10^6	50 (round)
HFPMT	R7525	Bialkali	25	5.0×10^5	490 (round)

Tests were performed at CERN H2 beamline [5]. Two different setups were used to test the muon response of the PMTs. In the first setup, PMTs were lined one after the other with their windows facing the beam direction. In the second setup , PMTs were placed side by side, their windows being parallel to the beam direction. Both PMT boxes were made light-tight and they were placed in front of wire chamber E (WCE) as the first test apparatus on the HF test table in order to perform precise position measurements with 1 mm resolution.

Two test stations were used to measure the differences in the response of the PMTs to Čerenkov light from electron showers : ~1 cm diameter bundle of regular HF quartz fibers (with 0.6 mm diameter) and the quartz fiber calorimeter. The fiber bundle was placed in ~3 cm diameter light guide and it splits into two parts at the readout end. The HFPMT was kept in place at one end of the fiber bundle while the candidate PMTs were interchanged at the other end. The portion of the fiber bundle that sees the beam was aligned to make ~45° with respect to the beam direction in order to maximize the Čerenkov light capture. The length of the bundle from the beam interaction to readout was ~1.5 m. WCE was used to select events with electrons that pass through the fiber bundle. The quartz fiber calorimeter consisted of an array of 6 mm diameter, 45 cm long steel rods in a 20 cm x 20 cm x 45 cm housing with quartz fibers (0.3 mm core diameter, 65 cm long) inserted in between the rods. The fibers were then bundled at the back of the calorimeter to form a single readout. The light guides at the readout end were 20 cm long with the same reflective material as HF light guide. Details of the readout can be found in [6].

3. Testing Muon Interactions with PMT Windows

Figure 1a shows the QIE charge profile as a function of wire chamber coordinates for the HFPMT. Using these profiles, a position cut with the correct dimensions at the PMT window edges was applied to study only the events where the charged particles traversed the PMT windows for all PMTs. The cut for the HFPMT window is shown in Fig. 1a. Figure 1b shows the responses of different types of PMTs when the muon beam hits the PMT window from the front. All responses are normalized to the gain of the HFPMT. Since HFPMT window is much thicker than that of the other PMTs, Čerenkov signal produced is much higher. It has a signal magnitude that is more than a factor of two compared to the single anode and four anode PMTs and a much larger spread with a long tail in the high end of the spectrum. The four anode and single anode PMTs both have ultra bialkali photocathodes and they exhibit the same behavior in the overall picture.

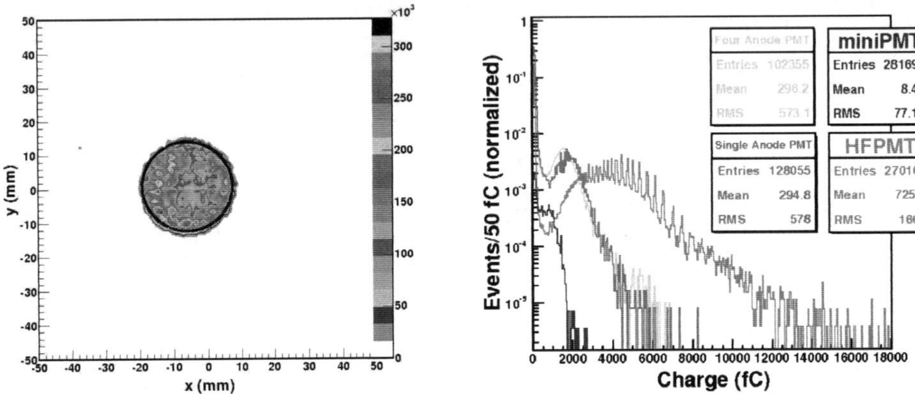

(a) (b)

Figure 1. (a) Integrated charge profile (in fC) as a function of WCE coordinates for HFPMT. The boundary of the PMT window event selection is shown as a black ring. (b) Charge distributions of the four anode PMT (R7600U-200-M4), the single anode PMT (R7600U-200), the miniPMT and the HFPMT produced by the front beam incidence.

With the PMT window cuts applied, PMT window event rates were also measured as the ratio of the number of events above pedestal to the total number of events: HFPMT 90%, four anode PMT 69%, single anode PMT 64% and miniPMT 51%. As expected, the window event rates were found to be correlated with the thicknesses of the PMT windows.

One of the four anode PMTs was used to study the dependence of the charge distribution on the angle of incidence of the beam. The PMT was exposed to beam with its window making an angle of 90° (front incidence), 70°, 50°, 30°, 10° and 0° (side incidence) with respect to the beam direction. Beam position cuts were applied around the PMT window using WCE information for each orientation. Figure 2a shows the charge distributions for 50°, 10° and 0°. As the angle is reduced from 90°, the thickness of the PMT window that is exposed to beam increases leading to an increase in signal magnitude (Fig. 2b). The results of the angular study are consistent with the results presented in [7] for bialkali photocathodes.

4. Testing Čerenkov Response of PMTs

Figure 3a shows the charge distributions for the PMTs reading out the fiber bundle signal. The distributions were normalized to the HFPMT gain. Single anode and four anode PMT signals both have a mean ~1.5 times the HFPMT signal.

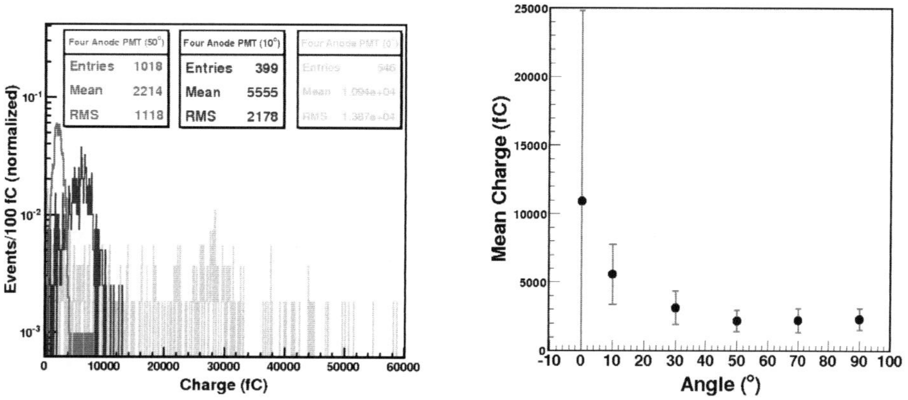

(a) (b)

Figure 2. (a) Charge distribution of the four anode PMT (R8900U-100-M4) for different angles of muon beam incidence. (b) Mean charge at all angles studied.

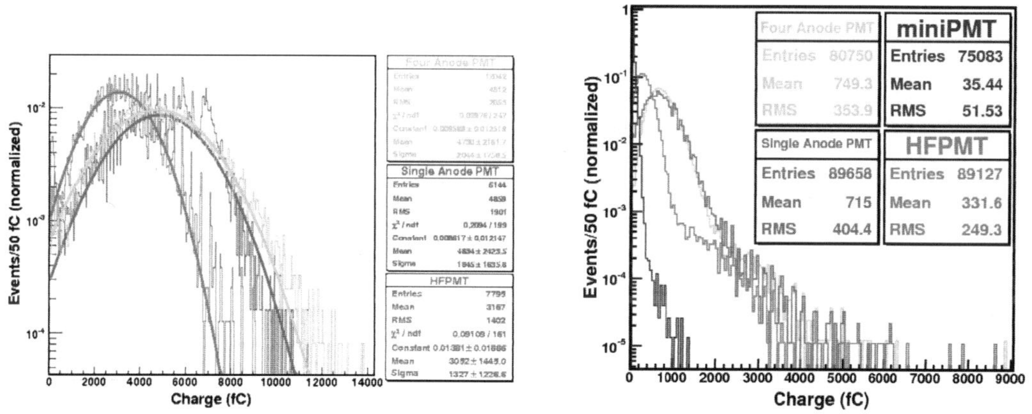

(a) (b)

Figure 3. (a) Fiber bundle charge distributions for the four anode PMT (R7600U-100-M4), the single anode PMT (R7600U-100) and the HFPMT. Gaussian fits are also shown. (b) Quartz fiber calorimeter charge distributions for four anode PMT (R7600U-200-M4), single anode PMT (R7600U-200), miniPMT and HFPMT.

Figure 3b shows the charge distributions of the four anode PMT, the single anode PMT, the miniPMT and the HFPMT when they were reading out the quartz fiber calorimeter. The distributions were normalized to the HFPMT gain. No PMT window area corrections were applied since proper light guides were utilized for each PMT window geometry.

5. Selecting PMT Window Events and Recovering the Signal with Four Anode PMT

A simple algorithm for identifying window events in the four anode PMTs has been developed with the assumption of using them in HF. Details of the algorithm is provided in [6]. Figure 4 shows the application of this PMT window event identification and signal recovery algorithm to a set of front muon incidence data. No pedestal subtraction has been applied to the data.

The algorithm successfully selects the PMT window events and recovers the signal back. The distribution is suppressed towards zero, slightly above pedestal - which is shown by the blue, crossed distribution - except for the events with muon hits around the center of the PMT window which correspond to ~2.5% of all the front-incident muon events.

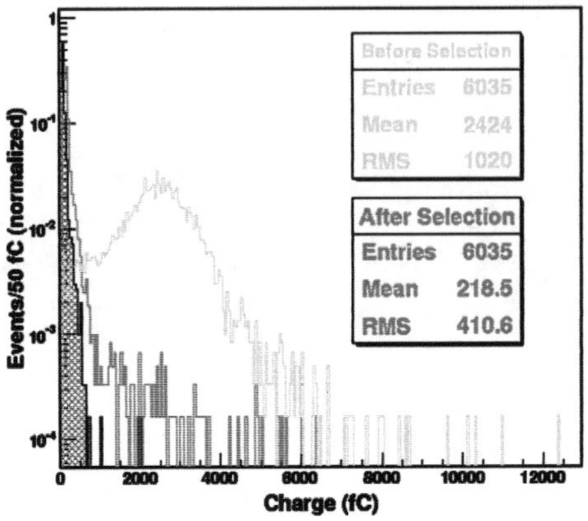

Figure 4. PMT window event selection and signal recovery for the four anode PMT with front incidence of muons. The blue, crossed area is the pedestal.

The algorithm was also applied to the data from the HF fiber bundle. The result is shown in Fig. 5a. The algorithm misidentifies about 6% of the signal events as PMT window events and reconstructs the signal suppressing it towards lower values. The misidentification is due to the algorithm itself; however, the suppression towards lower values is due to slight alignment variations that result in uneven illumination of the quadrants or real interference from particles scattered towards the four anode PMT or cosmic muons. For some of the runs, the fiber calorimeter was rotated by 60° or 90° to allow the electron shower to leak behind, and the fiber bundle PMTs were placed in the beamline at the back of the calorimeter. Although the probability for the PMTs to get real particle hits together with the bundle signal is too low, the algorithm was tested with this setup for its effectiveness in selecting PMT window events within a real signal. The result is shown in Fig. 5b. In the low signal region of the spectrum, the signal is suppressed towards lower values as in the case of the algorithm applied to pure signal. The most important aspect of this result is that the high tail in the original distribution, which clearly comes from PMT window events, disappears after the utilization of the algorithm.

6. Conclusions

In the search for a new photomultiplier tube for the CMS forward hadron calorimeters, candidate PMTs with different specifications were tested. The response of PMTs to relativistic charged particles traversing their windows at different incidence orientations was tested with 150 GeV/c muon beam. The HF calorimeter signal was also mimicked with two test setups in 80 GeV/c electron beam in order to compare PMT performances: around 1 cm-diameter bundle of HF quartz fibers and a 20 cm x 20 cm x 45 cm steel rod calorimeter with quartz fibers placed in between and alongside the rods. The fiber bundle was split into two at one end enabling two readout channels simultaneously.

At front incidence of muons on the PMT windows, the candidate PMTs exhibit significantly better performance with their reduced response magnitude and spread, and PMT window event

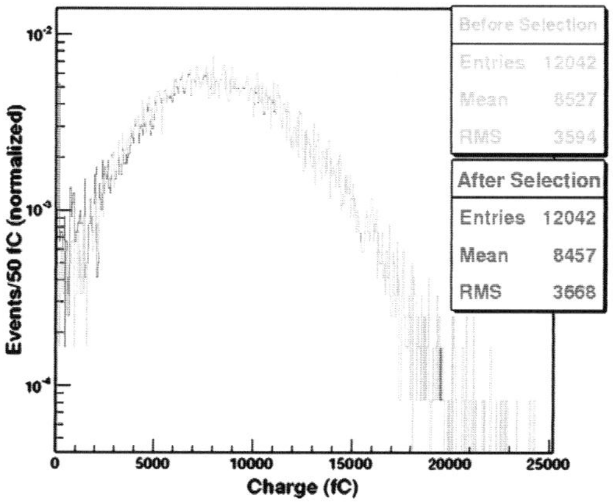

(a)

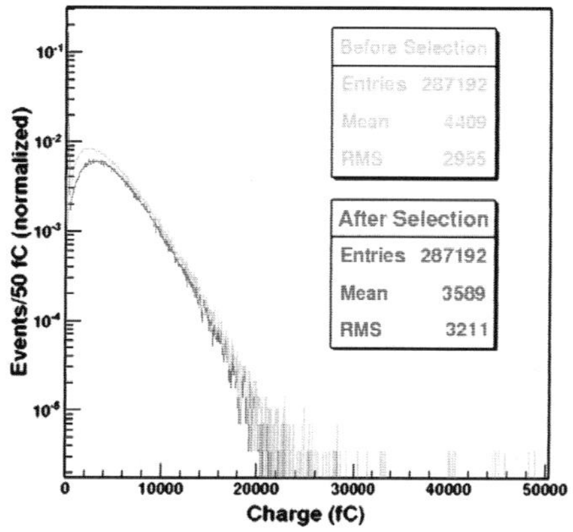

(b)

Figure 5. (a) Charge distributions before and after the algorithm for selection of PMT window events and signal recovery was applied to the four anode PMT of the HF fiber bundle. (b) The result of the application of selection and signal recovery algorithm to HF fiber bundle when the PMTs were in the beamline.

rates. Side muon incidence response is higher and has a wider spectrum when compared to front incidence response. Measurements with various angles of muon incidence also prove that the response becomes larger as the angle between the PMT window and the beam direction is decreased.

Measurements show that the candidate PMTs have superior performance over the HFPMT on detection of Čerenkov light as expected from their proposed quantum efficiencies.

For the four anode PMT, an effective and simple algorithm for selecting PMT window events and recovering the signal from the quadrants that do not show PMT window event signature is presented and implemented on muon incidence data as well as the fiber bundle data. This simple method proves to be an effective way of selecting PMT window events and recovering the signal with available information. The method can be utilized both for refining detector measurements and for the elimination of cosmic interference online or offline.

References

[1] Chatrchyan S et al. (CMS Collabration) 2008 "The CMS Experiment at the CERN LHC" *JINST* **3** S08004 145-149

[2] CMS Collaboration 2006 "CMS Physics TDR: Volume I, Detector Performance and Software" *CERN/LHCC* 2006-001

[3] Abdullin S et al. (CMS Collaboration) 2008 "Design, Performance, and Calibration of CMS Forward Calorimeter Wedges" *Eur. Phys. J.* **C53** 139-166

[4] http://www.hamamatsu.com/

[5] http://ab-div-atb-ea.web.cern.ch/ab-div-atb-ea/BeamsAndAreas/resultbeam.asp?beamline=H2

[6] CMS HCAL Collaboration 2010 "Study of Various Photomultiplier Tubes with Muon Beams And Čerenkov Light Produced in Electron Showers" *JINST* **5** P06002

[7] Motta D and Schonert S 2005 "Optical Properties of Bialkali Photocathodes" *Nucl. Instrum. Meth.* **A539** 217

Fast Neutron Induced Nuclear Counter Effect in Hamamatsu Silicon PIN Diodes and APDs

Rihua Mao, Liyuan Zhang and Ren-Yuan Zhu
256-48, HEP, Caltech, Pasadena, CA 91125, USA
E-mail: zhu@hep.caltech.edu

Abstract. Neutron induced nuclear counter effect in Hamamatsu silicon PIN diodes and APDs was measured by irradiating fast neutrons from a pair of ^{252}Cf sources directly to these devices. It was found that the entire kinetic energy of these neutrons may be converted into electron signals in these devices, leading to anomalous signals of up to a few million electrons in a single isolated calorimeter readout channel. Signals of such amplitude represent equivalent energy of several hundred GeV and a few GeV for PWO and LSO/LYSO crystals respectively assuming the corresponding light yields of 4 and 800 p.e./MeV. The overall rate of the neutron induced nuclear counter effect in APDs is found to be more than an order of magnitude less than that in PIN diodes. Increasing the APD gain was also found to reduce the neutron induced nuclear counter effect. An intelligent front-end chip capable of selecting un-contaminated signal is proposed to eliminate completely the nuclear counter effect without significant cost increase.

1. Introduction

Because of their immunity to magnetic fields silicon based readout devices, such as silicon PIN photo-diodes and avalanche photo-diodes (APDs), are widely used as photo-detectors for reading out scintillation photons. Modern crystal calorimeters, for example, use either PIN diodes or APDs as the readout device. These photo-detectors are located in the middle of particle showers during experimental data taking, where charged particles and neutrons are produced copiously. While the nuclear counter effect caused by minimum ionizing charged particles in PIN diodes [1] and APDs [2] is well understood it is less studied for neutrons. Thermal or fast neutrons are frequently used to irradiate various silicon based photo-detectors. The main purpose of these studies, however, is to investigate the neutron induced radiation damage effects in these devices, e.g. the degradation of the dark current etc. Fast neutron induced pulse height spectrum was reported for a Hamamatsu PIN diode S3590-02 [3] with a 200 μm thick depletion layer as a possible neutron sensor. This device, however, has a much reduced thickness as compared to the Hamamatsu S2744-08 PIN diodes with a 300 μm thick depletion layer which are widely used in high energy physics experiments. No previous publication was found for neutron induced pulse height spectrum in APDs.

GEANT simulations show that fast neutrons are produced copiously at hadronic colliders, such as the LHC, with an energy spectrum peaked at a few MeV [4]. A flux of about 10^5 cm^{-2} s^{-1} is expected for the Hamamatsu S8664-55 silicon APDs, which are used as the photo-detectors for the CMS barrel PWO electromagnetic calorimeter (ECAL), when the LHC is running at the designed LHC luminosity of 10^{34} cm^{-2} s^{-1} with 7 TeV beam energy [4]. Neutrons can also be produced through photo-nuclear reactions from bremsstrahlung photons via the Giant Dipole

Resonance (GDR) mechanism with an energy spectrum similar to and a rate of about 1% of that from hadrons [5].

Interactions between neutrons and photo-detector material are well understood. For fast neutrons, the dominant process with cross-section of up to ten barns is elastic neutron-nucleus collision through which a neutron transfers a fraction of its kinetic energy to a nucleus. The charged nucleus subsequently deposits its kinetic energy in the photo-detector through ionization and creates electron-hole pairs at a rate of 3.62 eV per pair. Up to a few million electrons may be induced when the entire kinetic energy of the fast neutron is converted to electron-hole pairs in a photo-detector. The corresponding equivalent energy of neutron induced electron signals in a calorimeter cell depends on crystal's light yield. Signals of a few million electrons correspond to several hundreds GeV for a PWO ECAL with typical light yield of 4 p.e./MeV for calorimeter size crystals with APD readout. Signals of the same amplitude would correspond to a few GeV for LSO/LYSO crystals with typical light yields of 800 p.e./MeV with APD readout. Since the upper energy scale is also about two hundred times different between the LHC and the proposed SuperB collider the impact of such signals to the physics output is comparable between the CMS PWO calorimeter at the LHC and the proposed SuperB forward LSO/LYSO calorimeter [6].

Since the beginning of LHC operations in late 2009 anomalous signals have been observed in single isolated readout channels in the CMS barrel PWO ECAL, where Hamamatsu S8664-55 silicon APDs are used as the photo-detector [7]. Contrary to the showering photons and electrons, these anomalous signals are characterized by little or no energy observed in their neighbors. They appeared at a rate of about 10^{-3} in minimum bias events [7]. Their amplitude was found to be up to several hundred GeV equivalent in the extreme cases. Strategies have been developed by CMS to reduce their impact on the physics output [7]. Effort was made to eliminate events contaminated by the anomalous signals using event topology, such as the spatial energy distribution and the event timing etc. For 2010 LHC runs at 3.5 TeV beam energy and about 10^{32} cm^{-2} s^{-1} luminosity the particular topology of these events allow a majority of the anomalous signals be removed at the trigger stage without affecting calorimeter performance for real physics. It, however, is clear that such rejection would be less effective when the beam energy and luminosity of the LHC increase because of the increased event energy and overlapping minimum bias and pile-up events. It is also clear that such effects can be completely eliminated by multiple photo-detectors with independent readout channels for a single calorimeter cell. Hardware solution would thus be preferred to make sure there was no contamination in physics analysis.

These anomalous signals observed by the CMS ECAL are caused by signals induced directly by shower particles in photo-detectors, conventionally called nuclear counter effect. Since the nuclear counter effect caused by minimum ionizing particles would not produce signals of such amplitude it was suspected that these signals were caused by heavily ionizing nuclei, which receive their kinetic energies through collisions with neutrons. To investigate neutron induced nuclear counter effect in silicon photo-detectors fast neutrons from a pair of ^{252}Cf sources were used to irradiate the following devices.

(i) A Hamamatsu S2744-08 PIN diode with an active area of 1×2 cm^2 and a 300 μm thick depletion layer.

(ii) A pair of Hamamatsu S8664-55 APDs with an active area of $2 \times 5 \times 5$ mm^2, a 5 μm thick active layer before the avalanche amplification and a 45 μm layer of silicon with a low gain behind the avalanche region [8].

Both photo-detector chips have a 500 μm thick epoxy coating at the top of the device surface, and are bounded to ceramic package with conductive epoxy of several tens of μm thick [9]. While Hamamatsu S2744-08 PIN diodes are used as photo-detectors for the *BaBar*, BELLE and BES III CsI(Tl) calorimeters, Hamamatsu S8664-55 APDs are used for the CMS PWO calorimeter.

Journal of Physics: Conference Series **293** (2011) 012012 doi:10.1088/1742-6596/293/1/012012

Both are candidate photo-detectors for the proposed SuperB LSO/LYSO calorimeter [6].

Fast neutron induced pulse height spectra in these photo-detectors were measured. Section 2 describes the experimental setup, the neutron sources used in this investigation and the calibrations of these photo-detectors for the absolute electron numbers and the APD gains. The neutron induced pulse height spectra measured by APDs and PIN diodes are presented in sections 3. Also presented in this section are the possible background pulse hight spectra by X-rays and low energy γ-rays. Section 4 describes a proposed hardware solution to eliminate the nuclear counter effect completely for future calorimeters using silicon photo-detectors. Section 5 provides a brief summary and a discussion of the cause of the neutron induced nuclear counter effect.

2. Experiment Setup and Calibrations

Figure 1 shows the experimental setup used to measure the neutron induced nuclear counter effect with a pair of ^{252}Cf neutron sources. Each of these two ^{252}Cf sources weighs 4.4 μg. As shown in the figure, these two sources were separated by 14 cm and were mounted at the top of a Paraffin wax shielding with dimension of $50 \times 10 \times 16$ cm (L$\times$W$\times$H). The photo-detectors were placed at the bottom of the Paraffin wax shielding at 8 cm from these sources. The neutron flux at the surface of photo-detectors was calculated to be 1.4×10^4 n/cm^2/s taking into account the ^{252}Cf constant 2.311×10^6 n/s/μg, the distances between the sources and the detector and the time since the delivery ($\tau_{1/2} = 2.6$ yr). A similar set-up using one ^{241}Am-Be neutron source was used for background study. In the ^{241}Am-Be setup the photo-detectors were placed 2 cm away from the source. The neutron flux at the surface of the photo-detectors was calculated to be 6×10^4 n/cm^2/s taking into account the NIST certified fast neutron flux (NIST Test No 846/257817-93), the distances between the source and the detector and the time since the calibration ($\tau_{1/2} = 432$ yr). The high flux of the ^{241}Am-Be neutron source allows a reduced data taking time for background runs.

Figure 2 shows the calibration setup for the Hamamatsu PIN diode and APDs. The absolute electron numbers in these devices were calibrated by using peaks of x-ray or low energy γ-ray

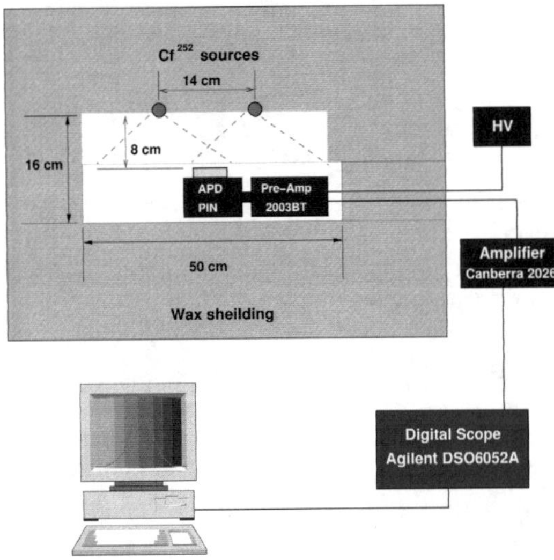

Figure 1. The experimental setup using a pair of ^{252}Cf sources.

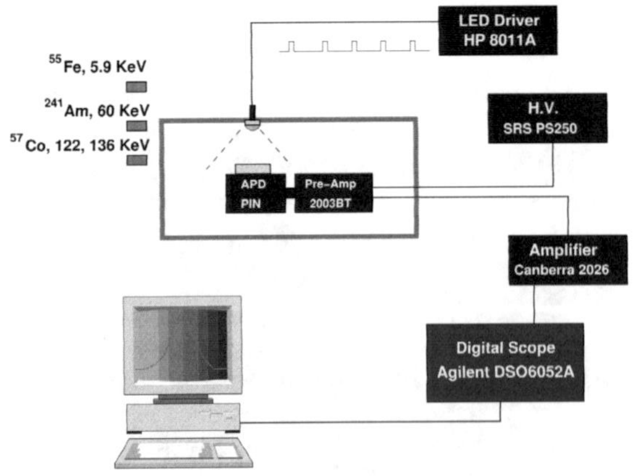

Figure 2. The photo-detector calibration setup using a pulsed blue LED as well as ^{55}Fe, ^{241}Am, and ^{57}Co x-ray sources.

sources: ^{55}Fe (5.9 KeV) for the APDs, and ^{241}Am (60 KeV) and ^{57}Co (122, 136 KeV) for the PIN diode.

Unless specified all irradiations were carried out with the front face of the photo-detectors facing the neutron sources. In both setups the signals from the photo-detectors were integrated by a Canberra 2003 BT pre-amplifier and shaped by a Canberra 2026 shaping amplifier before being sent to an Agilent 6052A digital storage oscilloscope. Data were taken with the scope operated in a self-trigger mode with threshold set at 1.5×10^5 electrons for APDs and 3.5×10^5 electrons for PIN diodes, allowing effective data taking to cover the high end of the pulse hight spectra. The silicon photo-detectors were reverse biased during the measurements, so that the PIN diode was fully depleted and the pair of APDs were set at a defined gain, e.g. 1, 10, 35, 100 and 200, while the nominal gain of these APDs in the CMS ECAL is 50. Because of the dark current and the large protection resistor in the Canberra 2003 BT pre-amplifier, the bias voltage from the power supply is different from that on the APDs. Unless specified, the bias voltage quoted in this text is the value applied to the pre-amplifier, which is larger than that on the APDs.

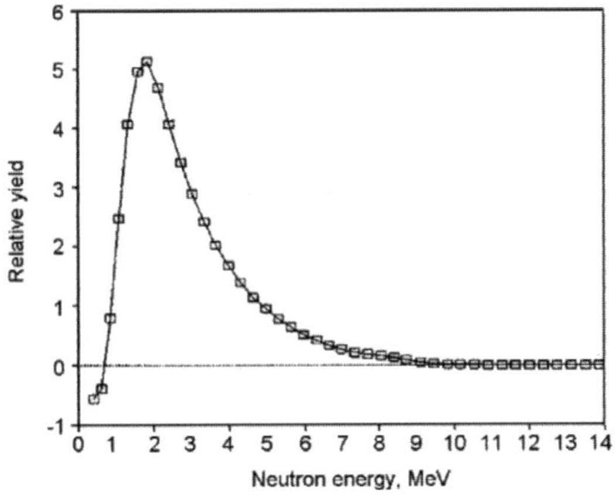

Figure 3. The neutron energy spectrum of the ^{252}Cf sources [10].

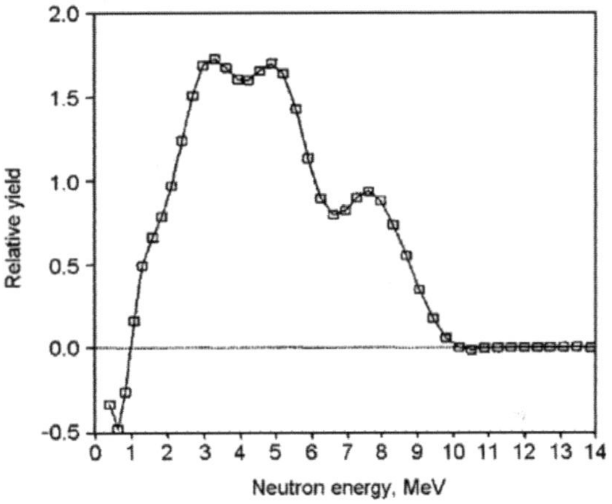

Figure 4. The neutron energy spectrum of the ^{241}Am-Be source [10].

Figure 3 and 4 show the energy spectra of neutrons from the ^{252}Cf source [10]. The ^{252}Cf sources provide fast neutrons with kinetic energy peaked at 2.2 MeV with a tail up to 10 MeV [10]. The peak kinetic energy of these neutrons match well with the fast neutrons produced in particle showers [4, 5]. The energy spectrum of the ^{241}Am-Be source has a broad distribution from 2 to 10 MeV with an average kinetic energy of about 4.5 MeV [10].

Calibration of the PIN diode is straight forward. The top plots of Figures 5 and 6 show pedestal distributions for the Hamamatsu S2744-08 photo-diode and the corresponding Gaussian fits. The bottom plots show the pulse height spectra obtained using a ^{241}Am source and a ^{57}Co source respectively. The corresponding peak ADC values for 60 keV, 122 keV and 136 keV shown in these figures were determined by Gaussian fits and with pedestal subtracted. Assuming 3.62 eV energy deposition is needed to create an electron-hole pair in the silicon device these peak values were used to extract the calibration constants of equivalent electron numbers per mV pulse height for the PIN diode. The calibrations constants obtained from these three peaks are consistent, as shown in Figure 7.

Calibrations for the APDs are more complicated. Figure 8 shows the APD gain as a function of the bias voltage applied to the Canberra 2003 BT pre-amplifier. As discussed before that this

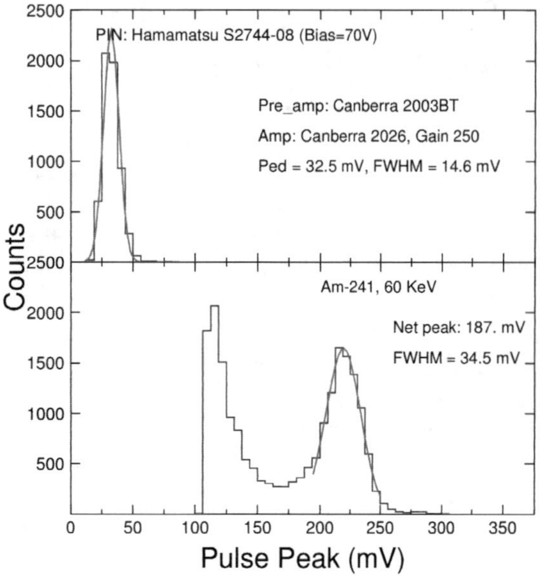

Figure 5. The pedestal (top) and the pulse height spectrum of 60 keV X-rays from a ^{241}Am source (bottom) are shown for the Hamamatsu S2744-08 PIN diode.

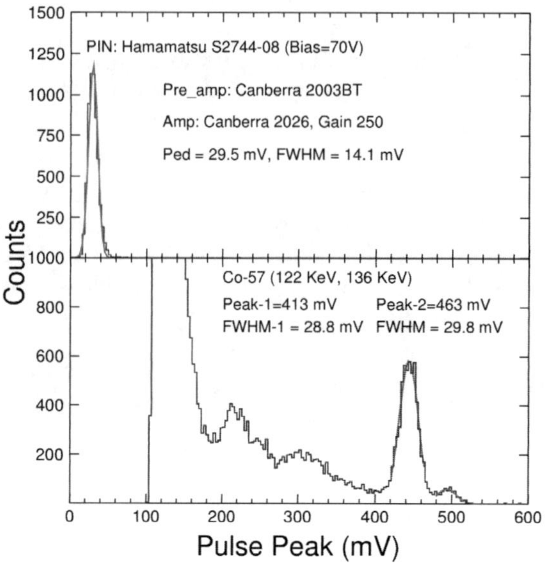

Figure 6. The pedestal (top) and the pulse height spectrum of 122 keV and 136 keV γ-rays from a ^{57}Co source (bottom) are shown for the Hamamatsu S2744-08 PIN diode.

bias voltage is larger than that on the APDs because of the dark current and the protection resistor in the pre-amplifier. The APD gain was measured with the pulsed blue LED light as

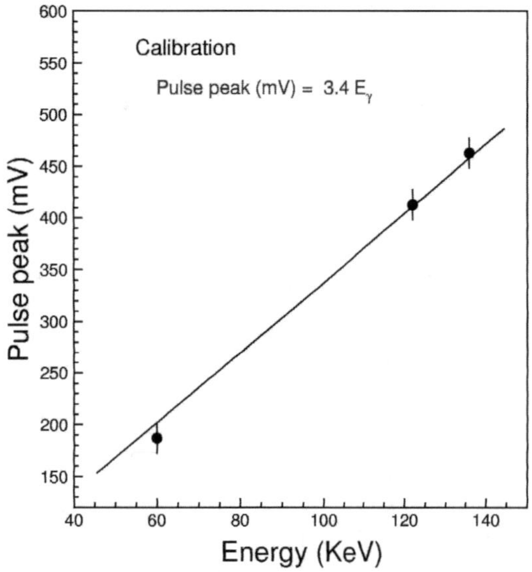

Figure 7. The peak values (in mV) of the pulse height spectra are shown as a function of the source energy (in keV) and a linear fit for the Hamamatsu S2744-08 PIN diode.

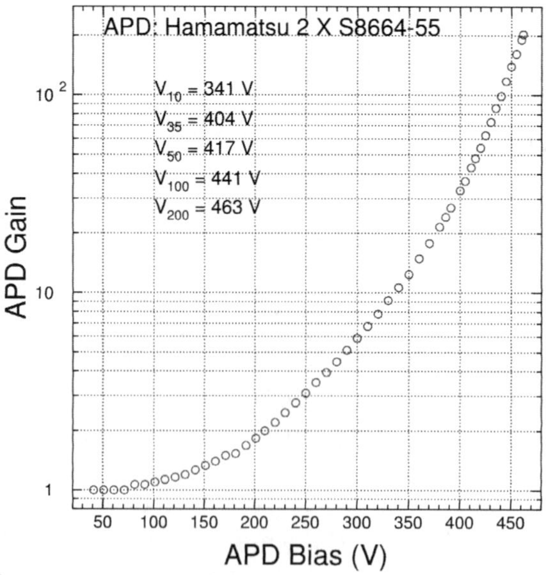

Figure 8. The APD gain is shown as a function of the bias applied to the Canberra 2003 BT pre-amplifier.

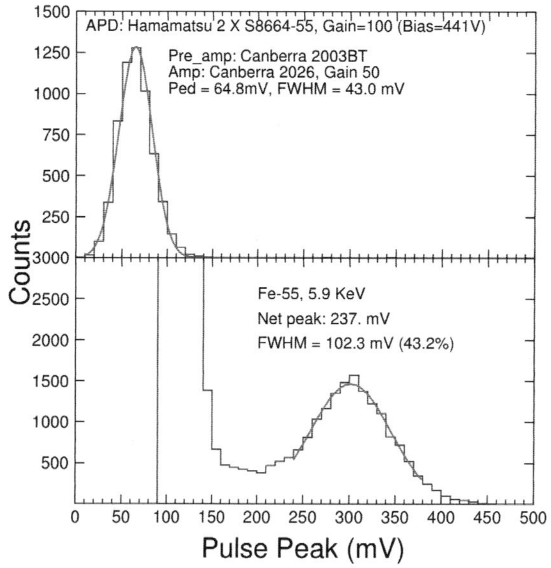

Figure 9. The pulse height spectra of the pedestal (top) and the 5.9 keV X-rays from a ^{55}Fe source (bottom) are shown for the Hamamatsu S8664-55 APDs.

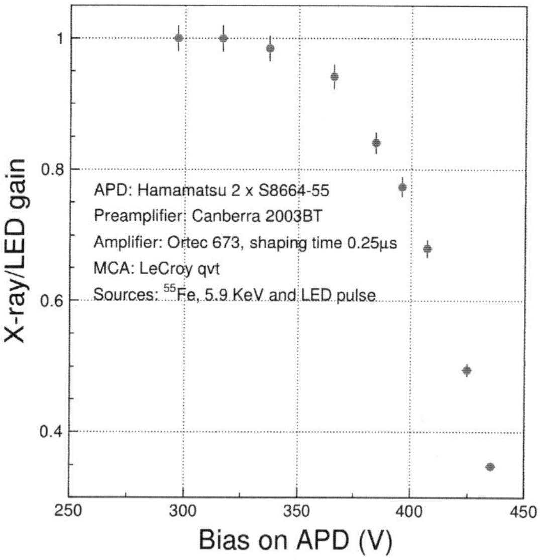

Figure 10. The correction factors used for the APD calibrations with 5.9 keV X-rays are shown as a function of the reverse bias voltage applied on the APDs.

shown in Figure 2. Since this is a relative measurement no specific knowledge of the photon numbers per LED pulse was required.

Figures 9 shows the pulse height spectra measured by the Hamamatsu S8644-55 APDs obtained by irradiating 5.9 keV X-ray from a ^{55}Fe source directly at the APDs. The APDs were reverse biased to set at a gain of 100. Similar calibration runs were carried out for the APD gains of 1, 10, 35 and 200. Like Figures 5 and 6, the top plot of Figures 9 shows a pedestal spectrum and a Gaussian fit. The bottom plot shows the spectrum taken with the ^{55}Fe source. It is known that when the reverse bias increases a fraction of the X-ray energies deposited inside the APDs does not receive the same amplification as scintillation photons would be [11]. Figure 10 shows the ratio between the APD responses to 5.9 KeV X-ray and to blue LED pulses as a function of the reverse bias applied to the APDs, which was used as the correction factors for the APD calibration at different gains. Once again no specific knowledge of the photon numbers per LED pulse was required in measuring these corrections since it is also a relative measurement.

3. Neutron Induced Nuclear Counter Effect in APDs and PIN Diodes

Figure 11 shows the pulse height spectra measured by two S8664-55 APDs for MeV neutrons from the ^{252}Cf sources. The horizontal scale is electron numbers determined according to the calibrations described in section 2. These electron numbers can be converted to equivalent energy deposition of scintillation photons according to the light output of the calorimeter, e.g. 4 p.e./MeV and 800 p.e./MeV respectively for PWO and LSO/LYSO crystals. Consistent spectra were observed in two runs lasting 91.5 h (blue dashes) and 62.1 h (green dots). The sum of the spectrum (solid red) represents a total run time of 154 h. Signals up to 2.2×10^6 electrons, corresponding to 8 MeV energy deposition in the sensitive layer of the APDs, are observed. A few ten overflow events caused by amplifier saturation are also noticed for these spectra taken with the APD gain of 35.

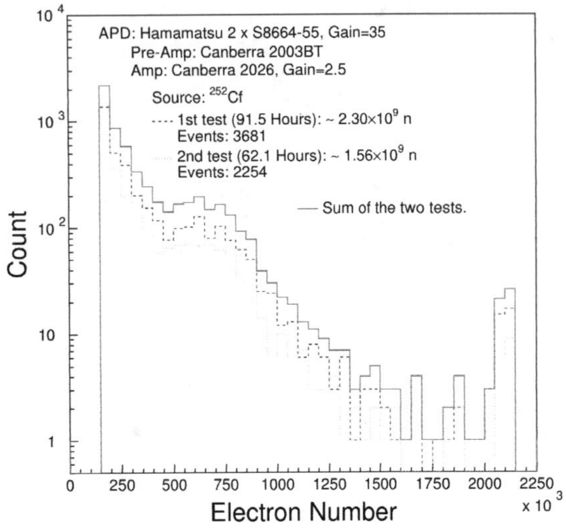

Figure 11. Two pulse height spectra (blue dashes and green dots) and their sum (solid red) observed by the APDs under irradiation with MeV neutrons from the ^{252}Cf source pair are shown as a function of electron numbers.

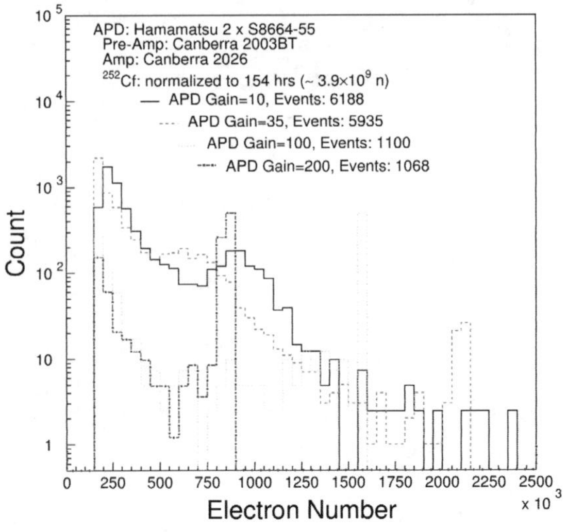

Figure 12. The pulse height spectra in electron numbers observed by the APDs under irradiation with MeV neutrons from the ^{252}Cf source pair are shown for APD gains of 10 (solid black), 35 (red dashes), 100 (green dots) and 200 (blue dotted dashes).

Figure 12 shows a comparison of the pulse height spectra measured by two S8664-55 APDs at various gains (solid black: 10, red dashes: 35, green dots: 100 and blue dotted dashes: 200), for fast neutrons from the ^{252}Cf source. Signals of up to 2.5×10^6 electrons are observed for the APD gain of 10 without overflow, indicating up to 9 MeV energy deposition. Overflow events caused by amplifier saturation are observed for gains of 35, 100 and 200. The corresponding fractions of events with signals of more than 1.5×10^5 electrons is 1.6, 1.5, 0.28 and 0.27×10^{-6} respectively for the APD gains of 10, 35, 100 and 200, indicating that a high APD gain reduces the neutron-induced nuclear counter effect in the APDs. This observation is easy to understand: when the gain increases the electron signals induced by neutrons deep in the APDs receive a reduced amplification as compared to scintillation photons.

Figure 13 compares the pulse height spectra measured by the pair of S8664-55 APDs (blue dashes) and the S2744-08 PIN diode (solid red) for fast neutrons from the ^{252}Cf source. The irradiations were carried out with the front face of the photo-detectors facing the neutron sources. For this test the gain of the S8664-55 APDs was set at one with 70 V bias. Signals up to 2.5 and 3.0×10^6 electrons are observed for APDs and PIN diodes respectively. The high endpoint is observed for the S2744-08 PIN diode since it has a significantly thicker active layer than the S8664-55 APDs. The fraction of events with signals of more than 3.5×10^5 electrons is 18 and 1.0×10^{-6} for the PIN diode and APDs respectively, indicating that the neutron-induced nuclear counter effect in APDs is more than an order of magnitude smaller than that in PIN diodes. It is clear that neutron induced nuclear counter effect can be reduced by using thin photo-detectors. In other words, a thick silicon photo-detector would be more effective as a neutron sensor.

It was suspected that the 500 μm thick epoxy resin at the front of these photo-detectors play an important role for the neutron induced nuclear counter effect since a low mass nucleus is an effective neutron moderator. It is well known that the maximum fraction of energy lost in a single elastic collision depends on the mass of the nucleus. It is ranged from 100% for ^1_1H to

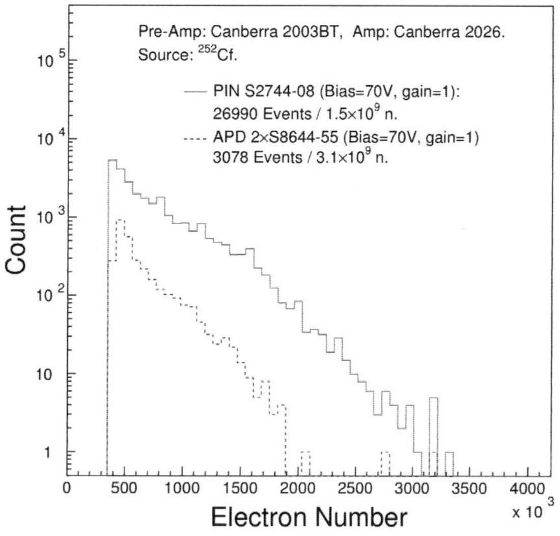

Figure 13. Pulse height spectra in electron numbers, observed in the Hamamatsu S2744-08 PIN diode (solid red) and the pair of Hamamatsu S8664-55 APDs (blue dashes), under irradiation by MeV neutrons from the ^{252}Cf source pair to the front face of the photo-detector.

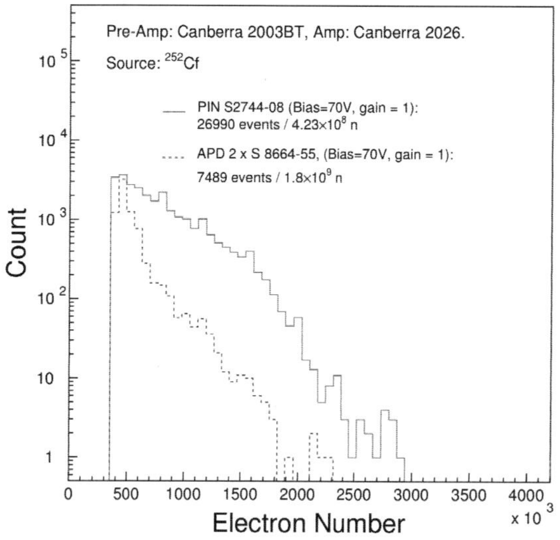

Figure 14. Pulse height spectra in electron numbers, observed in the Hamamatsu S2744-08 PIN diode (solid red) and the pair of Hamamatsu S8664-55 APDs (blue dashes), under irradiation by MeV neutrons from the ^{252}Cf source pair to the back face of the photo-detector.

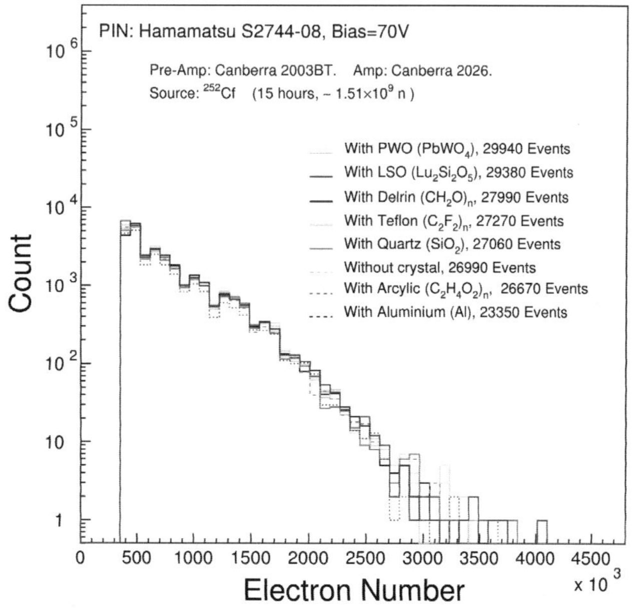

Figure 15. Pulse height spectra in electron numbers, observed in the Hamamatsu S2744-08 PIN diode under irradiation by MeV neutrons from the ^{252}Cf source pair with various materials inserted in front of the PIN diode.

1.7% for $^{238}_{92}U$. Figure 14 shows pulse height spectra similar to Figure 13 but with the back face of the photo-detectors facing the neutron sources. It is interesting to note that the fraction of events with signals of more than 3.5×10^5 electrons is 64 and 4.2×10^{-6} respectively for the PIN diode and the APDs, which is about a factor of four of what observed with the front face of the photo-detectors facing the sources. Since the thickness of the epoxy layers is 500 μm at the front but only a few ten μm at the back of these photo-detectors [9] this observation shows clearly that the 500 μm thick epoxy layer at the front of these devices do not enhance the neutron induced nuclear counter effect. A further measurement with an additional layer of epoxy of about 500 μm thick pasted at the front of a S2744-08 PIN diode also shows a reduction of the nuclear counter effect by about 10%.

To further investigate the effect of materials at the front of these photo-detectors various materials were inserted between these photo-detectors and the neutron source. Figure 15 shows the pulse height spectra of the Hamamatsu S2744-08 PIN diode for MeV neutrons from the ^{252}Cf source pair with various materials of about 1 to 2 cm thick inserted in front of the PIN diode. The observed spectra have overall consistent shape, indicating that 1 to 2 cm thick materials inserted have only a minor effect on fast neutrons as expected. The fraction of events with signals of more than 3.5×10^5 electrons, however, shows that the high Z materials, such as PWO and LSO, enhance slightly the neutron induced nuclear counter effect, while the low Z materials, such as H, C, O and Al, reduce slightly the effect. This observation is consistent with what was discussed above and reflects the fact that low Z material is a more effective moderator for fast neutrons.

1 MeV γ-rays from a ^{60}Co source and 60 keV X-rays from a ^{241}Am source were used to check possible background effect. Figures 16 and 17 show the pulse height spectra measured by the pair of S8664-55 APDs for fast neutrons from the ^{252}Cf source pair and the ^{241}Am-Be source respectively. Also shown in these figures are pulse height spectra measured by the APDs for the 1 MeV γ-rays from a ^{60}Co source and the 60 keV X-rays from a ^{241}Am source. While the fractions of events with signals of more than 1.5×10^5 electrons are 1.5 and 2.8×10^{-6} for the ^{252}Cf source and the ^{241}Am-Be source respectively, they are 13 and 5×10^{-11} for the

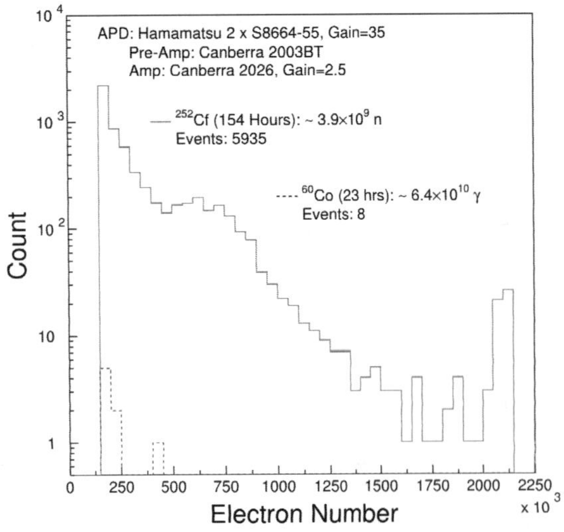

Figure 16. A comparison of the pulse height spectra in electron numbers, observed in the pair of Hamamatsu S8664-55 APDs under irradiation by MeV neutrons from the ^{252}Cf source (solid red) and MeV γ-rays from a ^{60}Co source (blue dashes).

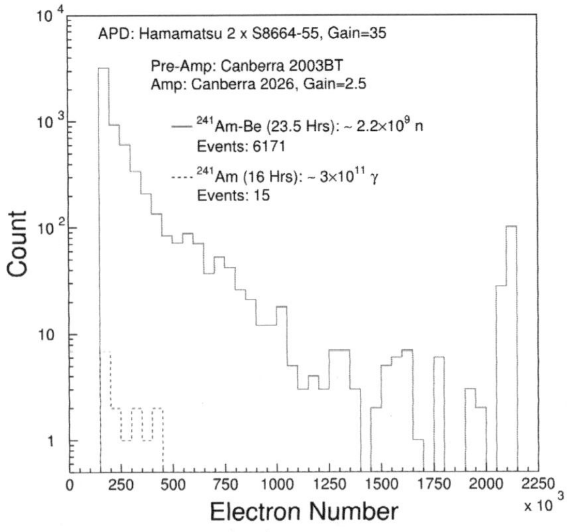

Figure 17. A comparison of the pulse height spectra in electron numbers, observed in the pair of Hamamatsu S8664-55 APDs under irradiation by MeV neutrons from a ^{241}Am-Be source (solid red) and 60 keV X-rays from a ^{241}Am source (blue dashes).

^{60}Co source and the ^{241}Am source respectively. It is clear contributions of X-rays and γ-rays are negligible in the spectra shown in Figures 11, 12, 13 and 15. Similarly, replacing photo-detectors with a capacitance showed no counting in pulse heigh spectrum indicating that these spectra are entirely produced by the photo-detectors with no contamination from the readout electronics.

4. Proposed Solution

As discussed in section 1, the contamination in calorimeter data by neutron induced nuclear counter effect may be reduced by using event topology based rejection, where isolation, shower shape and event timing may be used [7]. However, it is clear that a complete elimination would need multiple readout channels for a single calorimeter cell. Two photo-detectors mounted on a single calorimeter cell with independent readout, for example, would provide two signals of equal amplitude for normal events generated by scintillation photons. When a difference is observed between these two signals the signal with a larger amplitude can be identified as being contaminated by nuclear counter effect and can be discarded. Doing so would preserve valid data for the calorimeter cell and eliminate nuclear counter effect completely. Implementation of multiple readout channels for a single calorimeter cell, however, increases the overall calorimeter cost. Existing calorimeters do have multiple photo-detectors for single calorimeter cell, but do not have multiple independent readout channels to conserve channel counting.

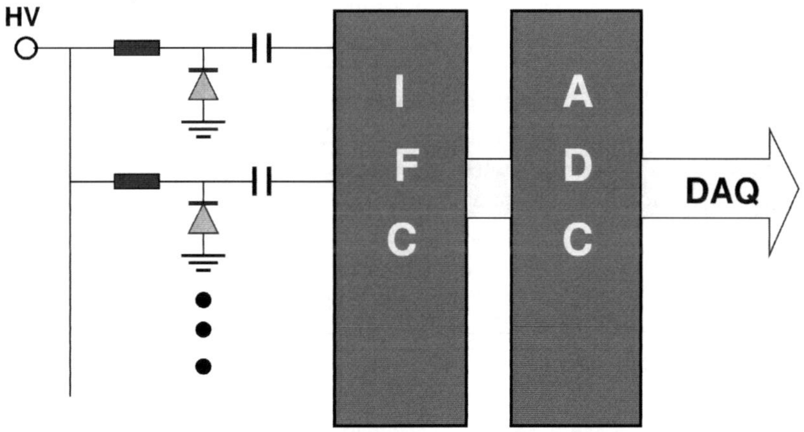

Figure 18. The proposed scheme to eliminate the neutron induced nuclear counter effect using multiple readout devices and an intelligent front end chip.

Figure 18 is a schematic showing a proposed readout scheme using multiple photo-detectors. In this scheme an intelligent front-end chip compares signals from different photo-detectors and feeds only the ones not contaminated for digitization. By using this scheme the nuclear counter effect would be eliminated completely without increasing the total channel counting. Intelligent front-end chips with such capacity are a standard in modern calorimeter readout design, where comparators and amplifiers are integrated in the chip to provide an extended dynamic range of 10^5. Integration of additional comparators and selected readouts in such a chip is straight forward. Solutions along this line are discussed in the proposed CMS upgrade technical proposal [12], and are also expected to be implemented in the design for the proposed SuperB LSO/LYSO endcap calorimeter [6].

5. Conclusion and Discussion

Fast neutrons from ^{252}Cf source induce signals of up to several million electrons in Hamamatsu silicon APDs and PIN diodes causing neutron induced nuclear counter effect. This effect is believed to be realized in a two step process in the bulk silicon of these photo-detectors. In the first step fast neutrons lose a fraction of their kinetic energy to nuclei through elastic neutron-nucleus collisions with a cross-section up to 10 barns. In the second step the charged nuclei deposit their kinetic energies in photo-detectors through ionization, which are converted into electron signals. The average number of elastic scattering collisions needed for a 2 MeV neutron to slow down to thermal energies, for example, ranges from 18 for ${}^{1}_{1}H$ to >2000 for ${}^{238}_{92}U$. Multiple collisions occur between a fast neutron and several nuclei in the photo-detectors, leading to the end point of the observed spectra. On the other hand, the absorption cross-sections, such as (n,p) and (n, α) are more than five orders of magnitude lower than the elastic collisions for fast neutrons. The absorption process would play a major role when the neutrons are slowed down to thermal energies. Our data show that the 500 μm thick epoxy coating at the front of these photo-detectors does not enhance the neutron induced nuclear counter effect. Our data also show that a low Z material at the front of these photo-detectors reduces the neutron induced nuclear counter effect.

The equivalent energy of the neutron induced nuclear counter effect in a crystal calorimeter depends on the crystal light yield. Anomalous signals of 2.5 million electrons, for example, correspond to 625 GeV and 3.1 GeV for PWO and LSO/LYSO crystals respectively assuming the light yields of 4 and 800 p.e./MeV for calorimeter size crystals. As discussed in section 1, the impact of such signals on the physics output is comparable between the CMS PWO calorimeter at the LHC and the proposed SuperB forward LSO/LYSO calorimeter at the proposed SuperB collider. The rate of the neutron induced nuclear counter effect in APDs is more than an order of magnitude lower than that in PIN diodes, indicating that APD is a preferred photo-detector in this aspect. Increasing the APD gain also reduces the effect. All these are due to the fact that only a portion of the neutron induced electron signals in the APDs receives full multiplication.

The neutron induced nuclear counter effect may be reduced by advanced data analysis by using event topology based selection. Its complete elimination requires multiple readout channels for a single calorimeter cell. An intelligent front-end chip with multiple photo-detectors is proposed to provide this function without requiring multiple readout channels for a single calorimeter cell, and thus leading to a significant cost saving for future calorimeters designed to be immune to the nuclear counter effect.

Acknowledgments
This work was supported in part by the U.S. Department of Energy under grant DE-FG03-92-ER-40701 and the U.S. National Science Foundation Award PHY-0612805. Many discussions with CMS and SuperB colleagues are acknowledged.

References
[1] A. Satpathy, K. Tamai, M. Fukushima, D.Y. Kim and M.H. Lee, "Nuclear Counter Effect of Silicon PIN photo-diode Used in CsI(Tl) Calorimeter," *Nucl. Instr. and Meth.* **A391** (1997) 423-426, and K. Ueno, S.K. Sahu, K.C. Peng W.S. Hou and C.H. Wang, "Detection of Minimum-Ionizing Particles and Nuclear Counter Effect with Pure BGO and BSO Crystals with Photo-diode Read-out," arXiv:physics/9704013v1 (1997) [physics.ins-det].

[2] Th. Kirn, Y. Musienko, Th. Flugel and D. Renker, "Performance of the Most Recent Hamamatsu Avalanche Photo-diode", *Nucl. Instr. and Meth.* **A387** (1997) 199-201, and D. Zurcher, "Nuclear Counter Effect and π-e Misidentification," *Nucl. Instr. and Meth.* **A442** (2000) 433-437.

[3] Y. Hosono, Sjafruddin, T. Iguchi and M. Nakazawa, "Fast Neutron Detector using PIN-type Silicon Photo-diode," *Nucl. Instr. and Meth.* **A361** (1995) 554-557.

[4] The CMS Collaboration, "The CMS Electromagnetic Calorimeter Project", CERN/LHCC 97-33 (1997).

[5] See, for example, discussions in the "Radioactivity and Radiation Protection," section 29 of the "Review of Particle Physics," *Phys. Lett.* **B667** (2008).

[6] C. Cecchi. "A LYSO Calorimeter for a SuperB Factory ," in *Proceedings of Fourteenth International Conference on Calorimetry in Particle Physics*, Journal of Physics Series (2010).

[7] The CMS Collaboration, "Electromagnetic calorimeter commissioning and first results with 7 TeV data," CMS Note 2010-012 in CERN Document System http://cdsweb.cern.ch/record/1278160?ln=en, and K. Theofilatos, "CMS ECAL Status and Performance with the First LHC Collisions," in *Proceedings of Fourteenth International Conference on Calorimetry in Particle Physics*, Journal of Physics Series (2010).

[8] K. Deiters, Y. Musienko, S. Nicol, B. Patel, D. Renker, S. Reucroft *et al.*, "Properties of the most recent avalanche photo-diodes for the CMS electromagnetic calorimeter", *Nucl. Instr. and Meth.* **A442** (2000) 193-197.

[9] T. Zarzycki of Hamamatsu, private communication.

[10] S.-T. Park, "Neutron Energy Spectra of ^{252}Cf, Am-Be source and of the D(d,n)^{3}He Reaction," *J. Rad. & Nucl. Chem.* **Vol 256** (2003) 163-166.

[11] J.M. Chen, R.H. Mao, L.Y. Zhang and R.-Y. Zhu, "Large Size LSO and LYSO Crystals for Future High Energy Physics Experiments," *IEEE Trans. Nucl. Sci.* **54** (2007) 718-724.

[12] The CMS Collaboration, "Technical Proposal for the Upgrade of the CMS Detector through 2020," **CMS U1TDR** (2011).

The High Dynamics Readout of PMT for BGO Calorimeter

Yunlong Zhang, Xiaolian Wang, Zizong Xu, Bing Li

Department of Modern Physics, University of Science and Technology of China, Hefei, Anhui, China, 230026

E-mail: ylzhang1@mail.ustc.edu.cn

Abstract. For one cell of BGO-EM energy deposition of electron/gamma with energy above TeV is between 0.5 MIPs and 10^5 MIPs, the readout system need to cover this high dynamic range.

1. Introduction

The role of calorimeter is used to measure the energy of incident particle. In order to measure the energy deposited in calorimeter, readout system should be designed carefully. In this paper, the current status of readout test is reported.

2. The Experiment Setup

The block diagram for the calibration measurement is shown in Fig.1. The LED was driven by pulse generator (GEN). Optical fibers were used to transmit light from the LED to the PMT which was coupled with a BGO crystal. The photon intensity incident on the photocathode was tuned by changing the amplitude of the supply voltage or the pulse width of GEN. The analog signal from PMT was amplified by VA32 which is a charged sensitive amplifier.

2.1. Response Function of PMT

The PMT response function could easily be described by eq. (1). [1-2]

$$\begin{aligned} S_{idear} &= P(n,\mu) \otimes G_n(x) \\ &= \sum_{n=0}^{n=\infty} \frac{\mu_n e^{-\mu}}{n!} \frac{1}{\sigma_1\sqrt{2\pi}} exp(-\frac{(x-nQ_1)^2}{2n\sigma_1^2}) \end{aligned} \quad (1)$$

Here the number of detected photoelectrons in PMT cathode follow a poisson distribution. The gaussian distribution describe the response of a multiplicative dynode system to a single photo-electron.

$$while \quad n = 0 \qquad \begin{aligned} Q_1 &= Q_0 \\ \sigma_1 &= \sigma_0 \end{aligned} \quad (2)$$

There are some parameters:

1. μ is the mean photon-electrons number detected by PMT,

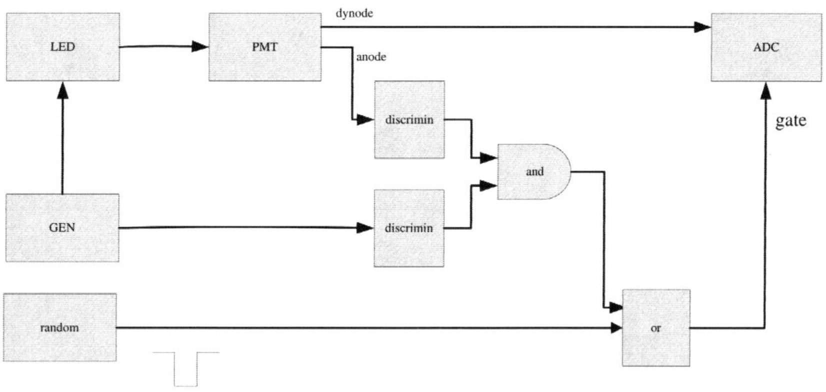

Figure 1. The block diagram for the calibration measurement.

Table 1. results of XP2262B spectrum			
μ	Q_0	Q_1	χ^2/ndf
1.6±0.06	72.2±5.5	339.5±9.2	252.9/247
2.6±0.08	78.2±7.3	325.0±8.7	260.4/265
10.0±0.2	70.0±4.1	328.1±7.3	306.1/259

Table 2. The Gain of XP2262B	
Mean Photon-electron Number (μ)	Gain(*10^6)
1.6±0.06	2.42±0.065
2.6±0.08	2.32±0.062
10.0±0.2	2.34±0.052

2. σ_1 is the single photon-electron spectrum's sigma,
3. Q_1 is the position of single photon-electron peak,
4. Q_0 is the pedestal position,
5. σ_0 is the pedestal width.

For more detailed expression, please see reference [1].

2.2. Single Photon-electron Spectrum

In order to test our PMT, we carried out several LED spectra that were measured by XP2262B PMT. Fig. 2 showed the output of XP2262B. we can see as the light intensity of LED increase, the mean photon-electron numbers increase too. The results of spectral processing are showed in Table 1. The spectrum were de-convoluted by means of a program based on the Minuit Minimization package using the eq. (1) as a fitting function.[2]

The position of XP2262B's single photon-electron peak was confirmed. So we can calculate the gain of PMT. The results were showed in Table 2.

Fig.3 shows the LED spectra measured by R5611 PMT. The fitting function is eq. (1) too. from the fitting result, we get the mean photon-electron number is 4.4. The position of single photon-electron peak of R5611 is 24.4 channel of ADC. We also could calculate the gain of R5611 is $1.7 * 10^5$ under 800 High Voltage.

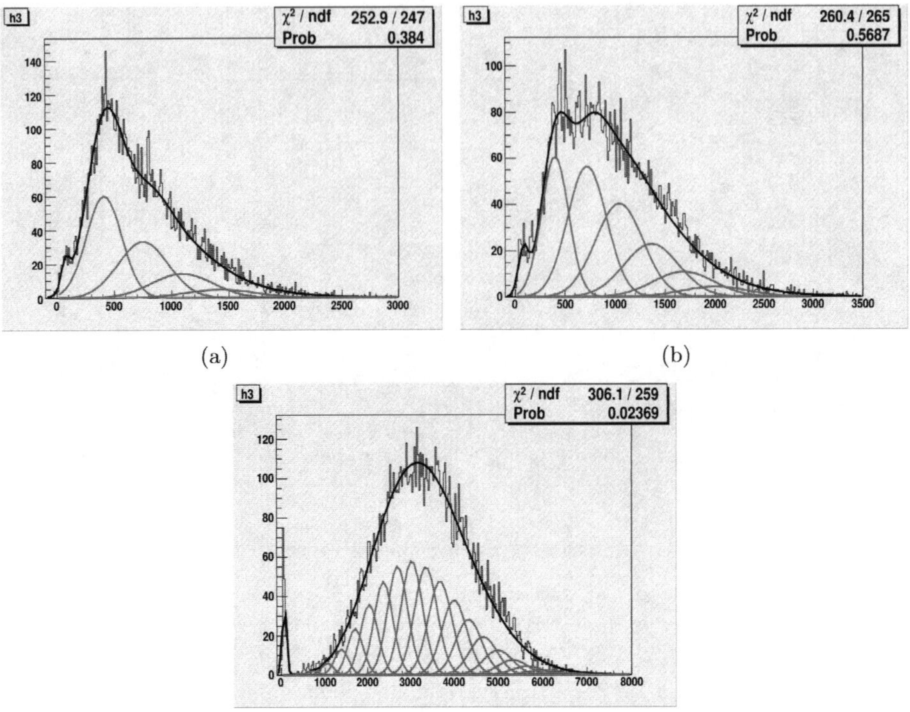

Figure 2. LED Spectrum Measured by XP2262B.

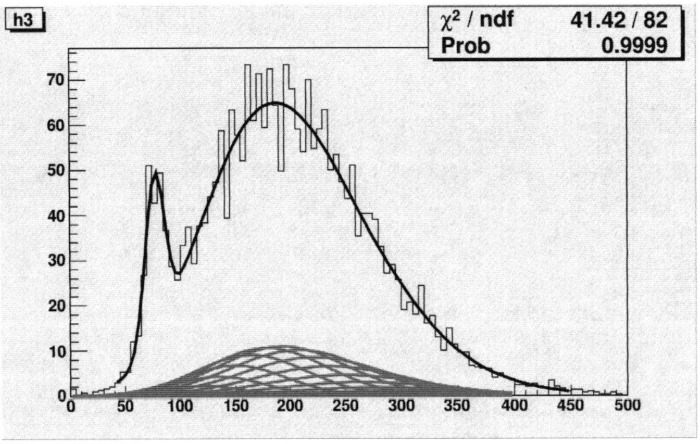

Figure 3. LED Spectrum Measured by R5611.

3. The Readout of PMT

In order to cover very high dynamic range (0.5 MIPs to 10^5 MIPs). We read signal from three dynodes of R5611. Fig. 4 shows the relationship between dynode 7 and dynode 4. As the light intensity of LED increase gradually, the signal of dynode 7 and dynode 4 increase too. The signal of dynode is about 20 times that of dynode 4. The range for each readout channel was tested using LED. Fig. 5 shows the output range of dynode 7 and dynode 4.
Fig. 5 shows the cosmic ray test system. The random trigger that we just want to know the pedestal position.The result of dy7 and dy4 output were showed in Fig. 6. The signal of dy7 is

Table 3. The Dynamic Range of Output Channel		
	Dy7	Dy4
Min(MIPs)	0.5	100
Max(MIPs)	250	2500

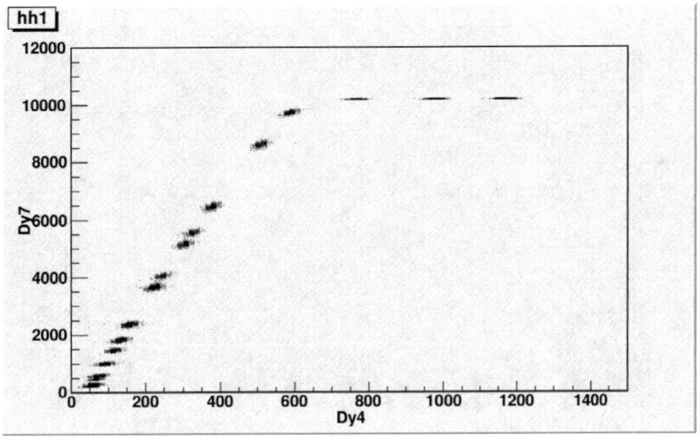

Figure 4. dynode 4 vs dynode 7.

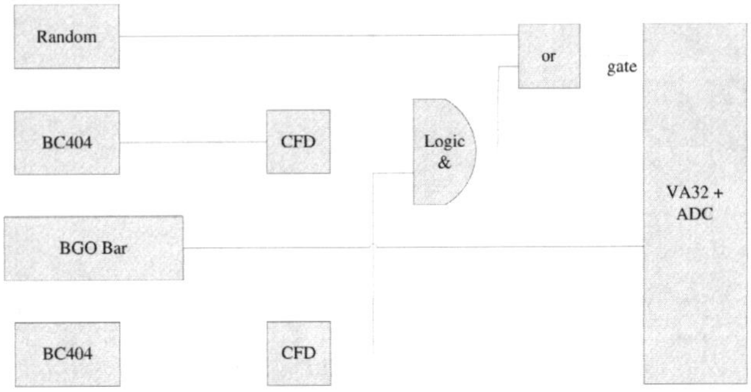

Figure 5. The block diagram for cosmic test.

about 3 pc and that of dy4 is about 150 fc.
We put an absorber between crystal and PMT window to attenuate photons (like Fig. 7), ensuring the response of Dy7 to 0.5 MIPs is about 40 channels. So the dynamic range of Dy7 is 0.5 MIPs to 125 MIPs. According the correlation between Dynode 7 and dynode 4. we can know the output range is from 100 MIPs to 2500 MIPs. The results were shown in Table 3.

4. Further Development

We should complete dynode readout studies, especially dynode 1. The "absorber" should test carefully. We will try to test another output way —- PD and APD.

5. Conclusion

We have presented response function of PMT. Use LED to test dynode output. Construct a cosmic ray test system to test the dynode output of muon.

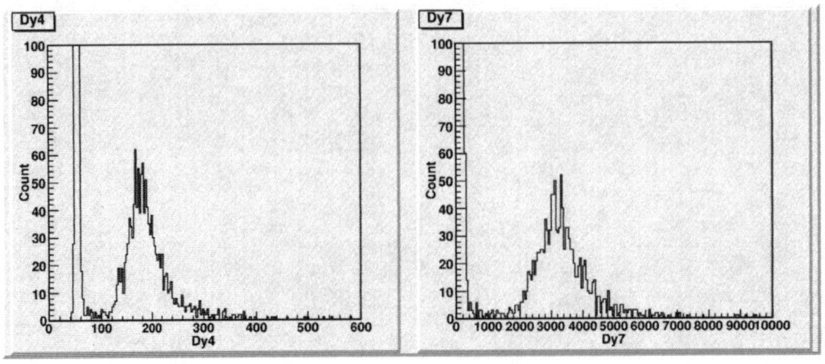

Figure 6. The result of cosmic ray test.

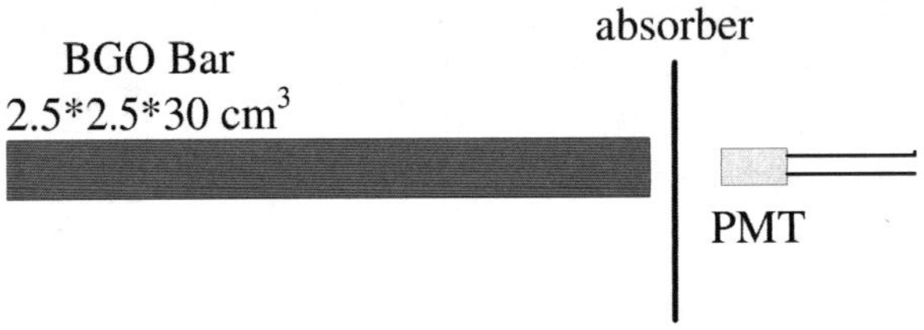

Figure 7. Absorber between crystal and PMT.

6. Acknowledgments

I appreciate the support and encouragement of Prof. Xu and Prof. Wang during the performance of this work. I would like to thank Bing Li, Qingchang Feng, Jianhua Guo for discussing some problems.

7. References

[1] E.H. Bellamy. Et al., Nucl. Instr. And Method., A339(1994) 468-476.
[2] S. TokSar, I. SSykora and M. Pikna., ATL-TILECAL-99-005.
[3] MINUIT Minimization and Error Analysis, Release 89.12j, CERN, Geneva, 1989 .

Journal of Physics: Conference Series **293** (2011) 012014 doi:10.1088/1742-6596/293/1/012014

Digital HCAL Electronics: Status of Production

Gary Drake and José Repond

Argonne National Laboratory

Corresponding author's e-mail address: drake@hep.anl.gov

Abstract. This is a status report of the production of the readout electronics for the Digital Hadron Calorimeter (DHCAL) prototype. The prototype will be equipped with Resistive Plate Chambers (RPCs), read out with 1 x 1 cm^2 pads. The readout of each channel is simplified to provide a yes or no (digital readout) within a time bin of 100 ns. Each detector layer with an area of 96 x 96 cm^2 contains close to 10,000 readout channels. The total channel count for the entire prototype calorimeter with 38 active layers is approximately 350,000.

1. The Digital Hadron Calorimeter

To fully exploit the physics potential of a future lepton collider, operating at centre-of-mass energies of 0.5 TeV and above, requires an unprecedented jet energy resolution, of the order of a factor of two better than previously attained. The preferred approach to achieve this performance is through the application of Particle Flow Algorithms (PFAs) [1]. PFAs utilize the tracker to measure the momenta of charged particles, the electromagnetic calorimeter to measure the energy of photons and the combined electromagnetic and hadronic calorimeters to measure the energy of neutral hadrons, i.e. the neutrons and K_L^0's. The major challenge of PFAs is the association of energy deposits in the calorimeter to the charged or neutral particles originating from the interaction point. This challenge is met with the development of highly segmented or so-called imaging calorimeters [2].

In this context we develop a Digital Hadron Calorimeter (DHCAL) utilizing Resistive Plate Chambers (RPCs) as active medium. Our RPCs are based on two different designs [3] with either two or only a single glass plate. In the latter case the readout board serves as anode and closes the gas volume. The chambers use 1.1 mm thick glass as resistive plates and are operated in avalanche mode. The gas gap measures 1.2 mm and is maintained with the help of fishing lines spaced 5 cm apart.

The chambers are read out with a readout board containing 1×1 cm^2 pads and a binary (or digital) electronic readout system. The latter is based on the DCAL chip [4] (developed by Argonne and Fermilab), which has been optimized for the readout of large number of channels.

The collaboration is currently assembling a large DHCAL prototype calorimeter with the aim of validating the technological approach and of measuring hadronic showers with unprecedented spatial resolution. This paper describes the status of the assembly and commissioning of the electronic readout system.

The project is being carried out by the DHCAL collaboration [5] including 36 people from Argonne National Laboratory, Boston University, Fermilab, Iowa University and the University of Texas at Arlington. It is part of the scientific program of the international CALICE collaboration [6], which develops finely segmented calorimetry for future lepton colliders.

2. Overview of the Electronic Readout System

The electronic readout system has been optimized for the readout of large number of channels. Figure 1 shows the block diagram of the system. The 1 x 1 cm^2 pads of the pad-board are read out by a custom front-end ASIC, the DCAL III chip. Each chip handles 64 channels and applies a common threshold to all channels. The chips can operate either in high gain (e.g. for the readout of GEMs) or low gain (for the readout of RPCs) mode. The data is acquired either in triggerless (noise measurement, cosmic ray data) or triggered (cosmic ray or test beam data) mode. Each front-end board contains 24 DCAL chips and is connected via dots of conductive glue to the pads of a pad-board. The latter cover an area of 32 x 48 cm^2 and contain 1536 pads.

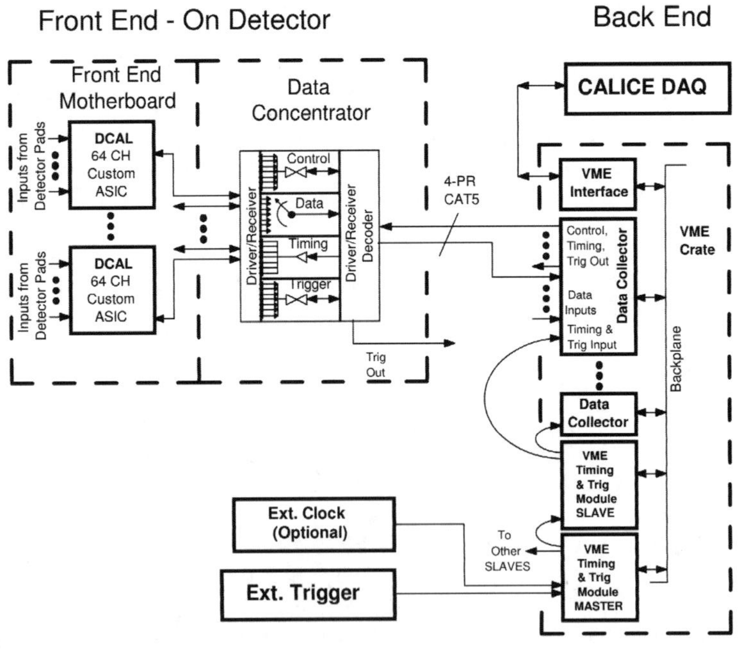

Figure 1. Block diagram of the DHCAL readout electronics.

The data from the 24 DCAL III chips are routed to a data concentrator (DCON), which is located on the edge of the front-end board. A VME-based data collector (DCOL) system receives the data from the DCONs, where each DCOL module connects to twelve DCONs or front-end boards. The entire system counts 20 DCOLs in two VME-crates. A Timing and Triggering Module (TTM) in each crate provides the necessary clock signals, resets and triggers for the front-end. An additional TTM serves as master to co-ordinate the (slave) TTMs in each crate.

3. Status of the Front-end Chip Production

The DCAL III front-end ASCIs were conceived in 0.25 μ CMOS technology. The output of the chips provides a timestamp with a 100 ns resolution and the corresponding hit pattern of its 64-channels. An engineering run produced eleven wafers with of the order of 10,300 parts. Of these a small number of chips were tested extensively on the bench. As an example of these tests Fig. 2 shows the S-curves for the 64 channels of a chip. Notice the sharp transition from 0 to 100 counts and the tight distribution of the curves. The bench tests failed to identify a single design flaw.

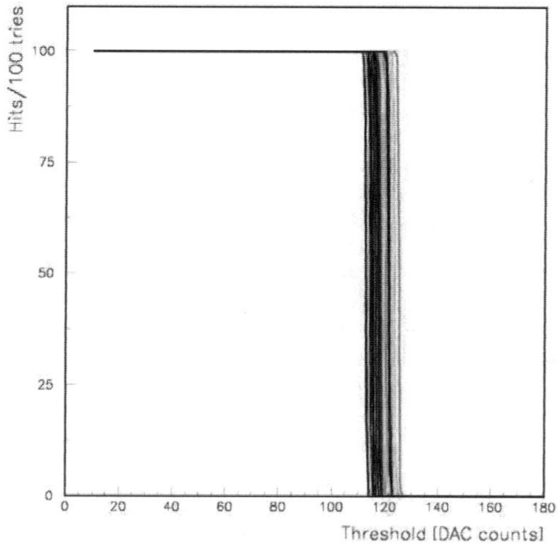

Figure 2. S-curves as measured with the DCAL III chips. The chip was operated in high gain mode and the charge was injected internally.

Before use in the assembly of the front-end boards, all chips were tested at Fermilab with the help of a robotic system, see Fig. 3. Of the initial 10,300 chips 8,644 passed the tests with zero errors, corresponding to an (average) yield of 84%.

Figure 3. Photograph of the robotic system testing the DCAL III chips.

4. Status of the Front-end Board Production

The front-end boards have been produced and are currently being assembled. Figure 4 shows a photograph of such a board. The darker area of the board indicates the data concentrator part. The firmware for the DCONs has been written and is entirely debugged. The boards have been operated without single malfunctioning for several weeks.

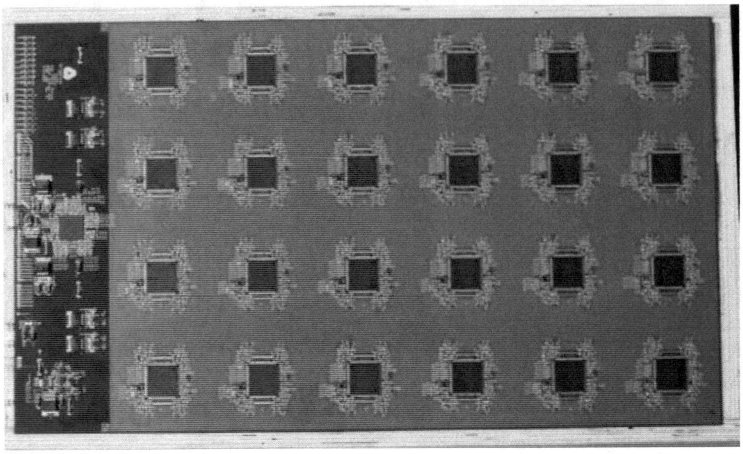

Figure 4. Photograph of a production front-end board containing 24 DCAL III chips.

The pad-boards are currently being fabricated. Since the boards do not contain any active components, no assembly is required.

A fixture for applying the glue dots to connect the front-end boards and pad-boards has been designed and built. To date of the order of 20/228 boards have been successfully glued. The operation takes approximately 55 minutes/board. With further improvements to the gluing fixture the required time is expected to be reduced by about a factor two.

The first cassette containing three RPCs and six front-end boards has been assembled. Figure 5 shows a photograph of the cassette with the top-cover removed to show the location of the front-end boards. The cassette has been operated for several weeks without problems of readout errors.

Figure 5. First detector cassette with six front-end boards. The top cover has been removed to show the location of the front-end boards.

5. Status of the Back-end Production

The data collector system has been fabricated, assembled and tested. The firmware is written and has been completely debugged. This work was carried out by Boston University.

The Timing and Triggering Modules (TTMs) required some modification compared to what was used in previous small scale tests [7]. In order to synchronize the clocks and the readout of separate VME-crates one TTM is configured to act as a master, fanning out the signals to slave TTMs, one per VME-crate. The modifications have been implemented into the design and the boards have been fabricated and (in part) assembled. The boards have been tested and perform as expected. The design work and the development of the firmware have been carried out by Fermilab.

6. Status of the Low-Voltage Power Supply

The operation of the front-end boards requires +5 Volts. The power is generated by Wiener power supplies and subsequently is distributed to the front-end boards. The distribution boxes have been designed and assembled. Figure 6 shows a photograph of one pair of power supply and distribution box. For safety reasons, each power line is separately fused.

Figure 6. Photograph of a Wiener power supply with the power distribution box located underneath.

7. Conclusions

The electronic readout system for a Digital Hadron Calorimeter (DHCAL) with Resistive Plate Chambers (RPCs) as active elements has been designed and developed. The system is optimized for the readout of large number of channels and utilizes the DCAL III chip for the front-end.

Production of the system for a large prototype calorimeter, with of the order of 350,000 readout channels, is ongoing. Particular care was devoted to thorough tests at every step in the prototyping process. Construction of the system is expected to be completed in early summer 2010 and will be followed by several running periods in the Fermilab test beam.

8. Acknowledgements

The authors would like to thank the organizers for an impeccably organized workshop and for the opportunity to present the work of the DHCAL collaboration to the calorimeter R&D community.

References

[1] M.A. Thomson, Particle Flow Calorimetry and the PandoraPFA Algorithm, Nucl. Instrum. Meth, **A**611:25 (2009) and references therein.
[2] J. Repond, Calorimetry at the International Linear Collider, Proceedings of the 10th Pisa Meeting on Advanced Detectors, La Biodola, Isola d'Elba, Italy, Nucl. Instrum. Meth, A572, 211 (2007).
[3] L. Xia, `Construction of the Digital Hadron Calorimeter Prototype', these proceedings.
[4] J. Hoff et al., A custom integrated circuit for calorimetry at the International Linear Collider, IEEE Nucl. Sci. Symp., Puerto Rico, 2005, FERMILAB-CONF-05-509.
[5] http://www.hep.anl.gov/repond/DHCAL_US.html
[6] https://twiki.cern.ch/twiki/bin/view/CALICE/WebHome
[7] J. Repond, 'Tests of a Digital Hadron Calorimeter', these proceedings.

Performance of the ATLAS missing E_T trigger with first $\sqrt{s} = 7$ TeV data

Pierre-Hugues Beauchemin for the ATLAS Collaboration
University of Oxford, Denys Wilkinson Building, Keble road, Oxford, OX1 3RH, UK
E-mail: h.beauchemin1@physics.ox.ac.uk

Abstract. We present an early study of the performance of the ATLAS trigger based on the vector ($E_{\mathrm{T}}^{\mathrm{miss}}$) and scalar ($\sum E_{\mathrm{T}}$) sums of the measured transverse energies, obtained with data collected during the first 2010 LHC runs, with proton-proton collisions at $\sqrt{s} = 7$ TeV. We demonstrate that because of the Level One configuration, calorimeter noise effects don't significantly affect $E_{\mathrm{T}}^{\mathrm{miss}}$ trigger rates for minbias events. We also show that there are good correlations between Event Filter and reconstructed offline $E_{\mathrm{T}}^{\mathrm{miss}}$ and $\sum E_{\mathrm{T}}$ quantities, and that the trigger efficiencies with respect to these offline quantities follow Monte Carlo expectations. From these studies, we conclude that $E_{\mathrm{T}}^{\mathrm{miss}}$ and $\sum E_{\mathrm{T}}$ HLT rejection can safely be enabled as soon as rates become too high for Level One rejection alone.

1. Introduction

Over the last few months, the ATLAS detector collected a sufficient number of 7 TeV LHC collision events to allow studying the performance of the ATLAS Trigger and Data Acquisition system (TDAQ). Such studies of the TDAQ system are critical for a successful physics program because they allow optimizing the rate and the efficiency at which interesting events are collected for thorough physics analyses. In particular, a large spectrum of physics results relies on the capacity of the ATLAS TDAQ system to efficiently collect events based on the estimate of the missing transverse energy ($E_{\mathrm{T}}^{\mathrm{miss}}$) contained in each event. The $E_{\mathrm{T}}^{\mathrm{miss}}$ trigger is, for example, the primary trigger to be used in new physics searches for processes involving new weakly interacting particles, which could account for the astronomically observed dark matter. In addition to discovery perspectives, the $E_{\mathrm{T}}^{\mathrm{miss}}$ trigger can also be used in combination with other triggers to control the rate of signatures involving low energy objects. For example, the $E_{\mathrm{T}}^{\mathrm{miss}}$ trigger is necessary in order to measure non-boosted Ws in the tau channel. Finally, the negligible statistical correlation with lepton triggers make it ideal to estimate the performance of those lepton triggers (and vice versa). It is currently being used to provide the first data-driven estimate of the electron trigger and identification efficiencies to be used in W and Z cross section measurements. The following performance studies of the $E_{\mathrm{T}}^{\mathrm{miss}}$ trigger will demonstrate that at the early stage of the detector commissioning the $E_{\mathrm{T}}^{\mathrm{miss}}$ trigger is already performing well enough to realise its physics potential.

Special triggers also exist, which set a minimum threshold to the scalar sum of transverse energies ($\sum E_{\mathrm{T}}$). Such triggers provide a good estimate of the overall calorimeter activity, and constitute a good complement to the $E_{\mathrm{T}}^{\mathrm{miss}}$ triggers for performance studies, as much as for physics purposes. The performance of such triggers will also be presented in the following.

1.1. Overview of the E_T^{miss} trigger system

The MET trigger requires that the magnitude of the vector sum of all transverse energies exceeds some thresholds. At Level One (L1) [1], only calorimeter information is used from the so-called "trigger towers" which correspond to the analog sums of all calorimeter cells in an (η, ϕ) range of 0.1×0.1 aligned along a given direction, formed by dedicated hardware. Next, the L1 pre-processor digitizes and applies noise subtraction to the formed trigger towers, and then produces jet elements, by summing the digital words of 4 adjacent trigger towers. Using all the jet elements in the event, the jet/energy processor then computes the calorimeter energy sums along the x-axis (E_x) and y-axis (E_y). Instead of computing MET from the quadratic sum of E_x and E_y, a look-up-table is used to accept/reject the events based on the (x, y) components of the measured energies. The scalar sum of the energy deposited in the calorimeter ($\sum E_{\mathrm{T}}$) is obtained similarly, without, however, the need for a look-up table to determine which thresholds the energy measurement satisfies.

The High Level Trigger (HLT) consists of a set of software-based algorithms which perform the $E_{\mathrm{T}}^{\mathrm{miss}}$ and $\sum E_{\mathrm{T}}$ reconstruction from detector input (Feature Extraction or FEX algorithms) or specify the selections which defined the various $E_{\mathrm{T}}^{\mathrm{miss}}$ or $\sum E_{\mathrm{T}}$ trigger chains to be included in the trigger menu (hypothesis algorithms) [2]. Because $E_{\mathrm{T}}^{\mathrm{miss}}$ and $\sum E_{\mathrm{T}}$ are global quantities, they only need to be computed once per event. This makes it possible to design many different hypothesis testing algorithms, for very specific physics needs, without noticeably increasing the execution time which is one of the major constraints at trigger level.

At Level Two (L2), the FEX algorithm simply fetches the L1 result and refines it by applying a correction taking into account the muons reconstructed at L2 (muons do not deposit much energy in the calorimeters and therefore constitute a source of fake $E_{\mathrm{T}}^{\mathrm{miss}}$). At the Event Filter (EF), the third and last of the trigger levels, contributions from both calorimeters and muon spectrometers are recomputed using the full granularity of the detectors. The 3D missing momentum vector is computed, although only its transverse projection is used to select events. In order to reduce the electronic noise contribution and therefore improve the resolution of the $E_{\mathrm{T}}^{\mathrm{miss}}$ measurement, only positive energy calorimeter cells above 3 noise RMS are considered in the $E_{\mathrm{T}}^{\mathrm{miss}}$ and $\sum E_{\mathrm{T}}$ reconstruction. This one-sided noise cut causes a shift in the $\sum E_{\mathrm{T}}$ measurement which can be somewhat accounted for by a similar shift in the $\sum E_{\mathrm{T}}$ thresholds defining the trigger chains. This resolution can be further improved by the application of a simple cell-based calibration, optimised to reproduce the offline calibrated Met and $\sum E_{\mathrm{T}}$ measurements.

1.2. Offline reference and configuration for commissioning period

The offline MET algorithms include corrections that cannot be implemented at trigger level. The reference selected for this study is the one provided by the offline MET_Topo, which is the best offline one still comparable to the EF algorithm described above. This algorithm accounts for the energy deposited in topological clusters [3] defined from seed calorimeter cells with an energy threshold (2-sided cuts) above 4 times the RMS of the signal expected from electronic noise, plus all cells above 2 RMS surrounding them and all adjacent cells. The $E_{\mathrm{T}}^{\mathrm{miss}}$ offline reconstruction obtained from MET_Topo is based only on uncalibrated calorimeter energy measurements. It provides the best offline performance for uncalibrated $E_{\mathrm{T}}^{\mathrm{miss}}$ measurements and therefore constitutes the targeted $E_{\mathrm{T}}^{\mathrm{miss}}$ reconstruction for trigger algorithms. The same topological clustering algorithm is also used for $\sum E_{\mathrm{T}}$ calculation, and is referred as MET_Topo $\sum E_{\mathrm{T}}$.

The ATLAS calorimeters are non-compensating, i.e. the correct hadronic energy scale is only recovered after applying an object-based calibration to the measured energy depositions. Such calibration is presently being validated for the full offline $E_{\mathrm{T}}^{\mathrm{miss}}$ reconstruction [4], and no reference is currently available to calibrate $E_{\mathrm{T}}^{\mathrm{miss}}$ trigger quantities (a simplified hadronic calibration has indeed been implemented, but is not yet used in trigger chains or monitoring histograms). For this reason we focus, for both trigger and full-reconstruction quantities during commissioning period, on the $E_{\mathrm{T}}^{\mathrm{miss}}$ performance with uncalibrated energies only (energies at the "electromagnetic" (EM) scale).

Finally, because the muon trigger is also separately in the process of being validated, muon corrections are ignored both at L2 and EF for the present performance studies. This means that the L2 results are identical to those at L1; hence only L1 and EF distributions will be shown. The effects of muon corrections on minimum-bias events are anyway negligible.

1.3. Data and preselection

The studies to be shown here have been obtained by taking the events in runs 152845 and 152878, all with stable LHC beam conditions, corresponding to an integrated luminosity of about 52 μb^{-1}. Such runs have been flagged as "good" by looking at the status of each subdetector and at the result of basic data quality checks. During these runs, the HLT was run in pass-through mode, meaning that it did not reject any events. However, the $E_{\mathrm{T}}^{\mathrm{miss}}$ trigger has been configured to process all events passing L1, which makes it possible to perform accurate debugging and better performance and efficiency studies.

A set of event preselections are applied in order to estimate the performance of the $E_{\mathrm{T}}^{\mathrm{miss}}$ and $\sum E_{\mathrm{T}}$ triggers on events of physics collision origin (as opposed to noise or cosmic events). Collision candidates are selected by requiring that the timing of the minimum-bias trigger scintillators is consistent with a collision (i.e. the time difference between scintillators installed at positive and negative pseudorapidities is consistent with zero). In addition a similar requirement has also been put on the timing of liquid-argon calorimeter cells. Furthermore, when explicitly mentioned, "good events" are selected by excluding those with badly measured jets. Bad jets are defined as those for which any of the following are true [5]:

- when less than 20% of the jet energy is deposited in the EM calorimeter, 90% of this jet energy is distributed over less than 6 calorimeter cells;
- when more than 95% of the jet energy is deposited in the EM calorimeter, more than 80% of the total jet energy comes from cells with an abnormal signal shape;
- the energy of the primary cells of the jet has been deposited more than 50 ns before or after the nominal proton-proton collision time.

We refer to these criteria as the jet clean-up cuts.

2. Results of performance studies

Figure 1 shows the correlation between L1 and offline $E_{\mathrm{T}}^{\mathrm{miss}}$, before and after removing events by applying the jet cleanup cuts described above. By comparing the upper and lower plots, we can see that the lowest L1 threshold on $E_{\mathrm{T}}^{\mathrm{miss}}$, which is L1_XE10, cuts away most of the events with badly measured or unphysical jets. This shows how the conservative L1 calorimeter-based trigger settings help to control $E_{\mathrm{T}}^{\mathrm{miss}}$ trigger rates. However, the L1 to offline correlation is not linear. This is because L1 is still using conservative commissioning-phase values, and future modifications of these settings are expected to improve L1 $E_{\mathrm{T}}^{\mathrm{miss}}$. For this reason, all current trigger chains based on $E_{\mathrm{T}}^{\mathrm{miss}}$ or $\sum E_{\mathrm{T}}$ start with a rather low L1 threshold compared to the EF threshold. The signal efficiency therefore currently depends mainly on the EF performance which is significantly better.

Figure 2 compares EF and offline $E_{\mathrm{T}}^{\mathrm{miss}}$ or $\sum E_{\mathrm{T}}$ for good events only. A clear linear relationship can be seen between trigger level and reconstructed offline quantities. This indicates that the selections operated by the EF $E_{\mathrm{T}}^{\mathrm{miss}}$ and $\sum E_{\mathrm{T}}$ algorithms don't significantly bias offline $E_{\mathrm{T}}^{\mathrm{miss}}$ and $\sum E_{\mathrm{T}}$ measurements and that such triggers will be highly efficient. The $\sim$15 GeV offset observed in the $\sum E_{\mathrm{T}}$ EF to offline scatter plot is an artefact of the 1-side noise suppression cut as explained in section 1.1.

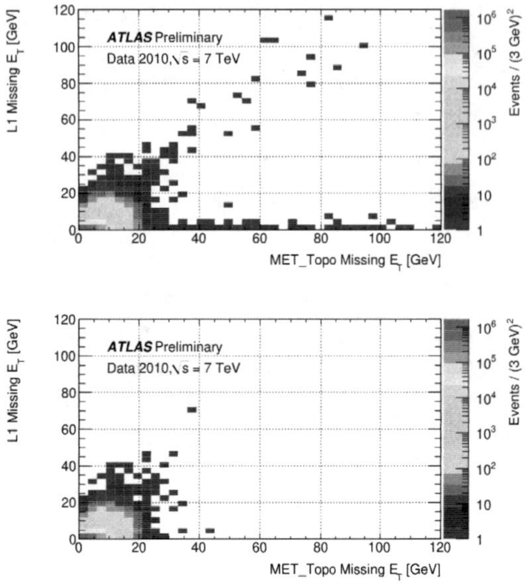

Figure 1. Correlation between the measured $E_{\mathrm{T}}^{\mathrm{miss}}$ at the L1 trigger level and the offline reconstructed $E_{\mathrm{T}}^{\mathrm{miss}}$ using topological clusters (MET_Topo $E_{\mathrm{T}}^{\mathrm{miss}}$) for collision candidates, before (top) and after (bottom) removing events by applying the jet cleanup cuts.

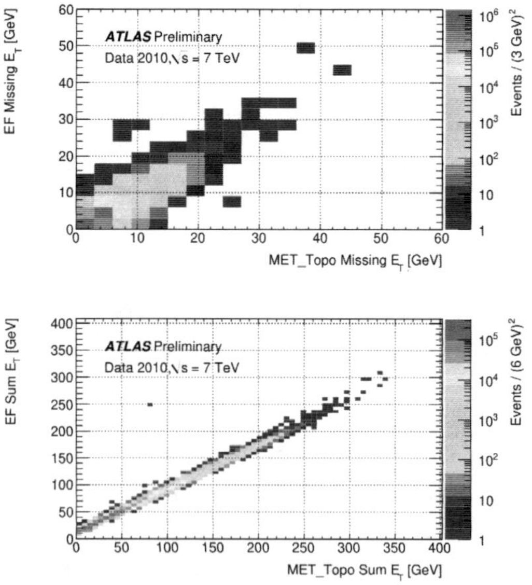

Figure 2. Correlation between the $E_{\mathrm{T}}^{\mathrm{miss}}$ (top) and $\sum E_{\mathrm{T}}$ (bottom) measured by the Event Filter (EF) trigger algorithm and by the offline algorithm using topological clusters (MET_Topo) for collision candidates, after removing events failing the standard ATLAS jet clean-up cuts.

This good correlation between EF and fully reconstructed quantities, however, also applies to events with unphysical jets, in contrast to what is observed at L1. As can be seen in figure 3, there is a difference between EF $E_{\mathrm{T}}^{\mathrm{miss}}$ distributions before and after applying jet clean-up cuts, showing that the EF $E_{\mathrm{T}}^{\mathrm{miss}}$ trigger alone will not eliminate events with bad jets affecting the EF $E_{\mathrm{T}}^{\mathrm{miss}}$ measurement. However, to pass the full trigger chain, both L1 and EF trigger decisions will be required, meaning that much of the bad jet contributions will be rejected by the L1 trigger and not survive to the final data sample. Figure 3 shows the effects of removing all events which do not pass L1_XE10 (the softest L1 $E_{\mathrm{T}}^{\mathrm{miss}}$ signature) or the ATLAS standard jet clean-up cuts. As one can see, only a few events at high $E_{\mathrm{T}}^{\mathrm{miss}}$ rejected by jet clean-up cuts survive the L1_XE10 trigger. Work is in progress to understand whether these bad jet events

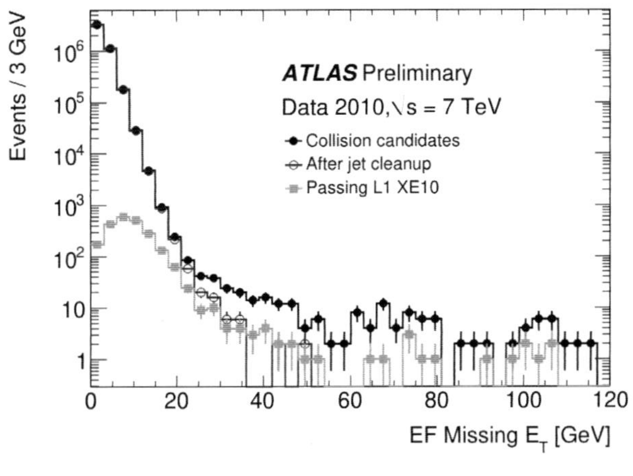

Figure 3. EF E_T^{miss} distribution for all collision candidates (full circles ●), after applying jet clean-up cuts (open circles ○), or passing a L1 Missing ET threshold of 10 GeV (full squares ■). The errors bars reflect statistical uncertainty only.

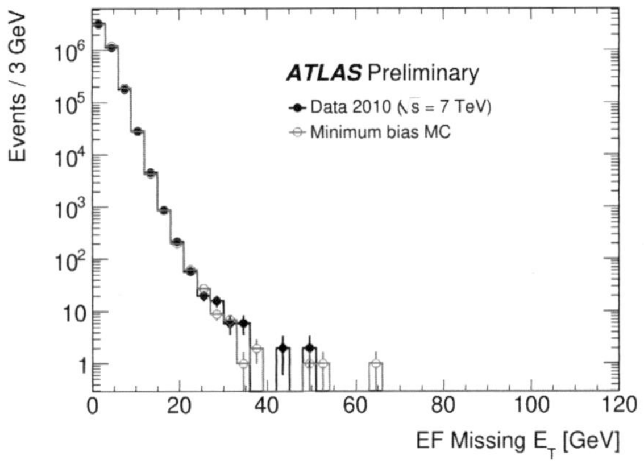

Figure 4. Event Filter (EF) trigger level E_T^{miss} distribution for real collision candidate data and Monte Carlo (MC) events, after applying the jet clean-up cuts. The MC distribution has been normalized to the data distribution (same integrated luminosity). The errors bars reflect statistical uncertainty only.

can be reduced at EF with additional checks.

Figure 4 shows the good agreement which is obtained by comparing the measured E_T^{miss} distribution at EF with the corresponding simulated distribution from minimum bias events, after applying the jet clean-up cuts. Figure 5 shows the trigger efficiency curves, the fraction of events which pass the EF trigger as a function of offline E_T^{miss} or $\sum E_T$, with no L1 trigger requirement. Good agreement is again seen between data and simulation. Figures 4 and 5 demonstrate that the ATLAS missing transverse energy HLT algorithms are well understood, and perform as expected on physics events. The E_T^{miss} and $\sum E_T$ trigger algorithms are therefore validated and ready for rejecting events (as soon as the L1 rate becomes too high). Moreover, the turn-on curves presented in figure 5 are steep: they reach an efficiency plateau only few GeV above the trigger threshold. This shows that the distortion of the offline E_T^{miss} or $\sum E_T$ distributions by trigger preselection is small. This demonstrates the high performance of the E_T^{miss} high-level trigger algorithms with respect to the offline E_T^{miss} quantities featuring the best performances (MET_Topo).

3. Summary and conclusion

The performance of the ATLAS triggers based on the scalar and vector sums of the transverse energies has been shown, for the first 2010 runs at $\sqrt{s} = 7$ TeV LHC proton-proton collisions. Because the hadronic calibration is still being validated for the full offline reconstruction, all plots here show the distributions obtained using the uncalibrated energy measurements.

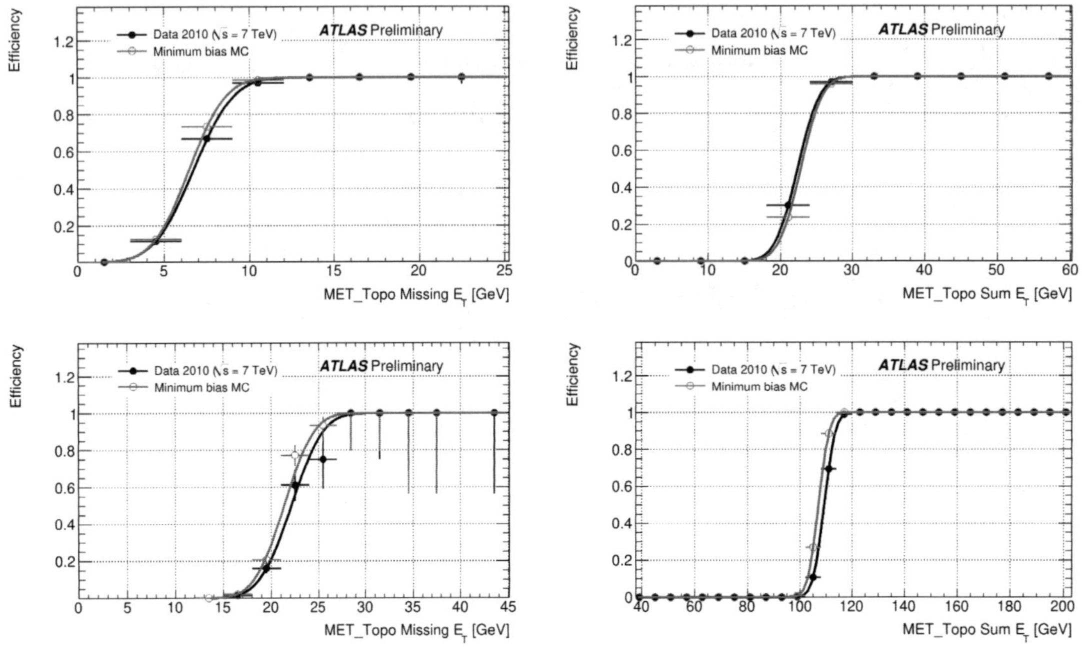

Figure 5. EF turn-on curves for $E_{\mathrm{T}}^{\mathrm{miss}}$ with threshold at 5 GeV (top left) and 20 GeV (bottom left), and for $\sum E_{\mathrm{T}}$ thresholds of 30 GeV (top right) and 100 GeV (bottom right). The simulation correctly reproduces the results obtained with real collision candidate data, after applying the jet clean-up cuts. The errors bars reflect statistical uncertainty only.

The results show:

(i) a good correlation between offline and EF trigger-level quantities ($E_{\mathrm{T}}^{\mathrm{miss}}$ and $\sum E_{\mathrm{T}}$),

(ii) bad jet contribution to trigger rates are controlled by L1 $E_{\mathrm{T}}^{\mathrm{miss}}$ preselection,

(iii) steep efficiency turn-on curves for EF $E_{\mathrm{T}}^{\mathrm{miss}}$ and $\sum E_{\mathrm{T}}$, and

(iv) good Monte Carlo to data agreement.

From these studies, we conclude that the HLT algorithms can be safely used to reject events, as soon as the L1 rate becomes too high. To optimise the signal efficiency, the current trigger chains based on $E_{\mathrm{T}}^{\mathrm{miss}}$ and $\sum E_{\mathrm{T}}$ all start with a rather low L1 threshold compared to the EF one, so that the final turn-on curve is very similar to what is obtained using EF alone. The $E_{\mathrm{T}}^{\mathrm{miss}}$ trigger is therefore in good shape to deliver its full physics potential.

References

[1] ATLAS Collaboration, 1998, Level 1 Trigger Technical Design Report, CERN/LHCC/98-14

[2] ATLAS Collaboration, 2003, High-Level Trigger, Data Acquisition and Controls Technical Design Report, CERN/LHCC/03-022

[3] Lampl W *et al.*, 2008, Calorimeter Clustering Algorithms: Description and Performance, Tech. Rep. ATL-LARG-PUB-2008-002, CERN, Geneva

[4] Cavalli D, Ma L, Yurkewicz A *et al.*, 2010, Performance of the Missing Transverse Energy Reconstruction and Calibration in Proton-Proton Collisions at a Center-of-Mass Energy of 7 TeV with the ATLAS Detector, Tech. Rep. ATL-COM-PHYS-2010-371, CERN, Geneva

[5] Baak M, Petteni M and Makovec N, 2010, Data-Quality Requirements and Event Cleaning for Jets and Missing Transverse Energy Reconstruction with the ATLAS Detector in Proton-Proton Collisions at a Center-of-Mass Energy of sqrts=7 TeV, Tech. Rep. ATLAS-COM-CONF-2010-038, CERN, Geneva

Journal of Physics: Conference Series **293** (2011) 012016 doi:10.1088/1742-6596/293/1/012016

Performance of 2nd generation CALICE/EUDET ASICs

C. de La Taille[1]
On behalf of the CALICE and EUDET collaborations
E-mail: taille@lal.in2p3.fr

Abstract. The paper reviews the performance of the three ASICs : HARDROC2, SPIROC2 and SKIROC2 developed to readout the ILC calorimeter prototypes. The chips integrate 36 to 64 channels of front-end, digitization and backend electronics in SiGe 0.35 μm technology. This second version was found mature enough to be produced in several hundreds to equip large scale technological prototypes and establish the feasibility of these highly granular "imaging" calorimeters as required for particle flow algorithms at the ILC. The low noise and low power sequential readout as well as power-pulsing operation at detector level and in magnetic field are proven.

1. Introduction

After the successful operation of "imaging" calorimeter physics prototypes for the ILC in the CALICE framework [1], a second generation of readout ASICs has been developed to establish the technological feasibility of large scale detectors, under the EUDET program [2]. These chips feature low-noise multi-channel readout with built-in auto-trigger, analog storage, digitization and readout. The design of these chips HARDROC, SPIROC and SKIROC that readout Silicon diodes, silicon photomultipliers or RPCs has been presented in the previous calorimetry conference [3]. The ASICs are also embedded inside the detector making ultra low power '~10-30 μW/channel) a key requirement to get rid of any cooling inside the detector. This low power is achieved through power pulsing, zero-suppression and low occupancy in the front-end chips.

2. HARDROC performance

The HARDOC (Hadronic Rpc Detector Read Out Chip) was designed to readout RPC semi-digital hadronic calorimeter prototype. The 1 cm2 pads are each connected to a variable gain current conveyor, fast shaper and 3 discriminators that record signals above ½ MIP (100 fC), few MIPs and several MIPs. The dispersion of trigger efficiency over the 64 channels is around a few percent as shown in Fig. The signals above threshold are stored in a large memory in the chip and are readout sequentially during the 200 ms inter-bunch period. The chip area is 16 mm^2 for 64 channels and is packaged in a thin (1.4mm) TQFP160 package, shown in Figure 1. The power dissipation is 1.5 mW/channel continuous or 7 μW with the 0.5% power pulsing cycle.

[1] To whom any correspondence should be addressed.

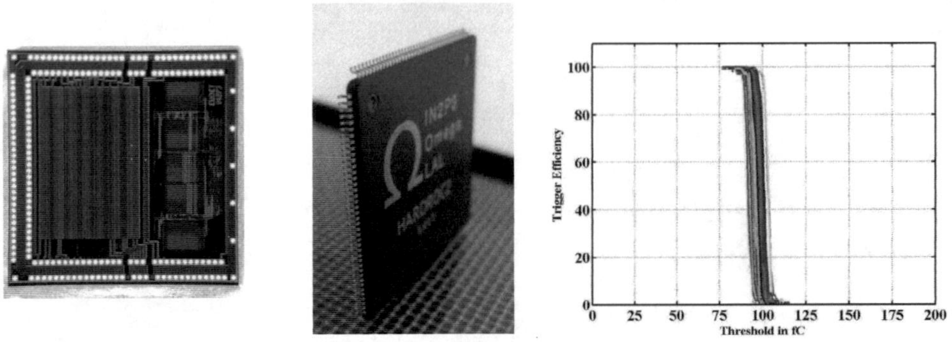

Figure 1 : Left : ASIC layout (die and package). Right : trigger efficiency for the 64 channels for a 100 fC input charge.

10,000 HARDROC have been produced in 2010 to equip a cubic meter size technological prototype, made of 40 layers of 1 square meter. The scalability of the detector has been an important issue and is displayed in Figure 2. where elementary boards of 32x48 cm^2 can be connected to make a larger detector read only on one side. The performance of such detector in cosmic tests or testbeam is very good with an efficiency above 95%, as shown in Ref. [4] from this conference.

Figure 2 : square meter RPC detector equipped with 144 HARDROC ASICs (courtesy of IPN Lyon)

3. SPIROC performance

The SPIROC (Silicon Photomultiplier Integrated Read Out Chip) ASIC was designed to readout the SiPMs (or MPPC) photodetectors of the analog hadronic calorimeter. Each SiPM is connected to a dual gain fast voltage amplifier with a gain of 1 and 10. Signals above thereshold are then shaped and stored for several ms in a 16-deep analog memory as well as the time of arrival thanks to a ramp TDC. After collisions, the stored values are digitized with a 12 bit 40 MHz Wilkinson ADC and the values are stored in a similar memory as for the HARDROC chip. Once the conversions are completed, each chip is readout sequentially with the same protocol as for the HARDROC chips, greatly simplifying the DAQ architecture. The chip contains 36 channels in an area of 25 mm^2 (Figure 3) and is packaged in a similar flat package TQFP208.

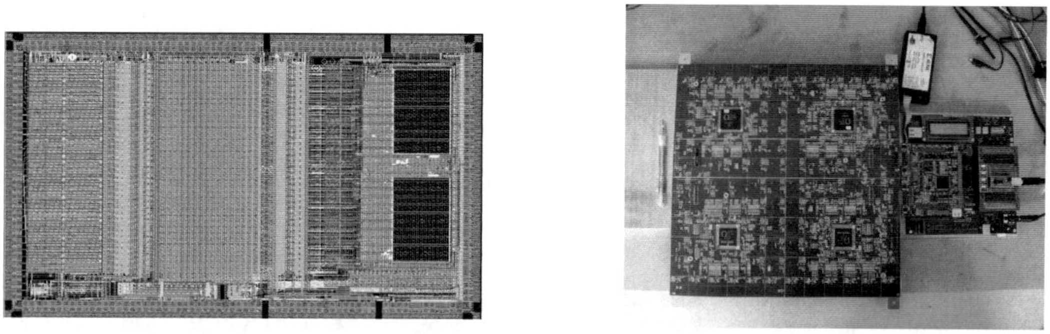

Figure 3 : (Left) layout of SPIROC. (Right) Front End Board of Analog Hadronic Calorimeter with 4 readout chips and 144 scintillating tiles and SiPM.

The first version of the chips has been produced in 2008 and extensively tested in Orsay and DESY. The analog front-end is derived from the previous chip version (FLC_SiPM) used in the CALICE physics prototype. The digitization stage is new, around a 12 bit Wilkinson ADC operated at 40MHz. The ramp is common to the 36 channels and carefully optimized to obtain good stability and linearity of +/- 0.025% (1 ADC count), as shown in Figure 4.

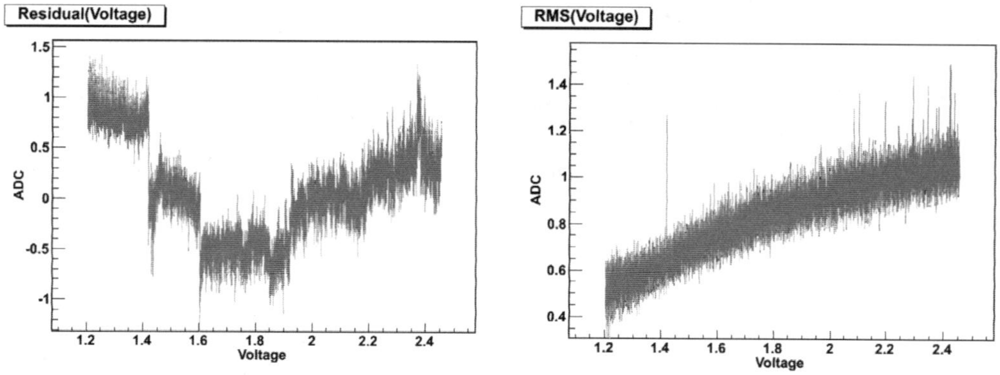

Figure 4 : (left) residuals to a linear fit of the 12 bit internal Wilkinson ADC. (right) ADC noise over the 36 ADC channels, corresponding to 1 ADC count = 250 µV.

The chips have been mounted on the HCAL front-end board displayed in Figure 3 and the performance measured with the photodetector as shown in Figure 5 and Ref [5] allowing a clean measurement of single photoelectron peaks and triggering on MIP signals.

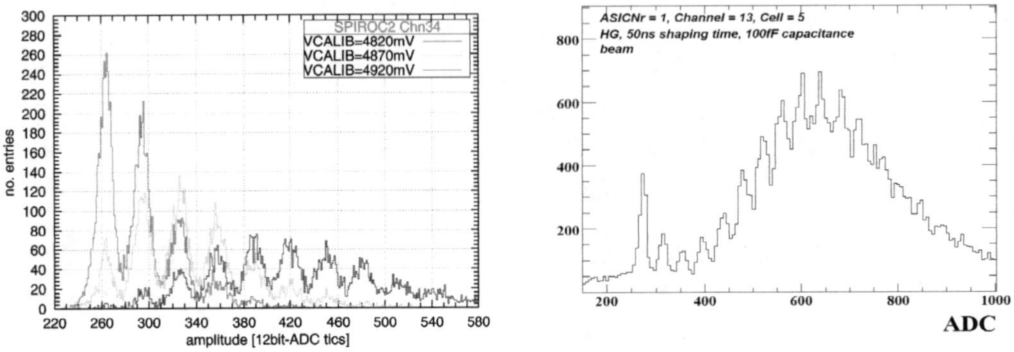

Figure 5 : (left) photoelectron peaks measured with the SPIROC2 chip and front-end board. (Right) MIP signal as measured with the same setup.

4. SKIROC design

The electromagnetic calorimeter exhibits similar requirements as the analog hadronic calorimeter in terms of dynamic range (15 bits) although the MIP signal is 200 times smaller (4 fC). An internal setting allows to emulate a SPIROC chip for this application, changing the input amplifier from voltage sensitive to charge sensitive. This allows to perform tests on the front-end boards and detector characterization, as shown in Figure 6. In parallel, a dedicated chip with 64 channels instead of 36 and a few specific ECAL optimizations has been developed and submitted, as shown in Figure 6.

Figure 6 : (left) ECAL front-end board (FEV7) equipped with 4 SPIROC chips in ECAL configuration. (Right) layout of the SKIROC2 ASIC.

5. Conclusion

The second generation readout ASICs are now ready to operate large scale calorimeter prototypes and establish the feasibility of a granular "imaging" calorimeter at the ILC, having validated the scalable sub-detector boards, readout scheme and power pulsing operation. The chips have been produced in a few thousand pieces in a dedicated production run and are described in detail in their data sheets given in Ref [6]. The detailed performance evaluation with the detector is ready to begin and will be continued in the AIDA FP7 framework [7].

6. References

[1] CALICE collaboration : https://twiki.cern.ch/twiki/bin/view/CALICE/WebHome
[2] EUDET FP6 framework web site : http://www.eudet.org
[3] C. De La Taille : proceedings of the 13th International conference on calorimetry, Pavia 2008.
[4] J. Blaha et al. proceedings of the 14th International conference on calorimetry, Beijing 2010
[5] E. Garutti et al. proceedings of the 14th International conference on calorimetry, Beijing 2010
[6] OMEGA web site : http://omega.in2p3.fr
[7] AIDA web site : https://espace.cern.ch/aida/default.aspx

ATLAS LAr calorimeters readout electronics upgrade R&D for sLHC

Hucheng Chen on behalf of the ATLAS Liquid Argon Calorimeter Group

Department of Physics, Brookhaven National Laboratory, Upton, NY 11973-5000, U.S.A.

E-mail: chc@bnl.gov

Abstract. The ATLAS Liquid Argon (LAr) calorimeters consist of an electromagnetic barrel calorimeter and two end-caps with electromagnetic, hadronic and forward calorimeters. A total of 182,468 signals are digitized and processed real-time on detector, to provide energy and time deposited in each detector element at every occurrence of the Level-1 trigger. A luminosity upgrade of the LHC will occur in the years ∼2020. The current readout electronics will need to be upgraded to sustain the higher radiation levels. A completely innovative readout scheme is being developed. The front-end readout will send out data continuously at each bunch crossing through high speed radiation resistant optical links, the data will be processed real-time with the possibility of implementing trigger algorithms. This article is an overview of the R&D activities and architectural studies the ATLAS LAr Calorimeter Group is developing.

1. Introduction

The ATLAS experiment [1] is a general-purpose detector designed to study the proton-proton collisions produced at the Large Hardron Collider (LHC) with a center-of-mass energy of 14 TeV and to explore the full physics potential of the LHC at CERN.

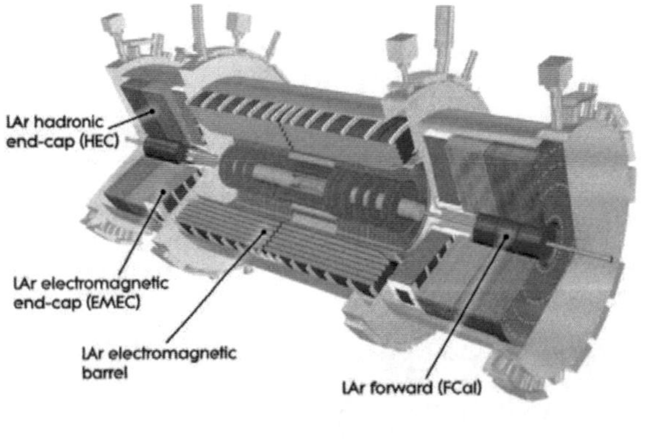

Figure 1. Schematic drawing of the ATLAS LAr calorimeters

Liquid Argon (LAr) calorimeters [2] in ATLAS are high precision, high sensitivity and high granularity detectors designed to provide precision measurements of electrons, photons, jets and missing transverse energy. The LAr calorimeters consist of four sub-detectors and are contained within three cryostats between the inner tracking detectors and the outer muon chambers as shown in Figure 1. The central cryostat houses the electromagnetic barrel calorimeter (EMB) which covers the pseudo-rapidity range $0 < |\eta| < 1.475$, while each end-cap cryostat contains an end-cap electromagnetic calorimeter (EMEC) which covers $1.375 < |\eta| < 3.2$, two hadronic end-cap wheels (HEC) covering $1.5 < |\eta| < 3.2$ and a three wheels forward calorimeter (FCAL) which covers the region very close to the beam axis ($3.1 < |\eta| < 4.9$) to achieve large acceptance.

2. Readout Electronics of LAr Calorimeters

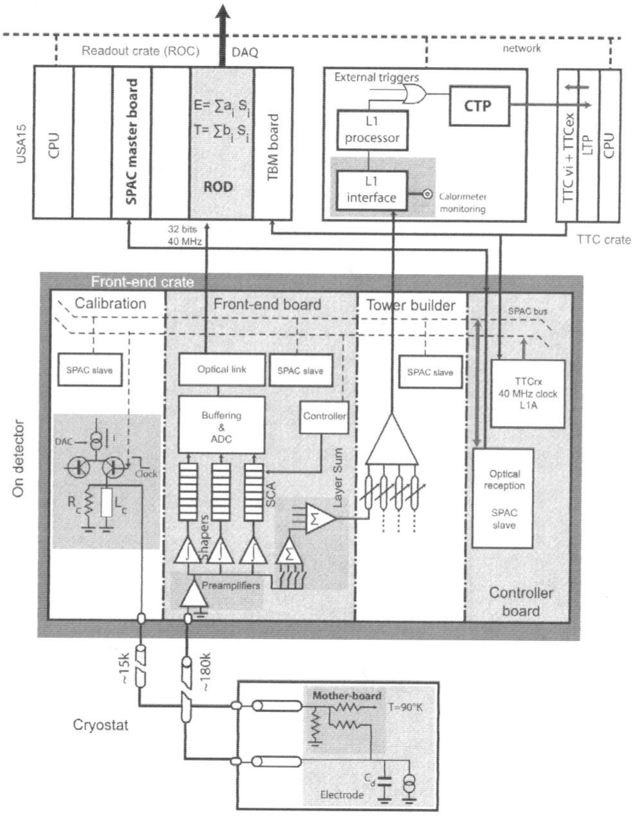

Figure 2. LAr calorimeters readout electronics architecture

The readout electronics architecture of LAr calorimeters is depicted in Figure 2. When a charged particle traverses and ionizes the liquid Argon in the gap between a LAr electrode and an absorber, an ionization current signal is preamplified, shaped, buffered and digitized by the front-end boards (FEB) [3] installed on the detector. The FEBs send the digitized pulse via optical links to the Readout Drivers (ROD) which are installed in a radiation free area (USA15) next to the detector cavern (UX15). In total 182,468 calorimeter cells are to be read out. Each ROD receives the data from up to 8 FEBs and calculates the energy deposited, the time of the deposition and a quality factor for cells with high energy deposit, using an optimal filtering algorithm [4]. RODs send data to the Readout Buffers (ROBs) hosted on PCs (ROS) through S-links. Analog trigger sum signals are formed on front-end electronics boards and sent to the receiver system through copper cables before interface to the Level-1 calorimeter trigger

(L1Calo) system. The LAr calorimeters front-end electronics system [5] is installed on detector and includes 58 Front End Crates (FEC) which house 1,524 FEBs and other electronics boards. The back-end electronics system [6] consists of 16 back end crates and 68 ROS PCs, and there are total 192 RODs plugged in back end crates.

3. Motivations of Readout Electronics Upgrade

LHC is expected to be upgraded in two phases. Phase 1 is planned in 2016 and phase 2 is planned in 2020 which is also known as super-LHC (sLHC). sLHC will have 10 times luminosity and the current LAr front-end electronics need to be upgraded to sustain the higher radiation levels [7]. The present front-end architecture is very complex. A FEB has 11 ASICs based on different technologies, some of which (e.g. DMILL) are already obsolete. The FEB is designed and produced for 10 years operation with limited number (~6%) of spares, it has also intrinsic limitations on Level-1 trigger acceptance. Therefore component level upgrade is not possible, and a new front-end architecture has been proposed as shown in Figure 3.

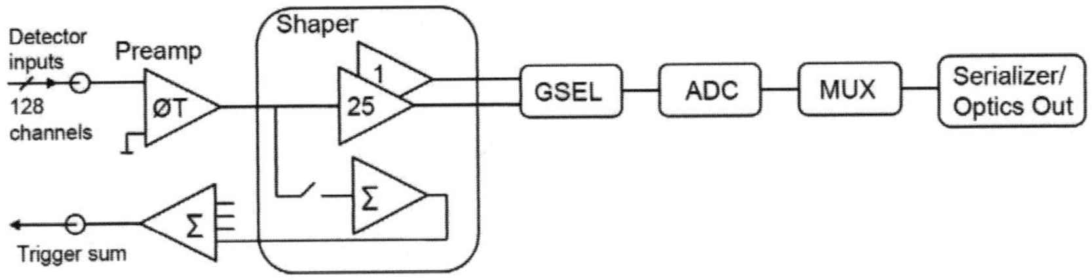

Figure 3. Proposed front-end readout architecture of LAr calorimeters

The new front-end readout will send out data continuously at each bunch crossing through high speed radiation resistant optical links. Large volume of data (~100Gbps each board) will be processed real-time with the possibility of implementing trigger algorithms for clusters and electron/photon identification at a finer granularity than what is currently implemented. The new architecture simplifies the system design while keeping many options open, such as pipeline design, shaping and gain settings etc. It requires many R&D efforts and the following section describes these R&D activities for LAr calorimeters readout electronics upgrade.

4. R&D Studies of Readout Electronics Upgrade

4.1. Front-end ASIC Design

Front-end electronics design for sLHC becomes more critical with requirements of both high performance and higher radiation levels. Three R&D efforts are going on which are focused on the analog front-end and mixed-signal front-end ASIC design.

4.1.1. Analog Front-end ASIC An analog front-end chip LAPAS (Liquid Argon PreAmplifier Shaper) has been designed and received in early 2009. It uses IBM 8WL 0.13 µm SiGe (silicon-germanium) BiCMOS technology. The preamplifier design of LAPAS chip is based on low noise line-terminating circuit topology presently used in LAr calorimeters. It has full 16-bit dynamic range and very low noise ($\sim 0.26nV/\sqrt{\text{Hz}}$). The shaper design of LAPAS chip is fully differential and based on $CR-(RC)^2$ topology with two gain settings (×1 and ×10). The uniformity is better than 5% across 17 tested ASICs, the INL (integral non-linearity) over full scale of two gains is less than 0.1%.

The LAPAS chip has been tested with hand wired prototype board, all measurements are as expected. The TID (total ionization dosage) test results show no significant concern. A new test print circuit board shown in Figure 4 is available. The full characterization is still in progress. The future plan is to explore SiGe technologies from different manufactures (e.g. IHP, AMS) and study the feasibility of CMOS only design.

Figure 4. Test print circuit board with LAPAS chip populated

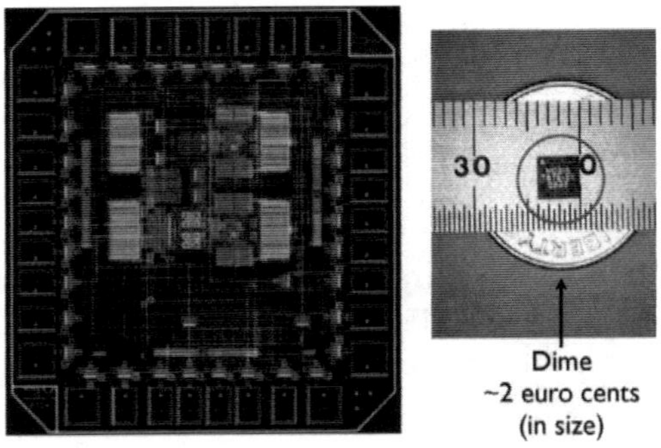

Figure 5. Layout and package of Nevis09 chip

4.1.2. Mixed-signal Front-end: ADC ADC is the most technologically challenging component in the new architecture due to the requirements of high dynamic range, radiation tolerance and SEE (single event effects) immunity. Some commercial ADCs (e.g. ADI-AD9259, ST-RHF1201 and TO-ADC5821) have been identified and planned to be tested in radiation environment, while a custom ADC is being developed.

The Nevis09 chip has been designed and received in late 2009. The layout and package of the chip is shown in Figure 5. It uses IBM 8RF 0.13 µm CMOS technology to implement 12-bit precision OTA (operational transconductance amplifier) with cascade of two T/H (track-and-hold) to achieve S/H (sample-and-hold) effect for testing. The preliminary test results show the circuit could reach 65dB, the full characterization test is still ongoing.

4.1.3. HEC Cold Electronics The HEC preamplifier and summing boards (PSB) are located at the perimeter of the HEC wheels inside the liquid Argon. The ASIC on PSB is designed and built by Gallium-Arsenide (GaAs) TriQuint QED-A 1 µm technology[8]. The present ATLAS requirements for the HEC PSB boards are $2 \times 10^{12}\,\mathrm{n/cm^2}$ per year. After 10 years of LHC operation, the present HEC cold electronics would be operated at its limit. It is therefore planned to develop a new ASIC that will be ten times more radiation hard against neutrons.

Table 1. Loss of gain of the transistors studied for a neutron fluence of $2 \times 10^{15}\mathrm{n/cm^2}$

Material	SiGe	SiGe	SiGe	Si	Si	Si	GaAs	GaAs
Transistor	Bipolar	Bipolar	Bipolar	CMOS FET	CMOS FET	CMOS FET	FET	FET
Foundry	IHP	IBM	AMS	IHP	IHP	AMS	Triquint	Sirenza
Type	npn	npn	npn	nmos	pmos	nmos	pHEMT	pHEMT
10 MHz	5%	5%	5%	4%	4%	3%	0%	4%
40 MHz	3%	2%	5%	2%	3%	3%	2%	2%

The radiation hardness against neutron irradiation has been studied for transistors of SiGe, Si and GaAs technologies. The loss of gain for the transistors for two different frequencies, studied at a neutron fluence of $2 \times 10^{15}\text{n/cm}^2$ is shown in Table 1. All technologies investigated show only a small degradation of the gain up to the irradiation level expected for sLHC. Based on these studies both options, bipolar SiGe as well as CMOS FET technologies, are sufficiently stable under neutron irradiation and are being investigated further. A first design of amplifier based on IHP SGB25V CMOS technology will be available for test in Sept 2010, while a similar design based on IBM 8RF CMOS technology is planned to be done in late 2010.

4.2. Radiation Resistance Optical Link

The new readout architecture requires high bandwidth radiation resistant optical link to transmit data out of the detector. The SMU_P1 chip based on 0.25 μm Silicon on Sapphire CMOS technology has been received in late 2009. It includes a 5Gbps 16:1 serializer, a 5GHz LC VCO based PLL, the CML driver, a divide-by-16 circuit, a varactor and an SRAM block.

The full characterization test is still ongoing. The preliminary test results show the serializer works in the range of 3.8 to 6.2Gbps with ~507mW power consumption. The tuning range of the LCPLL is 4.7 to 5GHz, the power consumption at 4.9GHz is ~121mW, the random jitter is less than 2.5ps and the deterministic jitter is less than 17ps.

4.3. Back-end High-speed Processing Unit

The ROD is the heart of the back-end electronics system, serving as the bridge between the front-end electronics system, the ROS and possibly the L1Calo system. With the new front-end architecture, ROD will have to be upgraded to process huge amount of data (~1.5Tbps per board) from the detector. A possible implementation, currently under study, of the ROD board is shown in Figure 6.

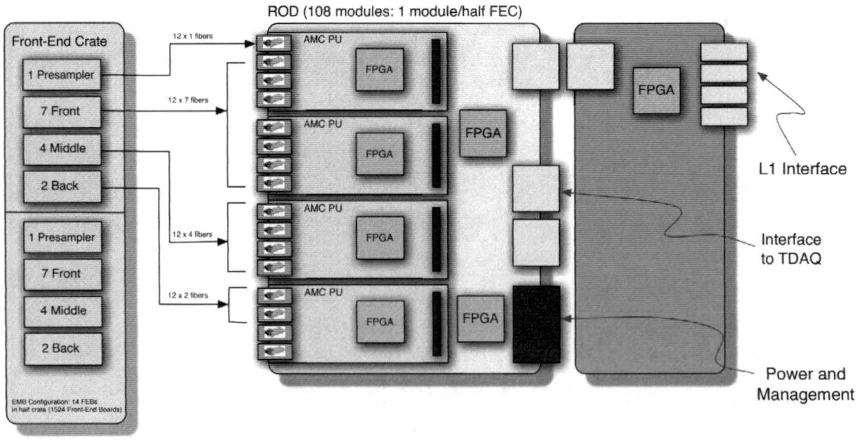

Figure 6. Schematic representation of the LAr signal dataflow

A SubROD and a SubROD Injector based on ATCA platform have been developed and shown in Figure 7 and 8, the integration test was performed in Oct 2009. They use commercial parallel optical transceiver running at 75Gbps with MPO®/MTP low profile connector, SERDES of Xilinx Virtex-5 FPGA and Altera Stratix II GX FPGA are tested successfully.

The feasibility of fast and large volume data preparation within fixed latency budget for Level-0 and Level-1 trigger upgrade is being investigated. The data volume is expected to be 40Tbps for L0Calo processing and 2Tbps for L1Calo processing. The other R&D studies related to ROD, such as IPMI (Intelligent Platform Management Interface) controller, interface to ROS over 10Gbit Ethernet and software development, are also being performed.

Figure 7. ATCA Sub-ROD module

Figure 8. ATCA Sub-ROD Injector module

4.4. Low Voltage Power Supply

The current front-end power supply system is of centralized architecture with each FEC having its own LVPS which takes 280V input from USA15 through ~100 m cable and provides seven different voltages. A LVPS has high power density (~3.2kW) and is water cooled. The new LVPS will have to sustain higher radiation levels, even though total power budget of front-end electronics will be kept same. The number and levels of voltages have to be rationalized, to reduce different voltages across devices. Two possible LVPS architectures, distributed power architecture and intermediate bus architecture, are being investigated.

5. Summary

Radiation levels and probably natural aging will require an upgrade of the ATLAS LAr calorimeters front-end electronics for sLHC. A new front-end readout architecture has been proposed and is being developed for the readout electronics upgrade. The upgrade R&D projects are progressing smoothly and more test results are expected in the near future. The feasibility of the readout architecture is expected to be demonstrated in next 2-3 years.

References

[1] G. Aad *et al.* [ATLAS Collaboration], "The ATLAS Experiment at the CERN Large Hadron Collider," JINST **3** (2008) S08003.

[2] [ATLAS Collaboration], "ATLAS liquid argon calorimeter: Technical design report," CERN-LHCC-96-41

[3] N. J. Buchanan *et al.*, "Design and implementation of the Front End Board for the readout of the ATLAS liquid argon calorimeters," JINST **3**, P03004 (2008).

[4] W.E. Cleland and E.G. Stern, "Signal processing considerations for liquid ionization calorimeters in a high rate environment," Nucl. Inst. Meth. A 338 (1994) 467

[5] N. J. Buchanan *et al.*, "ATLAS liquid argon calorimeter front end electronics," JINST **3**, P09003 (2008).

[6] A. Bazan *et al.* [Liquid Argon Back End Electronics Collaboration], "ATLAS Liquid Argon Calorimeter Back End Electronics," JINST **2** (2007) P06002.

[7] N. J. Buchanan *et al.*, "Radiation qualification of the front-end electronics for the readout of the ATLAS liquid argon calorimeters," JINST **3**, P10005 (2008).

[8] J. Ban *et al.*, "Cold Electronics For The Liquid Argon Hadronic End-Cap Calorimeter Of Atlas," Nucl. Instrum. Meth. A **556** (2006) 158.

Journal of Physics: Conference Series **293** (2011) 012018 doi:10.1088/1742-6596/293/1/012018

A Serializer ASIC at 5 Gbps for Detector Front-end Electronics Readout

Jingbo Ye, on behalf of the ATLAS Liquid Argon Calorimeter Group

Department of Physics, Southern Methodist University,
Dallas, Texas 75275, USA

E-mail:yejb@smu.edu

Abstract. Optical data links are used in detector front-end electronics readout systems of experiments in the Tevatron and the LHC. Optical links in high energy particle physics experiments usually have special requirements such as radiation tolerance, ultra high reliability and low power dissipation. These requirements are often not met by commercial components which are designed for applications in non-radiation, accessible (for maintenance) environment, and for multi-vendor systems so the parts must comply with certain standards. Future HEP experiments such as the upgrades for the sLHC call for optical links with ultra high data bandwidth, higher radiation tolerance and ultra low power dissipation. To meet these challenges and in particular those in the upgrade for the ATLAS Liquid Argon Calorimeter readout that calls for an optical link system of 100 Gbps for each front-end board, we adopted a full custom front-end electronics system design based on application specific integrated circuits. Reported here are the development and test results of two circuits: a 5 Gbps serializer and a 5 GHz phase locked loop, and a block diagram design for the 100 Gbps optical link.

1. Introduction

ATLAS Liquid Argon Calorimeter has 1524 front-end board (FEB) in its electronics readout system. Each FEB reads out 128 detector channels. Detector signals are amplified and split into two channels: the analog summation of detector layer information to build trigger towers for level 1 trigger, and shaping and digitization for precision energy measurements. In this channel, after a three gain shaper that works with a 12-bit ADC to cover the full dynamic range required by physics, signals are sampled at 40 MHz and stored in a switched capacitor array (SCA) awaiting level 1 trigger. Selected events are then digitized and formatted to be sent through a single channel optical link to the back-end electronics system for further processing [1, 2 and 3]. Because of the level 1 trigger selection, data rate over the fiber channel is 1.6 Gbps. A block diagram of this readout scheme is shown in figure 1a. A total of 11 ASICs and several commercial-off-the-shelf (COTS) components that are specially qualified to be radiation tolerant for this location in the ATLAS detector are used in the FEB, making a complex system that took a long development time [4]. This FEB is qualified for 10 years of operation at the LHC nominal luminosity of $10^{34}cm^{-2}s^{-1}$. Many of the COTS and ASIC technologies are obsolete making it impossible to upgrade this readout system adiabatically to meet a factor of 10 of luminosity increase (hence much higher radiation in the operation environment) in the sLHC. A new design will have to be adopted for the upgrade. A tentative FEB2 architecture, as depicted in figure 1b, streams data to the back-end electronic, greatly simplifies the front-end logic. The price to pay is mostly in the optical link and the back-end data processing.

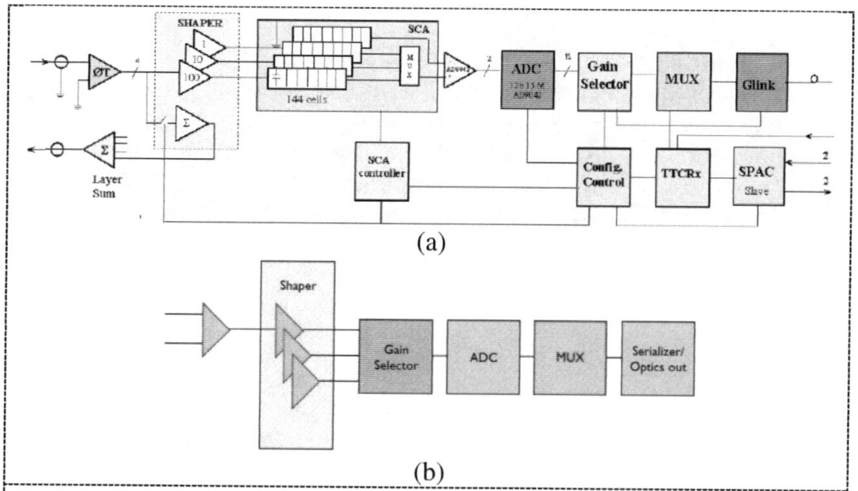

Figure 1. Block diagrams of the front-end board readout schematics of the current FEB (a), and the proposed upgrade FEB2 (b).

The Link-on-Chip (LOC) is a concept we proposed for the optical link upgrade which calls for a data bandwidth of 100 Gbps per FEB2, an increase of 62 times compared with the present system. The power dissipation of FEB2 is capped by the current cooling system that will not be upgraded, imposing a stringent power requirement on all ASICs for the front-end. We have identified an ASIC technology, a commercial 0.25 μm silicon-on-sapphire (SOS) CMOS process that meets the requirements on speed, power dissipation and radiation tolerance. A block diagram for the transmitting side of the link system is shown in figure 2. In this design a 6-lane serializer array called LOCs6 is the key component in the system. With a serializer data rate of 10 Gbps in each fiber channel, a 12-lane fiber ribbon will provide 100 Gbps data bandwidth plus 20% redundancy for system reliability. To check the serializer design, and to probe the highest speed possible with this technology, we submitted a prototype chip which we call SMU_P1. A picture of SMU_P1 is shown in figure 3. In this 3 mm × 3 mm tile, we have the following designs: LOCs1, a 5 Gbps 16:1 serializer; the LCPLL, a 5 GHz LC VCO based phase locked loop; The CML driver of LOCs1; A divide-by-16 circuit; A varactor, a voltage controlled capacitor; and an SRAM block, designed by INFN Milano.

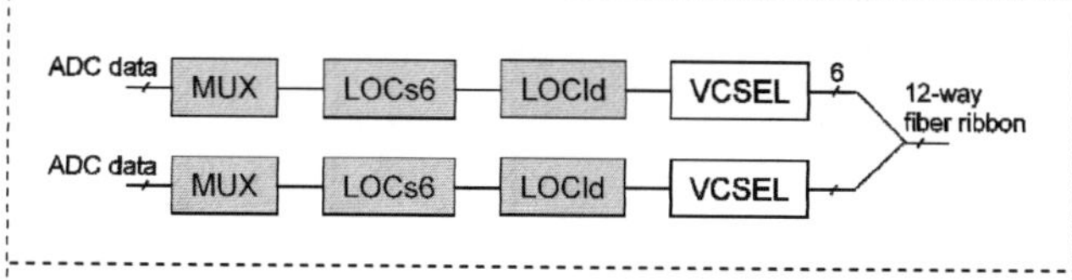

Figure 2. Block diagram for the transmitting side of the optical link upgrade to 100 Gbps per FEB.

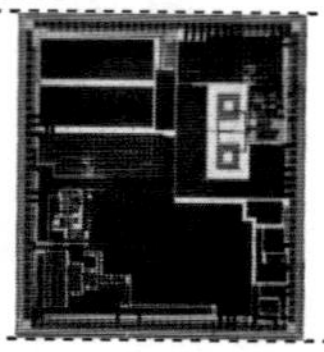

Figure 3. A picture of the SMU_P1 ASIC prototype.

SMU_P1 was submitted for fabrication in Aug. 2009. The chips arrived at SMU in Nov. 2009. Test setup was prepared during that time window. Test results of LOCs1 and the LCPLL are reported in section 2. A roadmap towards the 100 Gbps system is discussed in section 3.

2. Test results of LOCs1 and the LCPLL

Tests on the ASIC designs are in two categories: characteristic evaluations in lab and irradiation tests with a proton beam. Reported in this article are the former while results concerning the latter will be reported at TWEPP 2010.

Journal of Physics: Conference Series **293** (2011) 012018 doi:10.1088/1742-6596/293/1/012018

2.1. LOCs1.

LOCs1 is a 16:1 serializer based on a 2.5 GHz ring oscillator PLL. Using both edges of the clock, a serializing unit is constructed with a design speed of 5 Gbps, as shown in figure 4. Working with the power of 2 enables us to use the CMOS 2:1 multplexing unit (MUX 2:1) before the last stage where we trade power for speed by using a multiplexer driven by a complementary clock. In the future design of the 10 Gbps serializer, a full CML multiplexer will be used at this stage.

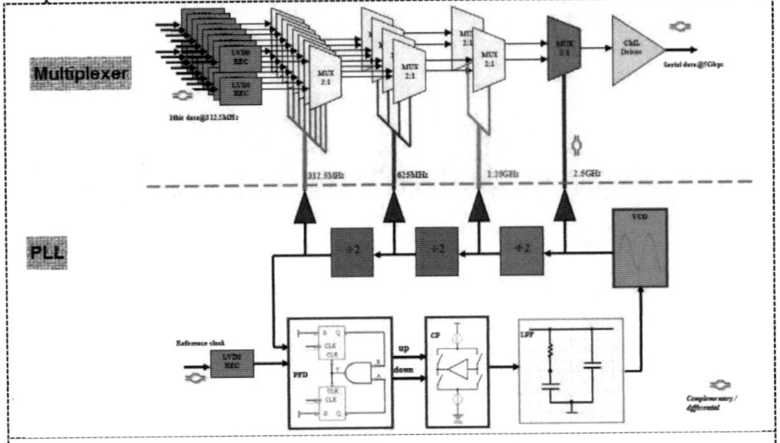

Figure 4. Block diagram of LOCs1.

There are two major measurements in the characteristic evaluations: the eye diagram and the eye mask test of the high speed output signal, and a stringent bit by bit check of the transmitted data in a bit-error-rate-tester (BERT), carried out in the form of the "bathtub scan" to measure the timing margin in the serial bit stream with a given error rate. To perform a BER test with LOCs1 under proton irradiation, a portable BERT has been developed in the ATLAS-CMS common project, the Versatile Link [5]. The VBERT, up to 4 channels in one FPGA, was successfully used in a recent irradiation test at the Indiana University Cyclotron Facility (IUCF) [6]. All tests are performed with an input data that generates a 2^7-1 pseudorandom bit stream. Shown in figure 5 is part of the test setup for LOCs1. A total of 12 carrier boards have been assembled.

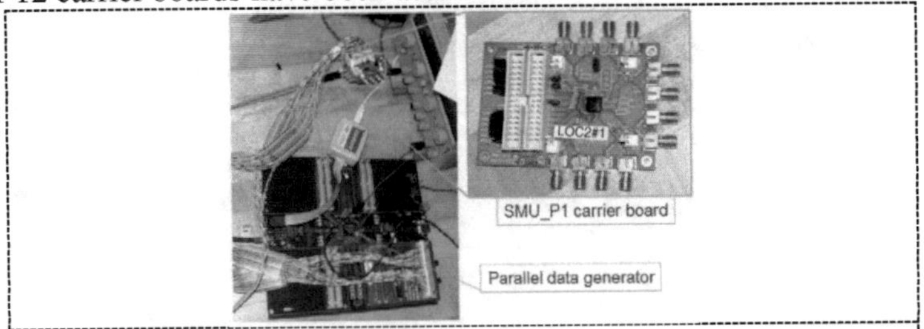

Figure 5. Pictures of part of the setup for testing LOCs1.

2.1.1. Eye diagram measurement and the eye mask test.

Eye diagram of the high speed output signal is measured to provide information on the rise and fall times of the signal, the amplitude and the timing jitter at the crossing pointing of the differential signal pair. An eye mask, adapted from Fiber Channel 4.25 standard and scaled up to 5 Gbps is imposed to the eye diagram which passes this eye mask test with a comfortable margin as shown in figure 6. Also shown in figure 6 is a simulated eye diagram with realistic output load to the last stage CML driver. As one can see, the measured eye diagram is slightly better than the simulated one, indicating that the conditions used in the simulation are conservative. The rise and fall times, signal amplitude and jitter are summarised in section *2.1.3*.

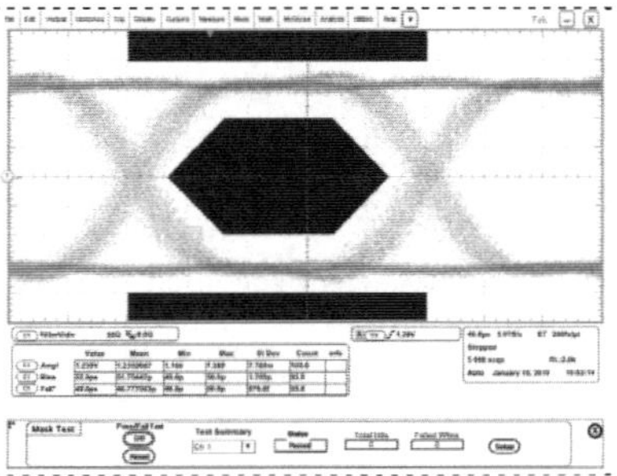

Figure 6a. The eye diagram measurement and eye mask test of LOCs1.

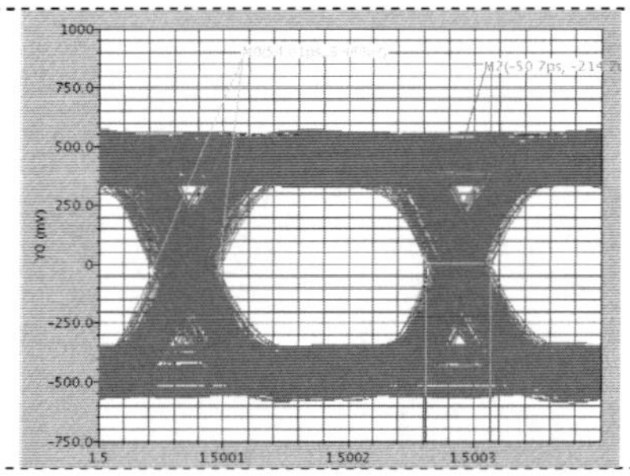

Figure 6b. The simulated eye diagram of LOCs1.

2.1.2. *The bathtub scan.*

Bathtub scans are performed at the design speed of 5 Gbps and at a higher speed of 5.8 Gbps. The results are shown in figure 7. At 5 Gbps there is an opening of 138 ps at a BER of 1×10^{-12}, comparable with the eye opening in the eye diagram measurement.

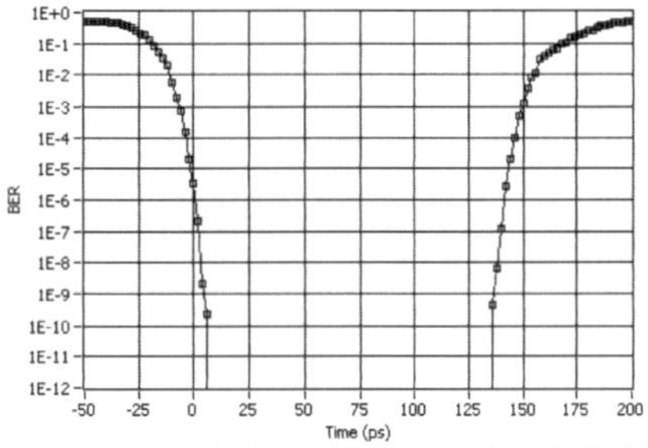

Figure 7a. Bathtub scan of LOCs1 at 5 Gbps.

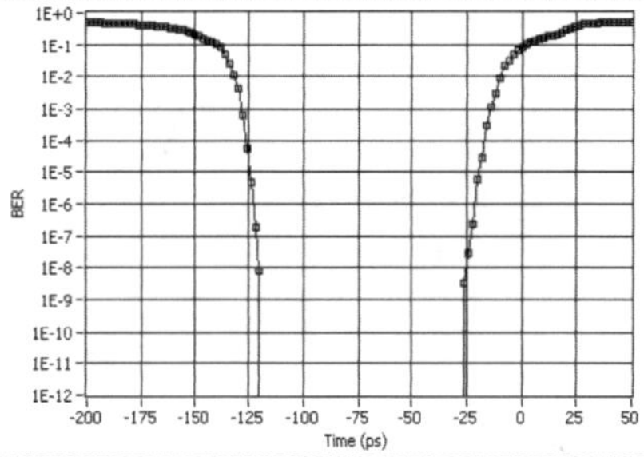

Figure 7b. Bathtub scan of LOCs1 at 5.8 Gbps.

2.1.3. *Summary of LOCs1 test results.*

A total of 12 LOCs1 were wire bonded to PCBs and tested. Five boards have problems: 2 Vcc shorts, 2 Vcc open, 1 with a stuck bit. They are all traced to the place where the chip is encapsulated, leaving unanswered questions to the problems: chip problems or wire bonding problems. Only tests on more chips will tell. For the 7 chips/PCBs that are working, the results are summarized below and in table 1.

- Range (Gbps): min: 3.8 – 4.0, max: 5.7 – 6.2.
- Power (based on 1 chip): 507 mW at 5 Gbps.

Table 1. Signal characteristics of LOCs1 at 5 Gbps

Amplitude (V)	1.16 ± 0.03
Rise time (ps)	52.0 ± 0.9
Fall time (ps)	51.9 ± 1.0
Random jitter (ps)	2.6 ± 0.6
Total deterministic jitter (ps)	33.4 ± 6.7
Total jitter at BER@1E-12 (ps)	61.6 ± 6.9
Eye opening at BER@1E-12	(69.3 ± 3.7)%
Bathtub curve opening at BER@1E-12 (ps)	122 ± 18

We consider the design goals of LOCs1 have been achieved.

2.2. The LCPLL

To reach a data bandwidth of 100 Gbps over a 10- or 12-lane fiber ribbon, we will need 8 to 10 Gbps transmission rate over a single fiber channel. The most crucial component in a serializer design is the PLL. In SMU_P1 we placed an LC based PLL design to work at 5 GHz. A block diagram of this PLL is shown in figure 8. The added CML driver is for test only.

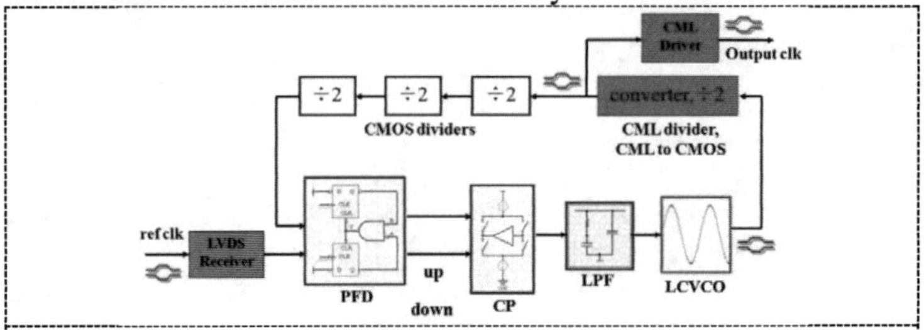

Figure 8. Block diagram of the LCPLL.

2.2.1. LCPLL measurement results

Four chips are measured and the results are listed in table 2. The tuning range of 4.7 to 5.0 GHz, although reaching the design goal of 5 GHz, is narrower than the simulated 3.8 to 5.0 GHz window. The power consumption of this PLL is 121 mW at 4.9 GHz, much lower than the ring oscillator based PLL used in LOCs1, which is 173 mW at 2.5 GHz, according to simulation.

Table 2. Parameters measured for the LCPLL

	Lower limit				Higher limit			
Board ID	1	2	3	4	1	2	3	4
Output frequency (GHz)	2.38	2.29	2.24	2.37	2.50	2.51	2.48	2.50
VCO frequency (GHz)	4.75	4.62	4.48	4.74	4.98	5.01	4.97	4.99
Rise time (ps)		66.2	47.0	67.1		67.2	42.1	66.1
Fall time (ps)		64.6	46.5	68.5		66.8	41.9	66.4
Random jitter RMS (ps)	1.28	2.49	2.05	1.81	1.26	1.06	1.08	1.83
Deterministic jitter (ps)	12.61	11.00	13.95	15.46	5.51	5.10	16.63	8.43

2.2.2. The narrow tuning range and the debugging

There two places that may cause the tuning range to reduce: the range of the varactor and the divider chain. Using the varactor in SMU_P1, we compared the measurement value against simulated values as shown in figure 9.

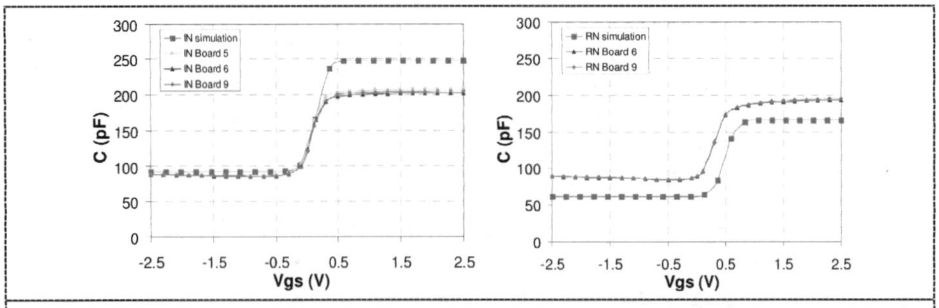

Figure 9. Varactor parameter measurements.

We use only RN type varators in the present design. Although there is a shift in absolute capacitance value, it is clear from the measurement that the tuning range should be the same as simulation.

Further investigation revealed a design mistake in the first stage divider that produces a glitch at low frequency end. This problem is understood and will be fixed in future designs.

3. Roadmap towards a 100 Gbps system

To achieve the required 100 Gbps per FEB data bandwidth for ATLAS LAr readout upgrade for the sLHC, the key development lies on the LOCs6 array serializer. A block diagram of LOCs6, together with the optical interface is shown in figure 10. A switch in the input data path provides a 20% redundancy for possible single channel failure. This design work is being carried out in the Department of Physics, Southern Methodist University.

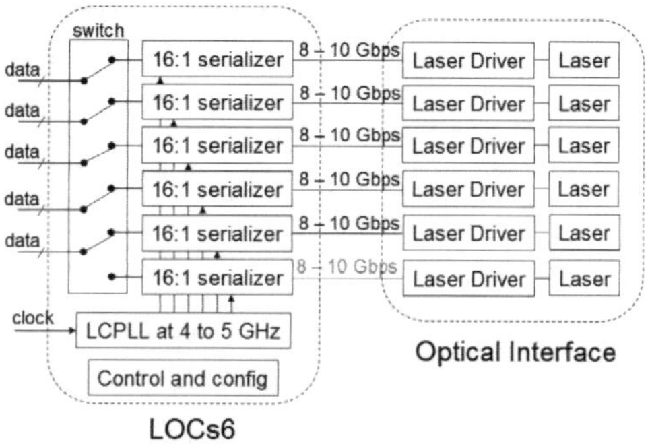

Figure 10. Block diagram of LOCs6 together with the optical interface.

We benefit from the ATLAS-CMS common project the Versatile Link on identification and irradiation evaluation on 10 Gbps lasers, fibers and passive components such as connectors and mechanical packages. We also adapt the system level specifications for optical links developed in the Versatile Link project.

Acknowledgements

This R&D work for the upgrade of the readout of the Liquid Argon Calorimeter of ATLAS is funded by US-ATLAS and is carried out by the team of Dr. D. Gong, Mr. C. Liu, Dr. T. Liu, and Dr. A. Xiang in the Department of Physics at SMU, Dallas, Texas. The team is deeply grateful to Paulo Moreira (CERN) for his kind help throughout the design of LOCs1. We also would like to thank Fukun Tang, Mauro Citterio, Francesco Lanni and many other collaborators in our community for their kind help and support in this project. We would like to thank Suen Hou and PK Teng from IPAS for helping us in the test setup.

References

[1] N.J.Buchanan et al., Design and implementation of the Front End Board for the readout of the ATLAS liquid argon calorimeters. JINST 3, P03004 (2008)
[2] A. Bazan et al., Atlas Liquid Argon Calorimeter Back End Electronics. JINST 2, P06002 (2007)
[3] N.J.Buchanan et al., Radiation Qualification of the Front-End Electronics for the readout of the ASTLAS Arogon Calorimeter. JINST 3, P10005 (2008)
[4] H.Abreu et al., Performance of the Electronic Readout of the ATLAS Liquid Argon Calorimeters. Accepted by JINST and posted at https://twiki.cern.ch/twiki/bin/view/LAr/LArPublications#LArg_Electronics
[5] For more information: https://espace.cern.ch/project-versatile-link/public/default.aspx
[6] For more information: http://www.iucf.indiana.edu/

XIV International Conference on Calorimetry in High Energy Physics (CALOR 2010) IOP Publishing
Journal of Physics: Conference Series **293** (2011) 012019 doi:10.1088/1742-6596/293/1/012019

Readout electronics of the ALICE photon spectrometer

Z B Yin[1], L J Liu[2], H Muller[3], D Röhrich[2], I Sibiryak[4], B Skaali[5], A Vinogradov[4], D Wang[1], Y P Wang[1], F Zhang[1], and D C Zhou[1]

[1] Institute of Particle Physics, Huazhong Normal University, Wuhan 430079, China
Key Laboratory of Quark & Lepton Physics (Huazhong Normal University), Ministry of Education, China
[2] Department of Physics and Technology, University of Bergen, 5007 Bergen, Norway
[3] CERN, PH Department, 1211 Geneva 23, Switzerland
[4] Russian Research Center Kurchatov Institute, Moscow, Russian Federation
[5] Department of Physics, University of Oslo, Blindern 0316, Norway

E-mail: zbyin@mail.ccnu.edu.cn

Abstract. The photon spectrometer (PHOS) in the ALICE experiment at LHC is a $PbWO_4$ crystal based electromagnetic calorimeter, dedicated to measuring photons, π^0's and η's over a broad p_T range from about 100 MeV/c to 100 GeV/c with the best possible energy and position resolution. The front-end electronics of the PHOS is thus required to cover a large dynamic range, to achieve a timing resolution better than $\sim$ 2 ns in order to discriminate against 1-2 GeV/c (anti-)neutrons, and to provide high p_T trigger to select rare high p_T events. In this paper, we present the full PHOS readout system, including the avalanche photo-diode, the low noise charge sensitive preamplifier, the 32 channel front-end electronics card, the trigger region unit and the trigger OR module. Results from PHOS commissioning with beam particles and cosmic rays will also be presented to address the performance of the PHOS readout electronics.

1. Introduction

ALICE [1, 2] is a general-purpose heavy-ion experiment designed to study the properties of the quark matter produced in ultra-relativistic heavy-ion collisions at LHC by comprehensive measurements of hadrons, leptons and photons. Photons produced throughout all stages of the collision are believed to carry important information on the evolution of the colliding system. The ALICE photon spectrometer (PHOS) [3] is designated to measure photons, π^0's and η's over a wide p_T range with the aim of exploring thermal and dynamical properties of the initial phase of the collisions through low p_T direct thermal photon measurements and studying parton energy losses through the measurements of high p_T π^0 and γ-jet correlations.

PHOS is a highly segmented electromagnetic calorimeter with individual detection cells made of very dense $PbWO_4$ (PWO) material with dimensions of $2.2\times2.2\times18$ cm^3. PHOS is subdivided into five PHOS modules and positioned at the bottom of the ALICE setup with a distance of 4.6 m to the nominal interaction point. A PHOS module contains 56×64 detection channels and subtends 0.24 in pseudo-rapidity and 20° in azimuth. Each crystal is optically coupled to an Avalanche Photo-Diode (APD), which is connected to a low-noise Charge-Sensitive Preamplifier (CSP). To increase significantly the light yield of the PWO crystal (by a factor of $\sim$ 3 compared

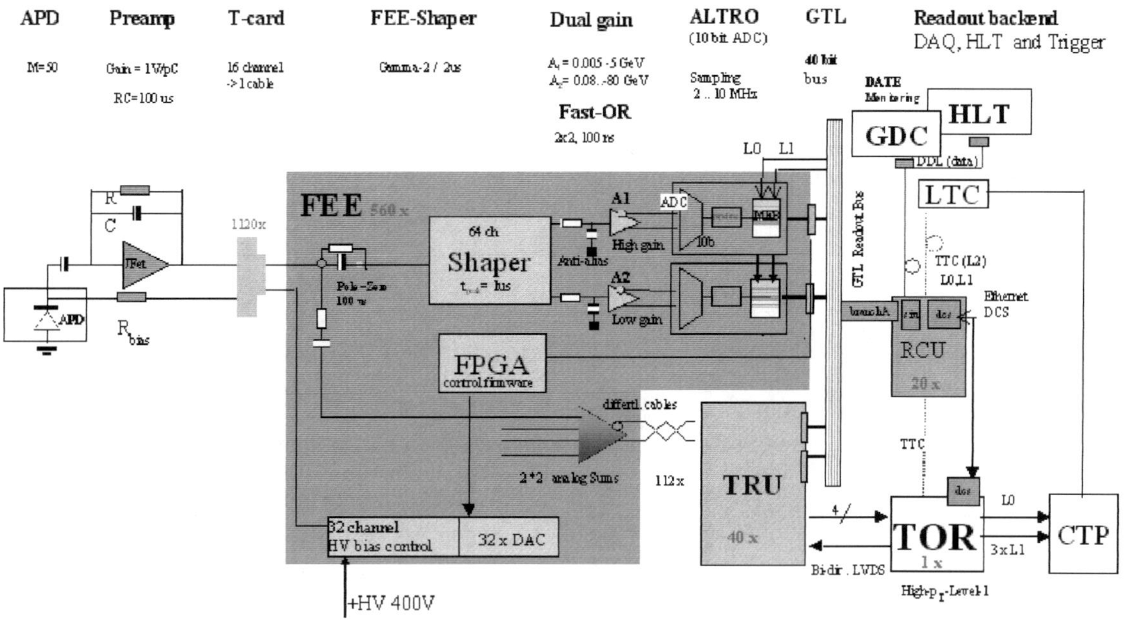

Figure 1. Schematic overview of the PHOS readout system.

to that at room temperature), the detector is located in a "cold" volume operated at -25 °C with a precision better than 0.3°C, whereas the readout electronics is separated by an insulation layer from the "cold" volume.

2. PHOS readout and trigger

The schematic overview of PHOS readout system is shown in Fig. 1. Light induced by incident particles is detected by APDs and the resulted charge is converted into a voltage signal by a CSP. 32 CSP output signals are readout via a PHOS FEE card [4], in which the signals are shaped in dual shapers with different gains, and then digitized with the 10 bit ALICE TPC Readout (ALTRO) chip [5] for a 14-bit effective dynamic range, covering a full energy range of 80 GeV with a least significant bit of 5 MeV.

The FEE also regulates the individual bias voltage between 210 and 400 V for each of the 32 APDs. The setup, control and readout of the FEE cards occurs over a custom GTL bus under mastership of a Readout and Control Unit (RCU) card [6]. Each RCU card masters 28 PHOS FEE cards. The RCU carries a Detector Control System (DCS) daughter card which provides control via a Linux operating system by Ethernet connection and provides the interface to the central trigger processor (CTP). The RCU carries an ALICE Source Interface Unit (SIU) card for transmission of data via an ALICE Detector Data Link (DDL) to the ALICE DAQ and high level trigger (HLT) system. Upon reception of a Level 1 (L1) trigger from LTU [7], the RCU sends via the GTL back-plane a strobe signal to all ALTRO chips in the FEE, which enables the storing of ADC samples of the triggered event in readout buffers. Data in these buffers are either transmitted via the RCU through a DDL to the DAQ and HLT system or discarded depending on whether L2a trigger arrives in ~ 88 μs.

PHOS also provides input to the Level 0 (L0) and L1 trigger decisions in ALICE. The signals of 2×2 detection channels are analog summed via a fast shaper in the FEE boards and transmitted via short cable to a local Trigger Region Unit (TRU) board, where trigger data are digitized and processed by the trigger algorithm implemented in FPGA to generate local L0 yes/no decision. Both L0 decision and the digitized trigger data are transmitted to the Trigger OR (TOR) board. TOR multiplexes the L0 decisions of all TRUs and generates global L1 trigger decisions.

The front-end electronics of one PHOS module consists of 112 FEE boards each processing the signal from 32 crystals, eight TRU boards each covering a region of 28 × 16 crystals, and 4 RCU boards each serving two branches of 14 FEE boards and 1 TRU board through GTL buses.

2.1. APD and charge sensitive preamplifier

PHOS uses the Hammamatsu S8148 (S8664-55) type APDs, which has an active area of 5 × 5 mm^2. The gain M of an APD increases with decreasing temperature, typically by a factor of 3 at −20°C compared to that at 20°C, with a bias voltage of 350 V. The voltage coefficient of the gain $\frac{1}{M}\frac{dM}{dV}$ increases with gain, which in turn increases with increasing bias voltage. Even with the same reverse-bias voltage, there is a considerable spread of the gain factors for production batches. Thus, to choose a nominal APD gain factor, a compromise is required between a high signal gain, noise and signal stability. To achieve best energy resolution, the APDs are decided to be operated at moderate gain of 50, which corresponds to a reverse-bias voltage of ∼ 300-390 V at operating temperature of −25°C. This choice has a lower excess noise which has a direct effect on the stochastic term of the energy resolution. More importantly, a low gain keeps the gain dependencies on temperature and voltage low. These dependencies have an important impact on the constant term of the energy resolution. Particularly, the programmable gain inter-calibration is limited if the gain is set too high. With a gain dependence of 3.3% per V at APD gain of 50, an inter-calibration precision of 0.66% is achievable with a bias voltage increment of 0.2 V/bit.

A charge-sensitive preamplifier [8] on a printed circuit board of area 19 × 19 mm^2 is mounted to the back side of the APD. The CSP produces an output voltage step which is proportional to the charge collected by APD on its input capacitance and inversely proportional to the feedback capacitor C_f. With effective C_f value of 1.2 pF, the amplifier gain is around 0.833 V/pC or 0.133 μV/e$^-$. Taking into account the detector surface and quantum efficiency of APD, about 200 electrons per MeV are generated in the APD by primary electromagnetic particles at the PHOS operating temperature of −25°C. Thus, the corresponding voltage signal at the preamplifier output is 26.7 μV/MeV.

2.2. Front-end electronics card

To extract a direct photon signal from a huge background of decay photons from neutral mesons (mainly π^0 and η), it is necessary for PHOS to achieve a good energy and position resolution. It is desired for PHOS FEE to have an electronic noise per channel of 5 MeV or less, and also a high inter-calibration precision of APD gain and uniformity across all channels.

Fig. 2 shows a photograph of the final 32-channel FEE card. To cover a large dynamic range with optimal energy resolution, 64 second-order (CR-RC2) signal shapers are implemented on a 10-layer board with a gain ratio of ∼ 16 for high and low gain, digitized separately with 10 bit ADCs contained in 4 ALTRO-16 chips. A shaping time of 1 μs has been adopted for the shapers in order to achieve a good timing resolution to discriminate against 1-2 GeV/c (anti-)neutrons. For each CSP input channel, a low noise buffer of gain 2 separates the combined RC differentiator/pole-zero network from the two active Bessel filters. The high gain shaper passes the low energy signals from 5 MeV to 5.12 GeV into the 1 V input range of the 10 bit ADC. With a total gain of 7.2, the low energy range shaper features an optimal RMS noise of 0.62 ADC counts with connected detector at APD gain M=50. The high energy range shaper applies an attenuation of 0.45 in order to fit high-energy CSP signal amplitudes of up to 2.4 V into the 1 V

Figure 2. Photograph of the 32 channel FEE card.

ADC window covering a energy range from 80 MeV to 81.92 GeV. The effective dynamic range is 14 bits. The semi-Gaussian pulse-shapes are digitized by the ALTRO-16 chip, which contains multiple buffers of up to 512 samples for each ADC channel. With a sampling frequency up to 10 MHz, a pipe-lined multi-event buffer inside the ALTRO is loaded with triggered events that can be readout via a custom GTL bus and used in off-line analysis to extract the energy and timing information by fitting the reconstructed pulse shape in time domain to a Γ-2 function [9].

The FEE card implements also 32 individual high-voltage bias controllers for gain control of all connected APDs. Each APD bias is set via a 10 bit digital value, converted by a 10 bit DAC with a bias voltage increment of 0.2 V/bit in the range of 210-400 V, giving a gain resolution of 0.66% at APD gain of 50. Each DAC is controlled via the Motorola SPI serial bus, mastered by the FEE board controller implemented in FPGA.

The FEE board controller, shown in Fig. 3, controls and monitors access to the FEE card resources. The RCU I^2C slave logic allows RCU address-mapped access to both the control and data sections of the FEE cards. The address spaces include ALTRO-chip internal registers, APD bias control registers, registers for monitoring voltages, currents and temperatures.

For trigger purposes, the sums of 2×2 analog signals are processed through a fast shaper amplifier on the FEE board and transferred into the TRU for further processing.

2.3. Trigger region unit

The TRU has been designed [10] to receive 112 fast OR signals from 14 FEE cards and to generate threshold triggers based on charge sums over 4×4 sliding windows over 448 channels. Octal, 12 bit 40 MHz Flash ADCs of 170 ns conversion time with serial outputs are used to digitize all 112 fast 2×2 analog sum signals and concentrate 112 Fast OR streams into a single Virtex FPGA. Fig. 4 is a block diagram of the TRU board controller. The trigger algorithm implemented in the Trigger Module of the FPGA calculates in parallel all 4×4 sliding window sums of the digitized energy signals from the trigger region and makes local L0 yes/no decision

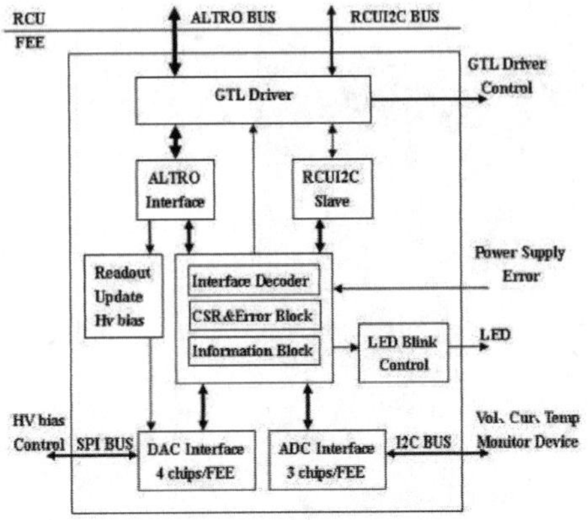

Figure 3. Block diagram of FEE board controller.

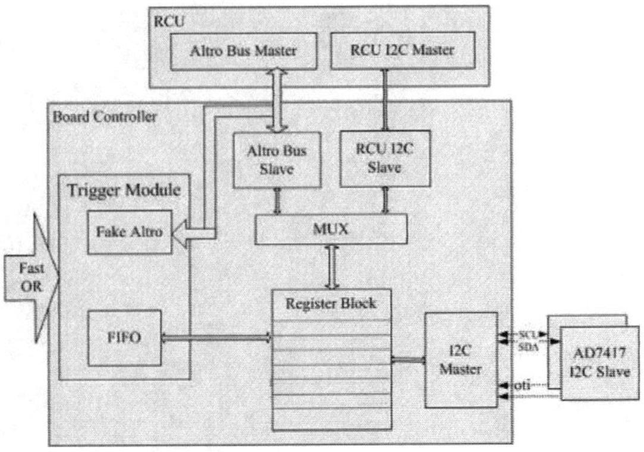

Figure 4. Block diagram of TRU board controller.

within 250 ns by searching for charge above threshold in 4×4 APD areas. Both L0 decision and the digitized 12-bit 2×2 sum trigger data for the global L1 processing are transmitted to TOR unit via LVDS cables. In addition, for offline analysis, raw 2×2 samples from the ADCs with the 2 least significant bit truncated can be readout similar to the FEE card via RCU and appended to the event stream through the "Fake ALTRO" mechanism [11] implemented in TRU board controller.

3. PHOS performance

The performance of PHOS FEE has been studied extensively with FEE prototypes and final products at laboratory with LED pulse and at T10 beam line of the CERN PS during the commissioning of PHOS module. The first PHOS module was assembled in 2006 for calibration at CERN PS. Calibration and measurements of the energy resolution were carried out at operating temperature of $-17°$C. Calibration of energy gain was obtained with 2 GeV electron beam for approximately 2/3 of the detection channels and it was found that the gains of the detection channels could be equalized to $\sim 4\%$ by adjusting the APD bias voltage. The response from different channels can be equalized to very high precision via offline calibration. Enough data were accumulated to estimate the energy resolution at 1, 2, 3, 4 and 5 GeV. The results are shown in Fig. 5 as a function of energy together with previous results from the PHOS prototype at $-25°$C. The measured energy resolution of the first PHOS module is consistent with previous results from prototype and fulfills the PHOS requirement.

While PHOS trigger commissioning is ongoing, cosmic data taken last year with PHOS L0 trigger show the energy distribution of all clusters (mainly single crystal clusters) in PHOS module 2 peak at the right energy position of 210 MeV, indicating that L0 trigger algorithm works as expected. Test on PHOS L0 trigger latency shows that PHOS L0 decisions arrive at the ALICE CTP within 800 ns. For detailed results on PHOS trigger commissioning, please refer to Ref. [12] in this proceedings.

4. Conclusion

We have presented in this paper the full readout electronics of the ALICE PHOS. The performance studies show that PHOS readout electronics are overall in very good shape. While PHOS trigger commissioning is ongoing, commissioning results have already shown that PHOS trigger makes correct L0 decision within ALICE L0 trigger latency.

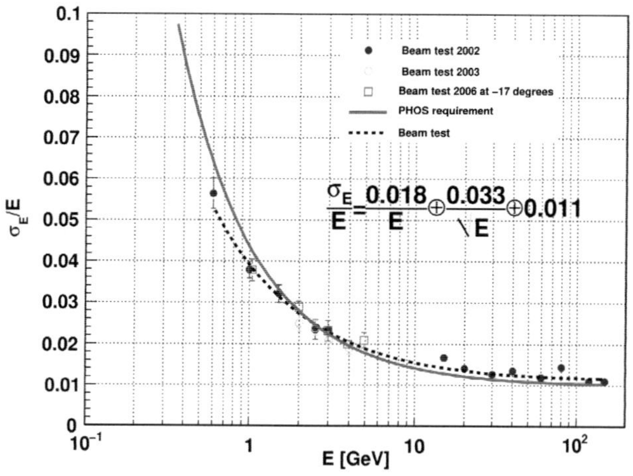

Figure 5. Energy resolution measured in 2006 with the first PHOS module together with previous results from the PHOS prototype.

Acknowledgments

This work is supported by the NSFC under Grant Nos. 10975061, 10875051 and 10635020, the Chinese Ministry of Education under Key Grant No. 306022, the Research Council of Norway (NFR), the Federal Agency of Science of the Ministry of Education and Science of Russian Federation, International Science and Technology Center, Russian Academy of Sciences, Russian Federal Agency of Atomic Energy, Russian Federal Agency for Science and Innovations, CERN-INTAS, and the ALICE CERN team.

References

[1] ALICE Collaboration 2004 *J. Phys.* G **30** 1517
[2] ALICE Collaboration 2006 *J. Phys.* G **32** 1295
[3] ALICE Collaboration 1999 *ALICE Technical Design Report* (Geneva:CERN)
[4] Muller H *et al* 2005 *Nucl. Instr. and Meth.* A **565** 768-83
[5] Esteve Bosch R, Jiménez de Parga A, Mota B and Musa L 2003 *IEEE Transactions on Nuclear Science* **50** 2460-69
[6] Lien J A 2005 *CERN THESIS* **013** (Bergen:University of Bergen)
[7] Evans D *et al* 2004 ALICE trigger system *Proc. 10th Workshop on Electronics for LHC and Future Experiments* (Boston) (Geneva:CERN) p 277-80
[8] Aleksandrov D V *et al* 2005 *Nucl. Instr. and Meth.* A **550** 169-84
[9] Sansen W M C, Chang Z W 1990 *IEEE Transaction on Circuits and Systems* **37** 1375
[10] Muller H *et al* 2005 *Proc. of the 11th Workshop on Electronics for LHC and Future Experiments* (Heidelberg) (Geneva:CERN) p 384
[11] Muller H *et al* 2010 *Nucl. Instr. and Meth.* A **617** 344-7
[12] Liu L 2010 *in this conference proceedings*

XIV International Conference on Calorimetry in High Energy Physics (CALOR 2010) IOP Publishing
Journal of Physics: Conference Series **293** (2011) 012020 doi:10.1088/1742-6596/293/1/012020

Front-End Electronics and Feature-Extraction Algorithm for the PANDA Electromagnetic Calorimeter

M. Kavatsyuk, E. Guliyev, P.J.J. Lemmens, H. Löhner, T.P. Poelman, G. Tambave for the PANDA collaboration

KVI, University of Groningen, The Netherlands

E-mail: M.Kavatsyuk@rug.nl

Abstract. The PANDA collaboration at FAIR, Germany, will employ antiproton annihilations to investigate yet undiscovered charm-mesons and glueballs aiming to unravel the origin of hadronic masses. A multi-purpose detector for tracking, calorimetry and particle identification is presently being developed to run at high luminosities providing up to $2 \cdot 10^7$ interactions/s. A trigger-less data-acquisition system will be employed with sub-detectors continuously providing data from incoming physics events. This paper describes readout electronics and the treatment of the digitised preamplifier signal for the Electromagnetic Calorimeter. The use of a Sampling ADC in the readout allows to achieve the design goals, namely a large dynamic range from 1 MeV to 10 GeV, a count-rate dependent low trigger threshold of about 1 - 3 MeV, and a time resolution better than 1 ns.

1. Introduction

1.1. PANDA Electromagnetic Calorimeter

PANDA is a general purpose hadron physics detector planned to be operated at the future Facility for Antiproton and Ion Research (FAIR) at Darmstadt, Germany. It will use cooled antiproton beams with a momentum between 1.5 GeV/c and 15 GeV/c and a momentum resolution up to $\Delta p/p=10^{-5}$. With hydrogen and various internal targets a peak luminosity of $2 \cdot 10^{32}$ cm^{-2}s^{-1} can be reached, allowing for up to $2 \cdot 10^7$ interactions/s. This, together with the high momentum resolution, allows to measure masses and widths of hadronic resonances with an accuracy 10 to 100 times better than achieved in any e^+e^--collider experiment. In addition states of all quantum numbers can be directly formed in antiproton-proton annihilations.

A sketch of the PANDA spectrometer is shown in figure 1. To obtain a good momentum resolution the detector is split into a target spectrometer at large angles based on a 2 T superconducting solenoid magnet surrounding the interaction point and a 2 Tm forward spectrometer based on a dipole magnet for small angle tracks. In both spectrometer parts tracking, charged particle identification, electromagnetic calorimetry and muon identification are available to allow to detect the complete spectrum of final states relevant for the PANDA physics objectives.

The experiment is focusing on hadron spectroscopy, in particular aiming to search for exotic states in the charmonium mass region, on the interaction of charmed hadrons with the

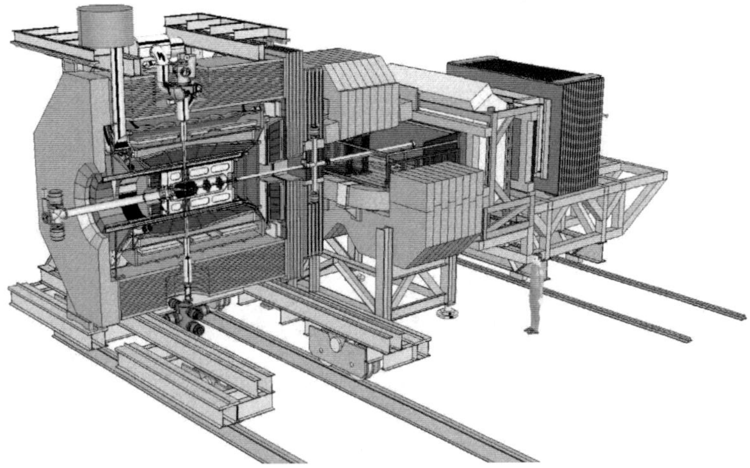

Figure 1. The PANDA detector with target spectrometer (within the solenoid magnet) and forward spectrometer. The target-spectrometer EMC is located inside the solenoid magnet and is depicted in violet colour.

nuclear medium, on double-hypernuclei to investigate the nuclear potential and hyperon-hyperon interactions as well as on electromagnetic processes to study various aspects of nucleon structure. These physics goals define the requirements for the PANDA-detector. For precision spectroscopy of charmonium states and exotic hadrons in the charmonium region it is extremely important to measure low-energy photons as final states with many photons can occur. Therefore, a low photon threshold of about 10 MeV is a central requirement for the ElectroMagnetic Calorimeter (EMC). This requires a threshold for individual crystals of about 3 MeV and correspondingly low noise levels of 1 MeV. Neutral decays of charmed mesons require the detection of a maximum photon energy deposition of 12 GeV per crystal at the given maximum beam energy. These requirements dictate the dynamic range for the readout electronics.

The PANDA EMC [1] will consist of $PbWO_4$ (PWO-II) crystals [2] arranged in the cylindrical barrel EMC with 11360 crystals, the forward endcap EMC with 3600 crystals and the backward endcap EMC with 592 crystals. In order to gain maximum light output from the PWO-II crystals, the calorimeter volume will be cooled to -25°C. The EMC will be placed inside the 2 T solenoid magnet of the target spectrometer. Therefore, the Large Area Avalanche PhotoDiodes (LAAPD) and Vacuum Photo Triodes/Tetrodes (VPT) were chosen as photo-sensors. The LAAPD will be employed for the backward endcap and the barrel parts of the EMC where the single-crystal hit-rate is expected to be in the range of 10 to 100 kHz. Two rectangular LAAPD with a sensitive area of 14×6.8 mm^2 each will be used to read each EMC crystal. For the forward endcap the VPT will be used because of higher expected single-crystal hit rates up to 500 kHz.

The gain of LAAPDs and VPTs is not sufficient to directly read out the output signals. Therefore, special low-power and low-noise preamplifiers were developed. The discrete component preamplifier (LNP) [3] will be used for the VPT readout, and the ASIC APFEL II [4] for the LAAPD readout. The ASIC has a built-in two-stage shaper and provides two output signals with high and low gains. The discrete preamplifier is a one-range resistor-reset type with decay constant of 25 μs. To achieve the best performance the preamplifiers will be placed near the photo-sensors and, therefore, will be kept at the same temperature as the EMC crystals, namely at -25°C.

For optimal conditions of event selection the PANDA experiment uses a triggerless data acquisition system. This requires that all sub-detectors have to provide all single-hit event information. The acquired data are sorted according to the time-stamps by the data acquisition

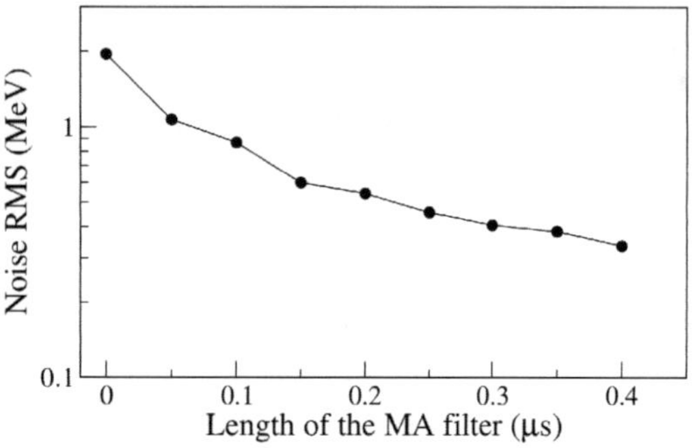

Figure 2. Noise level measured for different lengths of the moving averaging filter.

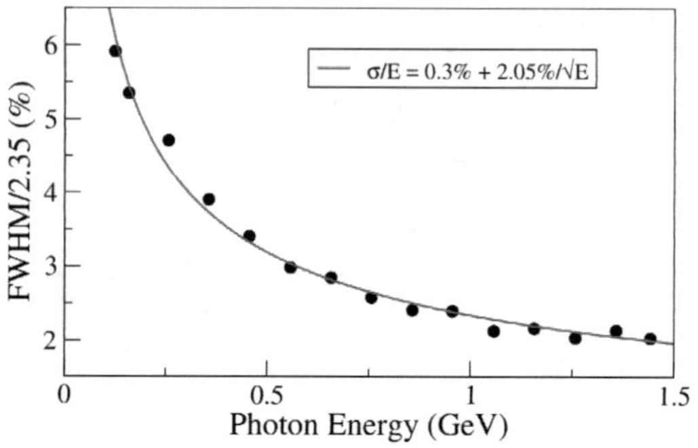

Figure 3. Cluster energy resolution as a function of tagged photon energy (points). The red solid line corresponds to the fitted curve with energy E given in GeV.

(DAQ) and the decision on the event selection is made based on the complete information about the primary interaction of the recorded event. To fulfil the requirement of the DAQ, the EMC-preamplifier signals are continuously digitised by sampling ADCs (SADC) and the data are processed on-line in FPGAs. The digitiser modules are placed in the EMC volume and kept at room temperature. The placement of the SADC system together with the on-line feature-extraction logic in the vicinity of the cold crystals restricts the allowed power consumption of the digitiser unit and, therefore, the complexity of the feature-extraction algorithm. Therefore, a simple and robust algorithm [5] was developed to process data from both the LNP and the ASIC EMC-preamplifiers.

1.2. PANDA-EMC prototype setup

Measurements with the prototype Proto60 [1] were carried out to estimate the expected performance of the PANDA EMC. The Proto60 comprises 60 PWO-II crystals cooled to $-25°$C and read out by one square LAAPD with a sensitive area of 1 cm^2, coupled to a LNP preamplifier. Measurements were done using tagged photons at the MAMI-C electron accelerator facility at Mainz, Germany. Fifteen tagger-defined energies in the range of 124 MeV – 1.4 GeV were selected. A matrix of 3×3 crystals was read out by the 100 MHz 16 bit SIS3302 SADC [6] operated in different runs at 100 MHz and at 50 MHz. The beam-spot diameter of the photon beam was about 1 mm, pointing to the centre of the middle crystal. The data acquisition was triggered by the external signal from the electron-tagger detectors. For each event 10 μs long traces were stored for all 9 channels. The obtained data were analysed offline, emulating the on-line feature-extraction algorithm. After optimisation the algorithm was implemented in VHDL for the SIS3302 SADC [6] enabling on-line data analysis for the future test experiments with the Proto60. The energy calibration of the individual channels of Proto60 was obtained from cosmic-ray data.

2. Performance of the EMC prototype

2.1. Noise level and energy resolution

To extract the information on the energy deposition from the digitised preamplifier signals the Moving Window Deoconvolution (MWD) [7] and Moving Average (MA) filters are used, see ref. [5] for more details. The noise level which determines the triggering threshold depends on the length of the MA filter (see figure 2). The length of the MA filter defines as well the

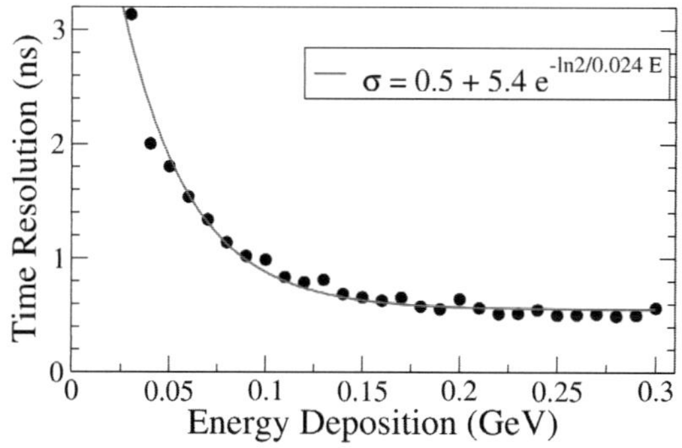

Figure 4. Measured time resolution RMS (black points) as a function of the energy deposition in the PWO-II crystal. The red solid line corresponds to the fitted curve with energy E given in GeV.

total width of the resulting pulse after the digital shaping and, therefore, the pile-up probability. Taking into account the single-crystal hit-rate between 100 and 500 kHz expected for the forward EMC-endcap the maximum tolerated pulse-length is about 100 ns. Therefore, the maximum MA-filter length is about 80 ns. As can be seen from figure 2, the achieved noise level is about 1 MeV in this case. The expected hit-rate for the barrel and backward endcap of the EMC is much lower and, therefore, longer shaping times can be used providing even better noise performance. To cover the required dynamic range up to 12 GeV, the resolution of the SADC should be 14 bits.

The cluster energy resolution measured with a 3×3 PWO-II crystal sub-matrix for the tagged photons is shown in figure 3. The measured energy resolution was fitted using the function $\sigma/E = A_0 + A_1/\sqrt{E/GeV}$ and the parameters $A_0 = 0.3\%$ and $A_1 = 2.05\%$ were found. In the Technical Design Report for the PANDA EMC [1] the limits $A_0 \leq 1\%$ and $A_1 \leq 2\%$ are required. The energy resolution of the EMC prototype almost fulfills these requirements with a marginally bigger stochastic term A_1. To allow for more than nine crystals per cluster would provide a better electromagnetic shower collection and, therefore, improve the energy resolution. As mentioned above in sect. 1.1, in the final configuration of the EMC each PWO-II crystal will be equipped with 2 LAAPD covering about 2 cm^2 of the crystal end-face. This larger coverage will lead to an improved photo-electron statistics and, according to test results, to an additional improvement of the resolution by almost a factor $\sqrt{2}$. Two independent readout channels per crystal will also increase the redundancy of the EMC and will allow to discriminate events contaminated by the nuclear-counter effect (i.e. detection of charged particles which penetrate the LAAPD).

2.2. Time resolution

To determine the precise time-stamp of signal pulses, the digital implementation of the constant-fraction discrimination is applied [5]. This method allows to achieve a timing precision much higher than the SADC sampling rate. To measure the time-resolution of the EMC-prototype, a special measurement was carried out. The tagged-photon beam was directed between two PWO-II crystals. During the data-analysis only events with about the same (±10 %) energy deposition in both crystals were selected. The time-resolution RMS for the single channel was deduced from the distribution of the time-difference between two crystals and is shown in figure 4 as a function of the energy deposition in the PWO-II crystal. For the energy deposition above 80 MeV the time resolution reaches below the level of 1 ns and improves to about 0.5 ns at 300 MeV. The achieved timing precision is much higher than the SADC sampling rate (50 MHz, see sect. 1.2) and is sufficient to relate hits in the PANDA-EMC to the primary-interaction events as well as suppress random coincidence background.

A sampling rate of about 40-50 MHz is optimal for the readout of the LNP preamplifier

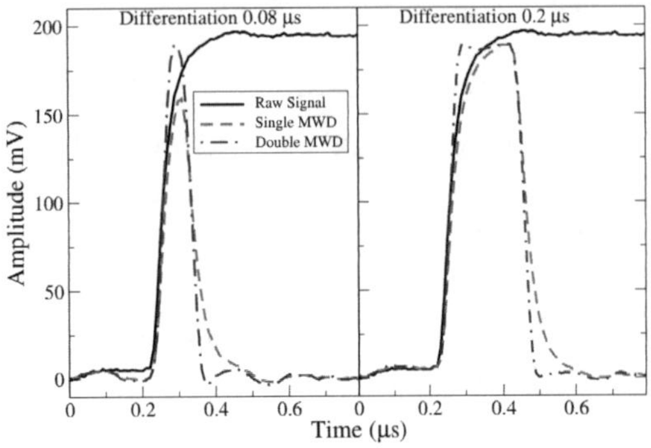

Figure 5. LNP-preamplifier pulse-shape before (black solid line) and after single (red dashed line) / double (blue dash-dotted line) Moving Window Deconvolution filtering for 80 ns (left panel) and 200 ns (right panel) differentiation time constants of the MWD filter.

signal. The usage of higher sampling rates does not improve the timing performance of the EMC and at the rate of about 25 MHz the time-resolution will worsen by about a factor of 2.

2.3. Performance at high rates

As mentioned above in the sect. 1.1 and 2.1, the forward endcap of the PANDA-EMC will operate at high single-hit rates, up to 500 kHz. To avoid pile-up of different events it is extremely important to keep the hit-response of the single-crystal detector as short as possible. The pulse-shape of the LNP preamplifier, used in the forward endcap, is shown in figure 5. The rise time of the pulse is about 100 ns and the fall time is 25 μs. The MWD digital-filter differentiates the incoming pulse and compensates for the exponential discharge of the integrating capacitor in the preamplifier. Therefore, the resulting pulse has the required width defined by the differentiating time-constant of the MWD filter. As can be seen in figure 5, the MWD filter does not distort the rising edge of the incoming pulse and the trailing edge repeats the shape of the rising one. For the LNP preamplifier this leads to the appearance of the exponential tail after the pulse, effectively increasing the total width of the resulting pulse by 100 ns. Therefore, the minimum achievable total pulse-width is about 300 ns. In addition, the application of the MWD filter with the differentiating time-constant shorter than the rising edge leads to a decreased amplitude of the resulting pulse while keeping the same noise level. This effectively increases the threshold for the hit-detection.

It was found that the resulting tail of the LNP-preamplifier pulse, shaped by the MWD filter, has an exponential behaviour. Therefore, such tail can be compensated applying a second MWD filter with the corresponding decay-constant. The resulting LNP-preamplifier pulse after double MWD shaping is shown in figure 5. The double MWD filtering allows to obtain much shorter pulses and recover its original amplitude for short differentiating time-constants without increasing the noise level. Figure 6 shows the recovery of the pulse amplitude after the double MWD filtering as a function of the differentiation time-constant of the first MWD filter. Only for the very short differentiation time-constant, below 80 ns, the pulse amplitude can not be completely recovered even though providing a much higher amplitude than the single MWD filtering. Such signal recovery technique allows to effectively achieve a low triggering threshold while reducing the pulse width.

The usage of the double MWD filtering does not influence the resulting energy resolution of the detector. As can be seen in figure 7 the cluster energy resolution for γ-rays is the same for the single and double MWD filtering. For the forward endcap of the PANDA-EMC the double MWD filtering with the differentiation time-constant of about 60 ns will provide the best performance in terms of low pile-up probability without compromising the energy resolution of the detector.

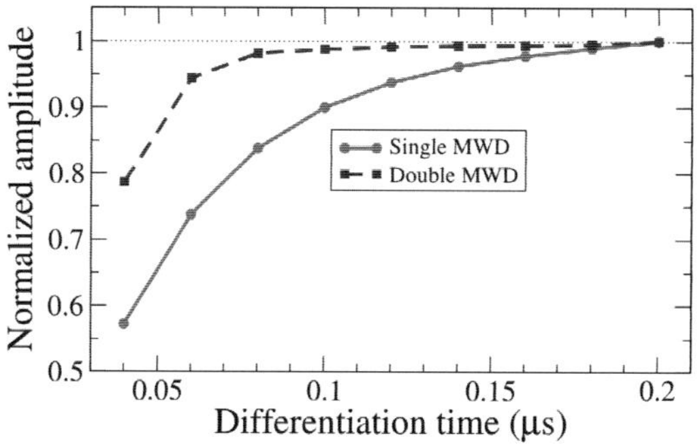

Figure 6. The pulse height after single (red solid line) and double (blue dashed line) MWD filtering as a function of the differentiation time-constant of the first MWD filter.

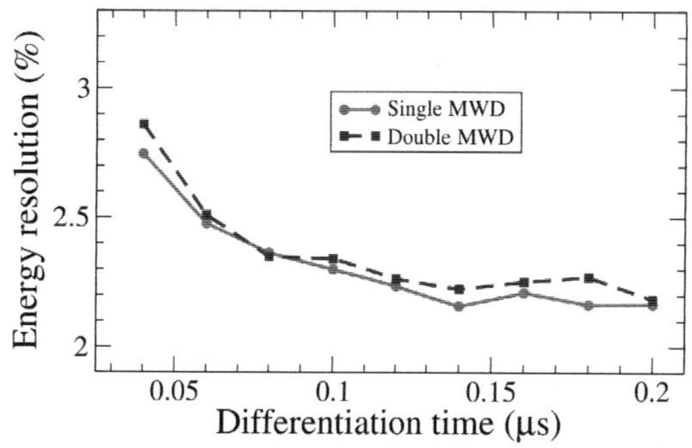

Figure 7. Measured cluster energy resolution for about 1 GeV photon energy obtained using the single (red solid line) and double (blue dashed line) MWD filtering method. The energy resolution is plotted as a function of the differentiation time-constant of the first MWD filter.

3. Conclusions

The measurements performed with the EMC prototype Proto60 using SADC readout demonstrated the performance close to the design goals of the PANDA-EMC. Namely it is possible to obtain a large dynamic range from 1 MeV to 10 GeV, a low trigger threshold of about 1 – 3 MeV, depending on the single detector hit rate, a time resolution better than 1 ns for an energy deposition above 80 MeV. The SADC sampling rate of about 40-50 MHz is sufficient to extract all the information from the LNP-preamplifier signals without loosing time and energy resolution. Such a low sampling rate and correspondingly moderate power consumption will allow to integrate the SADC into the front-end electronics at the calorimeter.

The digital filtering, namely double MWD shaping, allows to reduce the time span for the single-hit response of the EMC detector to about 160 ns. This reduces the hit pile-up probability.

The results presented in this paper are based on the off-line analysis of recorded signal traces. The used feature-extraction algorithm is implemented in VHDL for the FPGA logic of the SADC. Therefore, future tests of the PANDA-EMC prototypes will be performed under even more realistic conditions.

4. Acknowledgements

We thank the MAMI-C accelerator crew for providing the tagged photon beam and gratefully acknowledge the joint effort of the PANDA teams at Univ. Giessen and IPN Orsay for the Proto60 setup.

References

[1] PANDA EMC Technical Design Report, arXiv:0810.1216v1.
[2] Rainer W. Novotny, "High-Quality PWO Crystals for the PANDA-EMC", contribution to this conference.
[3] W. Erni, M. Steinacher, Univ. Basel, in PANDA Technical Progress Report, Feb. 2005.
[4] P. Wieczorek, H. Flemming, GSI report (2007) 30.
[5] M. Kavatsyuk et al., Conference Record, 2009 IEEE Nuclear Science Symposium, Orlando, Florida, USA, October 25-31 2009, N10-4; DOI 10.1109/NSSMIC.2009.5401798 (http://ieeexplore.ieee.org).
[6] SIS3302 SADC, http://www.struck.de/sis3302.htm.
[7] A. Georgiev, W. Gast, IEEE Trans. Nucl. Sci. NS-40 (1993) 770; J. Stein et al., Nucl. Instr. Meth. B 113 (1994) 141.

Particle Flow Calorimetry

Mark A. Thomson

University of Cambridge, Cavendish Laboratory, JJ Thomson Avenue, Cambridge CB3 0HE, UK

E-mail: thomson@hep.phy.cam.ac.uk

Abstract. The Particle Flow (PFlow) approach to calorimetry promises to deliver unprecedented jet energy resolution for experiments at future high energy colliders such as the proposed International Linear Collider (ILC). Since Calor 2008 there has been a significant improvement in understanding particle flow calorimetry and its potential. This contribution to the proceedings describes the current understanding of high granularity particle flow calorimetry in the context of the PandoraPFA algorithm and the ILD detector concept. It is shown that a jet energy resolution of $\sigma_E/E \lesssim 4\,\%$ is achievable for $40-400$ GeV jets, demonstrating that high granularity PFlow calorimetry can meet the challenging ILC jet energy resolution goals. The potential of high granularity PFlow calorimetry at a multi-TeV lepton collider, such as CLIC and a possible Muon Collider, is also discussed.

1. Introduction

In recent years the concept of high granularity Particle Flow calorimetry [1, 2] has been developed in the context of the proposed International Linear Collider (ILC) [4]. The goal for jet energy resolution at the ILC is that it is at least sufficient to cleanly separate W and Z hadronic decays; an invariant mass resolution comparable to the gauge boson widths, i.e. $\sigma_m/m = 2.7\,\% \approx \Gamma_W/m_W \approx \Gamma_Z/m_Z$, leads to an effective $2.5\,\sigma$ separation of the $\mathrm{W} \rightarrow \mathrm{q'\overline{q}}$ and $\mathrm{Z} \rightarrow \mathrm{q\overline{q}}$ mass peaks. It is only recently that it is been demonstrated that this level of performance can be achieved with Particle Flow Calorimetry [3].

In the traditional calorimetric approach, the jet energy is obtained from the sum of the energies deposited in the electromagnetic and hadronic calorimeters (ECAL and HCAL). This typically results in a jet energy resolution of the form

$$\frac{\sigma_E}{E} = \frac{\alpha}{\sqrt{\mathrm{E(GeV)}}} \oplus \beta.$$

The stochastic term, α, is usually greater than $\sim$50 % and the constant term, β, which encompasses a number of effects, is typically a few per cent. The stochastic term alone results in a contribution to the di-jet mass resolution of $\sigma_m/m \approx \alpha/\sqrt{E_{jj}}$, where E_{jj} is the energy of the di-jet system in GeV. At the ILC, operating at centre-of-mass energies $\sqrt{s} = 0.5 - 1.0\,\mathrm{TeV}$, the typical di-jet energies will be in the range $150 - 350\,\mathrm{GeV}$. Even if one neglects the constant term, to achieve the ILC goal of $\sigma_m/m = 2.7\,\%$, the stochastic term must be $\lesssim 30\,\%/\sqrt{\mathrm{E(GeV)}}$. This is unlikely to be achievable with a traditional approach to calorimetry.

1.1. The Particle Flow Approach to Calorimetry

On average, 62% of the jet energy is carried by charged particles (mainly hadrons), around 27% by photons, about 10% by long-lived neutral hadrons (e.g. n, $\overline{n}$ and K_{L}), and around 1.5 % by neutrinos. Hence, approximately 72 % of the jet energy is measured with the precision of the combined ECAL and HCAL for hadrons; the jet energy resolution is thus limited by the relatively poor hadronic energy resolution, typically $\gtrsim 50\,\%/\sqrt{\mathrm{E(GeV)}}$. A number of collider experiments (e.g. OPAL, H1 and D0) obtained improved jet energy resolution using the Energy Flow approach, whereby energy deposits in the calorimeters are removed according to the momentum of associated charged particle tracks. ALEPH [5] used particle flow techniques to attempt to reconstruct the four momenta of the particles in an event. However, due to the relatively low granularity of the calorimeters, energy depositions from neutral hadrons still had to be identified as a significant excesses of calorimetric energy compared to the associated charged particle tracks. Nevertheless, ALEPH achieved a jet energy resolution (for $\sqrt{s} = M_{\mathrm{Z}}$) equivalent to $\sigma_E = (59\,\%\sqrt{E/\mathrm{GeV}} + 0.6)$ GeV [5], roughly a factor two worse than required for the ILC even for these relatively low energy jets.

Within the ILC community it is believed that the most promising strategy for achieving the ILC jet energy goal is the Particle Flow (PFlow) approach to calorimetry using a *highly granular* detector. In contrast to a purely calorimetric measurement, PFlow calorimetry requires the reconstruction of the four-vectors of all visible particles in an event. The reconstructed jet energy is the sum of the energies of the individual particles. The momenta of charged particles are measured in the tracking detectors, while the energy measurements for photons and neutral hadrons are obtained from the calorimeters. In this manner, the HCAL is used to measure only $\sim 10\,\%$ of the energy in the jet. If one were to assume calorimeter resolutions of $\sigma_E/E = 0.15/\sqrt{\mathrm{E(GeV)}}$ for photons and $\sigma_E/E = 0.55\sqrt{\mathrm{E(GeV)}}$ for hadrons, a jet energy resolution of $0.19/\sqrt{\mathrm{E(GeV)}}$ would be obtained with the contributions from tracks, photons and neutral hadrons as given in Table 1. In practice, this level of performance can not be achieved as it is not possible to perfectly associate all energy deposits with the correct particles. For example, if the calorimeter hits from a photon are not resolved from a charged hadron shower, the photon energy is not accounted for. Similarly, if part of charged hadron shower is identified as a separate cluster, the energy is effectively double-counted as it is already accounted for by the track momentum. This *confusion* rather than calorimetric performance is the limiting factor in PFlow calorimetry. Thus, the crucial aspect of PFlow calorimetry is the ability to correctly assign calorimeter energy deposits to the correct reconstructed particles, placing stringent requirements on the granularity of the ECAL and HCAL. For the jet energy reconstruction, the sum of calorimeter energies is replaced by a complex pattern recognition problem, namely the Particle Flow reconstruction Algorithm (PFA). The jet energy resolution obtained is a combination of the intrinsic detector performance and the performance of the PFA software. The PandoraPFA algorithm[3] was developed to study PFlow calorimetry at the ILC and represents the state-of-the-art in high granularity PFlow reconstruction.

Component	**Detector**	**Energy Fract.**	**Energy Res.**	**Jet Energy Res.**
Charged Particles ($X^{\pm}$)	Tracker	$\sim 0.6\,E_j$	$10^{-4}\,E_{X^{\pm}}^2$	$< 3.6 \times 10^{-5}\,E_j^2$
Photons (γ)	ECAL	$\sim 0.3\,E_j$	$0.15\,\sqrt{E_\gamma}$	$0.08\,\sqrt{E_j}$
Neutral Hadrons (h^0)	HCAL	$\sim 0.1\,E_j$	$0.55\,\sqrt{E_{h^0}}$	$0.17\,\sqrt{E_j}$

Table 1. Contributions from the different particle components to the jet-energy resolution (all energies in GeV). The table lists the approximate fractions of charged particles, photons and neutral hadrons in a jet of energy, E_j, and the assumed single particle energy resolution.

It is important to realise that PFlow calorimetry is very different from the traditional approach, the aim is to use the energies measured in the calorimeters as little as possible. The accepted wisdom for calorimetry may not hold for particle flow calorimetry as it is a very different technique.

2. Overview a Particle Flow Detector

The studies presented in this contribution were performed in the context of the Geant4 [7] simulation of the ILD detector concept [8]. It consists of a vertex detector, tracking detectors, ECAL, HCAL and muon chambers. For the ECAL and HCAL the emphasis is on granularity, both longitudinal and transverse, rather than solely energy resolution. Suitable candidate technologies are being studied by the CALICE (calorimetry for the ILC) collaboration [9]. Amongst these are the Silicon-Tungsten ECAL and Steel-Scintillator HCAL designs assumed for the ILD reference detector simulation. The R&D and detector prototyping performed within the CALICE collaboration (described elsewhere in these proceedings) has reached the point where one can be confident that such a detector ultimately could be constructed.

In the ILD concept, both the ECAL and HCAL are located inside a solenoid which is taken to produce the 3.5 T magnetic field. The main tracking detector is simulated as a time projection chamber (TPC) with an active gas volume of half-length 2.25 m and inner and outer radii of 0.39 m and 1.74 m respectively. The ECAL is simulated as a Silicon-Tungsten sampling calorimeter consisting of 29 layers. The first 20 layers have 2.1 mm thick Tungsten and the last 9 layers have 4.2 mm thick Tungsten. The high resistivity Silicon is segmented into $5 \times 5\,\mathrm{mm}^2$ pixels. At normal incidence, the ECAL corresponds to 23 radiation lengths (X_0) and 0.8 nuclear interaction lengths (λ_I). The HCAL is simulated as a Steel-Scintillator sampling calorimeter comprising 48 layers of 20 mm thick Steel and 5 mm thick $3 \times 3\,\mathrm{cm}^2$ plastic scintillator tiles. At normal incidence the HCAL is $6\,\lambda_I$ thick.

The ECAL and HCAL in the ILD concept are well matched to the requirements of PFlow calorimetry. Tungsten is the ideal absorber material for the ECAL; it has a short radiation length and small Molière radius which leads to compact electromagnetic (EM) showers. It also has a large ratio of interaction length to radiation length which means that hadronic showers will tend to be longitudinally well separated from EM showers. The $5 \times 5\,\mathrm{mm}^2$ transverse segmentation takes full advantage of the small Molière radius. Steel is chosen as the HCAL absorber, primarily for its structural properties. The $3 \times 3\,\mathrm{cm}^2$ HCAL transverse segmentation is well matched to the requirements of PFlow calorimetry.

The performance of PFlow calorimetry depends strongly on the reconstruction software. For the results obtained to be meaningful, it is essential that both the detector simulation and the reconstruction chain are as realistic as possible. For this reason no Monte Carlo (MC) information is used at any stage in the reconstruction as this is likely to lead to an overly-optimistic evaluation of the potential performance of PFlow calorimetry. PandoraPFA runs in the MARLIN [6] C++ framework developed for the LDC and ILD detector concepts. Tracks in the TPC are reconstructed using TPC pattern recognition software [10] based on that used by ALEPH and track fitting software used by DELPHI [11]. PandoraPFA combines the tracking information with hits in the high granularity calorimeters to reconstruct the individual particles in the event. As an example of the information used in the reconstruction, Figure 1 shows a photon, a charged hadron (π^+) and a neutral hadron (K_L) as simulated in the ILD detector concept.

3. The PandoraPFA Particle Flow Algorithm

The PandoraPFA algorithm is described in detail in [3] and only an overview is presented here. It is a complex reconstruction programme with a number of distinct stages: *1) Track Selection/Topology:* track topologies such as kinks and decays of neutral particles in the detector

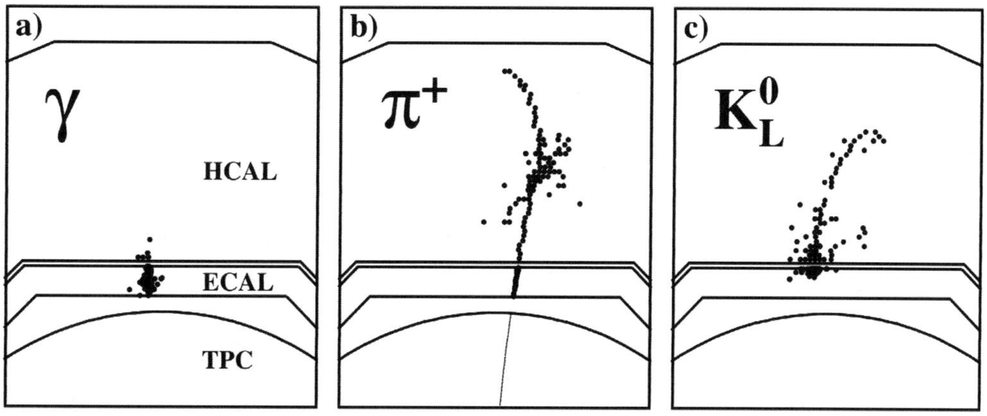

Figure 1. Example simulated single particle interactions in the ILD detector concept: a) a 10 GeV photon; b) a 10 GeV π^+ and c) a 10 GeV K_L. Hits in the TPC, ECAL and HCAL are shown. For the ECAL (HCAL) all hits with energy depositions $> 0.5\,(0.3)$ minimum ionising particle equivalent are displayed. Simulated TPC hits are digitised assuming 227 radial rows of readout pads.

volume (*e.g* $K_S \rightarrow \pi^+\pi^-$) are identified. *2) Calorimeter Hit Selection and Ordering:* isolated hits, defined on the basis of proximity to other hits, are removed from the initial clustering stage. The remaining hits are ordered into *pseudo-layers* and information related to the geometry and the surrounding hits are stored for use in the reconstruction. *3) Clustering:* the main clustering algorithm is a cone-based forward projective method [12] working from innermost to outermost pseudo-layer. The algorithm starts by seeding clusters using the projections of reconstructed tracks onto the front face of the ECAL. *3A) Photon Clustering:* PandoraPFA can be run in a mode where the above clustering algorithm is performed in two stages. In the first stage, only ECAL hits are considered with the aim of identifying energy deposits from photons. In the second stage the clustering algorithm is applied to the remaining hits. *4) Topological Cluster Merging:* by design the initial clustering stage errs on the side of splitting up true clusters rather than merging energy deposits from more than one particle into a single cluster. Clusters are then combined on the basis of clear topological signatures in the high granularity calorimeters. The topological cluster merging algorithms are only applied to clusters which have not been identified as photons. *5) Statistical Re-clustering:* The previous four stages of the algorithm are found to perform well for jets with energies of less than 50 GeV. For higher energy jets the performance degrades due to the increasing overlap between hadronic showers from different particles. Clusters which are likely to have been created from the merging of hits in showers from more than one particle are identified on the basis of the compatibility of the cluster energy, E_C, and the associated track momentum, p. In the case of an inconsistent energy-momentum match, attempts are made to re-cluster the hits by re-applying the clustering algorithm with different parameters, until the cluster splits to give a cluster energy consistent with the momentum of the associated track. *6) Photon Recovery and Identification:* A more sophisticated, shower-profile based, photon-identification algorithm is then applied to the clusters, improving the tagging of photons. It is also used to recover cases where a primary photon is merged with a hadronic shower from a charged particle. *7) Fragment Removal:* "neutral clusters" which are *fragments* of charged particle hadronic showers are identified. *8) Formation of Particle Flow Objects:* The final stage of the algorithm is to create the list of reconstructed particles, Particle Flow Objects (PFOs), and associated four-momenta.

4. Parametrising Particle Flow Performance: rms_{90}

Figure 2 shows the distribution of PFA reconstructed energy for simulated $(Z/\gamma)^* \to q\overline{q}$ events (light quarks only, i.e. q=u,d,s) generated at $\sqrt{s} = 200\,\mathrm{GeV}$ with the Z decaying at rest, termed "Z $\to$ uds" events. A cut on the polar angle of the generated $q\overline{q}$ system, $\theta_{q\overline{q}}$, is chosen to avoid the barrel/endcap overlap region, $|\cos\theta_{q\overline{q}}| < 0.7$. Only light quark decays are considered here as, currently, PandoraPFA does not include specific reconstruction algorithms to attempt to recover missing energy from semi-leptonic decays of heavy quarks.

It is essential to realise that the reconstructed energy distribution of Figure 2 is not Gaussian. This is not surprising; one might expect a Gaussian core for perfectly reconstructed events (where the calorimetric energy resolution for hadrons dominates), and tails corresponding to the population of events where confusion is significant. Quoting the rms, in this case 5.8 GeV, as a measure of the jet energy resolution makes little sense as it over-emphasises the importance of these tails. In order to quote the energy resolution a more robust estimator than the raw rms is required. The chosen measure, rms_{90}, is defined as the rms in the smallest range of reconstructed energy which contains 90 % of the events. For the reconstructed data shown in Figure 2, $\mathrm{rms}_{90} = 4.1\,\mathrm{GeV}$ (equivalent to a single jet energy resolution of 2.9 %). The advantage of using rms_{90} is that it is robust and is relatively insensitive to the tails of the distribution; it parametrises the resolution for the bulk of the data.

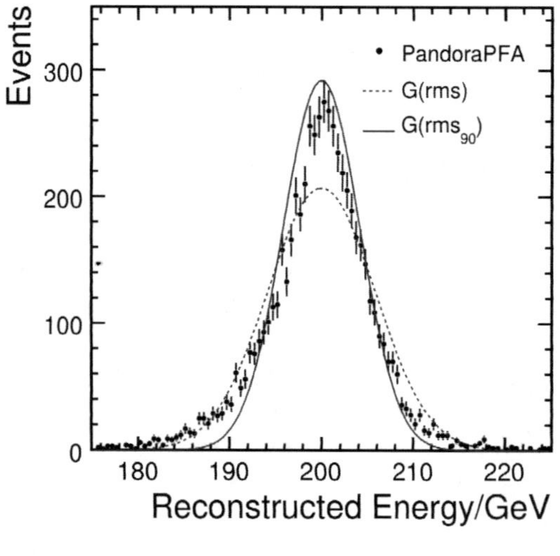

Figure 2. The total reconstructed energy from reconstructed PFOs in 200 GeV Z $\to$ uds events for initial quark directions within the polar angle acceptance $|\cos\theta_{q\overline{q}}| < 0.7$. The dotted line shows the best fit Gaussian distribution with an rms of 5.8 GeV. The solid line shows a Gaussian distribution, normalised to the same number of events, with standard deviation equal to rms_{90} (i.e. $\sigma = 4.1\,\mathrm{GeV}$).

A possible criticism [17] of the use of rms_{90} is that for a Gaussian distribution, rms_{90} is 21 % smaller than the true rms and thus significantly underestimates the true resolution. This misses the point. Unless the confusion term is eliminated, the PFA reconstructed jet energy distribution will never be Gaussian. For a non-Gaussian distribution the rms is not necessarily a good measure of the *analysing power* of a measurement. The equivalent Gaussian statistical analysing power, obtained from a MC study [3], shows that for the PandoraPFA reconstruction, rms_{90} is equivalent to a Gaussian resolution with $\sigma = 1.1 \times \mathrm{rms}_{90}$. Hence whilst rms_{90} does

Journal of Physics: Conference Series **293** (2011) 012021 doi:10.1088/1742-6596/293/1/012021

underestimate the resolution (in terms of equivalent analysing power), it is only a 10 % effect relative to a pure Gaussian distribution. It should be noted that quoting the raw rms of the PFlow reconstructed distribution over-estimates the equivalent analysing power by 30 %.

5. Particle Flow Performance

Jet Energy	rms	$\mathrm{rms}_{90}(E_{jj})$	$\mathrm{rms}_{90}(E_{jj})/\sqrt{E_{jj}}$	$\mathrm{rms}_{90}(E_j)/E_j$
45 GeV	3.4 GeV	2.4 GeV	25.2 %	(3.74 ± 0.05) %
100 GeV	5.8 GeV	4.1 GeV	29.2 %	(2.92 ± 0.04) %
180 GeV	11.6 GeV	7.6 GeV	40.3 %	(3.00 ± 0.04) %
250 GeV	16.4 GeV	11.0 GeV	49.3 %	(3.11 ± 0.05) %
375 GeV	29.1 GeV	19.2 GeV	81.4 %	(3.64 ± 0.05) %
500 GeV	43.3 GeV	28.6 GeV	91.6 %	(4.09 ± 0.07) %

Table 2. Jet energy resolution for Z →uds events with $|\cos\theta_{q\bar{q}}| < 0.7$, expressed as: i) the rms of the reconstructed di-jet energy distribution, E_{jj}; ii) rms_{90} for E_{jj}; iii) the effective constant α in $\mathrm{rms}_{90}(E_{jj})/E_{jj} = \alpha(E_{jj})/\sqrt{E_{jj}(\mathrm{GeV})}$; and iv) the fractional jet energy resolution for a single jet where $\mathrm{rms}_{90}(E_j) = \mathrm{rms}_{90}(E_{jj})/\sqrt{2}$.

The performance of the PandoraPFA algorithm with the ILD detector concept was studied using MC samples of approximately 10000 Z → uds generated with the Z decaying at rest with $E_Z = 91.2$, 200, 360, and 500 GeV. These jet energies are typical of those expected at the ILC operating at $\sqrt{s} = 0.5 - 1.0$ TeV. In addition, to study the performance at higher energies, events were generated with $E_Z = 750$ GeV and 1 TeV. Jet fragmentation and hadronisation was performed using the PYTHIA [13] program tuned to the fragmentation data from the OPAL experiment [14]. The events were passed through the MOKKA simulation of the ILD detector concept which is described in detail in [8]. The `LCPHYS`[16] Geant4 physics list was used for the modelling of hadronic showers. For each set of events, the total energy is reconstructed and the jet energy resolution is obtained by dividing the total energy resolution by $\sqrt{2}$. Figure 3 shows the jet energy resolution as a function of the polar angle of the quarks in Z → $q\bar{q}$ events. The energy resolution does not vary significantly in the region $|\cos\theta| < 0.975$. A small degradation in the energy resolution is seen for the barrel-endcap overlap region, $0.7 < |\cos\theta| < 0.8$. In addition, there is a small degradation in performance at $\cos\theta \approx 0$ due to the TPC central membrane and gaps between sections of the HCAL as simulated in the ILD detector model.

Table 2 summarises the current performance of the PandoraPFA algorithm applied to ILD detector simulation. For the typical ILC jet energy range, 45 − 250 GeV, the energy resolution is significantly better than the best resolution achieved at LEP, $\sigma_E/E \approx 0.65/\sqrt{E(\mathrm{GeV})}$. Table 2 also lists the single jet energy resolution. For jet energies in the range 45 − 375 GeV this is better than 3.8 %, which is necessary to resolve hadronic decays of W and Z bosons. These results clearly demonstrate the potential of PFlow calorimetry at the ILC; the jet energy resolution obtained is approximately a factor two better than might be achievable with a traditional calorimetric approach. Furthermore, it is expected that the performance of PandoraPFA will improve with future refinements to the algorithm.

It is worth noting, that for perfect PFlow reconstruction, the energy resolution would be described by $\sigma_E/E \approx \alpha/\sqrt{\mathrm{E(GeV)}}$, where α is a constant. The fact that this does not apply is no surprise; as the particle density increases it becomes harder to correctly associate the calorimetric energy deposits to the particles and the confusion term increases.

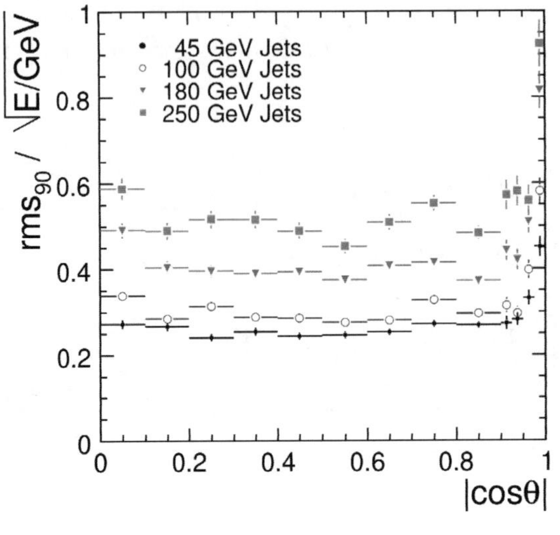

Figure 3. The jet energy resolution, defined as the α in $\sigma_E/E = \alpha/\sqrt{\mathrm{E(GeV)}}$, plotted versus $\cos\theta_{\mathrm{q\bar{q}}}$ for four different values of $\sqrt{s}$. The plot shows the resolution obtained from $(\mathrm{Z}/\gamma)^* \rightarrow \mathrm{q\bar{q}}$ events (q=u,d,s) generated at rest.

6. Understanding Particle Flow Performance

PandoraPFA is a fairly complex algorithm with a number of distinct stages which interact with each other in the sense that reconstruction failures in one part of the software can be corrected at a later stage. The contributions to the jet energy resolution have been estimated empirically by replacing different steps in PandoraPFA with algorithms which use MC information to perform: a) perfect reconstruction of photons as the first phase of the algorithm; b) perfect reconstruction of neutral hadrons; and c) perfect identification of fragments from charged hadrons. The jet energy resolutions obtained using these "perfect" algorithms enable the contributions from *confusion* to be estimated. In addition, studies using a deep HCAL enable the contribution from leakage to be estimated. Finally, MC information can be used to perform ideal track pattern recognition enabling the impact of imperfect track finding code to be assessed. For the current PandoraPFA algorithm, the contribution from the calorimetric energy resolution, $\approx 21\,\%/\sqrt{E}$, dominates the jet energy resolution for 45 GeV jets. For higher energy jets, the confusion term dominates. This behaviour is summarised in Figure 4. The contributions from resolution and confusion are roughly equal for 120 GeV jets. The most important contribution for high energy jets is confusion due to neutral hadrons being lost within charged hadron showers.

From the above study it is possible to obtain an semi-empirical parametrisation of the jet energy resolution:

$$\frac{\mathrm{rms}_{90}}{E} = \frac{21}{\sqrt{E}} \oplus 0.7 \oplus 0.004E \oplus 2.1\left(\frac{E}{100}\right)^{0.3} \%,$$

where E is the jet energy in GeV. The four terms in the expression respectively represent: the intrinsic calorimetric resolution; imperfect tracking; leakage and confusion. For a significant range of the jet energies relevant for the ILC, *high granularity* PFlow results in a jet energy resolution which is roughly a factor two better than the best achieved at LEP ($\sigma_E/E = 6.8\,\%$ at $\sqrt{s} = M_Z$). The ILC jet energy goal of $\sigma_E/E < 3.8\,\%$ is reached in the jet energy range 40 GeV – 420 GeV. Due to the confusion term, the performance of particle flow calorimetry, can not be

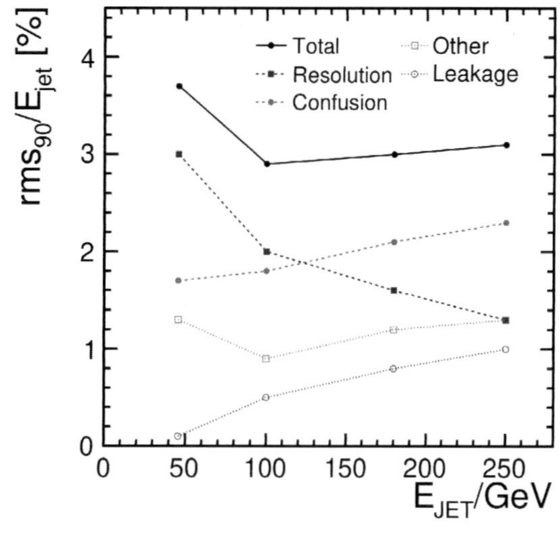

Figure 4. The contributions to the PFlow jet energy resolution obtained with PandoraPFA as a function of energy. The total is (approximately) the quadrature sum of the components.

expressed in the form $\sigma_E/E = \alpha/\sqrt{\mathrm{E(GeV)}} \oplus \beta$. Nevertheless, for comparison with traditional calorimetric approaches it is worth noting that for jets of energy less than 250 GeV a reasonable approximation of the *Gaussian equivalent* performance is:

$$\frac{\sigma}{E} < \frac{23\,\%}{\sqrt{E}} \oplus 3.0\,\%.$$

7. Dependence on Hadron Shower Modelling

Physics List	Jet Energy Resolution $r = \mathrm{rms}_{90}(E_j)/E_j$			
	45 GeV	100 GeV	180 GeV	250 GeV
LCPhys	$(3.74 \pm 0.05)\,\%$	$(2.92 \pm 0.04)\,\%$	$(3.00 \pm 0.04)\,\%$	$(3.11 \pm 0.05)\,\%$
QGSP_BERT	$(3.52 \pm 0.06)\,\%$	$(2.95 \pm 0.06)\,\%$	$(2.98 \pm 0.06)\,\%$	$(3.25 \pm 0.07)\,\%$
QGS_BIC	$(3.51 \pm 0.06)\,\%$	$(2.89 \pm 0.05)\,\%$	$(3.12 \pm 0.07)\,\%$	$(3.20 \pm 0.07)\,\%$
FTFP_BERT	$(3.68 \pm 0.08)\,\%$	$(3.10 \pm 0.06)\,\%$	$(3.24 \pm 0.06)\,\%$	$(3.26 \pm 0.08)\,\%$
LHEP	$(3.87 \pm 0.07)\,\%$	$(3.15 \pm 0.06)\,\%$	$(3.16 \pm 0.06)\,\%$	$(3.08 \pm 0.06)\,\%$
χ^2 (4 d.o.f)	23.3	17.8	16.0	6.3
rms/mean ($\sigma_r/\overline{r}$)	4.2 %	3.9 %	3.5 %	2.5 %

Table 3. Comparison of the jet energy resolution obtained using different hadronic shower physics lists. The χ^2 consistency of the different models for each jet energy are given as are the rms variations between the five models.

The results of the above studies rely on the accuracy of the MC simulation in describing EM and hadronic showers. The Geant4 MC provides a good description of EM showers as has been demonstrated in a series of test-beam experiments [15] using a Silicon-Tungsten ECAL of the type assumed for the ILD detector model. However, the uncertainties in the development of hadronic showers are much larger [18]. There are a number of possible effects which could affect

Journal of Physics: Conference Series **293** (2011) 012021 doi:10.1088/1742-6596/293/1/012021

PFlow performance: the hadronic energy resolution; the transverse development of hadronic showers which will affect the performance for higher energy jets where confusion is important; and the longitudinal development of the shower which will affect both the separation of hadronic and EM showers and the amount of leakage through the rear of the HCAL. To assess the sensitivity of PFlow reconstruction to hadronic shower modelling uncertainties, five Geant4 physics lists are compared: `QGSP_BERT`, `QGS_BIC`, `FTFP_BERT`, `LHEP` and `LCPhys`. These physics lists represent a wide range of models and result in significantly different predictions for total energy deposition, and the longitudinal and transverse shower profiles. For each Physics list, the calibration constants in PandoraPFA are re-tuned, but no attempt to re-optimise the algorithm is made. The jet energy resolutions obtained are given in Table 3. Whilst non-statistical differences are seen, the rms variations are relatively small, less than about 4 %. From this study it is concluded that, for 45 − 250 GeV jets, the jet energy resolution obtained from PFlow calorimetry as implemented in PandoraPFA does not depend strongly on the hadronic shower model.

8. Particle Flow for Multi-TeV Colliders

Given that the confusion term increases with energy, it is not *a priori* clear that PFlow calorimetry is suitable for higher energies such as $\sqrt{s} \sim 3\,\text{TeV}$ for the proposed CLIC collider. First studies of the potential of Particle Flow Calorimetry at CLIC were presented in [3]. For these studies the ILD concept, which is optimised for ILC energies, was modified; the HCAL thickness was increased from $6\,\lambda_I$ to $8\,\lambda_I$ and the magnetic field was increased from 3.5 T to 4.0 T. The jet energy resolution obtained for jets from $Z \rightarrow u\bar{u}, d\bar{d}, s\bar{s}$ decays at rest are listed in Table 4. For high energy jets, the effect of the increased HCAL thickness (the dominant effect) and increased magnetic field is significant. Despite the increased particle densities, the jet energy resolution (rms_{90}) for 500 GeV jets obtained from PFlow is 3.5 %. This is equivalent to $78\,\%/\sqrt{E(\text{GeV})}$. This is likely to be *at least* competitive with a traditional calorimetric approach, particularly when the constant term in Equation 1 and the contribution from non-containment are accounted for. Furthermore, it should be remembered that PandoraPFA has not been optimised for such high energy jets and improvements can be expected.

Jet Energy	$\mathbf{rms_{90}(E_{jj})/\sqrt{E_{jj}}}$		$\mathbf{rms_{90}(E_j)/E_j}$	
	3.5 T & $6\,\lambda_I$	4 T & $8\,\lambda_I$	3.5 T & $6\,\lambda_I$	4 T & $8\,\lambda_I$
45 GeV	25.2 %	25.2 %	$(3.74 \pm 0.05)\,\%$	$(3.74 \pm 0.05)\,\%$
100 GeV	29.2 %	28.7 %	$(2.92 \pm 0.04)\,\%$	$(2.87 \pm 0.04)\,\%$
180 GeV	40.3 %	37.5 %	$(3.00 \pm 0.04)\,\%$	$(2.80 \pm 0.04)\,\%$
250 GeV	49.3 %	44.7 %	$(3.11 \pm 0.05)\,\%$	$(2.83 \pm 0.05)\,\%$
375 GeV	81.4 %	71.7 %	$(3.64 \pm 0.05)\,\%$	$(3.21 \pm 0.05)\,\%$
500 GeV	91.6 %	78.0 %	$(4.09 \pm 0.07)\,\%$	$(3.49 \pm 0.07)\,\%$

Table 4. Comparisons of jet energy resolutions for two sets of detector parameters. This jet energy resolution shown is for $(Z/\gamma)^* \rightarrow$uds events with $|\cos\theta_{q\bar{q}}| < 0.7$. It is expressed as: i) the effective constant α in $\text{rms}_{90}(E_{jj})/E_{jj} = \alpha(E_{jj})/\sqrt{E_{jj}(\text{GeV})}$, where E_{jj} is the total reconstructed energy; and ii) the fractional jet energy resolution for a single jet where $\text{rms}_{90}(E_j) = \text{rms}_{90}(E_{jj})/\sqrt{2}$.

8.1. Gauge Boson Mass Reconstruction

The performance of PFlow calorimetry has, up to this point, been considered in terms of the jet energy resolution from particles decaying at rest. This is appropriate for the decays of particles

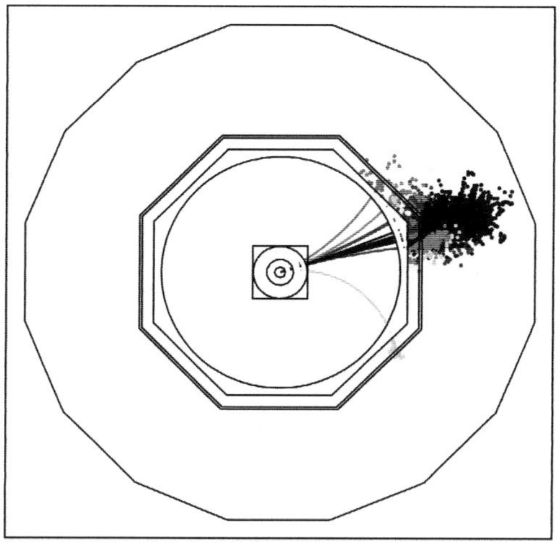

Figure 5. An example of a $\mathrm{Z} \to \mathrm{d}\overline{\mathrm{d}}$ decay with $E_Z = 1\,\mathrm{TeV}$ produced in a simulated $\mathrm{e^+e^-} \to \mathrm{ZZ} \to \nu\overline{\nu}\mathrm{d}\overline{\mathrm{d}}$ interaction in the ILD detector concept.

produced almost at rest. However, one of the goals for jet reconstruction at a future lepton collider is to be able to reconstruct gauge bosons which may be produced from the decays of BSM particles. In this case, the W/Z decays will not be at rest and the di-jet system will be boosted. At a multi-TeV lepton collider the boost may be significant. For PFlow calorimetry there are a number of effects associated with highly boosted jets: *(i)* The jet particle multiplicities are lower than those for jets of the same energy produced from decays at rest. This increases the average energy of the particles in the jet and, consequently, will result in less containment of the hadronic showers (greater leakage); *(ii)* The high jet boost decreases the average separation of the particles in the jet. This will tend to increase the confusion term; *(iii)* The two jets from the decay of a highly boosted gauge boson will tend to overlap to form a "mono-jet", as shown in Figure 5. The overlapping of jets has the potential to increase the confusion term.

Due to the likely increased confusion term, reconstructing the invariant mass of high energy gauge bosons presents a challenge for PFlow calorimetry. The PFlow reconstruction of boosted gauge bosons has been investigated by generating MC samples of $\mathrm{ZZ} \to \mathrm{d}\overline{\mathrm{d}}\nu\overline{\nu}$ and $\mathrm{W^+W^-} \to \mathrm{u}\overline{\mathrm{d}}\mu^-\overline{\nu}_\mu$ events at $\sqrt{s} = 0.25$, 0.5, 1.0 and 2.0 TeV. These final states give clean samples of single hadronic Z and W decays (the muons from the W decays are easy to identify and remove). The PFlow reconstructed W and Z invariant mass distributions are shown in Figure 6 and the corresponding mass resolutions are given in Table 5. A direct comparison with the jet energy resolutions of Table 4 is not straightforward due to the effects described above. However, the mass resolution (rms_{90}) of 2.8 GeV obtained from decays of gauge bosons with $E = 125\,\mathrm{GeV}$ is compatible with that expected from the jet energy resolution of Table 4 after accounting for the gauge boson width.

For the ILC operating at $\sqrt{s} = 0.5 - 1.0\,\mathrm{TeV}$, the typical energies of the gauge bosons of interest are likely to be in the range $E_{W/Z} = 125 - 250\,\mathrm{GeV}$. Here the reconstructed W and Z mass peaks are well resolved. The statistical separation, which is quantified in Table 5, is approximately 2.5σ, i.e. the separation between the two peaks is approximately 2.5 times greater than the effective mass resolution.

For CLIC operating at $\sqrt{s} = 3\,\mathrm{TeV}$, the relevant gauge boson energies are likely to be in the

$E_{W/Z}$	$\mathrm{rms}_{90}(m)$	σ_m/m	W/Z sep	ϵ
125 GeV	2.8 GeV	2.9 %	2.7σ	91 %
250 GeV	3.0 GeV	3.5 %	2.5σ	89 %
500 GeV	3.9 GeV	5.1 %	2.1σ	84 %
1000 GeV	6.4 GeV	7.0 %	1.5σ	78 %

Table 5. Invariant mass resolutions for the hadronic system in simulated ZZ $\rightarrow$ $\mathrm{d\bar{d}}\nu\bar{\nu}$ and $\mathrm{W^+W^-}$ $\rightarrow$ $\mathrm{u\bar{d}}\mu^-\bar{\nu}_\mu$ events in the ILD detector concept. The W/Z separation numbers, which take into account the tails, are defined such that a 2σ separation means that the optimal cut in the invariant mass distribution results in 15.8 % of events being mis-identified. The equivalent W/Z identification efficiencies, ϵ, are given in the final column. Even with infinitely good mass resolution, the best that can be achieved is 94 % due to the tails of the Breit-Wigner distribution and, thus, the possible range for ϵ is 50 – 94 %;

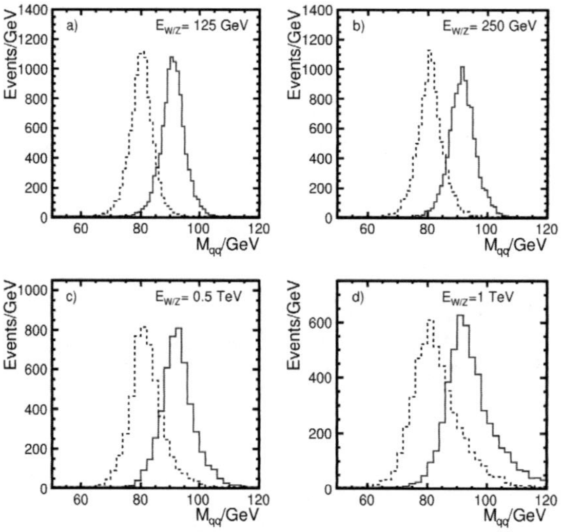

Figure 6. Reconstructed invariant mass distributions for the hadronic system in simulated ZZ $\rightarrow$ $\mathrm{d\bar{d}}\nu\bar{\nu}$ and $\mathrm{W^+W^-}$ $\rightarrow$ $\mathrm{u\bar{d}}\mu^-\bar{\nu}_\mu$ events as simulated in the modified ILD detector model.

range 0.5 – 1.0 TeV. At the low end of this range there is reasonable separation (2.1σ) between the W and Z peaks. Even for 1 TeV W/Z decays, where the events mostly appear as a single energetic mono-jet, the mass resolution achieved by the current version of PandoraPFA allows separation between W and Z decays at the 1.5σ level. However, it should be remembered that PandoraPFA has not been optimised for such high energy jets, and these results represent a lower bound on what can be achieved. Since Calor 2010 progress has been made in this area. From this study it is concluded that PFlow calorimetry is certainly not ruled out for a multi-TeV lepton collider. Particle flow reconstruction may be particularly well adapted to the determination of the masses of highly boosted system which appear as mono-jets, where the resolution of the structure within the jet is essential to obtain a good mass resolution.

9. Conclusions

A sophisticated particle flow reconstruction algorithm, PandoraPFA, has been developed to study the potential of high granularity Particle Flow calorimetry at a future linear collider.

The algorithm incorporates a number of techniques, e.g. topological clustering and statistical reclustering, which take advantage of the highly segmented calorimeters being considered for the ILC and beyond. Using this algorithm it has been demonstrated that Particle Flow Calorimetry can meet the ILC requirements for jet energy resolution. For jets in the energy range 40 – 400 GeV, the jet energy resolution, σ_E/E, is better than 4%. This conclusion does not depend strongly on the details of the modelling of hadronic showers. First studies of particle flow calorimetry for a multi-TeV collider such as CLIC show considerable promise. A jet energy resolution below 4 % is achievable for jets with energies less than approximately 600 GeV. Reasonable separation of the hadronic decays of W and Z bosons is achievable for W/Z energies of up to approximately 1 TeV.

In conclusion, since Calor 2008, our understanding of the potential of particle flow calorimetry has increased; it provides a technologically feasible route to achieving unprecedented jet energy resolution at a future lepton collider and as such form the current baseline for ILC and CLIC detector concepts. It should be remembered that the current PandoraPFA algorithm is still evolving and significant improvements in resolution are anticipated, particularly for higher energy jets.

10. Acknowledgements

I would like to thank the organisers of Calor 2010 for arranging such a stimulating and enjoyable conference. I would like to acknowledge the UK Science and Technology Facilities Council (STFC) for the continued support of this work. I would also like to thank colleagues in the CALICE collaboration and those working on the ILD detector concept for providing the high quality simulation and software frameworks used for the studies described in these proceedings.

References

[1] J.-C. Brient and H. Videau, "The calorimetry at a future e^+e^- linear collider", arXiv:hep-ex/0202004.
[2] V. L. Mogunov, "Energy-flow Method for Multi-jet Effective Mass Reconstruction in the Highly Granular TESLA Calorimeter", proceedings of the Snowmass Summer Study on the Future of Particle Physics, Snowmass, U.S.A. (2001).
[3] M. Thomson, Nucl. Instr. and Meth. **A611** (2009) 25.
[4] International Linear Collider reference design report. 1: Executive summary. 2: Physics at the ILC. 3: Accelerator. 4: Detectors, J. Brau (ed.), *et al.*, ILC-REPORT-2007-001 (2007).
[5] ALEPH Collaboration, D. Buskulic et al., Nucl. Instr. and Meth. **A360** (1995) 481.
[6] F. Gaede, Nucl. Instr. and Meth. **A559** (2006) 177.
[7] GEANT4 collaboration, S. Agostinelli *et al.*, Nucl. Instr. and Meth. **A506** (2003) 3; GEANT4 collaboration, J. Allison *et al.*, IEEE Trans. Nucl. Sci. 53 (2006) 1.
[8] ILD Letter of Intent, arXiv:1006.3396.
[9] J. C. Brient, *et al.*, "CALICE Report to the R&D Review Panel", ILC-DET-2007-024 (2007); arXiv:0707.1245.
[10] T. Behnke, *et al.*, "Track Reconstruction for a Detector at TESLA", LC-DET-2001-029 (2001), and references therein.
[11] DELPHI Collaboration, DELPHI Data Analysis Program (DELANA) User's Guide, DELPHI 89-44 PROG 137 (1989).
[12] M. A. Thomson, "Progress with Particle Flow Calorimetry", Proceedings of LCWS2007, DESY, Hamburg, June 2007. arXiv:0709.1360.
[13] T. Sjöstrand, Comp. Phys. Comm. **135** (2001) 238.
[14] OPAL Collaboration, G. Alexander *et al.*, Z. Phys. **C69** (1996) 543.
[15] CALICE Collaboration, C. Adloff *et al.*, accepted for publication by NIMA (2009); arXiv:0811.2354.
[16] "Linear Collider Physics List", http://www.slac.stanford.edu/comp/physics/geant/slac_physics_lists/ilc/ilc_physics_list.html
[17] R. Wigmans, Calor 2010.
[18] R. Wigmans, Proc. of HSW06, Fermilab, 2006, AIP Conf. Proc. 896 (2007) 123.
[19] G. Guignard (ed.), *et al.*, "A 3 TeV Linear Collider Based on CLIC Technology", CERN-2000-008 (2000).

Recent improvements on the description of hadronic interactions in Geant4

A. Dotti[1], J. Apostolakis[1], G. Folger[1], V. Grichine[1,2], V. Ivanchenko[1,3], M. Kosov[1,4], A. Ribon[1], V. Uzhinsky[1,5], D.H. Wright[6]

[1] CERN, Geneva
[2] P.N. Lebedev Institute of Physics (FIAN) - Russian Academy of Sciences
[3] Ecoanalytica, MSU, Moscow
[4] ITEP, Moscow
[5] LIT JINR, Dubna
[6] SLAC, Stanford

E-mail: andrea.dotti@cern.ch

Abstract.
We present an overview of recent improvements of hadronic models in Geant4 for the physics configurations (Physics Lists) relevant to applications in high energy experiments. During last year the improvements have concentrated on the study of unphysical discontinuities in calorimeter observables in the transition regions between the models used in Physics Lists. The microscopic origin of these have been investigated, and possible improvements of Geant4 code are currently under validation. In this paper we discuss the status of the latest version of Geant4 with emphasis on the most promising new developments, namely the Fritiof based and CHIPS Physics Lists.

1. Introduction

In modern high energy physics (HEP) experiments Monte Carlo programs are employed not only in the initial phase of the life time of an experiment to assist its design, but also for studying the performance of the reconstruction algorithms that will be used for the data analysis. In some cases simulations are used also to derive corrections to unfold detector and reconstruction effects.

Geant4 [1, 2] is a toolkit for the simulation of passage of particles through matter. Its areas of application include high energy, nuclear and accelerator physics, as well as studies in medical and space science. It includes the algorithms to track particles in complex geometries with or without the effect of magnetic fields, interfaces to analysis (*hits* and *digits*) or persistency frameworks and advanced visualization capabilities. Most important, a complete set of physics models to describe both electro-magnetic and hadronic interactions up to energies to hundreds of GeV is available.

QCD is the well-established theory of strong interactions, but most of the hadronic interactions that occur when particles cross matter (e.g. detectors) are happening in the non-perturbative region of the theory, where the theoretical predictions are not computable. Various phenomenological models exist to describe hadronic interactions in some limited kinematic regions, and to cover a wide energy range, from MeV to hundreds of GeV, various models have to be combined in a transportation code.

Geant4 provides a powerful framework to develop user specific algorithms describing hadronic interactions, and it comes with a variety of models with different performances in terms of physics precision and CPU needs, as well as different ranges of validity in terms of the energy and the type of the interacting particle. It is a user's responsibility to assemble the models to cover all particles and energies interested in an application. However, a set of *Physics Lists* are available with pre-assembled models to cover all energies and particles of typical modern HEP experiments. These Physics Lists are routinely validated by models developers and compared with test-beam data from past and present experiments.

Geant4 has been developed and tuned with the current generation LHC experiments as primary users. Strong requirements have been set [3, 4]. In the past years mainly the ATLAS and CMS experiments have extensively compared the simulation predictions with the collected test-beam data. A summary of the results can be found in [5].

In the past years, LHC experiments (in particular ATLAS and CMS) studied the physics performances of the different Physics Lists and converged towards the use of the so-called QGSP_BERT Physics List as the default one. It gives the best results in terms of describing the data collected with the calorimeters during the test-beam campaigns.
Before describing the models, used by this Physics List, it is important to remember that the models parameters are tuned with thin target data (single interaction data) rather than test-beam data. The simulation of calorimeters is used to verify, at a global level, the quality of a Physics List (for an overview of some results on the validation of models see: [6, 7, 8, 9]).

1.1. Main hadronic models

The Physics List QGSP_BERT which is used in production for ATLAS and CMS detector simulations, comprises the following physics models:

- Quark-Gluon String (QGS) model [10, 11, 12, 13, 14] for proton, neutron, pion and kaon interactions with nuclei at kinetic energies above 12 GeV, interfaced to Precompound (P) model [15] for the evaporation phase of the interaction.
- Low Energy Parameterized (LEP) model [16] for proton, neutron, pion and kaon interactions with nuclei at kinetic energies between 9.5 GeV and 25 GeV.
- Bertini cascade (BERT) model [17, 18], which includes intra-nuclear cascade, followed by precompound and evaporation phases of the residual nucleus, for proton, neutron, pion and kaon interactions with nuclei at kinetic energies below 9.9 GeV. Note that the Bertini model is not used for the simulation of secondary hadrons rescattering produced by the QGS model inside the nucleus.
- Parameterized (LEP + HEP) models for all remaining hadrons (i.e. hyperons and anti-baryons) interactions.
- Parameterized capture and fission for low-energy neutrons.
- CHIPS model [20, 21, 22, 23] for the nuclear capture of negatively charged particles at rest.
- CHIPS model for elastic scattering of neutrons and protons; for elastic scattering of all other hadrons a parametrized model below 1 GeV and a revised elastic scattering model above 1 GeV are used.
- Glauber model [19] for quasi-elastic scattering for proton, neutron, pion and kaon interactions with nuclei at kinetic energies above 12 GeV; in all other cases parametrized models.
- Standard electromagnetic processes [24].
- CHIPS model for gamma-nuclear and electron-nuclear interactions.

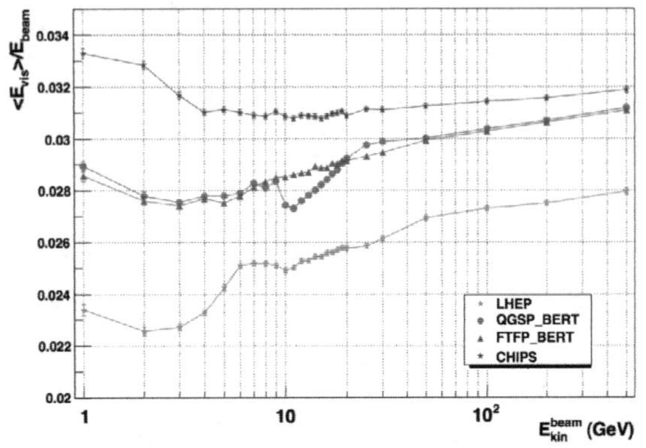

Figure 1. Simulated response in a simplified scintillator/iron sampling calorimeter for negatively charged pions as a function of primary kinetic energy. Different Geant4 Physics Lists are shown for comparison. Statistical and systematic errors are also shown, in many cases they are smaller than the symbol size.

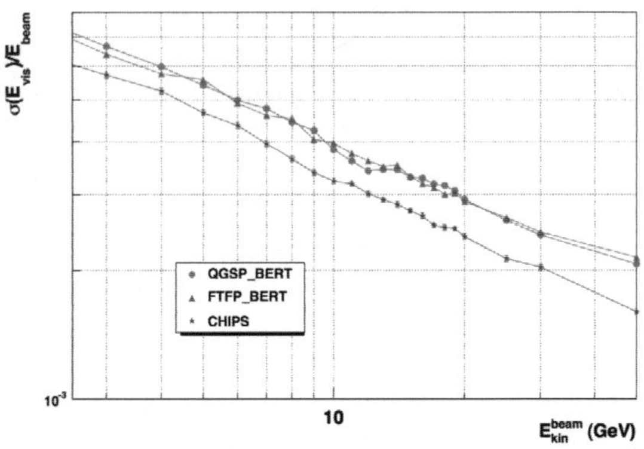

Figure 2. Simulated normalized response width in a simplified scintillator/iron sampling calorimeter for negatively charged pions as a function of primary kinetic energy (the interesting region at E_{beam} = 10 GeV is shown). Different Geant4 Physics Lists are shown for comparison.

- Parameterized model for muon-nuclear interaction.

There is a small transition region between BERT and LEP and a larger one between LEP and QGS. In these transition regions one of the models is chosen at random to compute the interaction with matter. The most promising alternatives to QGSP_BERT are the FTFP_BERT and CHIPS Physics Lists.

The FTFP_BERT Physics List is very similar to the QGSP_BERT one with the exception that it uses the well-known Fritiof model (FTF) instead of the QGS one to describe the interaction of highly energetic protons, neutrons, pions and kaons [25, 26]. While the QGS model is well-suited to describe interactions of projectile with kinetic energy E_{kin} > 10 GeV the FTF model can describe interactions starting from 4-5 GeV. Thus the string model can be coupled directly to the cascade model, that have a typical range of validity up to few GeV.

At high energy the CHIPS Physics List uses a one-dimensional parton multi-string model. The soft particles of a string fragment are absorbed by a target nucleus and handled by the low-energy model. A *quasmon* (parton plasma) is created and subsequently decayed. Finally, an evaporation model is used to de-excite the remnant nucleus. The advantage of the CHIPS Physics List is that it melds the high- and low-energy approaches in a coherent and unique theoretical framework. In addition, it can handle all types of projectiles, including kaons, anti-baryons and hyperons. While QGSP_BERT and FTFP_BERT share the same data sets for cross sections, CHIPS has its own data set for the interaction of kaons, hyperons and anti-particles; and for the capture of neutrons.

Finally all three Physics Lists share the same implementation of electromagnetic processes.

2. Status of the simulations of hadronic interactions for High Energy Physics experiments

Starting from 2006 both ATLAS and CMS experiments have decided to use QGSP_BERT (or its variants) as the default Physics List for the simulation of hadronic interactions. This list gives the best results when comparing with test-beam data:

- The response to hadrons, defined as the ratio between the measured energy and the beam energy, is described at the level of 3%.
- The energy resolution ($\sigma/\langle E\rangle$) is described by this list at the level of 10%, though if it is systematically smaller than measured data.
- With the addition of the Bertini intra-nuclear cascade, the shower shape description has substantially improved, however simulation still predicts shorter and narrower showers. Showers are shorter of about 10% (at a depth of $10\lambda_I$ in the calorimeter) for pion projectile, while for protons the agreement is at the level of 30% [27, 28].

2.1. Transition between models

The CMS experiment has found that the calorimeter energy response in its HCAL test-beam setup, as a function of the pion beam energy, presents an unphysical discontinuity around 9-10 GeV. The ATLAS experiment confirmed the same problem for its calorimeter test-beam setups.

The origins of this discontinuity, shown in figure 1, have been studied in detail in the past two years. It is now clear that the effect is caused by the use of the parametrized models for particle interactions in the energy range $9.5 < E_{kin} < 25$ GeV [29].

As a strategy to reduce the dependence on the parametrized models and to address the issue we have studied the performance of the FTFP_BERT Physics List, which has a reduced dependence on the parametrized models and it has different transition regions; and the CHIPS one, that does not depend at all on these parametrization and does not have, by construction, any strong transition. Predictions of these Physics Lists in a comparison with QGSP_BERT for simplified setups are discussed in the following section.

3. Results on simplified calorimeters

We have performed simulations of a $10\lambda_I$ depth and wide sampling calorimeter (100 periods made of a 16.8 mm thick iron slab followed by 4 mm thick slab of scintillator). Impinging pions of different kinetic energies (from 1 to 500 GeV) have been simulated. Given its importance for the response in scintillator-based calorimeters the Birks' attenuation effect has been implemented in the simulation with the parameters of [30]. Transportation of neutrons is stopped after 50 ns: this is the characteristic read-out time window in this kind of calorimeters[1]. A correction for front (albedo) and longitudinal leakage has been implemented. This correction varies with the primary particle's momentum, but it is typically of the order of 1% of the deposited energy. The systematic error associated with this correction is much smaller than the statistical error.

3.1. Energy response

Figure 1 shows the response as a function of primary energy for the considered Physics Lists. Red circles illustrate the response for the LHC experiments default Physics List (QGSP_BERT). The transition between models is clearly visible at around 10 and 25 GeV. Since these unphysical discontinuities are due to the use of the parametrized models, the results obtained with LHEP (a Physics List that uses parametrized models for all energies and all primary types) are also shown as reference. It becomes clear that the use of these models in QGSP_BERT for the intermediate energy region has the effect of reducing the simulation response and of producing the unphysical discontinuities.

The FTFP_BERT and CHIPS Physics Lists do not show these discontinuities and are much smoother. As discussed in section 2, the detailed simulations of the ATLAS and CMS calorimeters show that the response predicted by QGSP_BERT Physics List is in agreement with

[1] The response varies by about 3% increasing the neutron time-cut from 20 to 200 ns, this is the main systematic error in this analysis.

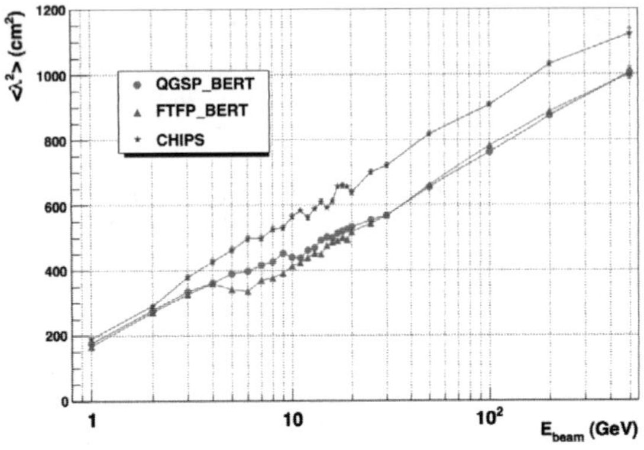

Figure 3. Longitudinal shower dimension as a function of the beam kinetic energy.

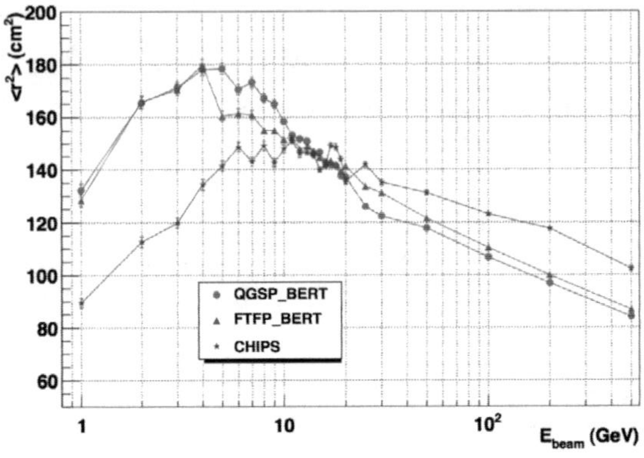

Figure 4. Lateral shower dimension as a function of the beam kinetic energy.

experimental data (typical test-beam have $E_{beam} > 20$ GeV, only few data points are available for lower energies). Thus, the FTFP_BERT Physics List agrees with QGSP_BERT which, in turn, agrees with the data. CHIPS response is too high (this Physics List is still considered as experimental one and its tuning with thin-target data is still ongoing).

3.2. Resolution

To study the smoothness of the energy resolution as a function of the beam energy, it is more convenient to show σ/E_{beam} instead of $\sigma/\langle E\rangle$ since the response of QGSP_BERT is not smooth. The results for the considered Physics Lists are shown in figure 2. Only in the region around the 10 GeV a small discontinuity is left. The transition effect is not particularly visible in this observable, and, once again, the FTFP_BERT combination gives similar results to the other string-based Physics List. CHIPS model predicts smaller widths. Data prefer simulations with larger resolution, thus more studies are needed for the CHIPS model to reach an agreement with data.

3.3. Shower shapes

We have examined the hadronic shower dimension as a function of the primary energy. In order to efficiently summarize the longitudinal ($\langle\lambda^2\rangle$) and lateral ($\langle r^2\rangle$) dimensions for each event, we have calculated the second moment[2] of the position of *read-out cells* with $E_c > 0$. The mean values of these quantities are shown in figures 3 and 4. For shower shape observables QGSP_BERT has irregularities around 10 and 25 GeV. CHIPS predicts longer showers, which is in the direction of a better agreement with test-beam data. However, for this Physics List, showers are more compact in the lateral dimensions at low energies, while they tend to be wider at higher energies in comparison with the other two Physics Lists. CHIPS also predicts stronger fluctuations as a function of beam energy, further development is needed for the CHIPS models.

In general FTFP_BERT Physics List gives similar results as the QGSP_BERT one, but a transition between the Bertini cascade model and the Fritiof string model between 4 and 5 GeV

[2] First we read-out the energy deposited in the detector in small *voxels* of 5x5x5 cm^3 of volume. For each event the characteristic shower dimensions have been calculated as:

$$\lambda^2 = \frac{\sum_{c\in\{cells\}} E_c\lambda_c^2}{\sum_{c\in\{cells\}} E_c} \ ; \ r^2 = \frac{\sum_{c\in\{cells\}} E_c r^2}{\sum_{c\in\{cells\}} E_c}$$

with λ_c and r_c being the center of cell c in polar coordinates with respect to the shower center and shower axis.

is visible.

4. Conclusions

In the past years the LHC experiments have compared in detail the prediction of the Geant4 simulations with test-beam data. The initial requirements set by LHC have been finally met. In particular the test beam experience has shown that at high energy the string based models describe reasonably well the experimental results. At low energies it has been shown that an intra-nuclear cascade code is a fundamental ingredient to improve the description of shower shapes.

New options have been created in Geant4 that are very promising. The FTFP_BERT Physics List, with a much reduced dependence on the parametrized models (responsible for the discontinuities), shows a smooth response as a function of the primary energy. The Physics List is in a good agreement with the QGSP_BERT predictions. The transition effect which is visible in FTFP_BERT at 4 GeV for the shower shape observables will be addressed in the future, however, its impact on global physics performance is expected to be small.

The CHIPS Physics List is a good alternative to string and cascade based Physics Lists. It uses a coherent treatment of hadronic inelastic interactions which results in a very smooth behavior in all observed observables. Though at the moment its response is higher, and the resolution is lower with respect to QGSP_BERT and FTFP_BERT, improvements are expected in the future since it is still undergoing tuning with thin-target data.

In conclusion, Geant4 is used with success by many HEP experiments. The quality of its simulation of hadronic interactions is constantly improved and the most widely used Physics List QGSP_BERT, has been stable since some years. Competitive models or new mixtures of models are emerging as promising alternatives. In most cases they give results similar of even better than QGSP_BERT Physics List.

References

[1] Agostinelli S et al. 2003 *Nuclear Instruments and Methods in Physics Research A* **506** 250-303
[2] Allison J et al. 2006 *IEEE Transactions on Nuclear Science* **53** No. 1 270-8
[3] Gianotti F et al. 2004 Simulation physics requirements from the LHC experiments *CERN-LCGAPP-2004-002*
[4] Gianotti F et al. 2004 Geant4 hadronic physics validation with LHC test-beam data: first conclusions *CERN-LCGAPP-2004-10*
[5] Ribon A et al. 2010 Status of Geant4 hadronic physics for the simulation of LHC experiments at the LHC physics program *CERN-LCGAPP-2010-002*
[6] Ribon A 2007 Third Benchmark Study for Validation of Hadronic Physics Simulations at LHC: Inclusive Charged Pion Production in Hadron-Nucleus Interactions at 100 and 320 GeV/c *CERN-LCGAPP-2007-01*
[7] Pokorski W 2004 In-Flight Pion Absorption: Second Benchmark Study for the Validation of Hadronic Physics Simulation at the LHC *CERN-LCGAPP-2004-11*
[8] Beringer J 2003 (p,xn) Production Cross Sections: A Benchmark Study for the Validation of Hadronic Physics Simulation at LHC *CERN-LCGAPP-2003-18*
[9] Banerjee S et al. 2008 Validation of Geant4 hadronic physics models at intermediate energies *Nuclear Science Symposium Conference Record* **NSS '08. IEEE** 2798-2800
[10] Amelin N S et al 1991 *Phys. Rev. Lett.* **67** 1523-26
[11] Amelin N S et al. 1992 *Nucl. Phys. A* **544** 463-6
[12] Bravina L V et al. 1994 *Nucl. Phys. A* **566** 461-4
[13] Bravina L V 1995 *Phys. Lett. B* **344** 49-54
[14] Folger G and Wellisch H P 2003 String parton models in GEANT4 *Preprints CHEP-2003-MOMT007, e-Print physics/0306007*
[15] GEANT4 Physics Reference Manual, Chapter 29: Precompound Model available at http://geant4.web.cern.ch/geant4/support/userdocuments.shtml
[16] Wright D H et al. 2006 *AIP Conf. Proc.* **867** 479
[17] Wright D H et al. 2007 *AIP Conf. Proc.* **896** 11
[18] Heikkinen A, Stepanov N and Wellisch H P 2003 Bertini intra-nuclear cascade implementation in GEANT4 *Preprint CHEP- 2003-MOMT008, e-Print physics/0306008*

[19] Glauber R J 1970 High Energy Physics and Nuclear Structure *Plenum Press, NY 1970* edited by S Devons
[20] Kosov M V 2000 Manual for the CHIPS event generator KEK internal report 2000-17, Feb. 2001 H/R
[21] Degtyarenko P V , Kosov P V and Wellisch H P 2000 *Eur. Phys. J. A* **8** 217-22
[22] Degtyarenko P V , Kosov P V and Wellisch H P 2001 *Eur. Phys. J. A* **9** 411-20
[23] Degtyarenko P V , Kosov P V and Wellisch H P 2001 *Eur. Phys. J. A* **9** 421-4
[24] Apostolakis J et al. 2008 The performance of the GEANT4 standard EM package for LHC and other applications *J. Phys.: Conf. Ser.* **119** 032004
[25] Andersson B 1987 *Nucl. Phys. B* **281** 289-309
[26] Nilsson-Almqvist B 1987 *Comp. Phys. Comm.* **43** 387-97
[27] Ribon A et al. 2008 Hadronic Shower Shape Studies in Geant4 *CERN-LCGAPP-2007-002*
[28] Ribon A et al. 2008 Hadronic shower shapes studies in Geant4: Update *CERN-LCGAPP-2008-001*
[29] Ribon A et al. 2009 Transition between hadronic models in Geant4 , Contribution at NSS-IEEE 2009 , Orlando, Florida, USA, 25-31 October 2009, Proceeding in Preparation
[30] Craun R L and Smith D L 1970 *Nucl. Instr. and Meth.* **80** 239-44

XIV International Conference on Calorimetry in High Energy Physics (CALOR 2010) IOP Publishing
Journal of Physics: Conference Series **293** (2011) 012023 doi:10.1088/1742-6596/293/1/012023

Gflash as a Parameterized Calorimeter Simulation for the CMS Experiment

Soon Yung Jun

Department of Physics, Carnegie Mellon University, Pittsburgh, PA, USA

E-mail: syjun@fnal.gov

Abstract. Gflash is a parameterized simulation of electromagnetic and hadronic showers. It uses parameterizations of longitudinal and lateral shower profiles in homogeneous or sampling calorimeters taking into account an individual shower fluctuation and correlation. Gflash can take over the Geant4 based tracking at the first inelastic interaction inside calorimeters. It can be tuned flexibly to data and speed up the calorimeter simulation significantly for high energy particle showers. In this paper, the Gflash implementation developed for the CMS calorimeter and tuned to the CMS test beam data is described. The performance of Gflash is also presented.

1. Introduction

The CMS experiment primarily uses the GEANT4 toolkit (Geant4) to simulate the passage of particles through the CMS detector. Geant4 provides various physics models that describe interactions of particles with all detector components. Physics processes included in a specific physics list govern to track all secondaries as well as the primary particle down to a certain minimum energy. In general, it requires much efforts to tune simulation of a general purposed high energy detector due to its complexity as well as various choices of options applied to a certain physics process. In addition, the computing time to simulate an event for the CMS experiment may takes several minutes due to energetic particles copiously produced by the high energy collision at LHC as well as multiple underlying interactions. Usually the calorimeter simulation is the primary contributor of the CPU time per event due to the large decay volume designed to produce particle showers by electromagnetic cascades or hadronic interactions.

Gflash is a parameterized simulation package for electromagnetic (EM) and hadronic showers. It was originally developed for the calorimeter simulation of the H1 experiment at HERA. Gflash uses parameterizations of longitudinal and lateral shower profiles in homogeneous or sampling calorimeters taking into account an individual shower fluctuation and correlations between parameters. The main advantage of using Gflash is to speed up the calorimeter simulation while it can be tuned flexibly to data. Details of the original program Gflash can be found elsewhere [1] [2].

The EM shower parameterization of Gflash can be applied to any homogeneous detector if one can use the energy in the unit of critical energy (E_c) and the length in the unit of the radiation length (X_o) according to the Rossi's approximation B [3]. Since the CMS EM calorimeter (ECAL) is homogeneous ($PbWO_4$) and relatively long ($25X_o$), the CMS EM shower parameterization adapt the original Gflash program without much modifications. However, both longitudinal and lateral parameterizations for the hadron shower are newly developed to take into account for the CMS specific detector structure, especially the presence of passive materials between ECAL and HCAL.

2. Particle Shower Parameterizations of Gflash

The simulation of particle showers in Gflash is divided into the spatial distribution of the deposited energy (E_{dp}) for a shower and the energy fraction of the deposited energy which is visible in the active medium. The spatial distribution of E_{dp} within the azimuthal symmetric calorimeter volume containing showers can be expressed as

$$dE_{dp}(\vec{r}) = \frac{E_{dp}}{2\pi} L(z)dzR(r)dr. \tag{1}$$

where $L(z)$ describes the longitudinal energy sub-profile along the shower depth (z) and $R(r)$ the lateral sub-profile to the radial direction from the shower axis (r). The energy fraction of the deposited energy which is visible in the active medium is determined by sampling fractions (or calibration scales) and their fluctuations.

2.1. Electromagnetic Shower Parameterization

The mean longitudinal energy profile of electromagnetic showers in Gflash is assumed to take a Γ–distribution,

$$L(x) = \frac{x^{\alpha-1}e^{-x}}{\Gamma(\alpha)} \tag{2}$$

where $x = \beta z$ and z is the shower depth in the unit of radiation length. The shape of showers is characterized by a correlated pair of parameter α and β which are corresponding to the first and second moments of the Γ–distribution. The distribution has a maximum at $z_{max} = (\alpha - 1)/\beta$ along the longitudinal axis $\hat{z}$.

The lateral profile of energy distribution for both electromagnetic and hadronic showers is described by a simple ansatz

$$R(r) = \frac{2rR_0^2}{(r^2 + R_0^2)^2} \tag{3}$$

where r in units of Moliére radius is the radial distance perpendicular to the shower direction and with origin in the center of the shower, and R_0 is a free parameter that is a function of the shower energy (E) and the shower depth (z). The mean of $R_0(E, z)$ is parameterized as

$$\langle R_0 \rangle (E, z) = [R_c + Q(E)z]^n$$

where n=1(2) for electromagnetic (hadronic) shower.

2.2. Hadronic Shower Parameterization Specialized for CMS

The longitudinal profile of hadron showers developed for the CMS calorimeter simulation is written as a combination of sub-profiles in two detector regions,

$$F = f_{\rm ecal}F_{\rm ecal} + f_{\rm hcal}F_{\rm hcal}, \tag{4}$$

where $F_{\rm ecal}$ and $F_{\rm hcal}$ represent the hadronic shower profile inside electromagnetic calorimeters (ECAL) and hadronic calorimeters (HCAL), respectively. In the CMS Gflash model for hadronic showers, both $F_{\rm ecal}$ and $F_{\rm hcal}$ taking the same functional hypothesis are the superposition of two Gamma distributions:

$$F_{\rm ecal}({\rm or} F_{\rm hcal}) = [cL(x_e; \alpha_e, \beta_e)dx_e + (1 - c)L(x_h; \alpha_h, \beta_h)dx_h], \tag{5}$$

$$L(x_i;\alpha_i,\beta_i) = \frac{x_i^{\alpha_i-1}e^{-x_i}}{\Gamma(\alpha_i)}, \qquad x_i = \frac{\beta_i z}{d_i} \qquad (i = e, h), \tag{6}$$

where c is the fraction of the deposited energy by any π^0 produced from the primary or secondary interactions, $L(x_e;\alpha_e,\beta_e)$ and $L(x_h;\alpha_h,\beta_h)$ are sub-profiles of the spatial energy distribution by π^0 and by any charged hadron tracks in the units of d_i ($d_e = X_o, d_h = \lambda_o$), respectively. We then parameterize the set of parameters describing the longitudinal profile. In particular, α and β of each Γ distribution can be semi–analytically estimated by using its mean (α/β) and variance (α/β^2) of the distribution. To build an initial set of parameters, $\vec{x} = (c, \alpha_e, \beta_e, \alpha_h, \beta_h)$, we use the CMS detector simulation with Geant4 to construct the statistical ansatz of the deposited hit energy distribution. The correlations between them are then parameterized as follow,

$$\vec{x} = \vec{\mu} + \vec{\sigma}\mathbf{C}\vec{z} \quad \text{with} \quad \rho = \mathbf{C}\mathbf{C}^T \tag{7}$$

where $\vec{\mu}$ is the vector of the means of x_i, $\vec{\sigma}$ is the vector of the standard deviation of x_i, $\mathbf{C}$ is the Cholesky decomposed matrix of the correlation matrix (ρ), and $\vec{z}$ is a vector of normal distributed random numbers. For the lateral profile, the mean and the variance of R_o are also newly parameterized depending on the position of the shower start to take into account the specific detector structure of the CMS calorimeters.

3. Tuning Shower Parameterizations of Gflash to the Test Beam Data

In order to tune shower shapes and responses of Gflash, we use CMS test beam data taken in 2006 for electrons and hadrons ($\pi^{\pm}$, $K^{\pm}, p, \bar{p}$). The version 9.2.p01 of Geant4 with the CMS test beam geometry is used to tune Gflash parameterizations. The default physics list of Geant4 used up to the first inelastic interaction includes the standard EM physics list and QGSP_BERT. The simulated energy response for the pure hadronic showers is calibrated with 50 GeV/c electron beams.

3.1. Tuning the EM parameteriszation

To tune the EM response of Gflash, the reference response of electron test beam data is used. For each electron beam energy (E_{beam} = 20, 30, 50, 80, 120, 150 GeV), the nominal energy response in $N \times N$ ($N = 1, 3, 5$) crystals is obtained from energy scan data. Optimized weights are used for the amplitude reconstruction and the crystals are inter-calibrated. The reference response of the detector is obtained by selecting only the electrons impacting close to the point of maximum containment which is determined for each crystal in order to disentangle the effects due to not fully containing the shower. Simulated data using Gflash are generated with the same set of configuration for simulation and reconstruction and they are then analyzed using the same analysis software as for the test beam data.

Since the absolute response may be sensitive to the calibration scale and the containment correct, the ratio of $N \times N$ crystal response (relative response) is a scale–independent measure of the energy response. The relative response is also a good measure for the lateral spread of the shower. Figure 1 shows an example of energy shape comparisons between Gflash and the test beam data for the 50 GeV electron beam. Both the absolute response ($N \times N$ crystal response) and the relative response (ratio of $N \times N$ responses) are in good agreement. Figure 2 shows the mean energy response and its resolution as the beam energy. Again, comparisons between Gflash and results from test beam data are in good agreement.

3.2. Tuning the Hadronic Parameterization

To tune hadronic response of Gflash, the reference response of the hadron test beam data is used. For the nominal energy response, the energy response of Gflash is compared to the sum

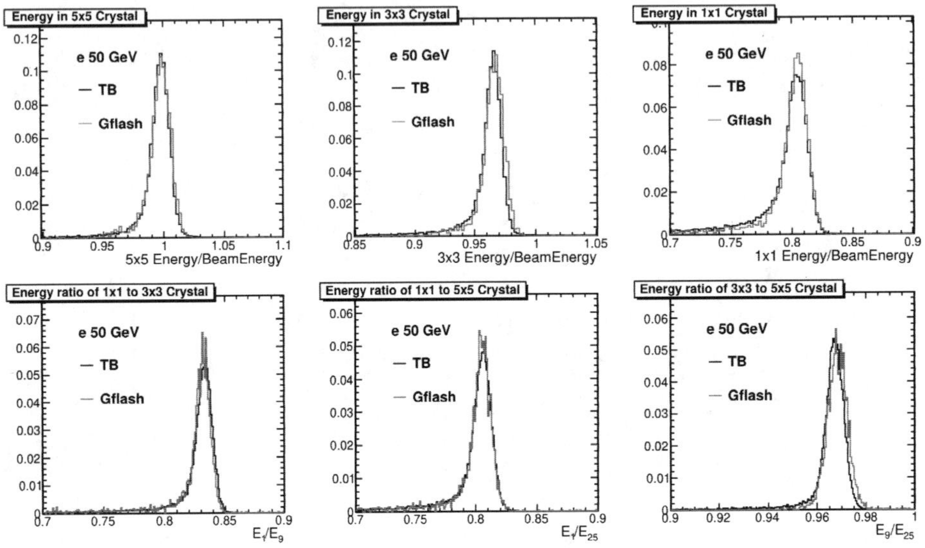

Figure 1. The energy response of 50 GeV e^- of Gflash (red) and the 2006 test beam data (black); the nominal energy response in $N \times N$ ($N = 1, 3, 5$) crystals (top) and the ratio of $N \times N$ crystal response (bottom)

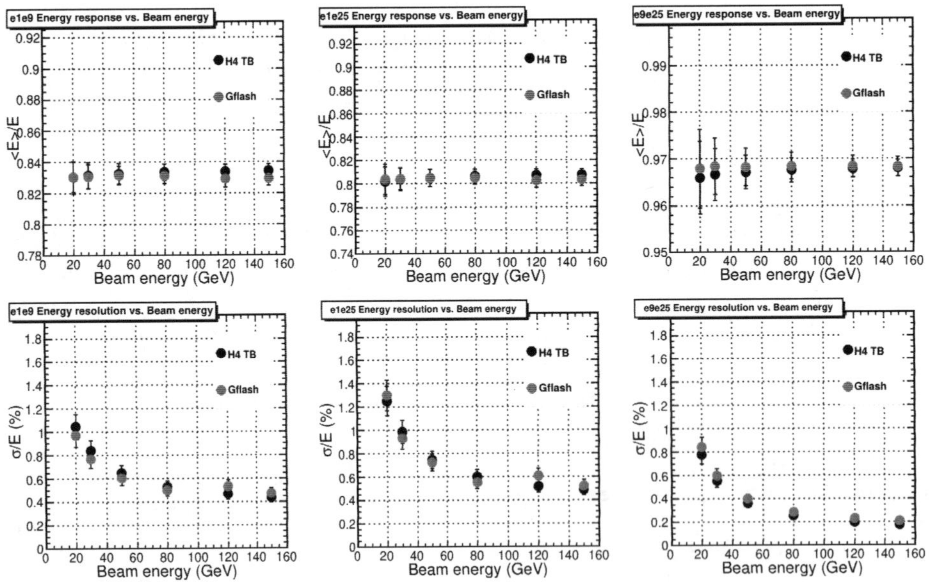

Figure 2. The mean energy response (top) and energy resolution (bottom) of e^- as the beam energy of Gflash (red) and the 2006 test beam data (black).

of energy in ECAL (7×7 matrix) and HCAL (4×3), and their correlations. For the lateral shower, the energy in each η-tower or ϕ-tower is compared using tower-by-tower energy matrix of ECAL (9×9) and HCAL (4×3).

For each energy of each particle type, the energy shape is tuned for 1) pure hadronic response (hadronic energy for $E_{ECAL} < 0.8$ GeV), 2) EM energy (ECAL) response 3) hadronic energy (HCAL) response. The total energy response is not attempted to be tuned, but used only as the final measure of tuning. Figure 3 shows examples of the energy shape tuning for 100 GeV π^- beam and 20 GeV p beam, respectively.

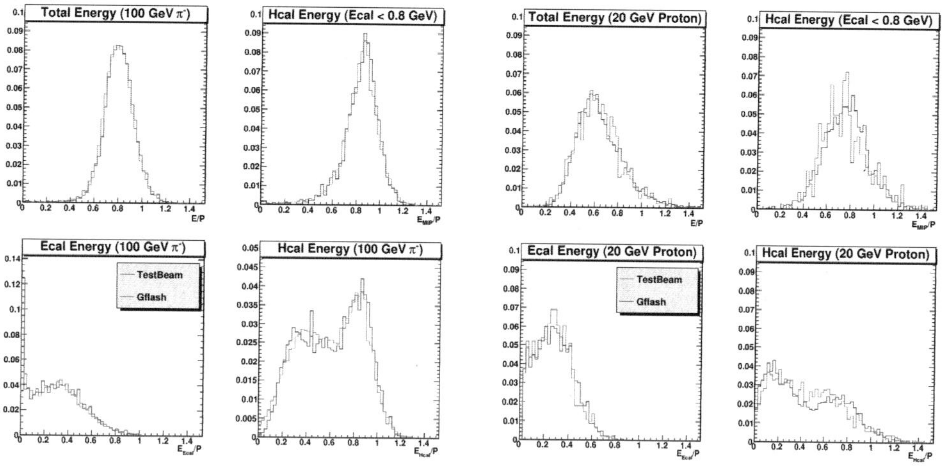

Figure 3. The energy shapes of 100 GeV π^- (left) and 20 GeV p (right) of Gflash (blue) compared to the 2006 test beam (red).

The measured mean energy as the beam energy ($\langle E\rangle/P$) for all available π^- and p beams are shown in Figure 4 and Figure 5, respectively. Compared to results from test beam data, both the average response and the resolution of Gflash are in good agreement for π^- and p.

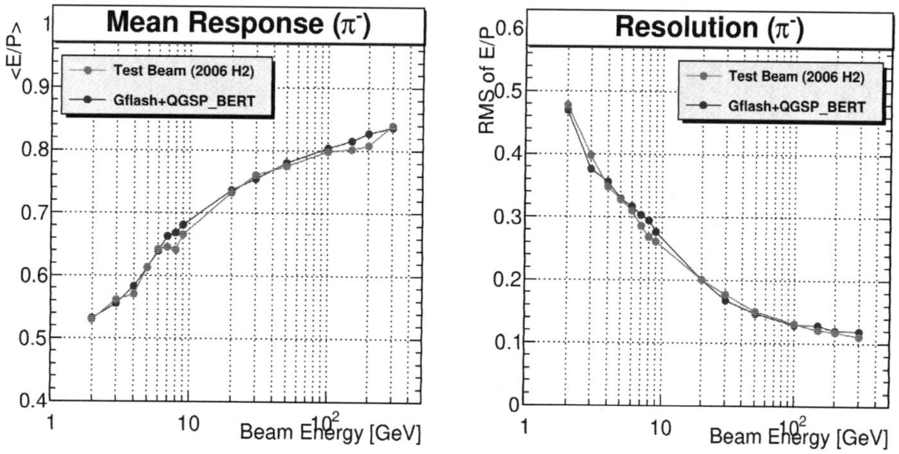

Figure 4. The energy response and resolution (π^-) as the beam energy of Gflash (blue) compared to those of the test beam data (red).

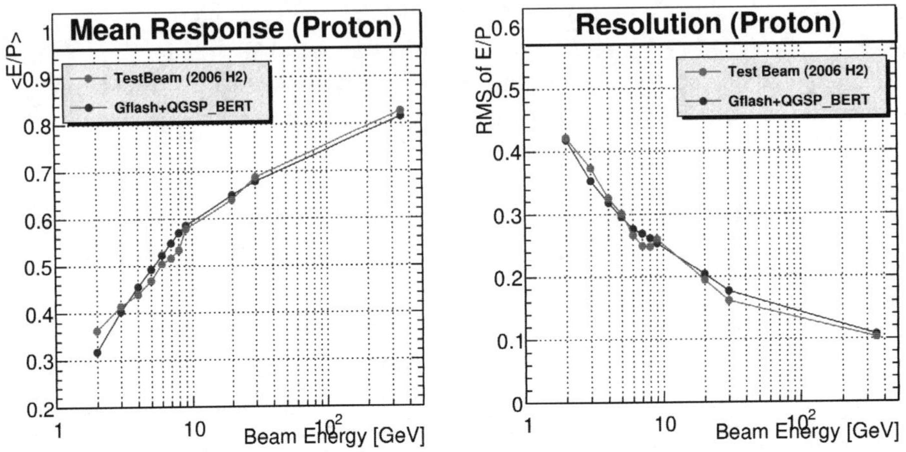

Figure 5. The energy response and resolution (proton) as the beam energy of Gflash (blue) compared to those from the test beam data (red).

4. Performance

Figure 6 shows the CPU time per event and the memory usage for first 100 $t\bar{t}$ events from Gflash and Geant4. The CPU gain using Gflash with respect to Geant4 is around factor 2 at low energy (2 GeV) and factor 10 at high energy (300 GeV). However, there is only marginal increment in the memory usage compared to Geant4.

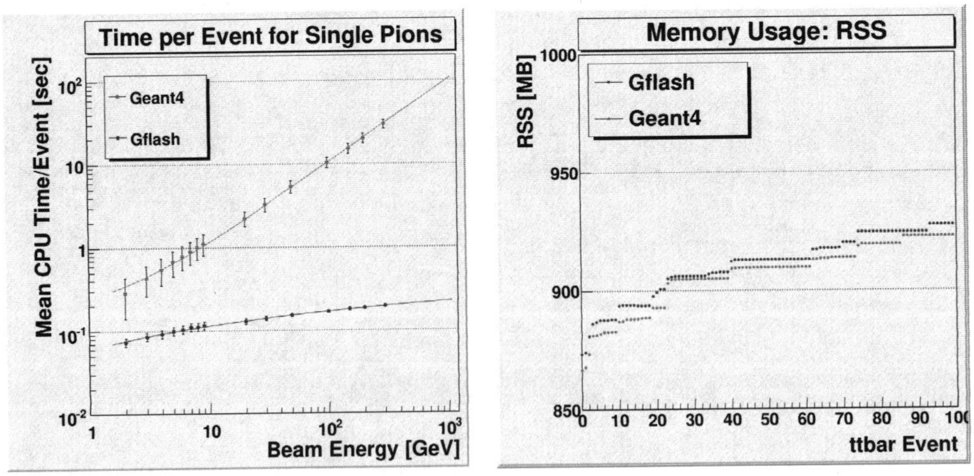

Figure 6. The CPU performance and the usage of memory for the first 100 $t\bar{t}$ events.

References

[1] Grindhammer G, Rudowicz M, and Peters S 1990 *Nucl. Instrum. and Methods* **A290** 469
[2] Grindhammer G and Peters S 2000 *Preprint* hep-ex/0001020
[3] Rossi B 1952 *High-Energy Particles* (New York: Prentice Hall)

XIV International Conference on Calorimetry in High Energy Physics (CALOR 2010) IOP Publishing
Journal of Physics: Conference Series **293** (2011) 012024 doi:10.1088/1742-6596/293/1/012024

Fast Simulation of Showers in the H1 SpaCal Calorimeter

Sasha Glazov

DESY, Notkestrasse 85, Hamburg, 22607 Germany

E-mail: glazov@mail.desy.de

Abstract. A method for the fast simulation of particle showers in the H1 lead/scintillating-fiber calorimeter is presented. The method uses a shower library technique in which the detector response is simulated based on a collection of stored showers for different particle types. The library is created using the GEANT programme. The fast simulation is compared to the data collected by the H1 experiment.

Simulation of high energy physics experiments requires large computational resources. A significant fraction of time is typically spent for the simulation of showers in calorimeters. Different methods have been developed to accelerate the simulation of showers. These include parametrisation methods [1], shower libraries [2] and modified shower libraries, so called frozen showers [3]. This note describes a "shower library" based simulation of the H1 calorimeter, more details of this method can be found in [4].

A "shower library" is a collection of GEANT [5] energies in the calorimeter for a given particle type, energy and location at the calorimeter surface. The structure of the shower library depends on the construction of the calorimeter and particle kinematics. The shower library provides description of the detector response approaching the quality of the GEANT simulation while the computing time is much reduced.

In the H1 detector [6], which is dedicated to the study of the deep inelastic scattering process using colliding electron and proton beams of 25.5 GeV and 920 GeV, respectively, from the HERA accelerator in Hamburg, the shower library is implemented for the lead/scintillating-fibre calorimeter (so called "Spaghetti calorimeter" - SpaCal) [7]. The SpaCal calorimeter contains electromagnetic and hadronic sections. The electromagnetic section of the SpaCal consists of 1192 cells with an active volume of $4.05\times4.05\times25$ cm^3 each. A transverse view of the calorimeter is given in figure 1. The cells are made of grooved lead plates and scintillating fibers with a diameter of 0.5 mm. The scintillation light of each cell is converted into an electric pulse using photomultiplier tubes. The active length of the electromagnetic SpaCal corresponds to 27.5 radiation lengths and about one hadronic interaction length. The Moliere radius of the SpaCal is 2.5 cm. The angular coverage of the calorimeter, which is measured with respect to the proton beam direction[1], is $153^\circ < \theta < 177.8^\circ$. The electromagnetic energy resolution can be parametrised as $7.5\%/\sqrt{E} \oplus 2\%$. The spatial resolution of the calorimeter in the transverse plane is about 3.5 mm. The amount of material from the interaction point to the calorimeter

[1] The z axis of the right-handed coordinate system used by H1 is defined to lie along the direction of the incident proton beam and the origin to be at the nominal ep interaction vertex.

surface varies between 0.6 and 1.0 radiation length.

The hadronic part of the SpaCal is made of 136 cells of $12 \times 12 \times 25$ cm^3 providing one nuclear interaction length. The fibers are of the same type as in the electromagnetic section but have a larger diameter of 1 mm.

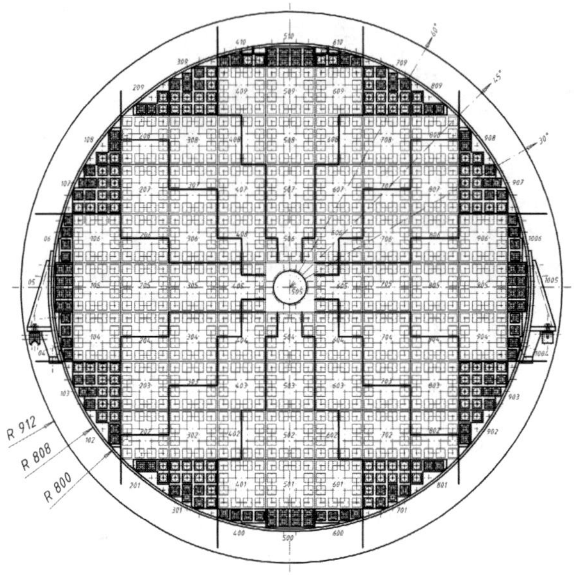

Figure 1. Transverse view of the SpaCal. Small boxes indicate individual cells. They are combined in 4 x 4 groups into super-modules.

In the simulation using the shower library a generated particle is traced through the detector components up to the calorimeter surface. At this point, if the impact position of the particle is far enough from the calorimeter boundaries, such that the shower is expected to be fully contained in the calorimeter, the shower library is applied: instead of using the GEANT package to simulate the calorimeter response, a suitable pre-simulated shower is selected from the shower library. The selected shower is then corrected for a difference between the actual particle kinematics and the kinematics of the particle used for the shower creation. Corrected energy deposits are added to the calorimeter response. The quality of the simulation based on the shower library therefore depends on how well it represents kinematics of simulated particles, how many showers are stored for each grid point and how large residual corrections are. Since the computer memory is limited, an additional important feature of the shower library is the packing scheme.

The shower library grid is binned in the particle energy, E, in the particle position with respect to the calorimeter surface, x and y, and in the particle direction angles, θ_x and θ_y, with respect to the lepton beam direction which is perpendicular to the calorimeter surface. The binning in energy is spaced uniformly in $\log E$, it should be fine enough compared to the varioation of the energy resolution as a function of energy. The grid should start from low energy, $\simeq$ 100 MeV, to describe showers starting before the calorimeter surface. The granularity of the position binning parallel to the calorimeter surface should be fine enough compared to the calorimeter spacial resolution. Symmetries of the calorimeter can be used to reduce the number of position bins. A reduction is possible for a large (vs shower size) calorimeter with translation symmetry. In this case it is sufficient to use binning inside the cell with the largest energy deposit ("hottest cell"). The granularity of the angular binning usually does not need to be too fine, if the angles are sufficiently small. A small angle $\theta_{x,y}$ leads to only a moderate (proportional to $\cos\theta_{x,y}$) increase of the shower size which is usually not very important. A larger effect is introduced by an effective shift of the shower x, y-position. This effect can be compensated for using an effective shower depth as described in [4].

The shower library is created using the GEANT simulation for particles selected with energies, positions and angles as defined by the shower library grid. For a given grid point, typically several showers are stored.

For simulation using the shower library, the bin which best fits the particle kinematics is selected. The energy bin is selected using the following algorithm:

(i) For a given particle energy, E_{gen}, find the nearest higher and lower energy grid points, E_h and E_l ($E_l < E_{gen} \leq E_h$).

(ii) With a probability proportional to the "distance" from a grid point to E_{gen}, randomly select an E_h or E_l bin. The "distance" can be measured linearly or logarithmically. For the logarithmic measure, which is used for the SpaCal shower library, the probability to select the lower grid point is:

$$P_l = \frac{\log \frac{E_h}{E_{gen}}}{\log \frac{E_h}{E_l}} \tag{1}$$

(iii) For the selected bin (h or l), take the shower from the shower library and, before adding to the calorimeter response, re-scale the energies of deposits as $E_{gen}/E_{l,h}$ ($E_{l,h}$ stands for E_l or E_h).

This procedure adequately reproduces the variation of the energy resolution as a function of energy correctly up to first order in $\log E_h/E_l$.

To obtain a good description of the calorimeter response and avoid statistical biases, the shower library should contain sufficient amount of showers. In order to avoid relatively slow disk access it is preferential to have the showers needed for all grid points buffered in the computer physical memory. In the case of multidimensional binning (in x, y, θ_x, θ_y, E and particle type) and fine calorimeter granularity (hundreds of energy deposits per shower), the total memory requirement can be rather high.

A usual strategy is to use buffers of packed showers. One buffer covers one complete shower library which covers all grid points and stores at least one shower per grid point. Several buffers are stored in a disk file. The showers from one buffer reside in memory, and are reused several times; a new buffer is read from the disk file after the repetition count exceeds some fixed value.

Electromagnetic showers are typically small in size and they can be stored using energy deposits in the cells in a fixed box around the hottest cell. To pack shower information effectively, it is usual to store the total energy of the shower as an uncompressed real number; the energy deposits are stored in terms of a bit packed fraction of the total energy. The fraction of energy of the hottest cell should be stored with the highest precision (e.g. with 24 bits) while for peripheral cells, just few bits can be sufficient. This packing scheme is illustrated in figure 2. For broad, asymmetric hadronic showers, many cells may contain no energy deposit. In this case, it is more efficient to store information only from the cells which contain energy.

The shower library for electromagnetic particles contains 12 bins in energy ranging logarithmically from 0.1 GeV up to 32 GeV, 10 bins in x and 10 bins in y which cover the front face of the hottest cell, 8 bins in θ_x and 8 bins in θ_y ranging from $-24.8°$ to $24.8°$ and two

5	5	6	8	6	5	5
5	10	10	12	10	10	5
6	10	20	20	20	10	6
8	12	20	24	20	12	8
6	10	20	20	20	10	6
5	10	10	12	10	10	5
5	5	6	8	6	5	5

Figure 2. Bit packing scheme for showers from the electromagnetic SpaCal shower libraries. The showers are stored using a fixed 7×7 cells template around the shower centre. Fraction of the shower energy contained in the cell at the shower centre is stored using 24 bits, while for outer cells 5 bits are used.

bins for particle type, for electron/positron[2] and photon. The total number of grid bins is

$$12 \cdot 10 \cdot 10 \cdot 8 \cdot 8 \cdot 2 = 153600,$$

which corresponds to one buffer of the shower library.

The shower library for hadrons contains 10 bins in energy from 0.1 GeV up to 20 GeV, 5 bins in x and 5 bins in y which cover the front face of the hottest cell, 4 bins in θ_x and 4 bins in θ_y from $-12.1°$ to $12.1°$ and 9 bins in particle type. The total number of grid bins is

$$10 \cdot 5 \cdot 5 \cdot 4 \cdot 4 \cdot 9 = 36000,$$

which corresponds to one buffer of the shower library.

The shower library for electromagnetic particles is built of showers containing the energy response of 49 cells from the electromagnetic section and 25 cells from the hadronic section. The shower library for hadrons is built of showers containing the energy responses from all cells in the electromagnetic and hadronic sections (1328) with non-zero energy. The packing of the energy information reduces the size for one buffer to about 8 MB.

The shower library for electromagnetic particles is used on average 30 times for a single electron from the interaction point. This is caused by the shower development in the material in front of the calorimeter. The repetition count for reading a new buffer is set to the total number of shower in the buffer, 153600, thus one buffer is used to simulate about 5000 electrons. There

[2] A detailed study of the showers produced by positrons and electrons shows that they are virtually identical; therefore no additional binning distinguishing electrons and positrons is introduced.

are in total 7 buffers used for the SpaCal simulation. Since the shower development up to the SpaCal is random and the resulting SpaCal shower is build from many showers from the shower library, the number of showers in the library is sufficient for the H1 analyses which study up to $\sim 10^7$ scattered electrons.

The main quantity used for particle identification in the SpaCal is the transverse size of the shower parametrised by a cluster radius. The determination of the cluster radius starts with the calculation of the cluster centre-of-gravity which is defined as a weighted average over the positions of the cluster cell centres:

$$X_{cluster} = \sum_i x_i w_i \qquad Y_{cluster} = \sum_i y_i w_i. \tag{2}$$

Here x_i, y_i are the x and y coordinates of the i-th cell centre and w_i defines the weight of cell i. Two definitions are used in H1 to calculate w_i, based on square root and logarithmic energy weighting:

$$w_{i,\mathrm{sqrt}} = \frac{\sqrt{E_i}}{\sum_j \sqrt{E_j}}, \tag{3}$$

$$w_{i,\log} = \frac{\max(0, w_{\mathrm{cut}} + \log(E_i/E_{\mathrm{cluster}}))}{\sum_j \max(0, w_{\mathrm{cut}} + \log(E_j/E_{\mathrm{cluster}}))}, \tag{4}$$

where E_i corresponds to the energy reconstructed in the cell i, and w_{cut} defines the logarithmic cut-off parameter for the SpaCal taken to be 4.85.

The cluster radius is then calculated as a weighted sum over the distances, R_i, between the centre of each cell, i, and the cluster centre-of-gravity. Corresponding to the two definitions of weighting, two cluster radius calculations are employed at H1. These are, so called,

$$ECRA = \sum_i R_i w_{i,\mathrm{sqrt}}, \tag{5}$$

and the logarithmic cluster radius,

$$R_{log} = \sqrt{\sum_i (R_i w_{i,\log})^2}. \tag{6}$$

SpaCal clusters corresponding to the scattered positrons, from e^+p interactions, are compared between the H1 data and simulation. For this comparison data collected by the H1 collaboration in the year 2000 are used. For these data, positrons with energy of 27.5 GeV collided with protons with energy of 920 GeV. The simulation of e^+p collisions uses the DJANGO event generator [8].

A pure sample of scattered positron clusters is obtained by requiring the energy of the cluster to exceed 15 GeV; this avoids hadronic background contamination. The selected cluster is required to be in the region of the SpaCal far from the inner and outer acceptance edges.

Figures 3 and 4 show comparisons between the H1 data and Monte Carlo simulation using the shower parametrisation [1] and between data and Monte Carlo simulation using the shower library. The comparisons are made for the R_{log} and ECRA estimators. Using the shower library a good description of the cluster radius is achieved.

Both simulation methods, using the shower library and the shower parametrisation, are about ten times faster than the full GEANT simulation for the scattered positron energies in the tested range.

An implementation of the shower library for the simulation of the H1 detector is presented. The software to generate, store and read back the shower library information is described. The shower library is created for the lead/scintillating-fibre calorimeter SpaCal.

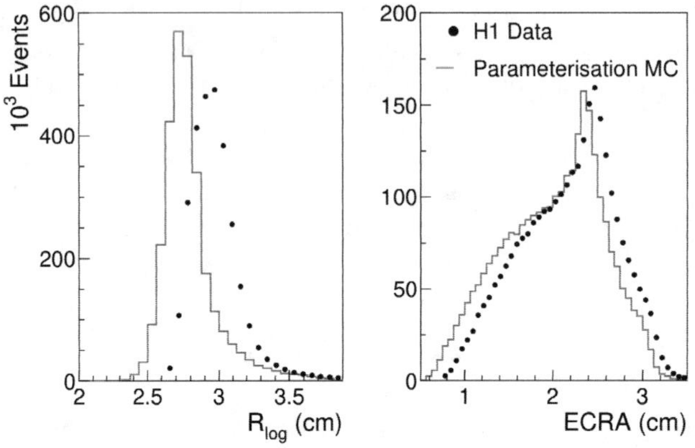

Figure 3. Comparison of H1 data (dots) with simulation using the shower parametrisation (histogram) for R_{log} (left) and ECRA (right).

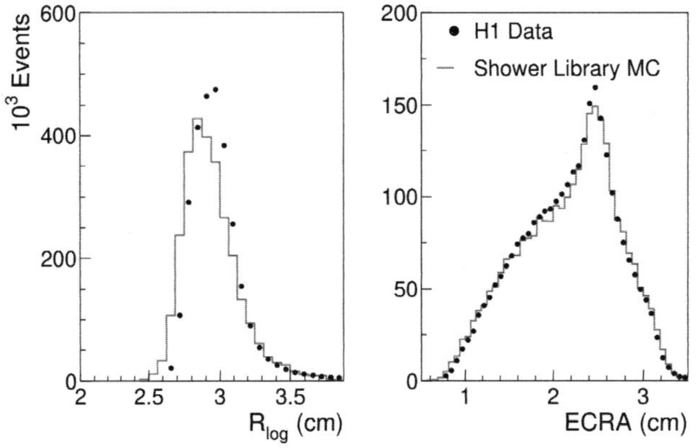

Figure 4. Comparison of H1 data (dots) with Monte Carlo simulation using the shower library (histogram) for R_{log} (left) and ECRA (right).

Monte Carlo simulation using the shower library provides good description of the cluster shapes observed in data. This reduces systematic uncertainty from electron identification efficiency which is important for the measurement of the structure function F_L [9]. The CPU time is reduced compared to the full GEANT simulation by about a factor of ten.

[1] G. Grindhammer, M. Rudowicz and S. Peters, Nucl. Inst. Methods A290 (1990) 469; G. Grindhammer and S. Peters, hep-ex/0001020.
[2] E. Turner Worcester, PhD Thesis, University of Chicago, Chicago, (2007).
[3] E. Barberio et al., J. Phys.: Conf. Ser. 160 (2009) 012082.
[4] A. Glazov, N. Raicevic, A. Zhokin, Comp.Phys.Comm, 181, 1008 (2010).
[5] R. Brun et al., GEANT3, CERN DD/EE/84-1 (1987); Application Software Group, GEANT Detector Description and Simulation Tool, CERN Program Library Long Writeup W5013, October (1994).
[6] I. Abt *et al.*, H1 Collaboration, Nucl. Instr. and Meth. A 386, 310 (1997); H1 Collaboration, Nucl. Instr. and Meth. A 386, 348 (1997).
[7] R.-D. Appuhn *et al.*, H1 SPACAL Group, Nucl. Instr. Meth. A 386 (1997) 397.
[8] G. A. Schuler and H. Spiesberger, Proc. Workshop on HERA Physics, Vol 3, eds. W. Buchmuller and G. Ingelman, Hamburg, DESY (1992), 1419; A. Kwiatkowski, H. Spiesberger and H. - J. Mohring, Comp. Phys. Comm. 69 (1992) 155.
[9] F.D. Aaron *et al.*, H1 Collaboration, Phys. Lett. B 665 (2008), 139.

Journal of Physics: Conference Series **293** (2011) 012025 doi:10.1088/1742-6596/293/1/012025

Simulation and reconstruction of the BESIII EMC

Miao He

Institute of High Energy Physics, Chinese Academy of Sciences, Beijing, China

E-mail: hem@ihep.ac.cn

Abstract. The simulation and reconstruction software of the BES III Electromagnetic Calorimeter (EMC), are developed based on the object-oriented language C++, in the framework of Gaudi[1]. Performance of EMC are studied with data and compared with Monte Carlo samples.

1. Introduction

The Beijing Spectrometer (BES) III [2][3] is a multi-purpose detector operated at the Beijing electron-positron collider (BEPC) II for physics at tau-charm energy region.

The BES III detector consists of four sub-detectors: the Drift Chamber (DC), Time-Of-Fight counters(TOF), the Electromagnetic Calorimeter (EMC), and Muon Counters(MUC). EMC is used to measure energies and positions of electrons and photons with high detection efficiency and good resolutions. The simulation software of EMC shall implement detector simulation and digitization, and reconstruction software shall process the raw data, correct all possible bias and reconstruct the energy and position of incident particle.

This paper introduces the design and the implementation of the EMC simulation and reconstruction software package, developed in the framework of BES III Offline Software System(BOSS)[4]. The configuration of the EMC is described firstly, followed by the discussion of basic simulation and reconstruction algorithms. The performance of EMC and comparison between data and simulation is shown at the end.

2. Construction of the BES III EMC

The BESIII electromagnetic calorimeter, made of CsI(Tl) crystals, is composed of one barrel and two endcaps. The barrel has an inner radius of 94 cm and a total of 44 rings of crystals along the z direction, each with 120 crystals. It covers the polar angle of $146.5° - 33.5° (|\cos\theta| = 0.83)$. The endcaps with inner radius of 50cm are placed at 138.5 cm from the collision point, each consists of 6 rings and is split into two half circles. It covers the polar angle of $32.5° - 21.3° (|\cos\theta| = 0.93)$. Hence the total acceptance of the barrel and endcaps is 93% of 4π, and The whole calorimeter has 6272 CsI(Tl) crystals with a total weight of about 24 tons. Each crystal has a length of 28 cm, a cross section of about $5.2\ cm \times 5.2\ cm$ in front face and $6.4\ cm \times 6.4\ cm$ in rear face. All crystals pointing to a position off from the interaction point by a few centimeters with a tilt angle of $1^o - 3^o$.

3. EMC Simulation

Simulation package is based on GEANT4, which contains the description of detector, physics list, tracking and digitization.

3.1. Detector geometry and materials definition

The barrel crystal is described by GEANT4's G4Trap volumn. While endcap cystal is not regular and can't be reproduced by any volumn in GEANT4 directly. Instead, we introduce a new volumn called G4IrregBox[5], which describes a general hexahedron and takes the coordinates of its 8 vertices as input parameters. Besides crystals, other materials like casing, photodiode, pre-amplifier box, cable, water pipe and support system are also included in detector description.

3.2. Physics list

A standard electro-magnetic interaction is provided by GEANT4 and used here for simulation. For hadronic interactions, several models have been tried and compared with BESIII data[6]. QGSP_BERT_HP gives best agreement between data and Monte Carlo.

3.3. Tracking and hits

The tracking of particles through the EMC is essentially done according to the GEANT4 standard. When a particle traverses a sensitive volume (a crystal or photodiode), the hit information, such as the particle position, energy deposit, etc., are saved for the digitization procedure which is discussed in the next section.

3.4. Digitization

The core component of digitization is the simulation of readout electronics. The response function of electronics can be calculated by forming an expression for the various components in the Laplace domain. The exponential decay constant is 1 μs for Thallium doped CsI, which gives a $1/(s+a)$ term in the Laplace domain, where a is the reciprocal of the decay time constant. The preamplifier is assumed to have an infinite time constant, giving a $1/s$ term. The differentiation stage of shaping circuitry gives a $s/(s+b)$ term ($b = 1/CR$) and the integration parts gives a $1/(s+c)^3$ contribution ($c = 1/RC$). Then the final expression is the product of above terms:

$$\frac{1}{s+a} \times \frac{1}{s} \times \frac{s}{s+b} \times \frac{1}{(s+c)^3}. \tag{1}$$

Since the time constant of differentiation and integration are 1 μs, the response function in time domain, i.e. waveform, is given by Inverse Laplace Transform:

$$h(t) = \frac{1}{24} t^4 e^{-t/\tau}, \tag{2}$$

in which τ=1 μs.

Based on the calibration constants, the waveform is easily scaled so the peak value equals energy deposit, i.e., energy sum of all hits in this crystal. Then this waveform is sampled at 50ns time interval starting with 1.5μs ahead of the peak, and lasts 3μs, i.e., 60 sampling points. The maximum amplitude of these sampling points is recorded as energy and the index of the maximum point is recorded as time.

3.5. Noise simulation

Beam related backgrounds are obtained from random trigger in real data, which contains energy and time information of each crystal caused by backgrounds. Each MC event is mixed with one of the background event which is randomly picked up from random trigger sample. If a crystal

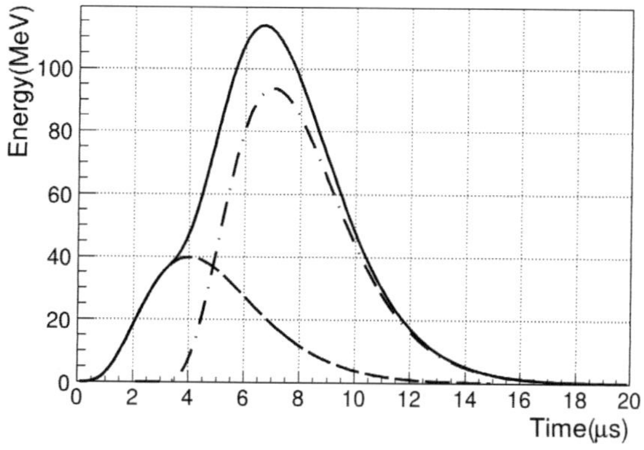

Figure 1. Sum of waveforms. ——Sum of waveforms, — · —signal, — — —background

has signal either from simulation or background, it will be kept. Otherwise the two signals in the same crystal will be added together by constructing two waveforms according to their energies and times, as shown in Fig.1. Obviously, the energy of mixed waveform is less than the sum of MC and background energies, and the time is between them.

Electronics noise is usually Gaussian distribution. A random noise is generated for each sampling point according to the measured noise and added with it. Fig.2 (a) shows ideal sampled waveform with 40 MeV energy and 4 μs time. While in Fig.2 (b), 0.5 MeV noise is added and the energy and time are both deviated.

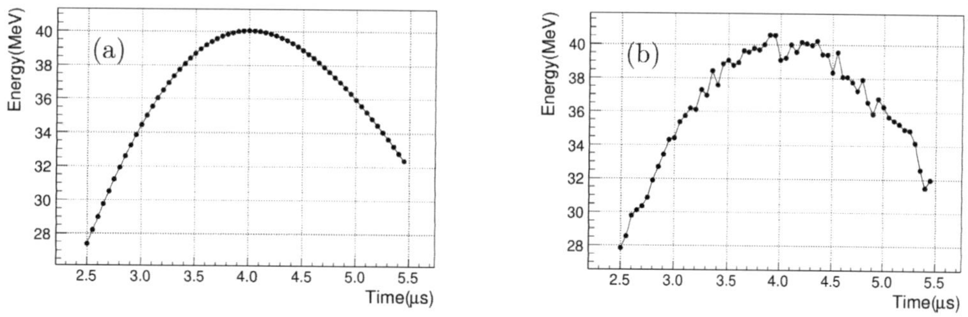

Figure 2. Sampled waveform (a) before and (b) after adding electronics noise.

4. EMC Reconstruction

When a photon or an electron hit the EMC, it interacts with CsI(Tl) crystals. The photon transforms to e^+e^- through gamma conversion, and e^+ or e^- releases photons through bremsstrahlung. These interactions continue alternately and form an electromagnetic shower. The shower develops laterally and longitudinally, losing energy in several connected crystals. The reconstruction algorithm searches these related crystals, adds their energy together, and calculates the hit position.

4.1. Clustering
The primary task of the reconstruction software of a crystal calorimeter is to identify connected regions, called clusters. There are several ways to define them for different arrangement of crystals. At BES III, the definition of a cluster is a contiguous group of crystals above a certain threshold. Its finding algorithm is to recursively search for a list of neighboring crystals above a certain energy threshold, and add them into a list. Crystals in the list are retrieved and its neighboring crystals are searched for again. This process is continued until no neighboring crystals are found. All the crystals in the list form a cluster. The same procedure is implemented for crystals not in the list, until no more clusters are found.

4.2. Cluster splitting
A cluster can be the result of energies deposited by one or more particles, corresponding to one or more showers. Each shower is recognized by a seed, which is the local maxima of energy deposit among its neighbors. If a cluster contains only one seed, only one shower, i.e., one particle is produced. If more seeds are found, for example, in a cluster caused by high momentum π^0, the cluster can be split to several showers. Every crystal in this multi-shower cluster will be assigned a weight from each seed, and contributes a corresponding shared energy to each shower.

4.3. Shower energy
Energy is one of the most important quantities for physics analysis. Although the sum over of all crystals in a shower seems to deliver the best energy resolution, the electronics noise and backgrounds may be included in the sum, hence deteriorate the resolution. Some of the common practices to avoid this problem is to define the shower energy as E3 $\times$ 3 or E5 $\times$ 5, a sum of 9 crystals or 25 crystals around the seed.

The distribution of the shower energy shows a non-Gaussian tail at the lower energy side, shown in Fig.3 and Fig.4, caused mainly by the front, rear and side leakage of energy. The energy loss in the Time-Of-Flight counters can be reconstructed and matched with EMC shower which is discussed in Ref.[9]. The energy peak then is corrected to expectation with π^0 samples or photon samples from BESIII data, discussed in Ref.[10].

4.4. Shower position
There are many different methods to obtain the shower position. The most simple and fast one is the weighted mean, as written in the following:

$$x_c = \frac{\Sigma_i W_i(E_i) x_i}{\Sigma_i W_i(E_i)}. \tag{3}$$

The weight, $W_i(E_i)$, is a function of energy deposited in the ith crystal, and x_i the coordinate of the crystal center at the front face. In Ref.[11] several weighting functions are carefully discussed. Two of them are listed here:

$$W_i^{(linear)}(E_i) = E_i, \tag{4}$$

$$W_i^{(log)}(E_i) = Max\{0, a_0 + ln(E_i) - ln(E_{total})\}. \tag{5}$$

They are called linear weighting function and logarithmic weighting function, respectively. Both of them need corrections because of the complexity of the EMC geometry. Bhabha event is chosen for this purpose, and the real hit position is supplied by the track extrapolation[12] from the drift chamber information.

5. Performance study

In this section, data with centre of mass system energy 3.686 GeV is taken from March, 2009 to April, 2009, and data at 3.097 GeV energy point is taken in June, 2009.

Fig.3 shows reconstructed energy spectrum (E5 $\times$ 5) of electrons and positrons from $e^+e^- \to e^+e^-$(bhabha events) at 3.686 GeV in barrel and endcap, respectively, which presents good agreement between data and simulation for high energy showers. Energy resolution in endcap is much more worse than in barrel because there are more dead materials in front of endcap than barrel. For low energy showers, they are also consistent, as shown in Fig.4, in which photons are generated in $\psi' \to \gamma\chi_{c1}$. A long tail is shown at the low energy side also because of the material effect in front of EMC.

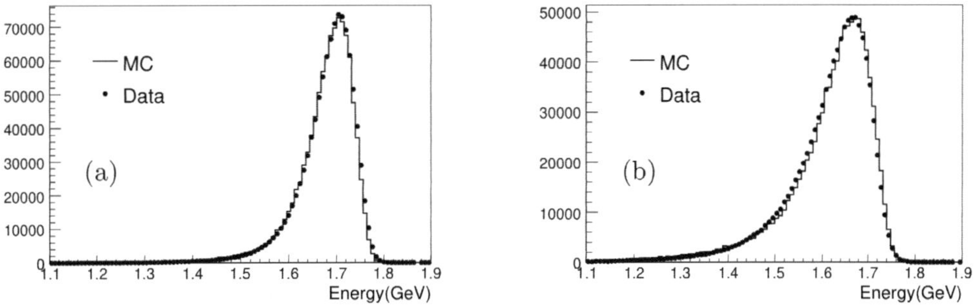

Figure 3. Energy spectrum of bhabha events in (a)barrel and (b)endcap

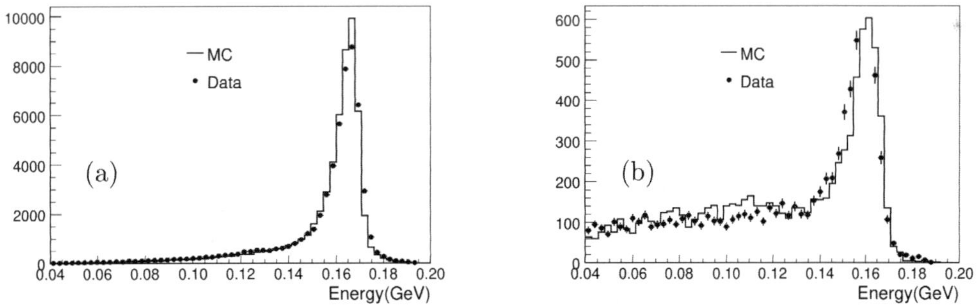

Figure 4. Energy spectrum of photon from $\psi' \to \gamma\chi_{c1}$ in (a)barrel and (b)endcap

Fig.5 shows the average energy peak and energy resolution of electrons and positrons from $e^+e^- \to e^+e^-$ at 3.097GeV as a function of crystal id in θ direction. Energy resolution is 2.3% in barrel and 4.1% in endcap for both data and simulation. While for photons from $e^+e^- \to \gamma\gamma$(diphoton events) at 3.097GeV, energy resolution is 2.7% in barrel and 4.2% in endcap, showing worse than bhabha events. The reason is photons interact with the materials in front of EMC by the production of electrons and positrons which introduce large fluctuation of energy loss, and make energy resolution worse. However, data and simulation show good agreement.

6. Conclusion

The simulation and reconstruction software packages for the BES III EMC have been developed and work normally during data taken. The performances are studied with data and basically

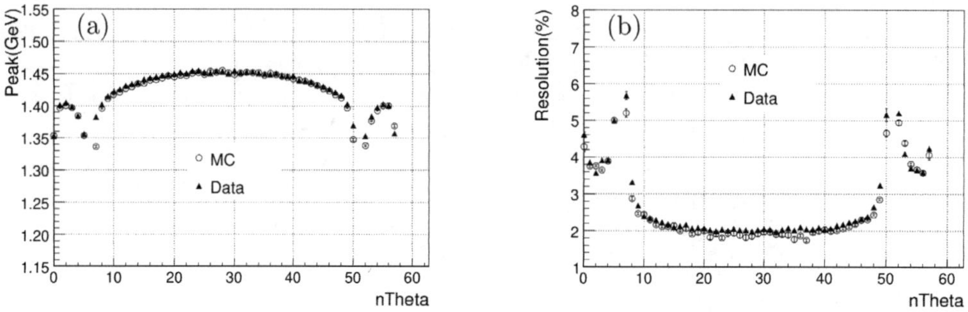

Figure 5. (a)Energy peak and (b)energy resolution of bhabha events

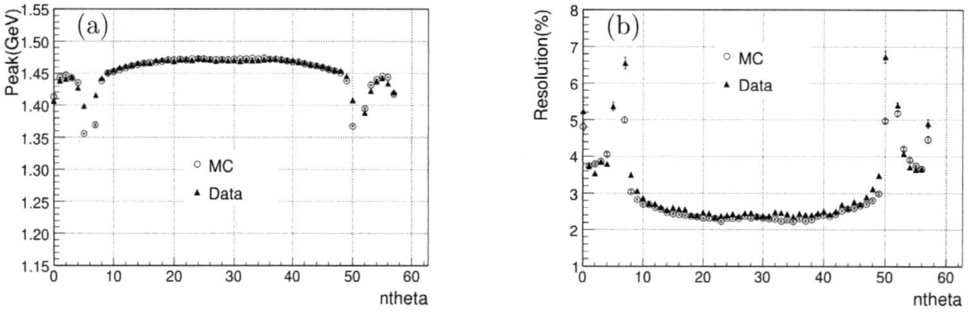

Figure 6. (a)Energy peak and (b)energy resolution of diphoton events

consistent with simulation.

References

[1] Barrand, G. et al., A software architecture and framework for building HEP data processing applications, Comput. Phys. Commun., 140: 45-55, 2001
[2] Y.F. Wang, The construction of the BESIII detectorInt. J. Mod. Phys. A 21(2006)5371.
[3] BESIII Collaboration, Nucl. Inst. Meth. A Volumn 614, Issue 3, 11 March 2010, Pages 345-399
[4] Weidong Li, Huaimin Liu et al., The Offline Software for the BESIII Experiment, February 2006, Mumbai, India.
[5] http://hypernews.slac.stanford.edu/HyperNews/geant4/get/geometry/305.html
[6] G.F.Cao et al. NIMA, Volume 606, Issue 3, 21 July 2009
[7] D.Antreasyan, "An algorithm for tagging photon and merged π showers in the crystal ball", Crystal Ball note 321, 1983
[8] R. Seitz, SLAC-BABAR-NOTE-294
[9] M.He, et al. Chinese Physics C (HEP & NP), 2008, 32(04): 269-274
[10] J.M.Bian, et al. Chinese Physics C (HEP & NP), 2010 34(01): 7277
[11] B.B.Brabson et al., Nucl. Instr. Meth., A332(1993), 419
[12] Liangliang, Wang et al., HEP&NP

Journal of Physics: Conference Series **293** (2011) 012026 doi:10.1088/1742-6596/293/1/012026

From raw data to physics results with ALICE PHOS

Renzhuo WAN [1,2] for the ALICE collaboration

[1]Key Laboratory of Quark & Lepton Physics (Huazhong Normal University), Ministry of Education, 430079, Wuhan, China

[2]Institut Pluridisciplinaire Hubert Curien, Université de Strasbourg, Strasbourg, France

E-mail: wanrz@iopp.ccnu.edu.cn

Abstract. A Large Ion Collider Experiment (ALICE) is designed to study the hot and dense QCD matter created in ultra-relativistic heavy-ion collisions and to study the phase transition between confined matter and the Quark-Gluon Plasma (QGP). Photons, produced throughout all stages of the collisions, is believed to carry important information on the hot dense system. In ALICE, photons can be precisely measured by the electromagnetic calorimeter PHOS (PHOton Spectrometer) over a wide momentum range. Here we introduce the PHOS detector and present the chain of data processing from raw data to physics results.

1. Introduction

Lattice QCD predicts the existence of a Quark Gluon Plasma (QGP) at extremely high temperature and energy density. A phase transition occurs from nuclear matter to a deconfined state of quarks and gluons at a critical temperature of ~170 MeV [1]. One of the major goals of high energy nuclear physics is to explore the phase diagram of strongly interacting matter, to study the QCD phase transition and the properties of the QGP. The ALICE experiment at the LHC (see Fig. 1) was designed to study the QCD matter by comprehensive measurements of hadrons, leptons, photons and jets in the high multiplicity environment of PbPb collisions at $\sqrt{s_{NN}}$=5.5 TeV [2]. Photons, emitted during the entire evolution of the collision, undergo no (strong) interactions with the medium and hence can reflect its state at the time they are produced. In ALICE, photons are measured with the photon spectrometer (PHOS) and the eletromagnetic calorimeter. The yield of direct photons are determined by subtracting from the inclusive photon yield the decay photon from light neutral mesons (π^0, η and $\omega(782)$ etc.), which can be reconstructed via two- or three-γ decay branch. The measurements of direct photon, π^0, η and $\omega(782)$ allow to test the perturbative QCD, to study the m_T and x_T scaling properties, and to deduce the thermal properties of the hot and dense medium. In addition, in back-to-back γ-jet events, the energy of high energy photons can be measured with high precision and serve as a reference to the jet energy, and thus provide a kinematic tag for the recoiling parton which might be affect by the hot and dense medium [3].

In this paper we concentrate on the photon measurement with PHOS and present the analysis procedure, from raw data to the physics results.

2. PHOS detector

The sketch of the PHOS assembly is illustrated in Fig. 2. PHOS [4] is a high resolution electromagnetic calorimeter placed at a radial distance of 460 cm from the interaction point. It consists

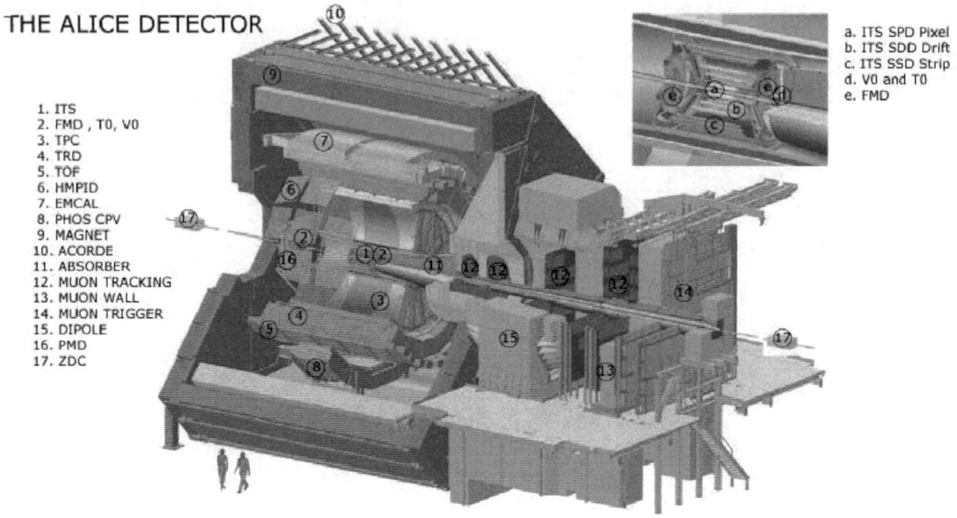

Figure 1. The ALICE detector at LHC.

of 5 identical modules and covers an azimuthal angle from 220° to 320° and a rapidity of $|\triangle\eta| < 0.12$ at mid-rapidity. Each module consists of a matrix of 56×64 $PbWO_4$ crystals [5]. The $PbWO_4$ is a fast scintillation crystal with a light emission of 400 nm to 500 nm. The size is of 22×22×180 mm^3, which corresponds to a $20X_0$ radiation length to fully contain high energy electromagnetic shower particles, and provides a fine granularity of $\Delta\phi\times\Delta\eta = 4.8{\cdot}10^{-3}\times 4.3{\cdot}10^{-3}$ to separate the different electromagnetic showers. The detector is operated at a temperature of $-25°$ in order to increase the PWO light yield by a factor of ~ 3 compared to the room temperature. S8664-55 APDs (Avalanche Photon Diode) with an active area of 5×5 mm^2 are mounted at the bottom of the crystals to detect scintillation light and the charge is collected by a low noise CSP (Charge Sensitive Preamplifier). Each 8×2 crystals are assembled in crystal strip units and two strip units are connected to a FEE (Front-End Electronics) [6, 7]. The FEE processes the signal from the CSP and send the data to data acquisition system via a Readout Control Unit. In addition, FEEs provide also fast 2×2 analog energy sum outputs to TRUs (Trigger Region Unit) for a L0/L1 trigger decision generation.

3. The calibration of PHOS

An energy calibration to a level of < 1% is necessary to achieve the physics goals of the PHOS detector. Calibration includes an absolute energy scale calibration and a relative gain calibration of individual detection channels. First, the gains of individual detection channels of PHOS modules were equalized to a precision of $\sim$ 20% by adjusting the APD bias voltage using the APD data sheet. Second, the first PHOS module was irradiated by a monochromatic electron beam at 2 GeV in the summer of 2006. Fig. 3 shows the obtained PHOS energy resolution, corresponding to $\frac{3\%}{\sqrt{E}} \oplus 1.1\%$. Finally, the channel-by-channel relative calibrations can be carried out by utilizing the energy deposition in each channel, and the distribution of energy deposited by MIP (Minimum Ionizing Particles). The absolute energy scale can be determined from the π^0 peak position and/or the electron E/p distribution. Note that a refined calibration will only be possible when there will be enough statistics accumulated (MIP, electrons, π^0).

It is crucial that all the calibration parameters, such as the ADC to GeV coefficient, the HG/LG ratio, the time offset and the bad channel map, are stored together with alignment data

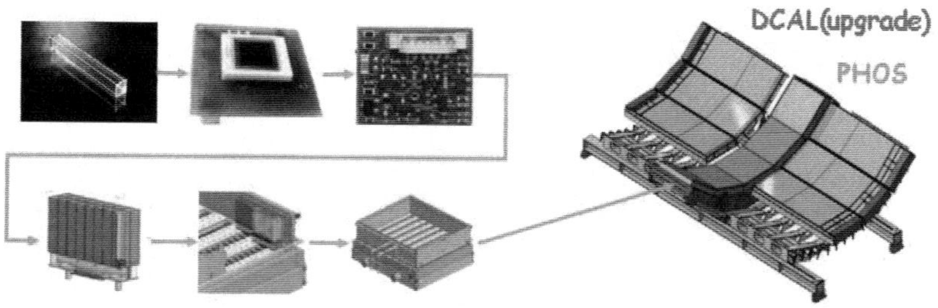

Figure 2. The assemble line from a PbWO4 to PHOS module (left). Right: The PHOS detector (yellow), and the upgrade detector DCAL (gray) which lies both side of PHOS with the same technology as ALICE EMCAL [8].

in the OCDB (Offline Conditions Data Base) to assure a correct event reconstruction.

4. From raw data to physics results

In this section, we focus on the analysis procedure, from the raw data reconstruction to the calculation of the particle kinematics in real data. AliRoot [9] (the ALICE offline framework build on the ROOT system) is our basic utensil. It is the software developed to perform simulations, reconstruct real & simulated events and perform the analysis.

4.1. Event reconstruction

The PHOS reconstruction procedure consists of four steps: (1) decoding raw data to extract the signal amplitude and timing information, (2) the clusterization and cluster unfolding to reconstruct the position and energy of the incident particles, (3) tracking by matching segments on TPC/CPV and EMC clusters, and (4) particle identification.

- Decoding raw data
  The energy and timing information are extracted by fitting the signal with a $\Gamma - 2$ function. The bad/hot channels causing a random electronic noise or operational defect are identified based on some stringent criteria on the signal shape and the typical behaviour of the energy deposition in the cells.
- The clusterizing
  A cluster, resulted in when a crystal is fired by an incident particle, is composed of a group of adjacent cells with an energy deposition above a threshold of ~ 5 MeV. Any cell with an energy larger than a minimal energy value (the cluster threshold) is considered as a seed of a new cluster. The local maximum of the cell energy is defined as the largest energy among the surrounding ones. The unfolding algorithm will deal with the clusters with multiple local maxima, which can occur in a high multiplicity environment. The detail of the algorithm can be found in reference [4]. The energy and position of the cluster in the (x, z) plane are calculated using the center of gravity algorithm $\bar{s} = \sum_i s_i w_i / \sum_i w_i$ with a logarithmic weight $w_i = max\ [0, w_0 + log(e_i/E)]$, and where the parameter w_0 is empirically determined. Then, the cluster energy is corrected for the angle of incidence of the incoming particle. Finally, to distinguish the electromagnetic and the hadronic showers, the shower topological parameters are calculated. They are the lateral dispersion in the (x, z) plane of PHOS, the two major axes (λ_1 and λ_2), and the cluster sphericity $S = |\lambda_1 - \lambda_2|/(\lambda_1 + \lambda_2)$. Additional quantities, such as the cell multiplicity, the core energy and the clusters timing can also be used to characterize the cluster.

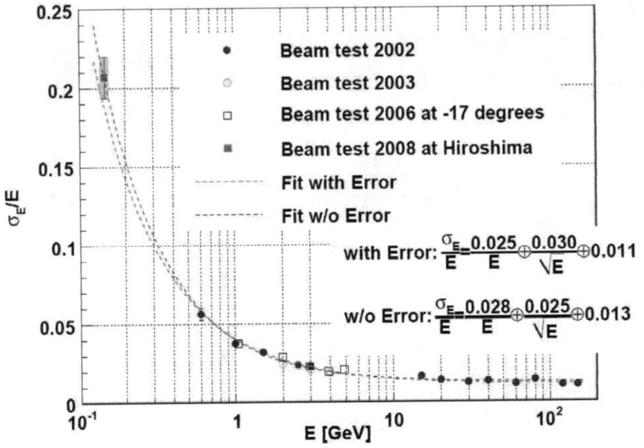

Figure 3. PHOS energy resolution measurement from beam test [10].

- Track segment reconstruction
  The track segments are reconstructed by matching the tracks position in the central tracking detectors and the clusters in the EMCs. Track matching with PHOS allows to discriminate the charged hadrons from neutral particles such as photons. The track matching performance between the TPC and PHOS with a realistic detector configuration has been studied with simulations [11]. The track matching efficiency reaches 80% for low charged particle density $dN/d\eta=400$ and 70% for $dN/d\eta=4000$ with a weak dependence on the particle transverse momentum.
- Particle identification
  Particle identification is based on: (1) time of flight information to reject the slow and heavy hadrons, (2) the CPV-EMC or TPC-EMC position matching to discriminate the charged hadrons, (3) the SSA (Shower Shape Analysis) to distinguish the electromagnetic and hadronic showers. Combing the above information on a statistical basis, the PID probability and purity can be calculated [12].
  Complementarily, the Bayesian method has been developed. It assigns a PID weight (or probability) to every reconstructed particle on an event-by-event basis. For PHOS, PID weight is calculated as:

$$W(i) = \frac{P(tof|i) \cdot P(d_{TE}^{xz}|i) \cdot P(dis|i) \cdot P(rec|i)}{\sum_s [P(tof|s) \cdot P(d_{TE}^{xz}|s) \cdot P(dis|s) \cdot P(rec|s)]}, \tag{1}$$

  where P is the density probability distribution for a given particle i and $s = \gamma, e^{\pm}, h^0, h^{\pm}$ and π^0. The PID probability of the different detectors can also be combined onto a global PID probability. The simulations indicate that, at low and intermediate p_T, PHOS PID efficiency is close to 100% for photons and electrons in a low-multiplicity environment, and about 80% to 90% in heavy-ion collisions. For the high-transverse-momentum π^0 ($p_T > 30$ GeV/c), the two decay photon clusters start to merge and can be misidentified as a single cluster. The Bayesian method does a good separation and obtains a π^0 PID efficiency up to 80% with a misidentification probability below 10% [13]. This is particularly interesting to maximize the high-p_T π^0 yield and minimize the systematical uncertainties for direct photon measurements.
  In addition, there is the isolation cut method [14, 15]. This method sits on the fact that direct photons are produced in the initial collisions and their dominant production processes

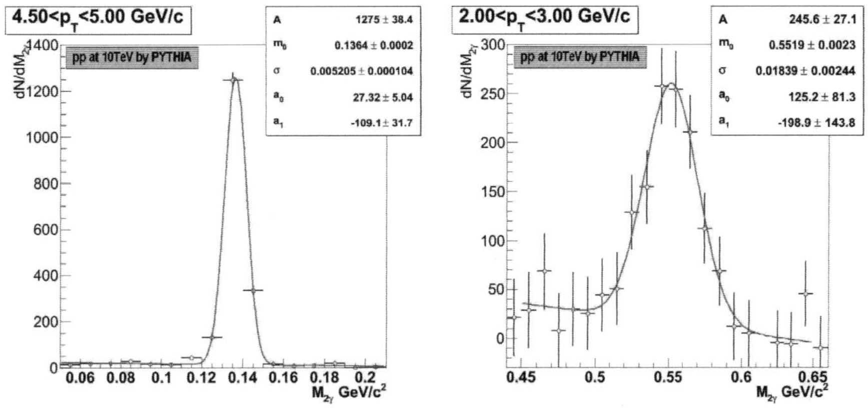

Figure 4. Invariant mass spectra for $\pi^0 \rightarrow 2\gamma$ at $4.5 < p_T < 5$ GeV/c (left) and $\eta \rightarrow 2\gamma$ at $2 < p_T < 3$ GeV/c (right) after the background subtraction by simulation with PYTHIA in pp@10 TeV.

are those of Compton scattering ($q(\overline{q}) + g \rightarrow \gamma + g$) and pair annihilation ($q + \overline{q} \rightarrow \gamma + g$) and are characterized by the (almost) back-to-back emission of the photon and the jet. In this picture, the photon is isolated, although the underlying event may vague this picture (specially in heavy-ion collisions). The idea is then to study the hadron and photon kinematics looking within a cone around the photon candidate defined as: $R = \sqrt{(\phi - \phi_\gamma)^2 + (\eta - \eta_\gamma)^2}$. The underlying event contribution is subtracted using the information from outside the cone. Simulations [16] show that direct photon identification efficiency would be improved using the isolation cut method, and that the overwhelming background from high p_T π^0 and η would be efficiently suppressed by using the SSA simultaneously. The potential sensitivity to the gluon PDF distribution at LHC energies has been studied in reference [17] using this method.

4.2. The analysis

The reconstructed particle properties and PID information are gathered in the analyses. They are handled with the same (flexible) code for simulations and real data (AliRoot framework). Fig. 4 shows the two-photon invariant mass distribution ($M_{\gamma\gamma}$) in pp collisions at 10 TeV by PYTHIA [18] simulations with a realistic detector configuration. We conclude that we could expect an invariant mass resolution is of 4-7 MeV/c^2 for π^0 and of 15-20 MeV/c^2 for η.

At present there was still not enough statistics to perform a detailed real data π^0 or η analysis. However, the expected particle production rate was estimated in reference [19]. The calculation considered $dN/dp_T = d\sigma/dp_T \cdot \mathcal{L}_{int} \cdot A \cdot \epsilon$ ($\mathcal{L}_{int}$ is the integrated luminosity, A is the geometrical acceptance and ϵ is the total correction factor). We concluded that we could reach a momentum range of up to 50 GeV/c for π^0, 40 GeV/c for η, 30 GeV/c for $\omega(782)$ and 20 GeV/c for direct photons with an $\mathcal{L}_{int} = 300\ nb^{-1}$ in pp collisions at 10 TeV. This would already allow us to test QCD in a wide momentum range.

Note that the early physics with PHOS focuses on the invariant cross section measurement of direct photons and light neutral mesons, via the formula quoted below:

$$E\frac{d^3\sigma}{d\mathbf{p}^3} = \frac{\triangle N(\triangle p_T)}{2\pi p_T \triangle p_T \triangle y} \cdot \frac{1}{\mathcal{L}_{int}} \cdot \frac{1}{\epsilon_{Acc.\times Rec.Eff.} \cdot \epsilon_{Conv.} \cdot \epsilon_{Trigger}} \tag{2}$$

Where the ϵ is the correction factor from geometrical acceptance, reconstruction efficiency, photon conversion, trigger efficiency, etc. The systematic uncertainties estimate will also be important to compare the measured distributions and the theoretical calculations.

5. Conclusion

PHOS is a high resolution electromagnetic calorimeter designed to measure photons over a large momentum range. It is partially installed (3/5), is operational and has joined the ALICE data-taking since 2009. Here we presented the data reconstruction and analysis procedure, from the raw data collection to the physics results. ALICE is presently collecting data, and soon the accumulated statistics will allow a fine PHOS calibration and measure the spectra of light neutral mesons. In summary, PHOS data analyses are in a good progress and promise exciting results soon.

Acknowledgement

Thank the ALICE collaboration and the ALICE funding agencies (the same as the acknowledgement in [20]). The work is supported partly by the NSFC (10875051, 10635020 and 10975061), the Key Project of Chinese Ministry of Education (306022 and IRT0624), and the Program of Introducing Talents of Discipline to Universities of China: B08033 and CCNU09C01002 (Key Project) of China.

References

[1] Karsch F., *Nucl. Phys. A* **698** (2002) 199c
[2] ALICE Collaboration, *JINST*, 3 S08002 (2008)
[3] Baier R. et al., *arXiv: hep-ph/0310274*
[4] ALICE Collaboration, *CERN-LHCC-99-4*
[5] Aleksandrov D. V. et al., *Nucl. Instrum. Meth. A* **550** (2005) 169
[6] Yin Z. and Zhou D., *HEP&NP*, Vol. **30**, No. 12 (2006) 1165
[7] Muller H. et al., *Nucl. Instrum. Meth. A* **518** (2004) 525
[8] ALICE Collaboration, *CERN-LHCC-2008-014*
[9] *http://aliweb.cern.ch/Offline/AliRoot/Releases.html*
[10] Maruyama Y. et al., *ALICE-INT-2009-013*
[11] Wan R. et al., *China Physics C*, Vol. **32**, No. 8 (2008) 620
[12] Conesa G. et al., *ALICE-INT-2005-053*
[13] Conesa G. et al., *ALICE-INT-2005-016*
[14] Aaltonen T. et al. CDF collaboration, *Phys. Rev. D* **80** (2009) 111106
[15] Abazov V. M. et al. D0 collaboration, *Phys. Lett. B* **639** (2006) 151
[16] Conesa G. et al., *Nucl. Instrum. Meth. A* **580** (2007) 1446
[17] Ichou R. and d'Enterria D., *arXiv: hep-ph/1005.4529*
[18] Sjostrand T., Mrenna S. and Skands P., *arXiv: hep-ph/0603175*
[19] Kharlov Y., Benhabib L. and Wan R., *POS* (2008LHC) 089
[20] K. Aamodt et. al ALICE collaboration, *Eur. Phys. J. C* **68** (2010) 89

Journal of Physics: Conference Series **293** (2011) 012027 doi:10.1088/1742-6596/293/1/012027

Commissioning of ATLAS electron and photon trigger selection

Valerio Dao for the ATLAS Collaboration
DPNC, Geneva University, Quai Ernest-Ansermet 24, CH-1211 Genève, CH
E-mail: `Valerio.Dao@unige.ch`

Abstract. First collisions at a center of mass energy of 7 TeV at the Large Hadron Collider were recorded by the ATLAS experiment at the beginning of 2010. Data selected by the hardware based Level-1 trigger were processed online by the software based High Level Trigger running without active rejection.

This paper gives an overview of the calorimeter based performance of the electron and photon triggers during the initial 2010 running period. These triggers will play a key role for the physics analyses in ATLAS. Examples of comparisons of the three trigger levels with the offline reconstruction will be presented, such as the Level-1 efficiency and the distributions of the calorimeter based selection variables used to distinguish between signal and background. This is an important step in the commissioning of the trigger system to ensure its correct functioning. Results from this first data are very encouraging. At the end a brief outlook on the commissioning steps and on the evolution of the e/γ trigger strategy with increasing luminosities will be given.

1. Introduction

The ATLAS experiment is one of the two general-purpose detectors at the LHC, the world's largest particle accelerator, built to study the Standard Model (SM) and to search for potential new physics scenarios. Proton collisions were collected at the end of 2009 at a centre of mass energy of 900 GeV, 2.36 TeV and since March 2010 at 7 TeV. With a 25 ns bunch spacing, the bunch crossing rate at the LHC design luminosity is approximately 40 MHz, while the event data recording is limited to 200 Hz. The requirement on the trigger system is thus an event rejection factor of at least $2 \cdot 10^5$ while retaining potentially interesting events. Maximum efficiency for selected physics processes is achieved by basing trigger selections on the identification of final state objects like electrons, photons, muons, taus, jets and missing transverse energy. Among these, events containing high-p_T electrons and photons in the final state play a very important role as they can be clearly identified in the detector and they provide signatures for many physics processes in the SM and beyond.

During the first months of 7 TeV collisions, data were selected using the first trigger level (L1) and subsequently processed online by the High Level Trigger (HLT). However, independent of the trigger decisions at the HLT all events were accepted as the L1 output rate was low enough. This particular running mode allows to evaluate the real time performance of the HLT triggers and compare them with the expectations before starting to actively reject events.

The first two sections of these proceedings contain a brief description of the ATLAS detector and its trigger system. This is followed by an overview of the implementation and performance of the electron and photon triggers. Finally the commissioning plans for the higher luminosity running period will be summarized.

2. The ATLAS detector

A detailed description of the ATLAS detector can be found in [1]. The most important components for electron and photon detection are the Inner Detector (ID) tracking system and the electromagnetic (EM) calorimeter.

The ID combines three subdetector with different technologies. From the inside to outside it consists of a three-layer silicon pixel detector, a four-layer semiconductor strip detector and a transition radiation tracker composed of straw-tube proportional chambers interleaved with a radiator material for electron/hadron discrimination. These devices are immersed in a 2 T solenoidal magnetic field and provide tracking for $|\eta| < 2.5$ with a p_T resolution of $\sigma_{p_T}/p_T = 0.05\% p_T \oplus 1\%$.

The electromagnetic calorimeter uses lead plates as absorber interleaved with liquid Argon(LAr)-filled gaps. An accordion geometry guarantees good uniformity with no cracks in the ϕ direction. It is longitudinally segmented into three layers and it is divided in a barrel part and two end-cap components. The first layer has a depth of $\approx 4.3X_0$ with a fine granularity cell structure in η direction ('strip' cells of size $\Delta\eta \times \Delta\phi = 0.003 \times 0.1$ for $|\eta| <1.8$, slightly larger size at larger η) for hadron rejection in the region $|\eta| < 2.5$. It is followed by a second layer of depth $\approx 16X_0$, where electromagnetic showers deposit most of their energy, with squared cells of size $\Delta\eta \times \Delta\phi = 0.025 \times 0.025$. The last layer of depth $2X_0$ is used for evaluating and correcting for the leakage into the hadronic calorimeter. The calorimeter is designed to achieve an energy resolution of $\sigma_E/E = 10\%/\sqrt{E} \oplus 0.7\%$, with E expressed in GeV, up to a pseudorapidity of 3.2; more details on the calorimeter performance are described in [2].

3. The ATLAS Trigger System

In order to achieve the desired data rate reduction, the ATLAS trigger system [3] is structured in three levels. The available detector granularity increases between L1 and HLT; so does the selection complexity at each level. This improves the resolution and the rejection power at the cost of longer decision time and thus lower processing rate.

The first trigger level (L1) is implemented with custom-made electronics and uses reduced-granularity information from the muon and the calorimeter subdetectors. The goal is to rapidly identify signatures from high-p_T muons, electrons/photons, jets, τ-leptons as well as events with large missing or total transverse energy. The decision needs to be taken within 2.5 μs while the information is still in the detector pipeline memories. The maximum L1 output that the system can handle is currently limited to 75 kHz.

The next two levels, Level-2 (L2) and Event-Filter (EF) are software-based and collectively referred to as the High Level Trigger. The L2 is seeded by the L1 through the 'Region-of-Interest' (RoI) mechanism. An RoI is a detector region around the η-ϕ direction identified by L1; the size of the RoI depends on the trigger signature (e/γ, τ, jets) and typically contains 2%-4% of the data in one event. When an event fulfill L1 selection, only full granularity detector information contained in the corresponding RoI is processed by L2. This approach guarantees a data reduction for network transfer and together with fast pattern recognition algorithms, ensure that the average L2 latency is within the allocated limits of 40 ms.

At the EF the full event information is available though currently the RoI approach is being used and offline algorithms are used for reconstruction; the average processing time is $\approx$ 4 s. The maximum EF output rate is 200 Hz due to constraints imposed by computing resources for offline event processing and storage.

3.1. Electron and Photon triggers

A detailed description of the electron and photon trigger signatures and their implementation in the ATLAS trigger strategy can be found in [4]. At L1, there is no distinction between electrons and photons since only calorimetric information is available: the granularity is given by the so-called 'Trigger Tower' (TT), with a size of $\Delta\eta \times \Delta\phi = 0.1 \times 0.1$, whose energy is computed summing up the signals from the cells along the full depth of the EM calorimeter. L1 clusters are created using a sliding window algorithm of 4×4 trigger towers. Events are selected if at least one (or two) clusters are identified above a configurable transverse energy threshold.

At the HLT, the trigger selection proceeds in a series of steps. At L2 a fast calorimeter reconstruction is performed and, in the case of electrons, fast pattern recognition algorithms are also executed for track reconstruction. After each of these steps, hypothesis algorithms are run to select electron and photon candidates and reject background coming predominantly from jets. The signal selection is based on discriminating variables mainly exploiting the differences in the EM shower profile. For electrons tracking information and cluster-track matching quantities can also be used to reject events containing hadron fakes. The EF uses the offline reconstruction algorithms but, due to timing constraints, some corrections like the bremsstrahlung and conversion recovery are not applied. For the signal identification the HLT relies on the same (or a slightly reduced) set of selection criteria used by the offline particle identification [5]. This implementation has been adopted in order to minimize potential sources of trigger inefficiencies on events selected by the offline analyses.

The various signatures that are implemented in the trigger differ mainly in the value of the transverse energy threshold and the identification criteria applied in the hypothesis algorithms.

4. e/γ Trigger commissioning

The following results are based on the analysis of $\approx 0.4\ nb^{-1}$ of stable beam proton collision data collected at $\sqrt{s} = 7$ TeV in April 2010. Given the relatively low peak luminosity (up to $5 \cdot 10^{27} cm^{-2}s^{-1}$), the main source of triggered events was provided by the Minimum Bias Trigger Scintillators (MBTS) [6] selecting collision through charged particles in the forward region ($2.09 < |\eta| < 3.89$). The first level calorimeter trigger was actively used to stream out events and provide an enriched sample of events containing electromagnetic clusters. After the neccesary stability checks the electron and photon HLT chains were activated in so-called *flagging* mode; the events were accepted independently of the HLT decision. The trigger algorithms were executed online and the decision of each chain and the reconstructed trigger objects used for making the decision were recorded into the data stream and could be analyzed offline.

This commissioning mode allows to evaluate the real time performance of the HLT and to crosscheck the trigger decisions with respect to the offline event selection, especially looking for misconfigurations or detector problems that could lead to an early rejection of events otherwise accepted by an offline analysis.

A detailed comparison of the trigger reconstruction performance with MC simulation expectations was also performed using a sample of non-diffractive minimum bias events generated with Pythia and processed with the full ATLAS simulation and reconstruction software.

4.1. L1 performance

The L1 calorimetric triggers proved a very good operational stability and reliability over the entire data taking period as reported in Figure 1. It shows the fraction of events collected with the MBTS trigger passing each of the three lower EM signatures, L1_EM2, L1_EM3 and L1_EM5 with a transverse energy (E_T) threshold approximatively at 3, 4 and 6 GeV[1]. At an

[1] transverse energy at L1 is the result of signal ADC conversion where 1bit corresponds to a deposited energy of $\approx$ 1GeV. The notation EM2 indicates that the bit count is required to be ≥ 3 GeV

Journal of Physics: Conference Series **293** (2011) 012027 doi:10.1088/1742-6596/293/1/012027

instantaneous luminosity of $10^{27}cm^{-2}s^{-1}$ the L1_EM2 was found to be 1.5 Hz. The visible step in Figure 1 is due to the update of the timing calibration constants performed to improve the signal reconstruction and energy resolution performance (see [7]).

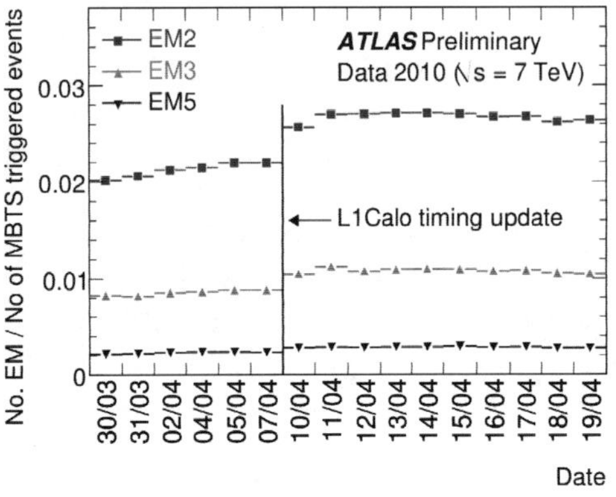

Figure 1. Fraction of events collected by the MBTS trigger satisfying the 3 lowest E_T L1 EM signatures as a function of time.

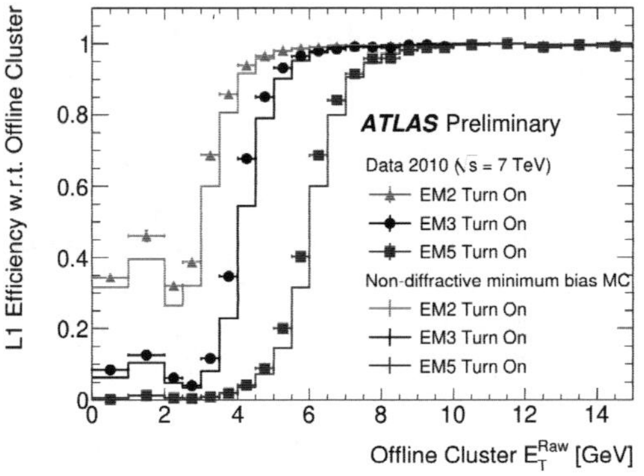

Figure 2. L1 efficiency for the 3 lowest E_T L1 EM signatures as a function of the raw offline cluster E_T.

The energy resolution has been evaluated by comparing the L1 electromagnetic clusters to the offline ones. Figure 2 shows the L1 efficiency as function of the raw[2] offline cluster E_T, computed requiring an angular matching criteria with $\Delta R = \sqrt{(\Delta\eta)^2 + (\Delta\phi)^2} < 0.15$ between the offline and the trigger objects. No selection on offline cluster shower shapes has been applied; the offline cluster is required to have $|\eta| < 2.5$ and not coming from the transition region between the barrel and end-cap calorimeters ($1.37 < |\eta| < 1.52$).

A good agreement in the shape of the trigger turn on is found for the 3 lowest threshold L1 triggers: the threshold position in MC is shifted towards higher value by less than 0.5 GeV. A few GeV above the threshold these triggers become fully efficient thus minimizing any bias on offline sample.

4.2. HLT performance

In order to evaluate the HLT performance, the level of agreement between offline and trigger selection variables needs to be assessed. For the following analysis the HLT electron and photon candidates were matched to an offline electron candidate, requiring a minimal distance of $\Delta R < 0.15$; the transition region of the calorimeters was again excluded. HLT chains were seeded by the L1 calorimetric trigger and the minimum E_T of the sample reflects the lower L1 threshold of 3 GeV. For the selected sample the distributions of kinematics and signal/background discriminating variables have been compared at the different trigger levels with the offline distributions as well as with the expectations from MC simulations. Since no identification criteria have been applied to offline reconstructed candidates, the distributions are still dominated by fakes. For the comparisons presented in the following, the Monte Carlo distributions are normalised to the number of events in the data sample.

As an example, Figure 3 and 4 show the distribution of two of the e/γ identification variables at EF for data and MC simulation. The first, called R_η is defined as the ratio of the energy

[2] The E_T^{Raw} is defined as the uncalibrated cluster transverse energy of the cluster before correction for losses in the upstream material that are not applied at L1.

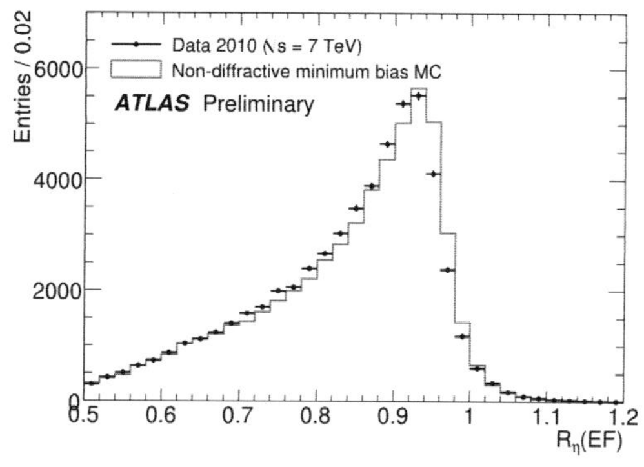

Figure 3. Data/MC comparison for the R_η identification variable computed at EF.

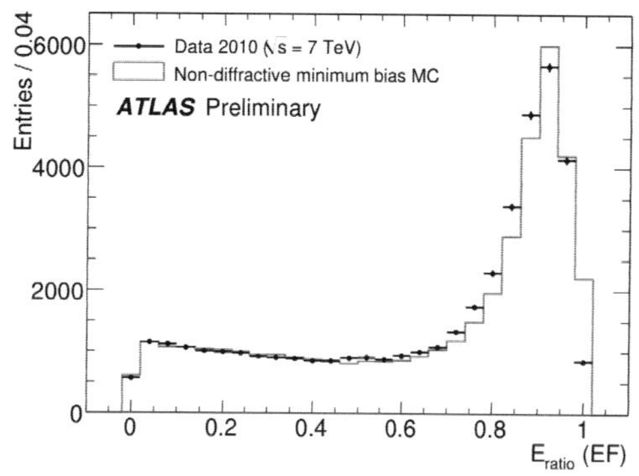

Figure 4. Data/MC comparison for the E_{ratio} identification variable computed at EF.

deposited in $\Delta\eta \times \Delta\phi = 3 \times 7$ cluster cells of the second layer of the EM calorimeter divided by the energy found in 7×7 cells. This quantity peaks towards one for electrons and photons which have showers well contained within 3×7 cells; hadronic showers have a broader spectrum[3]. A second variable, shown in Figure 4, is called E_{ratio} and it is computed as the difference between the highest and the second highest cell energy deposit in the first layer of the EM calorimeter divided by the sum of the two energy deposits. This variable peaks at one for single particle initiated showers and it tends towards lower values for showers initiated by more than one particle like a photon pair coming from π_0.

The figures show a reasonable agreement between data and simulations and similar behavior is seen for other variables. The small shift in the data to slightly lower values for R_η can partly be explained by the MC cross-talk modelling in the second sampling while the shift of the first layer variable are traced back to some missing material description in the current MC simulation. The same shifts are also observed for offline reconstructed objects and documented in [5].

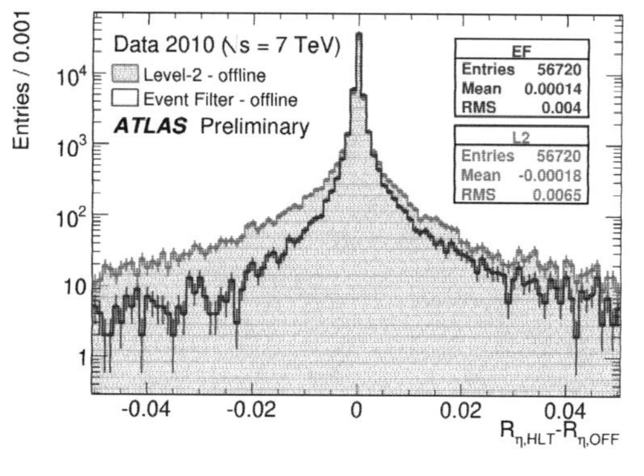

Figure 5. HLT trigger levels resolution on R_η variable with respect to offline reconstruction.

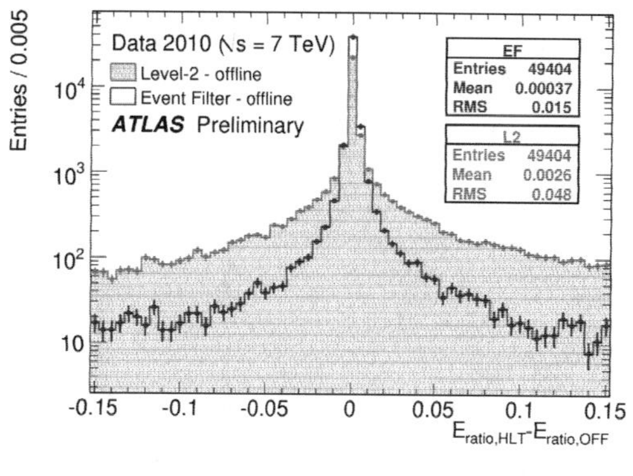

Figure 6. HLT trigger levels resolution on E_{ratio} variable with respect to offline reconstruction.

Another crucial check is the agreement between the different trigger levels and the offline reconstruction on the signal identification variables; this is a key ingredient in understanding

[3] Values of R_η above 1 are due to the noise contribution that can lead to negative signals in some cells.

the trigger effect on the offline analyses. Examples are provided in Figure 5 and 6 where the difference between offline and HLT levels on the R_η and E_{ratio} variables are shown for selected objects. For most of the clusters, trigger and offline calculations agree well and the broader resolution at L2 is due to the faster, and thus less refined, clustering algorithm and calibration. For both trigger levels the resolution is below few percent.

The presented results, for the example discussed above and also for the other selection variables not shown, provides evidence that the initial trigger selection criteria implementation developed using MC simulations are safe and robust for the HLT to be activated in rejection mode.

5. Conclusions

The calorimeter based performance of the ATLAS electron and photon triggers during the first months of LHC collisions at 7 TeV was presented in this paper. In the first phase of data taking at 7 TeV, data were collected with the L1 trigger since the event rate at the initial luminosity was low enough to be written directly to mass storage. For commissioning purposes the HLT was running online without rejecting any events.

The L1 e/γ triggers show good performance in terms of stability and high efficiency with respect to offline candidates above the trigger threshold. The selected events have been used for the first analysis concerning electrons and photons (W/Z, inclusive electrons and prompt photons). Studies on the performance of the software based trigger were crucial for preparing the HLT to start working in rejection mode for LHC luminosities above $10^{29}cm^{-2}s^{-1}$ when the output of the lowest L1 EM triggers needs to be limited in order to keep the data saving rate within the requirements. Data/MC comparisons showed that the distribution for the selection variables in the trigger are reasonably well modeled by the simulation. A good agreement between trigger and offline variables has been found showing that the e/γ triggers are working correctly; electron and photon offline candidates are selected by the HLT with minimal bias on offline reconstruction.

Based partly on the studies reported in this paper, the decision has been taken to activate the first low E_T HLT chains in rejection mode at the end of May 2010 when the LHC luminosity exceeded for the first time $10^{29}cm^{-2}s^{-1}$. With increasing luminosity more triggers will apply rejection at HLT in order to better fill the allocated bandwidth while still keeping good efficiency with respect to events containing low and medium energy electrons and photons. With higher integrated luminosity ($> 1pb^{-1}$) a sample of isolated electrons from $J/\Psi \to ee$, $W \to e\nu$ and $Z \to ee$ will be collected. This will allow to repeat the performance measurement described in this note on a very pure signal sample and evaluate with data-driven methods the trigger efficiencies to be used in physics measurements.

References

[1] The ATLAS Collaboration, G. Aad *et al.*, *The ATLAS Experiment at the CERN Large Hadron Collider*, JINST 3 (2008).
[2] The ATLAS Collaboration, H. Zhang, *The ATLAS Liquid Argon Calorimeter: overwiew and performance*, these proceedings.
[3] The ATLAS Collaboration, G. Aad *et al.*, *Expected performance of the ATLAS experiment: detector, trigger and physics*, e-Print: arXiv:0901.0512 [hep-ex], (2008).
[4] The ATLAS Collaboration, G. Aad *et al.*, *Performance of the Electron and Photon Trigger in p-p Collisions at $\sqrt{s}$=900 GeV*, ATLAS-CONF-2010-022, Geneva, (2010).
[5] The ATLAS Collaboration, D.Banfi , *Electron and photon reconstruction and identification with the ATLAS Detector*, these proceedings.
[6] The ATLAS Collaboration, G. Aad *et al.*, *Performance of the Minimum Bias Trigger in p-p Collisions at $\sqrt(s) = 900$ GeV*, ATLAS-COM-CONF-2010-025, Geneva, (2010).
[7] The ATLAS Collaboration, J. T. Childers, *ATLAS Level-1 Calorimeter Trigger Hardware: Initial Timing and Energy Calibration*, these proceedings.

Multiresolution Jet Reconstruction with FFTJet

Igor Volobouev

Texas Tech University, Department of Physics, Box 41051, Lubbock, Texas, USA 79409

E-mail: `i.volobouev@ttu.edu`

Abstract. This article introduces a new jet reconstruction method for particle physics. Event energy flow analysis is performed in two stages. Pattern recognition stage is applied first, utilizing multiresolution filtering techniques in the Fourier domain. Jet energy determination follows, conditional upon the choice of signal topology. The resulting algorithm is global, efficient, collinear and infrared safe, and allows the user to identify and avoid the event topology bifurcation points when energy reconstruction is performed. The performance of the new approach is better than that of commonly used jet algorithms such as k_{T}, anti-k_{T}, and seedless cone. An open source implementation is available at `http://projects.hepforge.org/fftjet/`.

1. Introduction

The necessity of improving pattern recognition capabilities of jet reconstruction algorithms has been recognized both in the context of multijet, high occupancy events and in the cases when massive particles decaying hadronically via electroweak interaction (W and Z bosons, top quarks) are sufficiently boosted so that the energy flow of their decay products can not be easily partitioned into well-separated jets. Due to software availability, LHC-targeted studies of particle physics processes of this kind utilized predominantly sequential recombination techniques [1, 2, 3, 4, 5]. However, sequential recombination algorithms are inherently more sensitive to fluctuations in the event energy flow than global methods, and clustering stability properties of sequential recombination algorithms are not fully understood. Changes in the recombination sequence at early clustering stages are not guaranteed to dissipate and may affect clusters obtained at later stages — this artifact led to the introduction of such concepts as "passive" and "active" jet areas which turn out to be distinct [6]. On the other hand, commonly used global reconstructions algorithms — such as seedless iterative cone and its various seeded variations — suffer from another important deficiency inherent in the cone-based jet reconstruction. This deficiency manifests itself as the pattern recognition ambiguity illustrated in Figure 1. Two energy deposits of similar magnitude separated by a distance larger than R but smaller than $2R$ produce three stable cone centers whose positions are shown with the arrows at the bottom of the figure. In various implementations of cone-based jet reconstruction procedures, this problem is usually addressed by the "split-merge" stage which happens after the stable cone locations are determined. During this stage, jets are merged if the energy which falls into the common region exceeds a predefined fraction of the energy of the jet with smaller magnitude. Even if the search for stable cones is performed in the infrared and collinear safe manner, the outcome of the split-merge stage is often unstable because the decision on whether to merge the two jets depends on the minute details of the energy deposition structure. Seeded

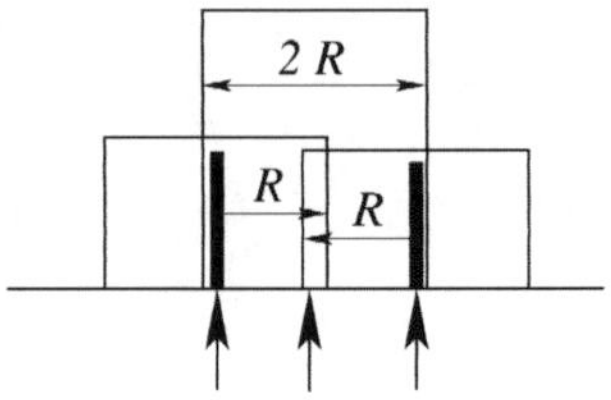

Figure 1. The locations of the three stable clusters reconstructed by the ideal iterative cone algorithm from two energy deposits are shown by the arrows at the bottom.

cone algorithms suffer from a variety of additional problems which lead to collinear and infrared instabilities of the reconstructed jet configurations [7].

2. Improved Pattern Recognition

The jet reconstruction method described here is based on a global technique which does not suffer from the problems mentioned in the previous section. This method was originally inspired by Refs. [8] and [9] and initially proposed in [10].[1] The study by Cheng [8] establishes an important connection between the iterative cone algorithm[2] and kernel density estimation (KDE) [16]. Cheng proves that the locations of stable cone centers correspond to modes (peaks) of the energy density built in the η-φ space using KDE with the Epanechnikov kernel. That is, all such centers can be found by convolving the empirical energy density

$$\rho_{\mathrm{emp}}(\eta, \varphi) = \sum_i \varepsilon_i \delta^2(\eta - \eta_i, \varphi - \varphi_i)$$

with the function

$$\mathrm{Epanechnikov}(\eta, \varphi) = \begin{cases} 1 - (\varphi^2 + \eta^2)/R^2, & \varphi^2 + \eta^2 < R^2 \\ 0, & \varphi^2 + \eta^2 \geq R^2 \end{cases}$$

and then funding all local maxima of the convolution. Here, ε is an energy variable which can stand for transverse momentum or transverse energy of a particle, calorimeter tower, *etc.* η and φ define the direction of the energy deposit as seen from the interaction point (depending on the choice of ε, η can stand for either rapidity or pseudorapidity, while φ is the azimuthal angle). $\delta^2(\eta, \varphi)$ is the two-dimensional delta function. The sum in the expression for ρ_{emp} is performed over all energy deposits expected to be jet constituents (typically, leptons and photons produced in the hard scattering process are excluded). R in the Epanechnikov kernel corresponds to the cone radius in the η-φ space.

Following Cheng, the reason for the pattern recognition ambiguity in the cone-based jet reconstruction can be explained as follows: it occurs because the sum of two Epanechnikov kernels placed at the locations of the deposits has a spurious peak in the middle, as shown in Figure 2 (left). Fortunately, there is a variety of kernels which do not suffer from this problem. In particular, the Gaussian kernel produces either two (narrow kernel) or one (wide kernel) peaks, as shown. Even though one still has to address the question of choosing the kernel width, the Gaussian kernel has a very important advantage: the whole split-merge stage is no longer necessary.

[1] Similar clustering and pattern recognition schemes have been introduced in various disciplines [11, 12, 13, 14].
[2] The iterative cone algorithm is known as the "mean shift" algorithm [15] in the pattern recognition literature.

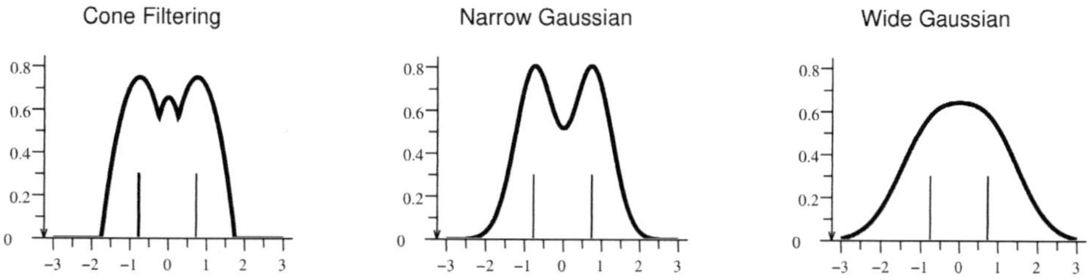

Figure 2. The convolution of two energy deposits with the Epanechnikov kernel (left) and with two Gaussian kernels of different width.

3. Multiresolution Description of the Event Energy Flow

An intelligent choice of the kernel width (or the R parameter in the k_{T} and cone algorithms) can not be performed until some assumptions are made about the expected jet shapes. In fact, optimal choice will be different for different signals. For example, a data analysis which searches for high energy dijet events with two well-separated jets is likely to make very different assumptions about jets from a data analysis which looks for $t\bar{t}$ events in the all-hadronic, 6-jet mode. Moreover, the optimal width is not necessarily the same for every jet in an event, as low momentum jets tend to have wider angular profiles, especially in the presence of magnetic field. Because of this, it is instructive to look at the jet structure of an event using a variety of kernel width, cone radius, *etc.* choices — all such parameters will be referred to by a general term "resolution scale" for the remainder of this paper. In the limit of continuous resolution scale one arrives at the description of event energy flow known as "mode tree" in the nonparametric statistics literature or "scale space image representation" in the computer vision theory.

Energy flow peaks found at different resolution scales in the η-φ space are used as "preclusters" which determine possible initial jet locations. In order to permit an effective use of the information contained in the multiresolution energy flow picture, preclusters are combined into a single hierarchical structure called "clustering tree". This tree is built using a distance function[3]. A precluster found at some resolution scale s_i is assigned a parent from the previous (larger) resolution scale s_{i-1} as follows: the distance between the precluster at the scale s_i is calculated to all preclusters at the scale s_{i-1}. The precluster at the scale s_{i-1} with the smallest such distance becomes the parent. This agglomeration strategy follows the approach of Ref. [9] and has the advantage that the obtained tree structure can also be utilized as a balltree [17]. A simple tree-like structure is obtained if the pattern recognition kernel is chosen in such a manner that the number of preclusters does not increase when the resolution scale increases, no matter how the event energy flow looks like. Although the 2-d Gaussian kernel is known to violate this condition on some rare occasions [18], it appears to perform very well for all practical purposes. An example clustering tree image can be found in Ref. [19].

Once the parent/daughter relationships are established between preclusters found at different resolution scales, dependence of various precluster characteristics on the scale parameter can be analyzed. Perhaps, the most useful such characteristic is the speed with which the precluster η-φ location changes as a function of scale (scale space drift). The drift speed becomes relatively high when two preclusters are about to merge. This indicates that the resolution scale is close to a bifurcation point — the scale at which two smaller preclusters form a bigger one. Near the bifurcation point the locations of the affected preclusters become very sensitive to small changes

[3] A simple distance is given by $d = \sqrt{\left(\frac{\Delta\varphi}{h_\varphi}\right)^2 + \left(\frac{\Delta\eta}{h_\eta}\right)^2}$, with some constant parameters h_η and h_φ. More complicated functions which take into account resolution scales and peak magnitudes can be constructed as well.

in the event energy flow.[4] This problem results in an increased uncertainty of jet energy and direction determination. The bifurcation points can be avoided by detecting them in the scale space with the clustering tree, and by using slightly modified resolution scales in case such points are found.

4. Jet Reconstruction Strategies

The information contained in the multiresolution energy flow description permits multiple strategies for jet reconstruction. The common technique of choosing a single best overall resolution scale according to some optimization criterion[5] can be easily improved upon by augmenting it with the bifurcation point detection method. A number of less traditional techniques can be employed as well:

- The jet configuration can be selected according to a clustering stability criterion. In the multiresolution picture, it is natural to assume that jet configuration which persists through a wide range of pattern recognition scales represents a salient feature of the event energy flow. Therefore, clustering stability can be expressed in terms of the configuration "lifetime" in the scale space. A reasonable lifetime definition is $J^\alpha \log(s_{\max}/s_{\min})$.[6] Here, J is the number of preclusters (perhaps, satisfying some imposed selection criteria) present in the clustering tree at each resolution scale s: $s_{\min} \le s \le s_{\max}$. Parameter α can be chosen empirically depending on the process under study (normally $0 < \alpha < 1$).
- A global resolution scale can be chosen on event-by-event basis in such a manner that the number of preclusters found in the clustering tree for that scale corresponds to the number of jets expected in the signal. The scale space lifetime of the expected jet configuration should be checked and events with unsatisfactory lifetimes should be discarded.
- The resolution scale can be adapted for each jet separately, in a manner that is consistent with the expected event topology. For example, if the process under study is expected to produce M jets at the leading order perturbation theory, the jet configuration can be chosen according to the principle by which the minimum transverse energy among M hardest jets is maximized over all possible cluster sets logically permitted by the clustering tree structure (*i.e.*, over all sets of M preclusters formed in such a way that no set member can be a clustering tree ancestor of another set member).
- Scale space differential blob detectors [23] can potentially be used for jet identification.
- Nontrivial clustering patterns can be identified in the signal, and similar patterns can be searched for in the clustering tree. For example, boosted resonances, such as W bosons or top quarks, are expected to produce one wide jet at higher resolution scales which has a prominent substructure at lower scales.

The best method will be process and detector-dependent. Perhaps, simple recipes for choosing an optimal jet reconstruction strategy will emerge as more experience with these techniques is accumulated.

5. The FFTJet Package

The FFTJet software package [19] implements the multiresolution pattern recognition ideas outlined in the previous sections. Package users are expected to reconstruct jets using the following sequence of steps:

[4] Bifurcation points are present in every jet reconstruction algorithm, including those which are normally considered to be infrared and collinear safe. Most algorithms do not have the capability to detect such points.

[5] For example, the fraction of events in which the number of reconstructed jets above certain p_T threshold equals the number of partons produced at the leading order perturbation theory can be maximized for the signal of interest.

[6] More powerful but significantly more complicated definition of the lifetime in the scale space is proposed in [22].

(i) The event energy flow is discretized using an equidistant grid in the η-φ space. The grid cell size is normally chosen in a manner consistent with the granularity of the detector calorimeter.

(ii) The discretized energy distribution is convolved with a kernel function $K(\eta, \varphi, s)$. The resolution scale parameter s determines the width and, possibly, the shape of the kernel. Many standard kernel functions are included in the FFTJet package, and user-defined kernels can be seamlessly added as well. The convolutions are efficiently performed by Discrete Fast Fourier Transform (DFFT).

(iii) The peaks of the convolved energy distributions are found. These are potential preclusters.

(iv) Preclusters with small magnitudes are eliminated in order to suppress the calorimeter noise.

(v) Steps ii through iv are repeated as many times as necessary using different values of s. The resulting preclusters are arranged in the clustering tree structure. Facilities for clustering tree sparsification and bifurcation point avoidance are provided.

(vi) Using the clustering tree information and assumptions about the signal spectrum, a decision is made about the event topology by choosing a set of preclusters. These preclusters are passed to the jet energy reconstruction stage.

(vii) Jet constituents are determined as follows. The event is viewed as a collection of energy deposits characterized by their direction (η, φ) and energy variable ε. Depending on the environment in which the code is used, these deposits can originate from detector calorimeter cells, reconstructed tracks, Monte Carlo particles, *etc.* A cluster membership function $M_j(\eta - \eta_j, \varphi - \varphi_j, \varepsilon, s_j)$ is associated with each precluster j at angular coordinates (η_j, φ_j) and scale s_j. There is also a membership function for the unclustered energy/underlying event. The cluster membership functions are evaluated for every energy deposit in the event. In the "crisp" clustering scenario, an energy deposit is assigned to the jet whose membership function for this deposit is the largest. In the "fuzzy" scenario, the deposit is split between all jets with weights proportional to their respective membership function values (the sum of all weights is normalized to 1 to ensure energy conservation).

(viii) Jet energies and directions are calculated according to one of the standard energy recombination schemes using weights determined in the previous step.

In order to further improve the jet energy resolution, the last two steps of the algorithm can be applied iteratively. In such a procedure, the jet directions (η_j, φ_j) and the membership function scales s_j are updated at each iteration using reconstructed jets from the previous iteration until some convergence criterion is satisfied.[7]

The computational complexity of the pattern recognition stage (steps i through v) is $O(SN \ln N)$, where N is on the order of the number of towers in the detector calorimeter and S is the user-selectable number of angular resolution scales (cone and k_{T} algorithms use only one resolution scale). This complexity is *independent from the detector occupancy* and therefore allows for predictable execution times which can be important for online use. The computational complexity of the jet energy reconstruction stage is $O(JMI)$, where J is the number of jets found, M is the number of objects (4-vectors) used to describe the event energy flow ($M \leq N$), and I is the number of times the last two steps of the algorithm are iterated (non-iterative version with $I = 1$ is often sufficient).

FFTJet permits arbitrary pattern recognition kernels and jet membership functions. The ability to use non-circular jet shapes translates into significant advantages in jet energy resolution (for the same expected instrumental noise and pile-up) when the method is used with calorimetric energy information provided by particle detectors which employ strong magnetic fields in their tracking systems. In such detectors, low-p_{T} jets experience significant widening in φ by the time

[7] This iterative procedure is known as "generalized mean shift" or "expectation maximization" algorithm [20, 21].

particles reach the calorimeter. As it was shown in [19], a simple replacement of Epanechnikov pattern recognition kernel by 2-d Gaussian and use of elliptical instead of circular cone results in a $\sim 30\%$ reduction of algorithm-related jet energy reconstruction uncertainty (compared to seedless cone algorithm) for a typical cone size of $R = 0.5$ in a detector with 3.8 T magnetic field. The gains over k_{T} and anti-k_{T} algorithms are similar. It is also apparent that development of more sophisticated cluster membership functions (which can, for example, mimic expected transverse energy profile of a jet) will result in further improvements in the method efficiency and jet energy resolution.

6. Acknowledgements

The author thanks the Jet Algorithms Group of the CMS collaboration for comments and discussions. The development of the FFTJet package was supported in part by the US Department of Energy grant DE-FG02-95ER40938.

References

[1] G.P. Salam, "Towards Jetography", The European Physical Journal C, Vol. 67, pp. 637-686, 2010.
[2] J. M. Butterworth, A.R. Davison, M. Rubin, and G.P. Salam, "Jet substructure as a new Higgs search channel at the LHC", Phys. Rev. Lett., Vol. 100, 242001 (2008).
[3] D. Krohn, J. Thaler and L. T. Wang, "Jet Trimming", J. High Energy Phys. JHEP 1002, 084 (2010).
[4] S.D. Ellis, C.K. Vermilion, and J.R. Walsh, "Techniques for improved heavy particle searches with jet substructure", Phys. Rev. D, Vol. 80, 051501 (2009).
[5] S.D. Ellis, C.K. Vermilion, and J.R. Walsh, "Recombination Algorithms and Jet Substructure: Pruning as a Tool for Heavy Particle Searches", Phys. Rev. D, Vol. 81, 094023 (2010).
[6] M. Cacciari, G.P. Salam, G. Soyez, "The catchment area of jets", J. High Energy Phys. JHEP04 (2008) 005.
[7] S.D. Ellis *et al.*, "Jets in hadron-hadron collisions", Progress in Particle and Nuclear Physics, Vol. 60, pp. 484-551, 2008.
[8] Y. Cheng, "Mean Shift, Mode Seeking, and Clustering", IEEE Trans. Pattern Analysis and Machine Intelligence, Vol. 17, pp. 790-799, 1995.
[9] M.C. Minnotte, D.W. Scott, "The Mode Tree: a Tool for Visualization of Nonparametric Density Features", J. Comp. Graph. Statist., Vol. 2, pp. 51-68, 1993.
[10] I. Volobouev, "Density-Based Clustering and Jet Reconstruction", presentation at the MC4LHC Workshop, CERN, July 2006.
[11] Y. Wong, "Clustering Data by Melting", Neural Computation, Vol. 5, pp. 89-104, 1993.
[12] S.J. Roberts, "Parametric and non-parametric unsupervised cluster analysis", Pattern Recognition, Vol. 30, pp. 261-272, 1997.
[13] E. Nakamura, N. Kehtarnavaz, "Determining number of clusters and prototype locations via multi-scale clustering", Patt. Rec. Lett., Vol. 19, pp. 1265-1283, 1998.
[14] Y. Leung, J.-S. Zhang, Z.-B. Xu, "Clustering by Scale-Space Filtering", IEEE Trans. Pattern Analysis and Machine Intelligence, Vol. 22, pp. 1396-1410, 2000.
[15] K. Fukunaga, L.D. Hostetler, "The Estimation of the Gradient of a Density Function, with Applications in Pattern Recognition", IEEE Trans. Information Theory, Vol. 21, pp. 32-40, 1975.
[16] D.W. Scott, "Multivariate Density Estimation: Theory, Practice, and Visualization", Wiley, 1992.
[17] S.M. Omohundro, "Five Balltree Construction Algorithms", ICSI Technical Report TR-89-063, 1989.
[18] M.A. Carreira-Perpinan, C.K.I. Williams, "An isotropic Gaussian mixture can have more modes than components", Technical Report EDI-INF-RR-0185, School of Informatics, University of Edinburgh, UK, 2003.
[19] I. Volobouev, "FFTJet: A Package for Multiresolution Particle Jet Reconstruction in the Fourier Domain", arXiv:0907.0270v1, 2009.
[20] W.H. Press, S.A. Teukolsky, W.T. Vetterling, and B.P. Flannery, "Numerical Recipes: the Art of Scientific Computing", 3rd ed., Cambridge University Press, 2007.
[21] M.A Carreira-Perpinan, "Gaussian Mean-Shift Is an EM Algorithm", IEEE Trans. Pattern Analysis and Machine Intelligence, Vol. 29, pp. 767-776, 2007.
[22] T. Lindeberg, "Effective Scale: A Natural Unit for Measuring Scale-Space Lifetime", IEEE Trans. Pattern Analysis and Machine Intelligence, Vol. 15, pp. 1068-1074, 1993.
[23] T. Lindeberg, "Feature Detection with Automatic Scale Selection", Int. J. of Computer Vision, Vol. 30, pp. 79-116, 1998.

Journal of Physics: Conference Series **293** (2011) 012029 doi:10.1088/1742-6596/293/1/012029

Simulation study for Forward Calorimater in LHC-ALICE experiment

Y.Hori, H.Hamagaki, T.Gunji

Center for Nuclear Study, Graduate School of Science, University of Tokyo, 7-3-1 Hongo, Bunkyo, Tokyo 113-0033, Japan

E-mail: yhori@cns.s.u-tokyo.ac.jp

Abstract. We propose Forward Calorimeter (FOCAL) as an upgrade plan of ALICE detector at LHC. FOCAL will contribute to small-x physics in p+Pb collisions and Quark Gluon Plasma (QGP) physics in Pb+Pb collisions. Our conceptual desgin of FOCAL consists of SiW sandwith-type calorimeter and Si Strip layers to measure prompt γ and $\pi^0 \rightarrow \gamma+\gamma$ up to 200 GeV energy, which corresponds to < 20 (10) GeV/c transverse momentum at $\eta = 3$ (4).

1. Introduction

The physics goal of FOCAL is the understanding of the parton structure inside the proton and nuclei at small-x, especially, the gluon saturation effect like Color Glass Condensate (CGC). For this purpose, FOCAL measures prompt γ and neutral mesons at the forward rapidity region in p+p and p+Pb collisions. FOCAL will also make an unique contribution for QGP physics by measuring inclusive γ, high p_T π^0 and jets at the forward rapidity region in Pb+Pb collisions.

Parton Distribution Functions (PDF) of the proton were precisely measured by the experiment at HERA and the rapid growth of the small-x gluon density was observed. The growth of the gluon density inside the hadron with decreasing x is ultimately expected to be balanced by the gluon-gluon fusion process. These nonliear effects lead to the gluon saturation at the small-x region [1]. This phenomenon is expected to occur when the area occupied by the gluons becomes equal to the area of transverse size of the hadron πR^2. This situation is expressed by the saturation momentum Q_s as $\Lambda^2_{QCD} \ll Q^2 < Q^2_s(x) \sim \alpha_s x G(x, Q^2)/\pi R^2 \propto A^{1/3} x^{-0.3}$. It is expected that Q^2_s reachs ~ 10 GeV^2 at LHC. The situation $\Lambda^2_{QCD} \ll Q^2_s$ will be realized in extremely high energy collisions at LHC.

2. Prompt γ measurements at the forward rapidity region

Experimentally, saturation effects on small-x gluons are expected to be revealed in measurements of forward particle production. For a $2 \rightarrow 2$ parton scattering, the minimum x probed in a process with a particle momentum p_T and its pseudo-rapidity η is given as

$$x_2^{min} = \frac{x_T e^{-\eta}}{(2 - x_T e^{\eta})}, x_1^{min} = \frac{x_2 x_T e^{-\eta}}{(2x_2 - x_T e^{\eta})} \tag{1}$$

where $x_T = 2p_T/\sqrt{s}$. Thus, x_2^{min} decreases by a factor of ~ 10 every 2 units of rapidity [2].

The production of prompt γ at forward rapidity is sensitive to the small-x gluon distribution.

The dominated production processes at LO level are compton scattering and annihilation process. Compton process is in particular interesting because small-x gluon and large-x (valence) quark are relevant if prompt γ is produced at the forward rapidity region. Figure 1 shows the accesible x range of the partons by the prompt γ measurement at $\eta = 3 \sim 3.5$ in 8.8 TeV collisions calculated by PYTHIA simulation. The x range that can be probed by prompt γ is $10^{-3} \sim 10^{-5}$, which has hardly been probed even for the proton. The left graph of Figure 2 shows accesible range of x and Q^2 by FOCAL at LHC and the other experiments.

The center and right graphs of Figure 2 show the cross section and annual yield of prompt

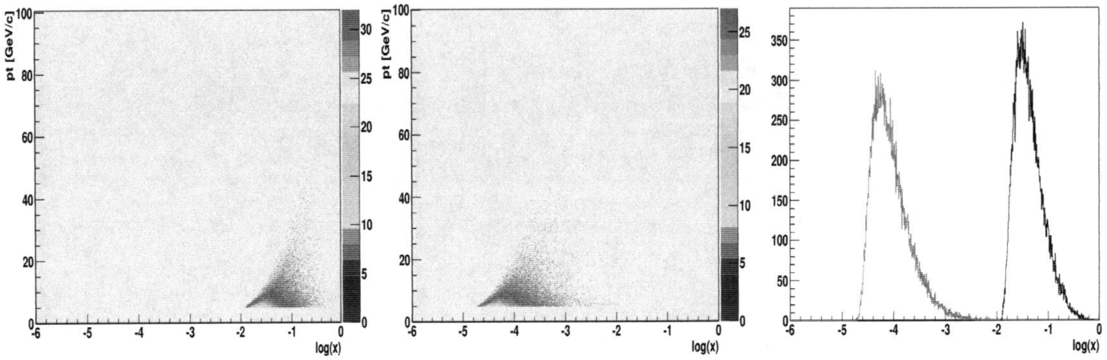

Figure 1. Left and Center: p_T and $x_1(x_2)$ range of relevant partons of γ production at $\eta = 3 \sim 3.5$, Right: Projection to the x-axis of the 2 left plots

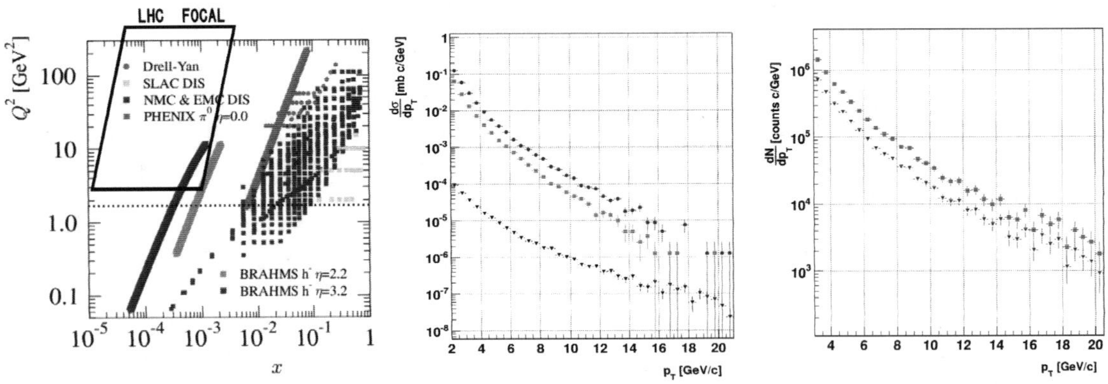

Figure 2. Left: Accesible range of x and Q^2 by FOCAL at LHC, Center: p_T spectrum at $\eta = 3 \sim 4$ in 8.8 TeV p+p collisions (blue - π^0, pink - γ from π^0, black - prompt γ), Right: Annual yield of prompt γ at $\eta = 3 \sim 4$ in p+p (blue) and p+Pb (red) collisions for a standard year of running at LHC, which does not include any nuclear effects

γ at $\eta = 3 \sim 4$ in 8.8 TeV p+p collisions. A width of p_T binning is 0.5 GeV/c and error bars are from simulation statistics. These graphs show that yields of prompt γ and π^0 are enough to be measured, but the ratio of prompt γ to γ from neutral mesons is quite small.

3. Conceptual design of Forward Calorimeter and Basic performance

Considering the physics requirements and the detector performance, the η coverage of FOCAL should be 3∼4 and the distance from interaction point should be about 4.5 m, where is between

ALICE L3 magnet wall and TPC. The layout of FOCAL is summarized in Table 1.

In order to measure γ and neutral mesons at the forward rapidity regions, FOCAL must

Distance from IP	4.5 m	η coverage	$3 \sim 4$
Outer Radius	45 cm	ϕ coverage	2π
Inner Radius	15 cm	Area	~ 0.6 m^2

Table 1. Layout of FOCAL

have a large dynamic range (γ energy < 200 GeV, which corresponds to $p_T < 20$ (10) GeV/c at $\eta = 3$ (4)) and highly granularity (multiplicity ~ 1 particle/cm^2 at $\eta = 4$ according to the HIJING simulation). FOCAL also has to have good separation ability of 2 γ from high energy π^0 decays. A minimum distance of 2 γ from 100 (200) GeV π^0 is $\sim$10 (6) mm.

These requirements can be fulfilled by a hybrid use of SiW sandwich-type calorimeter and Si

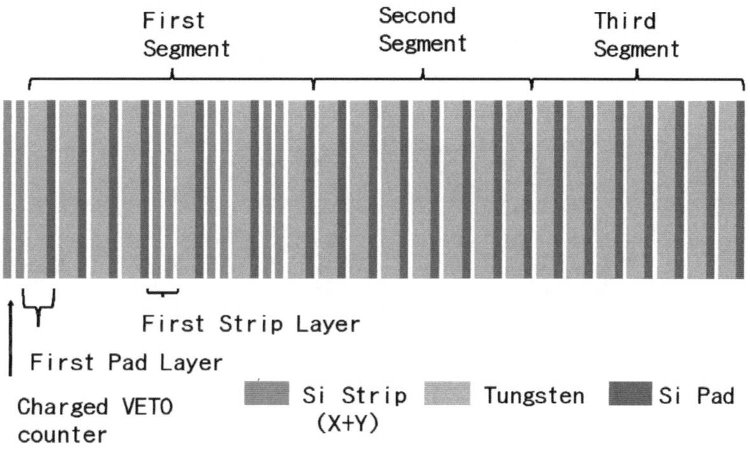

Figure 3. Conceptual design of FOCAL module

Strip layers as shown in Figure 3 [3]. The thickness of each Si Pad detector is 0.5 mm and that of W absorber is 3.5 mm. It has $21X_0$ depth and has 3 longitudinal segments. Each segment has 7 layers of Si Pad detector and W absorber. The Si Pad detector is laterally segmentized by 1 cm x 1 cm. A few layers of the Si Strip detector is used to separate 2 γ from high energy π^0. The thickness of Si Strip layer is 0.3 mm and the strip pitch is 0.5$\sim$1 mm.

Figure 4 shows the longitudinal shower shapes and sampling fraction at each segment calculated by GEANT 4 simulation. A shower maximum depth is about $5 \sim 9$ X_0 and total sampling fraction is almost constant up to 200 GeV γ. Figure 5 shows χ^2 distributions of longitudinal deposited energy of γ and π^+. γ shower is effectively identified by a cut for χ^2 distribution. Other basic performances are listed in Table.2.

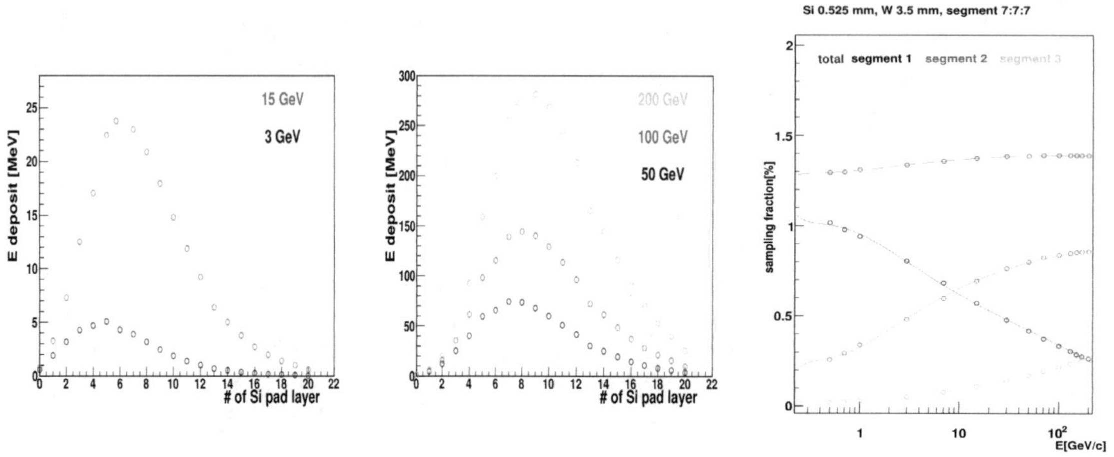

Figure 4. Left and Center: Longitudinal shape of γ shower with various energy, Right: Sampling fraction at each segment for γ shower

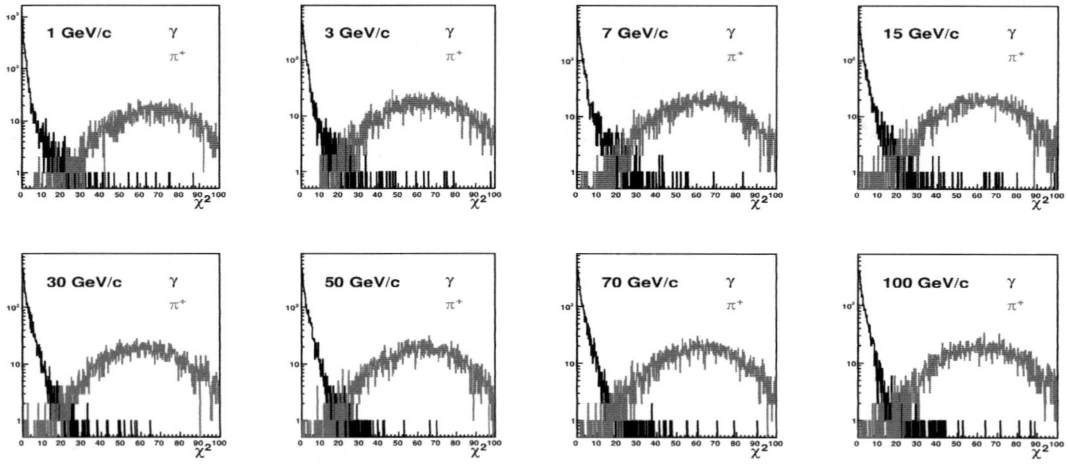

Figure 5. χ^2 distribution of γ and π^+ shower shape with various energy

Cell occupancy in p-Pb collisions	$< 10\%$
Detection efficiency	$> 95\%$
Linearity	$< 1\%$
Sampling fraction	1.4%
Energy resolution	$18/\sqrt{E}$ %
Position resolution at 10 GeV γ	1.5 mm
Charged Hadron rejection factor	250 at γ eff.90%

Table 2. Basic performance of Si Pad Detector

4. π^0 measurement

π^0 with 7 $\sim$ 70 GeV energy can be efficiently reconstructed by the invariant mass method using Si Pad detector and detection efficiency is $>$ 60% as shown in the left graph of Figure 6. The efficiency decreases above 60 GeV π^0 because 2 γ from π^0 are merged into same pad. The merging probabilities of 2γ from π^0 are shown in the right graph of Figure 6.

Figure 7 shows the invariant mass spectra at various p_T in $\eta = 3.0 \sim 3.5$ in 8.8 TeV p+Pb collisions calculated by HIJING event simulator. Energy and hit position are smeared according to the corresponding resolution listed in Table 2. Clear π^0 peak can be recognized.

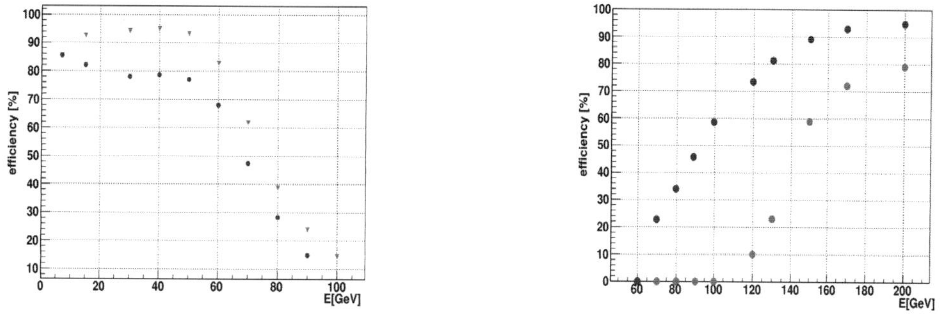

Figure 6. Left: Reconstruction efficiency of π^0 with invariant mass method (red - without the energy asymmetry cut, blue - with the energy asymmetry cut $<$ 0.8), Reight: 2 γ merging probalibity (red - 2 γ hit same pad, blue- 2 γ hit neighbor pad)

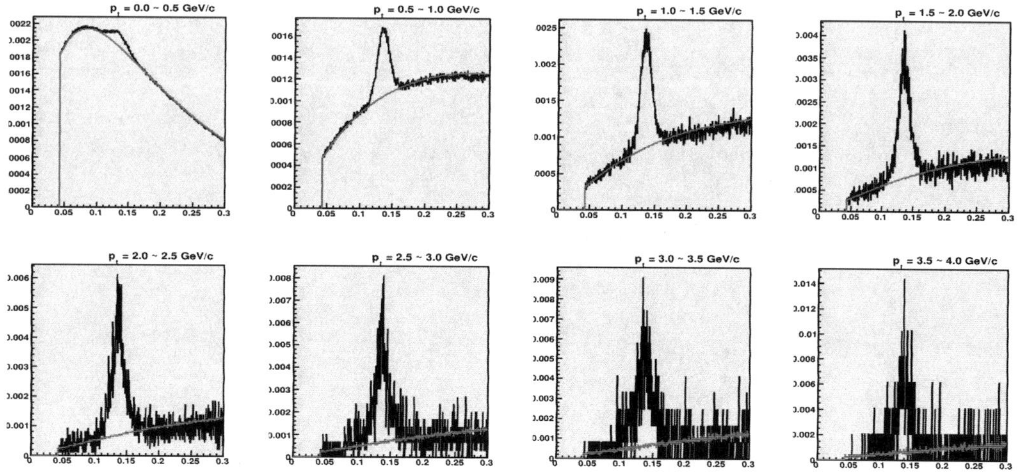

Figure 7. $\gamma\gamma$ invariant mass spectra in $\eta = 3.0 \sim 3.5$ in 8.8 TeV p+Pb collisions, Red line is estimated combinatorial background by event mixing method

By one layer of the Si Strip detector with 0.5 mm pitch strip, 100 $\sim$200 GeV π^0 can be detected and its efficiency is $\sim$ 50% with misidentification probability $<$ 1% as shown in Figure 8. The performance of the Si Strip detector with 1 mm pitch strip is not so different from that of 0.5 mm pitch strip. The position of the Si Strip detector should be the depth of 5 $\sim$ $7X_0$ or deeper, which corresponds to the shower maximum depth. It is also found that the use of multi Si Strip layers does not change the π^0 reconstruction efficiency drastically.

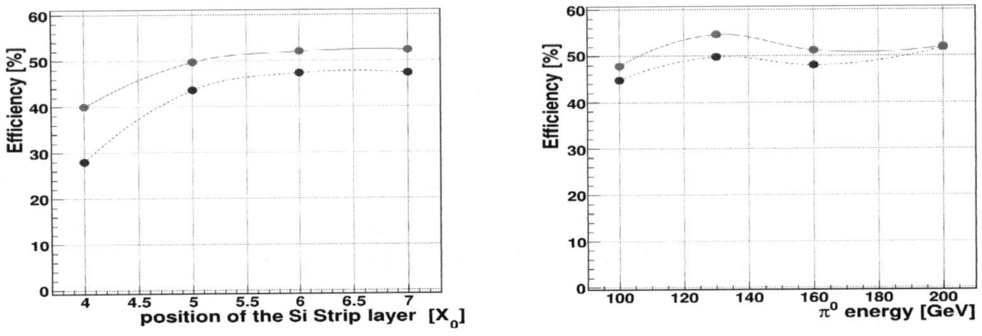

Figure 8. Left: Efficiency vs. the position of the Si Strip layer (red solid line - 200 GeV π^0, blue dot line - 100 GeV π^0), Right: Efficiency vs.π^0 energy (red solid line - 0.5 mm pitch strip, blue dot line - 1 mm pitch strip)

The annual yield of π^0 with $19.5 < p_T < 20$ GeV/c at $\eta = 3 \sim 4$ is $\sim 5 \times 10^4$ according to Figure 2. The accepatece for π^0 with p_T 20 GeV/c is almost 100 % and the reconstruction efficiency by the Si Strip detector is $\sim$ 50 %. Therefore, the annual count of π^0 with $19.5 < p_T$ < 20 GeV/c at $\eta = 3 \sim 4$ is $\sim 2.5 \times 10^4$.

5. Summary and future plan

We performed a dedicated simulation study for FOCAL in LHC-ALICE experiment. As a next step, we have to do full simulation and construct prototype of FOACL module including a readout electronics.

References

[1] L.D.McLerran and R.Venugopalan, Phys.Rev.D **49** (1994) 2233
[2] D. d'Enterria [CMS Collaboration], J.Phys.G:Nucl.Part.Phys.**34** (2007) S709
[3] Technical Design Report for a Nosecone Calorimeter for the PHENIX experiment.

A study on stochastic term of calorimetric energy resolution

Ilhan Tapan, Fatma Kocak

Uludag Universitesi, Fen Edebiyat Fakultesi, Fizik Bolumu, 16059, Bursa, Turkey

E-mail: ilhan@uludag.edu.tr

Abstract. The energy deposited in the active medium of the crystal fluctuates event by event. This fluctuations in the lateral shower containment contributes to the stochastic term. The lateral shower shape determines the distribution of the energy deposition in a cluster of crystals around the impact point. The contribution to the stochastic term coming from fluctuations in the lateral shower containment of calorimeter prototypes of $PbWO_4$ crystals have been simulated by GEANT4 for incident electrons at different energies.

1. Introduction

The particle detection using crystal–avalanche photodiode system is an important technique in research, medicine and industry. Also, for some selected applications, mainly in high-energy physics experiments. As an example, the CMS collaboration has chosen the $PbWO_4$ crystal for the electromagnetic calorimeter (ECAL) system. Compared to the other scintillating crystals, $PbWO_4$ crystal is an attractive detector material because of its high density and fast response. The drawbacks of this crystal are a high sensitivity to a temperature variation and poor light yield. Use of the $PbWO_4$ crystal for the CMS ECAL was given in reference [1] extensively.

2. Energy Resolution in Calorimeters

The Electromagnetic Calorimeter (ECAL) performance is expressed in terms of the energy resolution which is conventionally parameterized by the following quadratic sum,

$$\frac{\sigma(E)}{E} = \frac{a}{\sqrt{E}} \oplus b \oplus \frac{c}{E} \tag{1}$$

where denotes the square root of the quadratic sum, a is the stochastic term, b is the constant term, c is the noise term and E is the incident particle energy in GeV.

The stochastic term of the energy resolution for crystal-photodide combination is composed of a contribution from event to event fluctuations in the lateral shower containment ($a_{lateral}$) and a contribution from photo-electron statistics (a_{pe}) given in Eq.2.

$$a = \sqrt{a_{lateral}^{2} + a_{pe}^{2}} \tag{2}$$

The lateral shower shape determines the distribution of the energy deposition in a cluster of crystals around the impact point. The energy deposited in the active medium of the crystal fluctuates event by event. In order to compare the lateral shower development, it is possible to use the ratio of the energy contained into a single crystal (E1) over the energy contained into a 3x3 (E9) and a 5x5 (E25) crystal matrix centred around the hit crystal [2].

3. Simulation and Results

The contribution to the stochastic term coming from energy deposition fluctuations in the lateral shower containment of calorimeter prototypes of $PbWO_4$ crystals have been simulated by GEANT4 [3] for 0.2-100 GeV electrons. Electrons were injected into the central of the crystals and energy deposition and its fluctuation in the crystals has been obtained for the single crystal 1x1 (E1), 3x3 (E9) and 5x5 (E25) crystals matrices. The simulated size of the crystal is same size used in CMS ECAL. The crystal has a length of 23 cm ($25.8X_0$) and a truncated-pyramidal shape, resulting in a cross section that varied from 2.2x2.2 cm^2 on front side to 2.6x2.6 cm^2 to the rear side.

A simulation of the EM shower has been made for a 3x3 $PbWO_4$ crystal calorimeter. In the simulation the electrons at different energies were assumed to be injected in the center of the central crystal of the matrix. The energy of the incident particle is deposited in active medium of the crystals. Figure 1 shows the energy deposition spectra, at 1 GeV for example, obtained in the nine individual $PbWO_4$ crystals.

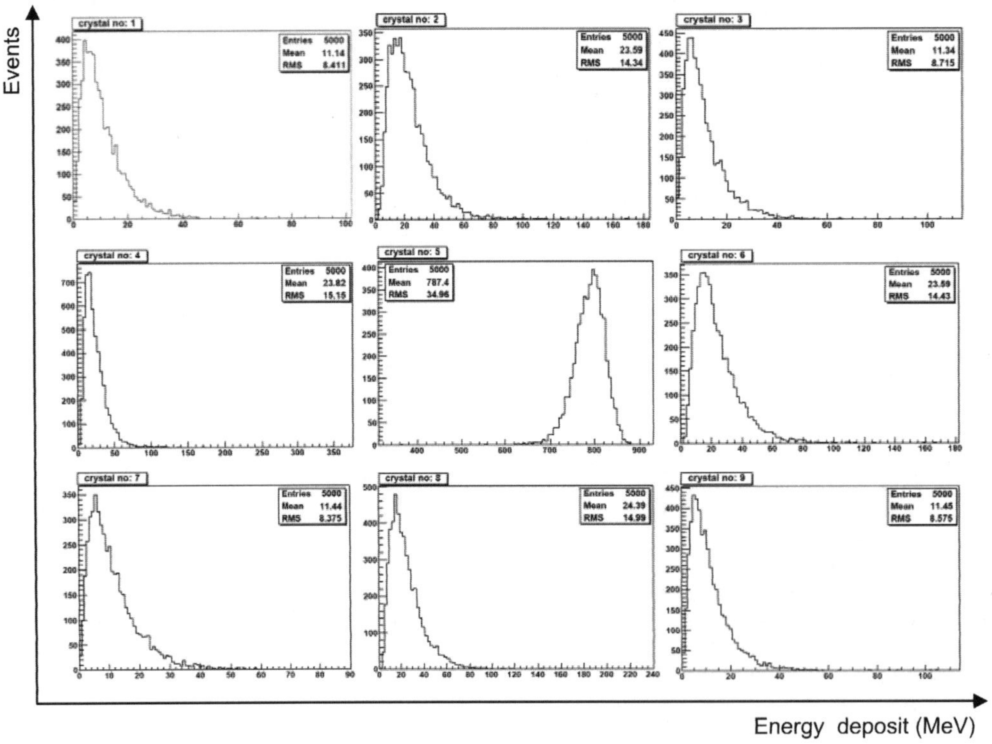

Figure 1. Energy deposits in 3x3 $PbWO_4$ crystals for 1 GeV electrons injected into the center of the central crystal.

As a result of that around 78% of the energy of the incident electron was deposited in the central crystal and that the total deposited energy in the nine $PbWO_4$ crystals was around 93%. Thus, the ratio of simulated energy deposition in the central crystal to that in all nine crystals was obtained around 84%. By summing up all the energies deposited in the nine crystal blocks, the energy spectra for the incident electrons at different energies were obtained as shown in figure 2. In order to compare GEANT4 results with the previous simulation results obtained by EGS4 [4], the simulation was performed in the order from 0.2 GeV to 1 GeV (figure 3).

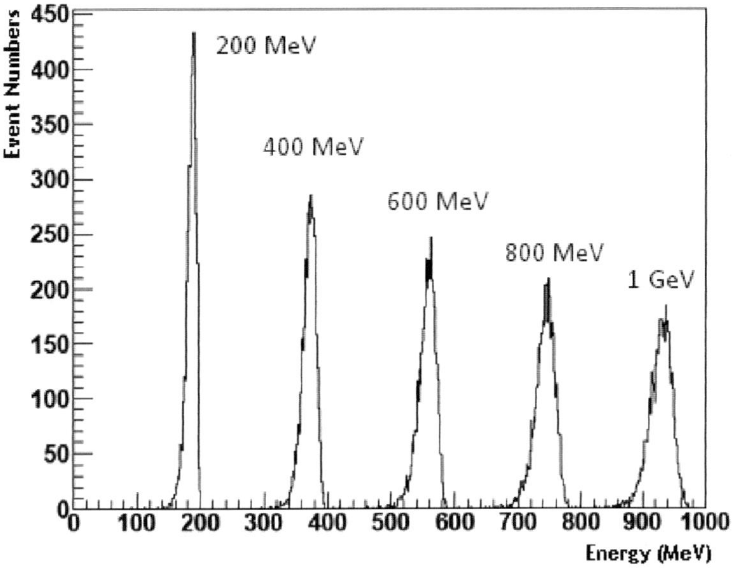

Figure 2. Energy spectra obtained by summing up all energy deposits in the nine crystals.

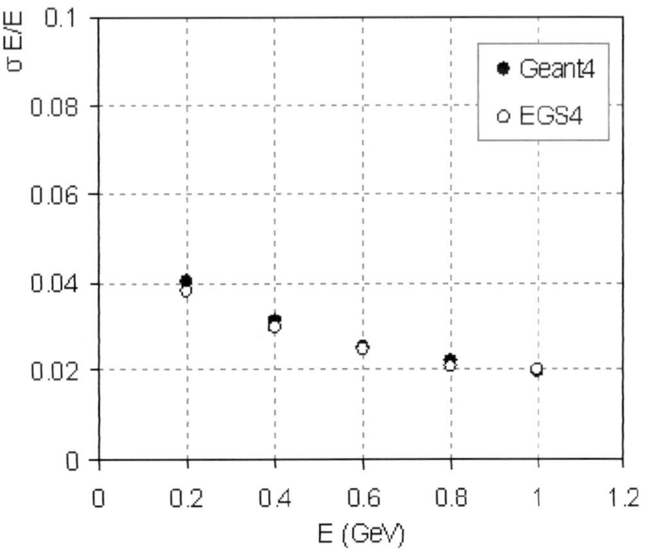

Figure 3. Energy resolution of the 3x3 crystal calorimeter for incident electrons of. 0.2, 0.4, 0.6, 0.8, and 1.0 GeV.

The simulation has been repeated for the calorimeter comprises 25 crystals arranged in a 5x5 matrix. In the simulation energy depositions and their fluctuations have been obtained up to 100 GeV incident electron energies. It turned out that the total deposited energy in the 25 $PbWO_4$ crystals was around 96%. Figure 4 shows the variation of energy deposition in the central crystal 1x1 (E1) and in the matrix of 3x3 (E9) and 5x5 (E25) crystals.

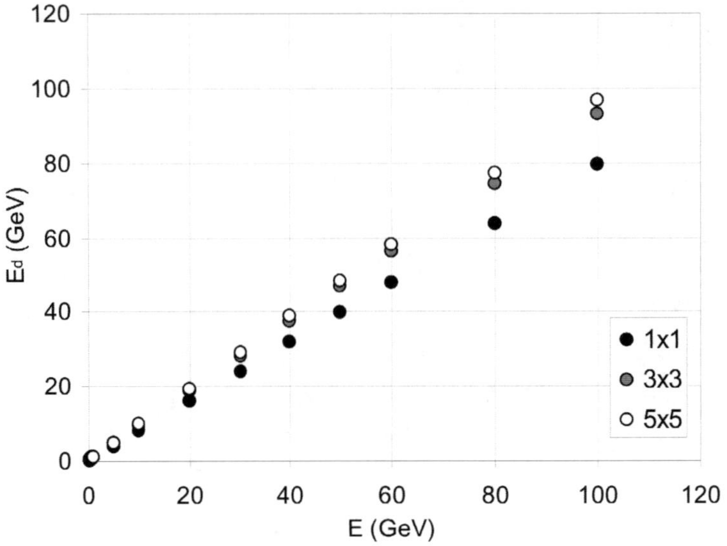

Figure 4. Deposited energy as a function of incident electron energy.

As can be seen from figure 5, the deposited energy fraction was fairly independent of the incident energy in the relevant energy region.

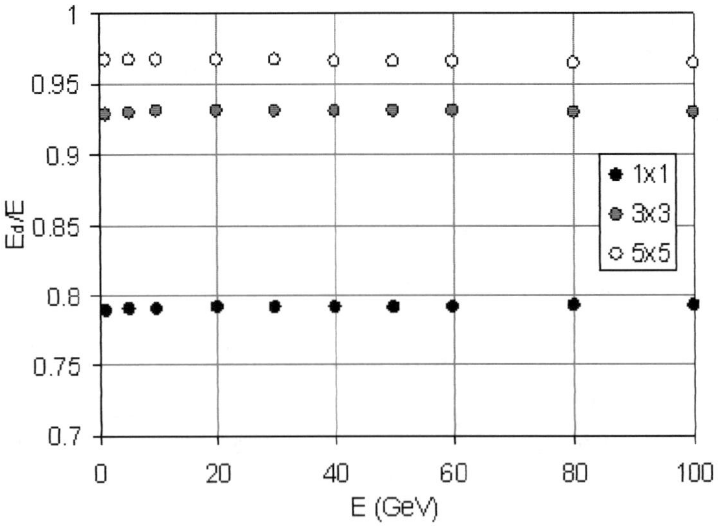

Figure 5. Deposited energy fraction as a function of incident electron energy.

As a result of this work, intrinsic energy resolution ($a_{lateral}$) for the 1x1, 3x3 and 5x5 crystal matrices have been obtained. Figure 6 shows intrinsic energy resolution for $PbWO_4$ crystal matrices as a function of incident electron energy.

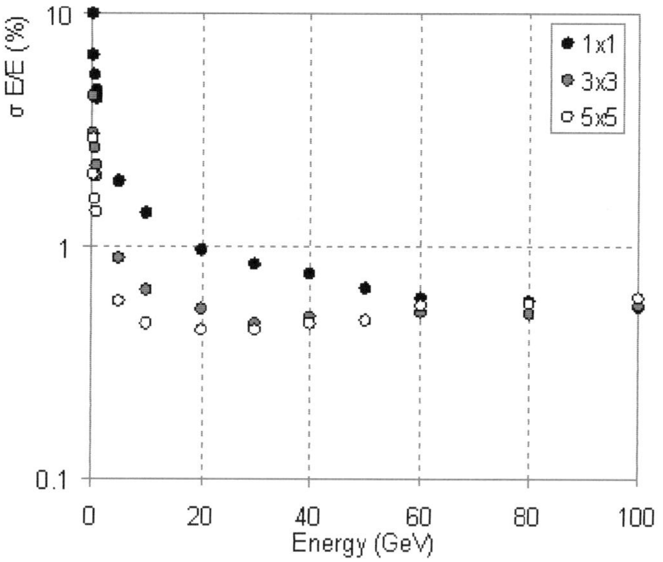

Figure 6. Intrinsic energy resolutions as a function of incident electron energies.

As can be seen from the variations, intrinsic energy resolutions decrease up to 60 GeV and then the resolutions are constant for all crystal matrices. When a particle with energy E creates a signal, then the relative precision of the calorimetric measurement of the energy $\sigma E/E$ will be related by the relative fluctuation in the photo-detector signal $\sigma S/S$. Decrease in the intrinsic energy resolution, $\sigma E/E$, correspond to a decrease in $\sigma S/S$.

Acknowledgment
This work is supported by Turkish State Planning Organization (DPT) under the Grant DPT2006K-120470. The numerical calculations reported in this paper were performed at TUBITAK ULAKBIM, High Performance and Grid Computing Center (TR-Grid e-Infrastructure).

References

[1] Zhu R Y 2004 *IEEE Trans. Nucl. Sci.*, **51** 1560–1567
[2] Meridiani P 2004 "Optimization of the discovery potential of the Higgs Boson in the decay channel $H \rightarrow ZZ^{(*)} \rightarrow 2e^{+}2e^{-}$ with the CMS detector" PhD Thesis, Universita Degli Studi Di Roma "La Sapienza"
[3] http://geant4.web.cern.ch/geant4
[4] Shimizu H et al. 2000 *Nucl. Instr. and Meth. in Phys. Res. A*, **447** 467-475

Journal of Physics: Conference Series **293** (2011) 012031 doi:10.1088/1742-6596/293/1/012031

Hadron response and shower development in the ATLAS calorimeters

Margar Simonyan
On behalf of the ATLAS Calorimeter Group, ATLAS Collaboration
Niels Bohr Institute, Copenhagen University.
Blegdamsvej 17, 2100 Copenhagen, Denmark

E-mail: margar.simonyan@cern.ch

Abstract. The response of the ATLAS Calorimeters to hadrons in the energy range between 2 to 350 GeV produced at CERN's Super Proton Synchrotron (SPS) H8 test-beam line has been measured. The article summarizes the results obtained from the combined test-beam, during which a full slice of the ATLAS detector was exposed to the beam. Results are also presented from a dedicated standalone hadron Tile Calorimeter test-beam where this sub-detector had a special configuration to allow measurements of the longitudinal development of pion and proton induced showers up to 20 nuclear interaction lengths. Data results are compared to GEANT4 simulations using several hadronic interaction models.

1. Introduction

ATLAS is a multi-purpose detector [1] built at the LHC collider at CERN. It consists of inner detector combining different tracking subsystems, sampling calorimeters and muon spectrometer. The ATLAS calorimeter system is divided into electromagnetic and hadronic sections. The electromagnetic calorimeter is lead-liquid argon (LAr) sampling detector with a fine granularity. In the central region the hadronic calorimeter is an iron-scintillator (Tile) sampling calorimeter. In this region the energy resolution for jets is about $50\%/\sqrt{E\ [GeV]} \oplus 3\%$ [1].

The energy for each calorimeter cell is reconstructed from digitized samples of the signal from each readout channel. The results presented in this paper are at the electromagnetic energy scale, i.e. no correction for non-compensation of the calorimeters or dead material energy losses are applied.

2. Monte Carlo Simulation

The simulation of the calorimeter modules was performed within the ATLAS software framework using the GEANT4 simulation toolkit [2]. Within the GEANT4 several models can be used to simulate the interaction of particles with matter. A "physics list" is a consistent collection of models that covers the interaction of all particles in the whole energy range from thermal energies to several TeV range. At high energies the Quark-Gluon String (QGS) and Fritiof (FTF) fragmentation models are used. In the intermediate energy region Bertini (BERT) intra-nuclear cascade model is applied. The Precompound (P) model is used at lower energies. Detector response simulation includes the emulation of Birks' law [3] and the simulation of the incoherent noise.

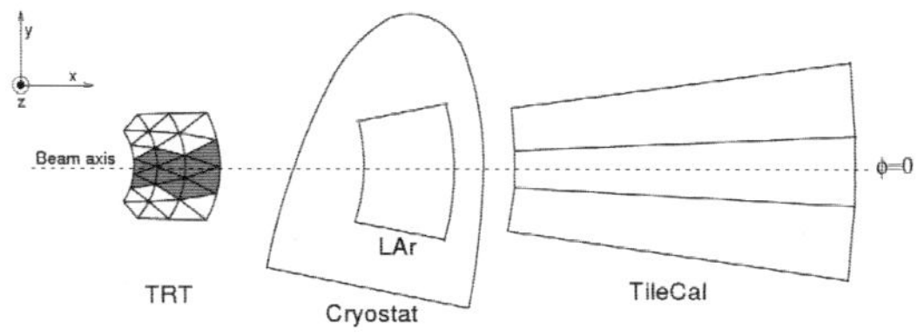

Figure 1. Layout of the calorimeter modules in the test-beam setup.

3. H8 beam line at the SPS at CERN

The H8 test-beam line at the SPS accelerator at CERN was instrumented with a set of beam detectors: scintillator trigger counters, wire chambers measuring the lateral coordinates of the beam particles and a helium filled Cherenkov counter. In order not to bias the calorimeter response, whenever possible, only minimal use of the calorimeter itself was made for particle identification. The signal in the scintillator counters was required to be compatible with single particle response. Events in the beam halo were rejected. A detailed description of all cuts used in the analysis can be found in [4], [5] and [6].

4. The combined test-beam measurements

The combined test-beam was an important step towards improving our understanding of the ATLAS detector. The drawing in Fig. 1 shows a simplified view of the detector setup in the combined test-beam. The transition radiation tracker (TRT) is used for particle identification. The electromagnetic (EM) calorimeter placed inside the cryostat and the hadronic Tile calorimeter (TileCal) located behind the EM section are shown. The beam energy ranged between 2 to 350 GeV. Data are compared to GEANT4 simulation version 9.1.

4.1. Pion response and resolution

A topological clustering algorithm [7] is applied to group neighboring cells into clusters and suppress the noise based on cell signal significance compared to the expected noise. The total energy response is defined as a sum of energies of all clusters.

The mean energy response and the energy resolution as a function of the beam momentum is presented in Fig. 2. The pion response (see Fig. 2 left) in the data is best described by the QGSP_BERT physics list. The data are described within 5% below 10 GeV and within 1% for higher momenta. The response for the QGSP physics list is about 5 - 10% lower than the one measured in the data. The Fritiof model is about 2 - 3% too low at high pion momenta, but 2–5% too high at low pion momenta. As in the case of the QGSP model adding the Bertini cascade increases the energy response, which is in better agreement with the data at high energy (+1%), but worse at low energies (5 - 10%). In general the Monte Carlo simulation predicts a better resolution than the one measured in the data (see Fig. 2 right). The resolution in the simulation is better by 10% below 10 GeV and about 5% at higher momenta. At high momenta adding the cascade model to the QGSP or FTFP models makes the resolution better which is in worse agreement with the data. However, the QGSP model gives a too broad resolution (by 5 - 10%) for pion momenta between 100 and 180 GeV. The QGSP_BERT model predicts a resolution that is lower by 5 - 10% than in the data for all pion momenta.

The response of the calorimeter to pions of 320 and 350 GeV energy as a function of pseudorapidity of beam impact point is shown in Fig. 3. The relative difference between MC and data results (Δ_E^{Comb}) is shown in the bottom part of the plots. The sum of the energies measured in the electromagnetic and hadronic calorimeters (E_{Comb}) is used. The MC simulation has the tendency to overestimate the response of the calorimeter. The response of the detector is uniform for data and MC across the calorimeter.

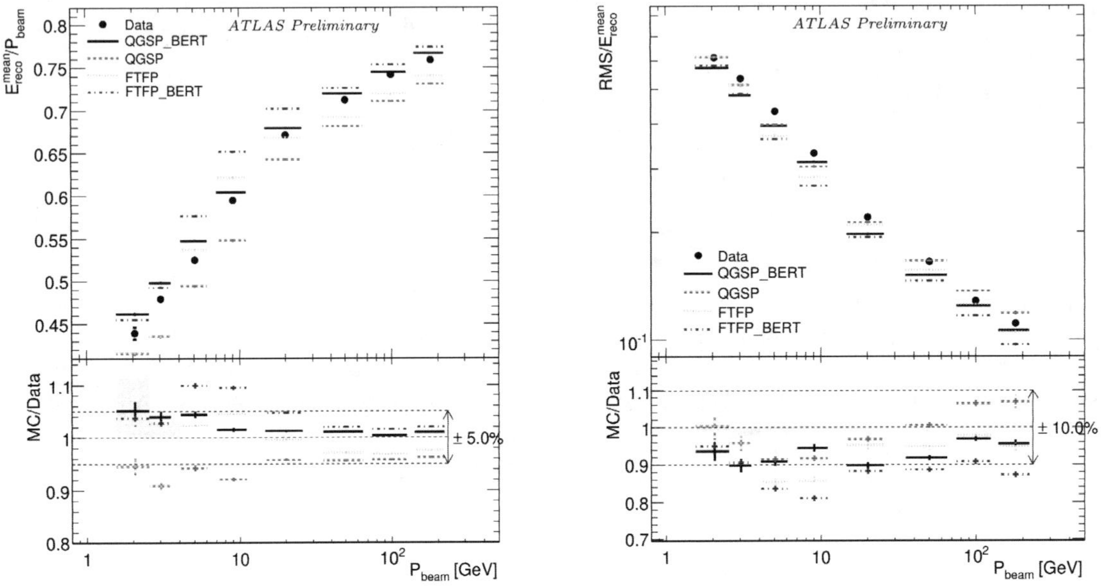

Figure 2. Mean energy (left) and resolution (right) for pions at beam momenta (P_{beam}) between 2 and 180 GeV. Only statistical uncertainties are shown. The yellow band indicates the uncertainty due to the particle identification cuts.

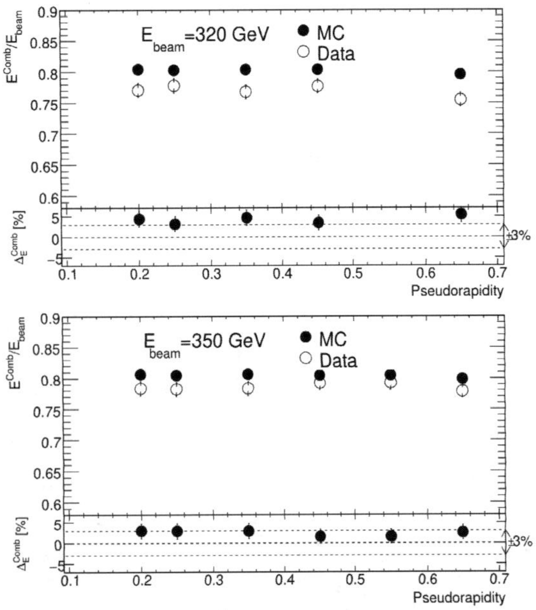

Figure 3. Calorimeter response as a function of pseudorapidity of beam incident point. The uncertainty includes statistical and systematic errors. QGSP_BERT physics list is used.

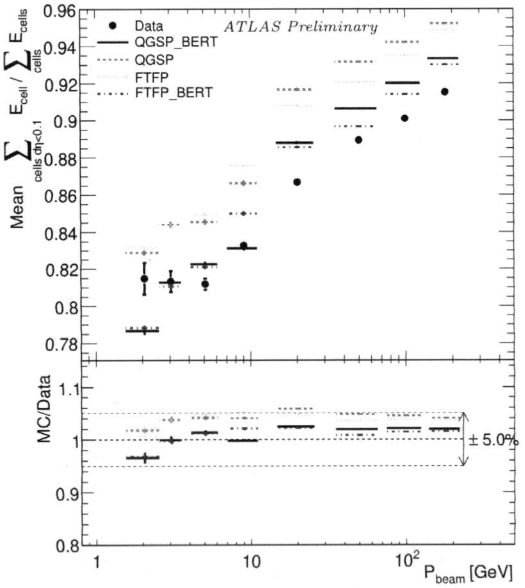

Figure 4. Mean energy fraction in the core of the shower as a function of beam momentum. The yellow band indicates the uncertainty due to the particle identification cuts.

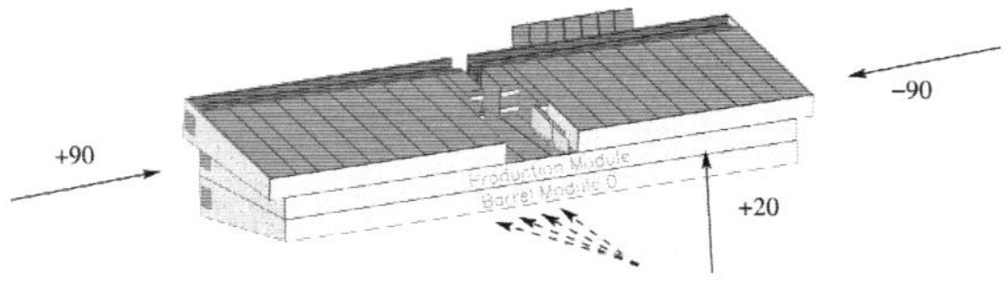

Figure 5. Layout of the calorimeter modules in the TileCal standalone test-beam setup. The arrows indicate the direction of the impinging particles.

4.2. Shower lateral spread

In order to study radial shower development we define the fraction of energy in narrow cone around the beam direction: $E_{core} = \Sigma_{cells\ with\ d\phi,d\eta<0.1} E_{cell} / \Sigma_{cells\ with\ d\phi,d\eta<0.3} E_{cell}$

The dependence of E_{core} on the beam momentum is shown in Fig. 4. The QGSP model produces showers which are too narrow for all pion momenta. The mean fraction is about 5% too high. The Fritiof model gives the same result for the beam momenta below 10 GeV, but produces showers that are a bit wider for high pion momenta (4%). Adding the Bertini cascade makes the shower wider for both the QGSP and the Fritiof models. In general, the shape of the data distribution is well described by these models.

5. The Tile Calorimeter standalone test-beam measurements

The layout of the Tile Calorimeter modules in the H8 test-beam line area is shown in Fig. 5. It consists of two Barrel Modules and two Extended Barrel modules. The beam impinges from the side (90°) of the middle located Barrel Module, perpendicularly to the scintillating tiles. The direction of the beam is defined to be the z-axis.

The Cherenkov counter is used to discriminate pions from protons event-by-event. At 20 GeV only negative beam was available. Electron and muon rejection has been done using calorimeter information. The resulting biases on the measurements are studied and they are found to be negligible. Data are compared to GEANT4 simulation version 9.3. No cut is applied on the cell energies.

5.1. Longitudinal shower profiles of pion and proton induced showers

The special test-beam configuration of the calorimeter modules made it possible to measure shower longitudinal profiles up to 20 nuclear interaction lengths[1] (λ).

The measured longitudinal profiles of pion and proton induced showers are presented in Fig. 6. The measurements extend up to 20 λ in depth. On average, hadron showers quickly deposit their energy and reach the maximum of the mean energy deposition within the first few nuclear interaction lengths in depth. The average energy deposition then exponentially decreases towards the end of the shower and is lower by approximately four orders of magnitudes at 20 λ.

The longitudinal profiles of pion induced showers obtained with MC simulation are compared to the measured data in Fig. 7. The distributions in the figure show that the simulation predicts shorter showers compared to the data for both pion (left plot) and proton (right plot) induced showers. The latter is worse described.

5.2. Impact of longitudinal leakage on the energy resolution

The large depth of the calorimeter in the test-beam allowed to study the influence of longitudinal leakage on the energy resolution. The depth dependence of the TileCal energy resolution is shown in Fig. 8 for pions. The resolution for the calorimeter having depth of 10.9 λ is used as reference. In the left plot the resolution is defined as the ratio of the RMS spread and the mean of the total energy distribution, while in the right plot the resolution is defined as the ratio of the standard

[1] We use throughout this analysis an effective nuclear interaction length for the Tile calorimeter $\lambda = 20.55$ cm, calculated using the known fraction of materials used in the detector construction and their nuclear interaction lengths.

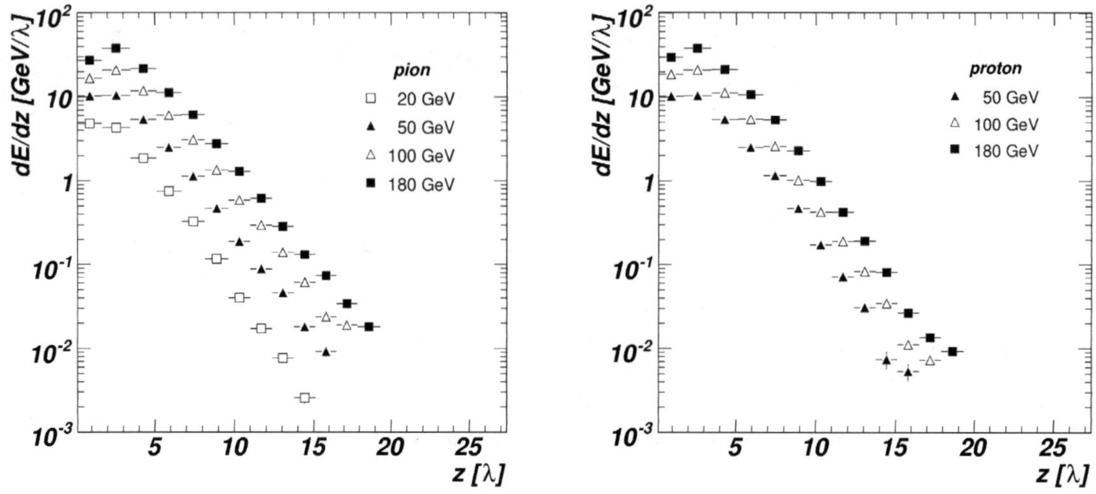

Figure 6. Measured longitudinal shower profiles of pion (left plot) and proton (right plot) induced showers as a function of the calorimeter depth (λ). The beam energy ranged between 20 to 180 GeV.

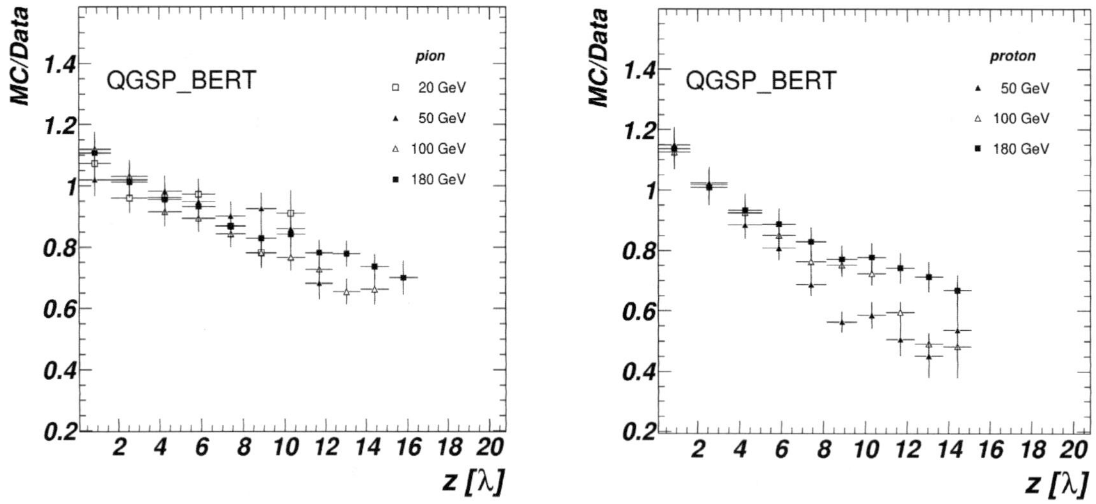

Figure 7. Data and MC simulation comparison of longitudinal profiles of pion (left plot) and proton (right plot) induced showers.

deviation σ and the peak of a Gaussian fit to the total energy distribution in the range -2σ to 2σ around the peak value. The worsening of the resolution (with the decrease of calorimeter depth) defined as RMS/mean ratio is significantly more than for the one defined as σ/peak ratio, since the first one takes into account the low energy tails in the total energy distribution caused by longitudinal leakage.

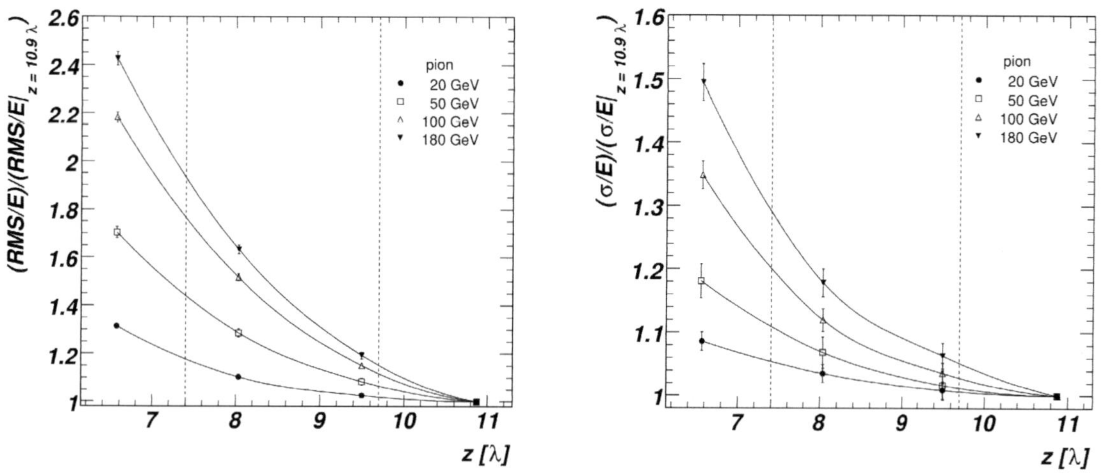

Figure 8. Dependence of the TileCal energy resolution on the depth of the calorimeter for pions at different energies. The resolution is defined as RMS/mean of the total energy distribution in the left plot and in the right one it is defined as σ/peak of a Gaussian fit to the distribution in the range -2σ to 2σ around the peak value. The vertical dashed lines denote the length of the hadronic and the full ATLAS calorimeter at $\eta = 0$.

Summary and Conclusions

The response of the ATLAS barrel calorimeters has been studied for hadrons within the energy range of 2 to 350 GeV in test-beam setup. Longitudinal profiles of pion and proton induced showers have been measured up to a depth of 20 nuclear interaction lengths.

The experimental data have been compared with the results of GEANT4 simulation, using FTFP and QGSP physics lists, as well as extensions with the Bertini intra-nuclear cascade model. The energy response of the ATLAS calorimeters to hadrons is described by the simulation within 5% in the energy range between 2 – 350 GeV, while the energy resolution is described with about 10% accuracy. Neither of those physics lists is able to satisfactorily reproduce the longitudinal and lateral developments of pion and proton induced showers in the whole energy range although the addition of the intra-nuclear cascade improves the description of data by the simulation.

References

[1] The ATLAS Collaboration, The ATLAS Experiment at the CERN Large Hadron Collider, *JINST (2008), 3, S08003*

[2] S. Agostinelli et al., GEANT4 – a simulation toolkit, *NIM A 506 (2003) pp. 250–303*

[3] Birks, John B., The Theory and practice of scintillation counting, International series of Monographs on Electronics and Instrumentation, v. 27, Macmillan, New York

[4] E. Abat et al., Response and Shower Topology of 2 to 180 GeV Pions Measured with the ATLAS Barrel Calorimeter at the CERN Test-beam and Comparison to Monte Carlo Simulations, *ATLAS Public note*

[5] E. Abat et al., Study of energy response and resolution of the ATLAS barrel calorimeter to hadrons of energies from 20 to 350 GeV, *NIM A 621 (2010) pp. 134–150*

[6] P. Adragna et al., Measurement of pion and proton response and longitudinal shower profiles up to 20 nuclear interaction lengths with the ATLAS Tile calorimeter, *NIM A, 615 (2010) pp. 158–181*

[7] W. Lampl et al., Calorimeter Clustering Algorithms: Description and Performance, ATLAS Note *ATL-LARG-PUB-2008-002*

Tests of Local Hadron Calibration Approaches in ATLAS Combined Beam Tests

Karl-Johan Grahn
Royal Institute of Technology, Stockholm, Sweden

Andrey Kiryunin, Guennadi Pospelov
Max-Planck-Institut für Physik, Werner-Heisenberg-Institut

for the ATLAS Calorimeter group

E-mail: guennadi.pospelov@cern.ch

Abstract. Three ATLAS calorimeters in the region of the forward crack at $|\eta| = 3.2$ in the nominal ATLAS setup and a typical section of the two barrel calorimeters at $|\eta| = 0.45$ of ATLAS have been exposed to combined beam tests with single electrons and pions. Detailed shower shape studies of electrons and pions with comparisons to various Geant4 based simulations utilizing different physics lists are presented for the endcap beam test. The local hadron calibration approach as used in the full Atlas setup has been applied to the endcap beam test data. An extension of it using layer correlations has been tested with the barrel test beam data. Both methods utilize modular correction steps based on shower shape variables to correct for invisible energy inside the reconstructed clusters in the calorimeters (compensation) and for lost energy deposits outside of the reconstructed clusters (dead material and out-of-cluster deposits). Results for both methods and comparisons to Monte Carlo simulations are presented.

1. Introduction

ATLAS [1] is a multi purpose detector that operates at the Large Hadron Collider at CERN aimed at the study of proton-proton collisions at a nominal center of mass energy of 14 TeV. One of the main goals of the experiment is the discovery of the Higgs boson.

The ATLAS calorimeters are sampling, non-compensating devices designed to address the large dynamic energy range requirement going from 30 MeV up to 3 TeV and provide hermetic coverage from -4.9 to $+4.9$ in units of pseudorapidity η. The electromagnetic (*em*) parts feature liquid argon (LAr) calorimeters with lead as absorber material in the central and endcap regions and copper in the forward region. The hadronic part consists of a steel-scintillator Tile calorimeter in the central region, LAr-copper in the endcap and LAr-tungsten in the forward regions respectively.

In the past, performance studies for electrons and hadrons in beam tests have been carried out for individual setups of the electromagnetic and hadronic calorimeter modules. Two combined tests in 2004 of the ATLAS barrel and endcap regions closed this extensive program:

- The 2004 Combined Beam Test was done in the endcap region ($\eta \sim 3.2$) in a particularly complex transition zone between the endcap and forward calorimeters.

- The 2004 Combined Beam Test in the barrel region ($\eta \sim 0.45$) included a full slice of the ATLAS Barrel region, including the LAr and Tile calorimeters.

These setups were kept as close as technically possible to the ATLAS geometry, both with respect to the calorimeter modules, the support structure and the dead material distribution.

The goal of this study is a validation with combined beam tests data of the ATLAS strategy to use hadron calibration based on simulations to reconstruct the correct energy of hadrons.

2. Hadron calibration technique

The detection of particles that are subject to the strong interaction is more sophisticated than the detection of particles that develop electromagnetic showers only. There are several factors which are responsible for non-linearity and degradation of energy resolution of hadrons.

The ATLAS calorimeter system is non-compensating, i.e. it generates smaller signals per unit of incoming energy for hadrons than for electrons ($e/h \sim 1.3$). Some energy losses (like binding energy losses in post-collision nuclear break-up) stay invisible and do not contribute to the signal. Additional imperfections in the reconstruction are imposed by energy deposits in the material outside of active zones of electromagnetic and hadronic calorimeters, i.e. in the dead material. These include cryostat walls, the magnetic coil, mechanical support structures, all material of the ATLAS Inner Detector and Muon System. Finally, some energy is deposited inside calorimeters but outside of any reconstructed object due to imperfect energy collection of the clustering algorithm.

In this study we consider two software compensation techniques to calibrate the detector response to hadrons: local hadron calibration (LHC) which is used in ATLAS, and layer correlation (LC). Both approaches rely on simulations, by establishing correlations between Monte Carlo truth energy deposits (energy in the dead material, invisible energy) with shower variables which can be measured in the experiment.

2.1. Local Hadron Calibration

The main goal of the local hadron calibration [2] is to provide jet algorithms with constituents — calibrated clusters with energies equal to their corresponding stable particle energies. The key feature of the approach is to factorize corrections in several sequential steps to disentangle detector effects of different types and to correct them independently.

The starting point of the calibration is the topological clustering [3] in the calorimeter cells which have been calibrated at the electromagnetic scale. Cluster shape variables are then used to classify clusters as having electromagnetic or hadronic origin. Hadronic clusters have smaller cell energy densities and larger depth in the calorimeter in comparison to electromagnetic ones. The hadron-like clusters are subject to a cell weighting procedure to compensate for the lower response of the calorimeter to hadronic energy deposits, while clusters classified as electromagnetic are kept at the original scale. In the next step out-of-cluster corrections are applied for the lost energy deposited in calorimeter cells outside of reconstructed clusters, i.e. in the tails of hadronic or electromagnetic showers rejected due to noise cuts. Finally dead material corrections are applied on the cluster level to account for energy deposits outside of active calorimeter volumes, e.g. in the cryostat, the magnetic coil and calorimeter inter-modular cracks.

2.2. Layer correlation method

The layer correlation technique [7] is an alternative approach to the standard ATLAS Local Hadron Calibration which has been used in the analysis of beam test data in the barrel region. It defines the total pion energy as the sum of clustered energy in 7 calorimeter layers of the combined electromagnetic and hadronic section. The event-by-event layer energy corrections

Journal of Physics: Conference Series **293** (2011) 012032 doi:10.1088/1742-6596/293/1/012032

are derived as a function of a specific pair of linear combinations of layer energies which account most for the shower fluctuation.

The selection of such combinations is done using a principal component analysis. An event is regarded as a point in the 7-dimensional vector space of calorimeter layer energy deposits. Its coordinates can be expressed in a new basis of eigenvectors of the covariance matrix between these layers. Eigenvectors are ordered by decreasing eigenvalue, meaning that the projections along the first few eigenvectors contain most of the information on event-by-event longitudinal shower fluctuations. These projections are used as input to build two-dimensional lookup tables, containing compensation weights to correct for the non-linear response of the calorimeters to hadrons and weights to correct for energy losses in the dead material.

3. Analysis of beam test data in the endcap region

The beam test in the particularly difficult forward region $2.5 < |\eta| < 4.0$ (the transition from the electromagnetic endcap calorimeter EMEC and hadronic endcap calorimeter HEC to the forward calorimeter FCal) was carried out in 2004 [4] in the CERN SPS H6 beam line.

The main elements of the setup are: beam instrumentation to measure the impact position and angle of beam particles, the liquid argon (LAr) cryostat with calorimeter modules and a tail-catcher to measure any leakage beyond the calorimeter modules. The load in the LAr cryostat consists of the inner section of one EMEC module (in ϕ 1/8 of the full EMEC wheel), eight front wheel HEC modules (8/32 of the full wheel), eight purpose-built rear wheel HEC modules and the FCal modules corresponding to the first 2 samplings of one quadrant.

In the two run periods more than 4000 data runs have been taken with electrons, pions or muons in the energy range $6\,\mathrm{GeV} \leq E \leq 200\,\mathrm{GeV}$ with about 80 million triggers in total. Energy scans have been taken at a standard set of impact points. In addition, horizontal and vertical scans have been done at fixed particle energies.

To compare data with Monte Carlo (MC) expectation the simulation code Geant4 [5], version 9.2 has been used. From the physics lists for hadronic shower simulations available in Geant4 the physics list QGSP_BERT and FTFP_BERT have been used. Reconstruction has been done using standard ATLAS software: 3d topological clustering for the reconstruction of the energy response at *em* scale, and local hadron calibration for the calibration to the final hadronic scale.

3.1. Energy response and shower shape studies.

Evaluation of relevant calorimeter performance parameters like the energy dependence of the response, resolution and shower shape in experiment and MC is crucial for any simulation based hadron calibration scheme.

Figure 1 (left) shows the average ratio of reconstructed energy to the beam energy as a function of the beam energy for electrons in the endcap area. The response is well described by MC. The final *em* scale is at the level of 98% of the beam energy (for $E > 50\,\mathrm{GeV}$) due to energy lost in dead material in front of the active calorimeter.

The response as function of energy has been studied for pions using energy deposits on electromagnetic scale (Fig.1, right). At the electromagnetic scale the energy is at the level of about 50-80% of the beam energy. This ratio increases with the beam energy due to the increasing electromagnetic fraction in hadronic showers. The expectations from simulation given by two physics lists are shown as well.

Additional studies show that the MC predicts a larger signal in the electromagnetic sections and a smaller one in the hadronic sections compared to the data. The FTFP_BERT physics list describes the sharing of energy between longitudinal layers slightly better than the QGSP_BERT list.

Several typical shower shape parameters, describing shower depth, length, width and average energy density, have been studied using charged pions at a fixed energy of 200 GeV (Fig. 2).

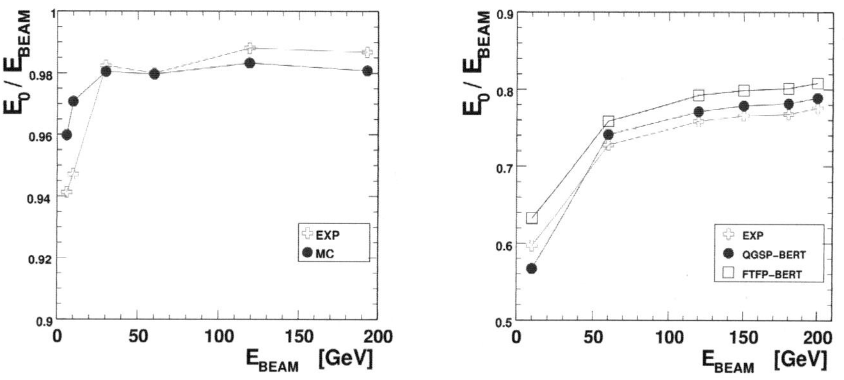

Figure 1. Energy response as a function of beam energy for electrons (left). Energy response at *em* scale as a function of beam energy for charged pions (right).

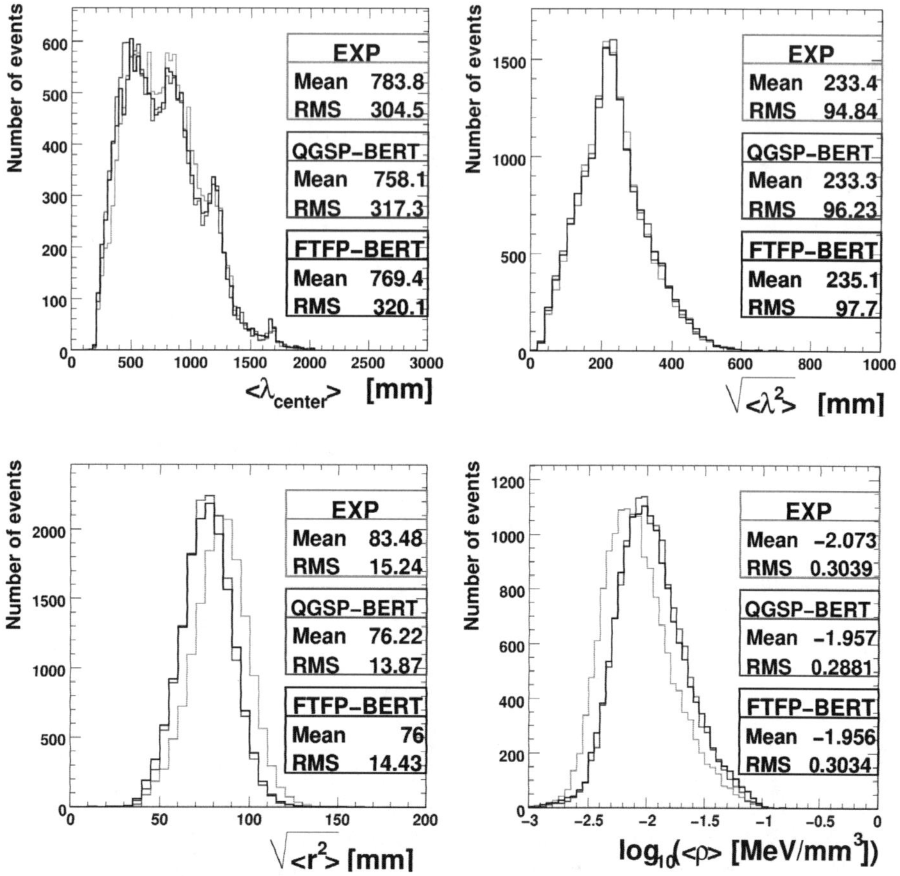

Figure 2. Shower shape parameters for 200 GeV charged pion: shower depth, length, width and energy density.

As results from the comparison of the simulations with the data we conclude:

- The total energy response is overestimated in MC: FTFP_BERT predicts +4%, QGSP_BERT predicts +2% more energy than the experimental data.
- Showers start earlier in MC, seen as larger energy deposition in electromagnetic calorimeter and smaller shower depth (FTFP_BERT describes data slightly better than QGSP_BERT).
- We observe more compact shower sizes in MC for both physics lists, seen as smaller shower width and larger average shower density.

3.2. Performance of local hadron calibration for pions.

The linearity in the endcap region as a function of pion energy after successively applying the correction steps is shown in Fig. 3 (left). Full markers indicate experimental data, while open markers show simulation results with the QGSP_BERT physics list. Each correction step — weighting, out-of-cluster, dead material corrections — makes the response more linear. After the final step, at hadronic scale, the linearity in the data and simulation is recovered within 2%, except at low energies. The energy resolution at *em* scale and at hadronic scale as a function of energy is shown in Fig. 3 (right). The simulation in comparison to data predicts a better resolution by about 20%. The resolution is improving on the level 5-10% after applying the hadron calibration in both, data and MC. Poor energy resolution for 10 GeV charged pions is explained by the relatively large contribution of electronics noise in reconstructed clusters. The moderate improvement in the resolution could be explained by the usage of standard ATLAS calibration constants rather than beam test specific ones, not accounting for limited acceptance and difference in dead material description.

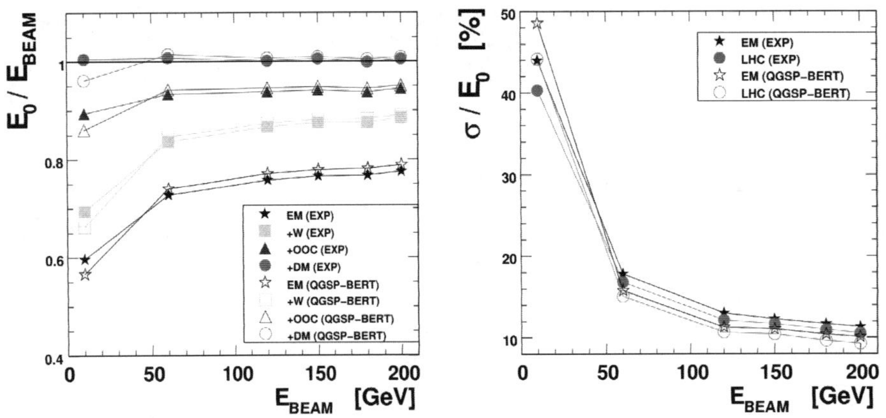

Figure 3. Energy dependence of the response for MC and data after each step of local hadron calibration (left). Energy resolution at *em* scale and after applying of LHC (right).

4. Analysis of beam test data in barrel region

The 2004 Barrel Combined Beam Test [6] was composed of a full slice of the ATLAS Barrel region, including the pixel detector, the silicon strip semiconductor tracker (SCT), the transition radiation tracker (TRT), the LAr and Tile calorimeters and the muon spectrometer. In addition, special beam-line detectors were installed to monitor the beam position and reject background events. The setup was exposed to beams of particles (pions, protons, electrons and muons) in the energy range 1 to 350 GeV in the CERN SPS H8 beam line. The calorimeters were positioned so that the beam impact angle corresponded to a pseudorapidity $\eta = 0.45$ of the ATLAS detector.

4.1. Linearity and resolution for the LC scheme

The performance for the fully corrected energy reconstruction is assessed in terms of linearity (Fig. 4, left) and the resolution (Fig. 4, right). The linearity and resolution are shown - first - at the electromagnetic scale - then - after applying the corrections: compensation weights, the LAr-Tile dead material correction, and finally after applying all corrections.

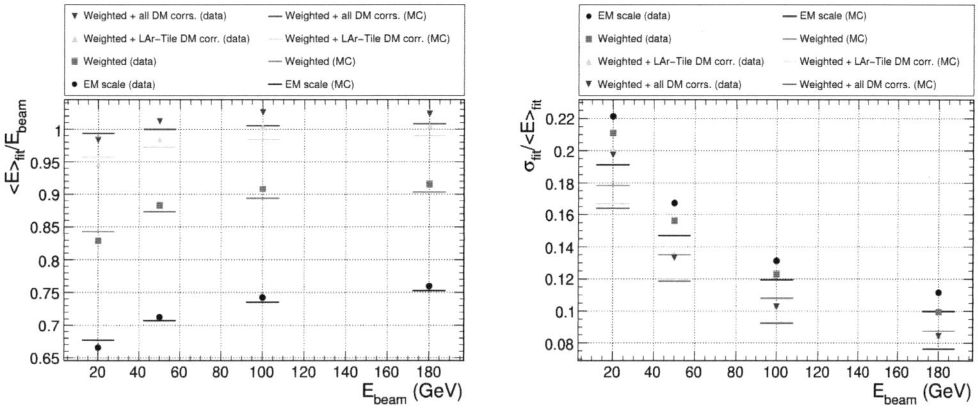

Figure 4. Linearity (left) and resolution (right) of reconstructed energy as a function of beam energy when successively applying the different parts of correction. Data (markers) and MC simulation (horizontal lines) are shown.

After all corrections, the linearity is recovered within 2% for beam energies above 50 GeV (3% for 20 GeV). The improvement in relative resolution when going from the electromagnetic scale to applying all corrections is about 17% to 20% for the data and 17% to 29% in simulation.

5. Conclusion

Two simulation based methods for the calibration of the calorimeter response to hadrons, local hadron calibration and layer correlation method, were studied with data of combined beam tests for the ATLAS endcap and barrel regions. Both methods were able to reconstruct the incoming pion energy within 3% in the energy range 20 – 200 GeV. The energy resolution is improved by about 20% in barrel region (LC method). In the endcap region (LHC method) it is improving by 10%, new beam test specific (rather than the ATLAS specific) correction constants have to be applied to reach the full performance.

References

[1] The ATLAS Collaboration, *G. Aad et al.*, 2008, The ATLAS Experiment at the CERN Large Hadron Collider, 2008 JINST 3 S08003.

[2] T. Barillari, *et al.*, Local Hadron Calibration, *ATLAS Note*, ATL-LARG-PUB-2009-001.

[3] Lampl W. *et al.* 2008, Calorimeter clustering algorithms: Description and Performance, *ATLAS Note* ATL-LARG-PUB-2008-002.

[4] Pinfold J. *et al.*, 2008, Performance of the ATLAS liquid argon endcap calorimeter in the pseudorapidity region $2.5 < |\eta| < 4.0$ in beam tests, *Nucl. Instr. and Meth.* A. **593** 323-342.

[5] S. Agostinelli et al., GEANT4 — a simulation toolkit, Nucl. Instr. and Meth. **A506** (2003) 250-303.

[6] Abat E. *et al.*, Response and Shower Topology of 2 to 180 GeV Pions Measured with the ATLAS Barrel Calorimeter at the CERN Test-beam and Comparison to Monte Carlo Simulations *ATLAS Note*, ATL-CAL-PUB-2010-001.

[7] Karl-Johan Grahn, A Layer Correlation Technique for Pion Energy Calibration at the 2004 ATLAS Combined Beam Test, arXiv:0911.2639.

PandoraPFA Tests using Overlaid Charged Pion Test Beam Data

O. Markin[1]
on behalf of the CALICE collaboration
[1]Institute for Theoretical and Experimental Physics ITEP, Moscow, Russia
E-mail: markin@itep.ru

Abstract. The test beam data obtained with CALICE calorimeter prototypes were used to test the PandoraPFA program. The program capability to recover a neutral hadron energy in the vicinity of a charged hadron was studied. The impact of overlapping of two hadron showers on energy resolution was investigated. The dependence of the confusion error on the distance between a 10 GeV neutral hadron and a charged pion was derived for pion energies of 10 and 30 GeV which are representative of a 100 GeV jet. The comparison of these test beam data results with Monte Carlo simulation using GEANT4 physics lists was performed.

1. Introduction

The Particle Flow Analysis (PFA) algorithm was implemented in the PandoraPFA program [1] as a part of the software for the future International Linear Collider (ILC) and was tested using Monte Carlo (MC) simulated jets. The capability of the PandoraPFA to recover a neutral hadron energy in the vicinity of a charged hadron is of crucial importance because the recovery mistake would degrade the energy resolution. This occurs in case the reconstructing program mixes up hits from showers created by charged and neutral hadrons as a result of shower overlapping. One more factor that degrades the energy resolution is an overlapping of a neutral hadron shower and a photon shower. However to resolve this confusion, contrary to the case of two hadron showers, one can use energy profiles of electromagnetic showers. In case of two hadron showers overlap the task for PFA becomes more complicated because the energy profiles are practically useless and only topological and energy criteria can help to disentangle showers.

The impact of the overlapping of showers on energy resolution for Monte Carlo simulated jets is shown in [1]. However it is known that different available physics lists give noticeably different predictions for hadron shower shapes, that might be important for resolving the overlapped showers. Moreover, the real detector could have not as good performance as that of the idealized MC model. The main goal of this study is to check the PandoraPFA performance using real test beam data and to compare the result with MC predictions. To investigate this issue we use test beam data collected at CERN in 2007 by the CALICE detector prototypes that are very similar to one of the detector technologies proposed for the future ILC and known as the Large Detector Concept (LDC) [2]. The detailed description of CALICE silicon-tungsten electromagnetic calorimeter (ECAL) and high granular analog hadronic calorimeter (HCAL) can be found in [3, 4]. For the first time the procedure of shower overlapping at the CALICE structure was described in [5].

To study the confusion error of overlapped hadron shower recovering, we pass to PandoraPFA two showers: one charged pion shower as it is and one emulating a neutral hadron (e.g. K_L^0 or neutron) shower. For the latter, all hits before the shower starting point were deleted. To confront test beam data with MC, GEANT4 simulation for two physics lists, LHEP and QGSP_BERT, was performed using beam profiles corresponding to the data runs. As the CALICE prototype was tested without magnetic field, the PandoraPFA processor has been adjusted for the case of straight tracks without modifications of the PandoraPFA shower separation algorithm. It is necessary to note that the magnetic field makes it easier for PandoraPFA to separate showers. Therefore our analysis gives a conservative estimate for the PandoraPFA performance.

2. Event preparation

For our study, positive and negative pion induced showers were used. The pion events in the energy range from 10 to 50 GeV were selected from the data collected in 2007 at CERN SPS test beam using complete CALICE setup [3, 4]. The information from the Čerenkov counter was taken into account to exclude beam contamination from electrons in the case of negative pion beam and protons in case of positive one. Besides this trigger, an additional selection procedure was used to purify the sample by excluding muons and trash events including multiparticle ones. The 0.5-MIP threshold was applied to all cell signals to reject noise. The events with MIP-like energy deposition in all calorimeter parts including tail catcher were considered as muons. The same selection procedure was applied to digitized MC samples.

The deposited energy distributions before and after selections are shown in Fig. 1. The blue histograms represent muons. The green histogram corresponds to an electron admixture to the negative pion beam identified by Čerenkov counter on the left plot and to a proton admixture to the positive pion beam on the right one. The resulting selected pion samples (red) are well fitted by Gaussian distribution. Trash and multiparticle contributions are not plotted here because of their low impact comparing to the shown components.

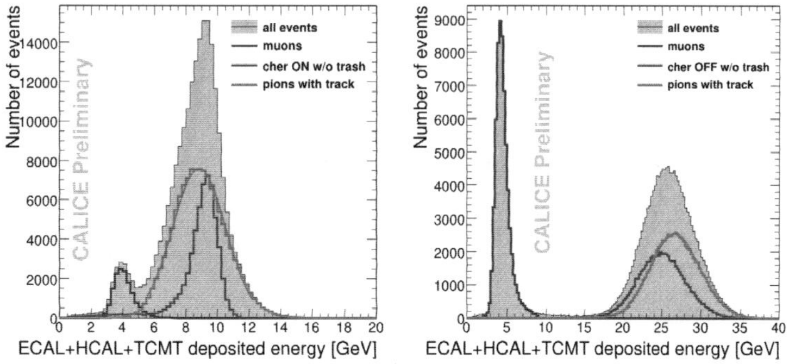

Figure 1. Deposited energy distributions before (shaded) and after selections for 10-GeV negative pion (left) and 30-GeV positive pion (right) beams. Red solid lines show Gaussian fits to resulting pion sample distributions. See text for details.

3. Mixing of showers

The selected pion events were overlaid and passed to PandoraPFA after an energy selection procedure: we selected for the analysis only events with overwhelming containment (more than 95% of deposited energy) in the ECAL plus HCAL, where the fine granularity allows us to

Journal of Physics: Conference Series **293** (2011) 012033 doi:10.1088/1742-6596/293/1/012033

retrieve the shower hit positions. At energies from 10 GeV to 30 GeV showers are not so long; their energy density drops e times after about 15 HCAL layers from the shower start. For this reason such a selection does not restrict physics too much. It means that we do not use showers which start in far HCAL layers and it is worth to note that such showers will be better separated due to magnetic field in a future detector and hence the confusion for them will be smaller.

We took two pion events from different runs and for one of them moved all hits by a certain transverse (w.r.t. beam direction) distance from their original position. This distance varied from 5 cm to 30 cm. For this displaced event we kept only hits behind the shower start, i.e. proper shower hits without incident track, and used this shower as an imitation of the neutral hadron shower. This imitated shower represents the main subject of our study. In what follows we will call the energy of this imitated hadron shower the *neutral hadron shower energy.*

The top row of Fig. 2 shows the energy distributions for the 30 GeV charged (left), 10 GeV charged (middle), and 10 GeV "neutral"(right) hadron events after our preparation procedure. These energies were chosen for the following analysis as being typical for 100-GeV jets.

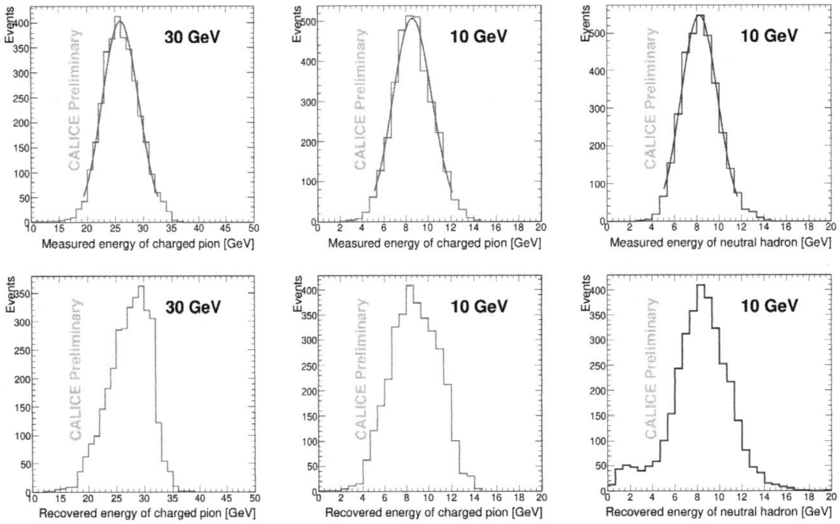

Figure 2. Top row: energy distributions for 30 GeV charged (left), 10 GeV charged (middle) and 10 GeV "neutral" (right) hadron events prepared from data runs for mixing of two showers. Solid lines correspond to Gaussian fit. Bottom row: the same energy distributions after their disentangling by PandoraPFA for the case of 15 cm distance between corresponding charged and "neutral" hadrons.

4. Recovering of showers

Due to shower overlapping in the calorimeter, the energy recovered by PandoraPFA for each of two showers is not accurate. The resulting energy distributions after their recovering by PandoraPFA are shown in the bottom row of Fig. 2. The overlapping is considerable in cases where the charged showering particle appears to be close to the neutral one. Since we know the neutral hadron energy from calorimetric measurements, we can compare it with the energy recovered by PandoraPFA and get the confusion error.

The maximum confusion takes place between the high energy charged hadron and the small energy neutral one (see left plot in Fig. 3). In such a case that high energy is given to the program as a reference point. So the PandoraPFA is to consider a significant number of neutral shower hits as charged shower hits while keeping the difference between the TPC track energy and the energy of charged hadron within the measurement error. Therefore for a large number of events PandoraPFA recovers the charged hadron energy higher than it actually was in the calorimeter, adding to it a significant part of the neutral hadron energy. That mostly happens for events, in which, due to intrinsic shower fluctuations, the difference between the measured charged hadron energy and the beam energy is comparable with the neutral hadron energy and PandoraPFA tries to cover up this difference at the expense of the neutral hadron energy. On the other hand, PandoraPFA can not add to the charged shower the energy much larger than the typical measurement error. Both effects together give a peak around -6 GeV in case of overlapping of 30 GeV charged and 10 GeV neutral hadrons (see Fig. 3).

The shoulder on the right slope of the zero difference peak appears due to the fact that the sufficiently large size of the 30 GeV charged hadron shower gives the PandoraPFA an opportunity to associate many hits of this shower with the neutral hadron and to recover its energy even higher than it actually was. At large distances this confusion vanishes and the mean value of the difference becomes positive since we include into the recovered neutral hadron shower all hits which PandoraPFA has not attached to the charged hadron shower, among them there are isolated hits and small clusters which actually belong to the charged hadron shower but were not recognized. Using the histograms shown in Fig. 3 one can extract a mean value of the difference between recovered energy and measured energy of the neutral hadron. The second characteristic used to estimate the confusion error is the RMS deviation value.

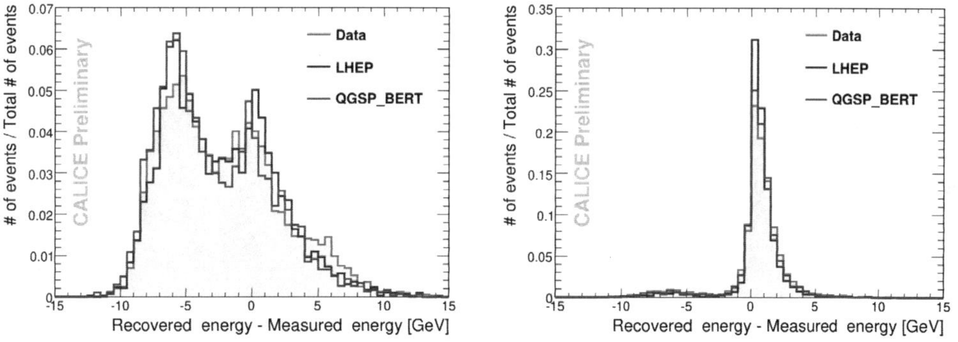

Figure 3. Difference between the recovered energy and the measured energy for the 10 GeV neutral hadron at 5 cm (left) and at 30 cm (right) from 30 GeV charged hadron. Data (red, yellow shaded) is compared to MC predictions for LHEP (blue) and QGSP_BERT (green) physics lists.

The mean value of the neutral hadron energy measured in the calorimeter exceeds the mean energy recovered by PandoraPFA at small distances between particles where shower overlapping is considerable (see Fig. 4). At large distance where confusion vanishes, the mean recovered energy of the neutral hadron becomes even higher than its mean measured energy. That happens because we include in the neutral hadron energy the energy of isolated hits and small clusters which in fact belong to the charged hadron shower but could not be associated with it because of remoteness.

The transverse shower size does not affect the situation appreciably whilst the showers are so close to each other that PandoraPFA tends to attach the small neutral particle shower to the big charged one. As the distance between hadrons grows, the transverse size naturally dominates the shower confusion (see Fig. 5). For this reason LHEP based simulation which gives smaller values for transverse shower sizes, predicts smaller confusion for the distances larger than 15 cm.

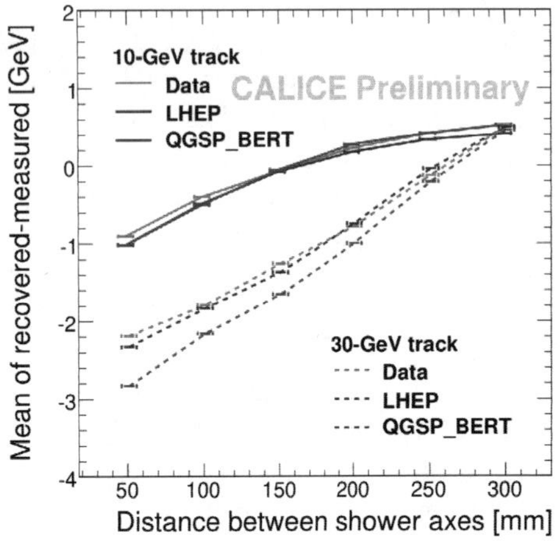

Figure 4. Mean difference between the recovered energy and the measured energy for the 10 GeV neutral hadron vs. the distance from the 10 GeV (solid) and 30 GeV (dashed) charged hadron.

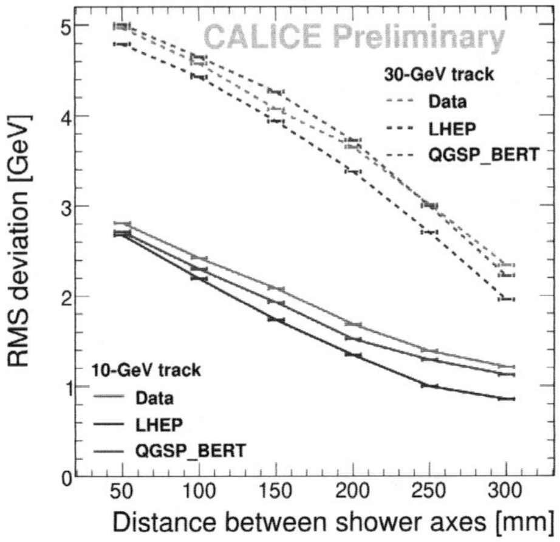

Figure 5. *RMS* deviation of the recovered energy of 10 GeV neutral hadron from its measured energy vs. the distance from 10 GeV (solid) and 30 GeV (dashed) charged hadron.

The RMS deviation of the recovered energy of a neutral hadron from its measured energy can be interpreted as a confusion error. It is particularly large for the 30 GeV charged and 10 GeV neutral hadrons overlapping. However, this does not affect the jet energy reconstruction accuracy too much because the probability to find a 30 GeV fragment in a 100 GeV jet is relatively low.

Fig. 6 shows the probability of recovering of the neutral hadron energy within 3 standard deviations from its real energy at different distances from 10 GeV and 30 GeV charged hadrons. For the 10 GeV neutral hadron we take the standard deviation equal to $0.6\sqrt{10 \times 0.86 \times 0.97}$ GeV, where 0.86 is the π/e ratio and the coefficient 0.97 takes account of track fragment loss for the imitated neutral shower. The latter coefficient is just the approximate ratio of the mean value of the right histogram in Fig. 2 to the mean value of the middle one.

If a charged hadron is situated in the vicinity of a neutral one with similar or exceeding energy, the confusion is not so large. The relative confusion is larger for small neutral hadron energy. This results in a lower probability of neutral hadron energy recovering for small neutral hadron energy (see Fig. 7).

5. Summary

To test the PandoraPFA algorithm, we have mapped pairs of CALICE test beam events shifted by the definite transverse distances from each other onto the LDC detector geometry. Then we modified the treatment of tracks in the PandoraPFA processor for the case of straight tracks. In this study we have investigated the hadron energy range for a 100 GeV jet. For jet fragment energies from 10 GeV to 30 GeV we estimated the confusion error for the recovered neutral hadron energy caused by the overlapping of showers. We have confronted our result for test beam data with the result of Monte Carlo simulations for LHEP and QGSP_BERT physics lists. The results for the data and MC are in a good agreement. This fact together with the successful PandoraPFA performance for simulated jets [1] allows us to consider the PandoraPFA program as a good reconstruction tool for a full-size experiment. All the way we were trying not to simplify the task for PandoraPFA to disentangle showers in order to get a conservative estimate

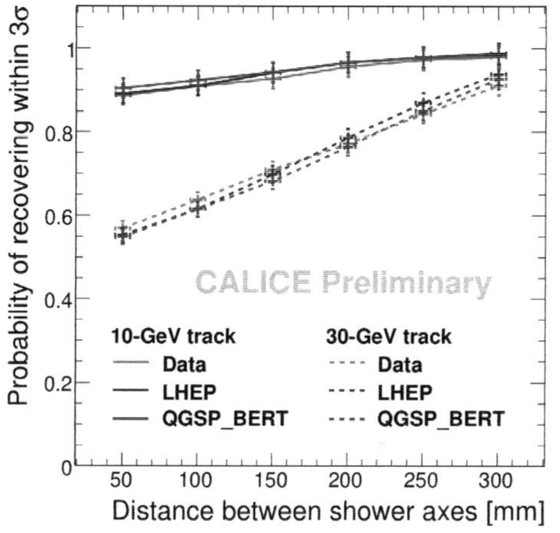

Figure 6. Probability of the 10 GeV neutral hadron energy recovering within 3 standard deviations from its real energy vs. the distance from 10 GeV (solid) and 30 GeV (dashed) charged hadron for data (red) and Monte Carlo simulations using LHEP (blue) and QGSP_BERT (green) physics lists.

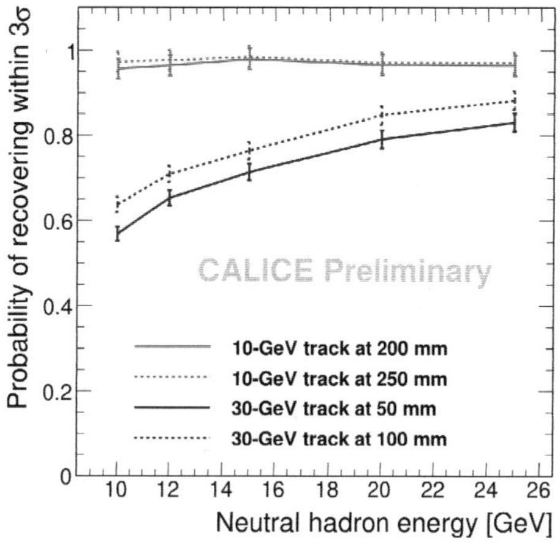

Figure 7. Probability of the neutral hadron energy recovering within 3 standard deviations vs. the neutral hadron energy in the vicinity of 10 GeV (red) and 30 GeV charged hadron (blue) for typical distances from the neutral hadron in a 4T magnetic field in LDC.

for the confusion error. In particular we assign all isolated hits to the neutral hadron shower and we also underestimate the separation of showers towards the end of the calorimter because, unlike a full detector, our testbeam apparatus has no magnetic field.

The agreement between the PandoraPFA performance achieved with real calorimeter prototype data and with the MC simulation demonstrates that the extrapolation to the complete detector is reliable. No hidden imperfections in the real data (non-perfect calibration, non-uniformity of tile response, cross talk between tiles, noise, dead or noisy channels) which could deteriorate the PFA performance were found. We point out that in our study LHEP physics list gives worse predictions for test beam data than the QGSP_BERT one.

This work was supported by grants of Russian National Educational Center 02.740.11.0239, Russian Scientific Schools 3270.2010.2, and Russian Foundation for Basic Research - Helmholtz Association 07-02-92281, 09-02-91321.

References

[1] Thomson M A 2009 Particle Flow Calorimetry and the PandoraPFA Algorithm *NIM A* **611** 25
[2] Behnke T, Damerell C, Jaros J and Miyamoto A 2007 *Detectors* (ILC Reference Design Report vol 4 DESY 07-046) p 25
[3] The CALICE collaboration 2008 Design and electronics commissioning of the physics prototype of a Si-W electromagnetic calorimeter for the International Linear Collider *JINST* **3** 08001
[4] The CALICE collaboration 2010 Construction and commissioning of the CALICE analog hadron calorimeter prototype *JINST* **5** 05004
[5] Morgunov V, Raspereza A 2004 *Novel 3-D clustering algorithm and two particle separation with tile HCAL* (Paris: Linear colliders vol 1) p 431

Journal of Physics: Conference Series **293** (2011) 012034 doi:10.1088/1742-6596/293/1/012034

Towards jet reconstruction in a realistic dual readout total absorption calorimeter

A Driutti[1], A Para[2], G Pauletta[1,#], N. Rodriguez Briones[2] and H Wenzel[2]

[1]Universities an INFN Trieste/Udine, [2]Fermi National Accelerator Laboratory

#Dept. of Physics, University of Udine, 33100 Udine, UD, Italy.

Abstract. Calorimeters implemented in future lepton colliders will inevitably suffer from leakage on account of space restrictions and will need to operate in strong magnetic fields. Both these circumstances will affect the reconstruction of jets and will give rise to the need for corrections. In dual readout calorimeters, these corrections are complicated by the need to deal with more than one signal. In this article we describe simulation studies of these corrections in a total absorption dual readout calorimeter.

1. Introduction

The chief purpose of high resolution calorimeters at future lepton colliders is likely to be di-jet and multi-jet spectroscopy. The primary figure of merit will be mass resolution and the main limitation will be hadron energy resolution.

Compensation by means of Dual Readout (DR) in totally active calorimeters [1] addresses this limitation but the design of collider experiments requires a careful optimization of various aspects: scientific, technical and financial. In particular, the requirements of a calorimeter inside the superconducting coil inevitably limits the thickness of a calorimeter and leads to leakage fluctuations while the magnetic field leads to magnetic corrections. Both these effects can limit the achievable resolution.

A longitudinally segmented total absorption calorimeter offers several possible ways of addressing these limitations and these have been the subject of simulation studies. The progress of this work is reviewed in the following sections. For particulars relating to the simulation package please refer to a paper presented by Hans Wenzel [2] in these proceedings Other related papers in these proceedings are refs. [3] and [4].

1.1. The calorimeter

A cylindrical calorimeter is filled with sensitive BGO-like material of uniform composition. Only the finer segmentation distinguishes the initial "EM" section from the remaining "HAD" volume. Different compositions and segmentations were investigated:

- EM calorimeter: 6-8 layers 5-3 cm thick with the same transverse segmentation.
- HAD calorimeter: 9-17 layers 10-6 cm thick, with the same transverse segmentation.
- The muon system/tail-catcher is implemented as a 48 layer sampling calorimeter.

Total calorimeter thickness: 120-126 cm, which corresponds to about 5.5 interaction lengths and is therefore subject to significant leakage. Magnetic fields up to 5T are catered for.

Journal of Physics: Conference Series **293** (2011) 012034 doi:10.1088/1742-6596/293/1/012034

2. Leakage and the Dual Readout compensation

Dual Readout compensation was first studied in detail for single hadron showers in the absence of magnetic field. The effects of leakage on the DR compensation (often referred to as the DR "correction") in these conditions were studied and different leakage correction methods devised and compared.

2.1. The DR correction

Data is simulated for incident pions and electrons. The ionization energy deposited is identified with the scintillation signal (S) and both ionization (S_i) and Čerenkov (C_i) energy depositions are recorded for each calorimeter element *i* and summed to form the total raw signals S_r and C_r, respectively. Assuming no energy loss for electrons, the corresponding raw signals are normalized to the incident electron energies and the corresponding normalization is applied to the pion signals: the normalized electron-energy distributions then peak around the average energy by construction whereas the average normalized pion energies manifest the expected losses due principally to nuclear breakup. Their S-distributions also manifest pronounced low-energy "tails" corresponding to leakage. Pions which pass through the calorimeter without hard interaction give rise to the usual minimum ionizing peak referred to as "punch-through". Whereas "punch-through" events are generally ignored in the calorimeter and delegated to the tracking elements, leakage must always be corrected for event-by-event in order to minimize the effects of its fluctuations on the calorimeter's energy resolution.

In the case of DR calorimeters leakage can also distort the correlation between the C and S which is used for the DR (compensation) correction as illustrated in figure 1. The correlation between S/E_π and C/S shown in Figure 1a clearly illustrates the effect of leakage on this correlation while Figure 1b illustrates how leakge is corrected for in one of a series of vertical slices of the correlation.

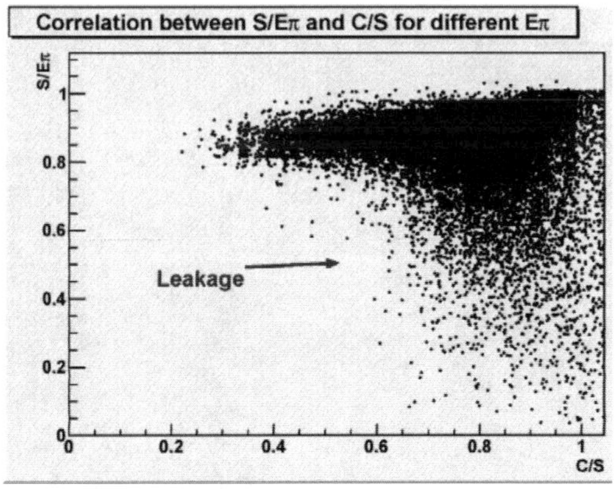

Figure 1a. S/E_π versus C/S illustrating effect of leakage on correlation

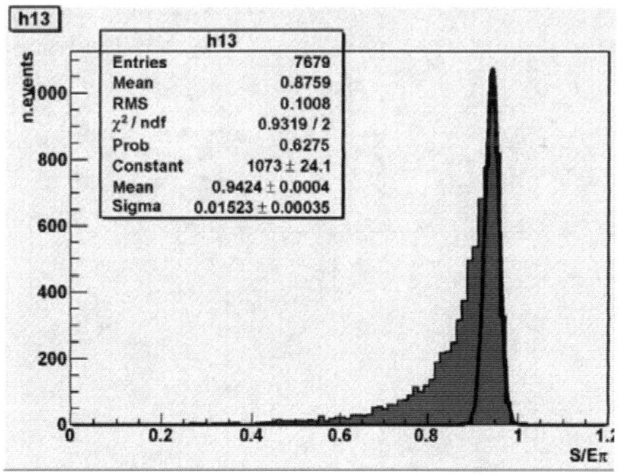

Figure 1b. A vertical slice of Figure 1a illustrating fit to correct effect of leakage

The resulting correlation is fitted with: $S/E_\pi = a(C/S)^2 + b(C/S) + c$, which is used to "correct" (DR correction) S:

$$S_{corr} = \frac{S}{a(C/S)^2 + b(C/S) + c}$$

Correction functions are are evaluated independently at each energy but are seen to manifest only a weak energy dependence (figure 2b). Distributions of S_{corr} peak at the correct energy as illustrated in Fig. 2a but retain the low energy tail corresponding to the leakage. The leakage energy fluctuates and the fractional fluctuation increases with energy until it exceeds the stochastic term and sets the limit on the achievable energy resolution. Leakage therefore needs to be corrected for event-by-event.

2.2. Leakage corrections

Two kinds of leakage correction are considered.

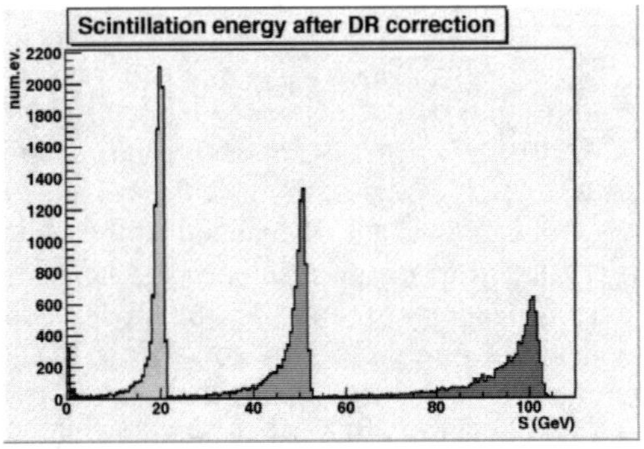

Figure 2a. Distributions of S_{corr}

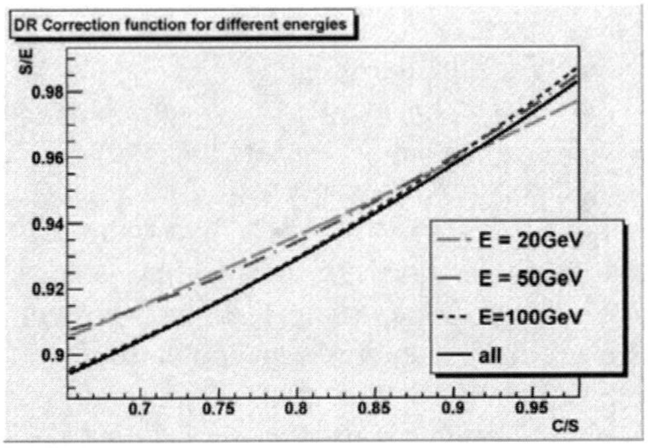

Figure 2b. .DR correlations at different energies

2.2.1. *Leakage corrections using the muon counters as "tail-catcher"*

For this correction, we use the correlation between the energy deposited in the calorimeter and the energy detected in the tail-catcher (figure 3a). The correction is applied before the DR correction and also partially corrects for punch-through.

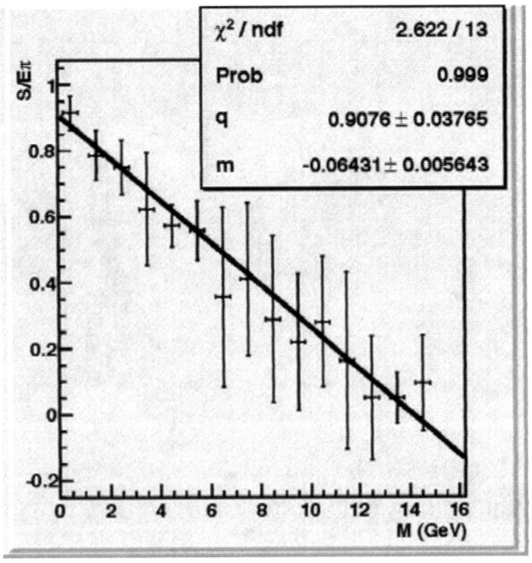

Figure 3a. Correlation between S/Eπ and the energy deposited in the tail-catcher

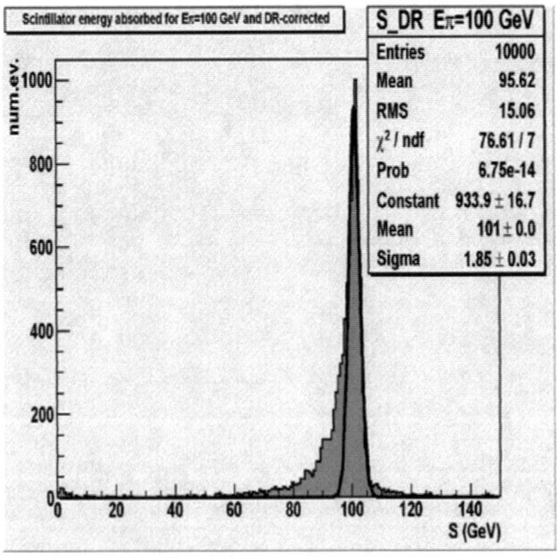

Figure 3b. .The distribution of S_{corr} after correction for leakage, using the correlation function in figure 3a, and the DR correction

Given that the tail-catcher is located outside the solenoid, the correction is expected to be under-estimated as would appear to be the case from figure 3b. Nevertheless, the resolution is ~0.2 GeV and the energy close to the correct one.

2.2.2. *Leakage corrections using longitudinal segmentation.*
Since leakage depends on shower evolution it longitudinal segmentation is expected to be useful for leakage correction. At first approximation, we use only the outermost layers: (S/E_p) was plotted against the fractional energy deposited in the two outermost layers. The resulting correlation function (fig. 4a) was then used to correct for the leakage event-by-event. Although we expect the correction to be over-estimated, the resulting distribution of S at 100GeV (figure 4b) is seen to be symmetric and centered close to the correct energy. The resolution is seen to be ~ 0.2 GeV.

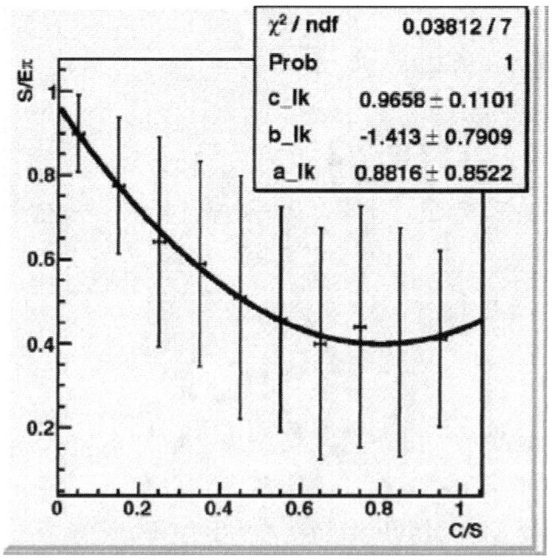

Figure 4a. Correlation between S/Eπ and the energy deposited in the last two layers of the calorimeter

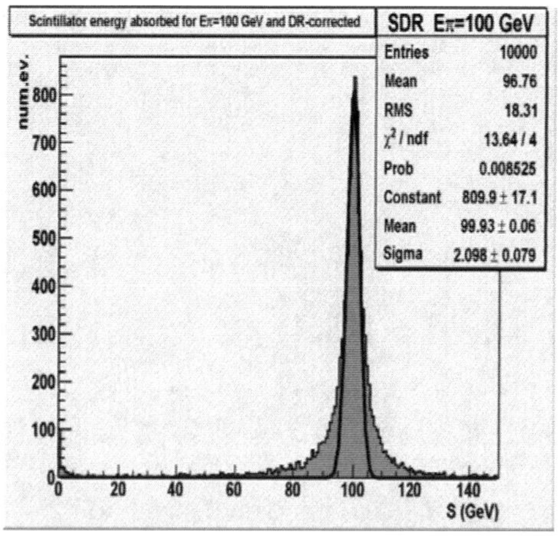

Figure 4b. The distribution of S_{corr} after correction for leakage using the correlation function in figure 3a

However, the above algorithm does not exploit all the information available: leakage fluctuations depend on the starting point of the hadron shower ("Interaction Depth or ID") and the extension of the shower, so it is expected that full use of the depth segmentation will improve the leakage corrections. S a first step in this direction, the data was therefore subdivided according to its ID and correlations for the DR and for the MLK corrections were evaluated separately for each ID.

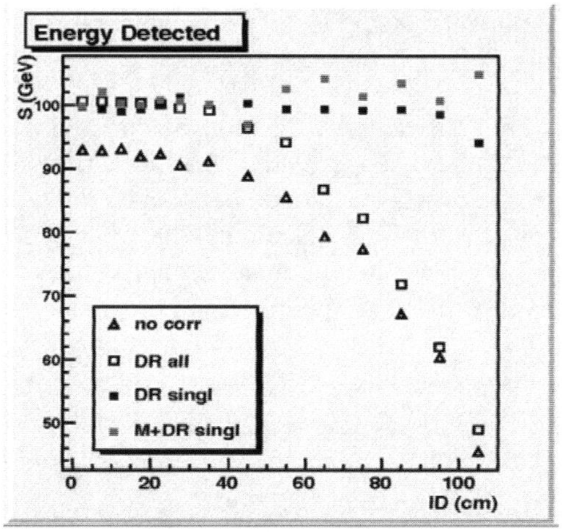

Figure 5a. The effect of these corrections are shown as a function ID. Significance increases with ID as expected

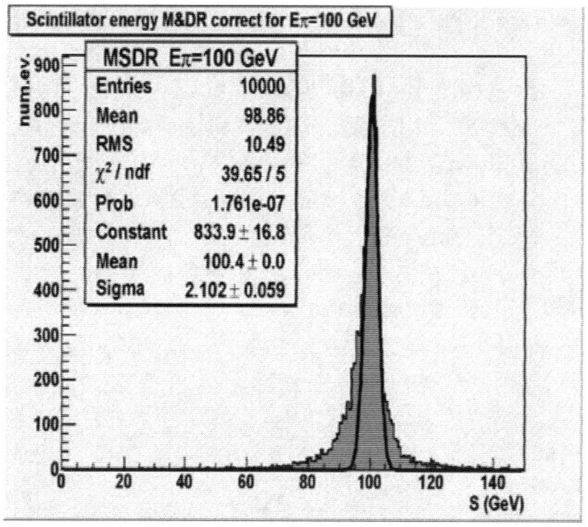

Figure 5b. Effect of corrections on the overall energy distribution

This led to a noteworthy reduction in the non-Gaussian tails observed in figure 3b, particularly at higher energies. Overcorrection is also evident for higher IDs. Punch-through is eliminated.

3. Applying corrections to jets

The processes are complicated by the mixed content and by jet reconstruction. The weak dependence.

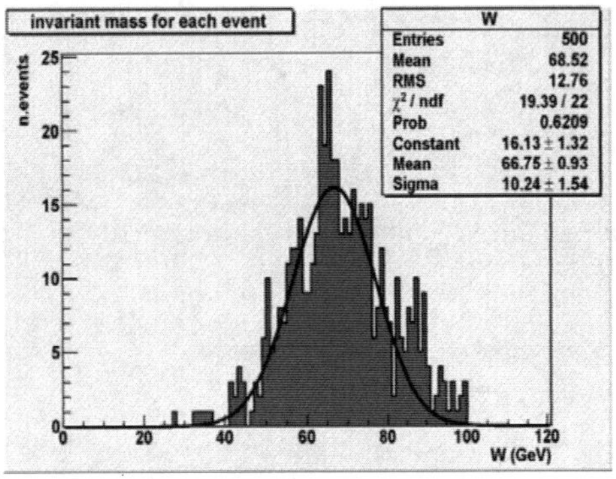

Figure 6a. W invariant mass distribution before corrections

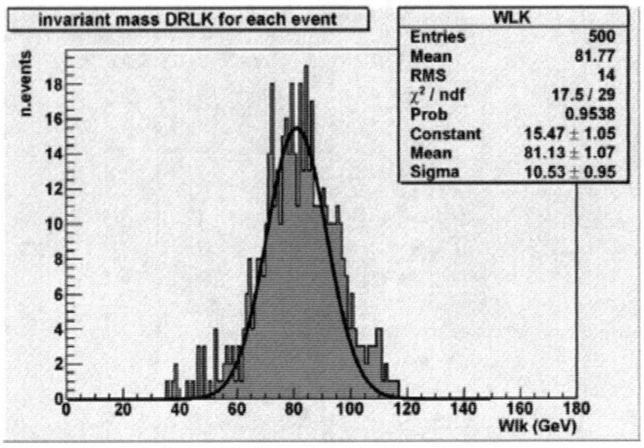

Figure 6b. W invariant mass distribution after corrections

of the DR correction should help and by restricting the study to jets from single W or W/Z to begin with, one can both try to implement corrections on reconstructed jets (as one must finally do in a realistic environment) and implement corrections independently of jet reconstruction by summing over the whole event, safe in the knowledge that the event contains only jets

In this spirit, hadron DR and leakage correction functions (from single pion data) are first applied to reconstructed jets (figure 6). New DR corrections are then evaluated for comparison by summing for S and C over all elements of the calorimeter (before jet reconstruction): they are smaller for all jet energies and this could be a reflection of the smaller average energy but also of a larger em content. However, no significant improvement is observed in the reconstructed W mass.

4. Correction for effect of Solenoid magnetic field

A magnetic field correction (ΔM) for every jet is based on the actual collection of charged tracks associated with the jet. The value of the correction reflects the change of the calculated mass of a calorimetric cluster due to the displacement of the energy deposited by charged particles. This change is derived using the momentum vector, as measured by the tracking detectors. The effect of this correction on the W invariant mass is shown in figure 7 for the maximum expected value of the solenoid field, both before and after the DR correction (applied using pion correction functions)
The residuals plotted in figure 7a are defined as follows:

$$R1 = \frac{(M_{corrected} - M_{true})}{M_{true}} \qquad R2 = \frac{M'_{corrected} - M_{true}}{M_{true}}$$

with

$$M'_{corrected} = M_{corrected} + \Delta M$$

where $M_{corrected}$ refers to the invariant mass after the DR correction and ΔM is the magnetic correction.

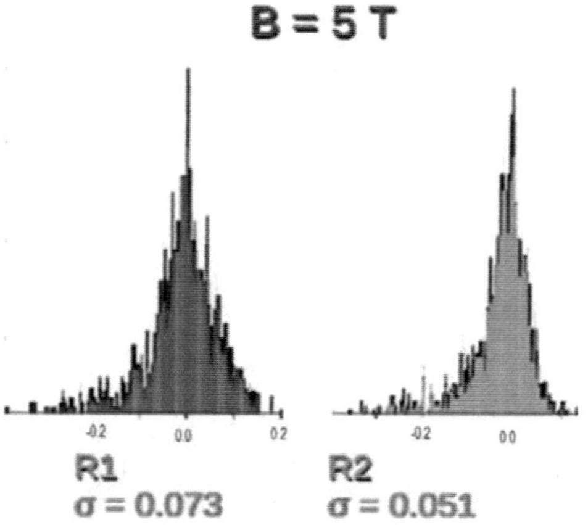

Figure 7a. Residuals (see text for definitions) illustrating effect of corrections on the reconstructed W mass at 5 Tesla

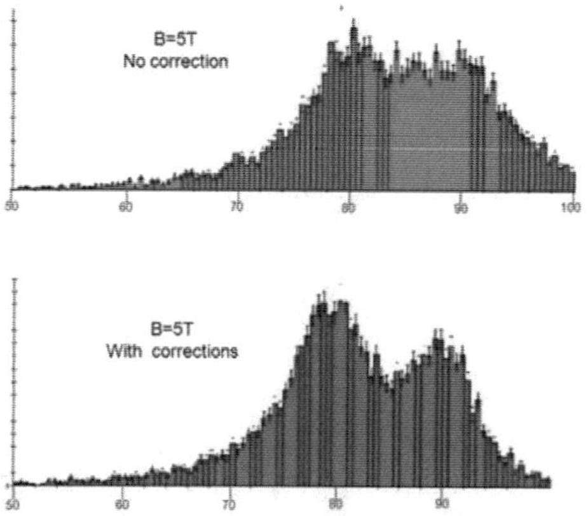

Figure 7b. Invariant mass distribution for a mixture of W and Z bosons before and after corrections

5. Conclusions

Corrections for leakage and magnetic field effects will be essential in future collider calorimeters – DR calorimeters are no exception. From a study involving single hadrons, on concludes that:

- In DR calorimeters, leakage must be accounted for in different stages: first in determining the DR correction and then in correcting for energy loss.
- Longitudinal segmentation is very effective to this end but does not correct for punch-through.
- Muon detector/tailcatcher assemblies can account for punch-through but tend to over-correct for leakage.

Despite the differences in content and energy distributions, one finds no great difference between single hadron DR correction functions and those obtained from simulated jet data. However leakage corrections are decidely smaller.

More work is needed to extend these studies to more realistic multijet environments.

References

[1] Para A 2010 *Total Absorption Dual Readout Hadron Calorimeter for future lepton colliders* (these proceedings)

[2] Wenzel H 2010 *Simulation of total absorption dual readout calorimetry: principles and performance for single hadrons and jets* (these proceedings)

[3] Wenzel H and Magill S 2010 *Enhancing the performance of a total absorption crystal calorimeter with dual readout (DR) by employing Particle Flow techniques* (these proceedings)

[4] Para A 2010 *Detailed Studies of hadron shower modeling in GEANT4* (these proceedings).

Journal of Physics: Conference Series **293** (2011) 012035 doi:10.1088/1742-6596/293/1/012035

The Status of the Telescope Array experiment

H. Tokuno[1], T. Abu-Zayyad[2], R. Aida[3], M. Allen[2], R. Azuma[1], E. Barcikowski[2], J. W. Belz[2], T. Benno[4], D. R. Bergman[5], S. A. Blake[2], O. Brusova[2], R. Cady[2], B. G. Cheon[6], J. Chiba[7], M. Chikawa[4], E. J. Cho[6], L. S. Cho[8], W. R. Cho[8], F. Cohen[9], K. Doura[4], C. Ebeling[2], H. Fujii[10], T. Fujii[11], T. Fukuda[1], M. Fukushima[9,22], D. Gorbunov[12], W. Hanlon[2], K. Hayashi[1], Y. Hayashi[11], N. Hayashida[9], K. Hibino[13], K. Hiyama[9], K. Honda[3], G. Hughes[5], T. Iguchi[1], D. Ikeda[9], K. Ikuta[3], S. J. J. Innemee[5], N. Inoue[14], T. Ishii[3], R. Ishimori[1], D. Ivanov[5], S. Iwamoto[3], C. C. H. Jui[2], K. Kadota[15], F. Kakimoto[1], O. Kalashev[12], T. Kanbe[3], H. Kang[16], K. Kasahara[17], H. Kawai[18], S. Kawakami[11], S. Kawana[14], E. Kido[9], B. G. Kim[19], H. B. Kim[6], J. H. Kim[6], J. H. Kim[20], A. Kitsugi[9], K. Kobayashi[7], H. Koers[21], Y. Kondo[9], V. Kuzmin[12], Y. J. Kwon[8], J. H. Lim[16], S. I. Lim[19], S. Machida[1], K. Martens[22], J. Martineau[2], T. Matsuda[10], T. Matsuyama[11], J. N. Matthews[2], M. Minamino[11], K. Miyata[7], H. Miyauchi[11], Y. Murano[1], T. Nakamura[23], S. W. Nam[19], T. Nonaka[9], S. Ogio[11], M. Ohnishi[9], H. Ohoka[9], T. Okuda[11], A. Oshima[11], S. Ozawa[17], I. H. Park[19], D. Rodriguez[2], S. Y. Roh[20], G. Rubtsov[12], D. Ryu[20], H. Sagawa[9], N. Sakurai[9], L. M. Scott[5], P. D. Shah[2], T. Shibata[9], H. Shimodaira[9], B. K. Shin[6], J. D. Smith[2], P. Sokolsky[2], T. J. Sonley[2], R. W. Springer[2], B. T. Stokes[5], S. R. Stratton[5], S. Suzuki[10], Y. Takahashi[9], M. Takeda[9], A. Taketa[9], M. Takita[9], Y. Tameda[1], H. Tanaka[11], K. Tanaka[24], M. Tanaka[10], J. R. Thomas[2], S. B. Thomas[2], G. B. Thomson[5], P. Tinyakov[12,21], I. Tkachev[12], T. Tomida[3], R. Torii[9], S. Troitsky[12], Y. Tsunesada[1], Y. Tsuyuguchi[3], Y. Uchihori[25], S. Udo[13], H. Ukai[3], B. VanKlaveren[2], Y. Wada[14], M. Wood[2], T. Yamakawa[9], Y. Yamakawa[9], H. Yamaoka[10], J. Yang[19], S. Yoshida[18], H. Yoshii[26], and Z. Zundel[2]

[1] Tokyo Institute of Technology, Meguro, Tokyo, Japan
[2] University of Utah, High Energy Astrophysics Institute, Salt Lake City, Utah, USA
[3] University of Yamanashi, Interdisciplinary Graduate School of Medicine and Engineering, Kofu, Yamanashi, Japan
[4] Kinki University, Higashi Osaka, Osaka, Japan
[5] Rutgers University, Piscataway, USA
[6] Hanyang University, Seongdong-gu, Seoul, Korea
[7] Tokyo University of Science, Noda, Chiba, Japan
[8] Yonsei University, Seodaemun-gu, Seoul, Korea
[9] Institute for Cosmic Ray Research, University of Tokyo, Kashiwa, Chiba, Japan
[10] Institute of Particle and Nuclear Studies, KEK, Tsukuba, Ibaraki, Japan
[11] Osaka City University, Osaka, Osaka, Japan
[12] Institute for Nuclear Research of the Russian Academy of Sciences, Moscow, Russia

[13] Kanagawa University, Yokohama, Kanagawa, Japan
[14] Saitama University, Saitama, Saitama, Japan
[15] Tokyo City University, Setagaya-ku, Tokyo, Japan
[16] Pusan National University, GeumJeong-gu, Busan, Korea
[17] Waseda University, Advanced Research Institute for Science and Engineering, Shinjuku-ku, Tokyo, Japan
[18] Chiba University, Chiba, Chiba, Japan
[19] Ewha Womans University, Seodaaemun-gu, Seoul, Korea
[20] Chungnam National University, Yuseong-gu, Daejeon, Korea
[21] University Libre de Bruxelles, Brussels, Belgium
[22] University of Tokyo, Institute for the Physics and Mathematics of the Universe, Kashiwa, Chiba, Japan
[23] Kochi University, Kochi, Kochi, Japan
[24] Hiroshima City University, Hiroshima, Hiroshima, Japan
[25] National Institute of Radiological Science, Chiba, Chiba, Japan
[26] Ehime University, Matsuyama, Ehime, Japan

E-mail: htokuno@cr.phys.titech.ac.jp

Abstract. The purpose of The Telescope Array experiment is to identify origin of the ultra high energy cosmic rays. The Telescope Array is a hybrid detector consists of a surface detector array and air fluorescence detectors. This hybrid detector is observing extensive air showers to measure the energy spectrum, anisotropy and composition of Ultra High Energy Cosmic Rays. The detector construction has been completed in March 2008, and the hybrid observation with the full configuration has been running since that time. In this talk, the status of observation and our prospects are described.

1. Introduction

The main purpose of the Telescope Array experiment is to identify sources of Ultra High Energy Cosmic Rays (UHECRs). UHECRs must interact with the cosmic microwave background, and a cutoff (GZK cutoff) in their energy spectrum is expected around its energy $10^{19.5}$ eV by Greisen, Zatsepin and Kuzmin [1, 2]. HiRes and Auger experiment reported that there is a cut-off structure in their observed energy spectrum [3, 4]. In contrast, the AGASA result shows that there are 11 events above 10^{20} eV [5]. These three experimental groups claim their systematic uncertainty in energy determination is about 20% respectively [3, 4, 5]. The rescaling energy spectra of AGASA, HiRes, and Auger within their uncertainties are in agree well below 10^{20} eV [6]. It seems that the inconsistency is due to the systematic uncertainties of these experiments in energy determination. To resolve this situation, we compare the response of AGASA type detector (plastic scintillation counter array) with that of HiRes type detector (air fluorescence light detector) directly using observed same EAS events.

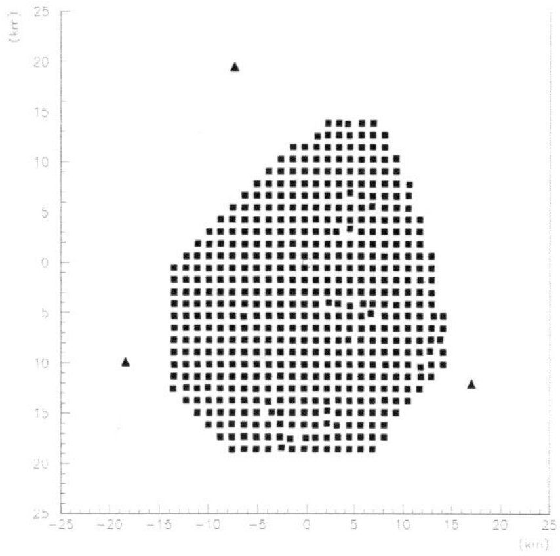

Figure 1. Layout of SD array (squares), and FD sites (triangles).

The Telescope Array is a hybrid detector consists of a surface detector (SD) array and fluorescence detectors (FDs) [7, 8]. This detector is located in Delta, Utah, USA (39.3 N, 112.9 W) with an

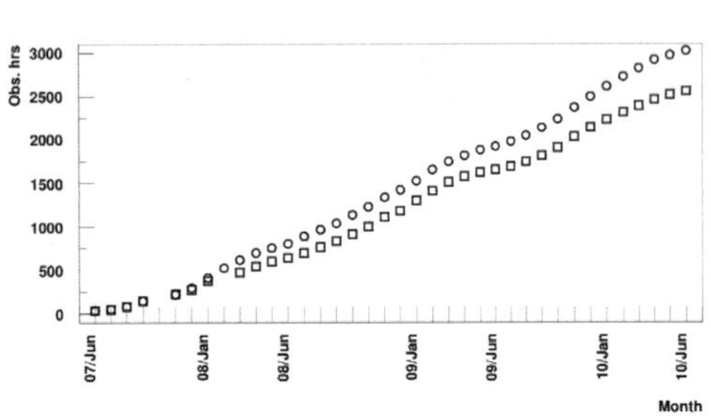

Figure 2. Total observation hours of FD from 2007/June to (circle: BR, square: LR).

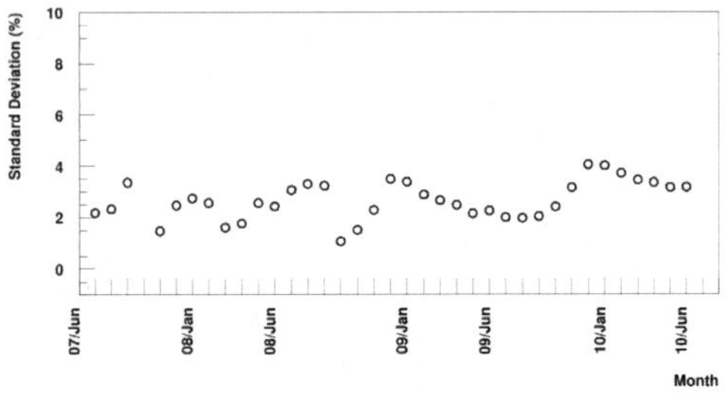

Figure 3. Standard deviation of relative PMT gains (Mean value of gain is equaled to 1 for each camera).

average altitude of 1400 m. The SD array has 507 detectors, each 3 m^2 in area, with 1.2 km spacing, and its total area is 700 km^2. Layout of SD array is shown in Fig. 1 (squares). Each SD consist of two layers of plastic scintillators with 1.2 cm thickness.

Three FD stations are surrounding the SD array, and observe the night sky above the SD array. Triangles in Fig. 1 show the location of FD site (South East: Black Rock Mesa, South West: Long Ridge, North West: Middle Drum). The each field of view (FOV) of these FD stations is 108 degree (114 degree for the third station) in azimuth and 3−31 degree in elevation. These FOVs in azimuth cover the whole region of SD array. More details of our detector configuration are described in our previous papers [7, 8]. The detector construction has been completed by March 2008, and the hybrid observation with the full configuration has been running since that time.

2. Observation Status

2.1. Fluorescence Detector

The first FD station (BR) and a half of the second station (LR) are running from May 2008, and the another half of the second station has been started from December 2008. The trigger rate for a station is about 2 Hz with 6% dead time. The total running time of the first and second station is 3020, 2560 hours respectively at 2010 June (Fig. 2). The running time of the second station is reduced for safety backup procedures, because the second FD station is operated remotely from the first station since May 2009.

FD performances (PMT gain, mirror reflectance, window transmittance, etc.) are monitoring continuously [9]. To monitor PMT gains, we employ absolute light pulsars on three PMTs each camera, and Xe-flashers for relative PMT gain monitoring mounted on each mirror. Fig. 3 shows the standard deviation of relative PMT gains (mean values of PMT gain at each camera are equaled to 1) through all the observation periods. The relative PMT gain are stable within 4% differences as shown in this figure. Using these monitor data, PMT gains are corrected in our analysis.

Atmospheric condition monitoring is important for FD analysis, because fluorescence light will be attenuated by aerosol components between generating point and FD. To monitor atmospheric conditions, we employ three special devices: a Central Laser Facility located on the same distance from the three FD stations (CLF location in Fig. 1: the circle), LIDAR (LIght Detection And Ranging), and an infrared camera for cloud monitoring at the first station ([10, 11], and Y. Tsunesada et al., T. Tomida et al., Proc. of 31th International Cosmic Ray Conference, 2009). LIDAR measures back scattered light of laser by air molecules and by aerosols in the air, and FDs measure side scattered lights from CLF laser. From CLF and LIDAR, atmospheric

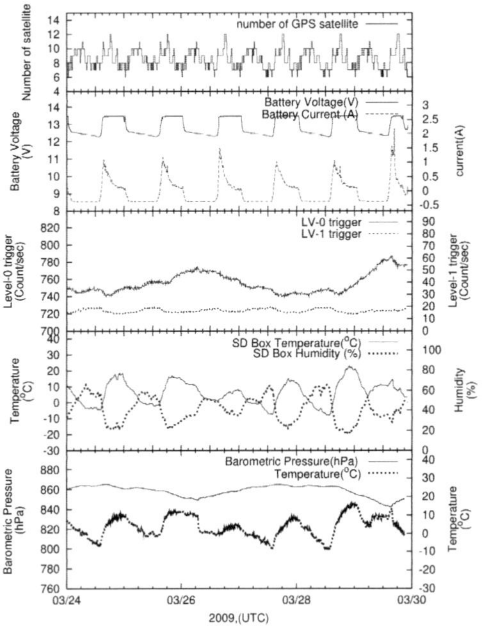

Figure 4. Plots of SD performance monitoring (status of GPS, Batteries, Low voltage power supply, Temperature, Humidity, Pressure).

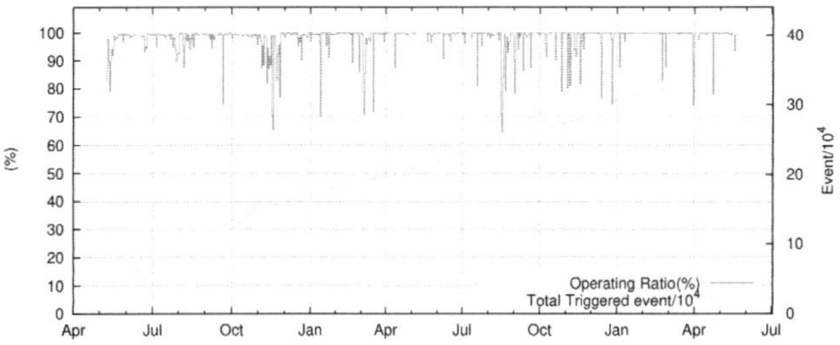

Figure 5. Plots of SD performance monitoring (Number of total Triggered event, Detector operating Ratio).

parameters are obtained, which include Vertical Aerosol Optical Depth (VAOD), extinction coefficients caused by Rayleigh scattering and Mie scattering. Clouds cause a strong attenuation of fluorescence light, therefore cloudiness in FOVs of FD is monitored by IR camera every 30 min.

The FDs at the third station were moved from HiRes-I, and this station have been running from December 2008. These FDs are calibrated and monitored by the same procedures of HiRes experiments and new additional equipments will provide for verification of their calibration and monitoring accuracies [12].

2.2. Surface Detector array

We installed 503 SDs in March 2008, and installed 4 additional SDs in November 2008. The trigger condition of the SD array is three adjacent detectors with more than three particle's equivalent signal within 8 usec. The average trigger rate is 20 triggers/hour under this condition. DAQ of SD was running with three sub-arrays from March 2008, and these three sub-arrays were integrated as a whole array in November 2008. This integration reduced array boundary, and increases number of useful events under our data quality cut in the analysis procedure. The total exposure will be comparable to that of AGASA with 13 years by 2010 Summer. SD performances are monitored every minutes independently of main DAQ. These information include the following items: Detector gain, accuracy of internal clock, charge level of battery, trigger rate, temperature and humidity inside the detectors, and accuracy of GPS clock (T. Nonaka et al., Proc. of 31th ICRC, 2009). Samples of the performance monitor are shown in Fig. 4,5. From this monitoring, the average number of available SD is more than 98 %, and down time of SD observation is less than 4% (Fig. 5). This on-line monitoring system help us to keep a stable DAQ running.

3. Prospects

We have obtained the following preliminary results: energy spectrum by FD + one SD analysis, photon limits, X_{max} distribution of air shower from FD experiments (see the presentation slides on the conference web page: http://bes3.ihep.ac.cn/conference/calor2010/). The FDs at the third station (MD) also provide a preliminary result of energy spectrum, which are consistent with HiRes-I experimental result (C.C.H. Jui et al. Proc. of 31th ICRC, 2009). In order to obtain energy spectrum from SD array, we have been studying its energy scale and robust procedures of energy determination by Monte Carlo simulation.

SD DAQ will be installed an additional trigger from FD in 2010 September. The SD trigger efficiencies is lower than 10% for 10^{18} eV from our simulation studies. In contrast, FD trigger efficiencies are higher than that for SD. However FD Mono analysis power for lower energetic events around 10^{18} eV is relatively lower. To improve analysis power, additional SD information is quite important. The FD mono analysis with SD information provides a better angular resolution and energy estimation accuracies than FD mono analysis. In this analysis, only information from one SD is enough to improve analysis power (D. Ikeda et al. 31th ICRC, 2009). For example, number of events which can be reconstructed by FD + one SD analysis will be 7 times larger than that of FD-Mono at 10^{18} eV. Estimated angler resolution of this analysis is also improved (e.g. better than 2° at 10^{18} eV).

Various equipments are applied to calibrate FDs optics and monitor for transparency of the atmosphere [10, 11, 12, 20]. Using these devices, we are monitoring FD performance, and atmospheric conditions continuously during observation periods. Estimation accuracy of FD aperture will also be studied carefully using these monitor data. We have obtained typical values of atmospheric parameters (e.g. VAOD, and extinction coefficient). We are studying the differences from typical values of these atmospheric parameters with various time scales, and their affect on FD analysis.

Uncertainty of fluorescence light yield is an issue of primary energy determination in air fluorescence experiments. Fluorescence light yield have been measured by some authors (e.g. [14, 15, 16, 17, 18]). However their experimental results cannot compare directly, because their experimental setting and units of results are different from each other. J. Rosado et al. [13] compared these results from different experiments, and their result suggests that there are 15% difference between the experiment results (Table 1 in [13]). Also some experimental results differ from estimated value by J. Rosado et al. about 10%. In order to obtain correct value of fluorescence yield, we need to understand these differences of experimental results, and the difference between experimental results and estimated values. To reduce this uncertainty, we have a plan to measure fluorescence light from controlled electrons in air using our FDs [19]. The module of electron light source will makes 10^9 electron bunch with their energy 40 MeV, and this pulse width is 2 us. This electron bunch will provide a standard candle of fluorescence light. This module has been installed at the front of the BR station with the distance 100 m between FD and the light axis. We are adjusting its setting to get stable electron bunches. This measurement will provide an End to End calibration which includes the energy loss rate of electron in air, fluorescence yield, and electronics performances of our FDs. From this measurement, our systematic uncertainties of primary energy estimation by FD will be reduced.

4. Summary

The hybrid detector of the Telescope Array has been measured EAS events since March 2008. Monitoring system of SD and FD performance and atmospheric condition is also running. The total exposure of SD array will be comparable to that of AGASA with 13 years by 2010 Summer. Some preliminary results were presented in this conference. In order to obtain robust analysis results, we are studying the energy scale comparison between SD and FD, FD-mono analysis comparison between the three station, etc. Also we will employ new devices: a standard candle of fluorescence light emitted from controlled electron bunch, an additional SD trigger from FD, etc in 2010.

Acknowledgments

The Telescope Array experiment is supported by the Ministry of Education, Culture, Sports, Science and Technology-Japan through Kakenhi grants on priority area (431) "Highest Energy Cosmic Rays", basic research awards 18204020(A), 18403004(B) and 20340057(B); by the U.S. National Science Foundation awards PHY-0307098, PHY-0601915, PHY-0703893, PHY-0758342, and PHY-0848320 (Utah) and PHY-0649681 (Rutgers); by the Korea Research Foundation (KRF-2007-341-C00020); by the Korean Science and Engineering Foundation (KOSEF, R01-2007-000-21088-0); by the National Research Foundation of Korea (NRF, 2010-0028071); by the Russian Academy of Sciences, RFBR grants 07-02-00820a and 09-07-00388a (INR), the FNRS contract 1.5.335.08, IISN and Belgian Science Policy under IUAP VI/11 (ULB). The foundations of Dr. Ezekiel R. and Edna Wattis Dumke, Willard L. Eccles and the George S. and Dolores Dore Eccles all helped with generous donations. The State of Utah supported the project through its Economic Development Board, and the University of Utah through the Office of the Vice President for Research. The experimental site became available through the cooperation of the Utah School and Institutional Trust Lands Administration (SITLA), U.S. Bureau of Land Management and the U.S. Air Force. We also wish to thank the people and the officials of Millard County, Utah, for their steadfast and warm supports. We gratefully acknowledge the contributions from the technical staffs of our home institutions and the University of Utah Center for High Performance Computing (CHPC).

References

[1] K.Greisen, 1966, Phys. Rev. Lett., v16, 748,
[2] T.Zatsepin and V.A.Kuzmin,1966, JETP Lett., v4, 78
[3] J. Abraham et al., 2008, Phys. Rev. Lett., v101, 061101
[4] R.U. Abbasi et al., 2008, Phys. Rev. Lett., v100, 101101
[5] M. Takeda et al., 2003, Astropart. Phys., v19, 447
[6] V. Berezinsky, 2009, Proc. of 30th ICRC, v6, 21
[7] H. Kawai et al., 2008, Nucl.Phys.B Proc.Suppl,v175-176, 221
[8] H. Kawai et al., 2009, J.Phys.Soc.Jpn. Suppl. A ,v78, 108
[9] H. Tokuno et al., 2009, NIM A, v601, 364
[10] S. Udo et al., 2008, Proc. of 30th ICRC, v5, 1021
[11] M. Chikawa et al., 2008, Proc. of 30th ICRC, v5, 1025
[12] J. N. Matthews et al., 2007, Proc. of 30th ICRC, v5, 1157
[13] J. Rosado et al., 2010, Astropart. Phys. in press (doi:10.1016/j.astropartphys.2010.06.007)
[14] F. Kakimoto et al., 1996, NIM A, v372, 527
[15] M. Nagano et al., 2004, Astropart. Phys. v22, 235
[16] R. Abbasi et al., 2008, Astropart. Phys. v29, 77
[17] M. Bohacova et al., 2009, Nucl. Phys. B Proc. Suppl. v190, 266
[18] P. Colin et al., 2007, Astropart. Phys. v27, 317
[19] T. Shibata et al., 2008, NIM A, v597, 61
[20] H. Tokuno et al., 2008, Proc. of 30th ICRC, v5, 1013

Journal of Physics: Conference Series **293** (2011) 012036 doi:10.1088/1742-6596/293/1/012036

The Fluorescence Detector of the Pierre Auger Observatory

Petr Nečesal for the Pierre Auger Collaboration

Fyzikální ústav AV ČR, v. v. i., Na Slovance 2, Praha 8, 182 21, Czech Republic and Observatorio Pierre Auger, Av. San Martin Norte 304, 5613 Malargüe, Argentina; http://www.auger.org/archive/authors_2010_08.html

E-mail: necesal@fzu.cz

Abstract. The Pierre Auger Observatory is a facility designed for the study of ultra-high energy cosmic rays. The Observatory combines two different types of detectors: a surface array of 1600 water Cherenkov stations placed on a 1.5 km triangular grid covering over 3000 km^2; and a fluorescence detector of 24 telescopes located in 4 buildings at the perimeter of the surface array. The fluorescence telescopes, each consisting of 440 photomultipliers, collect the ultraviolet light produced when the charged secondary particles in an air shower excite nitrogen molecules in the atmosphere. Because the intensity of the nitrogen fluorescence is proportional to the energy deposited in the atmosphere during the air shower, the air fluorescence measurements can be used to make a calorimetric measurement of the cosmic ray primary energy. Showers observed independently by the surface array and fluorescence telescopes, called hybrid events, are critical to the function of the Observatory, as they allow for a model-independent calibration of the surface detector. In this paper I describe the detector and the most important measurements.

1. Introduction

Ultra-high energy cosmic rays (UHECRs) are particles coming to the atmosphere of the Earth with energies that can exceed 10^{20} eV. While the first observation of cosmic rays (CRs) was made one hundred years ago, origins of these particles is not yet understood. Much of the reason is because of the very low flux of UHECRs at Earth. The differential flux of UHECRs is steeply falling with energy, and can be described by a power law

$$\frac{dN}{dE} \propto E^{-\gamma} \qquad (1)$$

with spectral index γ between 2.6 and 3.2. One can find two significant points in the energy spectrum – at 4×10^{15} eV there is so called 'knee' where γ changes from 2.7 to 3.1 [1] and known 'ankle' can be found at 4×10^{18} eV with another spectral index change from 3.3 to 2.6 [2]. Processes and mechanisms of achieving ultra-high energies are still undiscovered. So the main questions which people want to answer concern composition, energy spectrum and sources of UHECRs.

Primary CRs hitting the Earth's atmosphere interact with air molecules and fragment.The secondary particles produced in this collision will undergo further hadronic and electromagnetic interactions, ultimately producing an extensive air shower cascade. As the air shower develops in the atmosphere the number of secondary particles increases and their average energy falls.

Due to energy losses and decays, the shower will reach a maximum size, and then number of particles in the shower will begin to decrease. However, there are still enough energetic particles in a vertical air shower which can hit ground and can be detected by ground detectors. With increasing zenith angle of the primary particle an air shower must propagate through larger atmospheric mass and therefore only less particles survive at the ground level.

There are two possible types of UHECR detection – direct and indirect methods. Direct methods can be performed by means of satellites and balloons, but cannot considerably measure UHECR at energies larger than 10^{14-15} eV because of small statistics (Eq. 1). On the other hand indirect methods require sufficient energy ($E \gtrsim 10^{15}$) eV to develop shower and large detector area is necessary. Indirect methods are based on the measurement of secondary particles that reach the ground in array of Cherenkov detectors, scintillators or muon detectors or on the collection of the isotropic fluorescence light. Secondary charged particles in the shower excite nitrogen molecules, which then isotropically emit fluorescence light into several bands between 300 and 420 nm. These photons can be detected by fluorescence telescopes.

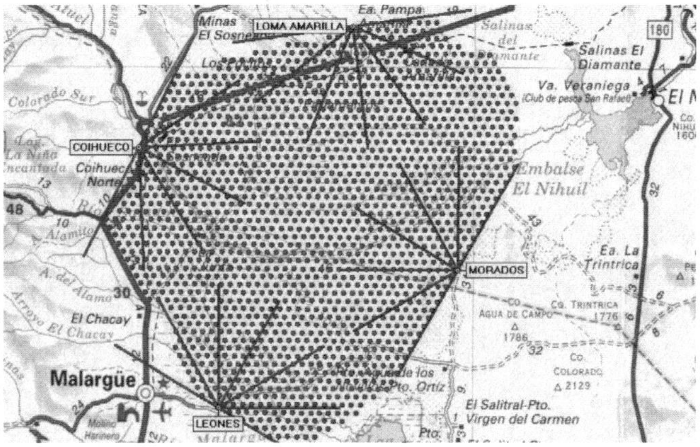

Figure 1. The detector layout of the Pierre Auger Observatory. Red dots indicate SD stations which are overviewed by 4 FD buildings. There are 6 fluorescence telescopes in each building. Green lines indicate field of view of each telescope.

Figure 2. A picture of an SD station and FD building. Each station has its own solar panel with battery and communication antenna. There is also the antenna nearby the FD building.

2. Pierre Auger Observatory

The Pierre Auger Observatory [3] is a facility studying UHECRs by means of two type of detectors: a surface detector (SD) which is composed of 1600 water Cherenkov stations and a fluorescence detector (FD) that comprises 24 fluorescence telescopes. The southern site of the Observatory is located in province Mendoza in Argentina near the city Malargüe (69° W, 35° S, 1420 m a.s.l.), the northern site is intended to be built in Colorado, USA. The layout of the southern site detectors is shown in Fig. 1, and the picture of an SD station and an FD building is shown in Fig. 2. Stations are deployed on a triangular grid with 1.5 km spacing. They cover an area of over 3000 km^2 and they have nearly 100 % duty cycle. There are 3 photomultipliers in pure water in each station.

The Pierre Auger Observatory is designed to measure UHECRs with energy above $\sim 10^{18}$ eV. However, there is also a low energy extension of SD station infill with plastic scintillators beneath the stations to detect muons (AMIGA [4]) and a set of 3 fluorescence telescopes with high elevation called HEAT [4]. An extensive program of atmospheric monitoring is essential part of the Observatory not only to control uncertainties concerning FD.

3. Fluorescence Detector

There are 4 buildings with 6 fluorescence telescopes each that overlook the area of SD stations (Fig. 1). Each telescope uses Schmidt optics and consists of a wide-angle, segmented spherical mirror, a spherical focal plane, a UV 300-410 nm passband filter, and a refractive corrector ring at the aperture of the telescope. The telescope field of view is 30° (in azimuth) x 28° (in elevation) so each building has 180° azimuth range (the 3 high-elevation telescopes observe 30° to 60° in elevation). There are 440 photomultipliers in the focal plane which collect reflected light. The longitudinal profile of a shower is thus measured as an image of active PMT pixels along the shower axis. The layout of the FD geometry is shown in Fig. 3.

There are several trigger levels. The first level digitizes signals from an analog board in each pixel at 10 MHz. The second level trigger is also an internal trigger to search for track segments of at least five active pixels. Every 100 ns, a scan over the full camera is performed and the triggered pixels are searched for track-like patterns. The third level trigger is a software algorithm that rejects events caused by lightning, muons which impact the focal plane, or randomly trigged pixels.

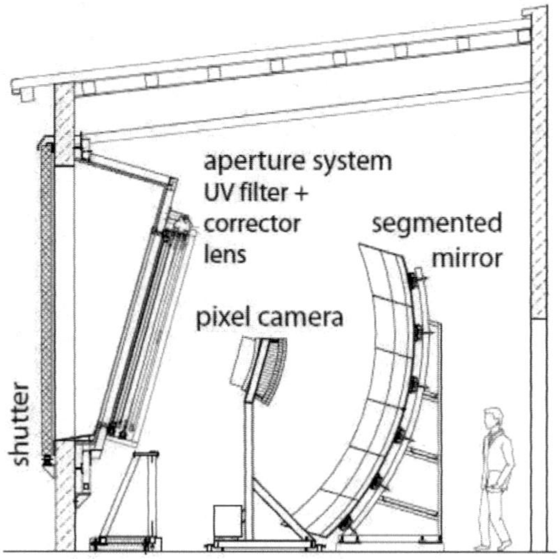

Figure 3. A scheme of telescope geometry. Light comes through the aperture, covered by a UV filter and a corrector ring, and it is reflected by a segmented mirror to the pixel camera with 440 photomultipliers.

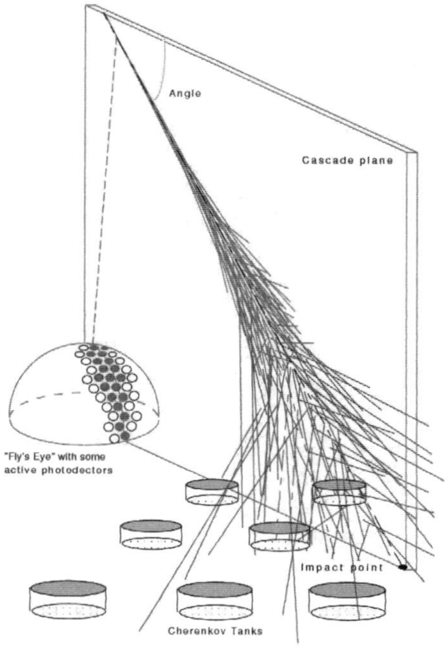

Figure 4. A cartoon of the hybrid detection technique. Charged particles at ground level are detected in SD stations while fluorescence light is collected in FD.

FD telescopes are calibrated to find the proper conversion between digitized counts and the true light flux (in photons). This calibration is performed in several steps using absolute and relative methods. The absolute calibration uses a calibrated light source (known as a 'drum') mounted at the telescope aperture [5]. Using the drum, the light flux at each pixel is known and the response is measured. In addition to the drum calibration, vertical laser shots at wavelengths 337 and 355 nm [6, 7] are used as an independent calibration method. Between absolute calibrations (which occur several times per year), the relative response of each fluorescence camera is recorded using light pulses from LEDs and xenon flashers. The relative calibration occurs before and after each night of observations.

3.1. Hybrid Detection Technique

The two measurement techniques in use at the Pierre Auger Observatory – surface detection and air fluorescence detection – are quite complementary. (Fig. 4). Whereas SD has 100% duty cycle time and acceptance can be calculated and it is model independent, FD has duty cycle of $\approx$ 13% [5] and acceptance depends on the atmospheric conditions (model dependent). SD measurement corresponds to one slice of a calorimeter at the last stage of shower development. To get the energy only from SD, Monte Carlo interaction models are needed in experiments without fluorescence detectors. On the other side FD collects photons emitted during development of the shower. It enables to sample a shower along its axis. Measured number of photons in different atmospheric slant depths can be converted to deposited energy to get shower longitudinal profile (Fig. 5). Then, the energy is measured directly with the atmosphere acting as a calorimeter for CR study. Events that are measured simultaneously by FD and SD are called hybrid.

Apart from the energy another very important characteristic of the shower development is the position of shower maximum. Shower maximum is the atmospheric depth where the shower longitudinal development reaches the maximum number of secondary particles.

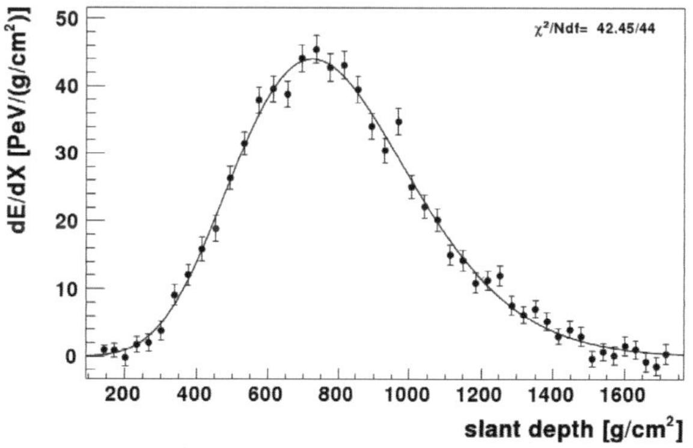

Figure 5. Longitudinal profile of energy deposit reconstructed from measured light by FD telescope. The line represents a Gaisser-Hillas function fitted to the measured profile.

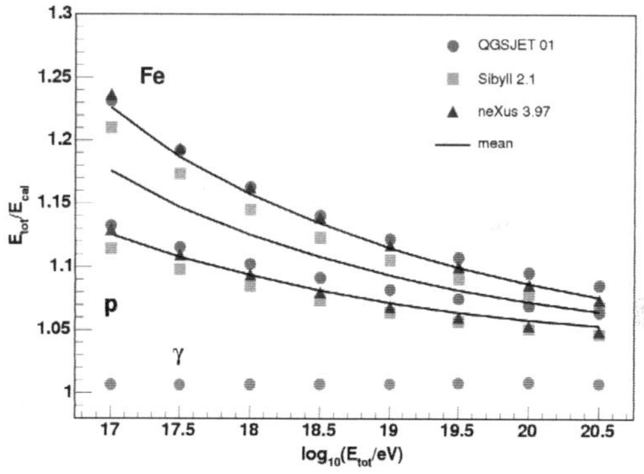

Figure 6. Correction factor for the missing energy calculated by several models (dots) with fitted curves [8] for proton and iron primaries.

3.2. FD Energy Measurement and Shower Reconstruction

Electromagnetic energy losses of charged particles are responsible for emitted fluorescence light. Therefore FD measures longitudinal shower profile. About 20 photons are emitted between 300 and 400 nm per 1 MeV loss. The total number of photons is proportional to the deposited energy of the air shower. Thus, the profile integral gives almost the whole shower energy ($\sim$ 90% of the total energy at 10^{19} eV):

$$E_{cal} = \int dX \frac{dE}{dX}. \quad (2)$$

The remaining part ($\sim$ 10% of the total energy at 10^{19} eV) of the shower energy cannot be directly measured by collecting emitted light. It concerns energy of neutrinos, muons and other particles which do not excite nitrogen molecules. This type of energy is called 'missing' and it is not included in the integral of the energy deposit derived from FD data. Nevertheless, one can calculate the correction factor that depends on the primary particle type and energy (Fig 6).

3.3. SD Energy Calibration

The biggest advantage of hybrid detection is the model independent energy calibration of the SD. The energies of hybrid events are determined from FD measurement, and then the signal from the SD can be calibrated. The SD records signals from electromagnetic and muonic shower component. One shower is typically large enough to hit many SD stations. Therefore one can derive (fit) the lateral distribution function of the signal at ground. The aim of such a procedure is to find the particle signal at a distance of 1000 m from the shower core (S_{1000}). The ground parameter S_{1000} depends on the zenith angle of the primary particle. The variable corrected for the zenith angle dependency is called S_{38} and corresponds to signal S_{1000} at 38° zenith angle. The calibration is based on finding the dependency of S_{38} on the energy measured by FD (Fig. 7).

3.4. Uncertainties on the Reconstructed Energy from FD

The use of the atmosphere as a calorimeter also determines the main uncertainties of the measurement. The total systematic uncertainty in the FD energy scale is 22%, with the major contributions due to uncertainties in the fluorescence yield (14%) and atmospheric transmission (7% for aerosol scattering, 1% for Rayleigh scattering). There are also 9% uncertainties in the telescope absolute calibration, and the longitudinal shower profile reconstruction contributes 10% to the total energy uncertainty. Missing energy gives 4% uncertainty.

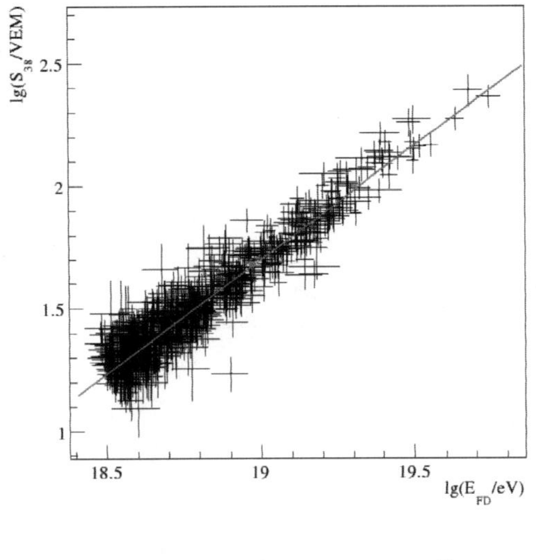

Figure 7. Ground parameter S_{38} as a function of the energy as measured with the FD for the 795 hybrid events [9].

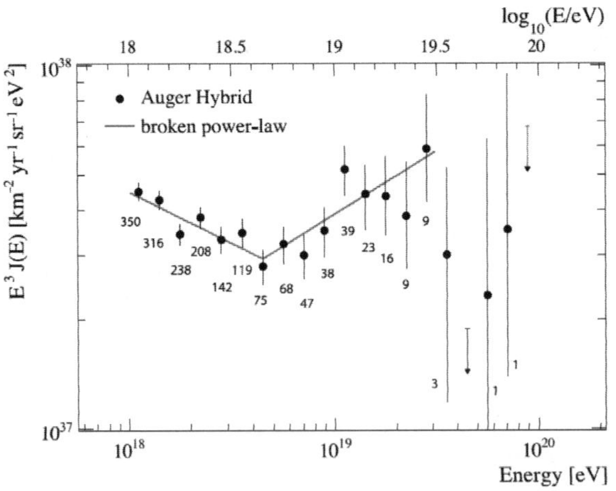

Figure 8. The energy spectrum of UHECR determined from hybrid measurements between November 2005 and May 2008 [2]. Statistical uncertainties are shown with a broken power law used to determine the position of the ankle [2], which is clearly visible.

Several devices have been installed to monitor atmospheric conditions, especially those for optical transmission determination. For example, the Central Laser Facility [10], Lidar stations [11], and the FRAM telescope [12] are used for atmospheric monitoring program.

The measurement of the shower maximum requires a conversion of geometrical altitude as observed by FD to atmospheric depth. The conversion depends on temperature, pressure and humidity, which vary with time and location. There are meteorological stations and a balloon launch facility to measure these thermodynamical parameters. Use of measured conditions at the site considerably improves the accuracy of the air shower reconstruction [13].

The fluorescence yield is the number of photons emitted in a given band per unit of energy loss by charged particles. It is an important quantity for the energy reconstruction. The Pierre Auger Observatory uses the absolute yield from Ref. [14].

4. Conclusions

The fluorescence detector of the Pierre Auger Observatory is an atmospheric calorimeter used for the hybrid detection of air showers. It enables a model-independent energy measurement and calibration of surface detector array. The first data taking started in late 2003. The Observatory has operated steadily at its full design size since July 2008. The fluorescence detector data have enabled for example precise measurements of the UHECR energy spectrum (Fig. 8), elongation rate [15] and CR composition.

5. Acknowledgements

The contribution is prepared with the support of Ministry of Education, Youth and Sports of the Czech Republic within the project LA08016 and with the support of the Charles University in Prague within the project 119810. Special thanks to my colleages from the Pierre Auger Collaboration for fruitful cooperation.

References

[1] Blümer J, Engel R, Hörandel J R 2009 Cosmic rays from the knee to the highest energies *Prog. Part. Nucl. Phys.* **63** 293 (*Preprint* arXiv:0904.0725 [astro-ph.HE]).
[2] Abraham J *et al.* (Pierre Auger Collaboration) 2010 Measurement of the energy of cosmic rays above 10^{18} eV using the Pierre Auger Observatory *Phys. Lett.* B **685** 239.
[3] Abraham J *et al.* (Pierre Auger Collaboration) 2004 *Nucl. Instrum. Meth.* A **523** 50.
[4] Abraham J *et al.* (Pierre Auger Collaboration) 2009 Operations of and Future Plans for the Pierre Auger Observatory *Proc. 31st Int. Cosmic Ray Conf. (Lodz, Poland)* (*Preprint* arXiv:0906.2354)
[5] Abraham J *et al.* (Pierre Auger Collaboration) 2010 The fluorescence detector of the Pierre Auger observatory *Nucl. Instr. and Meth. in Phys. Res.* A **620** 227.
[6] Roberts M D *et al.* (Pierre Auger Collaboration) 2003 Calibration of the Pierre Auger fluorescence detector *Proc. 28th Int. Cosmic Ray Conf. (Tsukuba, Japan)* (*Preprint* arXiv:astro-ph/0308410).
[7] Knapik R *et al.* (Pierre Auger Collaboration) 2007 The absolute, relative and multi-wavelength calibration of the Pierre Auger observatory fluorescence detectors *Proc. 30th Int. Cosmic Ray Conf. (Merida, Mexico)* (*Preprint* arXiv:0708.1924 [astro-ph]).
[8] Pierog T *et al.* 2005 *Proc. 29th Int. Cosmic Ray Conf. (Pune, India)* vol 7, p 103.
[9] Giulio C *et al.* (Pierre Auger Collaboration) 2009 The cosmic ray energy spectrum and related measurements with the Pierre Auger observatory *Proc. 31st Int. Cosmic Ray Conf. (Lodz, Poland)* (*Preprint* arXiv:0906.2189).
[10] Fick B *et al.* 2006 The central laser facility at the Pierre Auger observatory *JINST* **1** P11003.
[11] BenZvi S Y *et al.* 2007 The lidar system of the Pierre Auger observatory *Nucl. Instr. Meth.* A **574** 171.
[12] Prouza M, Jelínek M, Kubánek P, Ebr J, Trávníček P, Šmída R 2010 *Advances in Astronomy* vol 2010, Article ID 849382.
[13] Abraham J *et al.* (Pierre Auger Collaboration) 2010 *Astropart. Phys.* **33** 108.
[14] Nagano M *et al.* 2004 *Astropart. Phys.* **22** 235.
[15] Abraham J *et al.* (Pierre Auger Collaboration) 2010 Measurement of the depth maximum of extensive air showers above 10^{18} eV *Phys. Rev. Letters* **104** 091101.

Calorimeter R&D for the SuperNEMO Double Beta Decay Experiment

Anastasia Freshville on behalf of the SuperNEMO Collaboration
Department of Physics and Astronomy
University College London, Gower Street, London, UK, WC1E 6BT
E-mail: af@hep.ucl.ac.uk

Abstract.
SuperNEMO is a next generation double beta decay experiment, which will employ the successful tracker-calorimeter technique used in the currently running NEMO3 [1] experiment (located in the Laboratoire Souterrain de Modane (LSM)). SuperNEMO will study 100kg of the ^{82}Se isotope, reaching a sensitivity of a half-life greater than 10^{26} years for neutrinoless double beta decay ($0\nu\beta\beta$). This corresponds to a Majorana neutrino mass of $50-100$meV. One of the main goals and challenges of SuperNEMO R&D is to reach an unprecedented energy resolution of 4% FWHM at 3 MeV ($Q_{\beta\beta}$ value of ^{82}Se) for the calorimeter, which has been achieved by the collaboration.

1. Introduction

Neutrinoless double beta decay ($0\nu\beta\beta$), $2n \longrightarrow 2p + 2e^-$, is a lepton number violating process forbidden in the Standard Model which occurs via the exchange of a virtual neutrino. It is allowed only if the neutrino has a non-zero mass (as established by neutrino oscillation experiments) and if the neutrino is a Majorana particle (the neutrino and anti-neutrino are the same, as proposed by Ettore Majorana [2]) and can therefore provide further insight into the neutrino, such as its absolute mass and nature.

The experimental signature of $0\nu\beta\beta$ is two electrons coming from a common vertex with an energy sum equal to the $Q_{\beta\beta}$ value of the measured isotope. The half-life $T^{0\nu}_{1/2}$ measured from $0\nu\beta\beta$ experiments can be directly related to the electron neutrino mass $\langle m_{\nu_e}\rangle$, as shown in Equation 1, where $G^{0\nu}$ is the kinematic phase space ($G^{0\nu} \propto Q^5_{\beta\beta}$) and $M_{0\nu}$ is the nuclear matrix element. Equation 1 shows $0\nu\beta\beta$ via light neutrino exchange (the mass mechanism mode) as this mode requires minimal modifications to the Standard Model, however other modes also exist.

$$[T^{0\nu}_{1/2}]^{-1} = G^{0\nu}|M_{0\nu}|^2\langle m_{\nu_e}\rangle^2 \quad (1)$$

2. SuperNEMO Detector

SuperNEMO is based on the technology of the currently running NEMO3 experiment, using a tracker-calorimeter detection technique for particle identification. The detector will hold $\sim$ 100kg of double beta decay source isotope (^{82}Se is the 'baseline' design isotope, with ^{150}Nd and ^{48}Ca being considered depending on enrichment possibilities) to reach a projected sensitivity

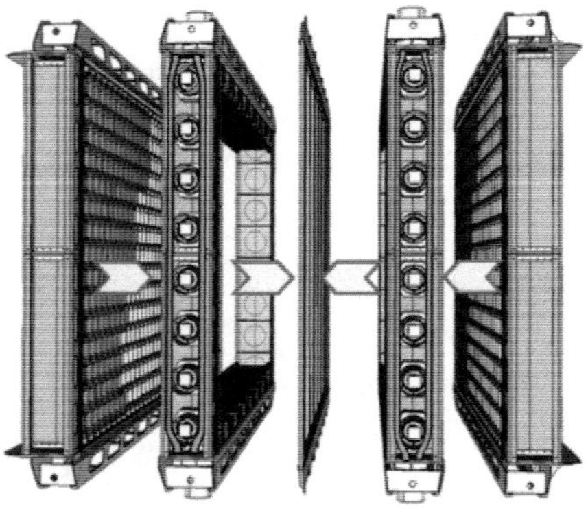

Figure 1: View of a SuperNEMO module's components: (from left to right) calorimeter wall, tracker, source foil, tracker and calorimeter wall.

of 10^{26} years (50 - 100 meV effective Majorana neutrino mass). The baseline design consists of 20 modules ($\sim$ 4m (height) x 6m (length) x 2m (width)), each containing 5kg of isotope source foil, surrounded by $\sim$ 2000 Geiger mode drift cells and enclosed by the calorimeter walls (containing $\sim$ 550 8" PMTs), as seen in Figure 1.

SuperNEMO is currently nearing the end of a three year R&D study, with four main areas of focus: source foil production, tracker development, calorimeter development and ultra-low background materials production and measurements. The goals of the R&D are summarised in Table 1.

Table 1: SuperNEMO Parameters and Goals

Parameters	Goals
Isotope	^{82}Se
Mass	100kg
$0\nu\beta\beta$ Detection Efficiency	30%
Energy Resolution FWHM at 3 MeV	4%
^{214}Bi Purity	$< 10\mu$Bq/kg
^{208}Tl Purity	$< 2\mu$Bq/kg
Operation Time	5 years
$T_{1/2}^{0\nu\beta\beta}$ Sensitivity	10^{26} years
Effective Majorana Mass $\langle m_{\nu_e} \rangle$	$50 - 100$meV

$$T_{1/2}^{0\nu} > \frac{ln2 \cdot N_A}{A} \cdot \frac{\epsilon}{\kappa_{CL}} \cdot \sqrt{\frac{M \cdot t}{N_{bkg} \cdot \Delta E}} \qquad (2)$$

The energy resolution of the calorimeter is of the utmost importance to SuperNEMO, as can be illustrated by the half-life sensitivity formula [4] (Equation 2), where the energy resolution ΔE is as significant as the isotope mass M, the run time t and the number of background events N_{bkg}. Factors N_A and A are Avogadro's number and the atomic mass of the isotope respectively, ϵ is the detector efficiency and κ_{CL} is the number of excluded $0\nu\beta\beta$ events at the chosen confidence level of the half-life sensitivity $T_{1/2}^{0\nu}$. One of the dominant backgrounds to

$0\nu\beta\beta$ is the irreducible $2\nu\beta\beta$ tail, making the energy resolution the determining factor of the detector's sensitivity to $0\nu\beta\beta$. This can be seen in Figure 2, where simulations carried out for ^{82}Se with a calorimeter energy resolution of 12% and 7% FWHM at 1MeV respectively and normalised to a 10^{26} year $0\nu\beta\beta$ half-life are shown. It can be seen that a much clearer separation between the $2\nu\beta\beta$ tail and the $0\nu\beta\beta$ peak is achieved with a 7% energy resolution.

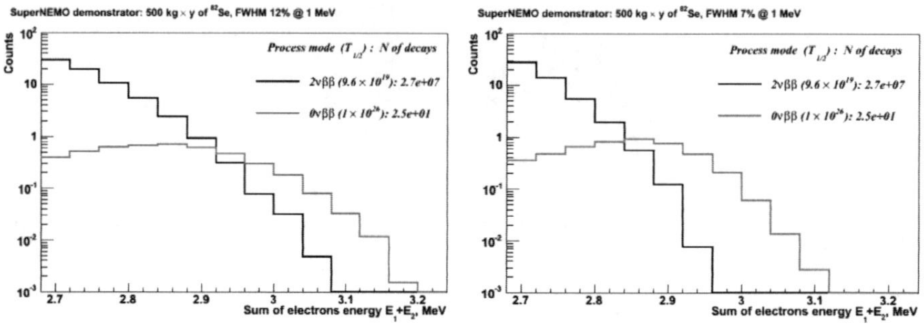

Figure 2: Calorimeter energy resolution simulations for ^{82}Se for 500kg·yr, with the $0\nu\beta\beta$ half-life (red) normalised to 10^{26} years. 12% (left) and 7% (right) FWHM at 1 MeV energy resolutions are shown, illustrating the overlap of the $2\nu\beta\beta$ (black) tail.

3. SuperNEMO Calorimeter Requirements

SuperNEMO has a need for a robust calorimeter (using reliable and time withstanding technology) with ease of manufacturing and assembly, providing unprecedented energy resolution results at a sensible cost. The calorimeter R&D for SuperNEMO consists of three main areas of study: energy and time resolution (see Section 3.1), calibration and photomultiplier tube (PMT) radio-purity.

During five years of data taking the gain and stability of $\sim 11,000$ PMTs must be monitored at a 1% level. The detector response must be linear, and any non-linear effects must be monitored at a 1% level up to $3-4$MeV (the region of interest for $0\nu\beta\beta$). The absolute calibration system will use ^{207}Bi sources (providing conversion electrons with single energies of 482keV and 976keV) inserted into the detector on a monthly basis. A UV-LED based light injection system is being developed for gain and linearity monitoring. Using low activity alpha sources embedded into the scintillator to monitor the gain is an additional possibility being studied. ^{60}Co (providing two coincident γs of 1.1 and 1.3 MeV) will occasionally be used for absolute time caiblration.

Specific to low background counting experiments ultra-pure materials must be used throughout the detector. One of the main sources of contamination comes from the PMTs, specifically from the cathode glass, which is closest to the active volume of the detector. The requirements for SuperNEMO have prompted development of low-background glass on a new scale and the collaboration is currently working closely with Hamamatsu to develop a radio-pure Barium salt free glass to reach the radio-purity requirements of the detector. Currently, the radioactivity limits reached with glass containing Barium salt are $^{40}\mathrm{K} < 0.6\mathrm{Bq/kg}$, $^{214}\mathrm{Bi} < 0.8\mathrm{Bq/kg}$ and $^{208}\mathrm{Tl} < 0.06\mathrm{Bq/kg}$.

3.1. Energy Resolution

In order to achieve a sensitivity of 10^{26} years SuperNEMO's calorimeter requires an energy resolution of $\frac{7\%}{\sqrt{E(MeV)}}$ (or 4% FWHM at 3MeV (the $Q_{\beta\beta}$ of ^{82}Se)). A high number of photo-electrons reduces the statistical error $1/\sqrt{N_{pe}}$ and leads to optimisation of the energy resolution, which can be simplified into three experimental objectives as seen in Equation 3.

$$\frac{N_{ph}}{E_e} \cdot \epsilon_{col}^{light} \cdot (QE^{PMT} \cdot \epsilon_{col}^{PMT}) = N_{pe} \quad (3)$$

$\frac{N_{ph}}{E_e}$ is the number of photons per unit energy and is determined by the scintillator light output. ϵ_{col}^{light} is the light collection efficiency and depends on the material, surface treatment and geometry of the scintillator, the reflector material and its efficiency and the optical coupling quality (gels, light guides etc). The scintillator must be a low Z material to minimise back scattering of low energy electrons and has to have a coincidence time resolution of $\sigma = 250$ps at 1MeV, as well as be radio-pure and cost effective. These requirements have led to a choice of plastic scintillator.

Intrinsic characteristics of the PMT include the quantum efficiency (QE) of the photo-cathode QE^{PMT}, the collection efficiency ϵ_{col}^{PMT} and, to a lesser extent, the gain of the first dynode. Significant breakthroughs have been made with new bi-alkali photo-cathodes developed by Hamamatsu and Photonis with a QE in the range of $35-43\%$ at the peak wavelength (compared to a previous QE of $\sim 25\%$) [5][6].

4. Calorimeter Design, R&D Setup and Measurements

Two possible designs for the calorimeter have been considered by the collaboration - the 'baseline' design, consisting of hexagonal 28cm diameter and 10cm minimum thickness polyvinyl-toluene (PVT) blocks and 8" PMTs (similar to that of NEMO3), and the 'bar' design, consisting of vertically placed 2m x 10cm x 2.5cm PVT bars and 3" PMTs at each end. The advantages of the block design are that a 7% FWHM at 1MeV energy resolution can be reached, good time resolution (250ps σ at 1MeV is anticipated) and the use of well understood technology from NEMO3 experience. $11,000$ block + PMT units would be required for the full detector, which increases the cost and the number of channels (and hence potentially the radioactivity). The advantages of the bar design are more efficient γ tagging (enabled by the layered design of the detector) and the reduction in the number of bar + PMT units to $5,000-7,000$, reducing the cost and the radioactivity. However, the 10% energy resolution and 400ps σ at 1 MeV time resolution achieved for the bar design do not meet the requirements of the SuperNEMO calorimeter. For this reason and due to the added difficulty of extrapolating the performance of the bar design to the full detector, the collaboration has chosen to focus on the baseline design.

4.1. Baseline Design R&D Setup and Measurements

The energy resolution measurement is carried out by exciting the scintillator with a flux of electrons of known energy and then analysing the resulting distribution. The monochromatic source of electrons approximates the delta function and therefore any smearing of the distribution that is seen is due to the light collection properties of the scintillator and PMT.

Many different scintillator, reflector and PMT combinations have been studied. The scintillators studied were liquid (toluene based) from CENBG, INR, ISM and JINR labs (consisting of 0.5% PPO and 0.0025% POPOP), liquid and solid hybrids, solid polystyrene (PST) from ISM and JINR labs (consisting of 1.5% PTP and 0.0175% POPOP) and solid PVT from Bicron (BC404 and BC408) and Eljen (EJ204 and EJ200). The specular and diffusive reflectors under study included Teflon, kapton, aluminised Mylar and Enhanced Specular Reflector (ESR) from Vikuiti and ReflechTech manufacturers. The three PMT competitors under study were Hamamatsu, Photonis and Electron Tubes Ltd. (ETL). Light-guide and optical gel studies to achieve the best optical contact between the scintillator and the PMT were also carried out.

Two different electron sources are used to measure the energy resolution. The first is a ^{207}Bi source, which is easy to implement in an experimental set up as the 976keV K-shell conversion electron (CE) is used. However, as the fitting function for ^{207}Bi needs to incorporate a convolution of X-rays, γs, L-shell and M-shell CEs the extraction of the energy resolution is complex. The second electron source is a ^{90}Sr β beam passed through a magnetic field to select monochromatic βs of known energy. The spectrum obtained from the beam can easily be fitted with a Gaussian function, however is more difficult to set up. Both methods have been used to cross check the energy resolution measurements acquired and have been found to be consistent (as shown in Figure 3).

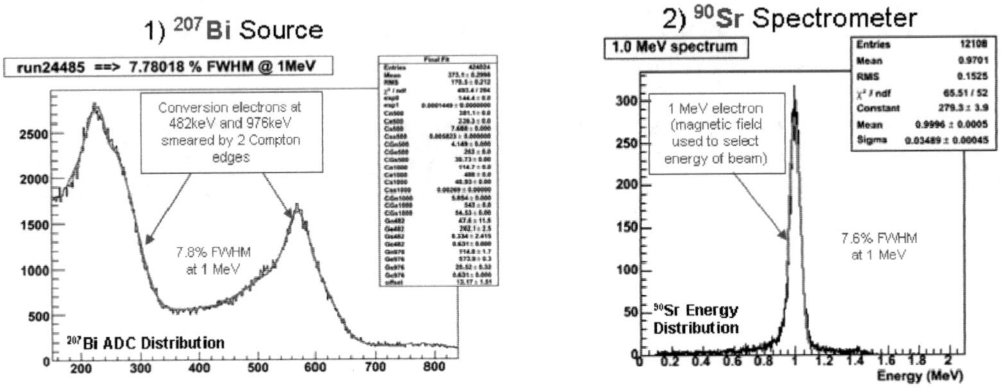

Figure 3: Two different fitting procedures for energy resolution acquisition using ^{207}Bi (left) and ^{90}Sr (right). The two methods are consistent with each other.

The target energy resolution of $\frac{7\%}{\sqrt{E(MeV)}}$ has been achieved with a solid PVT scintillator (EJ200) of a hexagonal shape, with the PMT looking directly at the hemispherical entry face of the block (the removal of the light-guide from the set up brought the breakthrough in achieving the target energy resolution) with a combination of aluminised Mylar wrapping on the entrance face and Teflon wrapping on the sides of the scintillator. An unprecedented energy resolution of $\frac{6.7\%}{\sqrt{E(MeV)}}$ has been achieved with an 8" high QE Photonis PMT, compared to that of $\frac{7.7\%}{\sqrt{E(MeV)}}$ with an 8" high QE Hamamatsu PMT. However, Photonis announced a full stop to all R&D and production in 2009 and since then the collaboration has been working closely with Hamamatsu to improve the collection efficiency of the 8" high QE PMT to improve the energy resolution further.

5. Light Collection Simulations with GEANT4

Full calorimeter simulations have been carried out with the GENBB event generator in GEANT4. The model used accounts for wavelength dependence of optical properties (most of which have been experimentally measured) of scintillators (self absorption and re-emission), reflective wrappings, photomultipliers (QE), optical coupling materials and refractive indexes of optical materials. The simulations were verified using data from the currently running NEMO3 detector and were shown to be accurate [7].

A SuperNEMO calorimeter unit of a hexagonal PVT (e.g. EJ200) block of 22.5cm radius wrapped in aluminised Mylar (92% reflectivity coefficient) around the sides and entrance face of the scintillator and Teflon on the top face of the scintillator coupled to a super-bi-alkali 8" hemispherical PMT (e.g. Hamamatsu R5912-MOD) with a QE of 33% was simulated. Using the symmetry of the block, each sextant of the hexagonal face was divided into 16 regions to check the uniformity of the energy resolution across the face of the block. The minimum energy resolution extracted was $\frac{7.14\pm0.5\%}{\sqrt{E(MeV)}}$, the maximum: $\frac{7.24\pm0.5\%}{\sqrt{E(MeV)}}$ and the mean: $\frac{7.19\pm0.5\%}{\sqrt{E(MeV)}}$, showing uniformity of the block and agreement between simulations and test bench measurements.

6. Block Shape Study

Details of the block design are still to be finalised, such as the exact shape of the block to be used. The effect of block geometry on the energy resolution has been studied with a PST block, which was machined to a new geometry after each measurement. Although PVT scintillator has given the best result a PST block was chosen for this study due to ease of availability and manufacturing. A study with a PVT block is currently under way.

All of the measurements were carried out under the same conditions with aluminised Mylar wrapping on the entrance face and Teflon wrapping on the sides of the block, an 8" Photonis standard QE (25%) PMT run at the same high voltage and alcohol to couple the block to the PMT. The source of electrons used was a 1MeV ^{90}Sr spectrometer beam injected at the centre of the entrance face. The block shapes studied can be seen in Figure 4 and the summary of the results in Table 2. The shape that will be chosen will be 4 or 5, depending on ease of manufacturing and cost.

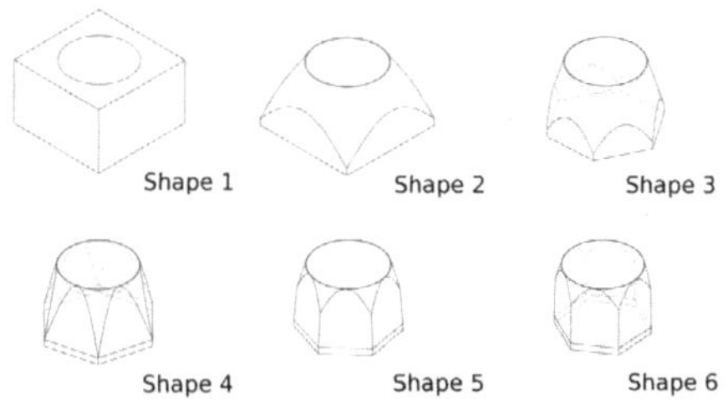

Figure 4: Block shapes studied.

Figure 5: Results for block shapes studied

Shape	FWHM (Measured)	FWHM (Simulated)
#1	10.0%	$10.1 \pm 0.1\%$
#2	9.6%	$9.9 \pm 0.1\%$
#3	9.5%	$9.5 \pm 0.1\%$
#4	9.3%	$9.3 \pm 0.1\%$
#5	8.9%	$8.8 \pm 0.1\%$
#6	8.8%	$8.6 \pm 0.1\%$

7. Conclusions and Future Plans

After three years of fruitful R&D an unprecedented energy resolution has been reached for low energy electrons in large volume low Z scintillators. The design energy resolution for the baseline (block) design has been reached at $\frac{7\%}{\sqrt{E(MeV)}}$, as well as the design resolution for the bar design at $\frac{10\%}{\sqrt{E(MeV)}}$. This is an increase of a factor of 2 in size from a NEMO3 block and an energy resolution improvement of a factor of 2. The baseline design has been chosen by the collaboration and the final details (such as the block shape) are close to being completed.

The first SuperNEMO module, known as the 'demonstrator', will begin construction in 2010 – 2011 and will be running with 7kg of ^{82}Se in 2013. The demonstrator sensitivity will reach a Majorana neutrino mass of 200 – 500meV. Full detector construction will begin in 2014 with physics running from 2014 – 2019 and a target sensitivity of 50 – 100meV.

References

[1] Arnold R. et al. Technical design and performance of the NEMO3 detector. *Nucl. Instrum. Meth.*, A **536**:79-122, 2005.
[2] Majorana E. Theory of the symmetry of electrons and positrons. *Nuovo Cim.*, **14**:171-84, 1937.
[3] Avignone III F.T., Elliot S.R. and Engel J. Double beta decay, Majorana neutrinos and neutrino mass. *arXiv*, nucl-ex:0708.1033, 2007.
[4] Avignone III F.T., King G.S. and Zdesenko Yu.G. Next generation double-beta decay experiments: Metrics for their evaluation. *New J. Phys.*, **7**:6, 2005.
[5] Hamamatsu SBA/UBA PMTs: http://jp.hamamatsu.com/resources/products/etd/eng/html/pmt_003.html
[6] Photonis High QE PMTs: pmm2.in2p3.fr/IMG/pdf/NNN08_Combettes-PHOTONIS.pdf
[7] Pahlka B et al. Spectral modeling of scintillator for the NEMO-3 and SuperNEMO detectors. arXiv:1004.3779v1[hep-ex]

Journal of Physics: Conference Series **293** (2011) 012038 doi:10.1088/1742-6596/293/1/012038

A Digital Calorimeter for Dark Matter Search in Space

X L Sun[1], J G Lu[1], T Hu[1], L Zhou[1], X Cai[1], B X Yu[1], J Fang[1], Y G Xie[1], Z H An[1], Z G Wang[1], A W Zhang[1], Z Xue[1]

[1]Institute of High Energy Physics, Beijing, China

E-mail:sunxl@ihep.ac.cn

Abstract. A novel digital imaging calorimeter is designed for the observation of high energy electrons and gamma rays on the Space Station of China. The MC data shows that the calorimeter has a capability of observing the electrons (without separation between e+ and e-) and the gamma-rays in 30GeV-5TeV with a high energy resolution of 4% over 500GeV, a good angular resolution of 0.5 degree over 500GeV, and a high proton rejection power of ~5x10^5(95% signal efficiency). The calorimeter has a large geometrical factor of ~3.3 m^2sr (5 sensitive faces) with ~1T weight. The excellent energy resolution make us can detect any change of the energy spectra and a line signature in the energy distribution, as expected from the dark matter.

1. Introduction

WMAP data reveals that dark matter comprises 23% of the universe [1]. This matter, different from atoms, does not emit or absorb light. It has only been detected indirectly by its gravity such as the rotation curves of galaxies, strong gravitational lenses and weak gravitational lenses. There are some candidates in particle physics: Stable weakly interacting massive particles (WIMPs) of mass 10GeV-10TeV, Massive neutrinos, Axions, Kaluza Klein particles, etc. WIMPs is preferred by most of theorist, which occur in several well-motivated theories beyond the Standard Model such as Supersymmetry, Inert Higgs models, etc. There are two detection mechanisms of WIMPs: Direct detection, detect signal of DM collisions with nuclei; Indirect detection, detect final decay products ($e^{\pm},\gamma,p,\bar{p},\nu$) of DM annihilation. Line signature from DM annihilation to double gamma-rays is the decisive evidence of DM. Dark matter search in space is an important method of indirect detection. Excellent capability of proton rejection with power ~10^5 is necessary for identifying electrons and gamma-rays in space because electron is about 10^{-2}-10^{-3} of proton and gamma-ray is about 10^{-3}-10^{-6} of proton between 30GeV-10TeV. Space Station of China will be built at 2020, a payload for dark matter search is proposed by IHEP. The detector should have capabilities:

- Energy range: 30GeV-5TeV (cover WIMP mass range)
- Energy resolution: < 5% (detect line signature requirement)
- Background rejection power: > 10^5 (suppress proton background requirement)
- Total power: < 500 Watts
- Total mass: < 1.5 Tons

2. Detector design

2.1 Detector concept design

The detector has two chief components: Anti-Coincidence Detector (ACD) and Digital Imaging Calorimeter (DIC). ACD is a positional sensitive plastic scintillator detector in order to reject charged particles for the gamma-rays observation. DIC is a 3D crystal array. Each crystal is a little cube with one face coupled with WLS fiber which collect and guild photons to ICCD camera. All separate cubes are glued together. "Digital" means hit mark '1', and non-hit mark '0' for each crystal. All the '01' describe the shower's image. We can get the information of particle energy, particle type and incident angle by the analysis of image.

2.2 ACD structure

ACD is composed of 1×1×80 cm plastic scintillator array along x and y direction, which covers more than 5/6 of the main body. Each scintillator has a WLSF inbuilt in a groove on surface. Reading out devoice is SiPM. ACD is positional sensitive, so the incident position of charged particle can be got associated with the shower axis in DIC.

2.3 Trigger structure

One fiber attached the surfaces of one string of crystals, and all fibers coupled with one PMT. When a high energy particle inject, we can get a fast signal (<100ns) which directed proportion the particle energy. So a threshold can be set for trigger.

2.4 Chief feature of the design

- Digital image means there is no problem of the amplitude saturation.
- 5 sensitive faces mean a large field of view.
- High level of granularity means good particle identify power and good angle resolution.
- CCD reading out means low power ~30 watts.
- ACD is positional sensitive, which means high suppress for backflash [2] charged particles.

3. MC simulation study on performance

3.1 Detector geometry

Total length is 60cm (~32 r.l.), and the side length of little crystal cube is 1.5-4.5cm. There is 1mm glue interval between two crystals. The code is written with GEANT4, physics list is QGSP. Digital energy cut is 12MeV (muon MIP deposit energy) for each crystal.

3.2 Relationship between incident energy and number of hit crystals

From the MC data we found the relationship is:

$$N_{hit}^{2} \propto E_{incident} \tag{1}$$

As can be seen from figure 1, linear relationship is good at high energy range and a little worse at low energy range.

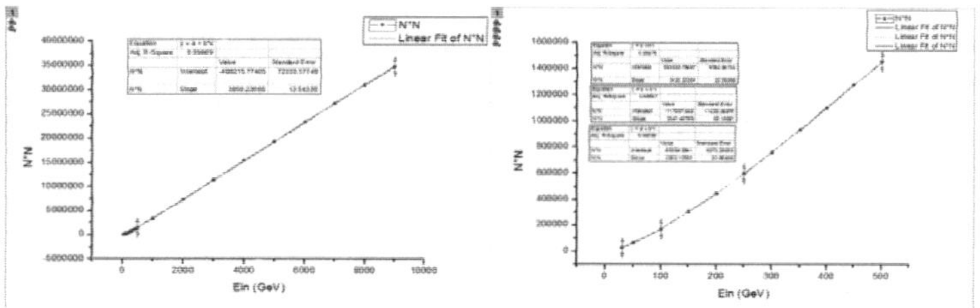

Figure 1. Relationship between incident energy and number of hit crystals.

3.3 Energy resolution

According to the relationship (1), we can get the energy reconstruction. Figure 2 is energy reconstruction of 1000GeV gamma-rays inject. The energy resolution is 3.27%. Figure 3 is energy resolution compare with different granularity 1.5-4.5 cm. About 4% can be reached above 500GeV.

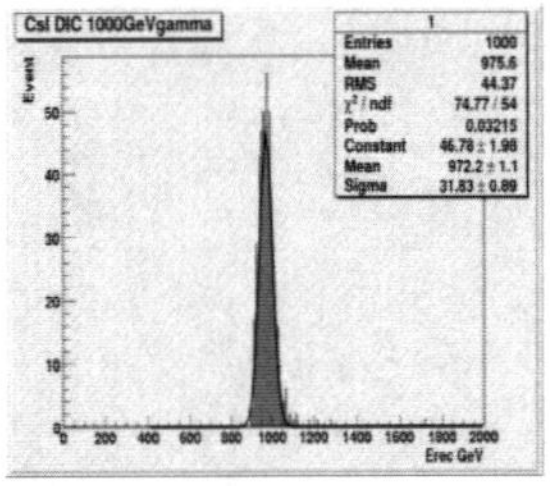

Figure 2. Energy reconstruction of 1000GeV gamma-rays.

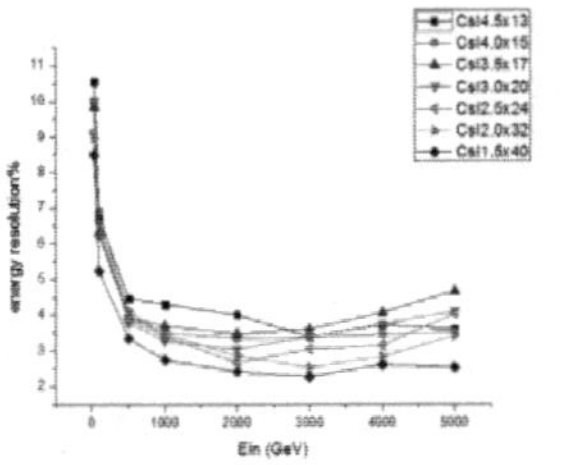

Figure 3. Energy resolution compare of different granularity 1.5-4.5 cm.

3.4 Angular resolution

Position and angular of incident particle can be got from shower axis reconstruction. Figure 4 shows shower axis reconstruction, figure 5 is position and angular distribution of incident particle, figure 6 is angular resolution compare with different angularity 1.5-4.5 cm. About 0.5° can be reached over 500GeV.

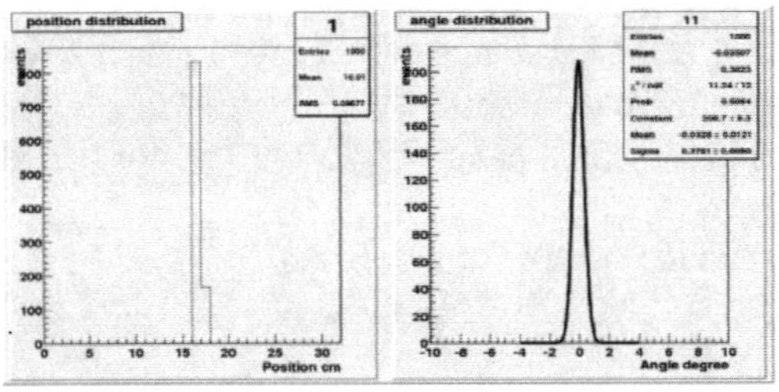

Figure 5. Position and angular distribution of incident particle.

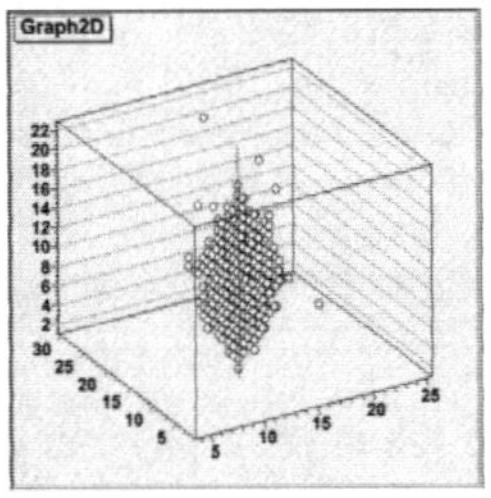

Figure 4. Shower axis reconstruction.

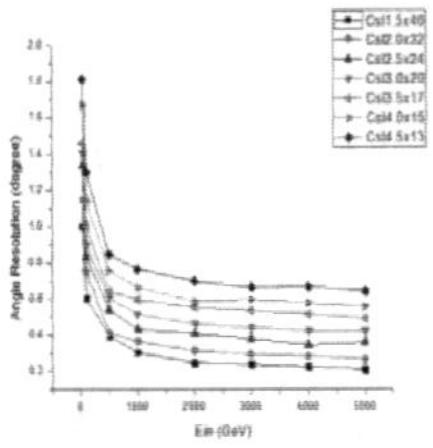

Figure 6. Angular resolution compare of different angularity 1.5-4.5 cm.

3.5 Particle identification

3.5.1 Gamma-ray proton separation There are two main differences between proton shower and gamma-ray shower:

- Proton induced showers are longitudinally wider than electron showers because of the spread of secondary particles.
- Gamma-ray induced showers will start and die off earlier than proton showers.

Figure 7 and figure 8 show the differences clearly.

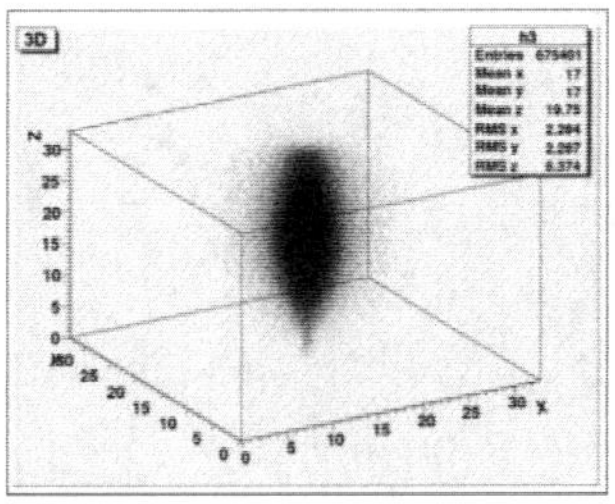

Figure 7. 3D image of electron shower.

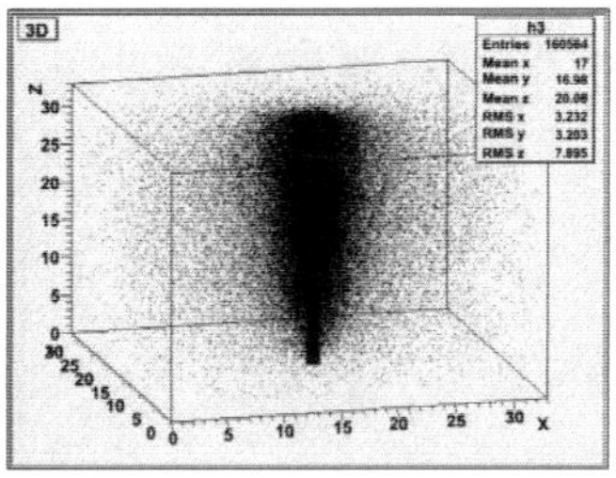

Figure 8. 3D image of proton shower.

A separation is done with TMVA tool in root. The input is:

- Signal: gamma-ray, 100GeV, 140000 events.
- Background: proton, 100GeV-5000GeV with power law 2.7, 1M events.

Figure 9 shows the output: proton rejection power is about 5×10^5, signal efficiency is 95% and BDT method has a better performance.

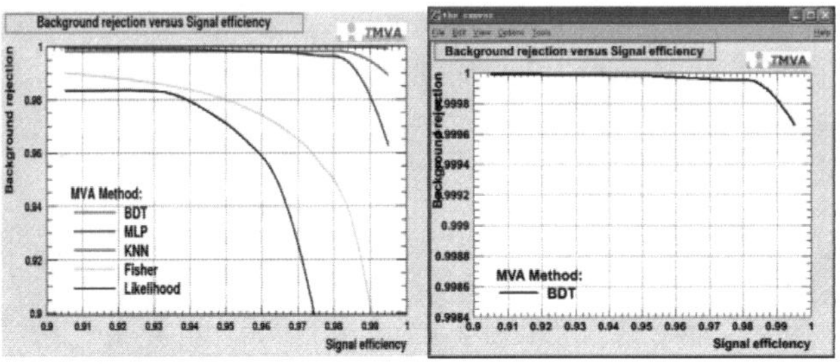

Figure 9. Background rejection versus signal efficiency.

3.5.2 Gamma-ray electron separation Difference is electron shower starts earlier than gamma-ray shower. A separation is done with TMVA, the input is:

- Signal: gamma-ray, 100GeV, 140000 events.
- Background: electron, 100GeV, 160000 events.

Figure 10 shows the output: electron rejection power is about 1×10^4,signal efficiency is 45% and BDT also has a better performance.

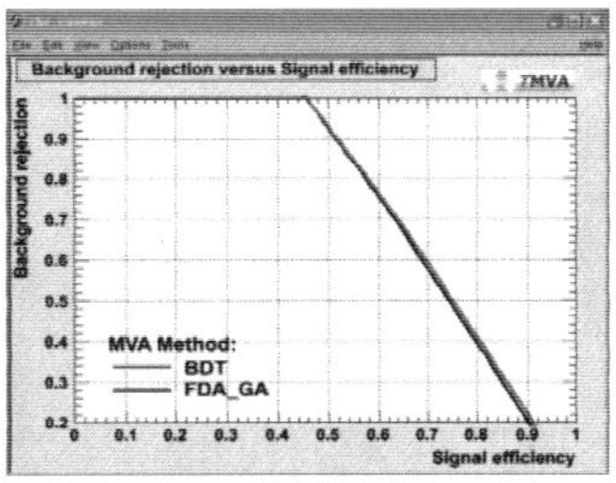

Figure 10. Background rejection versus signal efficiency.

4. Prototype study

4.1 Unit structure

The crystal side length is 2.5cm packing ESR, the WLS fiber plate attaching to one face of the crystal is the key structure for light reading out. Figure 11 is the unit structure.

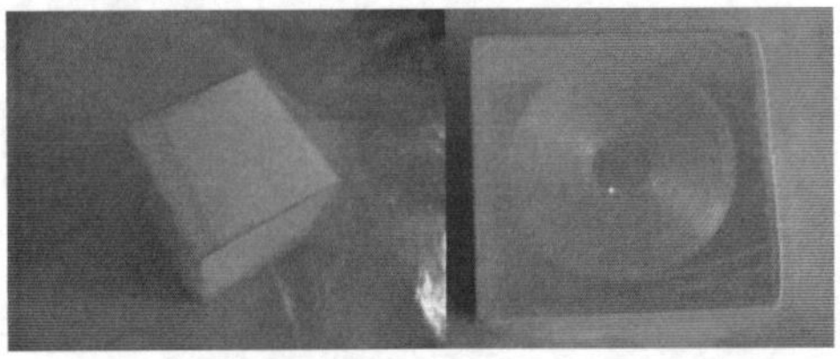

Figure 11. Unit structure.

Muon test is done for unit structure with different crystals. Figure 12 is view of the test system. We found CsI(Na) has the best light output about 188 photoelectrons for muon MIP signal. Figure 13 is the spectrum of muon MIP signal. Figure 14 is light output comparing with different crystals.

Figure 12. View of muon test system.

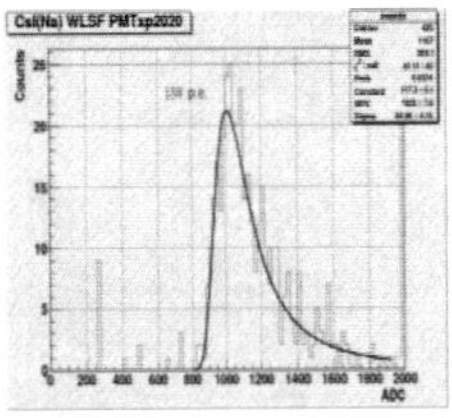

Figure 13. Spectrum of moun.

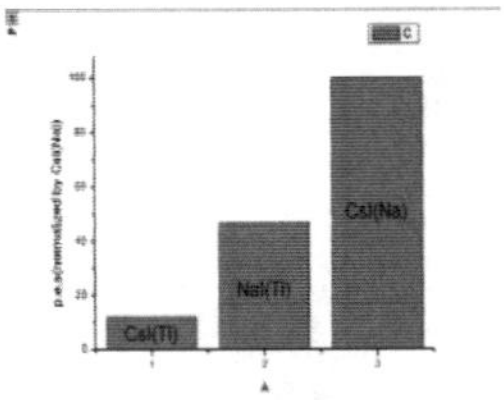

Figure 14. Light out of different crystals.

4.2 Prototype array

A prototype of 2×2×6 array is made for moun test. Figure 15 is schematic view of the muon test system.

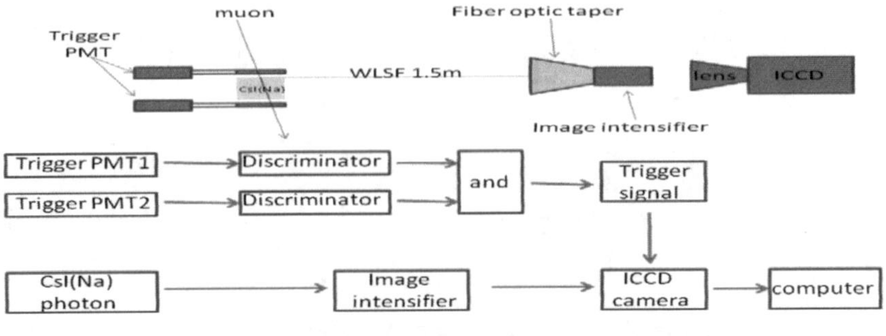

Figure 15. Schematic view of the muon test system.

Image intensifier is necessary, its key effective is: intensified image; photon delay for waiting trigger. Figure 16 is a ICCD photo of a muon event. The signal is clear and the ratio of signal to noise is about 8.

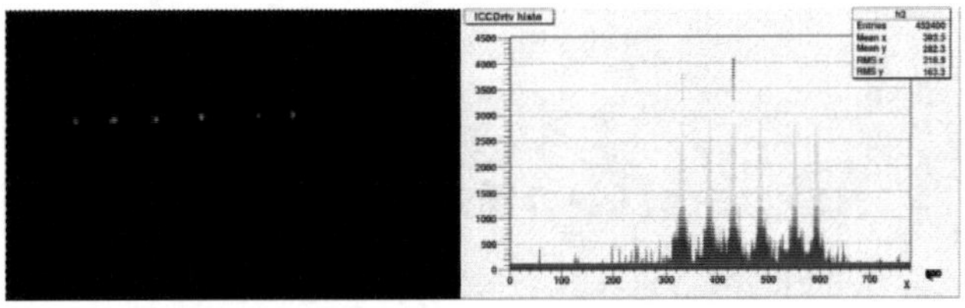

Figure 16. ICCD photo of a muon event.

4.3 Capacity analysis

Magnification of FOT and size of CCD are the key factors for capacity. For Φ 300um fiber with 100um interval and 450000 pixels CCD, the relationship between magnification and capacity is shown in table 1.

Table 1. Relationship between magnification and capacity.

Magnification	Capacity(fibers)
2:1	4500
3:1	10000
4:1	18000
5:1	28000

Capacity 28000 can cover a $30\times30\times30$ array, so only one CCD system needed for the whole calorimeter.

5. Summary

A scheme of 60cm cube CsI(Na) array is proposed: 5 faces sensitive, digital image, 32 radiation length, weight 1 ton, cover 30GeV to 5TeV energy range. The preliminary simulation is done: gamma proton separation power is 5×10^5 with 95% efficiency, electron proton separation power is 1×10^4 with 45% efficiency, energy resolution is 4% and angular resolution is 0.5°. The prototype has been built and tested and the system has a good performance.

References

[1] Spergel D N *et al.* 2007 *Astrophys. J. Suppl. Ser.* **170**, 377
[2] Moiseev A A, Ormes J F, Hartman R C, Johnson T E, Mitchell J W and Thompson D J 2004 *Astropart. Phys.* **22** 275

Journal of Physics: Conference Series **293** (2011) 012039 doi:10.1088/1742-6596/293/1/012039

NEUCAL, an innovative neutron detector for e/h discrimination: testbeam results

G Sguazzoni[1], O Adriani[2,1], L Bonechi[2,1], M Bongi[1], S Bottai[1], M Calamai[3,1], G Castellini[4], R D'Alessandro[2,1], M Grandi[1], P Papini[1], S Ricciarini[1], P Sona[2,1] and G Sorichetti[2]

[1] Istituto Nazionale di Fisica Nucleare, Sezione di Firenze, 50019, Sesto Fiorentino (FI), IT
[2] Dipartimento di Fisica, Università degli Studi di Firenze, 50019, Sesto Fiorentino (FI), IT
[3] Dipartimento di Fisica, Università degli Studi di Siena, 53100, Siena, IT
[4] IFAC-CNR Firenze, 50019, Sesto Fiorentino (FI), IT

E-mail: giacomo.sguazzoni@fi.infn.it

Abstract. An excellent hadron to electron discrimination is a crucial aspect of calorimeter-based experiments in astroparticle physics. Standard discrimination techniques require full shower development and fine granularity but in space detectors severe limitations exist due to constraints on dimensions, weight and power consumption. A possible approach is to exploit the different neutron yield of electromagnetic and hadronic showers. NEUCAL is a light and compact innovative neutron detector, to be used as an auxiliary complement of electromagnetic calorimeters. This new approach to neutron counting relies on scintillation detectors which are sensitive to the moderation phase of the neutron component. The NEUCAL prototype has been placed after a conventional calorimeter and tested with high energy beams of pions and positrons. The comparison of experimental data with a detailed Geant4 simulation and the encouraging results obtained are presented.

1. Introduction

Space experiments designed to study cosmic ray fluxes at very high energies must have, among other features, an excellent hadron to electron discrimination. Electromagnetic calorimeters have a very good intrinsic hadron rejection power based on the shower topology (depth and lateral shape). Unfortunately, to be fully exploited, this property requires calorimeters with a high granularity readout and sufficient thickness to ensure the full containment of the shower. These features cannot be fully implemented on satellites and airborne experiments in general due to weight constraints and power consumption limitations. The hadron to electron discrimination capabilities are consequently reduced.

A workaround solution was attempted in the successful PAMELA experiment [1]. A neutron detector, consisting of a moderator and a set of ^{3}He counters, was placed downstream of the calorimeter. The aim was to exploit the neutron component of the hadronic showers, much larger than in electromagnetic showers. This neutron counter, despite its very low efficiency, has been successfully used by the PAMELA collaboration for systematic effects evaluation and for measuring the calorimeter efficiency.

The NEUCAL project aims to further pursue and expand the use of neutron detectors coupled to calorimeters by introducing the new technique of *active moderation* [2], that consists in

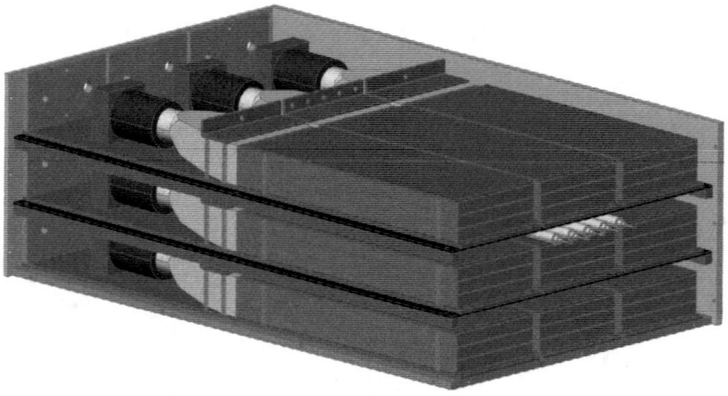

Figure 1. CAD open view of the NEUCAL detector prototype. The scintillators are in blue, and the five ^{3}He tubes (green) are placed on top of the central module.

detecting the signal of neutrons while their energy degrades within hydrogen-rich scintillators. The most promising time interval for the detection of the neutron moderation signal is between $\sim$ 10 ns to $\sim$ 100 ns after the shower core, when the peak of the neutron flux arrives. In the hadronic showers neutrons within this time window are produced by nuclear excitation processes and have energies of the order of 1 MeV.

2. NEUCAL prototype

The NEUCAL prototype [3, 4], sketched in Fig. 1, consists of a three by three matrix of identical modules placed in a light-tight aluminum box with three shelves. Each module is made up of three slabs (25 cm $\times$ 8 cm $\times$ 1 cm) of a fast polyvinyl-toluene scintillator (EJ-230 by Eljen Technology), coupled through a common Plexiglas light guide to a fast fine-mesh photomultiplier (R5946 by Hamamatsu Photonics). For the sake of flexibility, only optical grease was used in the optical couplings. Five ^{3}He tubes (12NH25/1 by Canberra) are also placed on top of the central module. The scintillators and the tubes were read out by fast digitizers capable of recording up to 10 ms of data samples. Two similar boards were used: one CAEN V1731 with 500 MS/s capability but only 8 bit range over an input dynamics of 1 Vpp; one CAEN V1720 with a more limited 250 MS/s capability but 12 bit range over an input dynamics of 2 Vpp. Each board has eight input channels. For a better timing precision one channel of each board was used to sample the trigger signal. Seven of the nine scintillator signals were sent to the 500 MS/s board, while the two remaining scintillator signals and the five ^{3}He tubes signals were sent to the 250 MS/s board. The readout was performed with a VME system.

3. Testbeam setup and simulation

In summer 2009 the NEUCAL prototype performances have been measured at CERN on the SPS line H4 during a test of the prototype of the the Cream-2 tungsten-scintillator calorimeter (*CalW*) [5]. Data were collected for negative pions (350 GeV), positrons (100 GeV and 150 GeV), and muons (150 GeV) which are not used for the present analysis. In all testbeam configurations, sketched in Fig. 2, NEUCAL was placed downstream CalW and all other possible devices and/or absorbers. In configuration a), used with positron beams, the upstream material corresponds to 16 X_0 and 0.6 λ_I; in configurations b) and c), used with pion beams, the upstream material corresponds to 29 X_0 and 1 λ_I, and 25 X_0 and 0.8 λ_I, respectively. In all configurations the NEUCAL active volume was symmetrically positioned and orthogonal with respect to the beam axis.

The NEUCAL prototype and the testbeam setup have been accurately modelled with Geant4 [6, 7] to validate the results against the Monte Carlo simulation. A sketch of the geometry of the configuration c), as implemented in Geant4, is visible in Fig. 3. The Geant4 simulation of NEUCAL has been also cross-checked with Fluka [8, 9] with respect to the single neutron response, as described in [3, 4].

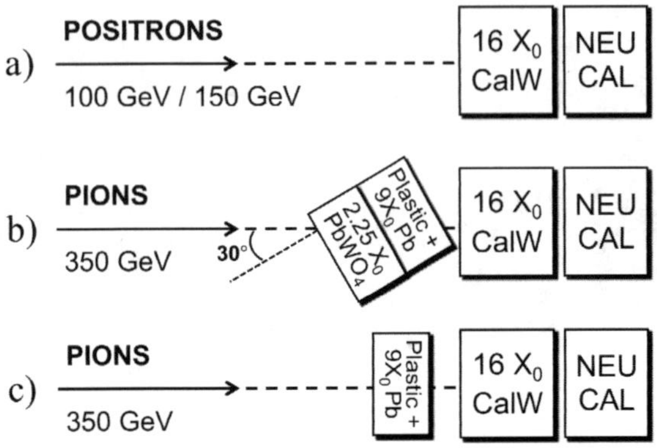

Figure 2. Sketch of the three testbeam configurations used for the present analyses of the NEUCAL data.

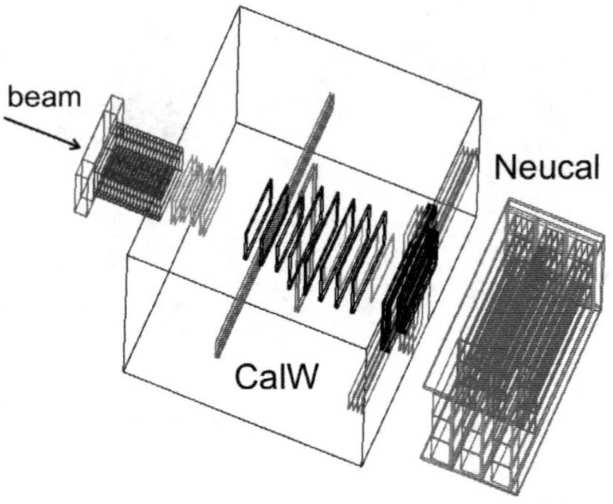

Figure 3. Testbeam geometry corresponding to configuration c) as implemented in Geant4.

Testbeam data comparison with Geant4 simulations are done only at the level of the energy deposition in scintillators and on the number of counts in ^{3}He tubes. In fact, no readout chain response is presently modelled in the simulation.

Two sets of simulated samples have been produced using different physics lists suitable for the NEUCAL use case [10]: `QGSP_BERT_HP` (*QBERT*), with Bertini model and Quark-Gluon String Precompound model to generate the final state for hadron inelastic scattering respectively below and above $\sim 10\,\text{GeV}$; `QGSP_BIN_HP` (*QBIC*), similar to QBERT but with the Binary Cascade model in place of the Bertini model for the final state generation below $\sim 10\,\text{GeV}$. Both physics lists include a high-precision model for low energy neutron transportation. The typical size of simulated samples is 20 thousand events for pions and 80 thousand events for positrons.

4. Results

By default data were taken in zero suppression mode, implemented in the digitizer boards, in order to reduce the event size and the readout time. The signals up to one millisecond after the initial charged particles shower were then reconstructed offline to search for delayed neutron interactions. The neutron signature is an isolated energy deposit in one of the scintillator modules or a pulse in the ^{3}He tubes, while a traversing charged particle will release energy on more than one scintillator module [4].

Unfortunately, the photomultipliers and the readout electronics often saturate because of the huge signal induced by the shower core, with different effects seen on data taken with and without zero suppression. In zero suppression mode the readout hardware spoilt the data in the time window close to the shower core as a consequence of reflections seen at the digitizer output lasting for a few hundred nanoseconds. In the few ten thousand events taken without zero suppression the board with limited ADC range (V1731) still showed slowly-recovering saturation effects. These issues have been understood and fixed in view of the next tests. Eventually, despite the difficult conditions, the samples of Table 1 have been used to produce the present results.

In the following, all plots report points corresponding to: pion sample data (-●-) and pion simulation predictions for QBERT (⊡) and QBIC (▩); positron sample data (-■-) and positron simulation predictions for QBERT (⊡) and QBIC (▩). All plots shown in this paper are normalized to the number of events which develop a shower in the upstream material: almost all of the positron and 45% to 60% of pion events, depending on the configuration.

The analysis of the traditional ^{3}He counters (see Table 1 for sample details) is an important

Table 1. Testbeam data samples used for each analysis. ZS stands for 'zero suppression'.

Analysis	Time Domain	Mode	e^+ sample	π^- sample
^{3}He counters	$0.1\,\mu$s – 0.1 ms	ZS	39k, 100 GeV; conf. a)	76k, 350 GeV; conf. b)
Scint. 'early'	$< 1\,\mu$s	no ZS	7k, 150 GeV; conf. a)	18k, 350 GeV; conf. c)
Scint. 'late'	$1\,\mu$s – 1 ms	ZS	39k, 100 GeV; conf. a)	76k, 350 GeV; conf. b)

validation of the simulation environment. The results are summarized in Fig. 4 where the number of pulses per event on all counters is plotted versus the logarithm of time, in microseconds. Both for data and simulation, the signal time is always referred to the arrival of the shower, taken as time zero. The behaviour is well reproduced by the simulation, but significant differences in the absolute prediction exist for pions between QBIC and QBERT, observable also in the scintillators analyses reported below. These discrepancies will be investigated further when new datasets from future tests will become available.

The analyses of the scintillator modules data search for neutron interaction candidates (*hits*) defined as single isolated signals exceeding 0.3 GeV in energy. This definition is also implemented at the simulation level. All scintillator modules were previously calibrated in energy with cosmic muons [11]. Details of the used samples can be found in Table 1. In particular, the analysis of the 'early' scintillator data ($< 1\,\mu$s) is performed on samples without zero suppression; 'late' domain analysis, from $1\,\mu$s to 1 ms, relies upon data with zero suppression.

Data from scintillators are contaminated by several background sources, not modelled in the simulation, for which specific rejection criteria have been deployed. Signal reflections in zero suppression mode are rejected by vetoing specific time delays between subsequent hits; the signals due to off-trigger beam particles can be reduced by asking no hits from different modules in coincidence; spurious effects due to saturation phenomena are reduced by rejecting signals with abnormally long duration.

An example of the results of the 'late' analysis, from $1\,\mu$s to 1 ms, is given in Fig. 5 where the hit energy per event is plotted as a function of the logarithm of time, in microseconds, for one

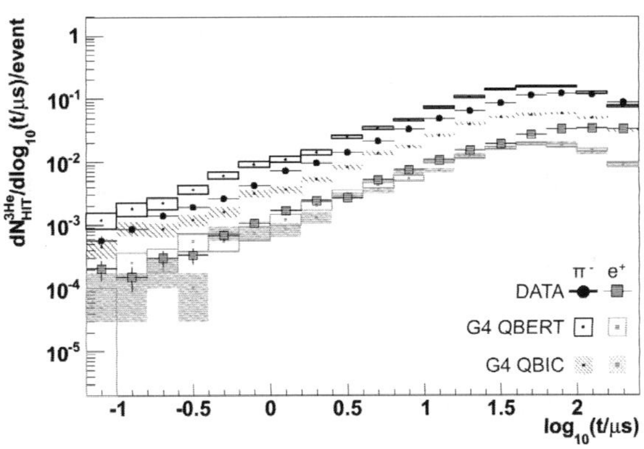

Figure 4. Number of pulses per event of all ^{3}He counters as a function of the logarithm of time, in microseconds.

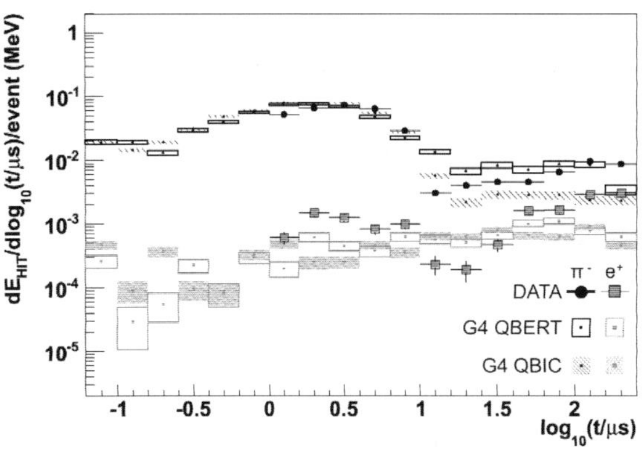

Figure 5. Hit energy per event of one top-side scintillator module as a function of the logarithm of time, in microseconds, 'late' domain.

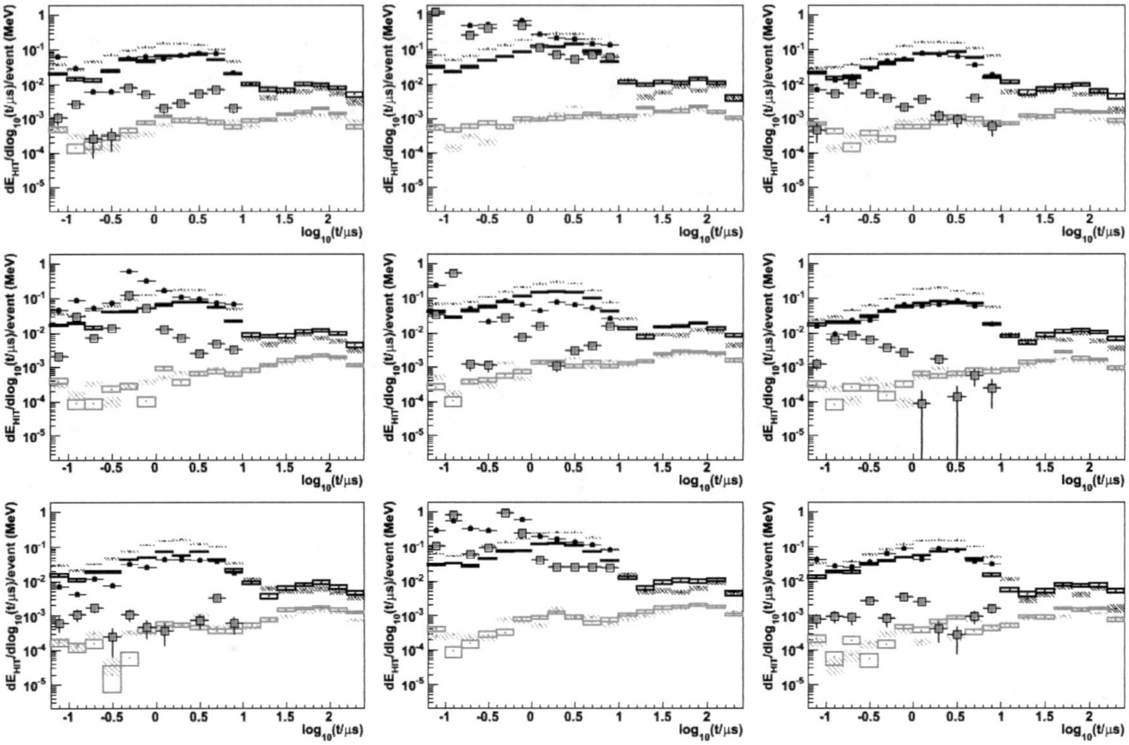

Figure 6. Hit energy per event as a function of the logarithm of time, in microseconds, for positrons and pions from the nine NEUCAL scintillator modules, 'early' time domain. Top row corresponds to the upstream module shelf with respect to the beam arrival direction. Same symbol conventions of Fig. 4 and Fig. 5 apply.

top-side module; all modules show similar behaviour. This range is populated by depositions resulting from neutron captures on nuclei and it is not directly interesting for the detection of the neutron moderation. Nevertheless, the good overall agreement between data and simulation is an indirect indication that neutron flux estimations are under control. Capture signals could be used as a complementary handle for neutron detection if they can be made numerically significant on the single event.

Off-trigger contribution due to beam contamination can be observed on the positron data for time greater than about $30\,\mu s$ in Fig. 5, and also in Fig. 4. This is due to the much greater intensity of the positron beam used in the test.

The 'early' time domain analysis ($< 1\,\mu s$), performed on data without zero suppression, is represented in Fig. 6 where the hit energy per event is reported versus the logarithm of time, in microseconds, for all nine modules. It is evident that central modules suffer the shower core contamination from pions and positrons, especially at short times. Side modules are less affected by this issue. Nevertheless the agreement between data and simulation is again not satisfactory for positrons, with the exclusion of the two bottom-side modules. This is a saturation effect in the modules read out by the V1731 boards due to the huge signal generated by electromagnetic showers not fully contained by the upstream calorimeter.

The analysis at times before 200 ns, where most of the neutron moderation signal is expected, is thus restricted to the bottom-side modules that are not directly flooded by the electromagnetic shower and that are read out by the board V1720, which has the dynamic range to withstand

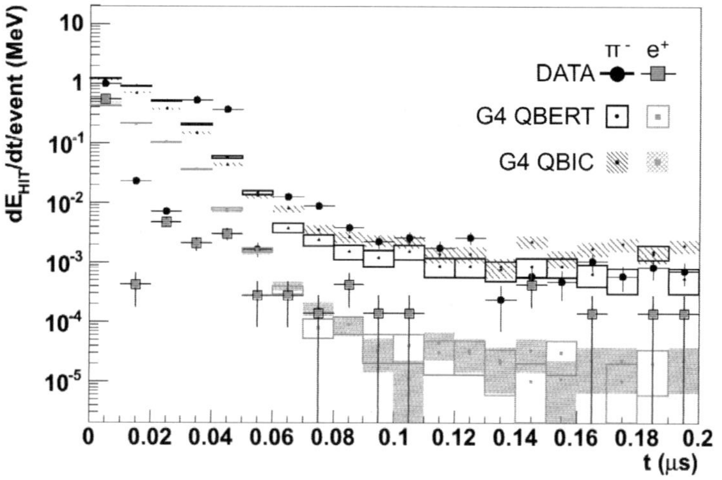

Figure 7. Hit energy per event as a function of time (linear scale) after the shower core.

the huge signal. The hit energy per event for the bottom-side modules as a function of time, now represented in linear scale, is shown in Fig. 7. The agreement with the simulation is satisfactory with the exception of the region below few tenth of nanoseconds where a significant shower contamination is still present and cannot be further reduced due to the limitations of the present setup. This plot demonstrates that the signal due to neutrons in pion data is significantly larger than in positron data in the time interval relevant for the NEUCAL application.

5. Conclusions

NEUCAL is a detector designed to exploit the innovative "active moderation" technique in the neutron counting. The final aim is the development of a light and compact device, suitable for space experiments, to help electromagnetic calorimeters in hadron to electron discrimination. The NEUCAL prototype has been tested with pion and positron showers, and data compared with an accurate Geant4 simulation. Results demonstrate in general good agreement between data and simulation and are very promising in view of the device application. Further confirmations are expected with upcoming tests that will profit from the experience gained so far and that will be organised with an improved and more performant setup.

Acknowledgments

The authors wish to express their gratitude to P S Marrocchesi, G Bigongiari, and P Maestro for their support during data taking at CERN and in the analysis of the data.

References

[1] Adriani O *et al.* (PAMELA) 2009 *Nature* **458** 607–609 (*Preprint* 0810.4995)
[2] Bottai S *et al.* 2010 *Nucl. Instrum. Meth.* **A617** 464 – 466
[3] Bonechi L *et al.* 2009 *Proceedings of ICATPP 2009 – 11th International Conference on Advanced Technology and Particle Physics* (Como, Italy) to be published (*Preprint* 1008.1886)
[4] D'Alessandro R *et al.* 2010 *Proceedings of VCI 2010 – 12th Vienna Conference on Instrumentation* (Vienna, Austria) doi:10.1016/j.nima.2010.06.347
[5] Bagliesi M G *et al.* 2003 *Nucl. Phys. Proc. Suppl.* **125** 358–362
[6] Agostinelli S *et al.* (GEANT4) 2003 *Nucl. Instrum. Meth.* **A506** 250–303
[7] Allison J *et al.* 2006 *IEEE Transaction on Nuclear Science* **53 No.1** 270
[8] Ferrari A, Sala P R, Fasso A and Ranft J 2005 CERN-2005-010
[9] Battistoni G *et al.* 2007 *AIP Conf. Proc.* **896** 31–49
[10] *Geant4 Physics Reference Manual* URL http://geant4.cern.ch/
[11] Tiberio A 2003 *Calibrazione del sistema di scintillatori Neucal* Master's thesis Università degli Studi di Firenze in italian

XIV International Conference on Calorimetry in High Energy Physics (CALOR 2010) IOP Publishing
Journal of Physics: Conference Series **293** (2011) 012040 doi:10.1088/1742-6596/293/1/012040

Commissioning of CMS zero degree calorimeter using LHC beam

O Grachov[1]**, M Murray, J Wood**
Department of Physics and Astronomy, University of Kansas
Lawrence, KS USA
Y Onel, S Sen, T Yetkin
Department of Physics and Astronomy, University of Iowa
Iowa City, IA USA
On behalf of HCAL-CMS collaboration

E-mail: grachov@ku.edu

Abstract. This paper reports on the commissioning and first running experience of the CMS Zero Degree Calorimeters during December 2009. All channels worked correctly. The ZDCs were timed into the data acquisition system using beam splash events. These data also allowed us to make a first estimate of channel-by-channel variations in gain.

1. Introduction

The CMS Zero Degree Calorimeters (ZDCs) [1] are located at the straight section of the LHC, between the two beam pipes at 140 m on each side of the interaction region. They are sampling calorimeters using tungsten (W), as radiator and quartz fibers (QF), as the active media. A significant advantage of this technology is that the calorimeter can be very compact, extremely fast and radiation hard. The Cherenkov light generated by charge particles passing through the fibers is brought to photomultiplier tubes (PMT). Schematic illustration of the ZDC is presented in Figure 1. Each ZDC includes electromagnetic (EM) and hadronic (HAD) section. Each section consists of a tungsten plate/ quartz fiber ribbon stack. The hadronic section is longitudinally segmented into 4 readout towers. The electromagnetic section segmented into 5 transverse towers with width of 15.86 mm. Both sections are 79.3 mm wide and 100 mm high. The main parameters of ZDC are presented in Table 1. The ZDC is fully functional and fully integrated into CMS [2] operations. Following very extensive local and global CMS running, commissioning was carried out first with splash events created by the circulating proton beams at 450 GeV impinging on a target. These splash events were useful for adjustment of read-out timing and to determine the relative timing of ZDC channels, as well as to validate relative gain calibration of the detector's towers.

[1] To whom any correspondence should be addressed.

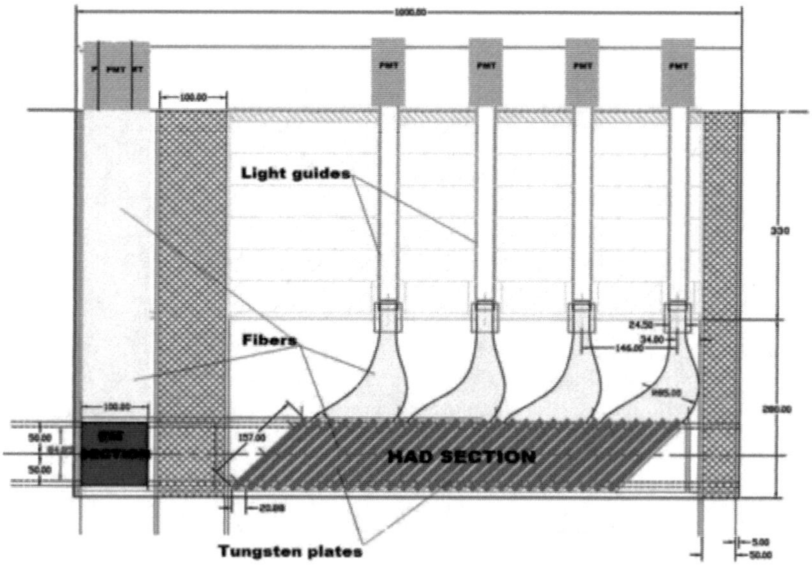

Figure 1. Schematic illustration of the ZDC (EM section and HAD section).

Table 1. Parameters of the ZDC

	Hadronic Section	Electromagnetic Section
Sampling Ratio	15.5 mm W/0.7 mm QF	2 mm W/0.7 mm QF
Number of Cells	24	33
Interaction (Rad.) Length	5.6λo	~1λo(19Xo)
Number of Channels	4 longitudinal segments	5 transverse towers

2. Detector's commissioning

2.1. Calibration with test beams

ZDC detectors were constructed at University of Kansas and assembled at CERN, and prior to installation were tested at CERN SPS test beam in 2006 and 2007 [3, 4]. The test beam line delivered positrons with energy between 10 GeV and 150 GeV and positive pions with energy 150 GeV, 300 GeV and 350 GeV. The energy resolution for positrons is $(\sigma/E)^2 = 0.49/E + 0.0064$. The linearity of the response as a function of positron beam momentum has been measured and the calorimeter is found to be linear over a range of from 10 GeV to 150 GeV to within 2%-3%. The energy resolution of the combined calorimeter (EM+HAD) for hadrons was obtained by a Landau fit and it is 21.5 % for 300 GeV positive pions. An extrapolation to energy 2.75 TeV will give the resolution of about 15%. Measurements show good linearity of detector in the positive pion energy range of 150GeV to 350 GeV.

2.2. Beam splash measurements

A total of 570 beam splash events for ZDC plus and 150 events for ZDC minus were recorded during the LHC 2009 commissioning exercises. The LHC beam with energy 450 GeV was steered into collimators located 150 m upstream in either direction of the CMS detector interaction point. Because of the large amount of material in the collimator only muons were able to penetrate the entire ZDC detector and form the signal in the calorimeter towers. A schematic representation of the geometry of the beam splash setup is shown in Figure 2.

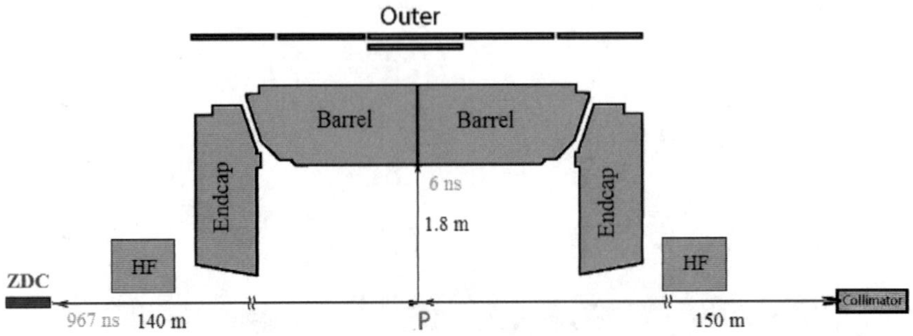

Figure 2. Beam splash measurement setup for one ZDC, showing the location of outer, barrel, two end caps and two HF hadron calorimeters, and ZDC calorimeter as well. ZDC located 140 m from CMS interaction point and 290 m from upstream collimator.

The splash events could not be used to study the absolute energy scale calibration because muons are minimum ionizing particles. Splash events were helpful for measuring the relative gain and delay between different channels of each ZDC and for overall adjustment of the timing of the calorimeter read-out.

2.3. Read-out

The read-out for one ZDC channel is shown in Figure 3. This scheme is the same as for read-out of HF (forward hadron) calorimeter [5].

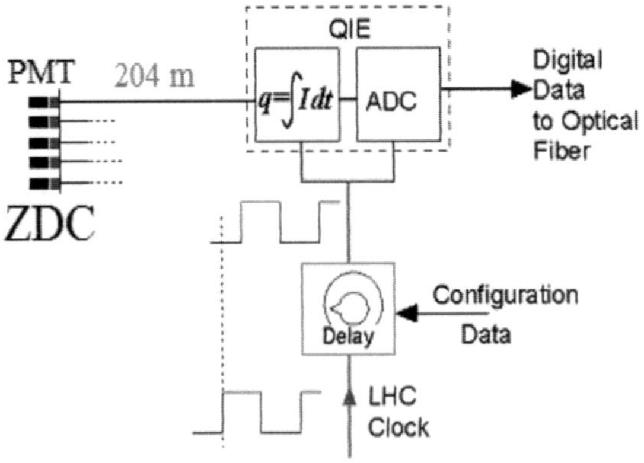

Figure 3. The read-out for one ZDC channel.

We are using 204 m length signal cable between phototubes and QIE (Q-charge, I-integrated, E-electronics). The QIE is a custom chip that contains the charge integrating electronics with an analog - to digital converter (ADC). The electric current collected from the PMT is integrated over each LHC clock period and then sampled. The integration clock can be delayed with respect to the LHC clock by programmable settings, which have a resolution of 1 ns.

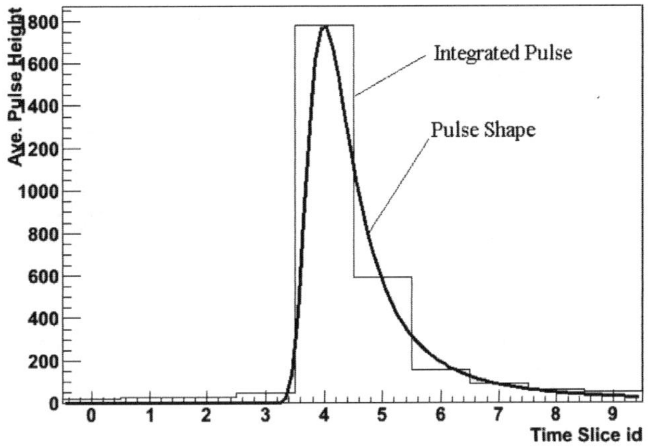

Figure 4. Integrated pulse and pulse shape. One time slice is 25 ns. The signal pulse is expected to start from the fourth sample and the baseline value can be estimated from the first three digitized samples (time slices 4,5 and 6).

Before the first beam, splash event's sampling delays were estimated from dead reckoning (time of flight, length of signal cable, electronic delay) and loaded into the ZDC front end electronics. These delay values were calculated to synchronize the ZDC with the rest of CMS. When beams arrived, events were selected by the CMS triggers system. For each event selected by the trigger the ten samples corresponding to the event were transmitted to the electronics for insertion into the CMS data stream. In order to estimate the amplitude of the signal and the timing for each channel, a fit was performed to the 10 digitized 25 ns time samples. The timing of each channel was calculated using charge-weighted algorithm based on two time slices around time slice with max energy deposition. The delay of the read-out pipeline is such that the signal pulse is expected to start from the fourth sample and the baseline value can be estimated from the first three digitized samples (time slices 4,5 and 6) (see Figure 4).

3. Results of measurement

3.1. Timing

The resulting distribution of the difference between the corrected time (after first splash beam) and predicted dead reckoning time is shown in Figure 5. This difference was corrected. Additional possible systematic effects, such as an offset in timing between the positive and negative ZDC, were studied with beam splash data from both directions. The timing for the ZDC plus before any corrections shows an average time differences that is 16 ns later than ZDC minus (see Figure 6). We corrected this difference and checked it using additional beam splash measurements.

The ZDC channel-by-channel synchronization was verified. Time variation is in the region of ±6 ns. For each individual EM and HAD channel, the signals generated by particles originating from the interaction point (IP) are registered with the different time, because of their different flight times and different intrinsic delays. The final off line time corrections are determined with collision data and expected to provide synchronization with a spread in the per channel mean times of 1 ns.

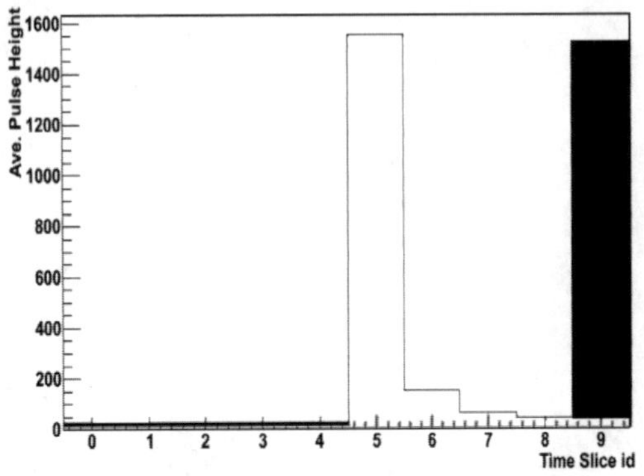

Figure 5. The resulting distribution of the difference between the corrected time (white) and predicted dead reckoning time (black) for positive ZDC. Data from LHC beam commissioning were used to validate the ZDC sampling delay settings derived from dead reckoning estimation.

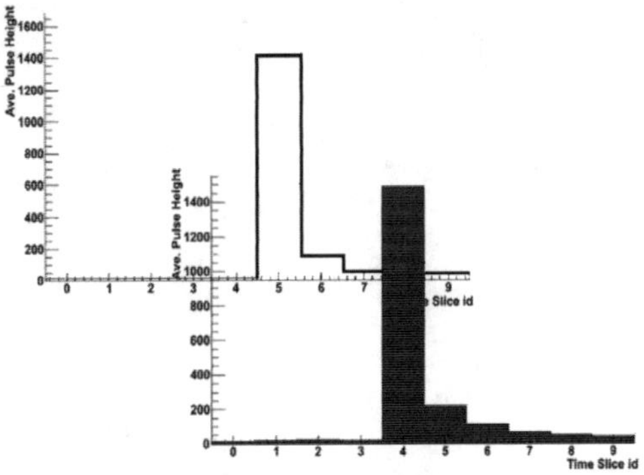

Figure 6. Offset in timing between the positive (upper plot) and negative (lower plot) ZDC, were studied with beam splash data from both directions. The timing for the ZDC plus before any corrections show an average time differences that is 16 ns later than ZDC minus.

3.2. Relative gain verification

The channel-to-channel response uniformity impacts on the energy resolution. Extensive data taking with splash beam muons resulted in rough verification of the tower's gain equalization. The muons produced from the dumping of the beam passed through the entire detector, depositing large signals in every tower of the ZDC. Typical pulse height distributions of the signal together with a Landau fit are shown in Figures 7 and 8.

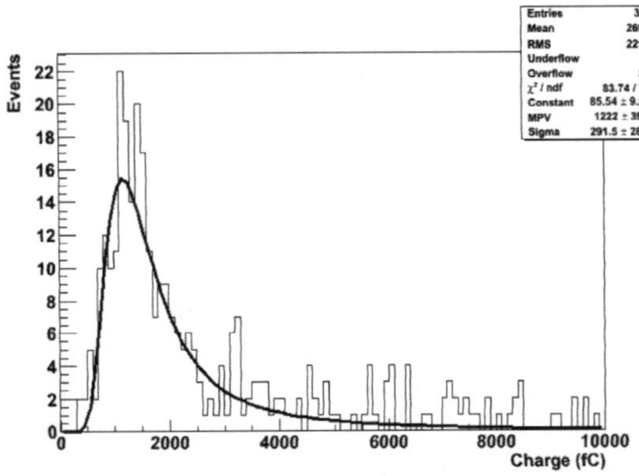

Figure 7. Pulse height distribution of energy deposition in one longitudinal readout tower of HAD section.

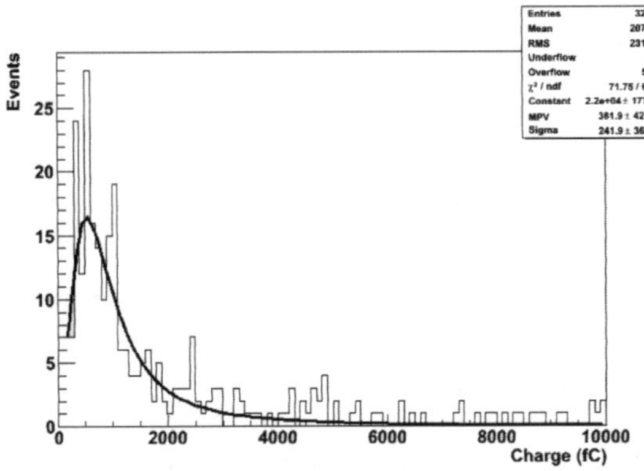

Figure 8. Pulse height distribution of energy deposition in one transverse tower of EM section.

Comparison of the most probably values of Landau fits of tower responses are used to estimate the channel-to-channel gain variation for EM (plus and minus) and HAD (plus and minus) sections separately. After this first set of measurements, we slightly corrected gain (changed of PMT high voltage) of two towers, one in EM and one in HAD. The second set of measurements showed that the relative gain variation (for EM and HAD sections separately) is in range of ~15%. More precise gain measurements and corrections will be done using events from beam-beam collisions.

4. Conclusion

The first running experience and commissioning of ZDC with splash events from the first LHC circulating beams were successful. All channels are working as designed. Initially dead - reckoning was used to set up the read-out timing. Beam splash data allowed this timing to be confirmed and optimized. The channel-to-channel gain variation was measured using beam splash events and corrected for uniform response.

References

[1] Grachov O A, Murray M J, Ayan A S, Debbins P, Norbeck E, Onel Y, d'Enteria D G 2006 *AIP Conf.Proc.867:258-265; e-Print:nucl-ex/0608052*
[2] By CMS Collaboration (Adolphi R et al) 2008 *JINST 3:S08004 361 pp*
[3] Grachov O A et al 2007 *Int.J.Mod.Phys.E16:2137-2142; e-Print:nucl-ex/0703001*
[4] Grachov O A et al 2009 *J.Phys.Conf.Ser.160:012059;e-Print arXiv:0807.0785[nucl-ex]*
[5] Chatrchyan S et al 2009 *J.Appl.Phys.83 826; e-Print arXiv:0911.4877[physics.ins-det]*

Journal of Physics: Conference Series **293** (2011) 012041 doi:10.1088/1742-6596/293/1/012041

Performance of the ATLAS Forward Calorimeters in First LHC Data

Dag Gillberg, on behalf of the ATLAS Liquid Argon Calorimeter Group
Department of Physics, Carleton University, Ottawa, ON K1S 5B6, Canada
E-mail: dag.gillberg@cern.ch

Abstract. The ATLAS Forward Calorimeters (FCal) cover the very forward region of $3.1 < |\eta| < 4.9$ and are hence designed to operate in a very high rate environment. This paper discusses the design of the FCal and presents studies of its performance in beam test data and LHC collision data. Measurements of the signal pulse shape, and the distribution of reconstructed cell energies are presented and compared with the expectation from simulations. A measurement of the location of the FCal is also performed in the data by studying the average deposited energy as a function of transverse distance to the beam axis.

1. The ATLAS Forward Calorimeters

The ATLAS detector is a multi-purpose detector built around one of the four collision points of the Large Hadron Collider (LHC)—the world's highest energy collider. The ATLAS Liquid Argon (LAr) Calorimeter is one of the major subsystems of ATLAS. An overview of the LAr calorimeters is shown in Figure 1.

The ATLAS forward calorimeters (FCal) are situated inside the endcap cryostats together with the electromagnetic endcap calorimeter (EMEC) and the hadronic endcap calorimeter (HEC). The FCal covers the very forward regions of $3.1 < |\eta| < 4.9$, where the energies and density of particles are very high. The main two motivations for the FCal are to improve the reconstruction of missing transverse energy $\not{E}_T$ (important for many physics analyses, e.g. SUSY

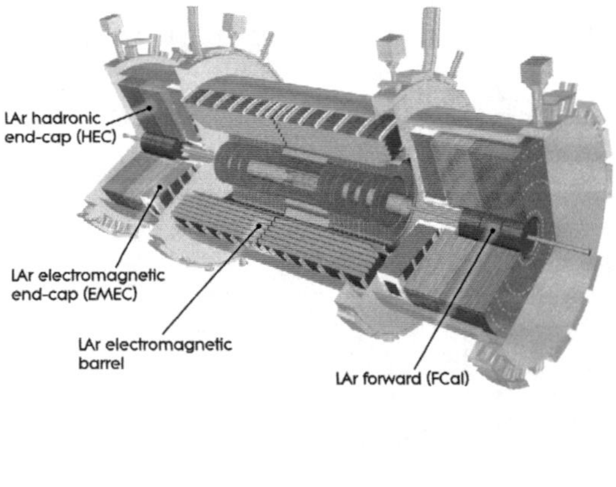

Figure 1. The ATLAS LAr calorimeters.

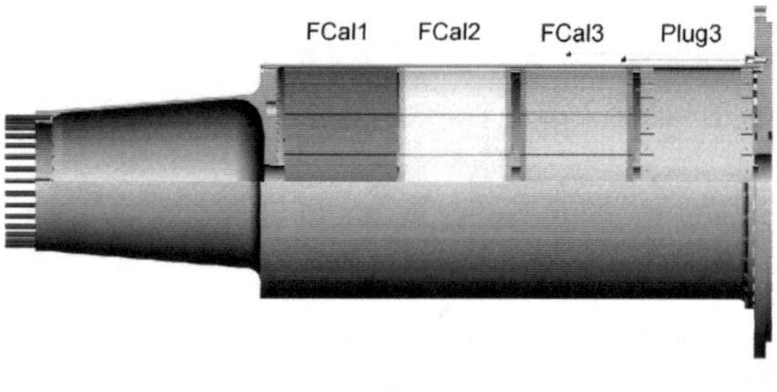

Figure 2. Cut-away view of the ATLAS forward calorimeter. Particles from the interaction point first traverse the electromagnetic FCal1 module (Cu/LAr-calorimeter) and thereafter the two hadronic FCal modules (W/LAr-calorimeters).

searches) and the identification and measurement of forward jets (for instance the forward jets that are produced in vector boson fusion processes).

The FCal consists of three layers. The layer closest to the interaction point, FCal1, is a Cu/LAr calorimeter designed for electromagnetic calorimetry. FCal2 and FCal3, are hadronic W/LAr calorimeters. Behind FCal3 is a passive layer of brass that absorbs hadronic shower remnants that punch through. Figure 2 shows a drawing of this arrangement.

Originally, the FCals were planned to be placed outside the endcaps, 15 m away from the interaction point (IP). Further studies favoured the current placement inside the endcap cryostat (5 m from the IP) for several reasons: this provides a smooth transition from the other calorimeters that reduces energy losses and improves the reconstruction of jets, electrons and $\not{E}_T$. It also provides shielding for the muon system.

In order to adequately reconstruct jets and missing transverse energy, the FCal design requirement for hadronic energy deposits was set to $\sigma_E/E = 1/\sqrt{E} + 0.1$, without significant non-Gaussian tails. The FCals are also required to have a fast response: $\mathcal{O}(25\text{ ns})$, to minimize the impact from pile-up, and to be radiation hard to cope with the design luminosity of the LHC: $\mathcal{L} = 10^{34}\text{ cm}^{-2}\text{s}^{-1}$.

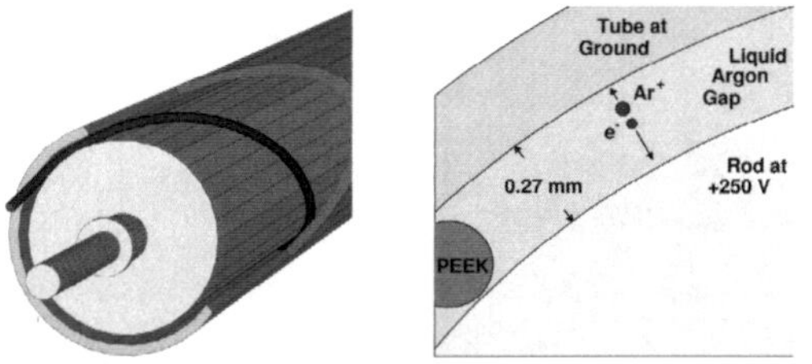

Figure 3. Close-up views of an electrode in the FCal1 module. FCal electrodes consists of a rod (anode), a liquid argon filled gap and a tube (cathode). The distance between the rod and the tupe are maintained by a insulating, helically wound PEEK fiber. Figure from [1].

A novel technology was developed to meet the design requirements. Each FCal module consists of an absorber matrix with cylindrical "electrodes" arranged parallel to the beam pipe, extending the full length of the module (~45 cm). As illustrated in Figure 3, each electrode consist of a "rod" (anode) inside a "tube" (cathode) with a narrow, LAr-filled, annular gap between the two. The LAr gap is significantly narrower than the 2 mm gap of a conventional LAr calorimeter: 0.27 mm, 0.38 mm and 0.50 mm for FCal1, 2 and 3 respectively. This provides a fast readout and avoids problems from ion buildup. The LAr gap between the rod and tube

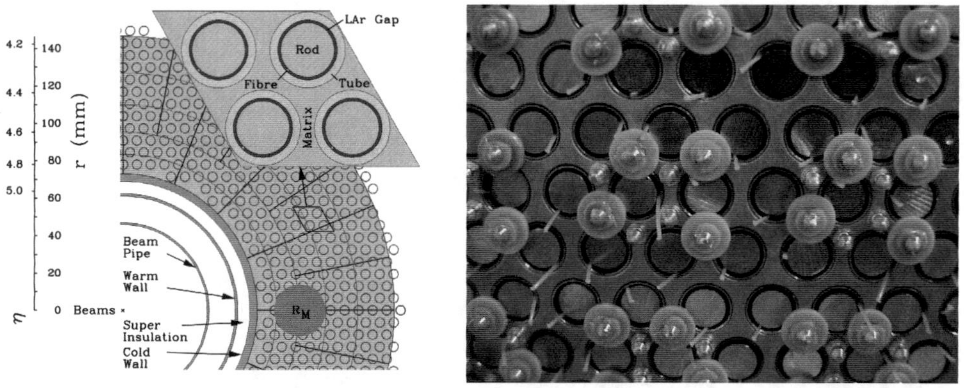

Figure 4. Left: Front-face view of the electrode arrangement of a portion of FCal1 close to the inner-edge. The Moliere radius R_M, the distance r to the beam, and the pseudorapidity η at the front face are also shown. Right: Photo of the front-face of a hadronic FCal3 module. The ends of the PEEK fibers can be seen between the rods and the tubes. Figures from [2].

is maintained by an insulating PEEK fiber that is wound around the rod.

A FCal readout cell consists of several neighbouring FCal electrodes read out together. Fewer electrodes are grouped close to the inner edge, which ensures an η-ϕ granularity of about 0.1×0.1 or better for most of the electromagnetic layer. The arrangement of electrodes is shown in Figure 4.

The choice of tungsten as absorber in the hadronic FCal2 and FCal3 modules limits the transverse hadronic shower spread, which is especially important for η determination in the forward region since a fixed distance here covers a larger η interval. This choice also provides a compact calorimeter that can be placed a bit deeper in the cryostat (see Fig. 1), which reduces neutron albedo back into the inner detector. FCal1 has a depth of $28X_0$, and all three FCal modules have a combined depth of 10λ.

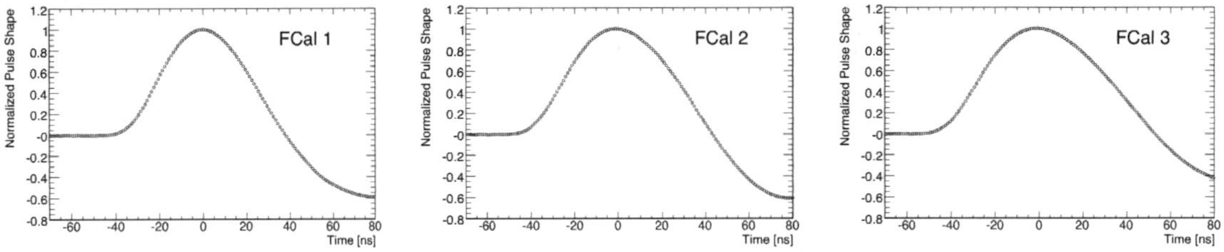

Figure 5. Expected FCal pulse shapes derived from beam test data [2].

2. Signal Pulse Shape

Figure 5 shows the FCal pulse shapes for each module derived from beam test data [2]. They have slightly different shapes due to the different LAr gap sizes, but are very similar for the channels within the same module. The pulse shapes have been studied *in-situ* using cosmic, beam splash and collision data. Figure 6 show the measured pulse shape in beam splash data for a channel in FCal3 compared with the expected pulse shape. The pulse shapes observed in data agree very well with the derived shapes from beam test data.

The energy of a channel is proportional to the amplitude of the pulse. The amplitude and timing of the pulse is reconstructed using the optimal filtering (OF) technique [3]. There are

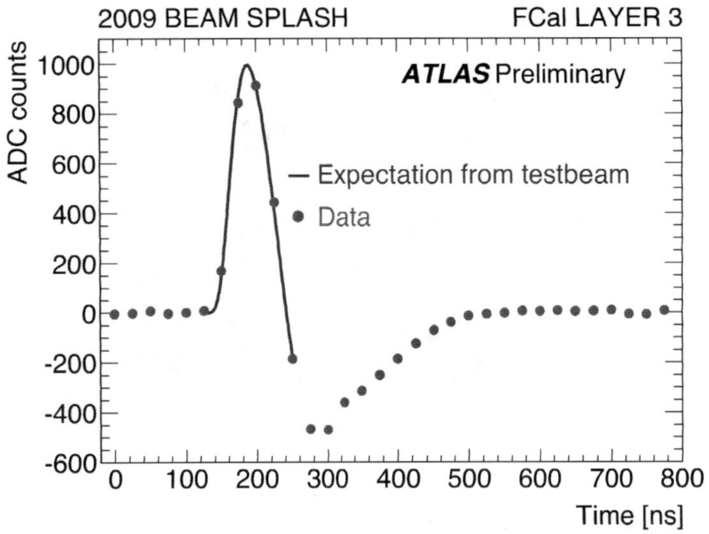

Figure 6. Pulse shape observed in 2009 beam splash data. The expected pulse shape derived from beam test data (see Figure 5) is shown as a solid blue line.

ongoing studies that try to find an optimal metric to determine the quality of the pulse, such that readout problems can be effectively identified.

3. Results from Beam Tests

The performance of the FCal was studied using beam tests in 2003. The purpose was to measure the response and the resolution, and to study the shower shape and energy losses near the inner edge. Beams of hadrons and electrons with fixed energies in the range 10-200 GeV were directed towards five different impact points on the FCal. The final C-side (negative z) FCal modules were used, with close-to-final electronics.

Figure 7 shows results from the electron beam tests. A good linearity was observed, and the intrinsic resolution was measured to $\sigma_E/E = 0.285/\sqrt{E/\text{GeV}} \oplus 0.035$ [2], which exceeds the design requirement of $\sigma_E/E = 0.35/\sqrt{E/\text{GeV}} \oplus 0.05$. Figure 8 shows the performance derived from hadron beam tests. When including the energy of cells in a cylinder with radius 8 cm around the impact point, the measured hadronic resolution is $\sigma_E/E = 0.95/\sqrt{E/\text{GeV}} \oplus 0.075$ [2], which is slightly better than the design criterion (see Section 1). For the measured results quoted above, the contribution from electronic noise was subtracted in quadrature from each measured resolution value. This was done since the electronic noise was not constant over the full set of runs used. Studies using more sophisticated clustering [4] and hadronic calibration schemes using radial weights have also been performed, and significantly improve the energy resolution [5].

These results were obtained from beam test with minimal upstream material and hence measure the intrinsic response and resolution. Beam tests were also performed with upstream material that provide more ATLAS-like conditions. Studies have also been performed of shower-shape variables that can be used to distinguish electromagnetic showers from hadronic ones (which can be used for electron identification), and of the energy losses down the beam pipe and other effects close to the inner edge of the FCal.

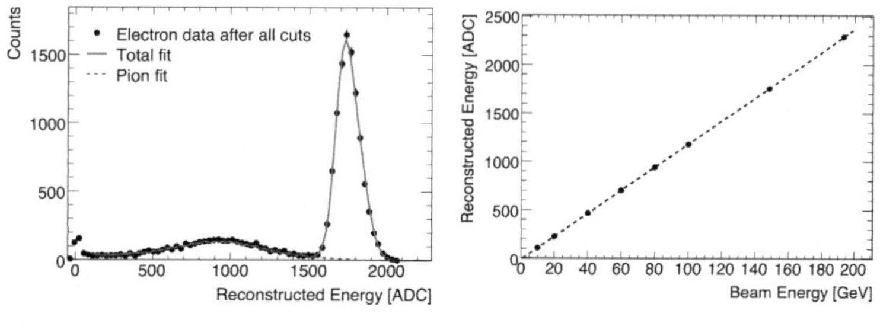

Figure 7. Left: Reconstructed energy in FCal1 for 150 GeV electron beam data. Energy from hadron contamination can be seen at lower energies. Right: Linearity of the FCal response for various electron beam energies.

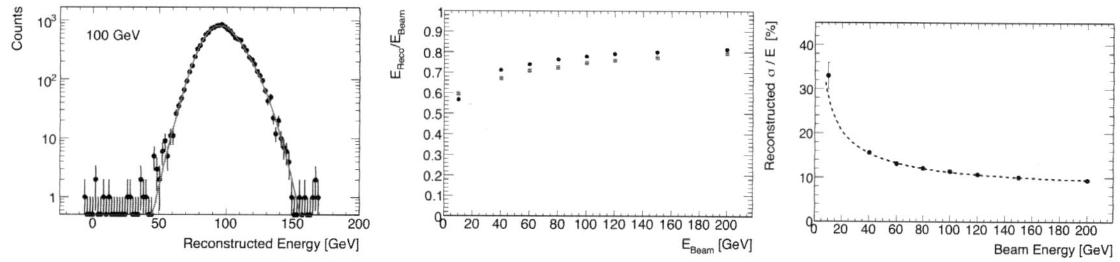

Figure 8. Left: Reconstructed energy from a 100 GeV hadron beam. Middle: Energy response at the EM scale versus beam energy for beam test data (black circles) and simulated beam test data (red squares). Right: Measured noise-subtracted resolution versus beam energy, for hadron beam tests data.

Journal of Physics: Conference Series **293** (2011) 012041 doi:10.1088/1742-6596/293/1/012041

4. Cell Energy Distribution

Figure 9 shows the measured energy for all FCal cells in $\sqrt{s} = 900$ GeV pp collision data recorded in late 2009, compared with the expected distribution from simulated non-diffractive minimum bias data. Reasonably good agreement is observed. Many other quantities such as jet energy, jet transverse width and fraction of energy deposited in the electromagnetic layer have been studied and also show quite good agreement with the expectations from simulations [6].

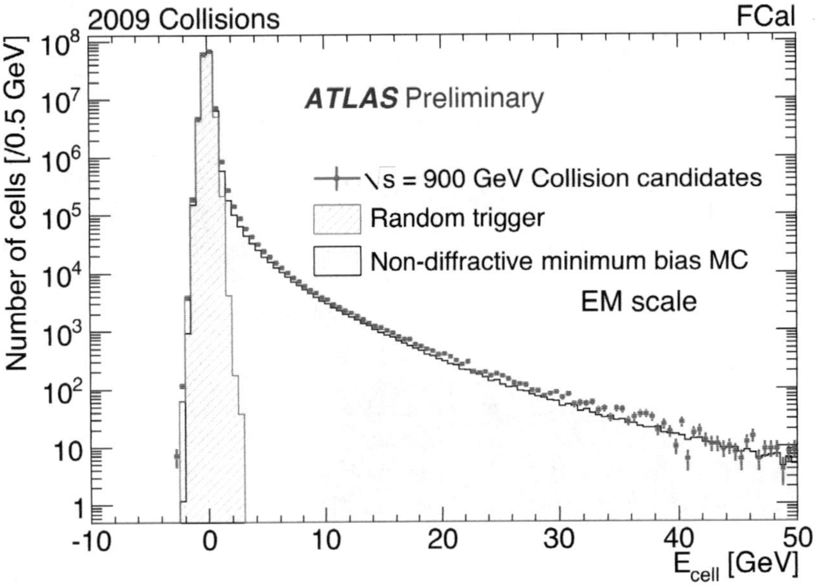

Figure 9. Energy distribution for the cells in the FCal for $\sqrt{s} = 900$ GeV pp collisions candidates (red squares) and randomly triggered data (hashed area). The predicted distribution from non-diffractive minimum bias simulation is shown in yellow.

5. Position Measurement

Since the flux and energy of particles increase steeply with η, and since this flux is symmetric around the beam spot, the energy density profile of a calorimeter can be used to determine its position relative to the beam axis. The FCals are very suitable for such measurements due to their position in the forward region close to the beam axis, and can be used both to determine its own position, and also to perform measurements of the beam spot location, complementary to the measurements performed by the inner tracking detectors.

Studies have shown that the energy density ρ_E measured in the FCals, can roughly be described by

$$\rho_E = k \cdot r^{-m}, \tag{1}$$

where r is the (transverse) distance to the beam spot, and k and m are constants obtained from a fit. The energy density for a channel can be calculated as $\rho_E = (\sum E_i)/(N_{\text{events}} \times N_{\text{electrodes}})$, where the sum is over all N_{events} events, and $N_{\text{electrodes}}$ is the number of electrodes of the channel. This quantity is proportional to the energy per unit area that hit the front face of a given FCal module, since each electrode has the same dimensions. Figure 10 shows a plot of the energy density for each cell in the electromagnetic layer of the C-side FCal measured in $\sqrt{s} = 7$ TeV collision data. One can see that the energy density increase at smaller transverse distances r.

Several methods that use the average cell energy to determine the position of the FCal have been tested. One of them plots the energy density for each channel against its x and y coordinates (as in Fig. 10), and then performs a two dimensional fit using Eq. 1, modified to introduce two new free parameters corresponding to the shift: $r = |(x, y)| \to |(x + \Delta x, y + \Delta y)|$. The vector $(\Delta x, \Delta y)$ obtained from this fit is then the measured shift of the FCal relative to the beam axis.

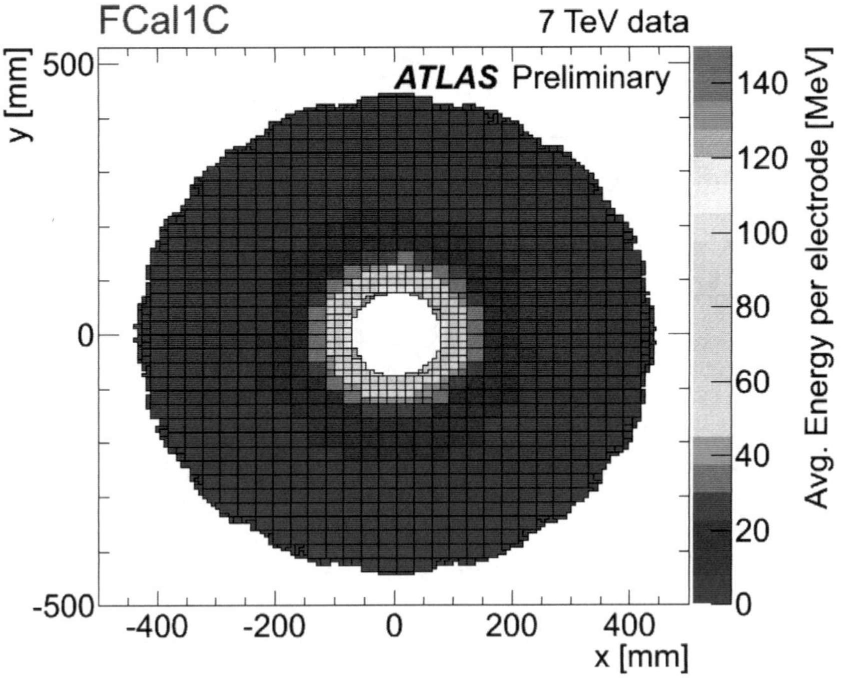

Figure 10. Average energy per event and electrode read out in the electromagnetic layer of the C-side (negative z) FCal in 7 TeV collision data. The 1006 readout cells each consist of 16 or 4 electrodes and are here enclosed with black lines. The upper half of this module sees more energy than the lower half, indicating that the FCal is located a bit low relative to the beam spot.

It should be noted that the simplified relation beween energy and distance given in Eq. 1 only give a reasonable fit for limited ranges of r. For instance, the energy density is not a smooth function of r close to $r = 210$ mm since there is more upstream material in this region due to the presence of a support structure. The method outlined above is therefore repeated several times for different segments of r, resulting in several measurements that later are combined. Channels with readout problems might bias the fit, and must be identified and removed.

The position of the FCal was measured in both 900 GeV and 7 TeV collision data using the method outlined above, and also using other methods. The measurements indicate that the geometrical axis of both A- (pos. z) and the C-side (neg. z) FCal lie about 2 mm below the centreline of ATLAS, with smaller measured shifts along the horizontal x-axis (+0.5 mm and +1.0 mm for the A- and C-sides respectively). The downward shift is consistent with the expected installed position of the FCal modules relative to the centreline of the endcap cryostats in combination with a possible sag in the support structure due to the weight of the FCals (12 tonnes per side).

Summary and Outlook

The ATLAS Forward Calorimeters use a novel design with tubular electrode readout and narrow LAr gaps in order to operate in the very high radiation environment of the forward region. The performance observed in beam tests meets the design requirements, and the pulse shapes in collision data agree very well with expectations. Distributions of cell energies and jet properties also agree quite well with the expectation from simulations. The position of the FCal has been measured in data.

References

[1] A. Artamonov *et al.* 2008, JINST **3** P02010
[2] J.P. Archambault *et al.* 2008, JINST **3** P02002
[3] W.E. Cleland and E.G. Stern, Nucl. Inst. Meth. A **338** (1994) 467.
[4] W. Lampl *et al.*, ATL-LARG-PUB-2008-002 (2008).
[5] Louise Heelan, J. Phys.: Conf. Ser. **160** 012058 (2009).
[6] The ATLAS Collaboration, ATLAS-CONF-2010-055 (2010).

XIV International Conference on Calorimetry in High Energy Physics (CALOR 2010) IOP Publishing
Journal of Physics: Conference Series **293** (2011) 012042 doi:10.1088/1742-6596/293/1/012042

CMS Electromagnetic Calorimeter status and performance with the first LHC collisions

Konstantinos Theofilatos (for the CMS ECAL collaboration)
ETH Zurich, Switzerland
E-mail: konstantinos.theofilatos@cern.ch

Abstract. The Electromagnetic Calorimeter (ECAL) of the Compact Muon Solenoid (CMS) detector will play a crucial role in new physics searches at the Large Hadron Collider (LHC) as well as in Standard Model precision measurements. The first LHC beams have been used to finalize the commissioning of ECAL read-out and trigger. The precision of the inter-channel synchronization and calibration has been verified and improved with in-situ data. The quality of the offline data reconstruction has been investigated with physics signals, in particular with di-photon states. A review of the CMS ECAL detector status and performance is presented in this note.

1. Introduction

The CMS detector is a multipurpose detector which will be used to perform precision Standard Model measurements and new physics searches [1]. The CMS ECAL plays a central role in the physics program of the CMS detector; ECAL was designed to have an outstanding electromagnetic energy resolution [2][3]. This is essential for particle searches such as the Higgs boson. For a low mass Higgs boson, $m_{\mathrm{H}} < 150$ GeV, a promising decay channel for its discovery is the $H \rightarrow \gamma\gamma$ channel which leads to a di-photon final state. A good di-photon mass energy resolution is essential to discriminate an expected signal with small cross section over a large and irreducible Standard Model background. The ECAL energy resolution was demonstrated to be $\Delta E/E < 0.5\%$ for $E > 100$ GeV using electron testbeams [4].

The CMS ECAL is a homogeneous crystal calorimeter and consists of a barrel (EB) section and two endcaps (EE) made in total of 75848 $PbWO_4$ crystals, with a 3 X_{o} lead - silicon strip preshower (ES) detector in the endcap section. The calorimeter is immersed within the 3.8 T magnetic field, produced by the CMS magnet. The scintillation light is read-out by a pair of avalanche photodiodes (APDs) for each EB crystal and a vacuum phototriode for each EE crystal. The small Moliere radius ($R_M \simeq 2.19$ cm) in combination with the large number of crystals results in a fine granularity for the lateral shower shape. In the forward region the granularity is further improved by the preshower detector (ES) $|1.65| < \eta < |2.6|$ which consists of 2 orthogonal planes of silicon strips. Each silicon sensor measures 63×63 mm^2, with an active area of 61×61 mm^2 divided into 32 strips (1.0 mm pitch).

2. Detector precalibration prior to LHC collisions

The ECAL was precalibrated prior to its installation inside the CMS detector. About 25% (3%) of the EB (EE) was intercalibrated with a precision of 0.5% using electron beams. The rest of the

ECAL detector was intercalibrated using a combination of precalibration data from laboratory measurements, cosmic stands and beam splash data, resulting in about $1-2\%$ precision in EB and about 5% in EE. The energy scale was set by the testbeam and validated in-situ by measuring the cosmic muon stopping power in the $PbWO_4$ crystals [5]. Apart from the obtained energy precalibrations, these extensive tests also helped to accumulate large operational experience with the detector. After the first LHC collision in November of 2009, only few hours were needed in order to observe the first π^0 candidates (Fig. 1). The $\pi^0 \to \gamma\gamma$ decay is now part of the CMS online data quality monitoring and provides and important stream for further improvement of the detector intercalibration in-situ.

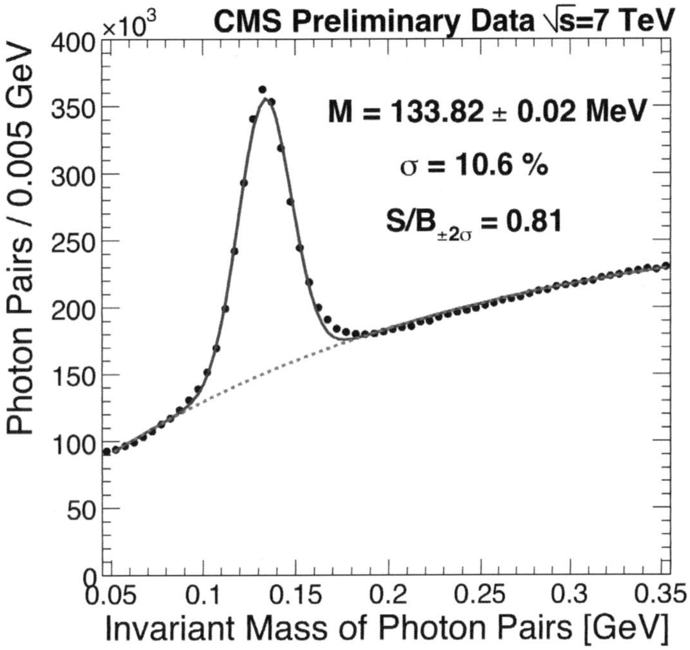

Figure 1. π^0 peak reconstructed in CMS data for 7 TeV center of mass energy collisions. About 1.5 M candidates are found for an integrated luminosity of about 0.43 nb^{-1}.

3. Channel timing synchronization

In September 2008 the LHC was running in single beam mode; the proton beam was stopped in a collimator about 150 m away from the CMS interaction point. These beam splashes produced an intense flux of muons crossing the CMS detector in the direction along the beam pipe. The response of the ECAL detector in LHC beam splashes was used to synchronize the detector. The hardware (online) synchronization was done with the granularity of a trigger tower (5×5 channels) in steps of 1.04 ns. The residual spread of the individual channels' timing within a trigger tower was further reduced offline. Overall, an RMS spread at the level of 300 ps is achieved (Fig. 2). The ECAL timing resolution was assessed by comparing the timing of neighboring crystals belonging to the same cluster. The timing resolution was found in collision data to be less than 1 ns for energies above $E > 1$ GeV (Fig. 3).

Journal of Physics: Conference Series **293** (2011) 012042 doi:10.1088/1742-6596/293/1/012042

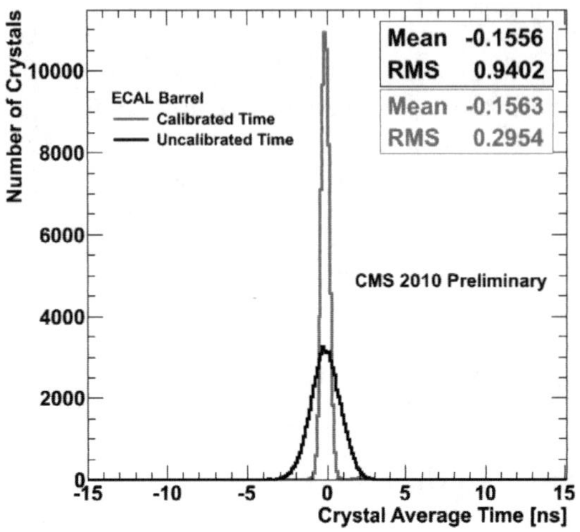

Figure 2. Mean time per crystal after alignment of the readout phase before time calibration (black) and after time calibration (red); preliminary offline calibrations derived from 2009 LHC beam splash data.

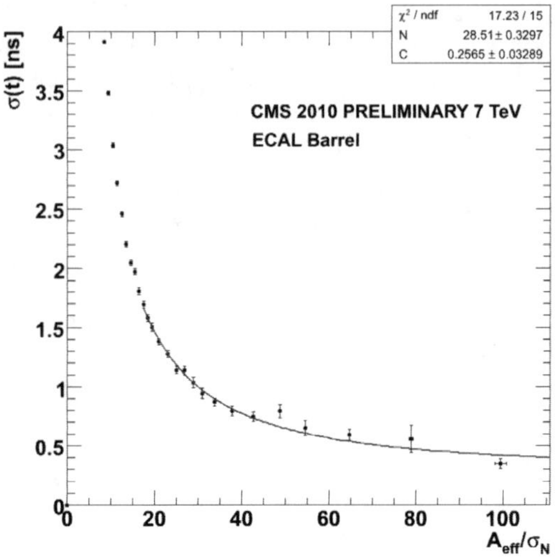

Figure 3. Time resolution as a function of the effective amplitude normalized in units of noise, derived by comparing the time in nearby crystals in the same cluster.

4. Status of the detector during 2008-2010 running

The ECAL detector was operating continuously during 2008-2010 taking calibration, cosmic and collision data whenever possible. In total more than 99% of channels are fully functional for physics. Typically the channels which are excluded in the read-out are seen as white regions in the occupancy plots (Fig. 4, Fig. 6 and Fig. 5). A fraction of those channels have a broken data-link. However most of them can be recovered using the information of the trigger primitives. In total only $\sim 0.15\%$ of channels have neither data nor trigger information and are accounted as non functional regions of the detector.

5. Comparison of collision data and Monte Carlo (MC) simulation

Establishing agreement between data and MC at the level of a single ECAL crystal is a necessary step towards understanding and predicting physics backgrounds with the help of MC simulation. The MC detector noise model was extensively studied and developed prior to the LHC start-up benefiting from the knowledge of the energy intercalibrations and was tuned to reproduce the noise measured in-situ using pedestal runs.

The energy spectrum of the individual channels matches very well what is expected from the simulation of minimum bias collisions (Fig. 7). Apart from the energy distribution very good agreement is also observed in the azimuthal occupancy of the channel with the highest reconstructed transverse energy (Fig. 8). Variations as a function of ϕ, fairly reproduced in MC, reflect modularity and the inhomogeneity of the energy equivalent noise in the ECAL.

The overall good agreement between data and MC reflects the level of understanding of the detector. In addition to the normal scintillation signals, some anomalous signals are observed in a small fraction of collision data with a rate about 1 per 1000 minimum bias events on 900 GeV collision data. These anomalous signals have a distinct pulse shape and are uniformly distributed in the EB while they are not seen in EE where the read-out is done by the VPTs. The origin of the anomalous signals appears to be mainly due to highly ionizing particles which give an

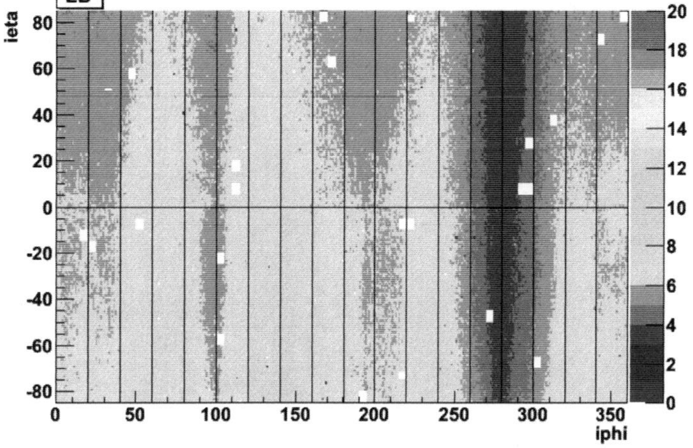

Figure 4. ECAL EB occupancy plots (mean energy per crystal) averaged over many LHC beam splash events. Energy modulations are a combination of the energy flow traversing CMS and geometry effects.

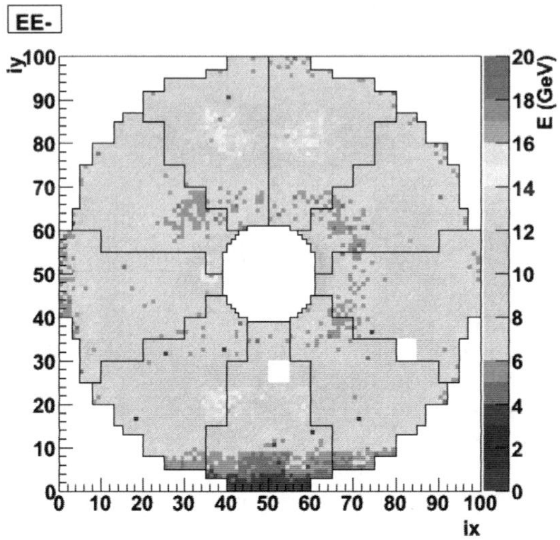

Figure 5. EE- occupancy plots (mean energy per crystal) averaged over many LHC beam splash events.

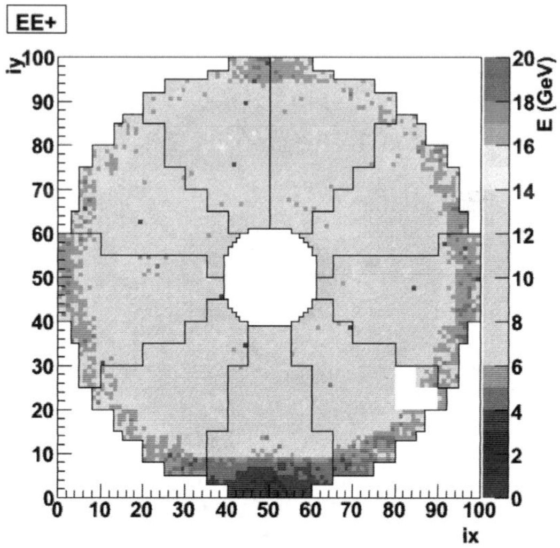

Figure 6. EE+ occupancy plots (mean energy per crystal) averaged over many LHC beam splash events.

instantaneous charge deposition in the pre-amplification region of the APDs. These signals are not accompanied by a shower generation in the crystal's body. They are efficiently suppressed by a quality selection such as the 1-E4/E1<0.95 where E1 is the energy of the seed crystal and E4 is the energy of the four crystals having a common face with the seeding crystal (Fig. 9).

Electron and photon objects used in physics analysis are reconstructed as clustered energy in the ECAL. An electromagnetic shower deposits its energy in several ECAL crystals; super-clusters (clusters of clusters) of crystals extended along ϕ (bending direction) are used to

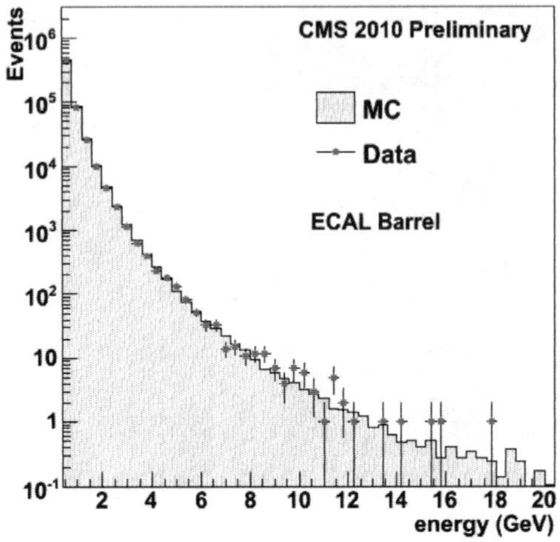

Figure 7. Energy spectrum in the individual EB channels. Data and MC correspond to the same integrated luminosity. Event quality and signal quality selections are applied.

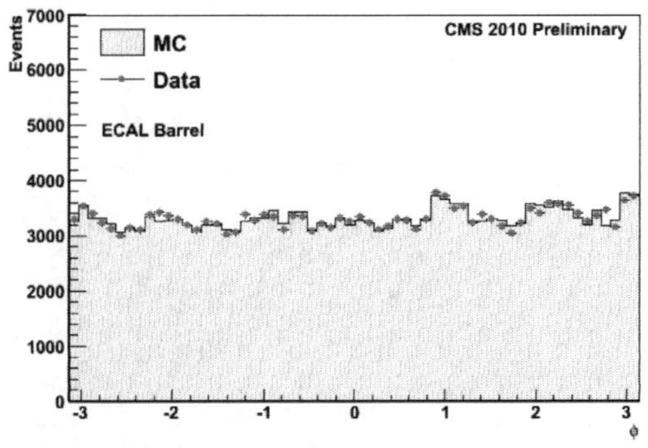

Figure 8. Azimuth distributions of the EB channel with highest reconstructed transverse energy (E_T).

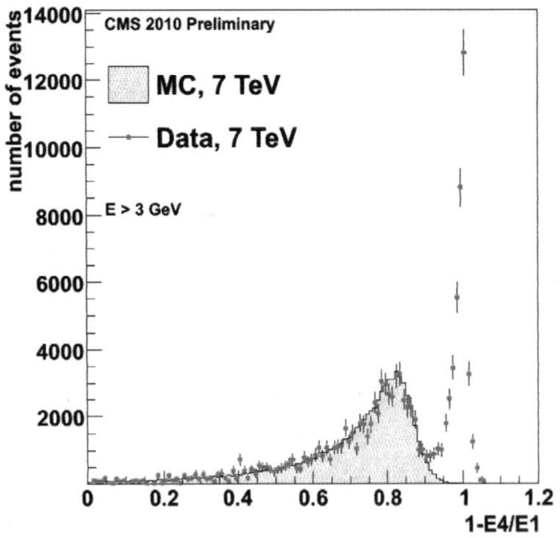

Figure 9. At the cluster level the anomalous signals appear as energy in a single crystal, while in electromagnetic showers the energy is typically shared between neighboring crystals. This fact is used to tag anomalous signals. E4/E1 : ratio of the energy deposit in one crystal (E1) over the energy in the four crystals (E4) having a common face with this crystal.

reconstruct the electromagnetic energy. The number of expected super-clusters and their transverse energy (E_T) are in very good agreement with the observed distribution in the data (Fig. 10 and Fig. 11). The good agreement between data and MC at the level of super-clusters is essential for the understanding of the electron and of the photon objects and controlling the expected physics background.

6. Conclusions

The CMS ECAL detector is fully functional and within the design specifications. More than 99% of the channels are operational and in good health for physics. The precalibrations have provided a good reference point for the start-up and in-situ π^0 calibration efforts are now ongoing. Apart

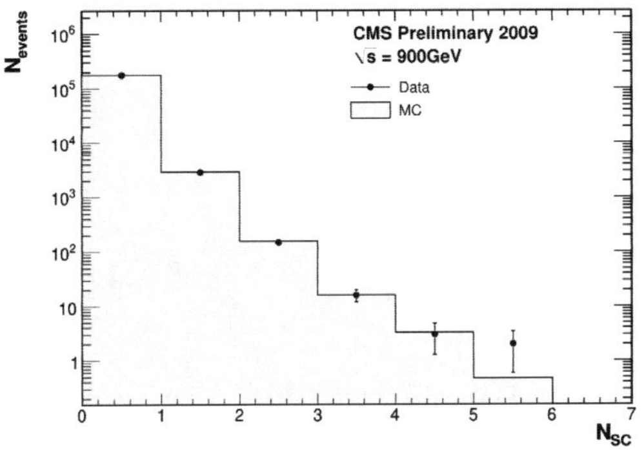

Figure 10. Number of ECAL super-clusters in 900 GeV collision data compared with the MC simulation.

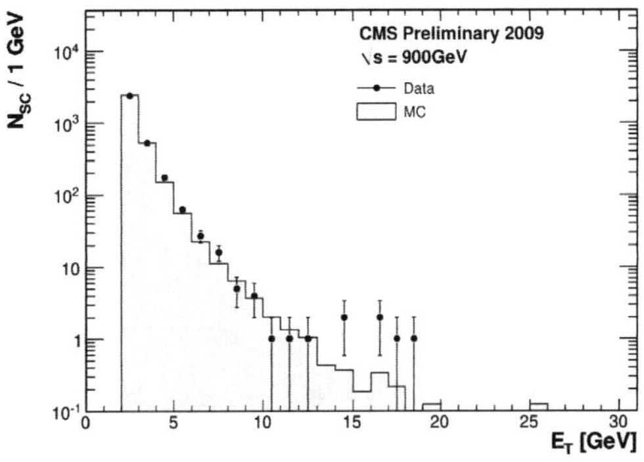

Figure 11. Transverse energy (E_T) of the super-clusters in 900 GeV collision data compared with the MC simulation

from a good energy calibration and resolution, CMS ECAL has also very good timing alignment and a timing resolution which is at the sub-nanosecond level. The comparison of the data versus the MC simulation shows a good overall understanding of the detector.

References

[1] CMS Collaboration, CMS Physics Technical Design Report, Volume 2: "Physics Performance" 2007,*J. Phys. G: Nucl. Part. Phys.*, **34** (2007) 995.
[2] CMS Collaboration, "The CMS experiment at the CERN LHC" 2008, *JINST* **3** S08004 (2008)
[3] CMS Collaboration, CMS Physics Technical Design Report , Volume 1: "Detector Performance and Software" 2006, *CERN/LHCC-2006-001; CMS-TDR-008-1*
[4] CMS Collaboration, "Energy resolution of the barrel of the CMS electromagnetic calorimeter" 2007, JINST 2 P04004 (2007)
[5] CMS Collaboration, "Measurement of the muon stopping power in lead tungstate" 2010, *J. Instrum.* **5** (2010) P03007 and *CMS-CFT-09-005*

Journal of Physics: Conference Series **293** (2011) 012043 doi:10.1088/1742-6596/293/1/012043

The ALICE Electromagnetic Calorimeter: EMCAL

A. Fantoni, on behalf of the ALICE collaboration

Istituto Nazionale Fisica Nucleare,
Laboratori Nazionali di Frascati
Via E. Fermi 40
00044 Frascati (Roma) - Italy

Alessandra.Fantoni@lnf.infn.it

Abstract. The ALICE Experiment (A Large Ion Collider Experiment) aims to study the properties of quark-gluon matter using Pb-Pb collisions at a center of mass energy (per nucleon pair) of $\sqrt{s_{NN}}$ = 5.5 TeV with the Large Hadron Collider (LHC) at CERN. The EMCAL detector is a large area electromagnetic calorimeter. ALICE's capability to perform jet reconstruction is greatly improved by measuring the neutral energy component of jets including photons and neutral pions. The calorimeter produces also a fast, high-p_T trigger providing an enhancement in high-p_T events, participates in the High Level Trigger (HLT), performing online reconstruction at the full data rates and generates trigger information based on reconstructed events. The performance of prototypes has been studied in test beam measurements at FNAL and CERN and are shown. Four EMCAL super-modules have been installed at CERN and are operational. First results from data taking are presented. Moreover, an extension of EMCAL for jet-jet and γ-jet physics will be discussed.

1. Introduction

ALICE (A Large Ion Collider Experiment) at the LHC is designed to carry out comprehensive measurements of high energy nucleus-nucleus collisions, in order to study QCD matter under extreme conditions and to explore the phase transition between confined matter and the Quark-Gluon Plasma (QGP) [1,2].

ALICE contains a wide array of detector systems for measuring hadrons, leptons and photons. The ALICE detector is described in detail in [3]. The large acceptance Electromagnetic Calorimeter (EMCAL), which is foreseen to be fully installed in 2011, significantly enhances ALICE's capabilities for jet measurements. EMCAL is designed to provide the following functions: efficient and unbiased fast level L0 and L1 trigger on high energy jets; measurement of the neutral portion of jet energy; improvement of jet energy resolution; measurement of high momentum photons, π^0 and electrons; γ/π^0 discrimination up to 30 GeV (considering invariant mass and shower shape techniques only); electron/hadron separation for momenta larger than 10 GeV/c; high uniformity of response for isolated electromagnetic clusters.

This paper presents a description of the EMCAL with its characteristics, the final results of the performance of prototype modules in test beam measurements at FNAL and at CERN, the present status of EMCAL, the first results from data taking and an upgrade for EMCAL, expanding the acceptance of EMCAL for di-jet and hadron-jet correlation measurements.

Journal of Physics: Conference Series **293** (2011) 012043 doi:10.1088/1742-6596/293/1/012043

2. EMCAL description and characteristics

The overall design of the EMCAL is heavily influenced by its integration within the ALICE [3] setup, which constrains the detector acceptance to a region of about 110° in the azimuthal angle ϕ, $0.7\leq\eta\leq0.7$ in pseudo-rapidity and 4.35 m < R_{EMCAL} < 4.7 m radial distance. The chosen technology is a layered Lead (Pb)-Scintillator (Scint) sampling calorimeter with wavelength shifting (WLS) fibers that run longitudinally through the Pb/Scint stack providing light collection (Shashlik) [4]. The basic building block is a module consisting of 2x2 optically isolated towers which are read out individually. All modules in the calorimeter are mechanically and dimensionally identical. Each module has a rectangular cross section in the ϕ direction and a trapezoidal cross section in η with a full taper of 1.5°, and spans $\Delta\eta x\Delta\phi$ = 0.014x0.014. A scheme of the module is shown in Fig.1, together with the dimensions in mm. White, acid free, bond paper serves as a diffuse reflector on the scintillator surfaces and provides friction between layers. The scintillator edges are treated with TiO_2 loaded reflector to improve the transverse optical uniformity within a single tower and to provide tower to tower optical isolation better than 99%. The requirement of a compact detector consistent with the EMCAL integration volume and the chosen detector thickness of about 20 radiation lengths, results in a lead to scintillator ratio by volume of about 1:1.22 corresponding to a sampling geometry of Pb (1.44 mm)/Scint (1.76 mm). The physical characteristics of the modules are summarized in Table 1.

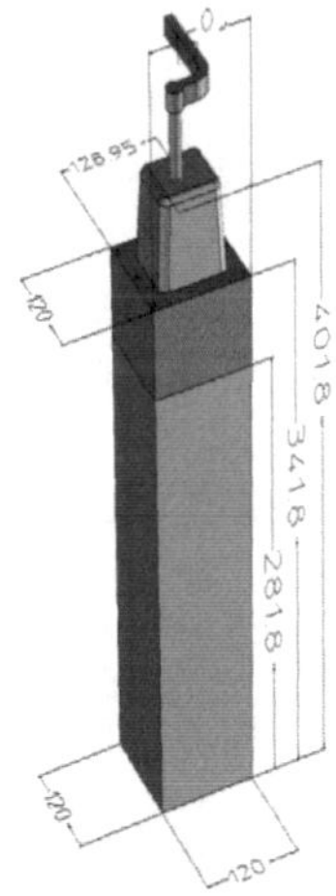

Parameter	Value
Tower size (at η=0)	~6.0 x ~6.0 x 24.6 cm^3
Tower size	$\Delta\phi$ x $\Delta\eta$ = 0.0143 x 0.0143
Sampling ratio	1.44 mm Pb / 1.76 mm Scint.
Layers	77
Scintillator	Polystyrene (BASF143E+ 1.5%pTP+0.04%POPOP)
Absorber	Natural lead
Effective RL X_0	12.3 mm
Effective MR R_M	3.20 mm
Effective density	5.68 g/cm^3
Sampling fraction	1/10.5
Radiation length (RL X_0)	20.1

Fig.1 (left): A single EMCAL module with the dimension in mm. Table 1 (right): EMCAL module physical parameters. Here, RL stands for Radiation Length and MR for the Moliere Radius.

Scintillation photons produced in each tower are captured by an array of 36 Kuraray Y-11 (200 M), double clad, wavelength shifting (WLS) fibers. Each fiber within a given tower terminates in an aluminized mirror at the front face of the module and is integrated into a polished, circular group of 36 fibers at the photo sensor end at the back of the module.

The 6.8 mm diameter fiber bundle from a given tower connects to the Avalanche Photodiode (APD) through a short light guide/diffuser. The selected photo sensor is the Hamamatsu S8664-55 avalanche photodiode chosen for operation in the high field inside the ALICE magnet. The APDs are operated at moderate gain for low noise and high gain stability in order to maximize energy and timing resolution. The number of primary electrons generated in the APD by an electromagnetic shower is ~4.4 electrons/MeV. The reverse bias voltage of the APDs are individually controlled to provide an electron multiplication factor (M) of 30 resulting in a charge output of ~132 electrons/MeV from the

APDs. All APDs used have been previously calibrated [5]. The charge output from the APD is integrated by a Charge Sensitive Preamplifier (CSP) with a short rise time of ~10 ns and a long decay time of ~130 μs, i.e., approximately a step pulse.
The readout electronics is similar to that of the Photon Spectrometer (PHOS): a detailed description is given in [6].
Twelve modules are assembled with a structural strongback to form a strip-module, a self-supporting unit like the module. A collection of 24 strip-modules forms a super-module. EMCAL will be formed by ten super-modules. Pictures of a strip-module and of a full super-module are shown in Fig.2, on the left and right panel, respectively.
More details on the mechanical assembly can be found in [7].

Fig.2: Single strip-module comprised of 12 EMCAL modules integrated onto a single strongback (left) and single super-module formed by 24 strip-modules (right).

3. Results from test beams

The performance of a 4x4 array of prototype modules and of a 4x4 array of final design modules for the ALICE EMCAL has been studied in test beam measurements at FNAL (2005) and CERN (2007), respectively. All towers were instrumented with the full electronics chain with shapers and APD gains operated as planned in ALICE. A LED calibration system was installed in order to monitor time-dependent gain changes. The readout of the front end electronics used the standard ALICE data aquisition system.
The final results are fully described in details in [8], showing an average light yield of (4.3±0.3) photoelectrons/MeV, a uniformity of the response within 1% for all towers and configurations; a good linearity of the response to electrons above 20 GeV, an only slightly deteriorated energy resolution when using the EMCAL default shaping time of 200 ns compared to 2 μs for PHOS, a position resolution described by 1.5 mm + 5.3 mm / √E (GeV) and a hadron rejection factor > 600 for an electron identification efficiency of 90%.
The energy resolution of $\sigma(E) / E = (1.7 \pm 0.3) + (11.1 \pm 0.4) / \sqrt{E} + (5.1 \pm 0.3) / E$ (where E is in GeV) is shown in Fig.3 (left panel), together with the x and y position resolution as a function of the energy deposit for electrons (right panel).
In conclusion, the performance of the tested EMCAL modules reaches all design criteria.

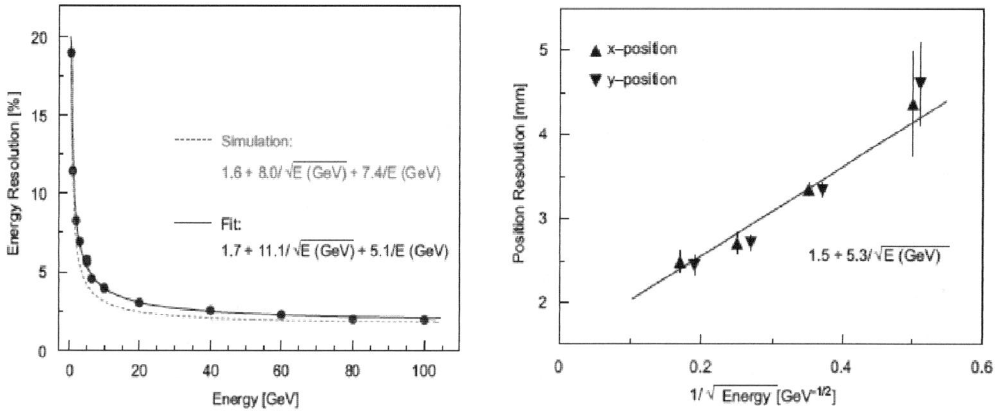

Fig.3: Left panel: energy resolution for electrons as a function of the incident beam energy. The beam energy spread was subtracted from the measured result and the dashed curve represents the resolution obtained from Monte Carlo simulations. Right panel: position resolution as a function of the deposited energy for electrons. The curve shows the best fit result.

4. The EMCAL status

The assembly of EMCAL modules started at the end of 2008. In 2009 four super-modules have been installed in ALICE and presently they are operational and taking data. The corresponding angular coverage is $\Delta\eta$ x $\Delta\phi$=1.4x1.05. Extensive commissioning of the readout electronic, L0, L1 and HLT triggering has been carried forward together with APD bias optimization and gain monitoring. First physics events have been successfully reconstructed. Presently four further super-modules are ready and under test and calibration. All details about test and calibration are reported in a dedicated proceeding of this Conference [9]. The modules for the last two super-modules are in preparation and the assembly is expected to be completed before the end of the summer; the corresponding cosmic calibration will be completed during the fall 2010. As soon as a long LHC shutdown will be available, the full EMCAL will be installed, presumably in 2011.

A picture of the four super-modules installed in ALICE is shown in Fig.4 (left side), where only the two in the front side are visible.

5. First results from data taking

After months of commissioning and running with cosmic rays, at the end of 2009 it was possible to see the response of EMCAL with the first proton-proton collisions at the center of mass energy $\sqrt{s}$ = 900 GeV and in 2010 at $\sqrt{s}$ = 7 TeV. Since the early commissioning phase of the LHC, it was possible to have the first events analyzed and displayed in the counting room by the offline reconstruction software and the online reconstruction software implemented in the high Level Trigger (HLT) analyzing the events in real time, indicating the start of the physics exploitation of the ALICE experiment. First results have been immediately obtained and published [10]. The four EMCAL super-modules are continuously present in the data taking. Invariant mass spectra of the π^0 are obtained in different energy bins, an example is shown on the right panel of Fig.4.

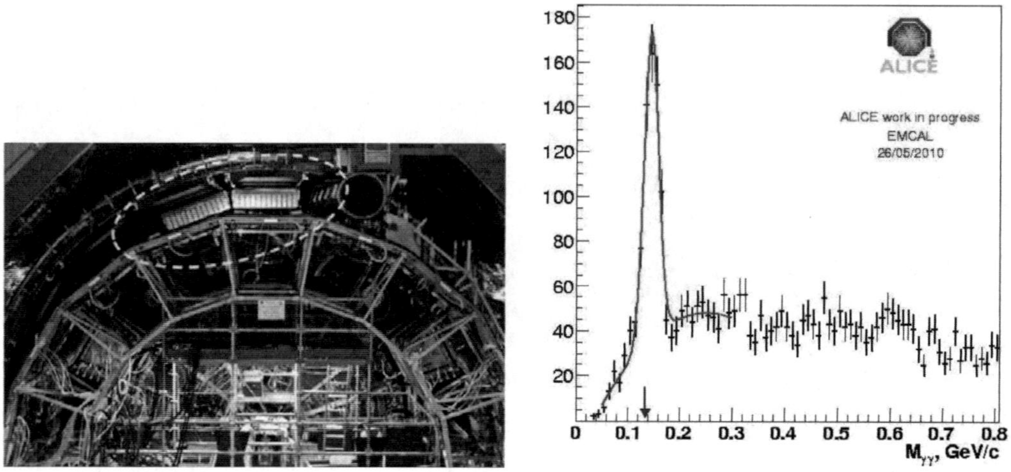

Fig.4 Left panel: four super-modules installed, only those two in front are visible. Right panel: π^0 mass reconstruction in the transverse momentum bin $5.0<p_T<5.5$ GeV.

6. The EMCAL upgrade: DCAL

Even though EMCAL was the last ALICE detector to be proposed, approved, assembled, and it is still only partially installed, the first upgrade approved by the ALICE collaboration (November 2009) is an extension of EMCAL, denominated DCAL (Di-jet Calorimeter) [11]. The DCAL expands the physics capabilities of the EMCAL by enabling back-to-back correlation measurements, which are impossible with the EMCAL alone, but are essential to obtain a complete picture of the physics addressed by the EMCAL. Together, the DCAL and EMCAL form a two-arm electromagnetic calorimeter. The EMCAL subtends 110° and the DCAL subtends 60° in ϕ, with both detectors covering $|\eta|<0.7$, thereby providing good acceptance for di-jets with radii $R\leq0.4$ up to transverse momenta $p_T \sim 150$ GeV/c. Simulation studies of the DCAL have been carried out and have verified that the technology originally developed for and implemented in the EMCAL meets all the needs of the DCAL project. As a consequence, from a technical perspective, DCAL is an extension of EMCAL. DCAL super-modules are built exactly as they are in EMCAL, out of strip-modules, but with reduced length in η: infact, each DCAL strip-module contains 16 modules instead of 24 present in EMCAL.
DCAL will be situated immediately adjacent to PHOS on both the ALICE A and C sides, causing unavoidably a small gap in η ($\delta\eta \sim 0.02$) between the sensitive volumes of the two detectors, due to the super-module structure. DCAL+PHOS can be considered as one integrated detector system for the study of jets, consequently all simulations done include PHOS as well as DCAL super-modules.
The assembly of the DCAL modules will start in summer 2010, after completion of EMCAL. Before the end of the year, all 6 DCAL super-modules will be assembled and tested, ready to be installed in ALICE as soon as a long shutdown will be available, presumably on 2011.
On the left panel of Fig.5 is shown a schematic view of the 6 DCAL super-modules with the PHOS super-modules in between and on the right panel the beam view of EMCAL and DCAL is illustrated.

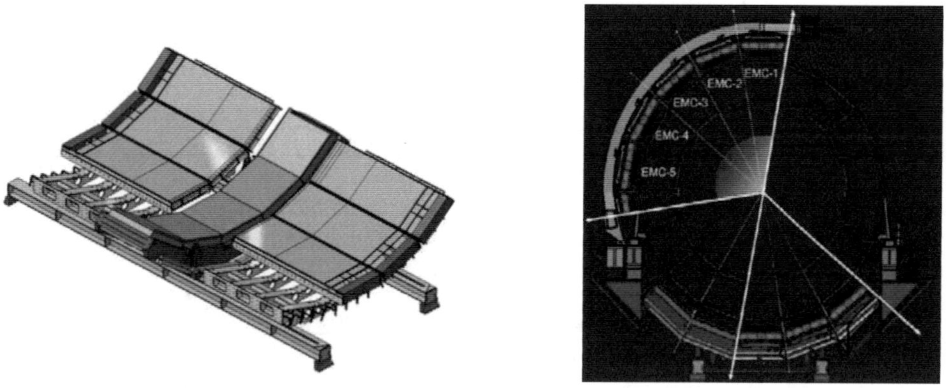

Fig.5: Left panel: 6 DCAL super-modules (in gray) with the PHOS super-modules (in orange) in between. Right panel: beam view of EMCAL and DCAL.

7. Conclusions

An accurate description of the ALICE electromagnetic calorimeter, with its characteristics has been given. Final results from measurements done in test beams in 2005 and 2007 have been reported, demonstrating that the performances of EMCAL satisfy all the design criteria. The status of the assembly and installation of EMCAL is illustrated: four super-modules are already installed in ALICE for π^0, γ and first jet physics. The first results obtained from data taking are promising and encouraging: of course some investigation and optimization is still going on.

The assembly of EMCAL is expected to be completed in summer 2010 and the corresponding calibration in the fall in order to be ready for the full EMCAL installation for the next long shutdown available.

An upgrade for EMCAL, denominated DCAL for di-jet and hadron-jet physics, is illustrated: the corresponding assembly will start in the next months and will be completed before the end of the year, with possible installation together with the rest of EMCAL super-modules.

References

[1] F. Carminati et al. (ALICE Collaboration), J. Phys. G30 1517 (2004).
[2] ALICE Collaboration, Physics Performace Report, Vol 2, J. Phys. G (2006).
[3] K. Aamodt et al. (ALICE Collaboration), JINST 3 S08002 (2008).
[4] ALICE Collaboration, ALICE Technical Design Report Electromagnetic Calorimeter, CERN LHCC-2008-014, 1 December 2008.
[5] A. Badalà et al., Nucl. Instrum. Methods A610, 200 (2009).
[6] PHOS Collaboration, ALICE Technical Design Report, CERN/LHCC 99-4, 5 (1999).
[7] F. Ronchetti et al., J. Phys. Conf. Series 160, 012012 (2009).
[8] J. Allen et al., Nucl. Instrum. Methods A615, 6 (2010).
[9] J. Faivre, this Volume.
[10] K. Aamodt et al. (ALICE Collaboration), Eur. Phys. J. C65, 11 (2010); K. Aamodt et al. (ALICE Collaboration), Eur. Phys. J. C67, 019 (2010).
[11] J. Allen et al. (ALICE Collaboration), ALICE DCal: An addendum to the EMCAL Technical Design Report, CERN LHCC-2010-011, 20 June 2010.

The ATLAS Liquid Argon Calorimeter: Overview and Performance

Huaqiao Zhang
on behalf of the ATLAS Liquid Argon Calorimeter Group
CPPM, Aix-Marseille Universite, CNRS/IN2P3, Marseille France
E-mail: huaqiao.zhang@cern.ch

Abstract.
The LHC is delivering collision data since November 2009 up to a 7 TeV center of mass energy world record. The ATLAS experiment has collected more than 1 nb^{-1} collision data until the end of April 2010. The ATLAS Liquid Argon (LAr) calorimeter system, operational since August 2006, has been carefully commissioned and checked using collision data as well as single beam data and cosmic data taken previously. This paper gives an overview of the ATLAS LAr calorimeters status and performance up to the latest 7 TeV collision data. Results on the LAr calorimeters high voltage, LAr temperature, purity and readout electronic stability as well as results on timing alignments and early collision performances are shown here. All indicates that the ATLAS LAr calorimeter system is well commissioned and shows good and promising performances for the physics at LHC era.

1. Introduction

The ATLAS experiment [1] at the Large Hadron Collider (LHC) is a general purpose detector aimed at Higgs and new physics searches as well as precision test of electroweak (EW) mechanisms. It includes an inner tracker, a calorimetry system and a muon spectrometer. The LAr Calorimeters [2], part of the ATLAS calorimetry system, are expected to play leading role in electron, photon, jet and missing E_t measurements and identifications. The test of the LAr calorimeter performances started during the construction. Electron and pion beam tests were performed on prototype and production modules between 1998 and 2004. The LAr calorimeters are operational in the collision cavern since August 2006. In-situ Cosmics data, as well as a short term of beam data in September 2008 and November 2009 were taken before the first LHC collision. Performances were studied and extrapolated to the collisions [3]. ATLAS has collected more than 1 nb^{-1} collision data up to April 2010. This paper reports an overview of the LAr calorimeter system status obtained from years of accumulated data and its in-situ performances from these first collision data.

2. The hardware status of the LAr calorimeters

The LAr calorimeters are sampling calorimeters with liquid argon as active material, full azimuthal angle (ϕ, defined in the transverse x,y plane) coverage and up to [-4.9,4.9] pseudo-rapidity (η) [1] coverage. The LAr calorimeter system includes the electromagnetic barrel

[1] $\eta = -ln(tan(\theta/2))$ where θ is the particle momentum direction versus beam axis polar angle

(EMB) calorimeter ($|\eta| < 1.475$, lead absorber) sitting in the barrel cryostat, and inside an end cap cryostat at each side of the barrel, an electromagnetic end cap (EMEC) calorimeter ($1.375 < |\eta| < 3.2$, lead absorber), a hadronic end cap (HEC) calorimeter ($1.5 < |\eta| < 3.2$, copper absorber), and a forward calorimeter (FCal, $3.1 < |\eta| < 4.9$, copper/tungsten absorber) as shown in Fig. 1. There are three or four longitudinal layers with different granularity in each of them. The EM calorimeter (EMB + EMEC) includes a highly granular accordion layer up to $|\eta| < 2.5$ and a presampler up to $|\eta| < 1.8$, providing information to trigger electron and photon objects (Fig. 2) and ensure proper electron and photon identification and measurement.

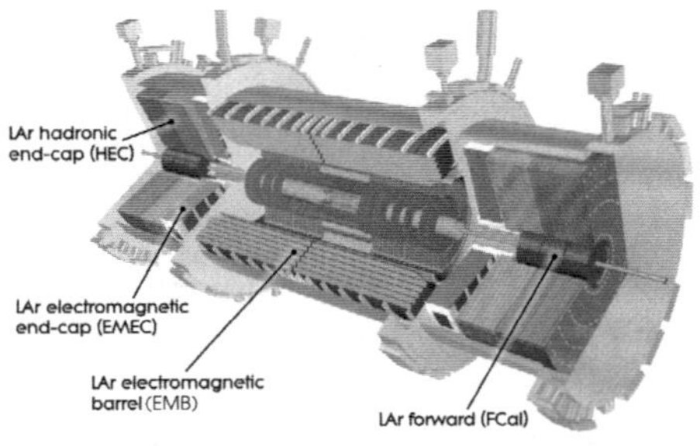

Figure 1. ATLAS LAr calorimeters.

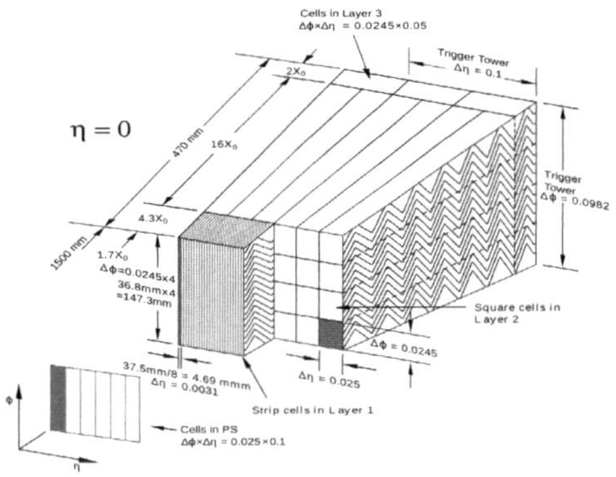

Figure 2. EM calorimeter layers.

The LAr calorimeter system has, in total, 182,468 readout channels/cells, of which 98.6% are operational (4th of May 2010). There are 1.4% dead readout channels, including 20 dead optical transmitters (EMB+EMEC 1.5%, HEC 0.1%, FCal 0%), that will be recovered at a next long shut down. Only 0.0186% electrode channels are unrecoverable.

2.1. High voltage status and liquid argon temperature and purity

High energy particles passing through the LAr calorimeters produce, by interaction with absorbers and LAr, secondary particles traveling in LAr and creating ionized charges. Those charges are drifted by the high voltage (HV) in the liquid argon gap and creates an electric signal on the electrodes. The collected signal depends on the HV and the temperature and purity of the liquid argon.

- HV: LAr calorimeter cells are set to constant HV in $\delta\eta$ x $\delta\phi$ regions of size 0.2*0.2. Only 6% of those HV regions are working at reduced value [4]. Signals in reduced HV regions are recovered with only a small loss of accuracy by correction factors.
- LAr temperature: The LAr signal sensitivity to LAr temperature is -2%/K (density: -0.45%/K , Velocity: -1.55%/K) [5].A homogeneity and stability better than 100 mK is required. The LAr temperature is monitored by probes immersed in the cryostat. The achieved homogeneity, within each cryostat, is 59 mK with 1.5 mK RMS for each probe.
- LAr purity: Gas such as O_2 can capture electrons and reduce signal collection. An impurity level better than 1000 ppb O_2 equivalent is required. Purity monitors [6] in each cryostat show that the impurity level in the LAr is in the range of 200±100 ppb O_2 equivalent.

2.2. LAr calorimeter pulse shape and drift time

Cell collected signals are then amplified, shaped and passed through a switched capacitor array which samples the signal every 25 ns. LAr cells pulse shapes can be predicted from calibration pulses by several models such as Response Transformation Method (RTM) and First Principles Method (FPM). Both model requires that the drift time be measured. Studies of EM calorimeter drift time use special 32 samplings (25 ns per sampling, 32 samplings covering the full signal pulse region) cosmics runs. The measured drift time is around the expected 460 ns in the barrel and varies with cell gap size in the end cap region. The cosmics results show that the maximum differences of the pulse shape and the predictions are within 2% [7].

2.3. Electronics readout stability

The energy reconstruction of LAr cells is based on an Optimal Filtering algorithm:

$$E_{cell} = F_{\mu A \to MeV} . F_{DAC \to \mu A} . \frac{1}{\frac{M_{phys}}{M_{cali}}} . R . [\sum_{j=1}^{N_{samples}} a_j (s_j - p)], \qquad (1)$$

where the pedestals (p), the ramps (R), the Optimal Filtering Coefficients (a_j) and the correction factor for the difference between the calibration and the physics pulse($\frac{1}{\frac{M_{phys}}{M_{cali}}}$) of each cell are monitored by an automatic electronic calibration procedure and stored in a data base that is updated weekly or once significant variations are observed. The calibration run frequency can be up to every LHC fill thus any small deviation of the constants could be corrected within a short time. A study of the calibration constant variations over a six months period shows in Fig. 3 the good stability of the electronic calibration constants. For example, the RMS of EM calorimeter pedestals is less than 0.03 ADC count(typical pedestal is ~1000 ADC counts).

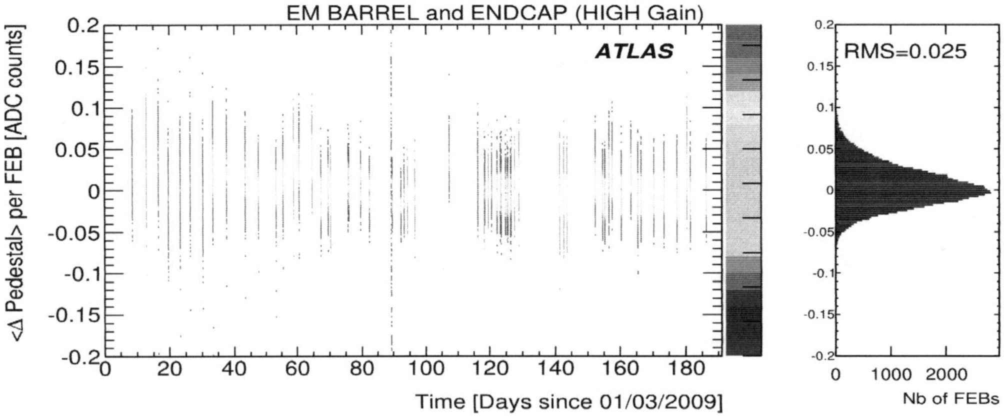

Figure 3. Pedestal stability over six month

2.4. LAr EM calorimeter uniformity measurement with cosmics

Besides the test-beam results, the LAr EM calorimeter response uniformity has also been estimated by comparing cosmics data and MC simulations that did not emulate any non-uniformity. Although this method is limited by muon event properties and statistics (it is hard to extrapolate to electron/photon cases in real collisions), it gives first in-situ uniformity of the ATLAS EM calorimeter. The studies show that the EM barrel second layer non-uniformity is at the level of 1%(as shown in Fig. 4).

3. The LAr calorimeter performances

ATLAS started to record LHC beam data since end of November 2009. These data, first single beam data, then collisions at 900 GeV then at 2.36 TeV and finally at 7 TeV center of mass energy, allowed extensive performance studies of the LAr calorimeter system performances.

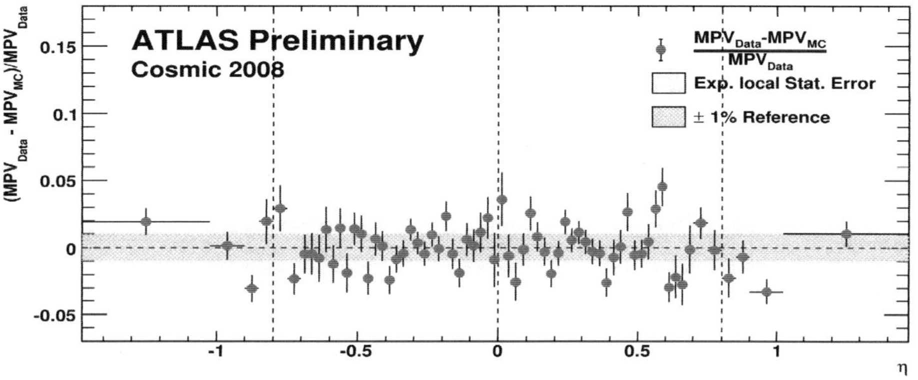

Figure 4. Uniformity of EMB estimated in cosmics data.

3.1. Timing alignment

The so called beam splash events, issued from single beam collisions on a target/collimator 149m far upstream of the ATLAS detector, were used to align timing constants for both the Level one calorimeter trigger and the main LAr calorimeter readout. These events, with both side single beams, created plenty of secondary particles, and fired almost all the ATLAS calorimeters. The LAr Level one (L1Calo) trigger timing constants for all EM trigger towers are now aligned to an accuracy of ± 5 ns [9]. The LAr calorimeter main readout timing constants were tuned from the average timing of each readout front end board in collision data. An accuracy about 2 ns for all LAr partitions has been achieved for now. With the future increase of collision data statistics, these timing constants will be tuned further at the channel level both for the the L1Calo trigger and the LAr calorimeter systems.

3.2. Level one EM calorimeter trigger performance

The L1Calo trigger and the LAr main readout use the same detector, but different granularity and readout path. An analogue sum is made of all four (or three) layers of LAr EM cells connected to a trigger tower, typically 0.1 x 0.1 in $\delta\eta$ x $\delta\phi$ up to $|\eta| < 2.5$ and grosser trigger tower in higher η region (Fig. 2). The EM L1Calo trigger performance was checked firstly looking at the good correlation of transverse energy (E_T^{L1}) measured by the trigger tower versus the sum E_T^{LAr} of all offline cells connected to this trigger tower as shows in Fig. 5. Then the EM L1Calo trigger E_T resolution with respect to offline sum E_T was measured to be less than 3% at about $E_T > 30GeV$ as shown in Fig. 6, well within the designed 5% E_T resolution. Finally turn on curves of the L1Calo trigger system, such as for the EM5 item, which requires level one EM cluster E_T greater than 5 GeV, have expected performances, as can be seen for EM5 in Fig. 7, and provide electon/photon triggers for offline analysis.

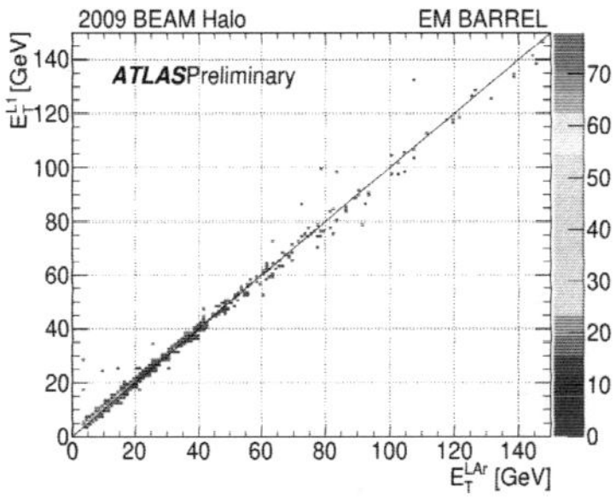

Figure 5. E_T^{L1} versus E_T^{LAr}.

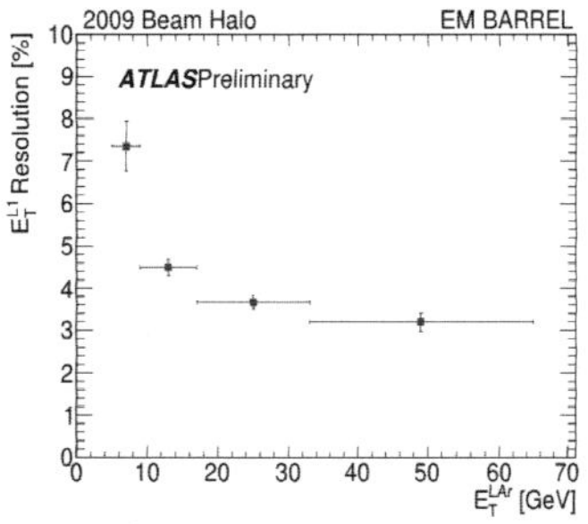

Figure 6. E_T^{L1} resolution versus E_T^{LAr}.

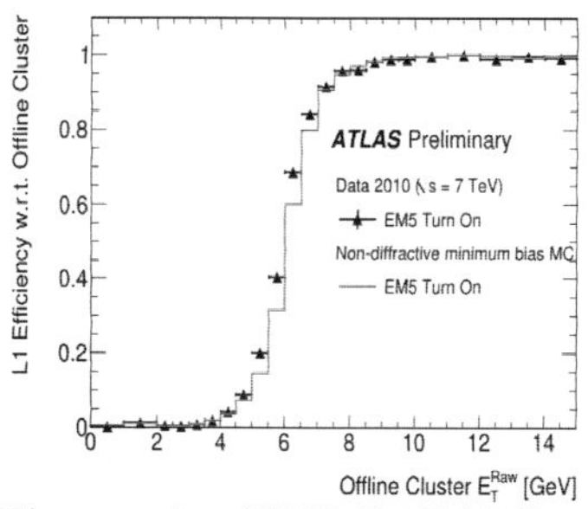

Figure 7. EM5 L1CALO Trigger Turn on.

3.3. LAr noise and missing transverse energy in random events

Fig. 8 shows the noise measured with random trigger events at the cell level as a function of η for all LAr layers. It is in agreement with expectations. The noise of single cells is constantly checked by the online monitoring system. The noise can also be checked through global cell quantities such as missing transverse energy $E_T^{\rm miss}$ in random events. The definition is:

$$E_T^{\rm miss} = \sqrt{(\sum E_x^{\rm miss})^2 + (\sum E_y^{\rm miss})^2} \tag{2}$$

Summing up all the LAr cells above two sigma of noise, the obtained LAr calorimeters $E_T^{\rm miss}$ agrees well with the gaussian model. However, as shown in Fig. 9, better noise suppression is achieved with a dynamical topological cluster summed $E_T^{\rm miss}$, which is used as the default physics $E_T^{\rm miss}$ algorithm.

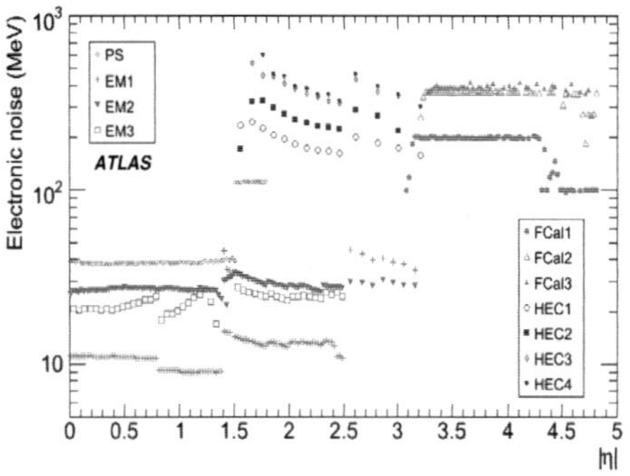

Figure 8. LAr noise at cell level

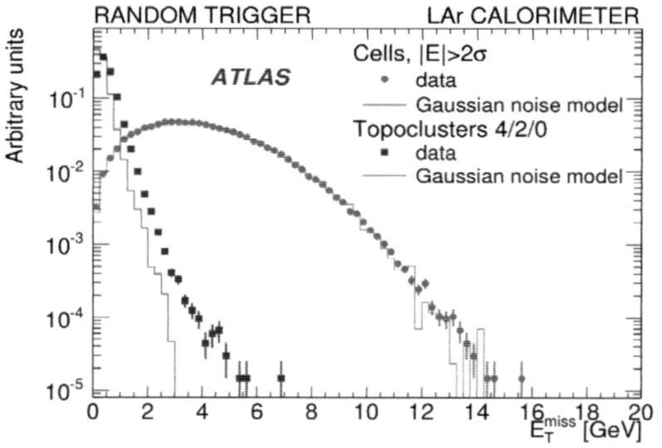

Figure 9. $E_T^{\rm miss}$ in random events

3.4. LAr calorimeter energy flow

The energy deposited in LAr cells by collision data is compared to Non-Diffractive minimum bias Monte Carlo simulations and to random trigger events. The collision signal tails is well visible in all LAr calorimeters and the simulation agrees well with the data as shown for the endcap electromagntic calorimeter in Fig. 10. A more detailed check of LAr calorimeter energy flow is done by looking at the occupancy map of each calorimeter layer. This map is obtained for each layer by summing up, for all selected minimum bias 7 TeV events, all the cells with energy above $5\sigma_{\rm noise}$. The occupancy convolutes physics and detector performance problems. Each hot/cold zone in the map shown in Fig. 11 is identified as detector defects, such as for example the blank areas which are dead readout regions or the green squares which are non nominal HV regions, and have been taken care off.

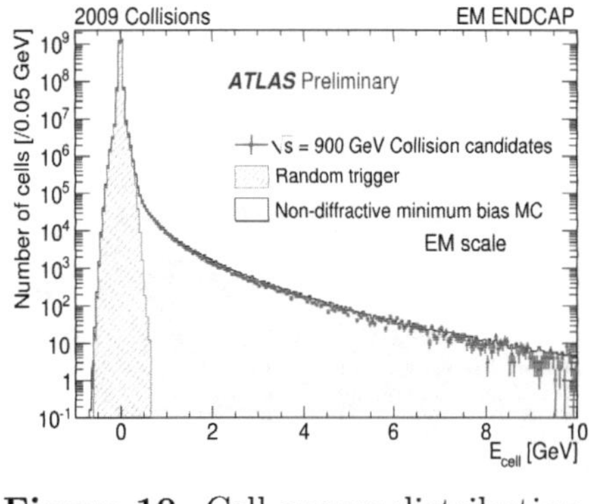

Figure 10. Cell energy distribution

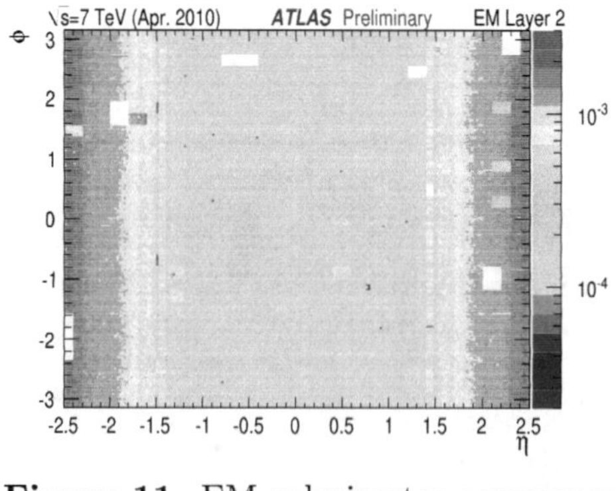

Figure 11. EM calorimeter occupancy.

3.5. Electron, photon, jet, E_T^{miss}

After the check of the LAr calorimeters global performances, the study of electrons, photons, jets and E_T^{miss} [10] has been made with the first 900 GeV and 7 TeV collision data. As an example, the E_T^{miss} distribution from a topological cluster algorithm is shown in Fig. 14. The di-jet $\delta\phi$ distribution from Anti-k_t jet algorithm is shown in Fig. 13. And after a careful selection of photon candidates, a reconstructed invariant $\gamma\gamma$ mass spectrum is produced, as shown in Fig. 12, with visible π^0 and η^0 signal peaks. For all of those distributions, the agreement between the simulation and the data is good at 900 GeV as well as at 7 TeV, illustrating the excellent performances of the LAr calorimeter system.

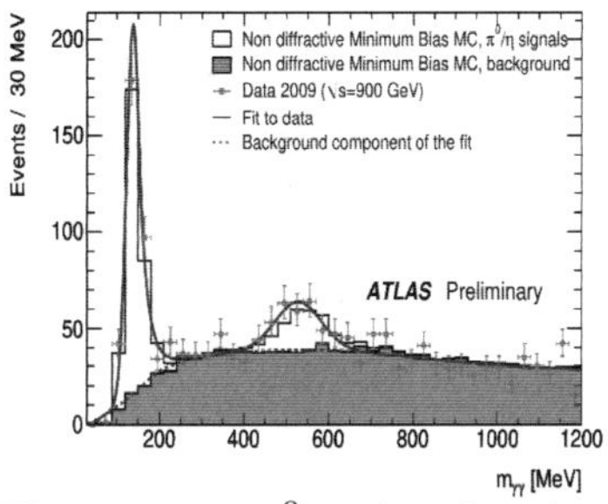

Figure 12. π^0 and η signal

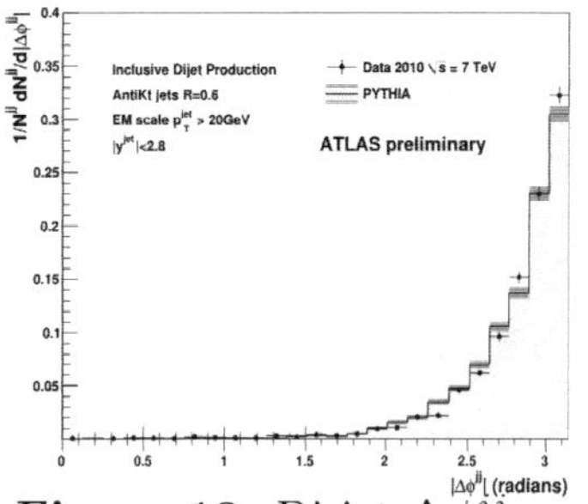

Figure 13. Di-jet $\Delta\phi^{JJ}$.

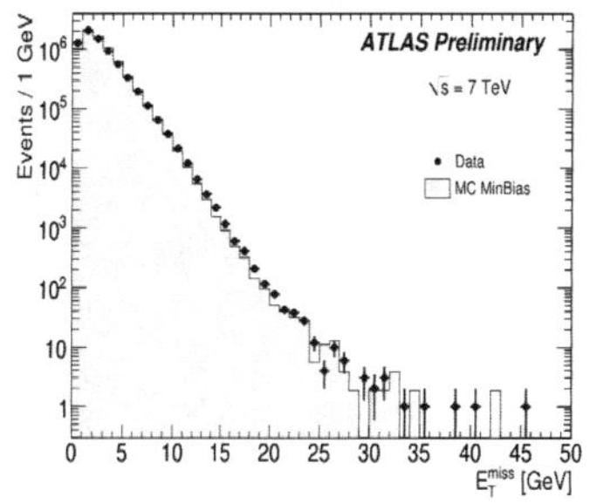

Figure 14. Missing E_t.

4. Conclusion

The studies of recent collision data as well as years of test-beam, cosmics and single beam data recorded by the ATLAS detector, assess the excellent performance of the Liquid Argon Calorimeters. Remarkable stability of the LAr temperature, purity, and electronics is achieved well within the designed requirements. Less than 1% overall energy response non-uniformity even in the cosmics data, $\pm$ 2 ns overall LAr timing in the first collision data and a well modeled and monitored noise. The first performance measurements of the LAr trigger system, that's to say electron, photon, jet and E_T^{miss} reconstruction using collision data show very promising results and allow the first observation of particle peaks, such as π^0 and η^0. The LAr calorimeter system is now ready for the long ongoing data campaign.

References

[1] G Aad et al, The ATLAS Experiment at the CERN Large Hadron Collider, The ATLAS Collaboration JINST 3 (2008) S08003
[2] ATLAS liquid argon calorimeter: Technical design report, The ATLAS Collaboration, CERN-LHCC-96-41 (1996)
[3] G.Aad et al., Readiness of the ATLAS Liquid Argon Calorimeter for LHC Collisions, The ATLAS Collaboration, Accepted by EPJC
[4] B.Belhorma et al., Evaporating short-circuits in the ATLAS liquid argon barrel presampler, ATL-LARG-PUB-2005-003
[5] C. de la Taille, Temperature dependence of the ATLAS electromagnetic calorimeter signal. Preliminary drift time measurement. ATL-LARG-95-029 (1995)
[6] M.Adams et al., A purity monitoring system for liquid argon calorimeters, Nucl. Inst.Meth. A 545 (2000) 613-623
[7] G.Aad et al., Drift Time Measurement in the ATLAS Liquid Argon Electromagnetic Calorimeter using Cosmic Muons, The ATLAS Collaboration, Accepted by EPJC
[8] H. Abreu et al., Performance of the Electronic Readout of the ATLAS Liquid Argon Calorimeters, Accepted by JINST
[9] ATLAS Level-1 Calorimeter Trigger: Timing Calibration with 2009 LHC Beam Splashes, The ATLAS Collaboration, ATL-DAQ-PUB-2010-001
[10] The ATLAS collaboration, ATLAS-CONF-2010-001, ATLAS-CONF-2010-005, ATLAS-CONF-2010-006, ATLAS-CONF-2010-008

Calibration of CMS Electromagnetic Calorimeter at LHC startup

Riccardo Paramatti

Abstract. The first 7 TeV LHC collisions recorded with the CMS detector have been used to perform a channel-by-channel calibration of the electromagnetic calorimeter (ECAL). Decays of π^0 and η into two photons as well as the azimuthal symmetry of the average energy deposition at a given pseudorapidity are utilized to equalize the response of the individual channels. The ECAL comprises a central *barrel* section and two *endcaps*. Based on an integrated luminosity of $\sim$ 100 nb^{-1}, a channel-by-channel *in-situ* calibration precision of 1.15% has been achieved in the barrel ECAL in the pseudorapidity region $|\eta| < 0.8$. The energy scale of the ECAL has been investigated and found to agree with the simulation to within 1% in the barrel and 3% in the endcaps.

1. Introduction

The electromagnetic calorimeter (ECAL) of the Compact Muon Solenoid (CMS) experiment consists of a barrel section (EB) and two endcaps (EE). The EB contains 61 200 lead tungstate crystals arranged in 170 $\eta-$rings of 360 $PbWO_4$ crystals each (one crystal covers 1 degree in ϕ). It is subdivided into 36 supermodules each containing 1700 crystals and provides coverage in pseudorapidity up to $|\eta| < 1.5$. The two ECAL Endcap (EE) subsystems consist of four half-disk Dees, each containing 3662 crystals. The EE provides a coverage of $1.5 < |\eta| < 3.0$. The scintillation light is detected by avalanche photodiodes (APDs) in the EB and vacuum phototriodes (VPTs) in the EE. The EE is additionally instrumented with a preshower detector (ES) on its front faces, consisting of a two layer lead-silicon sampling calorimeter. An extended description of the CMS ECAL is provided in [1]. The general performance and commissioning of the ECAL with the first 7 TeV data is described in [2]. A detailed description of the CMS experiment can be found in [3].

In the environment of CMS, for unconverted photons with energies in the range of interest for physics analyses, $E_{shower} \sim 100$ GeV, the energy resolution will be dominated by the constant term. As a consequence, the performance of the CMS ECAL at the LHC will strongly depend on the quality of the calibration and monitoring. Achieving the design-goal inter-calibration precision of 0.5% will be particularly important for a discovery of the Higgs boson in the decay channel $H \rightarrow \gamma\gamma$, one of the primary goals of the LHC physics program.

This article describes results of the first crystal-by-crystal calibrations and global energy response measurements of the CMS ECAL at the LHC, performed during the first months of operation. The performance of the specialized ECAL calibration data-taking streams is also discussed.

Journal of Physics: Conference Series **293** (2011) 012045 doi:10.1088/1742-6596/293/1/012045

2. The CMS ECAL calibration strategy

The ECAL was precalibrated before installation using several methods (section 2.1). The final calibration will be made *in-situ* at the LHC using collision data.

The estimated particle energy, obtained from the ECAL, can be expressed as:

$$E = F \cdot \sum_{cluster\ crystals} G(GeV/ADC) \cdot C_i \cdot A_i \qquad (1)$$

where the sum is over the crystals in a *cluster*. A_i are the reconstructed amplitudes in ADC counts. C_i is the inter-calibration constant while G is the ECAL energy scale. The factor F is defined as an additional energy correction which depends on the type of the particle, its energy and pseudorapidity and in particular takes into account shower leakage and bremsstrahlung losses for electrons. This article describe the analysis focused on the determination of C_i and G.

For the calibration in-situ at the LHC, the following strategies have been explored:

- ϕ-symmetry inter-calibration (Section 3) is expected to provide a fast inter-calibration exploiting invariance around the beam axis of energy flow in minimum bias events [4]. This method inter-calibrates crystals at the same pseudorapidity, so that other methods are needed to inter-calibrate regions at different pseudorapidity.
- π^0 and η calibration (Section 4) uses the photon pairs selected as $\pi^0(\eta) \to \gamma\gamma$ candidates [5]. At the startup, this method is also used to investigate the ECAL energy scale (Section 4.3).
- isolated electrons from $W \to e\nu$ and $Z \to e^+e^-$decays can be used to compare the energy measured in ECAL to the track momentum measured in the silicon tracker. This method is expected to become the main channel-by-channel calibration tool when the integrated luminosity reaches several fb^{-1} [6]. The integrated luminosity accumulated so far is not sufficient to use this method. $Z \to e^+e^-$ dacays will be used also to monitor and correct the absolute ECAL energy scale [7].

In addition, variations in the crystal light output and transparency are monitored with a laser-based system [8]. With the present luminosity, the crystal transparency changes are negligible (and then they have no effect on ECAL response).

2.1. The pre-calibration

The CMS ECAL has been pre-calibrated prior to installation with laboratory measurements of crystal light yield and photo-detector gain during the construction phase (all EB and EE channels), with test beam electrons (9 EB supermodules and about 500 EE crystals) and with cosmic ray muons (all EB channels). After installation in the LHC, circulating beams were stopped in collimators 150 m away from CMS in September 2008 and November 2009. The resulting *beam dump* events have been used to improve pre-calibration precision.

Inter-calibration constants coming from different sources are combined as a weighted average. The precision of a given calibration method is determined empirically by comparing the new inter-calibration constants to those derived at the test beams. The comparison for the EE constants (derived from the combination of laboratory measurements and beam dump calibration) is shown in Figure 1 for a subset of the crystals exposed to beam. The results of these pre-calibrations have been reported in [9] and [1].

The channel-by-channel calibration precision at the start-up at 7 TeV is estimated to be:

- EB: about 0.5% for the 9 supermodules calibrated with test beams and 1.5% to 2.2%, depending on pseudorapidity, for the other 27 supermodules;
- EE: below 1% in the $\sim$ 500 crystals calibrated with beam and about 5% elsewhere;

In the rest of the article, a crystal-by-crystal calibration constant is defined as a multiplicative factor c_i relative to the precalibration constant derived for this crystal: $C_i = C_i^{precalib.} \cdot c_i$.

Journal of Physics: Conference Series **293** (2011) 012045 doi:10.1088/1742-6596/293/1/012045

3. Calibration with the ϕ-symmetry method

The ϕ-symmetry method is based on the expectation that for a large sample of minimum bias events the total deposited transverse energy (E_T) should be the same for all crystals in a ring at fixed pseudorapidity. Inter-calibration in ϕ is performed by comparing the total transverse energy (ΣE_T) deposited in one crystal with the mean of the total ΣE_T collected by crystals at the same absolute value of η.

In the determination of the transverse energy sum, only deposits with energies between lower and upper thresholds are considered. The former is applied to remove the noise contribution and is derived by studying the noise spectrum in randomly triggered events. The latter is applied to avoid a possible bias from very high E_T deposits (e.g. from electrons originating from W or Z decays).

In order to compare the relative supermodule scale in the EB with test beam and with other calibration methods, we derive the inter-calibration constants from a data sample obtained with an integrated luminosity of 8.8 nb^{-1}, corresponding to about $1.6{\cdot}10^8$ minimum bias events. For each of the 36 supermodules we study the distribution of the 1700 constants. We use the mean of the Gaussian fit to this distribution as our definition of SM scale. To estimate the systematic error, we have derived the distribution of the relative scale for those supermodules which were calibrated at the test beam, with a precision of several times better than expected from our method with the available data. The width of this distribution, about 0.5%, is the convolution of the precision of the test beam inter-calibration and the ϕ-symmetry inter-calibration, and poses therefore an upper limit to our systematic error. In Figure 2, we show a comparison of the SM scale derived from 2009 data ($\sim$ 0.01 nb^{-1} at $\sqrt{s} = 900$ GeV) and 2010 data ($\sqrt{s} = 7$ TeV).

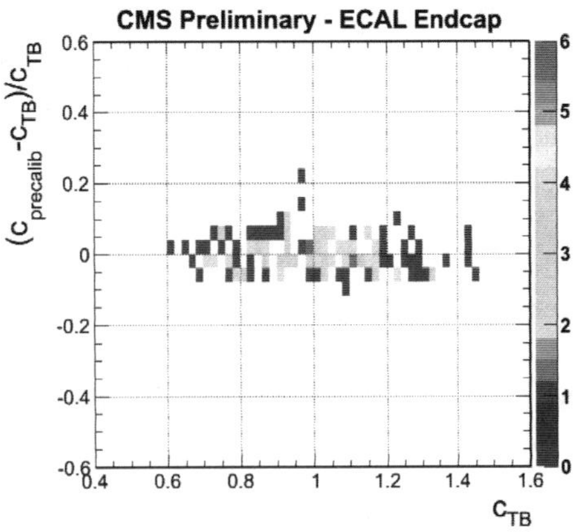

Figure 1. Relative difference between the inter-calibration coefficients measured at the test beam and the coefficients derived from the combination of all other available measurements (laboratory, 2008 and 2009 beam dump data) versus the inter-calibration coefficients from test beam.

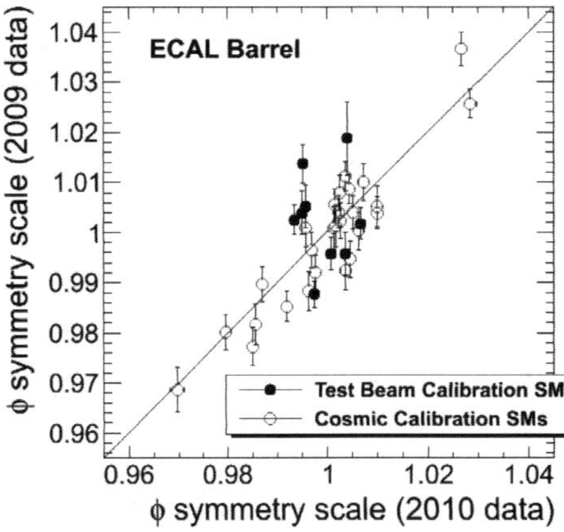

Figure 2. Comparison of relative supermodule scale with ϕ-symmetry method between $\sqrt{s} = 900$ GeV (2009) and $\sqrt{s} = 7$ TeV (2010) data. Only statistical errors shown. The systematic uncertainty of the method is estimated to be $\pm 0.5\%$.

3.1. Crystal-by-crystal calibration

We applied the ϕ-symmetry inter-calibration procedure on $1.6{\cdot}10^8$ minimum bias events and derived inter-calibration constants c_i for all crystals in the EB. To estimate the precision of our set of constants, we derived the distribution of the c_i for each ring of constant η. The width of the distribution gives the convolution of the precision of the pre-calibration with the precision of our method. For the nine test-beam calibrated supermodules, the expected pre-calibration precision is about 0.5%, several times better than expected for the ϕ-symmetry method. Therefore, the width of distribution of the c_i limited to those supermodules gives an upper bound of the inter-calibration precision achieved with the ϕ-symmetry method. The width of the c_i distribution is estimated by performing a Gaussian fit and extracting the sigma. In Figure 3, we show the estimated precision as a function of ring index. The pattern is due to the amount of material in front of ECAL, which is not constant in pseudorapidity.

4. Calibration with π^0 and η decays

An inter-calibration technique based on reconstructing $\pi^0 \rightarrow \gamma\gamma$ decays has been studied using both the simulations and a dedicated test-beam study [5]. In this section the performance of the specialized π^0 and η calibration streams and the first crystal-by-crystal calibration results are reported.

4.1. Performance of the calibration stream

In order to take advantage of the high rate of π^0 decays, a specialized data-taking stream has been developed. In this stream, the candidate di-photon decays are selected directly from events passing single-e/γ and single-jet L1 triggers. After selection, only information about a limited region of ECAL (energy deposits in 20 to 40 individual crystals) near the π^0 candidates is stored for the actual calibration.

In the EB region, the photon candidates are reconstructed using a simple 3×3 window clustering algorithm that requires that the energy deposited in a seed crystal is above 0.5 GeV and no crystals are allowed to belong to more than one cluster. The cluster energy is then computed as the sum of nine crystal energies. Thus, the constructed clusters can comprise at most nine crystals with non-zero energy deposits. The shape of the cluster energy deposition should be consistent with that of an electromagnetic shower produced by an unconverted photon. In order to take into account the energy losses due to the noise suppression algorithm and shower leakage, we apply an energy correction derived as a function of the cluster energy and pseudorapidity (factor F in eq. 1). The correction is derived using the simulation and applied both to the data and simulation, in the same manner.

To be considered in the combinatorial selection of di-photon pairs, a photon candidate is required to have a transverse energy above 0.8 GeV. The π^0 candidates are then selected by requiring their transverse energy to be above 1.6 GeV. The $\pi^0 \rightarrow \gamma\gamma$ candidate should also be isolated from other significant energy deposits. In addition, the π^0 candidates are required to have an invariant mass below 0.25 GeV/c^2. Figure 4 shows the invariant mass distributions of the selected π^0 candidates after all selection cuts both for data and simulated minimum bias events, for the EB. The distributions for data are obtained with 18.7 nb^{-1}. The number of events in the simulation distribution is normalized to that for the data.

To estimate the signal rate, the invariant mass distributions are fitted to a sum of a Gaussian and fourth order polynomial background. The signal-to-background ratio (S/B) is then estimated to be the ratio of the integral of the fitted signal and background functions in the mass window, $|M_{inv} - M_{fit}| < 2 \cdot \sigma_{fit}$. The π^0 peak width is found to be 8.2% in data and 8.1% for simulation. The S/B is measured to be 2.3 and 2.4 in data and simulation, respectively.

A separate calibration stream has been implemented to select $\eta \rightarrow \gamma\gamma$ decays. The same selection and reconstruction methods are used except that the photon pair transverse energy cut is tightened to $E_T^{\gamma\gamma} > 3$ GeV. In addition, a veto cut is applied to reject photons coming from di-photon candidates within the 3σ window around the fitted π^0 peak.

In the EE, the on-line stream has also been implemented selecting π^0 and η di-photon candidates using the same methods.

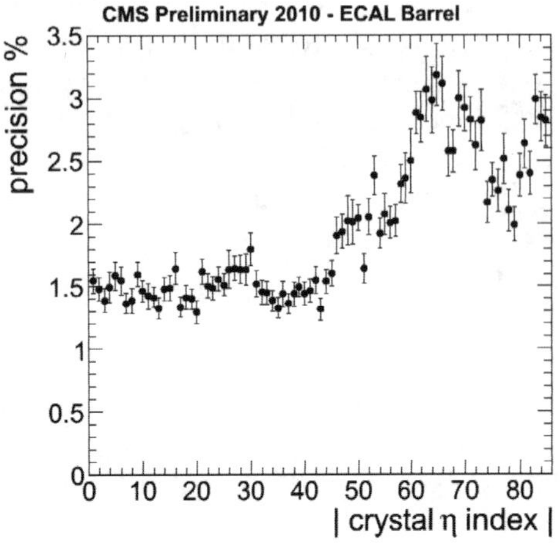

Figure 3. Precision of the inter-calibration constants derived from ϕ-symmetry as a function of ring index in EB.

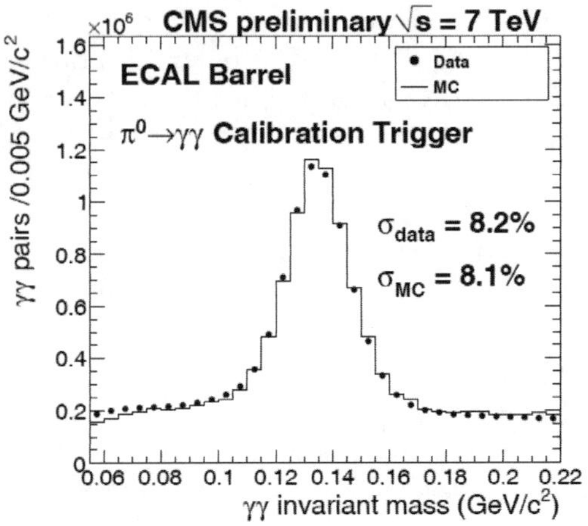

Figure 4. Invariant mass distributions for photon pairs accepted by the on-line π^0 stream for the data and simulations in the EB.

4.2. Inter-calibration Results

To investigate the potential of the $\pi^0 \to \gamma\gamma$ inter-calibration, we use two independent calibration algorithms [5]. Both algorithms are based on an iterative procedure where the inter-calibration constants are updated after each iteration step.

Nine supermodules in this region have been pre-calibrated to a precision of about 0.5% in the test-beams. Crystals in those supermodules are used to estimate the in-situ inter-calibration precision in the same way as for the $\phi-$symmetry method, as shown in Figure 5. In the region $|$crystal η index$| \leq 45$, the crystal-by-crystal calibration precision is found to be 1.4% which is consistent with the expectation from Monte Carlo studies which give $1.3\pm0.3\%$. The systematic limit on the precision of this method has not yet been reached and the current inter-calibration accuracy is dominated by the statistical uncertainty.

4.3. The ECAL energy scale with π^0 and η decays

The absolute scale of the ECAL, the G factor defined in eq. 1, has been measured during the test beam campaign for EB and EE separately. In-situ, the ECAL energy scale will be determined by reconstructing di-electron and di-photon invariant mass peaks. While the $Z \to e^+e^-$samples collected so far are not sufficient to measure the energy scale, a first measurement can be done using π^0 and η reconstructed decays.

The energy scale is derived by measuring the ratio of the reconstructed invariant mass peak position between data and Monte Carlo. Assuming perfect simulation of the material in front of ECAL and alignment of the detector, this ratio provides the correction to be applied to the ECAL scale. More conservatively, we consider this measurement as an indication of the current ECAL scale accuracy.

Systematic uncertainties have been investigated separately for EB and EE. The error sources considered for the mass shift are the systematics due to the pseudorapidity cut (0.5% in EB and 1.3% in EE), due to the transverse energy cut variation (0.6% in EB and 1.7% in EE) and due to the energy corrections (0.4% in EB and 0.5% in EE).

The EB scale is found to agree with the simulation at a level of 1%, with the measured mass shift (data vs simulation) of $-$ 0.7% $\pm$ 0.02% (stat.) $\pm$ 0.9% (syst.). In the EE, the scale is

Journal of Physics: Conference Series **293** (2011) 012045 doi:10.1088/1742-6596/293/1/012045

found to agree with the simulation to about 3%. The average mass shift is measured to be + 2.5% ± 0.2% (stat.) ± 2.2% (syst.).

5. Combination of the π^0 and ϕ-symmetry inter-calibration results

The inter-calibrations with the π^0 and ϕ-symmetry methods are performed independently and a comparison between their results provides a useful cross-check. We combine the crystal-by-crystal calibration constants obtained with the two methods. As shown in Figure 6, for the central barrel region (|crystal η index| $\leq$ 45) the combined inter-calibration precision is found to be 1.15% which is in good agreement with the expectation of 1.1% coming from the estimated precisions of the two methods (see Figures 3 and 5).

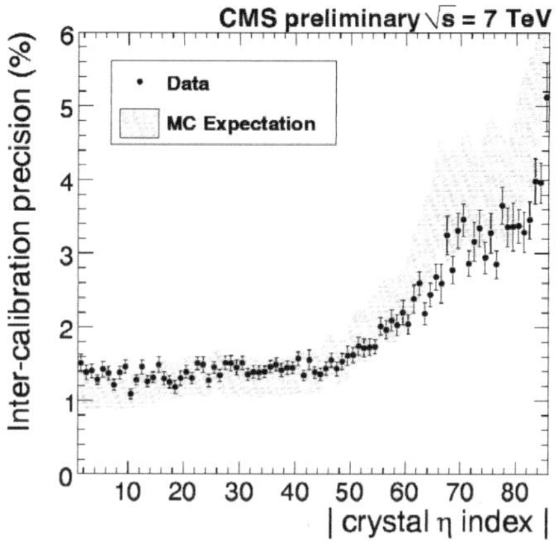

Figure 5. Inter-calibration precision as a function of η-index. The expected precision estimated from simulation studies is also shown.

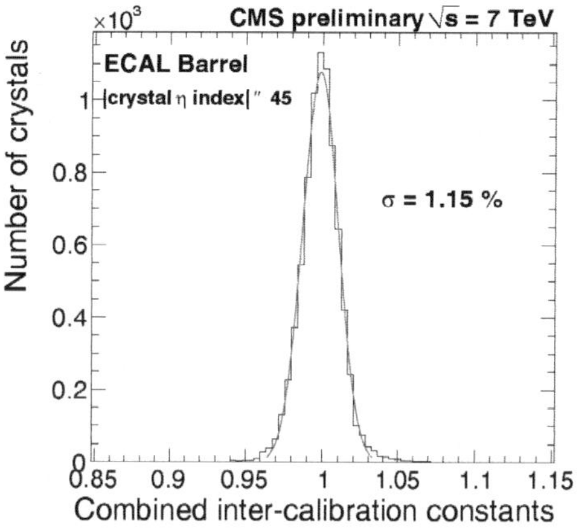

Figure 6. Distribution of the combined inter-calibration constants in the central region |crystal η index| $\leq$ 45.

6. Conclusions

Using the first $\sim$ 100 nb^{-1} collected with the CMS detector in 2010 at a center of mass energy of 7 TeV, studies of the CMS ECAL calibration methods and procedures are carried out. The achieved channel-to-channel calibration precision in the central barrel with the sample size currently available is 1.15%, in good agreement with the expectation from Monte Carlo studies.

The global energy response scale of the ECAL is also studied, and found to agree with the expectation to within about 1% in the barrel and 3% in the endcaps. In addition, the performance online calibration streams is found to be consistent with the simulation. The calibration of the CMS ECAL will continue with the upcoming LHC data, with collection of large samples of neutral pion decays as well as W and Z bosons decaying into energetic and isolated electrons.

References

[1] CMS Collaboration, "Performance and operation of the CMS electromagnetic calorimeter", JINST **5** (2010) T03010.
[2] CMS Collaboration, "Performance and operation of the CMS electromagnetic calorimeter", CMS Note **2010/012**.
[3] CMS Collaboration, "The CMS experiment at the CERN LHC", JINST **0803** (2008) S08004.
[4] D. Futyan and C. Seez, "Intercalibration of ECAL crystals in ϕ Using Symmetry of Energy Deposition", CMS Note **2002/031**.

[5] V. Litvin, "Inter-calibration of the CMS barrel electromagnetic calorimeter using $\pi^0 \rightarrow \gamma\gamma$ decays", CMS Conference Report **2007/069**.
[6] L. Agostino *et al.*, "Inter-calibration of the CMS electromagnetic calorimeter with isolated electrons", J. Phys. **G33** (2007) N67-N84.
[7] P. Meridiani and R. Paramatti, "Use of $Z \rightarrow e^+e^-$ events for ECAL calibration", CMS Note **2006/039**.
[8] M. Anfreville *et al.*, "Laser monitoring system for the CMS lead tungstate crystal calorimeter", CMS Note **2007/028**.
[9] P. Adzic *et al.*, "Intercalibration of the barrel electromagnetic calorimeter of the CMS experiment at start-up", JINST **3** (2008) P10007.

Journal of Physics: Conference Series **293** (2011) 012046 doi:10.1088/1742-6596/293/1/012046

Relative pre-calibration of the ALICE electromagnetic calorimeter EMCAL

J Faivre for the ALICE collaboration

Laboratoire de Physique Subatomique et de Cosmologie (LPSC), Université Joseph Fourier, CNRS-IN2P3, Institut Polytechnique de Grenoble, Grenoble, France

E-mail: julien.faivre@lpsc.in2p3.fr

Abstract. Several experiments showed that calorimetry is an essential tool for the Quark-Gluon Plasma study. ALICE, on the LHC collider, therefore integrated two electromagnetic calorimeters in its design. We will explain how the relative pre-calibration of one of them, the EMCAL, is performed thanks to cosmic muons, through an iterative procedure which consists in varying the high-voltage associated to each channel to ajust its gain. After 3 iterations of high-voltage tuning, of which each requires a day of data-taking, the dispersion (RMS over mean) of the channel amplitude of the response to a muon at minimum ionisation is found to be less than 2 %. We also show that more iterations allow to lower this value down to 1 %.

The results obtained at the RHIC ultra-relativistic heavy-ion collider in the domain of the Quark-Gluon Plasma study showed the importance of jets, photons and heavy flavor (charm and beauty) particles to probe the partonic medium created in such collisions. ALICE is the experiment dedicated to the study of the Quark-Gluon Plasma at the CERN LHC collider. Since jets, photons and heavy flavor particles are very well studied using an electromagnetic calorimeter in conjunction with a TPC, the ALICE experiment integrated two electromagnetic calorimeters in its design : PHOS and EMCAL. After an introduction to the physics motivations and a brief presentation of the EMCAL, the method and results of the relative pre-calibration of the channels before insertion of the detector in ALICE will be detailed.

1. Physics motivations

The strong interaction, one of the four known fundamental forces, can be studied by measuring the properties of the analog of the electromagnetic plasma for the strong force. In the past years, much progress has been done both on the experimental and theoretical sides. For example, new experimental results have constantly been validating or constraining calculations carried out with the theory of the strong interaction, Quantum Chromodynamics (QCD). The particles which are sensitive to the strong interaction, quarks and gluons (partons), carry a strong force charge, or color charge. As for the electromagnetic interaction, increasing the color charge density screens the strong force potential down to a value for which color deconfinement is reached. Such a medium is called a Quark-Gluon Plasma (QGP).

To achieve this large color charge density, ultrarelativistic heavy ion collisions are done. The studies ultimately performed with the RHIC collider have shown that a QGP is undoubtedly produced and has interesting properties [1]. The next large experiment dedicated to the QGP study is ALICE, at the CERN LHC collider. p-p, p-Pb and Pb-Pb collisions will be looked at, and their unprecedented energy will allow for the first time for a quantitative study of the QGP.

Many probes carry information about the QGP phase of the collision, but results obtained at the RHIC collider have pointed to the importance of calorimetry. Photons (unaffected by the strong interaction), jets and azimuthal correlations between jets and hadrons or photons, and heavy flavor decaying leptonically, will allow to probe the temperature, density, equation of state of the medium, as well as to test various QCD predictions. They will furthermore provide data to constrain the nucleon and nuclei parton content, and can also be triggered for selecting particularly interesting collisions. For these reasons, two of the ALICE subdetectors are electromagnetic calorimeters. The expected ALICE Physics performance can be found in [2; 3].

2. The ALICE electromagnetic calorimeter EMCAL

The EMCAL electromagnetic calorimeter is the ALICE calorimeter which has the largest acceptance ; it spans 1.4 units of pseudo-rapidity η and 100 degrees in azimuth φ. It divides into 10 supermodules and 11 520 towers of size $\Delta\varphi \times \Delta\eta = 0.0143 \times 0.0143$. Each supermodule in turn divides into 24 stripmodules along the beam direction, of 2×24 towers each. For the towers to be close to a projective geometry, a given stripmodule is tilted by $1.5\,^\circ$ with respect to the neighboring stripmodule in the direction of the experiment center (*cf.* figure 1).

The chosen calorimeter technology is Shashlik, for which the towers consist of a 20.1 radiation lengths lead-scintillator sandwich crossed by wavelength shifting fibers, which guide the emitted photons towards an Avalanche PhotoDiode (APD). The APD are operated at room temperature, with electron multiplication factors around 30. Beam tests performed with an 8×8 tower prototype with electron beams from 0.5 to 100 GeV allowed to measure an EMCAL energy resolution of $\frac{\sigma_E}{E}(\%) = 1.7 + \frac{11.1}{\sqrt{E\,(GeV)}} + \frac{5.1}{E\,(GeV)}$ [4]. More details on the detector design and properties can be found in [5; 6].

Because various towers don't have strictly identical properties (scintillator or lead thickness, light attenuation, etc...), a given energy deposit does not lead to an identical response from two different towers. A relative calibration must therefore be done, and consists in changing, tower per tower, the APD gains in a way that they compensate for the tower-to-tower variations of the response, and eventually achieve a uniformity in the response of the supermodule.

A pre-calibration with cosmic muons, before insertion of the supermodules in the ALICE experiment, was expected to provide a dispersion (RMS divided by the mean) of the average amplitude of the towers'response better than 10 %. The next section will present the procedure and results of this calibration. Since the postponement of the LHC startup allowed to have 40 % of the EMCAL installed for the first Physics runs, i.e. 4 out of the 10 supermodules, this will soon be improved by an on-site relative and absolute calibration thanks to e.g. π^0 decays to photons (which can be associated with photon electroconversion), minimum ionizing particles coming from the proton-proton collisions, or electrons identified in the tracking detectors.

3. Relative pre-calibration

3.1. Relative pre-calibration procedure

Cosmic muons were chosen as a permanent and free source of minimum ionizing particles (MIP). Simulations of their energy distribution have shown that roughly two thirds of them cross the whole calorimeter thickness and behave as MIP, with a mean deposited energy in the active part (scintillator) of about 28 MeV, which corresponds to a $\simeq 300$ MeV incoming electron.

Since the vast majority of the APD have undergone test bench measurements [7], the value of their gain G as a function of the high voltage V applied to the APD is known, and the relative calibration is therefore performed through individual APD high-voltage tuning. The parametrization used is $G = A_i + B_i e^{k_i V}$, where A_i, B_i and k_i are the parameters of a fit performed over the data taken with APD i on the test bench. For the few APD of which the gain dependence as a function of the high voltage is not known, average parameters $A_i = \alpha(V_i^{30})$, $B_i = \beta(V_i^{30})$, $k_i = \kappa(V_i^{30})$ are calculated, where V^{30} stands for the voltage defined as $G(V^{30}) = A + Be^{kV^{30}} = 30A$, and α, β, κ are fits over a significant part of the APD database. These average parameters showed to be good enough to tune the APD gains when no measurement of these parameters was available.

The relative calibration procedure therefore consists in calculating a new high voltage for each APD by measuring the average amplitude from cosmic muons data taking and choosing a reference value to be achieved. The procedure is repeated, and results in a lowering of the towers'response dispersion after each iteration.

3.2. Experimental bench

The measurements were carried out on an experimental bench schematized on figure 1. A supermodule is calibrated by thirds, of 8 stripmodules each. The experimental bench consists in 16 scintillator paddles placed by pairs above and under the stripmodules. The size, position and orientation of the paddles are such that a particle which crosses both paddles of a pair necessarily crosses the associated stripmodule, and only that one.

On both ends of a paddle, the photons created in the scintillating medium are guided to a photomultiplier tube. When all four photomultipliers of a pair of top and bottom scintillator paddles see a signal, a trigger is issued and the whole 1/3 supermodule data is read. The paddle pair which triggered is identified, and the times at which the signal has been seen by each of the four photomultipliers are also registered.

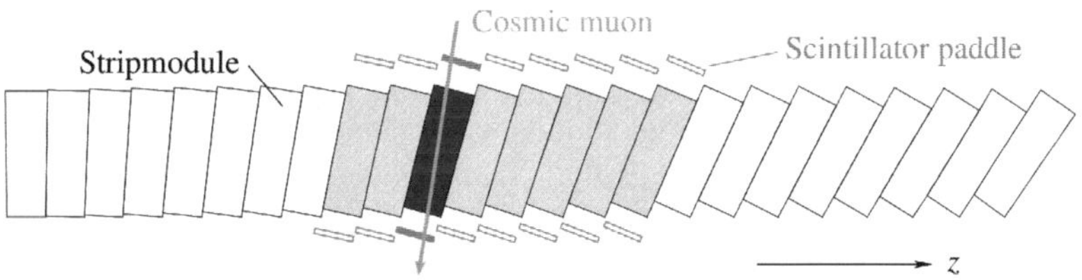

Figure 1. Schematic view of the experimental bench along the beam direction z : sketched are the stripmodules, a third of which being placed between scintillator paddles. A pair of paddles is fired by a cosmic muon.

3.3. Cosmics signal selection

Because of the simple configuration of the scintillator paddles, the cosmic muons which issue a trigger don't necessarily cross a single tower. The distribution of the length of the muons' paths through a tower is therefore broad, and that of the energy deposit is consequently wide as well. Since a reasonable precision on the average MIP signal amplitude requires a narrower deposited energy distribution, the muons which cross more than one tower must be discarded. This is achieved by the means of two cuts described below. Figure 2 shows the distribution of the signal amplitudes for a given tower without cuts and with all cuts applied.

The isolation cut consists in discarding a cosmic muon when a tower neighboring the tower of interest also has some signal. Because of the noise, the isolation cut level has been set to the minimal value of 3 ADC, which corresponds to a 5 - 6 MeV energy deposit.

The second cut uses the time information of the signal arrival to the photomultipliers. For a given paddle, the time difference between the "left" and "right" arrival times gives the position along the paddle of the muon trajectory with an accuracy of 2 to 3 centimeters. Figure 3 shows the distribution, for 12 out of 24 towers, of this time difference for the muons which gave a signal in a selected tower. Figure 4 shows the linear behavior of the means of these distributions as a function of the tower number, which is bijectively related to the distance along the strip.

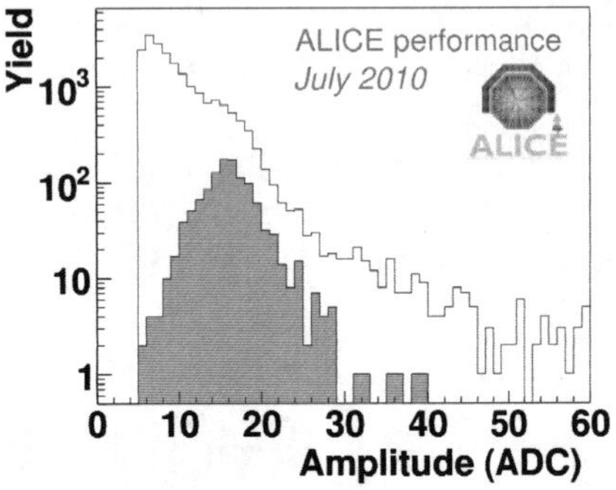

Figure 2. Distribution of the signal amplitude for the collected muons without cuts (line), and with all selection cuts applied (filled histogram).

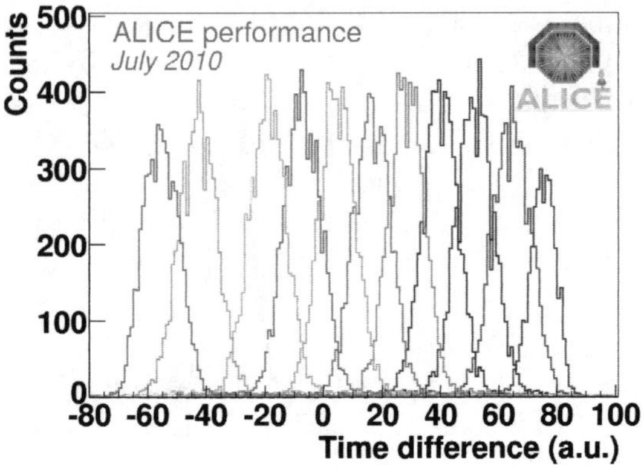

Figure 3. Time difference for muons which gave a signal in a given tower, every two tower.

Studies showed that both cuts applied together strongly reduce the contribution of oblique muons. The shift of the average MIP signal amplitude to lower values induced by a loosening of the isolation cut from 3 to 5 ADC is negligible. This shift is of a fraction of ADC when the time difference cut is not applied.

3.4. Tower MIP *signal amplitude*

The distribution of the signal amplitude for the collected muons after the two selection cuts is fitted with a gaussian shape for every tower i, with a resulting χ^2/ndf between 0.5 and 3.0. The mean μ_i of the gaussian fit is then taken as the average MIP signal amplitude for the considered tower. The desired average signal amplitude chosen as a target for the convergence process is around 16 ADC (the chosen value depended on the test hall temperature).

The RMS σ_i of the gaussian fit was such that the relative width σ_i/μ_i approximatively ranged from 14 to 17 % for the 1/3 supermodule looked at in this study. We estimated the magnitude of various contributions :

- A simulation showed that the fluctuations in the energy deposited in the scintillators range from 12 to 16 % ;
- The calorimeter intrinsic resolution due to the statistics of photoelectrons is about 4 % ;
- The digitization leads to a 1.8 % uncertainty ;
- The smearing due to temperature fluctuations during data taking for the results presented here was estimated to be between 1.5 and 3 %. This result includes the APD-to-APD variation of the gain dependence on the temperature ($\frac{1}{G}\frac{dG}{dT} = -1.5$ to $-4\ \%/K$), and the uncertainty on the measurement of the temperature variations ($\simeq 0.2\,^\circ$).
- Other contributions like the electronic noise, the signal fit or the residual oblique muons still have to be estimated, but only the electronic noise is expected to be somewhat significant.

The present status of the study does not allow to draw strict conclusions, but the data shows that the deposited energy fluctuations are the main contribution to the variance of the signal amplitude distributions, and does not leave much space to significant unknown contributions.

The statistical uncertainty on the measured average MIP signal amplitude for each tower is then calculated as $\sigma_{\mu,i} = \sigma_i/\sqrt{N_i}$, with N_i the number of signal entries for tower i. 24 hours of data taking showed to be sufficient to reach a statistical relative uncertainty on the average MIP signal amplitude below the 1 % level for almost all the towers ; this sets the limit on the reachable level of relative calibration with cosmic muons over these timescales. Figure 5 shows the distribution of this relative uncertainty for a supermodule. The statistics has been accumulated for 1354 minutes for one third. The results for the two other thirds, with data taking times of 975 and 910 minutes, were scaled to 1354 minutes of data taking. The asymetric queue to larger relative uncertainties results from two contributions. The first one is shown as the grey histogram, and comes from the towers located on both ends of the stripmodules along the scintillator paddles'direction : as the paddles have the same length as the stripmodules, not all the cosmic muons which cross these towers will issue a trigger. The second contribution is due to the towers which belong to the most tilted stripmodules and originates from the cosine squared dependence of the cosmic muons rate on the azimutal angle.

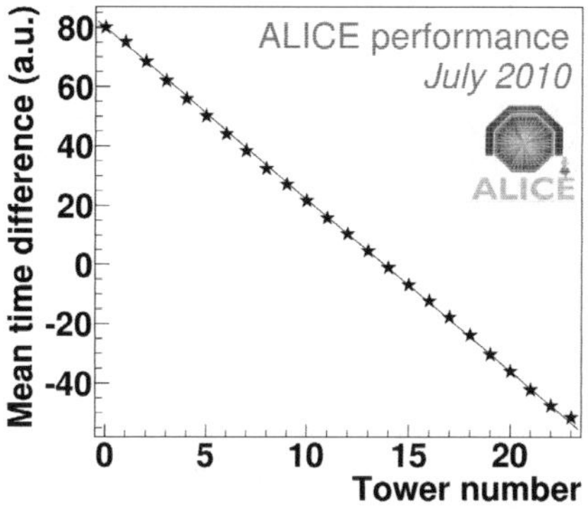

Figure 4. Mean time difference as a function of the tower number.

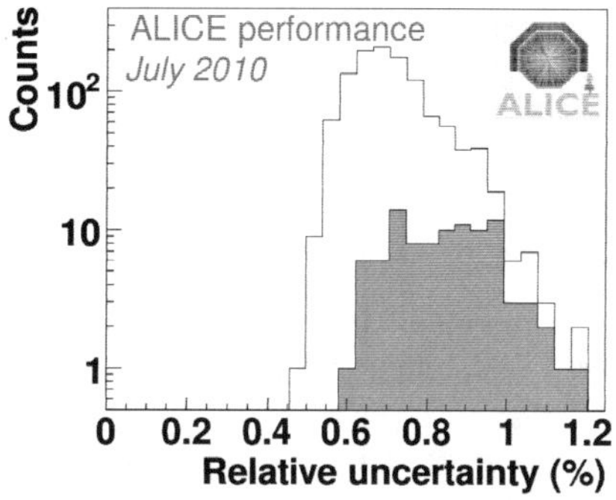

Figure 5. Typical distribution of the relative uncertainty on the average MIP signal amplitudes for a supermodule for a $\simeq$ 24-hour data taking period. The grey histogram shows the contribution of the towers located on both ends of the stripmodules only.

3.5. Relative calibration results

As explained earlier, the average MIP signal amplitude is adjusted by tuning the APD high voltages. As shown in figure 6, the 16×24 measured average MIP signal amplitudes of the 1/3 supermodule used for these tests progressively converge towards the desired value, with a steadily decreasing dispersion even up to the 5th iteration, down to the 1 % level. The data taking time was chosen to be sufficient to have statistical errors significantly lower than the dispersion value. We simultaneously controlled the RMS of the distribution of the APD high voltages difference between two consecutive iterations $\mathtt{HV}_n - \mathtt{HV}_{n-1}$: as expected, this RMS decreases, except for the 4th iteration because of a significant change of the temperature in the experimental hall. We stopped at the 5th iteration, as this RMS was getting close to the high voltage digitization value of 0.2 V/bit, which corresponds to about 0.5 % on the relative gain.

Figure 7 shows the initial and final average MIP signal amplitude distributions for the 1/3 supermodule on which these studies have been performed. Most third supermodules have been observed to behave similarly, with initial dispersions worse than 10 % and a value better than 2 % reached after 2 or 3 iterations, i.e. a number well below the requirements. The steadily dropping dispersion of the average MIP signal amplitudes up to the 5th iteration indicates that the relative calibration, usually provided with 3 iterations, is reliable.

4. Conclusions

As the various towers of a calorimeter don't have strictly identical geometrical, physical, chemical or electronical properties, they generate different signals as a response to the same deposited energy, unless a relative calibration is performed. The EMCAL, one of the two electromagnetic

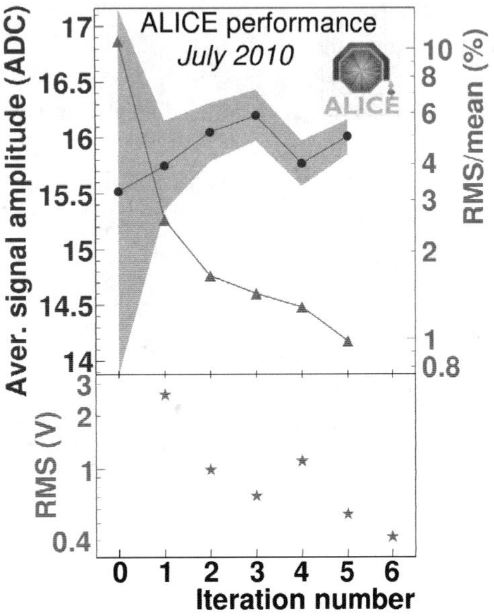

Figure 6. As a function of the iteration number : evolution of the mean (•) and dispersion (▲) of the distribution of the average MIP signal amplitudes (top), and of the RMS (★) of the distribution of the high voltages difference between two consecutive iterations (bottom).

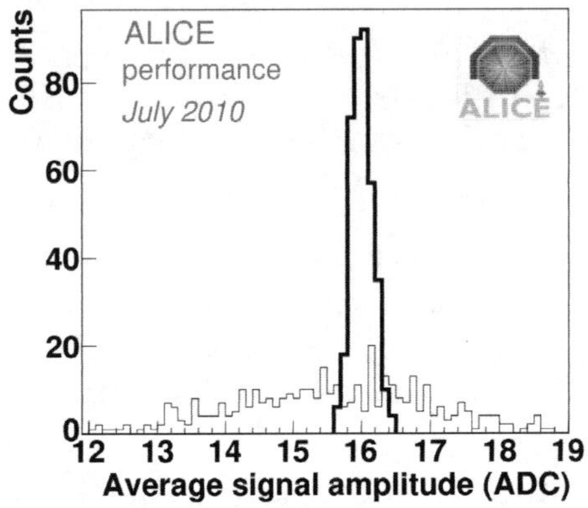

Figure 7. Initial (thin) and final (bold) average MIP signal amplitude distributions.

calorimeters of the ALICE experiment, undergoes a relative pre-calibration before insertion in the experiment, by measuring the response to minimum ionizing cosmic muons and adjusting the channel gains so that the calorimeter's response is uniform over all the channels.

While ongoing simulation studies aim at better quantifying the systematic uncertainties and the bias (towers on the edges, most tilted stripmodules), we already showed that three one-day periods of cosmic muons data taking were sufficient to perform the relative calibration of a 1/3 supermodule down to a dispersion of 2 %, well below the 10 % requirement. This pre-calibration will be used as a starting point for refined relative and absolute calibration thanks to e.g. particle decays into photons or electrons. The target relative calibration value is 1 % and will essentially be limited by the statistics collected.

These very procedure and experimental bench will also be used for the relative pre-calibration of the third ALICE electromagnetic calorimeter, DCAL, which has identical technology and properties to the EMCAL and a slightly smaller acceptance.

References

[1] Arsene I *et al.* (BRAHMS), Back B *et al.* (PHOBOS), Adams J *et al.* (STAR), Adcox K *et al.* (PHENIX) 2005 *Nucl. Phys.* A **757** 1–27, 28–101, 102–183, 184–283

[2] Alessandro G (Ed.) *et al.* (ALICE) 2006 *J. Phys.* G **32** 1295–2040

[3] Bellwied R *et al.* (ALICE EMCal) 2010 ALICE EMCal Physics Performance Report (*Preprint* arXiv:1008.0413)

[4] Allen J *et al.* (ALICE EMCal) 2010 *Nucl. Instrum. Meth.* A **615** 6–13

[5] Fantoni A These proceedings

[6] Cortese P *et al.* (ALICE) 2008 ALICE Electromagnetic Calorimeter Technical Design Report CERN-LHCC-2008-014

[7] Badalà A *et al.* 2009 *Nucl. Instrum. Meth.* A **610** 200–203

Absolute energy calibration and the use of timing information of the BESIII EMC

Jianming Bian
(For the BESIII Collaboration)
Institute of High Energy Physics, Chinese Academy of Sciences, Beijing 100049, China
E-mail: bianjm@ihep.ac.cn

Abstract. We introduce the absolute energy calibration of the electromagnetic calorimeter at BESIII. The precision after calibration is about 0.5% and the resolution difference between data and the simulation is about 5%. We also present a method to veto beam-gas background using the timing information of the EMC, which is crucial for physics analyses involving low energy photons at BESIII.

1. Introduction

Beijing Electron Positron Collider II and Beijing Spectrometer III (BEPCII/BESIII)[1] is a major upgrade of BEPC/BESII[2]. BESIII is designed to study τ-charm physics[3]. The BESIII detector is equipped with a CsI(Tl) crystal electro-magnetic calorimeter (EMC) which provides much improved energy measurement. The design energy resolution of BESIII EMC is 2.5% at 1GeV, which is about 10 times better than BESII. In this talk, we present an introduction to the absolute energy calibration, as well as the performance after calibration. As BESIII works in a relatively low energy region with high luminosity, the beam-gas background is a major consideration for physics involving low-energy photons, such as radiative charmonium transitions and reconstruction of low momentum π^0s. We also present a method of using timing information of EMC to reject beam-gas background at BESIII.

2. BESIII electro-magnetic calorimeter (EMC)

The electromagnetic calorimeter is designed to precisely measure the energies of photons above 20 MeV and to provide trigger signals. It has a good e/π discrimination capability for momenta higher than 200 MeV/c. The structure of EMC is described in Ref [1], it consists of 6,240 large CsI(Tl) crystals located outside of the TOF counters and inside the coil of the solenoid. Its performance is greatly enhanced compared to the sampling EM calorimeter of BES/BESII which was made from streamer tubes sandwiched in between lead plates. The inner radius of the BESIII EMC is 94 cm. The length of the crystals is 28 cm or 15 radiation lengths (X_0) with a front face of 5.2 cm $\times$ 5.2 cm. The depth and segmentation of the EMC are chosen to optimize the detector performance and to control the cost of the detector. The total weight of the crystals is approximately 24 tons. The design energy resolution of EM showers is $\sigma = 2.5\%$ and the position resolution is $\sigma = 0.6$ cm E (E in GeV) at 1 GeV. The angular coverage of the barrel EMC is $144.7^o < \theta < 33.5^o$ ($|cos(\theta)| < 0.83$). The two end caps cover $21.3^o < \theta < 34.5^o$ and $145.4^o < \theta < 158.7^o$ ($0.85 < |cos(\theta)| < 0.95$). Performance checks using radiative Bhabha and Bhabha events indicate the precision of the energy measurement is better than 0.5% and the relative resolution reaches the design value of 2.5%.

3. Absolute energy calibration

The off-line calibration at BESIII goes through two steps. The first one is the calibration with Bhabha events which determines the gain factor for each crystal that converts ADC counts to deposited energy. However, due to the material in front of the EMC and the transverse and longitudinal shower energy leakage, the reconstructed energy during this step is usually not equal to the incident energy. Therefore, the second step, absolute energy calibration, is needed to determine the absolute energy scale which corrects deposited shower energies (E_{shower}) to incident energies(E_{true}). The main idea of the absolute energy calibration is to use photons of known energy to calculate the correction factor.

3.1. MC calibration

To calibrate the reconstructed shower energy in the MC simulation, single photon samples with known energies and directions were generated. These samples are subjected to the reconstruction process. Then the ratios of the peak value of the reconstructed energy to the value of the true energy for each energy-position (polar angle) combination are extracted as the correction factors in MC and stored in a 2-D table. When applying the correction, factors are obtained from the 2-D table via interpolation. The MC correction can improve the precision of the energy reconstruction in MC to better than 0.1%. The geometry-depended energy leakage can also be corrected in this process and the remaining discrepancy as a function of energy will be further corrected using a data-driven method.

3.2. DATA calibration

$\pi^0 \to \gamma\gamma$ and $e^+e^- \to \gamma\gamma$ events are used to perform the absolute energy calibration in the data. Because of the energy leakage out of crystals and interaction in the material in front of the EMC, the line shape of photon energy deposits does not follow a Gaussian distribution. We parameterize the asymmetric distribution of photon energies by a Novosibirsk function:

$$f_{Nov}(m) = A\exp\left(\frac{-\ln^2(1+t\frac{\sinh(t\sqrt{\ln 4})}{t\sqrt{\ln 4}}\frac{m-m_0}{\sigma})}{2t^2} - \frac{t^2}{2}\right), \tag{1}$$

where A is the normalization constant, $\sigma(> 0)$ is the resolution, m_0 is the mean value and $t(< 0)$ parameterizes the tail. This function can be seen as a Gaussian distribution with an asymmetric tail.

In the $\pi^0 \to \gamma\gamma$ process, the invariant mass of the π^0 can be calculated as follows:

$$m_{\gamma\gamma}^{exp} = \sqrt{2E_{low}E_{high}(1-\cos\theta_{\gamma\gamma})}, \tag{2}$$

where E_{low} and E_{high} are the energies of the low and high momentum photon from the π^0 decay and $\theta_{\gamma\gamma}$ is the angle between them. If either photon from the π^0 decay loses a part of its energy, the reconstructed mass of the π^0 will be lower than its nominal value. The line shape of the invariant mass of the π^0 can also be parameterized by a Novosibirsk function.

The correction factor E_{shower}/E_{true} is determined by adjusting the invariant mass of photon pairs to the nominal π^0 mass, where E_{shower} is the deposited shower energy in a 5×5 crystal cluster and E_{true} is the true energy of the incident photon. To extract correction factors, the reconstructed photons are grouped according to their energies. Thus for each π^0 daughter photon pair(E_{low} and E_{high}), one photon has an energy in the i-th bin and the second one has an energy in the j-th bin. Then invariant masses of photon pairs in the data are obtained and fit for these energy bin combinations. Each spectra is fitted with a Novosibirsk function plus a 2nd order polynomial background term.

Due to the asymmetric shape of the deposited photon energies, the peak value of the photon pairs invariant mass distribution is not equal to the π^0 nominal mass ($0.135\text{GeV}/c^2$) even after shifting the photon energy peak to its true value[4][5]. For this reason, the expected π^0 mass for each energy bin combination is not simply equal to the nominal mass but needs to be obtained by MC simulation. The needed shift C_{ij} is determined by:

$$C_{ij} = \ln m_{\gamma\gamma}^{data} - \ln m_{\gamma\gamma}^{exp}, \tag{3}$$

where $m_{\gamma\gamma}^{exp}$ and $m_{\gamma\gamma}^{data}$ are π^0 masses for Monte Carlo simulation and data. The errors of $\ln m_{\gamma\gamma}^{exp}$ and $\ln m_{\gamma\gamma}^{data}$ are $\sigma_{ij}(\ln m_{\gamma\gamma}^{exp})$ and $\sigma_{ij}(\ln m_{\gamma\gamma}^{data})$. Then the error of C_{ij} can be evaluated:

$$\sigma_{ij}^2(C_{ij}) = \sigma_{ij}^2(\ln m_{\gamma\gamma}^{data}) + \sigma_{ij}^2(\ln m_{\gamma\gamma}^{exp}). \tag{4}$$

With about 10 M events in the 106 M ψ' data events collected by BESIII, we performed the absolute energy calibration. The correction factor at E=1.843GeV($\ln E/1GeV$=0.611) is obtained using $e^+e^- \to \gamma\gamma$ events and the others are obtained from π^0 calibration. A 3rd order polynomial is fit to the factors as the calibration function.

3.3. Performance

Radiative photon in $\psi(2S) \to \gamma\chi_{cJ}(1P)$ and radiative Bhabha events, as well as π^0s in the $\psi' \to \pi^0\pi^0 J/\psi$ process are used to check the performance of the absolute photon energy calibration. As shown in Fig. 1 and 2, the difference between the data and MC is found to be 0.5% in energy scale, 5% in resolution.

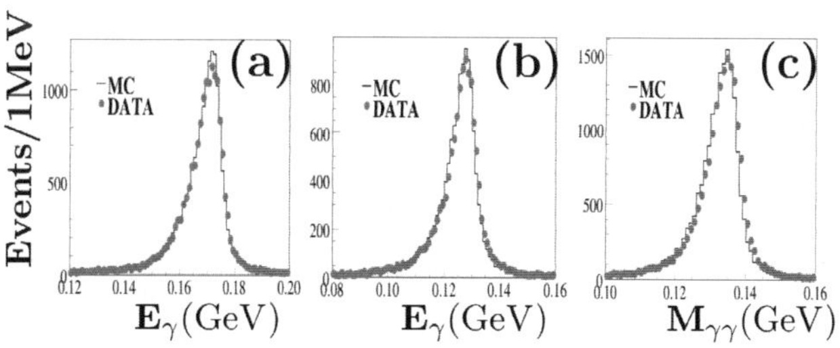

Figure 1. Comparisons between MC (lines) and data (dots): (a) energy distribution of the radiative photon in $\psi' \to \gamma\chi_{c1}, \chi_{c1} \to \gamma J/\psi, J/\psi \to l^+l^-$; (b) energy distribution of the radiative photon in $\psi' \to \gamma\chi_{c2}, \chi_{c2} \to \gamma J/\psi, J/\psi \to l^+l^-$; (c) invariant mass distribution of π^0 in $\psi' \to \pi^0\pi^0 J/\psi, J/\psi \to l^+l^-$.

4. Using of EMC timing information at BESIII

Physics analyses at BESIII which involve low energy photons such as $\psi' \to \pi^0 h_c$ ($p(\pi^0) =$ 84MeV), $\psi' \to \gamma\eta_c'$ ($E_\gamma = 47$MeV) require good photon identification in the low energy region. However, as the luminosity is much higher than at previous experiments, BESIII suffers from much higher beam-gas background. This is a substantial problem when the photon energy is less than 200 MeV. Because the time distribution of physics photons is Gaussian-like while beam-gas background photons are randomly distributed, a peaking time of the energy deposit in the EMC is reconstructed with a precision of 50 ns to discriminate them.

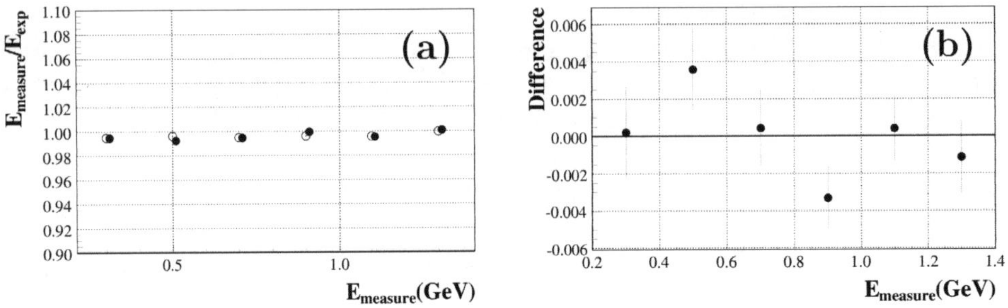

Figure 2. Peaks of the ratio of measured to expected energies ($E_{measure}/E_{exp}$) vs. the measured energy of the photon: (a)Comparison of the data(circle) and MC simulation(dot); (b) The difference between the data and MC simulation.

4.1. Timing information in a crystal

The BESIII-CsI(Tl) Calorimeter has 6240 detection units. Each unit consists of a CsI(Tl) crystal and two photodiodes. The main purpose of the read-out electronics of the detection unit is to measure the energy, timing information and to provide a fast energy trigger. Signals from the two photodiodes are amplified by two charge sensitive preamplifiers independently. These signals are summed and coupled to the main amplifier circuit. Then the combined signal is subject to three parallel 10 bit FADC channels with different full scale ranges corresponding to about 0−78, 78−625 and 625−2500 MeV. These FADCs, which are capable of operating at 20.8 MHz with 15 bit equivalent resolution, are realized by three 10 bit FADC channels with three different gains: $\times 0.5$, $\times 2$ and $\times 16$. The data from FADCs are sent to a digital pipeline and the peak finding circuit scans the peak of the signal waveform within 3 μs after the L1 trigger arrives. The step length of the scanning is 50 ns. The peak value is compared with the threshold stored in the threshold register, and signals smaller than the threshold are discarded. The amplitude of the peak is stored as the read-out energy information and the timing step number of the peak is stored as the time information. The precision of the time information is equal to the scanning step length, 50 ns. For the same trigger type (L1 time), a hit of a physics photon will be a bump in the 3-μs time scanning window while hits caused by beam-gas background are distributed randomly. The time resolution is determined by the fluctuation of the peaking time of crystals and uncertainty of the trigger time. Fig. 3 shows the crystal time information of physics photons with the same trigger time in the three energy (FADC) regions. The time resolutions are about 150 ns.

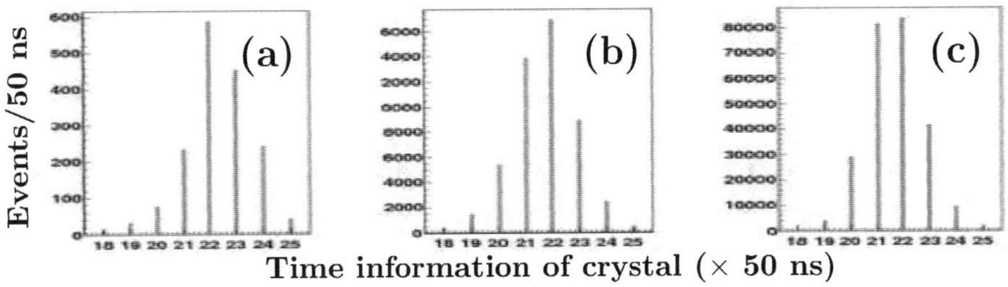

Figure 3. Time information of a crystal in different energy region in the data: (a) 0−78 MeV (b) 78−625 MeV (c) 625−2500 MeV.

4.2. Time information in a shower

The time information of the seed crystal in a shower will be reconstructed as the EMC time information of this shower. This raw EMC time is shifted by the trigger type and affected by the resolution of the trigger time. We use EMC time minus event start time in a event to cancel these effects. The subtracted EMC time is used to distinguish physic photon and beam-gas background. Using $\psi' \to \pi^0\pi^0 J/\psi, J/\psi \to e^+e^-, \mu^+\mu^-$ events, we obtain the the EMC shower time before subtract event start time, event start times for different triggers and the subtracted EMC time used for physics analysis, as shown in Fig.4. From the third column of Fig.4 we find that the subtracted shower time information has little effect arising from the trigger.

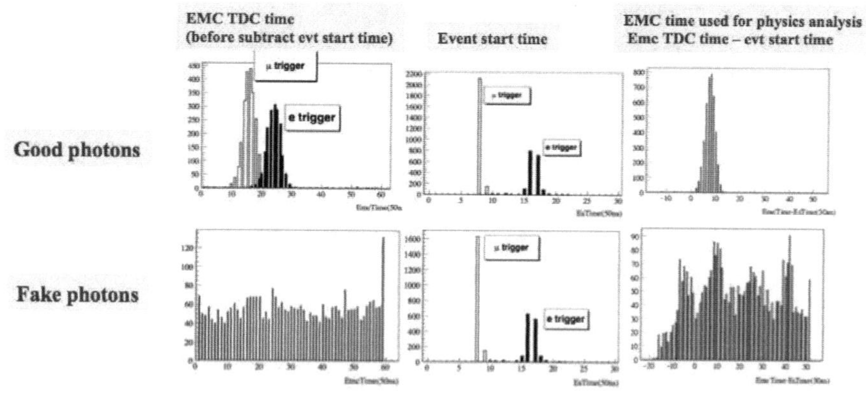

Figure 4. Time information of EMC shower.

4.3. Use of EMC timing information

The cut imposed on the shower time is set to 0−14 ($\times$ 50 ns) because this value is safe enough to retain signal and veto beam-gas background (veto ratio 70%), as shown in Fig.5. The mis-combination background of low-energy π^0 can also be significantly suppressed by this technique, as shown in Fig. 6. With the help of the time information, we have performed the inclusive analysis of $\psi' \to \pi^0 h_c$ [6] which includes a very low momentum π^0 ($p(\pi^0)$ 84MeV). Timing information from the EMC is also playing an important role in background rejection in analyses of exclusive decays of $\psi' \to \pi^0 h_c$ and $\psi' \to \gamma\eta_c(2S)$ (E_γ=47 MeV), which are ongoing at BESIII.

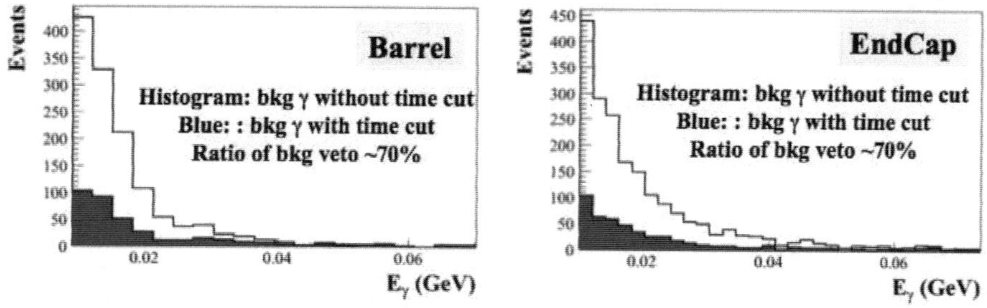

Figure 5. Performance of using time information to reject background for photon.

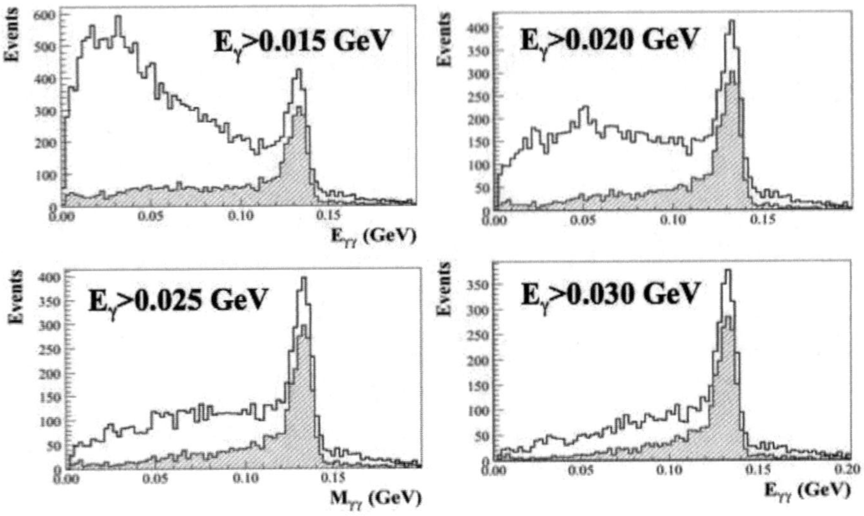

Figure 6. π^0 signals with different lower limits of photon energy before (lines) and after (shades) cuting time information.

5. Summary

In summary, the absolute energy calibration of the EMC works well at BESIII and the precision of the energy measurement reaches 0.5%. Energy scales and lineshapes of photons and π^0s have good DATA/MC agreements after calibration. Timing information is very useful as a beam-gas background veto at BESIII.

References

[1] M. Ablikim, *et al.* "Design and construction of the BESIII detector", arXiv:0911.4960, accepted by Nucl. Instrum. Meth. A

[2] J. Z. Bai *et al.* (BES Collaboration), Nucl. Instrum. Methods Phys. Res., Sect. A 458, 627 (2001); 344, 319 (1994).

[3] "Physics at BESIII", Edited by K. T. Chao and Y. F. Wang, Int. J. Mod. Phys. **A 24**, No.1(2009) supp.

[4] A. Bukin and H. Marsiske, BaBar Note 433(1996).

[5] S. Li & B. Heltsley, CBX 02-17.

[6] M. Ablikim *et al.* (BESIII Collaboration), Phys. Rev. Lett. **104**, 132002 (2010)

Journal of Physics: Conference Series **293** (2011) 012048 doi:10.1088/1742-6596/293/1/012048

Energy calibration of the EMC with Bhabha events at BESIII

ChunXiu Liu

Institute of High Energy Physics, 19B Yuquanlu Road, Shijingshan district, Beijing 100049

liucx@ihep.ac.cn

Abstract. In this report we present the calibration of energy in the electromagnetic calorimeter using Bhabha events collected with BESIII detector. Since Bhabha events have advantage of high statistics and known kinematics, absolute counter-by-counter calibrations have been performed by matrix conversion method. We have achieved the energy resolution of 2.3% at the beam energy 1.843GeV.

1. Introduction

The BESIII detector[1] is working at the interaction point of the BEPCII e^+e^- collider[2]. The Electro-Magnetic Calorimeter (EMC) plays an important role in the BESIII detector, whose primary function is to measure precisely energies and positions of electrons and photons. Detecting direct photons and photons from decays of π^0, η, ρ, etc. produced in ψ', ψ'', τ and D decays is essential for BESIII physics. In radiative decay processes, such as $J/\psi\to\gamma\pi\pi$, $J/\psi\to\gamma KK$, $J/\psi\to\gamma\eta\eta$, direct photons must be precisely measured and separated from photons from π^0, η decays. Also the EMC plays an important role in electron identification.

The desired photon energy detection range for BESIII is from ~20MeV to the full beam energy of 2.3GeV for studying the $e^+e^- \to \gamma\gamma$ process. Accurately measuring photon energies and their hit coordinates is crucial for reconstructing physics processes and for new discoveries. In order to get exact energy, two steps of calibration are required. First, the energy of each counter is calibrated and then the calibration of shower energy after clustering is performed. So several channels are used such as cosmic-ray, Bhabha, radiative Bhabha, gamma-gamma, π^0 and so on, are used. Each channel has different usage and purpose. For counter-by-counter calibration, Bhabha channels are very useful due to high statistics and known kinematics. The calibration of shower energy is considered in details in the report[3] at this conference.

In this report, we describe the crystal counter energy calibration with Bhabha events to achieve good performance to match physics requirements.

2. CsI(Tl) crystal calorimeter

The BESIII EMC is composed of one barrel and two end cap sections. The entire calorimeter have 6240 CsI(*Tl*) crystals. The length of the crystals is 28 cm or 15.1 X_0. All crystals are arranged as 56 rings (44 rings in the barrel and 2×6 rings in the end caps). Each crystal covers an angle of about 3° in both polar and azimuth directions.

The 5280 barrel crystals are divided into 44 rings. A barrel ring has 120 crystals. All crystals point to the interaction region with a small tilt of 1.5° in the ϕ directions and 1.5° to 3° in the θ directions (i.e., ±5 cm from IP in the beam direction), in order to avoid photons from the interaction point escaping through cracks between crystals.

The each end cap has 6 rings. The number of crystals in 6 rings is 96, 96, 80, 80, 64, 64. Each of the 6 rings are split into two half circles and can be opened for accessing the drift chamber. The end caps with inner radius of 50cm are placed at $z = \pm 138$ cm from the collision point.

In order to minimize the dead areas and gaps between the crystals, the crystals are suspended from the support girder. In Ref.1, the details about the calorimeter construction are described.

3. Energy calibration

From the digitized signal of i-th counter (ADC_i), the energy deposit (E_i) is obtained by the following formula:

$$E_i = \frac{ADC_i - PED_i}{e_i} \times c_i, \tag{1}$$

where ADC_i is the ADC value, PED_i is the pedestal value, e_i is the gain of the electronics chain and c_i is the energy conversion constant. The values of ADC_i and PED_i are experimentally measured. The pedestal and electronics gain constants are obtained from online. The digitized signal read out from the offline raw data is subtracted pedestal and calibrated by electronics, i.e. $(ADC_i - PED_i)/e_i$. The energy conversion constant c_i that relates the incident photon energy to the input charge of the preamplifier is the most difficult calibration constant to obtain.

Before installation into the container, every CsI(*Tl*) crystal was tested in the lab. The scintillation light obtained with ^{137}Cs source read out by a PMT was measured. Cosmic ray tests in which the crystals were read out by photodiodes and preamplifiers were performed. The energy conversion constant database established by these tests is used in the BESIII EMC crystal calibrations as initial values. These constants, however, can only be used to determine the relative performance of the crystals. So the initial c_i was obtained using the cosmic rays data without magnetic field which are taken during the BESIII detector joint debugging. It ensured the quick start of the calorimeter with good performance at the initial stage.

But the light output can be changed in each counter under the container due to several reasons: the pressure of structure, the geometry of calorimeter, radiation damage and so on. To achieve more accurate c_i, counter-by-counter calibration is executed using Bhabha events.

3.1. Cosmic ray calibration

During the BESIII detector joint debugging, the cosmic ray data without magnetic field are taken. We select the events with the back-to-back condition and only two crystals with the energy larger than 10MeV. The cosmic ray data is mainly μ lepton in BESIII detector, and its momentum is basically lager than 1GeV, so we may suppose that the deposited energy is proportional to the depth through CsI crystal. We check the ADC (MeV) distribution for each crystal. First each crystal was calibrated by the relative light yield, the calibration constant is $C_i^{LY} = LY_{average} / LY_i$, where $LY_{average}$ is the average of the relative light yield, LY_i is the relative light yield of the i-th crystal. Then the comparison with the MC was made, the Monte Carlo constant is $C_i^{MC} = E_i^{average}(MC) / E_i^{average}(data)$. Finally the cosmic ray calibration constant for each crystal is $C_i^{\cos mic} = C_i^{LY} \times C_i^{MC}$.

Fig.1 shows the performance of Bhabha events at a center-of-mass energy of 3.686GeV in Oct 2008, i.e. in the early stage of BESIII running. The triangle dots are the results after the cosmic calibration. The square dots are the results of the first Bhabha calibration. We can see that the cosmic calibration is very rough, but as the input of Bhabha calibration it ensured the initial digi-calibration successfully accomplished.

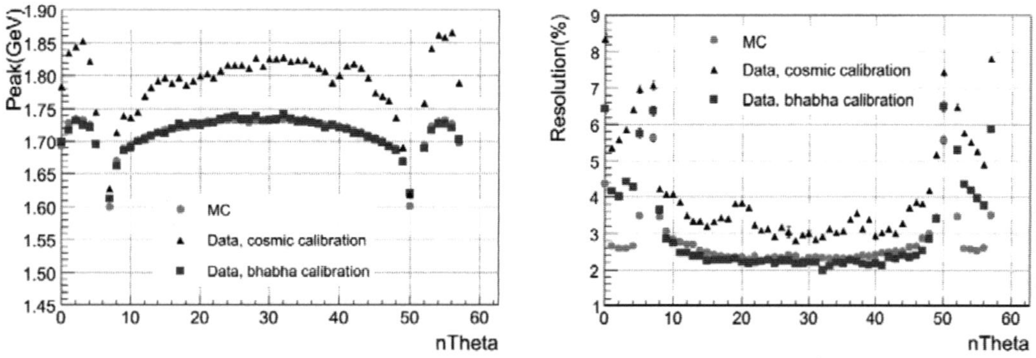

(a) (b)

Fig.1 (a) and (b) are the energy peak and resolution versus ThetaID of crystal counter, respectively.

3.2. Bhabha Calibration

3.2.1. Bhabha Events

We use the online Bhabha events or offline PreSelected Bhabha events. The criteria conditions of Bhabha sample are in the following:

$$1GeV < E_{sh1} < 4GeV\,,\ 0.4GeV < E_{sh2} < 4GeV\,,$$

$$|\theta_1 - \theta_2| < 3^\circ\ ,\ -25^\circ < \phi_1 - \phi_2 < -4^\circ \,\|\, 2^\circ < \phi_1 - \phi_2 < 20^\circ,\ \text{Mdc\_hit>20}.$$

Event selection is very important for Bhabha calibration. Because the light yield and uniformity change with time through the radiation damage, and the status of every crystal is different, we select the shower with the energy in the range of energy Peak(ixtal) $\pm$ 1σ(Iθ) for every crystal.

3.2.2. Method

We wish to find the parameters (g_i) that minimize the discrepancy between the expected shower energy and the measured one. The χ^2 is given by

$$\chi^2 = \sum_{k=1}^{N} \frac{E_{\exp}^k - \sum_i^{5\times5} g_i E_i}{\sigma(\theta,\phi)}\ , \tag{2}$$

$$E_{\exp}^k = E_e(\theta,\phi) f(E_e,\theta,\phi)\ . \tag{3}$$

Where $E_{\exp}^k$ is the expected shower energy, E_i is the measured energy in i-th counter and is summed to shower range up to 5×5 around maximum energy counter and g_i is weight value for each counter, i.e., Bhabha calibration constant which represent the energy response in i-th counter. The expected shower energy $E_{\exp}^k$ is written by Eq.(3). $E_e(\theta,\phi)$ is the $e^\pm$ kinematics energy determined from incident direction but is not real deposited energy due to passive material and energy leakage. So those effects are supplemented by the function $f(E_e,\theta,\phi)$ which is the energy leakage from MC.

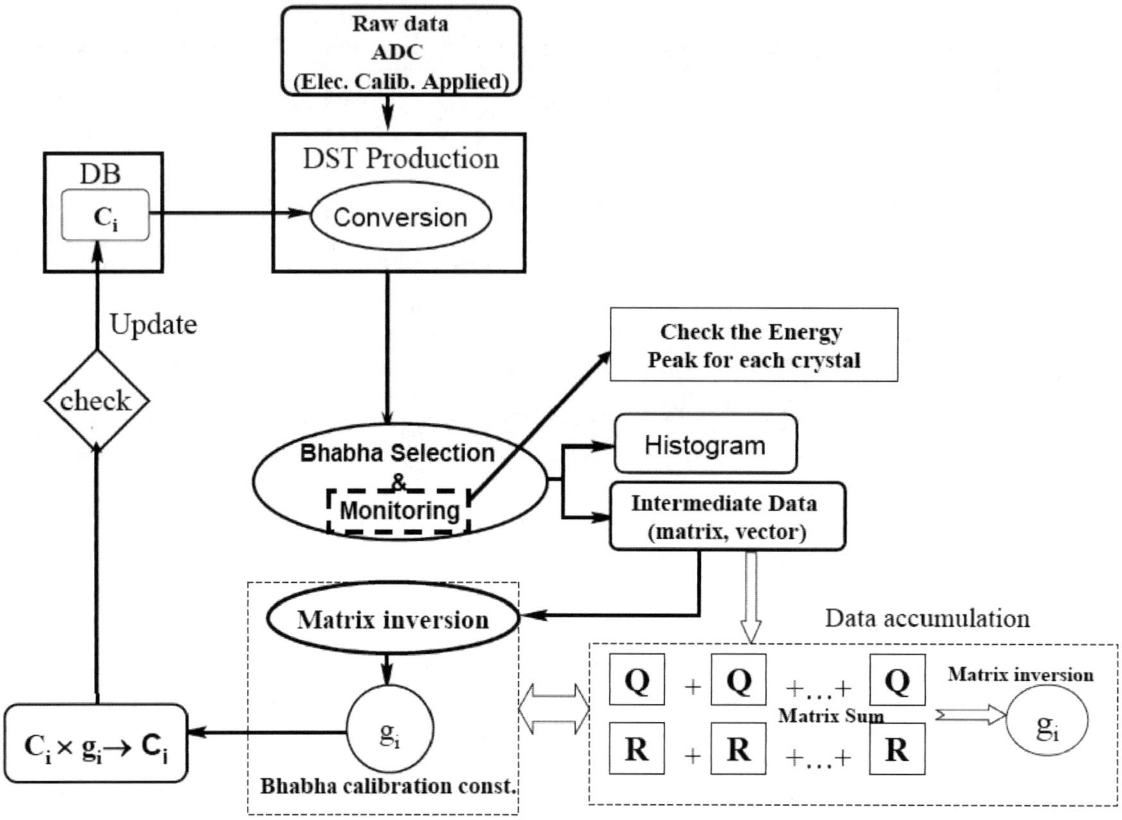

Fig.2 Bhabha calibration procedure

By minimizing the χ^2 defined in Eq.(2), matrix equation is extracted.

$$\sum_j g_j Q_{ij} = R_i \quad , \tag{4}$$

$$Q_{ij} = \sum_{k=1}^{N} \frac{E_{ik} E_{jk}}{\sigma_k^2} \quad , \tag{5}$$

$$R_i = \sum_{k=1}^{N} \frac{E_{ik} E_{\exp}^k}{\sigma_k^2} \quad . \tag{6}$$

Where Q is matrix with order 6240. Matrix Q is sparse because Bhabha event hit specific region in which counters have correction. All g_i are decided simultaneously by inverting matrix equation. Matrix conversion is main part in Bhabha calibration and the difficulty result from the inverse of matrix Q. Sparse matrix package (SLAP) [4] is used to invert matrix equation. Bhabha calibration procedure is shown in Fig.2.

If there is no difference between counter in energy response, the value of all g_i is unity. Bhabha calibration constant g_i distribution is shown in fig.3. Finally original energy conversion constant (c_i) is corrected by Bhabha calibration constant (g_i) i.e., $c_i \times g_i \rightarrow c_i$.

After Bhabha calibration we check the calibration constants via checking the performances of E5x5cms energy peak and resolution versus theta-index, E5x5cms energy peak and resolution versus phi-index in barrel, and E5x5cms energy versus ixtal profile and so on. Those checks are shown in Fig.4.

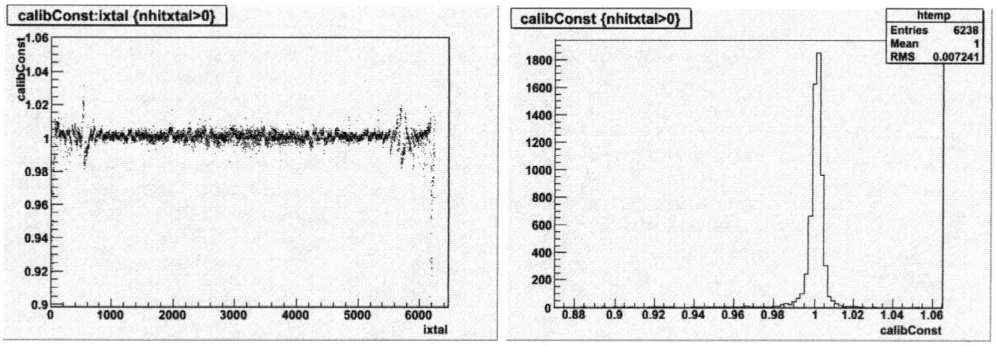

(a) (b)

Fig.3 Bhabha calibration constant in second calibration version.

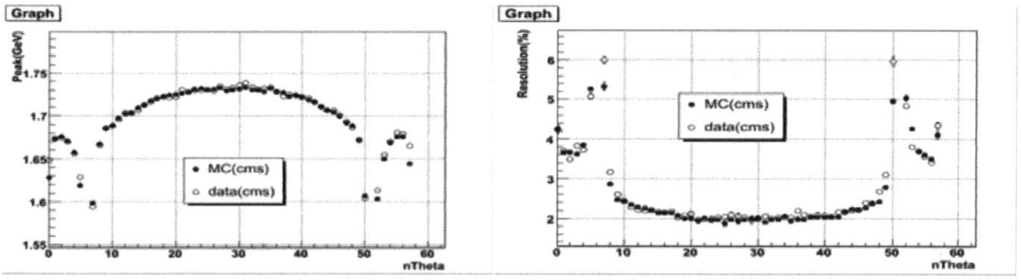

(a) (b)

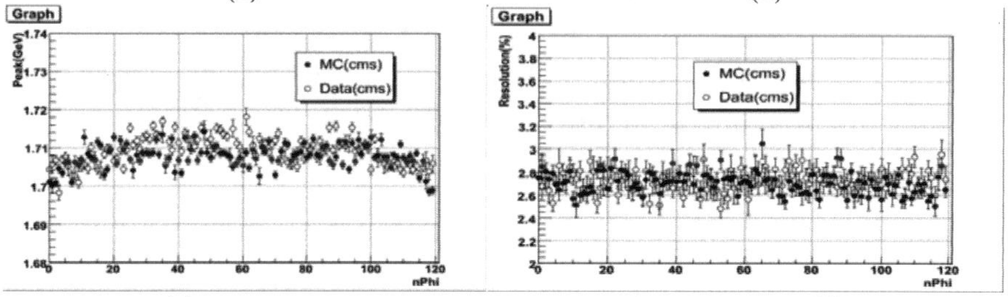

(c) (d)

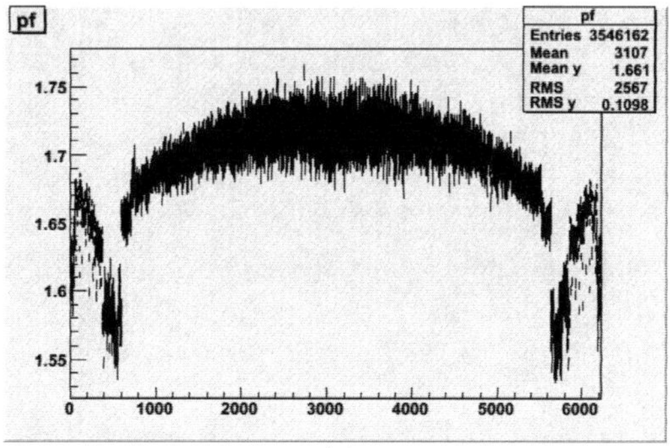

(e)

Fig.4 (a) and (b) are the E5x5cms energy peak and resolution versus theta-index, respectively (c) and (d) are the E5x5cms energy peak and resolution versus phi-index in barrel, respectively (e) is the E5x5cms energy versus ixtal profile.

Journal of Physics: Conference Series **293** (2011) 012048 doi:10.1088/1742-6596/293/1/012048

3.2.3. EMC Performance after Bhabha calibration

After Bhabha calibration EMC performance are shown in Fig.5. In the center-of-mass system, energy resolutions for electrons (from $e^+e^- \rightarrow e^+e^-$) are ~2.3% in barrel and ~4.1% in end cap, for photons (from $e^+e^- \rightarrow \gamma\gamma$) ~2.7% in barrel and ~4.2% in end cap. Data and MC also consist very well.

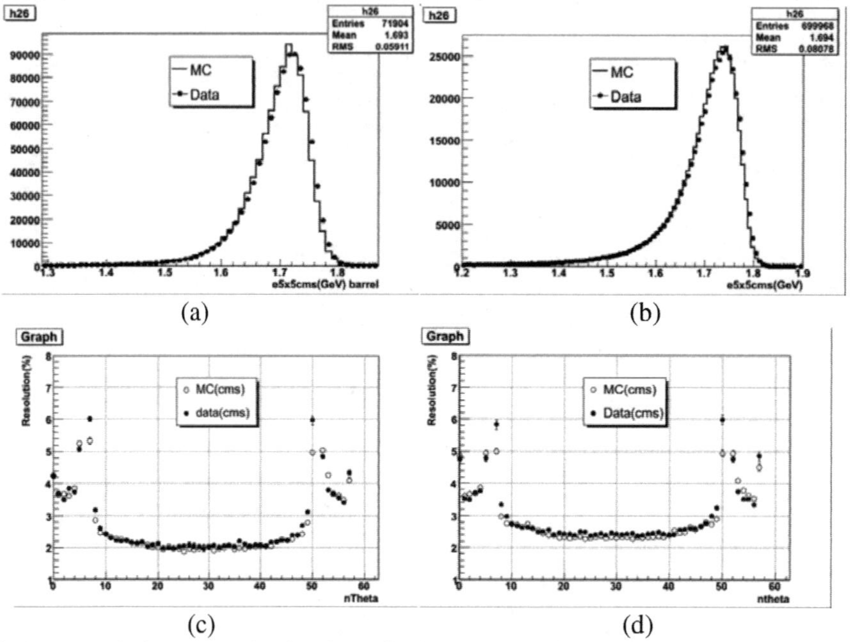

Fig.5 (a) and (b) are e5x5 energy distribution of Bhabha events and digamma events at the center-of-mass energy 3.686GeV , respectively. (c) and (d) are the energy resolution versus the thetaID of Bhabha events and digamma events, respectively.

4. Conclusion

The BESIII CsI(Tl) calorimeter has been working quite well for two years. All the 6240 CsI(Tl) counters are alive. The single crystal energy calibration is very good. Data and MC consist very well after the calibration with Bhabha. The energy resolution for $e^+e^- \rightarrow e^+e^-$ is ~2.3% at the center-of-mass energy 3.686GeV. Current Bhabha calibration method shows the stable and reliable picture and EMC all crystal counters are monitored by Bhabha calibration.

References

[1] M. Ablikim et al. (BESIII Collaboration), NIM A 614 (2010) 345－399
[2] C. Zhang for BEPC & BEPCII Teams, Performance of the BEPC and progress of the BEPCII, in: Proceedings of APAC, 2004, pp.15-19, Gyeongju, Korea.
[3] JianMing Bian, "Absolute energy calibration and the use of time information of BESIII EMC", talk at this conference.
[4] Mark K. Seager et al. Sparse Linear Algebra Package（SLAP）Version 2.0, 1988, http://www.netlib.org/slap/.

XIV International Conference on Calorimetry in High Energy Physics (CALOR 2010) IOP Publishing
Journal of Physics: Conference Series **293** (2011) 012049 doi:10.1088/1742-6596/293/1/012049

Measurement of the Muon Stopping Power in Lead Tungstate with the Electromagnetic Calorimeter in CMS

A D Benaglia[1] on behalf of the CMS Collaboration
[1] Università degli Studi di Milano - Bicocca and INFN - Piazza della Scienza 3, Milano, Italy
E-mail: andrea.benaglia@mib.infn.it

Abstract. A large sample of cosmic ray events collected by the CMS detector has been exploited to measure the muon stopping power in the lead tungstate ($PbWO_4$) of the electromagnetic calorimeter. The events were recorded in October-November 2008, during commissioning runs of the CMS detector with the solenoid at the nominal field strength of 3.8 T. The measurement spans a momentum range from 5 to 1000 GeV/c. The results are consistent with the expectations over the entire range. A comparison of collision losses with radiative losses allowed for a first experimental determination of muon critical energy in lead tungstate, measured to be $160^{+5}_{-6}\,(\text{stat.}) \pm 8\,(\text{syst.})$ GeV, in agreement with expectations.

1. Introduction

During the commissioning phase conducted with cosmic ray events [1], the electromagnetic calorimeter (ECAL) of the Compact Muon Solenoid (CMS) experiment has been exploited to measure the muon stopping power in the lead tungstate of its crystals.

The stopping power for muons in the energy range considered can be conveniently written as

$$f(E) = \left\langle -\frac{\mathrm{d}E}{\mathrm{d}x} \right\rangle = a(E) + b(E)E\,,$$

where E is the total muon energy, x is the thickness of the traversed material, commonly measured in mass per unit surface, $a(E)$ is the stopping power due to collisions with atomic electrons, and $b(E)$ is due to radiative processes: bremsstrahlung, direct pair production, and photonuclear interactions; $a(E)$ and $b(E)$ are slowly varying functions of E at energies where radiative contributions are important [2, 3].

Numerical values of stopping power and related quantities quoted throughout this report are taken from tables in [4, 5]. The theoretical uncertainties are everywhere smaller than the statistical precision of the results in this analysis.

The definition of critical energy as the energy at which the average rates of energy loss through collision and radiation are equal [2, 3] is adopted here. Analysis of data in the full momentum range permits a comparison of collision losses with radiative losses, thus leading to the first experimental determination of muon critical energy.

See [6] for a detailed description of the CMS detector.

2. Measurement of physical variables and event selection

The physical variables to be measured for each muon are the momentum (p), the path length in ECAL (Δx), and the energy lost in ECAL (ΔE) [7].

The muon momentum is measured from the track fit performed in the inner tracking system [8].

The path length in ECAL is estimated by propagating the measured muon track inside the calorimeter taking into account bending due to the magnetic field and expected energy loss. The average path length in ECAL for the selected muon sample is 22.0 cm, from which it follows that the typical energy loss for a minimum ionizing muon is about 300 MeV.

The energy deposited in ECAL barrel[1] by cosmic ray muons is measured by means of a dedicated algorithm. Online data reduction is based on zero suppression (ZS), that is, a readout of channels above a ZS threshold of about 20 MeV, and on a full readout of selected regions (SR) of high interest, with a threshold set to about 170 MeV [9]. When the condition for SR is met, a matrix of 3×3 trigger towers is read out. Each tower is comprised of 5×5 crystals and the matrix is centered around the tower with energy deposit above the SR threshold.
The energy reconstruction algorithm then identifies clusters starting from a central crystal (seed) with an energy deposit of at least 139.5 MeV (or from the most energetic of a pair of adjacent crystals with at least 46.5 MeV each), and adding all the channels above the clustering threshold (18.5 MeV) belonging to a 5×5 matrix of crystals centered on the seed. Contiguous clusters are eventually merged. The final cluster energy is obtained after application of channel-to-channel relative calibration constants and of an absolute energy scale factor, set with data taken with a 120 GeV/c electron beam [10].

The initial sample of useful CMS triggers is reduced by the combined requirements of a muon at the trigger level, of a single muon track reconstructed by the tracker [8] in the range 5-1000 GeV/c, and of an energy deposit associated with the track in the lower ECAL barrel, since the muon momentum has to be measured upstream of the energy release in the calorimeter. After these and further quality selections applied to the event sample, 2.5×10^5 muons are left.

The measured muon spectrum, after the selections, is shown in Fig. 1(a). The spectrum of energy deposited in crystals is shown in Fig. 1(b). Figure 2 displays the distributions of $\mathrm{d}E/\mathrm{d}x$ for muons with momentum below 10 GeV/c, where collision losses dominate (a), and for muons with momentum above 300 GeV/c, where sizable radiation losses are expected (b).

3. Data analysis and systematic correction

3.1. Instrumental effects

The relevant instrumental effects are related to the online data reduction and to the energy reconstruction procedure described in Section 2. The presence of thresholds introduces a bias in energy reconstruction: noise fluctuations above threshold contribute a positive bias, while energy deposits below threshold contribute a negative bias.

When the muon direction is close to the crystal axis (angle between muon and crystal axis $\lesssim 0.1$ radians), the condition for SR is met in the majority of cases. The main positive bias then arises from the probability of noise fluctuation above the clustering threshold. A 14.7 MeV bias is estimated by means of a dedicated analysis of the noise spectrum.

At larger muon angles to the crystal axis, the crystal multiplicity in cluster increases and the average energy deposit per crystal decreases correspondingly. Therefore conditions for ZS readout occur more frequently, with the higher ZS threshold reducing the noise bias. Moreover, the probability that the energy deposited in a single crystal is below clustering threshold increases, thus contributing a negative bias. Bias is then expected to decrease with increasing angles.

[1] Because of the angular distribution of the cosmic ray muon flux and the lower sensitivity of the ECAL end-caps to the energy released by mip muons, only the barrel part of ECAL is used in the present analysis.

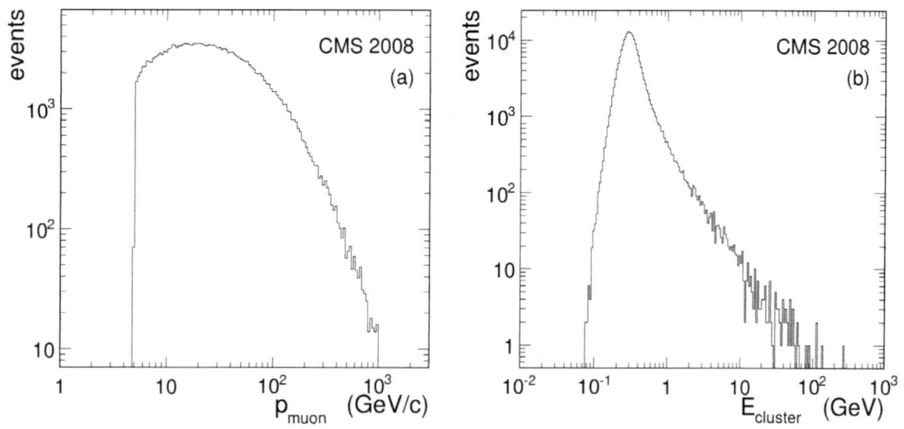

Figure 1. (a) Momentum spectrum of the muons passing the selections; (b) spectrum of the reconstructed energy in the lower ECAL hemisphere. A logarithmic binning is used.

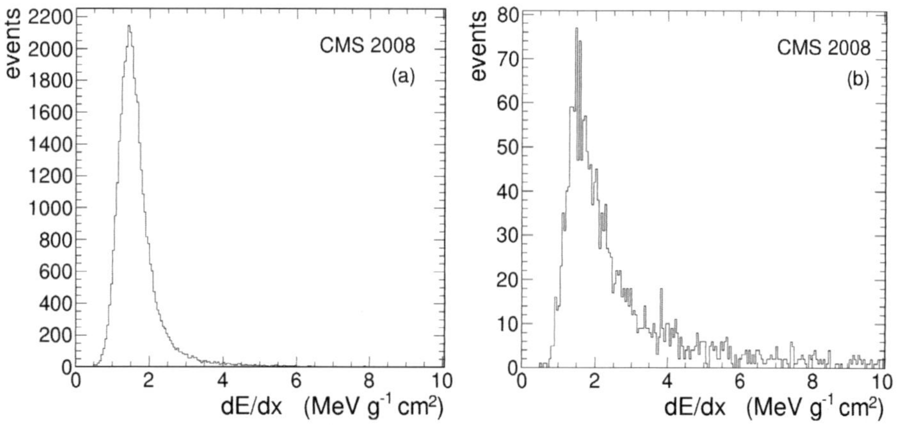

Figure 2. Measured distributions of $\mathrm{d}E/\mathrm{d}x$ in ECAL; (a) for muon momenta below 10 GeV/c; (b) for muon momenta above 300 GeV/c; the fraction of events with $\mathrm{d}E/\mathrm{d}x > 10\ \mathrm{MeV\,g^{-1}\,cm^2}$ is 1.3×10^{-3} and 8×10^{-2} in (a) and (b) respectively.

This bias dependence on angle for non-radiating muons is observed in experimental data, as shown in Fig. 3, where the raw measured $\mathrm{d}E/\mathrm{d}x$ is displayed versus angle for muons below 10 GeV/c.

A correction is thus applied to the estimated collision component of the stopping power by applying the following procedure. At small angles, a bias of 14.7 MeV is subtracted. At larger angles, the bias to be subtracted is estimated from a fit to the data shown in Fig. 3. The uncertainty in the evaluation of the 14.7 MeV bias and the statistical uncertainty of the fit contribute to the systematic uncertainty of the bias correction, yielding an overall systematic uncertainty of 3.5 MeV, which corresponds to about 1.2 % of the average energy deposited by a low-momentum muon.

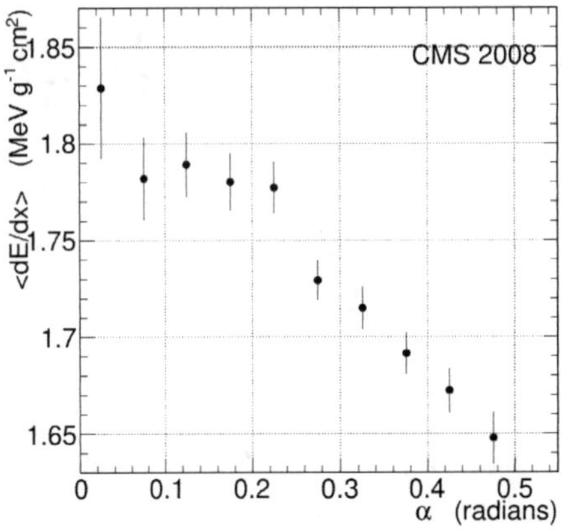

Figure 3. Dependence of the raw $\langle \mathrm{d}E/\mathrm{d}x \rangle$ on the angle α between the muon direction and the crystal axis, for muon momentum between 5 and 10 GeV/c. Vertical bars represent statistical errors.

3.2. Containment corrections

An issue to be taken into account is that energy ΔE lost by a muon traversing a given volume is in general different from the energy deposited in that same volume, which is the first output of the measurement procedure. For the present experimental setup, it can be shown that there is always a non-negligible flow of energy carried by secondaries leaving the calorimeter volume from the rear crystal face, since secondaries are produced along the entire muon path.

To estimate the difference between energy lost and energy deposited in ECAL, a model based on the Geant4 [11] simulation toolkit, with a simplified description of the detector and materials surrounding it, was developed. Since energy leakage and the effects of upstream and downstream materials on it are quantitatively different for collisions and radiative processes, the two different regimes have been treated separately.

For collisional processes, according to the simulation, the energy carried by secondaries flowing out of the rear detector surface is, on average, 3 % of the energy lost in the crystals for a muon with momentum 15 GeV/c. This represents an upper limit to containment corrections in the low-momentum region, since the rear leakage can be compensated by secondaries produced in upstream material entering the front detector surface. A comparison of the deposits in the upper and lower hemispheres of the ECAL barrel, for muons with momentum between 5 and 10 GeV/c, shows no difference in $\mathrm{d}E/\mathrm{d}x$, with a sensitivity better than 1 % [12]. Given this result, in spite of the asymmetric distribution of the outer and inner material around ECAL, containment correction has been neglected, and 1 % systematic uncertainty on the null correction has been assumed for collision losses.

Energy containment corrections for radiative processes have been derived from dedicated simulations as the mean value of two extreme cases: with no material in front of ECAL and with the whole material budget of the tracker concentrated just in front of ECAL, representing the upper and the lower limit to the containment correction. Half of the difference between the results of the two simulations is taken as the associated systematic error. The correction is maximal at 1 TeV/c and corresponds to (28 ± 5) % of the average energy lost, while at 170 GeV/c it reduces to (14.5 ± 2.5) %. Such an uncertainty is small compared to the statistical uncertainty in the measurement of the stopping power in radiative region.

4. Experimental results

In Fig. 4(a), the specific muon energy loss resulting after corrections is compared to expectations as a function of the muon momentum [4]. Fig. 4(b) shows the ratio of experimental measurements to expected values. Two regions are indicated in Fig. 4: the expected 68 % probability central interval (grey shaded area), and the minimum interval corresponding to 68 % probability (region delimited by continuous curves). The reasons for considering both intervals and the procedure adopted for their estimate are summarized below.

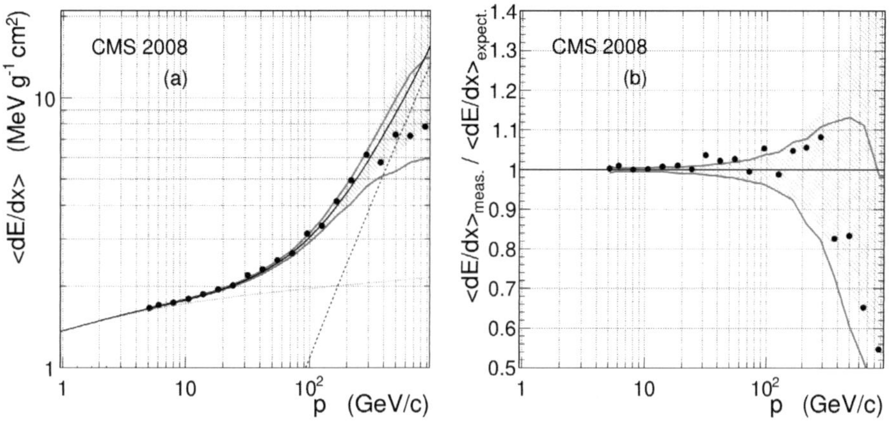

Figure 4. (a) Muon stopping power measured in $PbWO_4$ (dots) as a function of muon momentum compared to expectations [4] (continuous line). The expected contributions from collision and radiative processes are plotted as well (dotted line and dashed line respectively). (b) Ratio of the measured and the expected values of the muon stopping power, as a function of muon momentum. The shaded grey area indicates the expected 68 % probability central interval, while the continuous curves delimit the minimum interval containing 68 % of the expected results.

Radiative losses are rare events characterized by large energy releases. As a consequence, the probability density functions (pdf) of energy loss for single events are positively skewed distributions, with long tails at high energy. The skewness of the single event pdf increases with increasing muon momentum, owing to the increasing contribution of radiative processes with momentum. At the same time, the number of events per bin in the selected data sample rapidly decreases with momentum for $p > 50$ GeV/c (Fig. 1(a)). It is thus expected that, due to the combination of these two effects, the pdf of the sample mean of $\mathrm{d}E/\mathrm{d}x$ cannot be approximated by a Gaussian in the higher momentum bins. Moreover, when this condition occurs, sample mean and sample variance are highly correlated. It follows that the widely used "experimental error", namely the sample rms divided by the square root of the sample population, is not a good measure for the consistency of the sample mean with the expected value.

Therefore the uncertainty of the result has been estimated by means of a numerical technique based on the simulation of the expected results with Geant4. Ten thousand pseudo-experiments per momentum bin, each with the same statistics as the actual bin population, were simulated. The expected pdf of the mean was then obtained for each bin as the distribution of the mean values of the stopping power from the different experiments. For each pdf two 68 % probability intervals for the expected result were derived: the central interval, obtained by discarding 16 % of the results on each tail of the pdf, and the interval of minimum width containing 68 % of the results; this last interval always contains the most probable value.

The curve $(\mathrm{d}E/\mathrm{d}x)_{meas} = \alpha \left[\left(\frac{\mathrm{d}E}{\mathrm{d}x}\right)_{coll} + \beta \times \left(\frac{\mathrm{d}E}{\mathrm{d}x}\right)_{rad} \right]$, where *coll* and *rad* label the predicted energy losses in $PbWO_4$ due to collisions with atomic electrons and radiative processes respectively [4], is fitted to experimental stopping power data using a binned maximum likelihood and the pdf described above. The parameters α and β account for the overall normalization of the energy scale and for the relative normalization of radiation and collision losses. With the adopted parameterization the β parameter, from which the critical energy is measured, is independent of the overall energy scale. The fit results in:

$$\alpha = 1.004^{+0.002}_{-0.003}\,(\text{stat.}) \pm 0.016\,(\text{syst.})$$
$$\beta = 1.07^{+0.05}_{-0.04}\,(\text{stat.}) \pm 0.6\,(\text{syst.}).$$

Adding statistical and systematic contributions in quadrature, it may be concluded from the above results that the energy scale is consistent with expectations within a systematic uncertainty of 1.6 %, with a 1.2 % contribution from the uncertainty on the energy scale dependence on the angle and on the clustering, and a 1.0 % contribution from uncertainty in containment corrections. This result is mainly determined by the precision of the measurements in the momentum region below 20 GeV/c, where radiation losses are marginal.

From the fit results, a muon critical energy of $160^{+5}_{-6}\,(\text{stat.}) \pm 8\,(\text{syst.})$ GeV is obtained, in agreement with the computed value of 169.5 GeV for $PbWO_4$ [4]. The systematic uncertainty includes a contribution of 4.5 GeV from the uncertainty of the containment corrections, dominated by the limited knowledge of the correction for radiation losses, and a contribution of 6 GeV due to the stability of the result against bias subtraction and event selection.

5. Conclusions

The muon stopping power in $PbWO_4$ has been measured over the momentum range from 5 GeV/c to 1000 GeV/c. In the region corresponding to muon momenta less than 20 GeV/c, where collision losses dominate, the average energy deposited in the crystals is of order 300 MeV. Thus the agreement (to within 1-2 %) between the measured stopping power and the calculated values confirms that the energy calibration of the detector, previously determined with 120 GeV/c electrons, remains valid down to the sub-GeV scale.

From a comparison of the rate of collision losses with that of radiative losses, a first experimental determination of the muon critical energy has been obtained. The experimental value $160^{+5}_{-6}\,(\text{stat.}) \pm 8.0\,(\text{syst.})$ GeV is in agreement with numerical calculations for this material.

References

[1] CMS Collaboration 2010 Commissioning of the CMS experiment and the cosmic run at four tesla *JINST* **5** T03001
[2] Particle Data Group, Amsler C et al. 2008 Review of particle physics *Phys. Lett.* B **667** 1
[3] Groom Donald E, Mokhov Nikolai V and Striganov Sergei I 2001 Muon stopping power and range tables 10-MeV to 100-TeV *Atom. Data Nucl. Data Tabl.* **78** 183-356
[4] Particle Data Group 2009 Atomic and Nuclear Properties of Materials, tables for $PbWO_4$ *web pages at: http://pdg.lbl.gov/2009/AtomicNuclearProperties/*
[5] Ivanov D, Kuraev E A, Schiller A and Serbo V G 1998 Production of e+ e- pairs to all orders in Z alpha for collisions of high-energy muons with heavy nuclei *Phys. Lett.* B **442** 453-8
[6] CMS Collaboration 2008 The CMS experiment at the CERN LHC *JINST* **3** S08004
[7] CMS Collaboration 2010 Measurement of the muon stopping power in lead tungstate *JINST* **5** P03007
[8] CMS Collaboration 2010 Performance of CMS muon reconstruction in cosmic-ray events *JINST* **5** T03022
[9] Almeida N et al. Data filtering in the readout of the CMS electromagnetic calorimeter 2008 *JINST* **3** P02011
[10] Adzic P et al. 2008 Intercalibration of the barrel electromagnetic calorimeter of the CMS experiment at start-up *JINST* **3** P10007
[11] GEANT4 Collaboration, Agostinelli S et al. 2003 GEANT4: A simulation toolkit *Nucl. Instrum. Meth.* A **506** 250-303
[12] CMS Collaboration 2010 Performance and Operation of the CMS Crystal Electromagnetic Calorimeter *JINST* **5** T03010

XIV International Conference on Calorimetry in High Energy Physics (CALOR 2010) IOP Publishing
Journal of Physics: Conference Series **293** (2011) 012050 doi:10.1088/1742-6596/293/1/012050

Drift Time measurement in the ATLAS Liquid Argon electromagnetic calorimeter using cosmic muons

Carolina Gabaldón, on behalf of the ATLAS Liquid Argon Calorimeter Group
CERN, *CH* – 1211 Geneve 23, Switzerland
E-mail: carolina.gabaldon.ruiz@cern.ch

Abstract.
In this article, measurements of the drift time (T_{drift}) of the ionization electrons in the ATLAS ElectroMagnetic (EM) calorimeter are presented, using 32-samples cosmic muon data. In particular, the drift time measurement is used to assess the intrinsic uniformity of the calorimeter gaps and estimate its impact on the constant term of the energy resolution. This analysis leads to an estimated contribution to the constant term of $(0.29^{+0.05}_{-0.04})$ % in the barrel and $(0.54^{+0.06}_{-0.04})$ % in the endcaps. The same data are used to measure the drift velocity of ionization electrons in liquid argon, which is found to be 4.61 ± 0.07 mm/μs at 88.5 K and 1 kV/mm.

1. Introduction

The ATLAS liquid argon (LAr) calorimeter [1] is composed of sampling detectors with full azimuthal symmetry. The EM calorimeter is a device based on an accordion shape geometry consisting of lead absorbers and electrodes coupled to a fast electronic integration. The EM calorimeter is divided into one barrel, covering $|\eta| < 1.475$ and two endcaps, $1.375 < |\eta| < 3.2$. In depth, three very granular layers (denoted as layer 1,2,3) allow a detailed measurement of the electrons and photons, crucial during the operation of the LHC.

The required precision and discovery reach places stringent requirements on the performance of the calorimeter. The uniformity of the calorimeter response over a large acceptance is particularly important for the overall resolution. Non-uniformities due to intrinsic characteristics of the ionization gaps in the liquid argon system have to be measured. The electron drift time in LAr, which can be obtained from the signal pulse shape resulting from ionizing particles that deposit sufficient energy above the intrinsic noise level in a calorimeter cell, is a powerful monitoring tool.

Cosmic muon runs are recorded regularly since august 2006 to test *in situ* the detector performance [2]. At the end of the summer and during autumn of 2008 stable cosmic muon runs were taken with the detector fully operational and using various trigger menus. The main challenge was to extract the electron drift time in LAr to measure the asymmetries and intrinsic non-uniformities of the calorimeter.

2. Inonization signal in the calorimeter

Electrons and photons hitting the detector induce electromagnetic showers. The fraction of the shower energy lost by ionization in the LAr gaps generates the device signal: the ionization electrons drift in the field produced across the LAr gap by the high voltage, generating a current with an amplitude

Journal of Physics: Conference Series **293** (2011) 012050 doi:10.1088/1742-6596/293/1/012050

proportional to the released energy. This current is then amplified, shaped, sampled at the LHC bunch-crossing frequency (25 ns) and digitized.

In the straight section of the LAr gaps the ionization current has a typical ionization-chamber triangular shape with a fast rise time less than 1 ns followed by a linear decay for the duration of the maximum drift time:

$$T_{drift} = w_{gap}/V_{drift}, \tag{1}$$

where w_{gap} is the LAr gap width and V_{drift} the electron drift velocity. For a fixed value of the high-voltage HV applied to a LAr gap, the corresponding V_{drift} depends on the electrical field as:

$$V_{drift} = V_{ref} \cdot \left[\frac{HV}{HV_0} \cdot \frac{w_{gap0}}{w_{gap}}\right]^{\alpha} \tag{2}$$

where $\alpha = 0.3^{+0.04}_{-0.02}$ [3]. The drift time is approximately four times more sensitive (Equation 1 and 2) than the peak current in a LAr cell [1] to a variation of the gap width.

Figure 1 shows two typical digitized signal shapes, one for the barrel and the other for the endcap. The data samples in each plot correspond to a single cosmic muon event in a single cell, and fluctuations of the amplitude in each sample due to noise can be observed. The prediction relies on the modeling of each readout cell and on the description of the signal propagation. Two methods have been developed to predict the signal: Response Transformation Method (RTM) [4] and First Principles Method (FPM) [5] (for the EM barrel). The drift time measurements can be performed by fitting a pulse shape prediction to the physics signal from data, taking into account the contribution to the drift time coming from the straight and the bend regions on the accordion electrodes (see Figure 2) and the nominal values for the W_{gap} and the high voltage (see Figure 3).

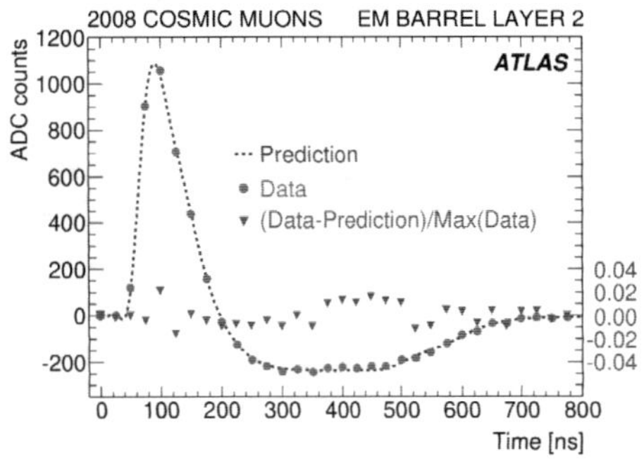

(a) Barrel

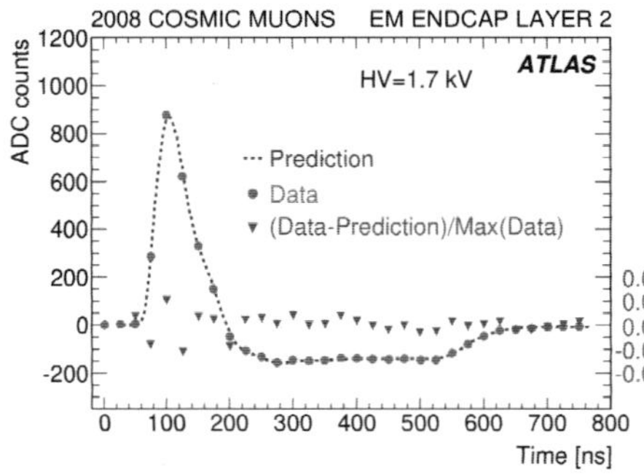

(b) Endcap

Figure 1. Typical single ionization pulse in a cell of layer 2 of the barrel (a) and endcap (b). The large red dots show the data samples, the small blue dots the prediction and the grey triangles the relative difference (data (S) - prediction (g))/S_{max}.

3. Drift time measurement and response uniformity

In the barrel, the LAr gap size is designed to be constant (2.09 mm) independent of η. The drift time is then expected to be constant, except for the presence of local non-uniformities. Figure 4(a) presents the drift time T_{drift} averaged over the azimuth (ϕ) as a function of η for the barrel. The results of the two methods differ by 0.1 ns on average with an RMS of 1.3 ns. The full purple line illustrates the prediction

[1] $I_0 = \rho \cdot V_{drift}$, with ρ the negative linear charge density along the direction perpendicular to the readout electrode.

Journal of Physics: Conference Series **293** (2011) 012050 doi:10.1088/1742-6596/293/1/012050

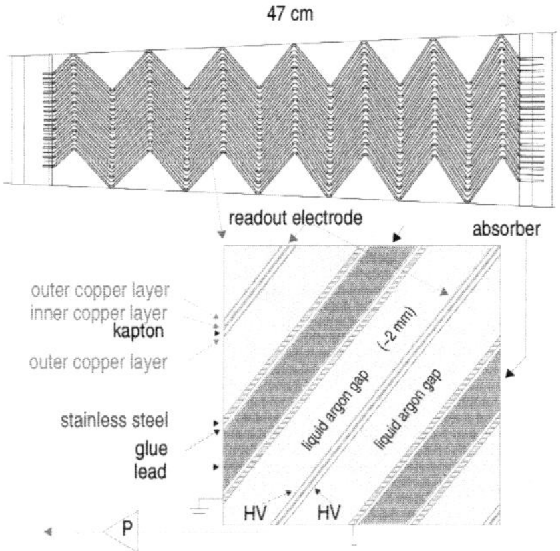

Figure 2. Accordion structure of the barrel.

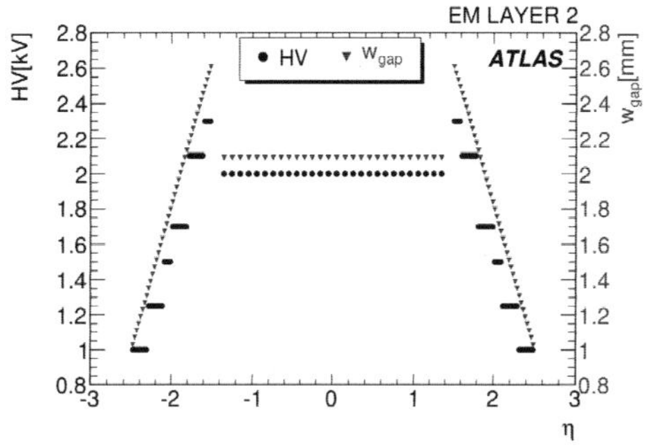

Figure 3. Nominal HV (black dots) and nominal gap width w_{gap} (blue triangles) versus η in the 2nd layer of the EM calorimeter.

from absorber thickness measurements made during the calorimeter construction [6]. There is a good agreement between the prediction and the data, except in the transition regions around $\eta = 0$, ± 0.8 and -1.4, where the lower field due to edge effects at the end of the electrodes induces a larger T_{drift}.

The measured drift time versus η for the two endcap wheels is shown in Figure 4(b). A general decrease of T_{drift} with increasing pseudorapidity is observed, as expected from the corresponding reduction of the design gap size (Figure 3). In addition, regular steps are found, corresponding to the locations of the boundaries between high voltage regions. The resulted T_{drift} is compared to the Monte Carlo (MC) calculation using 10 GeV electromagnetic showers. There is a good agreement between data and MC at the level of 1 %.

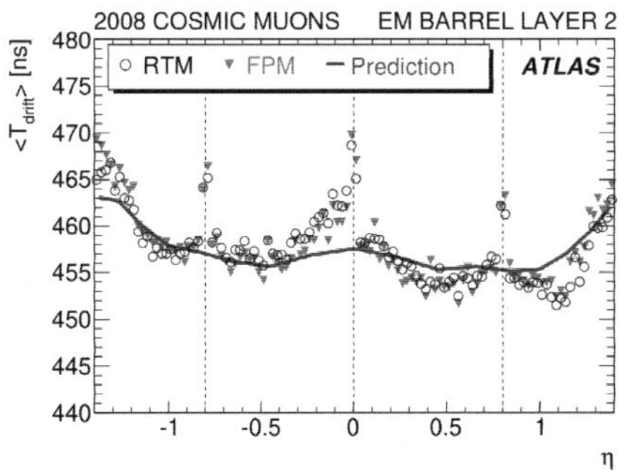

(a) Barrel

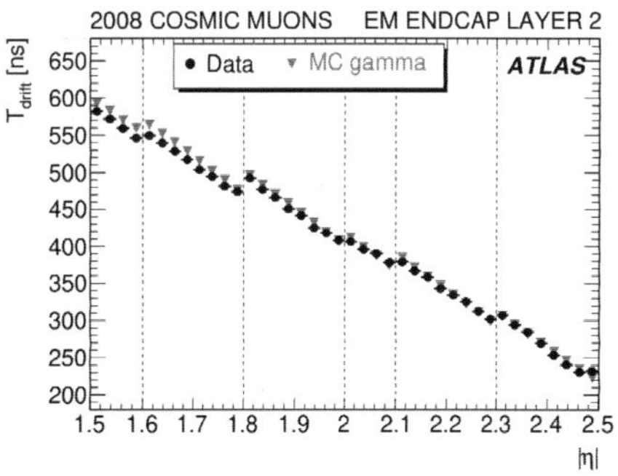

(b) Endcap

Figure 4. (a) T_{drift} as a function of η in layer 2 of the barrel: using the RTM method (open dots), the FPM method (red triangles) and the prediction (purple line). (b) T_{drift} as a function of η in layer 2 of the endcap: Black points are the data and red triangles MC predictions for photons.

The reconstructed value of the energy deposited in the calorimeter by an electron or photon should be independent of the position of its impact on the calorimeter. The non-uniformity coming from local variations of the response due to gap fluctuations can be determined using the drift time measurements. Figure 5 shows the distribution of the drift time averaged over groups of 4×4 cells corresponding to an area of 0.1×0.1 in $\Delta\eta \times \Delta\phi$ plane for the barrel (a) and for the endcap (b). In the endcap, the normalization $< T_0 >$ cancels out the change of the drift time due to the nominal design gap size variation with η. From this figure, the drift time uniformity is extracted for the barrel, $(1.28 \pm 0.03)\%$, and for the endcap, $(2.8 \pm 0.1)\%$ [2].

From the relation between the drift time and the drift velocity (Equation 1), it follows, that the drift time uniformity leads to a dispersion of the response due to intrinsic gap variations of $(0.29^{+0.05}_{-0.04})$ % for the barrel and $(0.54^{+0.06}_{-0.04})$ % for the endcap.

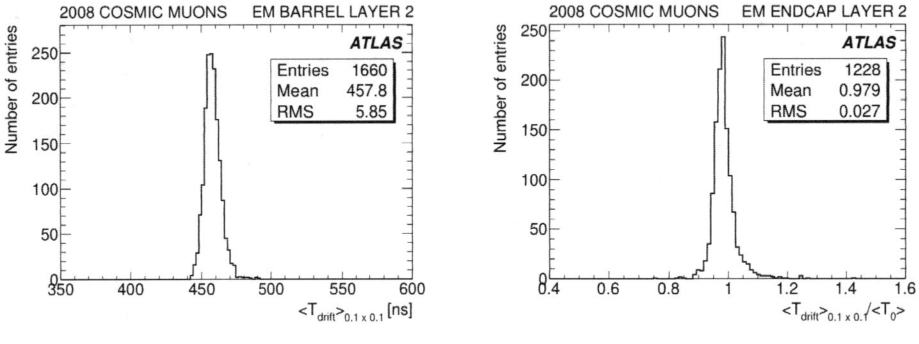

(a) Barrel (b) Endcap

Figure 5. Drift time uniformity between groups of 4×4 cells ($\Delta\eta \times \Delta\phi = 0.1 \times 0.1$) for barrel layer 2 (a) and for endcap layer 2 (b).

4. Drift velocity measurement

The drift velocity can be obtained from the drift time measurement using Equation 1 where the nominal values for w_{gap} are used (see Figure 3). While w_{gap} is designed to be constant for the barrel, it depends on η for the endcap where in addition there are different high voltage regions. In order to compare V_{drift} for barrel and endcap, the actual drift velocity can be scaled to a reference field of 1 kV/mm (Equation 2).

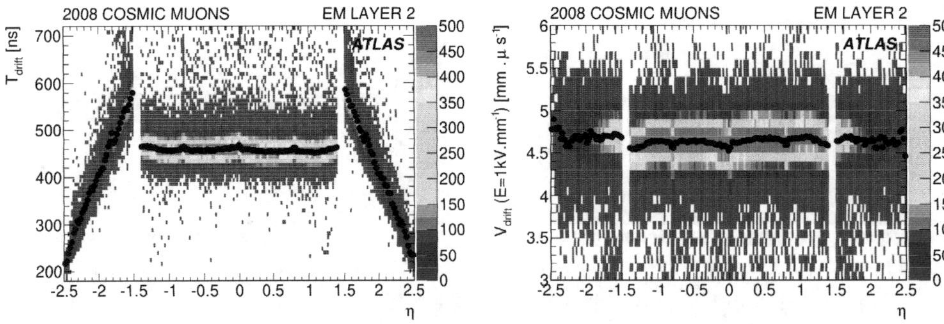

(a) T_{drift} (b) V_{drift}

Figure 6. Drift time (a) and Drift velocity (b) (at $E = 1$ kV/mm) vs η in layer 2.

[2] The contribution from pure statistical fluctuations is not neglitible for the endcap, (1.5 ± 0.1) % and is quadratically subtracted.

Figure 6(a) shows the distribution of the drift time values and the behavior of their mean as a function of η. As expected, variation in the endcap is observed due to the decrease of the gap size as $|\eta|$ increases. Using those drift time values, the drift velocity for a given field is extracted as a function of $|\eta|$ for the whole electromagnetic calorimeter (Figure 6(b)). A rather flat behavior is observed except at the transition regions in the barrel where the field is lower. In addition, endcap C ($\eta < 0$) is higher than endcap A ($\eta > 0$), which is consistent with the difference of temperature between both wheels (0.5 %).

Figure 7 shows the comparison of V_{drift} for the different layers of the barrel (a) and endcaps (c). The drift velocity mean values are given in Table 1, together with the uncertainties explained further in [3]. All the results are compatible within errors, although the barrel presampler is below the average. Averaging over the significant layers (presampler and layer 2) gives as the final result for the reference field of 1 kV/mm: $V_{ref} = (4.61 \pm 0.07)$ mm/μs ($T = 88.5$ K). This value can be compared with the measurements from [7] which give (4.65 ± 0.12) mm/μs for a LAr temperature of 88.5 K and provides good agreement with the present measurement.

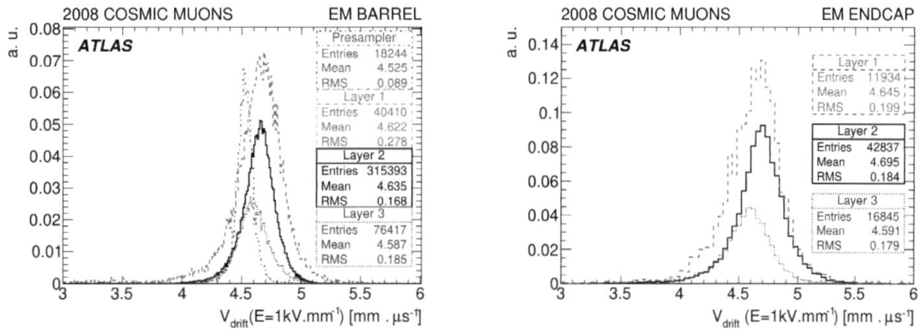

(a) Barrel (b) Endcap

Figure 7. Drift velocity distribution for the barrel (a) and endcap (b).

	Layer	Drift velocity (in mm/μs at 1 kV/mm)
Barrel	Presampler	4.52 ± 0.001 (stat) $^{+0.11}_{-0.07}$ (syst)
	Layer 1	4.62 ± 0.003 (stat) $^{+0.06}_{-0.14}$ (syst)
	Layer 2	4.63 ± 0.002 (stat) $^{+0.06}_{-0.14}$ (syst)
	Layer 3	4.59 ± 0.002 (stat) $^{+0.06}_{-0.14}$ (syst)
Endcap	Layer 1	4.65 ± 0.002 (stat) $^{+0.10}_{-0.14}$ (syst)
	Layer 2	4.69 ± 0.001 (stat) $^{+0.10}_{-0.14}$ (syst)
	Layer 3	4.59 ± 0.002 (stat) $^{+0.10}_{-0.14}$ (syst)

Table 1. Drift velocity at $E = 1$ kV/mm in the different layers of the calorimeter.

5. Direct determination of local gap

Using Equation 1 and Equation 2 it is possible to express the gap size as a function of the measured T_{drift}, namely:

$$w_{gap} = [A \cdot T_{drift}]^{1/(1+\alpha)} \quad (3)$$

with $A = V_{ref} \cdot \left[\frac{HV}{HV_0}\right]^{\alpha} \cdot w^{\alpha}_{gap0}$. The gap size should not depend on the reference, which has been taken as

$w_{gap0} = 2$ mm, $HV_0 = 2$ kV and $V_{ref} = 4.61$ mm$/\mu$s for both Barrel and Endcaps. Data for the endcaps have been corrected for the temperature difference, and rescaled to 88.5 K.

The additional information yielded by this analysis shows directly how the ratio of the measured gap to the designed gap varies as a function of position in the detector. Figure 8 shows the relative difference between the calculated gap sizes from previous formula and the design values as a function of pseudorapidity. One can see that the ratio between calculated and design values, spanning a gap range between 1 mm and 2.5 mm, has an RMS of 0.83%, i.e. typically 16 μm. In general, a reasonable agreement is found apart from some local effects. In the barrel part one recognizes the systematic effects associated with the slight bulging of the absorbers, and the "transition regions" at $\eta = 0$, ± 0.8 and ± 1.4.

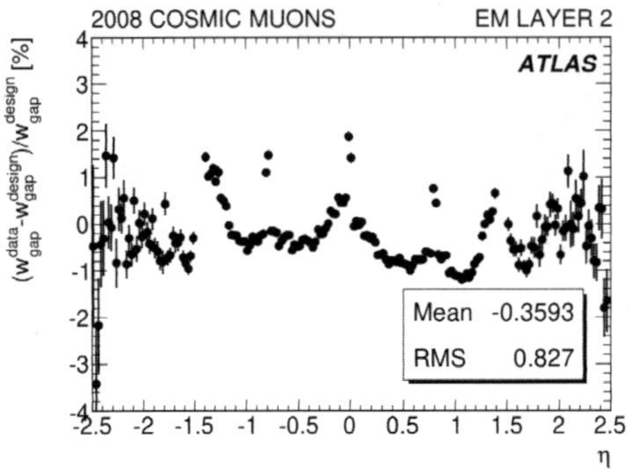

Figure 8. Relative difference between the design gap values and the values extracted from T_{drift} measurements.

6. Conclusions

The average electron drift time in each cell of the highly granular LAr electromagnetic calorimeter of ATLAS, is measured from the samples of shaped ionization pulses induced by high energy deposits from cosmic muons recorded during the 2008 commissioning run. The pulses were recorded in a wide time window (32 samples) to assure the required precision of the measurement.

The drift time measurement has shown to be a powerful tool to detect asymmetries and intrinsic non-uniformities along the calorimeter. Taking advantage of these facts, an estimate of the calorimeter non-uniformity of response due to gap size variations has been inferred, namely $(0.29^{+0.05}_{-0.04})$ % and $(0.54^{+0.06}_{-0.04})$ % respectively for the barrel and the endcaps.

The drift velocity at 1 kV/mm is obtained from the drift time measurements resulting in (4.61 ± 0.07) mm$/\mu$s. This value is compatible with previously published measurements at the same operating temperature of 88.5 K.

References

[1] ATLAS Collaboration, *The ATLAS Experiment at the CERN Large Hadron Collider*, JINST 3 (2008) S08003
[2] ATLAS Collaboration, *Readiness of the ATLAS Liquid Argon Calorimeter for LHC Collisions*, Accepted by EPJC.
[3] ATLAS Collaboration, *Drift Time Measurement in the ATLAS Liquid Argon Electromagnetic Calorimeter using Cosmic Muons, The ATLAS Collaboration*, Accepted by EPJC.
[4] D. Banfi, M. Delmastro and M. Fanti, *Cell response equalization of the ATLAS electromagnetic calorimeter without the direct knowledge of the ionization signals*, SN-ATLAS-2005-054, J. Inst **1** (2006) P08001.
[5] C. Collard *et al.*, *Prediction of signal amplitude and shape for the ATLAS electromagnetic calorimeter*, ATL-LARG-PUB-2007-010.
[6] B. Aubert et al. *Construction,assembly and tests of the ATLAS electromagnetic barrel calorimeter*, NIM A 558 (2006) 388.
[7] W. Walkowiak, *Drift Velocity of Free Electrons in Liquid Argon*, NIM A 449 (2000) 288-294.

Monitoring the stability of the CMS electromagnetic calorimeter

Federico Ferri[1]
DSM/IRFU, CEA/Saclay
91191 Gif-sur-Yvette, France

E-mail: `federico.ferri@cern.ch`

Abstract. The lead tungstate electromagnetic calorimeter of the CMS experiment has been proved to achieve an excellent energy resolution during the commissioning phase of the detector. The uniformity of the relative response of individual channels and the energy scale for electrons and photons are constrained by the several calibration procedures. The stability of the detector in time is constantly monitored throughout the LHC operation by means of dedicated runs and data taken at 100 Hz in the 3 μs abort gap at the end of each 89 μs beam cycle. A laser monitoring system is used to track the single channel response variations with time, as changes of the crystal transparency caused by irradiation. The stability of crucial detector parameters such as high voltage, temperature and electronic noise and the performance of the operation of the light monitoring system are shown to fulfill the requirements needed to achieve the target resolution of 0.5% at high energies.

1. Introduction

The Compact Muon Solenoid (CMS) detector [1] is a general purpose detector installed at the CERN Large Hadron Collider (LHC). It consists of a silicon central tracking device surrounded by the electromagnetic and hadron calorimetry, all immersed in a magnetic field of about 4 T, and by a muon detector placed in the return yoke of the magnet. The Electromagnetic Calorimeter (ECAL) of CMS [2] is a hermetic homogeneous calorimeter made of about 76000 lead tungstate ($PbWO_4$) crystals, equipped with avalanche photo-diodes (APDs) in the "barrel" part and vacuum photo-triodes (VPTs) in the end-caps for the collection of the scintillation light. The barrel calorimeter is divided into 36 super-modules, each containing 1700 crystal arranged in four modules, and covers a pseudo-rapidity (η) region up to $|\eta| < 1.48$. It is closed at each end by an end-cap consisting of two *dees* of 3662 crystal each and extending the pseudo-rapidity coverage up to $|\eta| < 3$. The design of ECAL has been optimized for the search of the Higgs boson via its electromagnetic decay $H \to \gamma\gamma$. To fully exploit its discovery potential, the electromagnetic calorimeter must achieve and maintain the target resolution of 0.5% at high energies. The resolution is affected by the uniformity of the detector response, determined by the accuracy of the channel-to-channel calibration, whose ultimate precision will be reached *in-situ* using events from the LHC collisions. Collision events also allow to determine the energy scale and linearity of the ECAL response to electromagnetic particles. In fact, in the Tracker material in front of ECAL electrons may loose energy through bremsstrahlung and photons may be converted into

[1] On behalf of the CMS Electromagnetic Calorimeter Group

electron pairs. These effects may affect the energy measurement and resolution, thus they also need to be calibrated.
An essential part of the calibration procedure is the time stability of the calorimeter response. The main sources of variation in the ECAL response are due to the dependence of the crystal transparency on the absorbed radiation dose-rate, to the temperature dependence of the scintillation process, and to the temperature and high-voltage dependence of the APD gain. The electronic noise is also playing an important role, as it affects directly the energy resolution, though it becomes negligible at high energies.
The following sections will give an overview of the stability of the main parameters affecting the operation of the electromagnetic calorimeter. More detail can be found e.g. in [3].

2. ECAL readout and monitoring procedures

The signals from the scintillation light of the lead tungstate crystals are pre-amplified and shaped, and then amplified by three amplifiers with nominal gains of 1, 6 and 12. The signals are then digitized into 10 samples by a 40 MHz, 12-bit ADC. An integrated logic provides for each sample the highest non-saturated value among the three available gains. The readout phase of the detector is adjusted so to have the sixth sample on the maximum of the electronic signal, with the first three samples providing a measurement of the baseline. Digital filtering techniques are applied to reconstruct the amplitude and the time of the signal maximum with respect to the readout phase.
In order to monitor the main parameters affecting the ECAL operations, a so called "calibration sequence" run continuously during the data taking using $\mathcal{O}(1\%)$ of the LHC beam abort gaps, a period of 3 μs at the end of each 89 μs of beam cycle. This allows to monitor the crystal transparency via the laser system described in the following sections, the electronic noise via pedestals, and the electronic response stability via a fixed charged injection in the readout chain (Test Pulse). In addition to that, the on-board electronics gives a continuous readout of parameters such as the temperature of the thermistors, high voltage, low voltage, APD dark currents. Finally, dedicated runs can be taken to monitor specific quantities.

3. Noise stability

During the Cosmic Run At Four Tesla (CRAFT), a one month period of extensive cosmic rays data taking that took place in October-November 2008 [3], the electronic noise has been monitored via dedicated pedestal runs. The electronic noise, defined as the r.m.s. of the reconstructed signal amplitude, has been evaluated to be for the barrel (end-caps) 1.06 (1.96) ADC count with less than 0.1% of the barrel (end-caps) channels with a noise greater than 2 (3) ADC count[2]. These values are consistent with those measured during the test beam campaigns (see e.g. [5]) and with the specification of the readout electronics [6].

4. High voltage stability

High voltage is supplied to the barrel APDs and end-cap VPTs via custom power supplies developed in collaboration with CAEN (model SY1527).
The APD gain $\mathcal{G}$ (nominal value of 50) is very sensitive to the bias voltage, showing a variation of $1/\mathcal{G}(\partial\mathcal{G}/\partial V) \approx 3\%/\text{V}$. The nominal operating voltage, which is between 340 and 430 V, has to be kept stable to better than 60 mV in order to provide a negligible contribution to the energy resolution of the calorimeter. During CRAFT all the channels have been proved to be stable to better than 10 mV with an average fluctuation of about 2.1 mV (r.m.s.) and 97% of the channels below 5 mV. It is worth mentioning that the measurements performed during the CRAFT data taking showed APD dark currents below the measurable threshold for all the

[2] The average energy equivalent noise can be evaluated to be roughly 40 (140) MeV for the barrel (end-caps) .

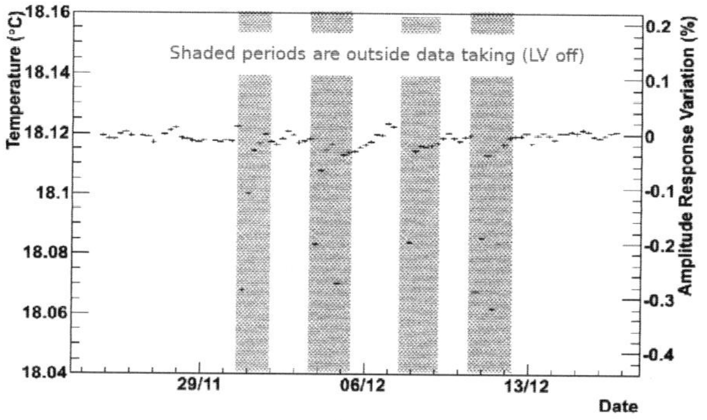

(a)

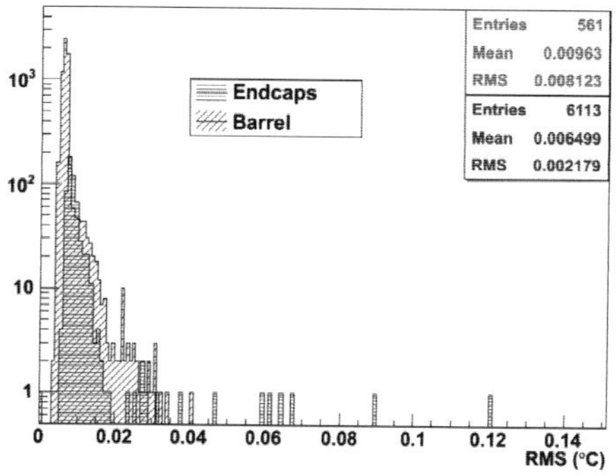

(b)

Figure 1. (a) Average temperature of the ECAL barrel over a period of one month of data taking at the start of the LHC operations in 2009. The shaded regions correspond to periods outside the data taking in which the Low Voltage (LV) was off. (b) Histogram of the corresponding temperature stability measured by each single thermistor for the barrel (blue diagonal lines) and the two end-caps (red horizontal lines).

channels but 11 (0.02%).
At the operating bias used in CMS, the VPT gain is close to saturation. Therefore, the voltages for the end-caps do not need to be controlled very precisely, as the gain dependence on high voltage is less than 0.1%/V [7].

5. Temperature stability

Fluctuations in the temperature of ECAL directly affects the scintillation process (the temperature light yield of the crystals is approximately -2%/°C) and the gain $\mathcal{G}$ of the APDs as $1/\mathcal{G}(\partial\mathcal{G}/\partial T) \approx -2.3\%/°\text{C}$ [8]. In the end-caps, the temperature variation of the VPT relative response is assumed to be negligible with respect to the temperature sensitivity of the crystal light yield [9, 10]. According to these considerations, to provide a negligible contribution to the energy resolution the temperature of the ECAL barrel is required to be stable within 0.05 °C, while a less stringent requirement of 0.1 °C is assumed for the end-caps.
The nominal operating temperature of ECAL is 18 °C. A cooling system utilising water flow [4, 2] is used to regulate the temperature of the barrel and end-cap crystals. As a figure of merit, the power dissipated by the electronics of 400 crystals is of about 5 kW.
The temperature is measured by two independent group of sensors. Precision temperature thermistors manufactured by EPCOS (10 per supermodule, 24 per *dee*) monitor the temperature on each side of the crystal volume and on the incoming and outgoing cooling water. In addition, thermistors are fixed to the back of each 5×2 (5×5) array of crystal in the barrel (end-cap). Figure 4(a) shows the average temperature of the ECAL barrel over one month of data taking at the start of the LHC operations in 2009. The histogram of the corresponding temperature stability measured by each single thermistor for barrel and end-cap is shown in figure 4(b) and for both barrel and end-caps has proven to be much better than the required 0.05 °C.
It is worth noting that eventual local in-homogeneities of the temperature are irrelevant, provided they are stable in time, as they are absorbed into the definition of the channel-to-channel calibration constants.

6. Crystal transparency monitoring

The lead tungstate crystals composing the calorimeter show a response variation dependent on the absorbed radiation dose-rate. This is due to the formation of colour centres which absorb a

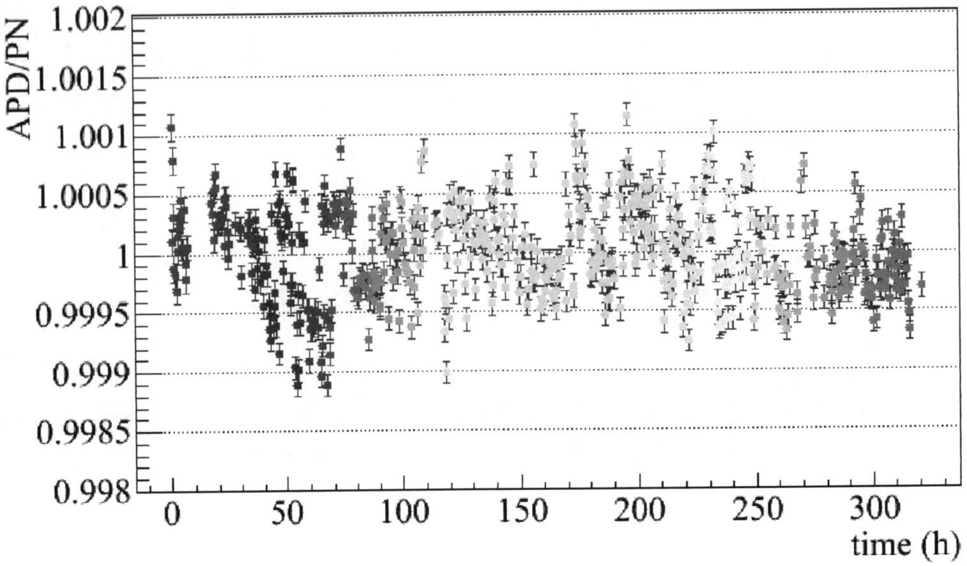

Figure 2. Typical channel stability, defined as the r.m.s of the amplitude measured by the APD normalised to the PN amplitude, after applying all the corrections mentioned in the text. The point colours (if visible) please the eye over the about 350 h of LHC data taking, which shows a stability better than 0.04% (to be compared with the required 0.2%).

fraction of the transmitted light, thus modifying the crystal transparency. At the ECAL working temperature of 18 °C, the colour centres anneal and the crystal progressively recover its initial properties. Given the nominal LHC operations, these changes take place on a time scale of hours and cause a transparency variation of a few percent at the design luminosity.
In order to measure the evolution of crystal transparency, laser pulses are injected into each single crystal via a system of optical fibres. The crystal response is normalised to the laser pulse magnitude, measured using silicon PN photo-diodes. Before being injected in the light distribution system, the laser signals are sampled at 1 GHz by a fast CAEN digitizer (model V1729).

Two lasers are used, both manufactured by Quantronix (Nd:YLF 527DQ-S Q-switched and Ti:Sapphire, custom made), to provide sources of two different wavelengths. A blue one ($\lambda = 440$ nm), very close to the scintillation emission peak, probes the crystal transparency. A red one ($\lambda = 790$ nm), far from the scintillation peak, is very little affected by the changes in transparency and is used to verify the stability of the elements in the system. The lasers are operated at 100 Hz and exploit the 3 μs of the LHC beam abort gaps in order not to interfere with the normal data taking. The transparency of each crystal is measured every 20 to 30 minutes. For more detail as well as for the complete technical specifications of the laser system see e.g. [11].

To reach the ECAL target resolution, the laser monitoring system is required to monitor the transparency changes of a single crystal with a precision of 0.2%. With such a target, several second order effects cannot be neglected. In particular, the non linearities of the silicon PN photodiodes are measured and corrected for via a highly precise charge injection system (PN Test Pulse). In addition to that, the difference in shaping time of APD (VPT) and PN is also taken into account, by using the APD (VPT) Single Pulse Response of each individual channel and PN, convoluted with the fast-digitized laser signal in order to improve the estimate of the

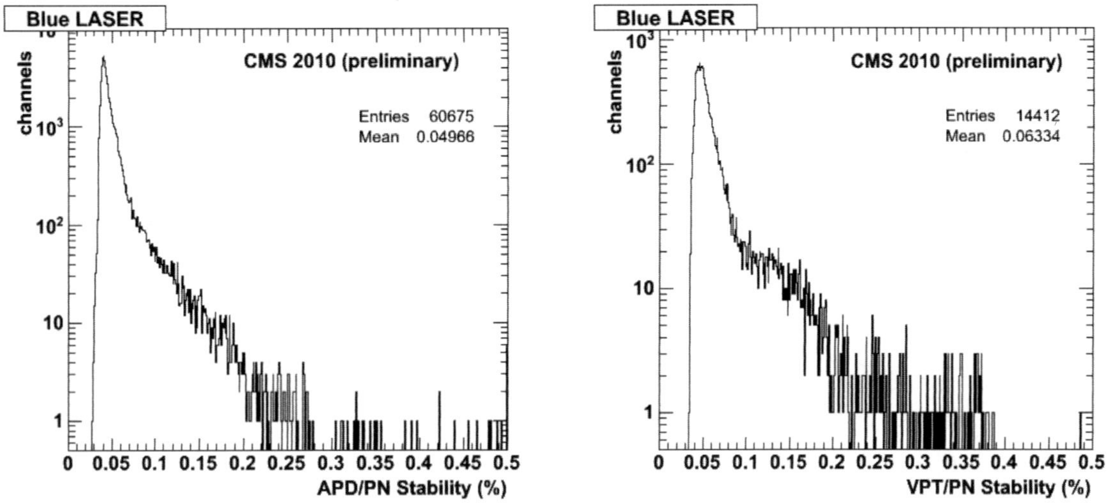

(a) (b)

Figure 3. Stabilities for the whole ECAL barrel (a) and end-caps (b) over about 350 h of LHC data taking. The stability for the barrel (end-caps) is defined as the r.m.s of the amplitude measured by the APD (VPT) normalised to the PN amplitude, after applying all the corrections mentioned in the text. The shoulder visible in (b) that still meets the requirements is due to a region of crystals for which the PN system was temporarily not fully operational.

measured amplitudes.
The capability of this system to allow corrections for transparency changes was proved with test beam data [4]. With no significant irradiation for the ECAL crystals, the 2010 LHC data taking period has given the possibility to demonstrate the stability of the laser monitoring system for the whole ECAL over a much longer period of time and using the nominal procedures in the calibration sequence.
An example of a typical channel is given in figure 6 for a period of 350 h, and shows a stability better than 0.04%. Figure 6 shows the stability of the blue laser for the whole barrel (a) and end-caps (b) over the same 350 h of LHC data taking. The average stabilities of the whole ECAL barrel and end-caps are of about 0.05% and 0.06% respectively, and are much better than the required 0.2%.

7. Summary and conclusions

The operation of the CMS electromagnetic calorimeter is able to meet the stringent requirements imposed by reaching the target resolution of 0.5% at high energies. For the whole data taking period starting from the cosmic rays campaigns till the ongoing LHC collisions, all the detector parameters have been constantly monitored and proven to be extremely stable during the CMS operations. In particular, the temperature stability of the detector is much better than the requirements, and so is the stability of the ECAL laser monitoring system.

References

[1] CMS Collaboration, *The Compact Muon Solenoid Technical Proposal, CERN/LHCC 9438* (1994).
[2] CMS collaboration, *The Electromagnetic Calorimeter Project: Technical Design Report, CERN-LHCC-97-033* (1997).
[3] CMS Collaboration, *Performance of the CMS Electromagnetic Calorimeter*, 2010, *JINST 5 T03010*
[4] CMS collaboration, *The CMS experiment at the CERN LHC*, 2008 *JINST* **3** S08004.

[5] P. Adzic et al., *Reconstruction of the signal amplitude of the CMS electromagnetic calorimeter*, *Eur. Phys. J.* **C46S1** (2006) 23.

[6] M. Raymond et al., *The MGPA Electromagnetic Readout Chip for CMS*, *CERN-LHCC-2003-055* (2003).

[7] K.W. Bell et al., *Vacuum phototriodes for the CMS electromagnetic calorimeter endcap*, *IEEE Trans. Nucl. Sci.* **51** (2004) 2284.

[8] D. Renker, *Properties of avalanche photodiodes for applications in high energy physics, astrophysics and medical imaging*, Nucl. Instrum. Meth. **A 486** (2002) 164.

[9] See, for example, the technical reprint from ET Enterprises Ltd, *The determination of photomultiplier temperature coefcients for gain and spectral sensitivity using the photon counting technique*, http://www.electrontubes.com/pdf/rp081colour.pdf.

[10] See, for example, figure 29 of the brochure from ET Enterprises Ltd, *Understanding Photomultipliers*, http://www.electrontubes.com/Photomultipliers/Understanding.pdf.

[11] M. Anfreville et al., *Laser monitoring system for the CMS lead tungstate crystal calorimeter*, *CMS-NOTE-2007-028*, http://cms.cern.ch/iCMS/jsp/openle.jsp?type=NOTE&year=2007&les=NOTE2007_028.pdf.

Current status and performance of the LHCb electromagnetic and hadron calorimeters

Irina Machikhiliyan for the LHCb calorimeter group
Laboratoire d'Annecy-le-Vieux de Physique des Particules (LAPP), 9 Chemin de Bellevue, F-74941 Annecy-le-Vieux Cedex, France
E-mail: Irina.Machikhiliyan@cern.ch

Abstract.
As a part of the calorimeter system of the LHCb experiment both the Electromagnetic Calorimeter (ECAL) and the Hadron Calorimeter (HCAL) play a key role in the LHCb trigger system providing it with energetic electron, photon and hadron candidates. The article discusses the designs of both calorimeters and reports their present performance as it was obtained from the analysis of early data collected in the spring of 2010.

1. Introduction

LHCb [1] is one of six experiments operating at Large Hadron Collider (LHC) at CERN. Its physics program is focused on b-physics; in particular it covers such topics as the studies of CP-violation and the indirect searches of new physics via rare B-decays. Experimental set-up is the single arm spectrometer which includes the vertex detector, the tracking system supplemented with dipole magnet, two RICH detectors, the calorimeter system and several layers of muon chambers. LHCb trigger system should be able to reduce the events rate from original 40 MHz down to 2 kHz in the stage of data storage. The selection scheme adopts two-level structure. On the first step (Zero Level Trigger or L0) the fast hardware triggering is performed on events containing secondary particles of high transverse momenta which is the indication of b-events. On the second step (High Level Trigger) the selection is refined further by dedicated software which analyzes partially reconstructed data.

One of major purposes of the LHCb calorimeter system [2] is to provide fast (within 4 μs) L0 trigger on events containing energetic e^+/e^-, photon, π^0 or hadron. The main detectors of this system are the Electromagnetic Calorimeter and the Hadron Calorimeter supplemented by the preceding Scintillator Pad Detector (SPD) and Preshower Detector (PS). SPD and PS are two single layers of scintillator tiles separated by lead sheet 2.5 X_0 in depth. They are vital for the discriminating between photons, electrons and hadrons in the stage of L0 decision; detailed discussion of their design can be found elsewhere [2]. Apart from being involved in the trigger ECAL is also used for the precise reconstruction of the kinematical parameters of neutral particles and for the identification of electrons and positrons. HCAL on the contrary does not play any significant role in offline analysis since is not able to ensure the full containment of energetic hadronic showers on account of the insufficient depth. Therefore it is only employed in L0 where it is expected to provide up to 70% of triggers. Both ECAL and HCAL are the subjects of the present article which discusses their design, describes the procedures used for

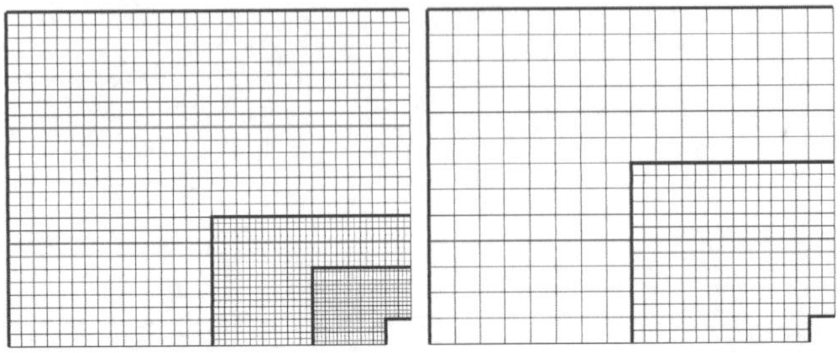

(a) (b)

Figure 1. Detector layout (1/4): (a) ECAL (b) HCAL

time-alignment, calibration and monitoring as well as presents current performance of both calorimeters.

2. Design overview

In many aspects designs of ECAL and HCAL are similar. The detector bodies are rectangular walls with solid angle coverage 300×250 mrad excluding central 30 mrad cut-out for the LHC beam pipe. The calorimeter media is the interleaving structure of absorber plates and scintillator tiles readout by WLS fibers. LHCb ECAL is the calorimeter of shashlik type: layers of lead and scintillator, orientated perpendicularly to the beam direction, are pierced by optical fibers. HCAL design is similar to the one of ATLAS TileCal [3]: plates of iron and scintillator are arranged parallel to the beam axis and the light is collected from the top and the bottom tile edges. WLS fibers from square areas are reunited in bundles forming readout cells; cell size increases from the center of detectors to their periphery (see figure 1). In general ECAL comprises three sections - Inner, Middle and Outer - of different transverse segmentation. In case of HCAL the subdivision into two sections - Inner and Outer - is enough.

The light collected over each cell is readout by photomultiplier (PM). The typical duration of the PM anode pulse is compatible with LHC bunch spacing 25 ns and the trailing edge of the pulse extends partially outside the 25 ns time slot. This small tail is canceled in the first stage of the signal processing by means of the dedicated clipping circuit embedded in the PM power supply. The shaped pulse is transmitted then to the input connector of Front-End Board (FEB) for further treatment which includes integration, digitization by means of analog-to-

Table 1. Selected parameters of ECAL and HCAL. $\sin(\theta)=\sqrt{(x^2+y^2)/(x^2+y^2+z^2)}$, where (x,y,z) are coordinates of the center of the detector cell; $N_{p.e.}$ is the number of photoelectrons.

	ECAL			HCAL	
	Inner	Middle	Outer	Inner	Outer
Cell size, cm	4.04	6.06	12.12	13.1	26.2
# of cells	1536	1792	2688	880	608
Lateral size, m×m	1.9×1.5	3.9×2.4	7.8×6.3	4.2×3.7	8.4×6.8
Distance to the interaction point, m		12.5		13.3	
Depth		25 X_0, 1.1 λ_I		5.6 λ_I	
Upper limit of dynamic range, GeV		7+10/sin(θ)		15/sin(θ)	
Energy resolution, % (E is in GeV)		$(8\div10)/\sqrt{E}\oplus0.9$		$69/\sqrt{E}\oplus9$	
Average light yield, $N_{p.e.}$/GeV	3077	3516	2569	105	

digital converter (ADC) and the following subtraction of the ADC pedestal. Each FEB is also equipped with delay chips for precise (1 ns accuracy) adjustment of the relative timing of the ADC clock and the integrated PM signal in order to compensate for different PM signal arrival times (see next section). One FEB is able to process data from 32 calorimeter cells thus hosting 32 readout channels. ADC readings are stored in the pipeline deep enough to keep the data for the latency of L0 decision. In parallel they are used by the trigger part of the FEB which locates trivial cluster candidate with highest transverse energy in the detector region served by given board. A list of such clusters serves as an input for the L0 trigger level.

To monitor the functionality of the readout chain described above and the stability of its characteristics both ECAL and HCAL are equipped with monitoring systems on the basis of light emitting diodes (LEDs). Each photomultiplier is illuminated by light flashes of fixed intensity, so the value of average PM response can be used to follow up the behaviour of each readout channel (mostly defined by the stability of the PM gain). In order to eliminate the effects, coming from possible variations of the intensity of LED flash, LEDs themselves are monitored by means of very stable silicon PIN photodiodes.

Selected parameters of both calorimeters are presented in table 1.

3. Time-alignment

For both ECAL and HCAL the time of the arrival of PM signal at the FEB input depends on the cell position. The main contribution comes from two factors: the typical particle time-of-flight increasing from the center of detector to its periphery and the different signal collection time inside photomultiplier decreasing with the distance from the beam axis. In total the spread of timing over all detector cells is on the level of $4\div5$ ns. In order to ensure the fine mutual synchronization of readout channels of each detector and also the correct time alignment of different detectors with respect to the accelerator cycle, LHCb data acquisition system offers special Time-Alignment Events which can contain up to ±3 consecutive time slots of 25 ns around the one under interest. If the timing is optimal the signal fraction registered in the time slots adjacent to the central one is minimal. In the particular case of ECAL and HCAL the asymmetry-like variable $R = (E_{cent} - E_{next})/(E_{cent} + E_{next})$ is used to estimate the size of the misalignment in time δt in each readout channel, where E_{cent} and E_{next} are average energy depositions in the central time slot and the next one after it respectively. It was shown that $R \sim \delta t$ if (i) the timing is already not very far from the optimal one and (ii) the measurement of R is done on the condition that an artificial shift of timing of about 12.5 ns is introduced in the readout system. It allows calculating δt directly from one measurement avoiding multi-step time scan in 25 ns range. This method was used for the fine adjustment of the timing for both ECAL and HCAL at the end of 2009 when LHC delivered first collisions. After the procedure was completed a small time scan in few ns range was performed in order to verify the optimal sampling of the PM integrated signals. The same exercise was also repeated in the spring of 2010 as soon as LHC resumed its operation after long technical stop. It was shown that for the vast majority of ECAL and HCAL cells the timing was set with the precision ±1 ns.

4. Stability of detector response

The monitoring of ECAL and HCAL responses is performed online in parallel with data taking. LEDs flash with the frequency of about 50 Hz synchronously with one of accelerator empty buckets which are not filled with protons. The stability of ECAL and HCAL was studied using data collected during the period from 6 April to 19 July 2010. For each data sample the deviation of response on LED in each cell i from its value averaged over all data sets was calculated: $\Delta_i = (A_i - \overline{A_i})/\overline{A_i}$. Figure 2 represents the value of Δ_i averaged over all operational detector cells (>99% of all cells) as function of time. It is seen that the average stability of the response of both ECAL and HCAL is well within 2 %.

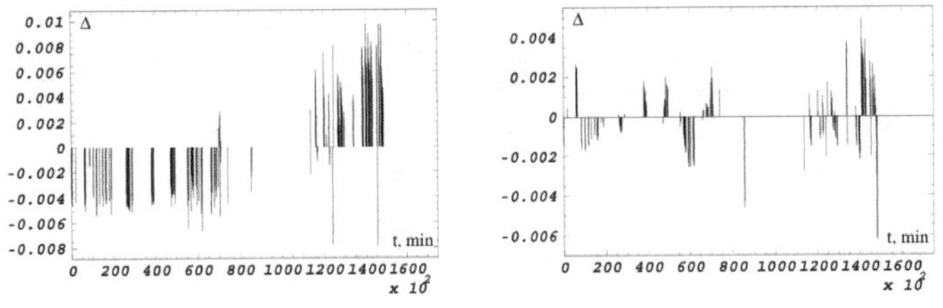

(a) (b)

Figure 2. Average relative variation of detector response in time: (a) ECAL (b) HCAL. Time scale corresponds the period from 6 Apr to 19 Jul 2010

5. Calibration of ECAL

The calibration of ECAL was performed in three consecutive steps. During detector commissioning gains of photomultipliers were pre-calibrated with the help of LEDs. With the first LHC collisions the calibration of ECAL was improved using energy flow method. Finally the fine calibration of the detector was done using $\pi^0 \to \gamma\gamma$ decays.

Pre-calibration of PM gains. Major factors defining the accuracy of the initial adjustment of the ECAL energy scale were:

(i) the cell-to-cell variation of light yield value Y. Average values of Y for each ECAL section were measured with the help of cosmic particles and on the test beam (see table 1). These studies showed also that the r.m.s. of the distribution of Y-values is on the level of 8%;

(ii) the dispersion in the sensitivity s_{ADC} of readout ADCs. Permissible variation for s_{ADC} value was set to $\pm 5\%$ limit; only ADCs fulfilling this requirement were allowed to be used in FEBs;

(iii) the dispersion in the quantum efficiency of PM photocathodes. Related values of green sensitivity index q_{gr} were provided by the manufacturer for each PM individually;

(iv) the precision of the adjustment of PM gains.

To minimize the influence of the last factor all photomultipliers were pre-calibrated in situ with the help of the monitoring system. The PM gain G was calculated as $G = K \times (\sigma^2_{LED} - \sigma^2_{ped})/(A_{LED} - A_{ped})$, where A_{LED}/A_{ped} - positions of Gaussian peaks associated with PM response on LED / ADC pedestal, $\sigma_{LED}/\sigma_{ped}$ - widths of these peaks and K - known parameter defined by hardware properties. The formula is valid on the condition that the main contribution to the width of LED peak comes from photo statistics, i.e. $\sigma_{LED} \sim \sqrt{N_{p.e.}}$, where $N_{p.e.}$ is the number of photoelectrons emitted from PM photocathode. To fulfill this requirement the LED spectra for the measurement of G were collected in the reference point at LED flash intensity as low as possible. In order to fit small PM signals into ADC dynamic range the high voltages (HV) supplied to PMs were significantly increased. To cover the nominal range of ECAL gains the relative dependence $G(HV)$ with respect to the reference point was also measured for each PM following the relative change of PM response with HV at fixed LED flash intensity. The outcome of the procedure described above was the set of regulation curves $G(HV)$, one per each detector channel. The statistical error of the measurement was 3÷4% and the accuracy of the method itself was found to be within 8%.

Finally individual regulation curves, individual q_{gr} values as well as average values of Y and s_{ADC} were used to produce initial HV settings for all ECAL PMs. The overall average accuracy of initial channel-to-channel inter-calibration was estimated to be within 13%. Such precision was sufficient to observe the peak corresponding to decay $\pi^0 \to \gamma\gamma$ immediately after first collisions were delivered by LHC.

Calibration with energy flow. Energy flow method allows improving cell-to-cell inter-calibration up to 4% level on the basis of relatively small statistics (few millions of minimum

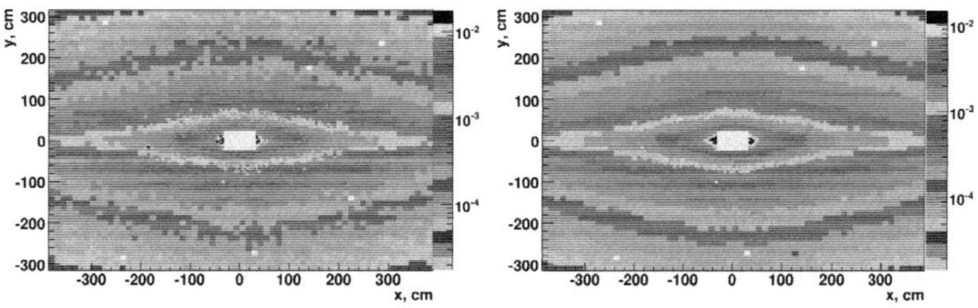

(a) (b)

Figure 3. ECAL cluster occupancy: (a) before (b) after energy flow calibration

bias events). Its basic idea is the smoothing of the map of transverse energy depositions over detector surface. For each cell correction factor is produced with respect to the mean deposit over eight neighboring cells. The method does not require any information from other sub-detectors and uses only the raw energy deposits in detector cells, i.e. it does not depend on Monte Carlo (MC) based parameters employed in the reconstruction software. All these advantages make the method very profitable in the early stage of detector operation. However it does not provide absolute calibration so for the normalization the positions of the net peaks associated with decays $\pi^0 \to \gamma\gamma$ and $\eta \to \gamma\gamma$ were used. Figure 3(a) and figure 3(b) show the maps of ECAL cluster occupancies before and after energy flow calibration respectively. The resulting smoothing is clearly seen. Corresponding improvement of the net π^0 peak relative width was from 9.1% to 7.5% while η peak width decreased from 6.1% to 5.4%.

Calibration with neutral pions. The fine calibration of ECAL with π^0 decays is now in progress. This is an iterative procedure which uses a set of the invariant mass distributions for combinations of two photon candidates. Such distributions are constructed for each detector cell in order to detect the π^0 peak. The shift of its position with respect to the 135 MeV/c^2 is used to correct the adjustment of energy scale of given cell. The required amount of statistics is rather high: for the most peripheral cells it numbers up to 250 millions of interactions.

The specific of LHCb ECAL is that it can not be calibrated by π^0 method in standalone mode since the measured parameters of particles are constructs based on the energy depositions and

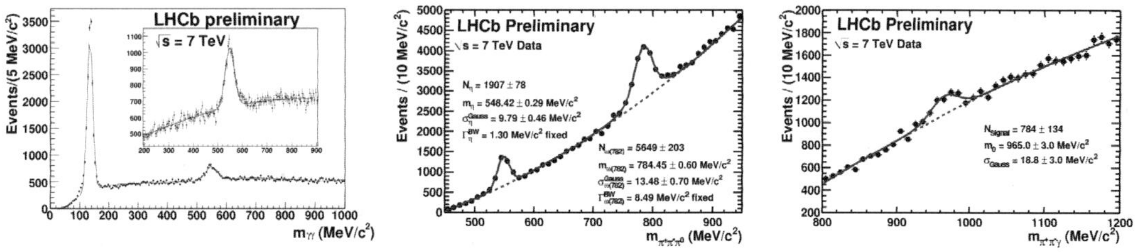

(a) (b) (c)

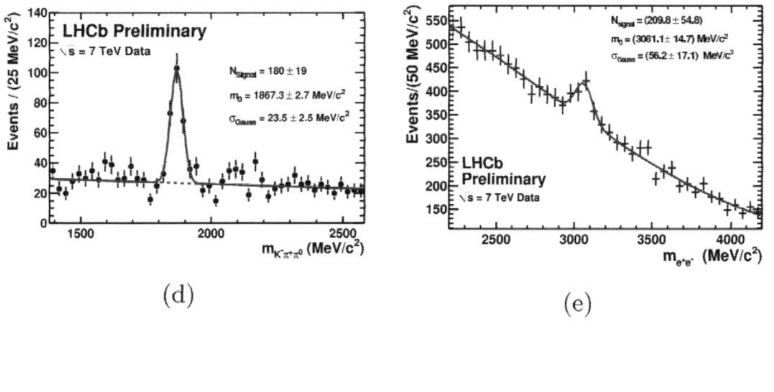

(d) (e)

Figure 4. Particle signals reconstructed with the participation of the ECAL: (a) $\pi^0/\eta \to \gamma\gamma$ (b) $\eta/\omega \to \pi^+\pi^-\pi^0$ (c) $\eta' \to \rho^0(\pi^+\pi^-)\gamma$ (d) $D^0 \to K^-\pi^+\pi^0$ (e) $J/\Psi \to e^+e^-$

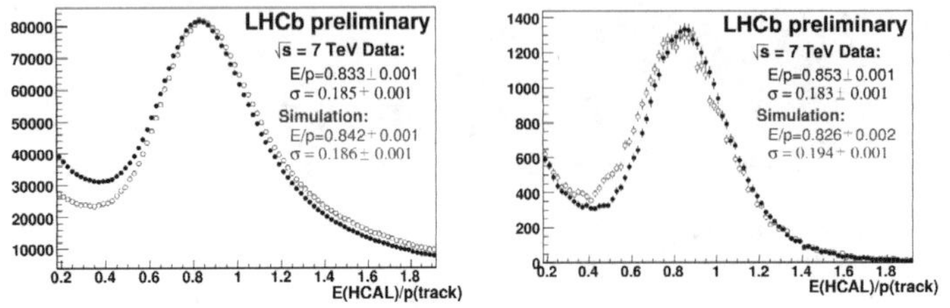

(a) (b)

Figure 5. HCAL, distribution of E/p: (a) Inner (b) Outer section. ● represents data, ○ shows MC simulations

the coordinates of impact points both in PS and ECAL detectors. Also in case of photons both SPD and tracking system are used to suppress conversions in front of the calorimeter system. PS is now calibrated with the precision better than 5% [4]. Also various weighting and correction factors employed in the reconstruction software have been verified and tuned on the first data. With regard to all these improvements the first cycles of π^0 calibration were performed recently. The accuracy of ECAL calibration which is presently achieved is on the level of 2÷2.5%.

As the demonstration of present ECAL performance figure 4 shows signals reconstructed with the participation of ECAL for the following particles: π^0, η, ω and η' in various decay modes as well as D^0 and J/Ψ. For $\pi^0 \to \gamma\gamma$ and $\eta \to \gamma\gamma$ decays where the resolution is defined mostly by ECAL performance the relative widths of the peaks are 7.1% and 3.4% respectively.

6. Calibration of HCAL

The HCAL calibration is performed on the regular basis with the help of embedded cesium radioactive source calibration system originally developed for ATLAS TileCal. Two ~10 mCi ^{137}Cs sources (one per each detector half) are driven by hydraulic system so they traverse each scintillator tile. The PM response is measured with the help of dedicated system of current integrators. Relation factors between the value of PM anode current and the measured particle energy are known from test beam studies, which showed also that this method allows achieving cell-to-cell inter-calibration level 2÷3%. Such precision is more than enough to fulfill all HCAL objectives in the LHCb experiment. The source calibration was also cross-checked with energy flow method which proved that HCAL is currently inter-calibrated with the accuracy better than 4%. As concerns the adjustment of absolute energy scale, its precision is defined mostly by the uncertainty in the values of source activities which are known with the accuracy of about 10%. In order to verify the global normalization the basic physics signals are used such as the position of hadron peak on the distribution of ratios of energy E deposited in HCAL to the particle momentum p measured by the tracking system. Examples of E/p distributions for HCAL Inner and Outer section are shown in figure 5(a) and figure 5(b) respectively (full circles). Open circles represent E/p distributions for MC events. The difference between simulations and data does not exceed 3%.

Acknowledgments

Author is mostly grateful to M. N. Minard and P. Perret for their comments and useful discussion. Author would also like to thank CALOR2010 organizers for their hospitality.

References

[1] Alves A *et al* 2008 *Journal of Instrumentation* **3** S08005
[2] LHCb Collaboration 2000 LHCb technical design report: calorimeters *Preprint* CERN/LHCC/2000-0036
[3] Aad G *et al* 2008 *Journal of Instrumentation* **3** S08003
[4] Niess V These proceedings

Commissioning of the ATLAS Tile Hadronic Calorimeter with cosmic muons, single beams and first collisions

Valerio Rossetti
on behalf of the ATLAS Tile Calorimeter Group, ATLAS Collaboration

Institut de Fisica d' Altes Energies (IFAE).
Edifici Cn, Universitat Autonoma de Barcelona (UAB), E-08193 Bellaterra (Barcelona), Spain.

E-mail: valerio.rossetti@cern.ch

Abstract. Until December 2009 the ATLAS detector has undergone a long period of commissioning with cosmic rays. The LHC collider delivered to the ATLAS detector brief periods of single beams and from December 2009 months of proton-proton collisions at different center of mass energy (900 GeV, 2.36 TeV and 7 TeV). The focus of these running periods was to commission the ATLAS detector for the 2010 collision run during which physics results were expected. In this paper, we will show how the data from cosmic muons, from single beam events and early collisions has been used to commission and to calibrate the ATLAS Tile Hadronic Calorimeter.

1. The Tile Hadron Calorimeter

The ATLAS Tile Calorimeter (TileCal) [1] is the barrel hadronic calorimeter of the ATLAS experiment [2] at the CERN Large Hadron Collider [3]. A cut-away drawing of the calorimeters and the inner detector is shown in Fig.1. The main goal of TileCal is the precise measurements of hadrons, jets, taus and the missing transverse energy (E_T^{miss}).

TileCal is a sampling calorimeter made of iron plates as absorber and plastic scintillator as active material. It is divided into three cylindrical sections, whose axes coincide with the colliding beam axis (z axis). The cylindrical sections are referred to as the long barrel (LB) and extended barrels (EB).

Each of the three sections is composed of 64 azimuthal segments, referred to as modules, subtending $\Delta\phi = 2\pi/64 \simeq 0.1$. Radially, TileCal extends from an inner radius of 2.28 m to an outer radius of 4.25 m. In the direction perpendicular to the beam axis the calorimeter is composed of 3 radial layers. The typical cell dimensions are $\Delta\phi_{cell} \times \Delta\eta_{cell} = 0.1 \times 0.1$ (0.1×0.2 in the last radial layer). This segmentation defines a quasi-projective tower structure covering the region $|\eta| < 1.7$. The deviations from perfect projectivity are small compared to the typical angular extent of hadronic jets.

A schematic drawing of a module is given in Fig.2. The TileCal plates made of iron and scintillating material, are placed perpendicular to the colliding beam axis and are radially staggered in depth.

Journal of Physics: Conference Series **293** (2011) 012053 doi:10.1088/1742-6596/293/1/012053

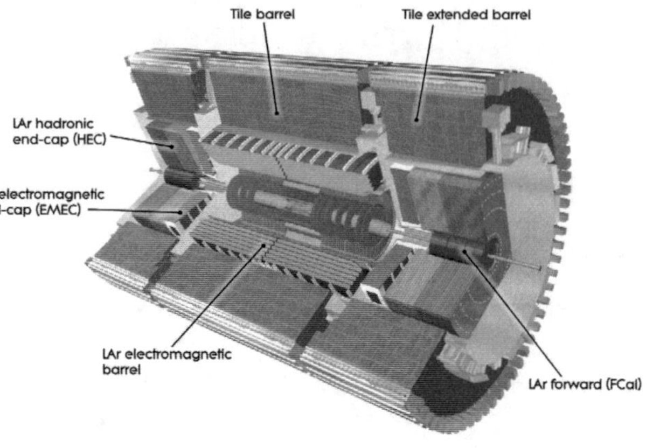

Figure 1. A cut-away drawing of the ATLAS inner detector and calorimeter systems. The TileCal consists of one barrel and two extended barrel cylindrical sections.

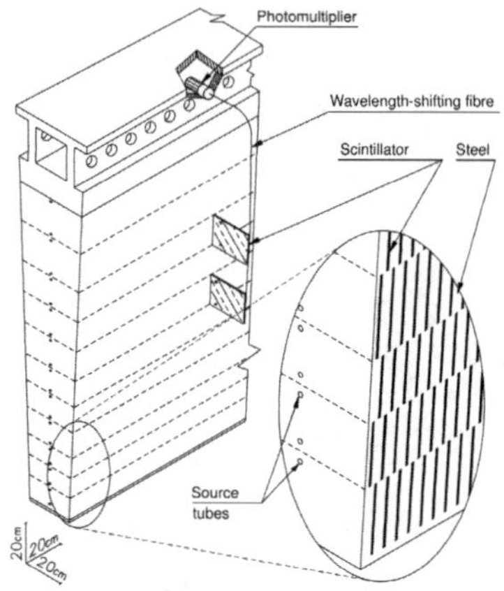

Figure 2. Schematic drawing of the mechanical assembly and optical readout of the TileCal, corresponding to a ϕ wedge.

2. Calorimeter response to cosmic muons

Since the interaction of muons with matter is well understood and well modeled, cosmic rays measurements were used to provide information about the response of the detector.

The cosmic ray events used in this analysis have been collected with the ATLAS experiment between September and October 2008. The selection of the events relies on the RPC and TGC systems of the Muon Spectrometer [4]. The analysis is based on a total of about 1M triggered events. The tracks are reconstructed using the Inner Detector system [5] information and are extrapolated through the volume of the calorimeters taking into account of material and magnetic field. The track path length $\mathrm{d}x$ is then evaluated as the distance between the entrance and the exit points for every cell crossed by the muon.

The events are required to have a reconstructed track in the Inner Detector with at least 8 hits in the Pixel and SCT systems. The tracking requirement introduce some cut-off in the distributions of transverse and longitudinal impact parameters. These are $|d_0| < 380$ mm and $|z_0| < 800$ mm respectively. The events are further required to have the track momentum in the window 10 GeV $< p <$ 30 GeV. The lower limit was applied to minimize effects of multiple scattering. The upper cut restricts the muon radiative energy losses which otherwise cause considerable fluctuations in the deposited energy. A small fraction of events has the cell response compatible with the pedestal level although the cells should be hit by the muon. This happens because the particle actually crosses a neighboring cell and it is consistent with the expected deviation from a linear muon trajectory due to multiple scattering. In order to limit this effect, only tracks that are well within a module were selectected by applying the cuts $|\Delta\phi_{in}| = |\phi_{cell} - \phi_{in}| < 0.045$ and $|\Delta\phi_{out}| = |\phi_{cell} - \phi_{out}| < 0.045$, where $\phi_{in}(\phi_{out})$ is the azimuthal angle of the entrance (exit) point of the muon trajectory in the cell. In order to remove residual noise contribution, a low cut $\mathrm{d}E_{cut} = 60\mathrm{MeV}$ was applied on the measured cell energy $\mathrm{d}E$. Deposits of energy in a short path have a larger variation of the sampling fraction. In order to limit this effect only tracks with path length $\mathrm{d}x > 200$ mm were considered. The energy deposited by a muon track with a trajectory close to the vertical direction is badly measured in TileCal due to the strong sampling fraction variation caused by the vertical orientation of the scintillating tiles. To ensure more stable results, tracks are required to have a minimum cell path z component. The cut $z = |z_{in} - z_{out}| > 6\mathrm{cm}$ was applied, where z_{in} and z_{out} are the crossing points at the z cell surfaces.

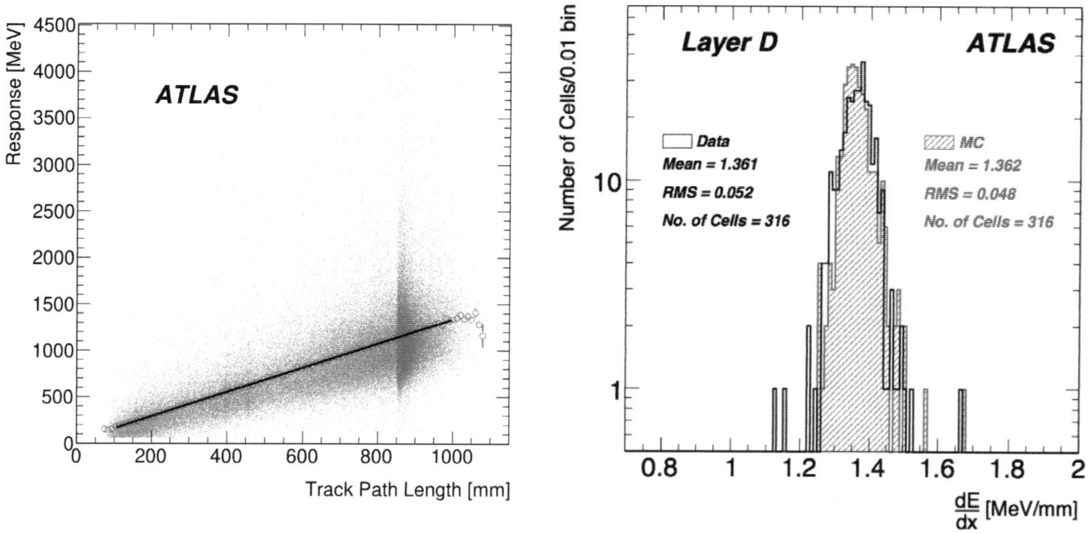

Figure 3. On the left the response of the barrel module BC cells as a function of track path length is shown. A linear fit to the corresponding distribution of mean values is superimposed. The excess of events around the track path length of 840 mm (radial size of the BC cells of the barrel module) is a purely statistical effect, since most of the cosmic ray muons enter the calorimeter at small zenith angle. On the right the distribution of the truncated mean $\mathrm{d}E/\mathrm{d}x$ per cell, shown for the radial layer D for data and MC.

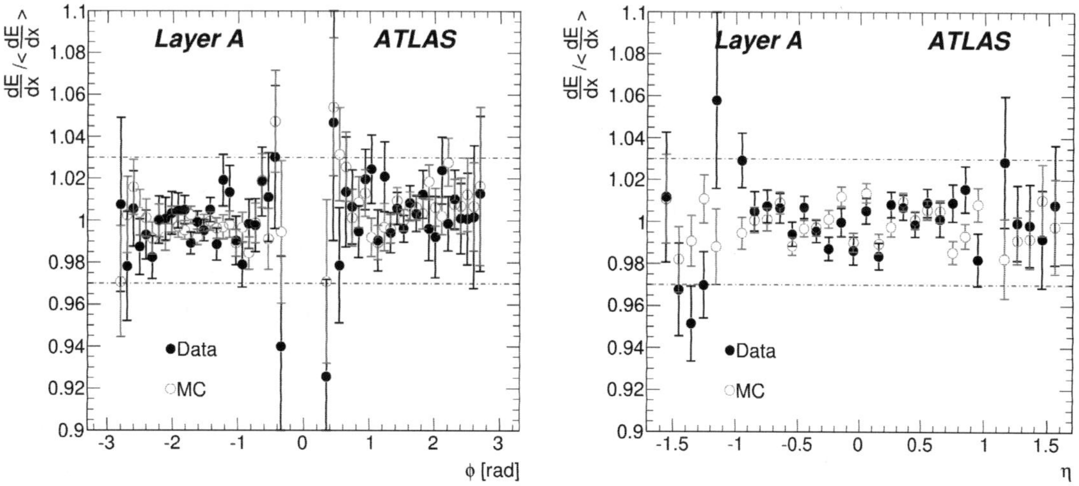

Figure 4. Uniformity of the A layer cell response to cosmic muons, expressed in terms of normalised truncated mean of $\mathrm{d}E/\mathrm{d}x$, as a function of azimuthal angle ϕ (left) and the pseudorapidity η (right).

At energies in the range between 10 GeV and 30 GeV the dominant energy loss process is ionization and the energy lost is essentially proportional to the track path length (Fig. 3-left). The response of the detector was then studied determining the ratio between the energy deposited in a calorimeter cell ($\mathrm{d}E$) and the length of the path of the track in the cell ($\mathrm{d}x$). Our estimator for the muon response is the truncated mean of $\mathrm{d}E/\mathrm{d}x$, defined as the mean in which 1% of the events in the high-energy tails of the distribution are removed. This quantity is less affected by the higher energy loss processess (bremsstralung, δ-rays...) that can cause fluctuations on the mean $\mathrm{d}E/\mathrm{d}x$.

The uniformity of the cell response in the layer D is shown in Fig. 3-right. The cells considered are required to have at least 100 selected muon deposits. For the D layer 316 cells fullfill this requirement and represent 38% of the total number of cells in this layer. The distribution is compared with Monte Carlo (MC) simulated cosmic events. Since the MC shows an RMS compatible with that of data, it indicates that cells are well intercalibrated within the layer. Fig. 4 shows the cell response, expressed in terms of truncated mean of $\mathrm{d}E/\mathrm{d}x$, as a function of ϕ and η for the layer A cells. The $\mathrm{d}E/\mathrm{d}x$ values are normalized on the average $\mathrm{d}E/\mathrm{d}x$ of all the cells of the layer. Data and MC show similar pattern and the dispersion of the data points is below 3% for the explored range.

3. Timing calibration with "splash" events

To allow for optimal reconstruction of the energy deposited in the calorimeter the timing of the readout channels must be adjusted with a precision of $\sim$ 1 ns. These time offsets can be measured using the laser calibration system. The timing precision for channels in the same module achieved with laser is 0.6 ns for 99% of the Tile Calorimeter readout channels.

The inter-partition timing and global timing with respect to the rest of ATLAS were coarsely set using cosmic-ray data and more accurately using the so called "splash" events. In these events the LHC proton beam hits a completely closed collimator placed on the beam line 140 m away from ATLAS interaction point. The effect is that $O(10^5)$ particles (mainly muons) arrive simultaneously in the ATLAS detector, depositing in TileCal a total energy of $\sim 10^3$TeV.

Fig. 5 (a) shows the cell time measured in beam splash events, averaged over the full range of the azimuthal angle ϕ for all cells with the same z-coordinate of ATLAS (along the beam axis). The visible discontinuities at $z = 0$ and $z = \pm 3000$ mm for the 2008 data are due to the uncorrected time differences between the four TileCal partitions. These were calculated using the 2008 data and adjusted for the 2009 running period. After the muon time-of-flight corrections (b), the timing shows an almost flat distribution within 2 ns in each partition, confirming a good intercalibration between modules with the laser system. In 2009, the TOF-corrected timing distribution (c) shows an timing intercalibration between the partitions of $\pm$ 5 ns. In preparation for the 2010 run, the 2009 single beam results were used to provide the offsets for all cells and, as is shown in Fig. 5 (d) for the 2010 single beam results, all remaining non-uniformities were corrected for. The spread of the TileCal cell timing distribution at the start of the 7 TeV collisions is of order 0.5 ns [1].

4. Calorimeter performance in collision events

In November 2009 the Large Hadron Collision entered a tuning phase to reach the optimal operating conditions. From November 2009 up to now the center of mass energy of the proton-proton collisions was increased from 900 GeV, corresponding to protons colliding at the injection energy, to 2.36 TeV and finally up to 7 TeV. In Fig.6 the cell response spectrum is shown for collision events at different $\sqrt{s}$. The same distribution for MC simulated events at $\sqrt{s}$ = 7 TeV is shown for comparison. The events were triggered using the MBTS system (Minimum Bias Trigger Scintillators) [1] both in data and MC. In order to compare the signal spectrum with the

[1] This value takes into account 97% of the TileCal channels.

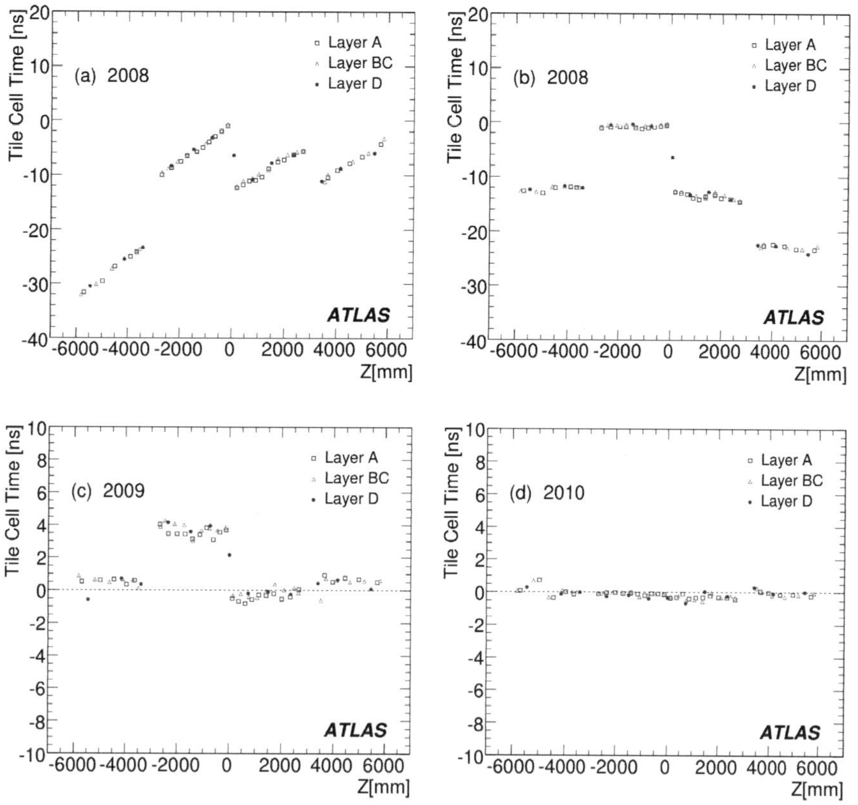

Figure 5. Timing of TileCal signals recorded with single beam data in September 2008 (a and b), November 2009 (c) and February 2010 (d). The time is averaged over the full range of the azimuthal angle ϕ for all cells with the same z-coordinate (along beam axis), shown separately for the three radial layers. Corrections for the muon time-of-flight along the z axis are applied in the b), c) and d) figures, but not on the top left (a).

noise the distribution resulting from random triggered events are also shown. The results show that the MC provides a good description of the response spectrum both for noise and signal regions. The response as a function of η is shown in Fig.7. In order to suppress the noise only the cells with a signal $>$ 500 MeV are used. The agreement is within 2% in the central region, while is worse at large eta ($\eta > 1.2$) where also the statistics is poorer.

References

[1] ATLAS Tile Calorimeter Collaboration, *Tile Calorimeter Technical Design Report*, CERN/LHCC **96-42** (1996)

[2] ATLAS Collaboration, G. Aad et al., *The ATLAS experiment at the CERN Large Hadron Collider*, **JINST 3** (2008) S08003

[3] L. Evans et al., *LHC Machine*, **JINST 3** (2008) S08001

[4] ATLAS Collaboration, *Commissioning of the ATLAS Muon Spectrometer with Cosmic Rays*, **http://arxiv.org/abs/1006.4384** (2010)

[5] ATLAS Collaboration, *The ATLAS Inner Detector commissioning and calibration*, **http://arxiv.org/abs/1004.5293** (2010)

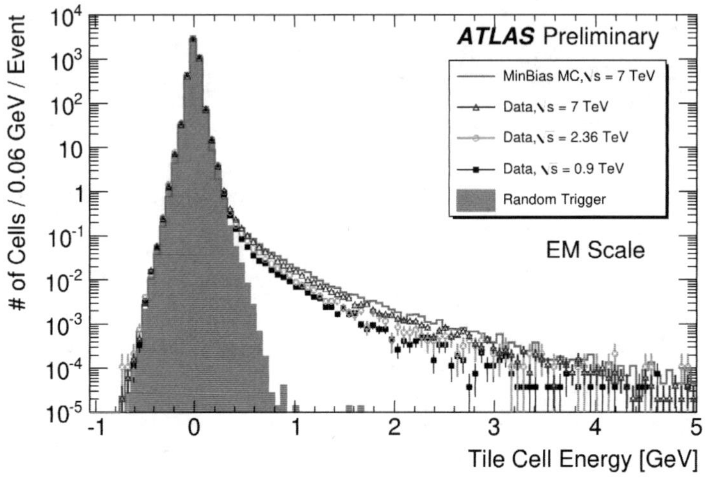

Figure 6. Energy response of the TileCal cells. The distributions from collision data at 7 TeV, 2.36 TeV, and 0.9 TeV are superimposed with non-diffractive minimum bias Pythia MC and randomly triggered events.

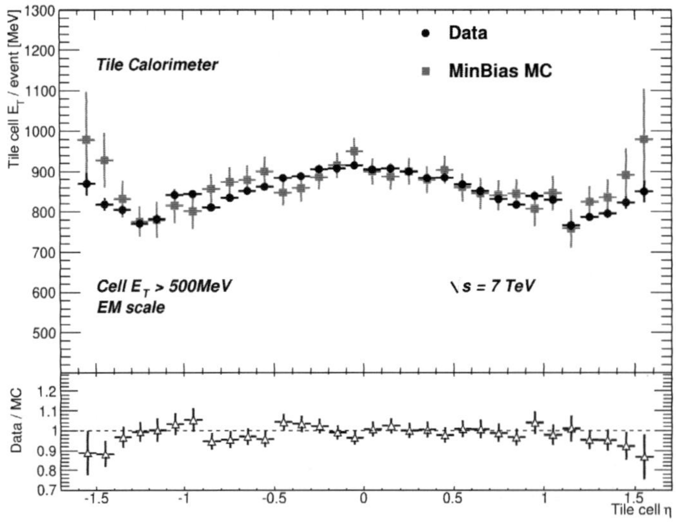

Figure 7. Transverse Energy (E_T) of TileCal cell as a function of eta in collision candidate events at $\sqrt{s} = 7$ TeV. Only cells with transverse energy above 500 MeV are considered. Non-diffractive minimum bias Pythia MC events with the same energy cut are superimposed with the collision candidate events. The cut is chosen at 500 MeV in order to have negligible electronic noise contribution.

Journal of Physics: Conference Series **293** (2011) 012054 doi:10.1088/1742-6596/293/1/012054

Operational experience with the CMS hadronic calorimeter system

Taylan Yetkin
(On behalf of the CMS Collaboration)
University of Iowa, Iowa City, IA, USA
E-mail: taylan-yetkin@uiowa.edu

Abstract. The hadronic calorimeter (HCAL) of CMS was commissioned before and during the initial proton collisions in Large Hadron Collider. Various phases of HCAL commissioning were used to gain operational experience and prepare the detector for physics. In this note we briefly summarize the activities and outcomes from the the commissioning studies.

1. Introduction

The CMS detector at CERN Larger Hadron Collider is built for Higgs particle and new physics searches by using various objects such as muons, electrons, photons, taus, jets, and missing transverse energy ($\not{E}_T$) [1, 2]. Hadronic calorimeter (HCAL), together with the electromagnetic calorimeter, is responsible from jet and $\not{E}_T$ measurements, and it will play very crucial role for new physics discovery program in CMS. HCAL is composed of various parts: The barrel (HB) and endcap (HE) calorimeters are outside the tracker and electromagnetic calorimeters, all of which are inside the magnet solenoid. They are made of brass absorber with scintillation tiles and wavelength shifting fibers. The outer hadron calorimeter (HO) situated behind the solenoid and have similar scintillation tile and wavelength shifter fibers. The forward hadron calorimeters (HF) are positioned 11.2 m away from the interaction point in both sides and they use quartz fiber technology with steel absorber. More details about the HCAL can be found in Ref.[1].

With the data from first 7 TeV collisions, the experiments will re-confirm Standard Model to demonstrate that the detectors are well understood. In parallel to that, tools for background estimations based on data driven methods will be refined and developed further. Before reaching its readiness, CMS detector has undergone various commissioning phases. In the last couple of years CMS experiment collected data from millions of cosmic ray muons. With this period we learnt how to operate HCAL as a part of the CMS experiment, and made different measurements which varies from timing adjustments to calibration of sub-detectors. Also, first circulating beams and beam shots on collimators in 2008 gave us different opportunities to prepare HCAL for physics. After the repair period in LHC, which was due to the electrical failure in LHC magnets, the protons were collided with 900 GeV and 2.36 TeV center of mass energies. On March 2010, a new world record was announced with 7 TeV collisions in LHC experiments, and at the end of April 2010 CMS collaboration announced that more than 1 nb^{-1} of data were collected.

In all these milestones, HCAL collaboration made sure that the sub-detectors were operated in their highest performance, and they are ready for physics. In this report I will focus on what

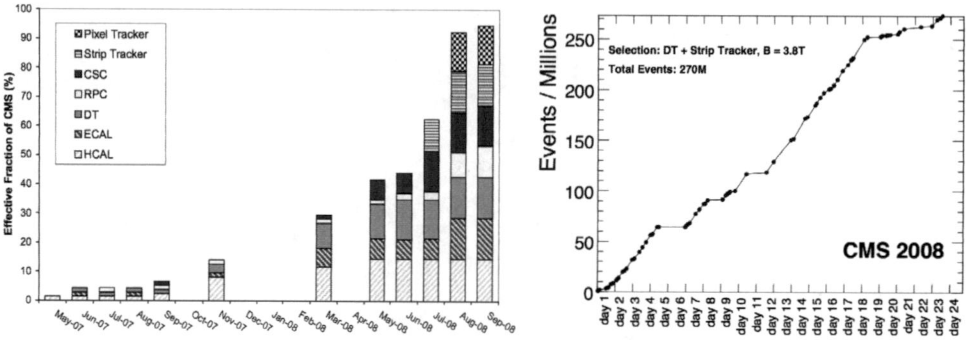

Figure 1: The fraction of participation of seven sub-detectors in global runs during commissioning versus time. The unit of fraction is $\frac{1}{7}$.100% (left). The number of collected events triggered by cosmic ray muons versus days in CMS global runs (right)[3].

we learnt from cosmic muons, beam splashes and first collisions to operate HCAL to make it a better device for physics.

2. Monitoring HCAL status

HCAL is equipped with high-voltage, low-voltage, laser, LED, and radiation damage monitoring systems. High voltage and low voltage are continuously monitored through detector control system which is controlled by several PCs in the CMS control room. Relevant alarms and warnings are also installed to provoke actions to recover failures. The pedestal and timing parameters are downloaded to front-end electronics through database and control system. The pedestal values are periodically checked by using dedicated pedestal runs; actions are taken at the HCAL channel level if drifts from loaded pedestal values are observed. The timing of the HCAL channels are checked by using laser runs. For the detector noise studies, dedicated self-triggered events are also collected. These monitoring runs (pedestal, laser, and self-triggered) are taken during local activities of HCAL. During the collisions, however, local time is limited. Therefore, monitoring and calibration events are also collected in orbits where no proton bunches exist (also known as abort gap). The gain stability of the HCAL is monitored by using LED runs. The local and abort gap monitoring data are processed by quality control jobs to produce histograms and tables for overall HCAL status. In case of problems, HCAL operations experts intervene and correct the behavior. During the commissioning and operation of HCAL, the development of the tools continued, and an expert team was formed to help operation of HCAL.

3. Performance studies with cosmic ray muons

Data taken with cosmic ray muons played crucial role in commissioning of CMS detectors. All sub-detectors were included and commissioned to act as a combined, unified system throughout cosmic ray muon tests[3]. As shown in Figure 1, left, by the end of commissioning phase all sub-detectors were successfully included in global runs. HCAL was one of the few sub-detectors whose commissioning completed earlier than others. Another very important goal was to collect more than 300 million events for performance studies. As shown in Figure 1, right, in about 25 days this goal was achieved and performance studies from these data resulted in much better understood detector.

HCAL group had specific goals for the cosmic ray muon runs. Among them was to participate to global runs with all components, measure the impact of magnetic field on HCAL response, measure the muon energy in a wide muon momentum range to confirm the absolute energy scale,

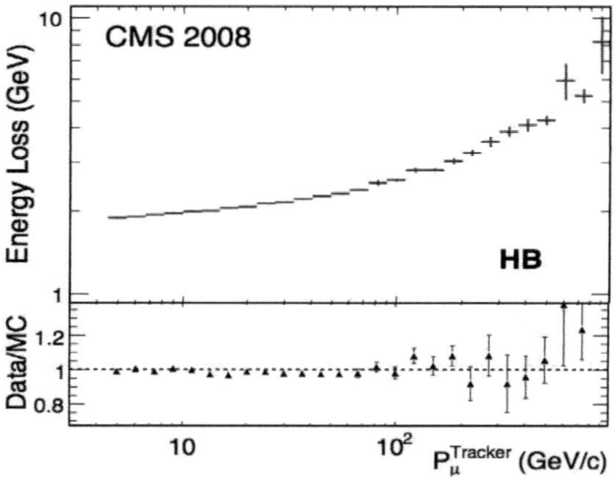

Figure 2: The energy loss of cosmic ray muons in HB versus momentum of the muons measured by the tracker (top). The ratio of data and Monte Carlo simulation predictions versus momentum of the muons measured by the tracker (bottom)[9].

confirm initial calibrations and if possible improve them, and measure the electronics noise rates and effect of magnetic field on the rate.

The active material in HB and HE are plastic scintillators and about 10% increase in response is expected with the magnetic field[4–8]. Of 10%, 5-8% is due the intrinsic brightening of the scintillator[8], 1-2% is due to the increased path length of charged particles inside scintillator and 2% is due to the cross-talk between hybrid photo-diode (HPD) pixels due to small misalignments[9]. On average 9.5% increase in HB and 8.4% increase in HE was observed[9].

One of the benefits of cosmic rays and having a working tracker system was the ability to validate absolute energy scale of HCAL. Muons with momentum in the range of ~100-200 GeV/c were used in the analysis[9]. Independent checks were also done for data and MC agreement, and it was shown that they agreed each other reasonably (Figure 2, bottom). The muon energy loss in calorimeter was measured in beam tests by using 150 GeV/c muons and found as 2.80±0.03 GeV in HB[10]. By using cosmic ray muons in the range of 135 and 170 GeV/c the energy deposition was found to be 2.85±0.02 GeV/c (Figure 2, top) [9].

The calibration constants in HCAL were determined by using particle beams of known momentum in beam tests and by using radioactive source measurements. For the barrel and endcap calorimeters absolute energy scale was set using the response to 50 GeV/c charged pions of the parts of the detector exposed to the beam[9]. Since these parts correspond to small fraction of readout segments (in ϕ), ϕ to ϕ variations in the response are expected, and they are mostly due to HPD gain differences. The variations were partially corrected by using cosmic ray muons as well[9].

The splash events taken in 2008 were used to adjust timing in η, where we expect variations due to cable length and readout box positioning. Since the position of the collimator is known, the timing was able to be calculated and converted into time measured from interaction point to particular HCAL cell[11]. At first, delay settings for barrel came from test beam measurements, whereas the delay settings for other sections were set to zero. Figure 3, left, shows the difference of predicted and measured times for different parts of the detector and for different readout depths. Measured time matches to predicted within 1 ns for barrel section. After calculating the delays by using splash events the synchronization in η is achieved within 2 ns as shown in Figure 3, right[11]. The splash events were also used to find corrections for calibration constants.

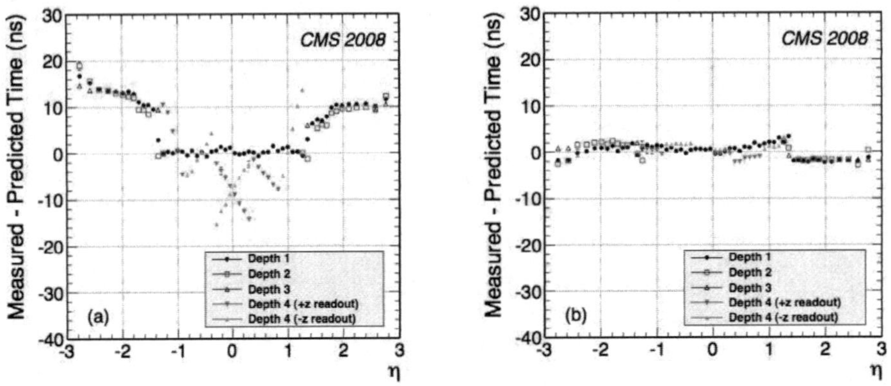

Figure 3: Difference between measured and predicted time as a function of η before (left) and after (right) adjusting delay settings by using beam splash events[11].

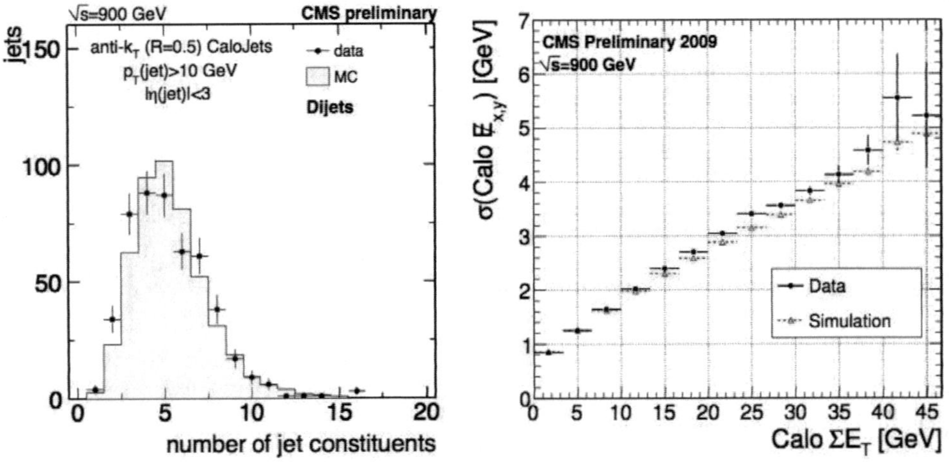

Figure 4: The number of jet constituents in 900 GeV collisions for jet algorithm that uses calorimeter towers (left)[12]. Resolution of x, y components of $\not{E}_T$ versus scalar sum of calorimeter E_T[13].

These corrections were used to smooth the response in ϕ for barrel, endcap, and outer parts.

Since December 2009, LHC is colliding protons in the center of experiments. The collisions started with first in 900 GeV and 2.36 TeV center of mass energies. The CMS experiment has been collecting data with very high efficiency to enable first physics studies. The reconstructed objects which uses HCAL and their comparison with Monte Carlo reflect the understanding and performance of HCAL. For example, jet and $\not{E}_T$ can be reconstructed by using calorimeter readout towers for which understanding of shower shapes and energy scale is very important. Figure 4 shows number of jet constituents (i.e., calorimeter towers) and $\not{E}_T$ resolution for 900 GeV collision data[12, 13]. Although there is a long way for more precise measurements, the Monte Carlo description of first data is remarkable.

Studies during beam tests and cosmic ray muons helped to understand and categorize noise sources in HCAL [14]. In HCAL there are two major sources for noise: the noise from HPDs and photo-multiplier tubes, the former is used for barrel, endcap, and outer calorimeter, the latter is used for forward calorimeter. Both noise classes are studied and rates were measured during cosmic runs and collisions. Overlap rate of HPD noise with physics event is less than 10^{-5}[14]. For the PMT hit events, the rates are around 6×10^{-3} per event[14]. There are filtering algorithms developed both at trigger and analysis level to reject noisy events. Further improvements on the noise and R&D studies with new photo-devices for the upgrade is ongoing.

4. Conclusion

7 TeV collisions at LHC in March 2010 has marked the beginning of new era in particle physics. Experiments will exploit the first collision data in the coming months to re-confirm Standard Model to demonstrate that the detectors are well understood. HCAL, as being the device for energy measurements of strongly interacting particles, will play very crucial role for new physics discovery program in CMS. In the last couple of years CMS detectors commissioned fully and collected millions of cosmic ray muons. With this period we learnt how to operate HCAL as a part of the CMS experiment. As of this conference, LHC was delivering beams with 10^{10} protons per bunch, and CMS experiment collected more than 1 nb^{-1} of data with these bunches. During this period, the efficiency of HCAL, i.e. participation percentage, was more than 99%, and HCAL had more than 99% live channels. Consequently, HCAL is operational and the experience gained during the commissioning is paying well, as HCAL participates jet and $\not{E}_{\mathrm{T}}$ measurements with very high efficiency.

References

[1] Chatrchyan S *et al.* (CMS Collaboration) 2008 *JINST* **3** S08004
[2] Bayatian G L *et al.* (CMS Collaboration) 2007 *J. Phys. G* **34**
[3] Chatrchyan S *et al.* (CMS Collaboration) 2010 *JINST* **5** T03001
[4] Bertolucci S *et al.* 1987 *Nucl. Instrum. Methods Phys. Res., Sect. A* **254** 561 – 562
[5] Cumalat J P *et al.* 1990 *Nucl. Instrum. Methods Phys. Res., Sect. A* **293** 606 – 614
[6] Blmker D *et al.* 1992 *Nucl. Instrum. Methods Phys. Res., Sect. A* **311** 505 – 511
[7] J M *et al.* 1992 *Nucl. Instrum. Methods Phys. Res., Sect. A* **312** 451 – 456
[8] Bertoldi M *et al.* 1997 *Nucl. Instrum. Methods Phys. Res., Sect. A* **386** 301 – 306
[9] Chatrchyan S *et al.* (CMS Collaboration) 2010 *JINST* **5** T03012
[10] Abdullin S *et al.* (USCMS Collaboration) 2009 *Eur. J. Phys. C* **60**(3) 359–373
[11] Chatrchyan S *et al.* (CMS Collaboration) 2010 *JINST* **5** T03013
[12] URL `http://cdsweb.cern.ch/record/1248210/files/JME-10-001-pas.pdf`
[13] URL `http://cdsweb.cern.ch/record/1247385/files/JME-10-002-pas.pdf`
[14] Chatrchyan S *et al.* (CMS Collaboration) 2010 *JINST* **5** T03014

Journal of Physics: Conference Series **293** (2011) 012055 doi:10.1088/1742-6596/293/1/012055

Response of single isolated hadrons in the ATLAS calorimeter

Margar Simonyan
On behalf of the ATLAS Collaboration
Niels Bohr Institute, Copenhagen University.
Blegdamsvej 17, 2100 Copenhagen, Denmark

E-mail: margar.simonyan@cern.ch

Abstract. Data produced in proton-proton collisions at a center-of-mass energy of $\sqrt{s} = 900$ GeV collected by the ATLAS experiment at CERN are used to select isolated tracks and to probe the calorimeter response associated to them. The ratio E/P between the energy E deposited in the calorimeter around the track impact point, and the track momentum P is measured over a wide rapidity range for particles with momentum range from 0.5 to 10 GeV/c. The average measured E/P is compared to the predictions of the Monte Carlo simulation.

1. Introduction

The ATLAS detector is a large collider detector built to be used for LHC collisions at CERN. It consists of inner detector combining different tracking subsystems, sampling calorimeters and muon spectrometer.

Single isolated tracks can be used to assess the calorimeter response to single particles by comparing the track momentum to the energy measured in a cone around the track. For the particle momenta produced in 900 GeV proton-proton collisions, the track momentum measurement is precise and accurate. The transverse momentum resolution of the inner detector to charged track is $0.05\% P_T \; [GeV/c] \oplus 1\%$. In the central region the typical energy resolution of the ATLAS calorimeter to jets is about $50\%/\sqrt{E} \; [GeV] \oplus 3\%$. The ratio E/P is an observable that allows to assess the calorimeter response to hadrons.

The basic idea of the analysis is to associate a calorimeter energy deposit to an isolated track and compare the energy in the calorimeter to the track momentum as measured by the inner detector. A cone of a given size is used around the extrapolated impact point of the track at the calorimeter entrance, and the energy which is deposited in the cone is measured. However, there are two sources of contamination that must be accounted for when comparing to the single particle energy deposit. First, the shower associated to a track can be superimposed with showers associated to other close-by tracks. This contamination can be largely reduced with an isolation cut on the track, such that a track is accepted for analysis only if it is far enough from any other track. A track is accepted for the analysis only if the track under study is far enough from any other track. Second, isolated tracks can be accompanied by neutral particles whose energy deposit can overlap to that of the charged particle. There is no obvious way to eliminate this contamination. Its contribution to the final result has been estimated in Sec. 4 and Sec. 7.

2. Event and track selection

All the events have been collected in the LHC commissioning period of December 2009. The runs considered for the analysis have been selected requiring that the Pixel and Silicon Microstrip tracker (SCT) detectors, the ATLAS solenoid and the ATLAS calorimeters [1] were operational. Events have been triggered with the requirement of at least 1 hit on either side of the Minimum Bias Trigger Scintillators, which cover the pseudo-rapidity range, $2.09 < |\eta| < 3.84$ and are located in front of the end-cap electromagnetic calorimeters. Collision candidate events have been selected by requiring the presence of one reconstructed vertex with at least two tracks associated to it.

For each event, a search for isolated tracks is performed. For each track candidate i, its $\eta - \phi$[1] position is extrapolated to the second longitudinal layer of the electromagnetic calorimeter $(\eta^i_{EM2}, \phi^i_{EM2})$. Then, for each of the other track candidates j, the extrapolated position $(\eta^j_{EM2}, \phi^j_{EM2})$ is also computed. If

$$\Delta R_{ij} = \sqrt{(\eta^i_{EM2} - \eta^j_{EM2})^2 + (\phi^i_{EM2} - \phi^j_{EM2})^2} > 0.4 \tag{1}$$

for all j, then the track candidate is defined to be isolated and it is processed further in the analysis.

The following selection has been applied on the resulting isolated track candidates:

- a transverse track momentum of $P_T > 500$ MeV/c.
- at least 1 hit in the Pixel detector and 6 hits in the SCT.
- the transverse and longitudinal impact parameters computed with respect to the primary vertex, $|d_0| < 1.5$ mm, $|z_0| \sin\theta < 1.5$ mm.

This selection is the same as that applied in [2]. In this study tracks with absolute value of pseudo-rapidity smaller than 2.3 have been used.

3. Monte Carlo simulation

The Monte Carlo (MC) simulation used in this paper consists of a sample of about one million events of proton-proton collisions at $\sqrt{s} = 900$ GeV produced by non–diffractive processes. The event generation was done with Pythia [5] 6.4.21. The phenomenological model to describe proton-proton collisions was tuned to reproduce minimum bias and underlying event data at a center-of-mass energy range between 200 GeV and 1.96 TeV. The ATLAS detector simulation software based on GEANT4 [3] has been used to process the generated events.

4. Definition of the E/P Observable

The sum of the energy deposits in all the calorimeter longitudinal layers associated to a track (E) is the observable that will be compared to the track momentum P. It is computed making use of topological clusters at the electromagnetic scale, i.e., without applying any correction for the calorimeter non-compensation and for the energy loss in dead material. The purpose of the topological clustering algorithm is to identify areas of connected energy deposits in the calorimeter and suppress the noise based on the significance of the energy deposits in cells with respect to the expected noise level. A topological cluster is initiated by cells with an energy deposit $|E_{cell}| > 4\sigma_{noise}$, where σ_{noise} comes from the Gaussian fit of the electronic noise in the cell. Then, iteratively, the cluster expands adding all neighboring cells with $|E_{cell}| > 2\sigma_{noise}$.

[1] The ATLAS reference system is a cartesian right-handed coordinate system, with the nominal collision point at the origin. The anti–clockwise beam direction defines the positive z-axis, while the positive x axis is defined as pointing from the nominal interaction point to the center of the LHC ring. ϕ is the polar angle with respect to the x axis, θ is the azimuthal angle with respect to the z axis. The pseudo-rapidity is defined as $\eta = -\ln\ \tan\theta/2$.

Journal of Physics: Conference Series **293** (2011) 012055 doi:10.1088/1742-6596/293/1/012055

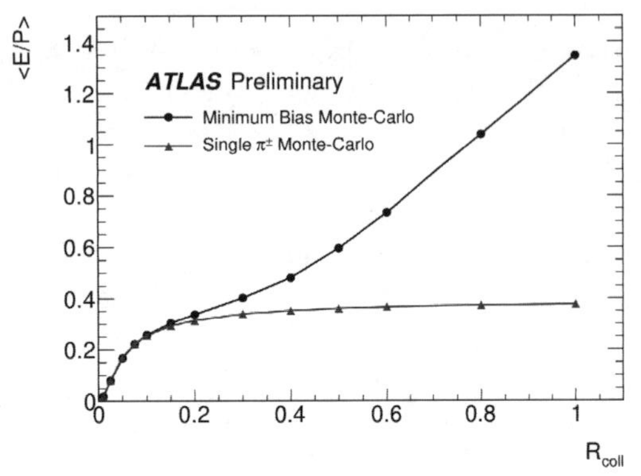

Figure 1. Mean value of the E/P as a function of the opening of the collection cone for single pions (blue triangles) and for a non–diffractive minimum bias sample (black dots).

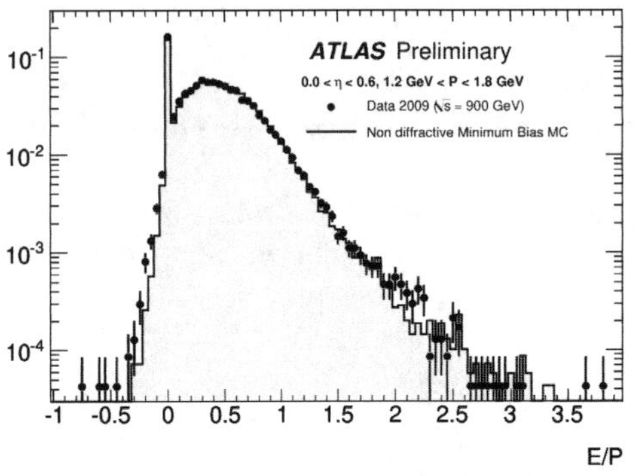

Figure 2. E/P distribution for isolated tracks with an impact point in the region $0 < |\eta| < 0.6$ and with a momentum in the range $1.2\ GeV/c < P < 1.8\ GeV/c$.

Finally, the cells surrounding the resulting cluster are added, regardless of their energy. The $\eta-\phi$ position of a cluster in a given longitudinal calorimeter layer is computed as the energy–weighted position of the cells in the layer.

For each longitudinal layer, the layer cluster energy E_j is associated to the good track k, if the distance between the extrapolated track position $\eta_{tr}^{kj}, \phi_{tr}^{kj}$ (obtained using the direction of the track at the impact point on the calorimeter) and the cluster position is smaller than a given value R_{coll}. That is, if

$$\sqrt{(\eta_{tr}^{kj} - \eta_{cl}^{ij})^2 + (\phi_{tr}^{kj} - \phi_{cl}^{ij})^2} < R_{coll} \tag{2}$$

then the cluster energy in the layer E_j is associated to the track. Subsequently the cluster energies in all the calorimeter layers is summed

$$E = \sum_j E_j \tag{3}$$

The parameter R_{coll} has been chosen based on a trade-off between maximizing the particle shower containment and minimizing the background contribution coming from neutral particles produced close to the track.

Fig. 1 shows the comparison of the arithmetic mean value $\langle E/P \rangle$ as a function of the size of R_{coll}. The momentum and pseudo–rapidity spectra of the single pions have been weighted such to reproduce those of the charged particles in the non–diffractive minimum bias sample. The energy associated to a track in the minimum bias sample is equivalent to the single pion energy, provided that R_{coll} is small enough. The $\langle E/P \rangle$ ratios start to differ for collection cone opening angles larger than $\Delta R = 0.2$, which correspond to a radial shower containment of about 90%. A cone of $R_{coll} = 0.2$ is chosen as the best compromise between shower containment and background contribution.

5. The E/P Distribution

As an example the E/P distribution for tracks with momentum in the range from 1.2 to 1.8 GeV/c is shown in Fig. 2. The simulation reproduces the data well.

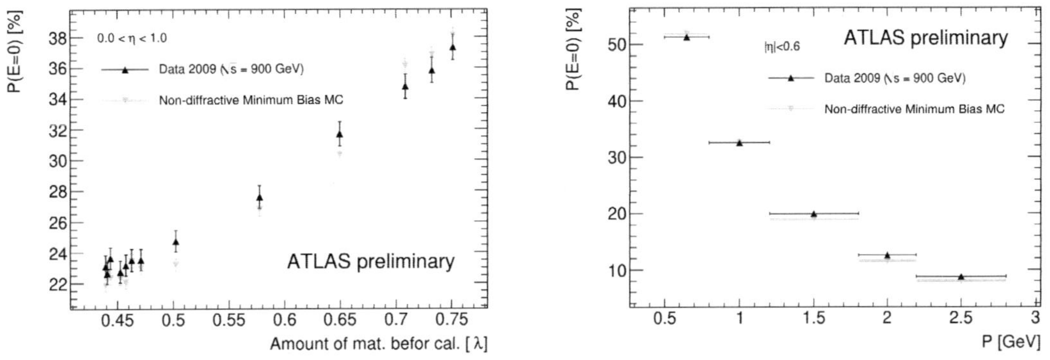

(a) (b)

Figure 3. Probability to get no calorimeter response $P(E=0)$ as a function of the amount of material in nuclear interaction lengths in front of active volume of the calorimeter in $0.0 < \eta < 1.0$ region (a). $P(E=0)$ as a function of track momentum (b).

The large number of entries with $E/P = 0$ corresponds to isolated tracks that have no cluster associated in the calorimeter. The following effects are responsible for that:

- Particles can undergo hadronic interactions in the material in front of the calorimeter. This can happen in the inner detector volume, in the cryostat or in the solenoid placed in front of the calorimeter. Such particles can change their direction, or produce a large number of low momentum particles emerging from the hadronic interaction.
- As discussed in Sec. 4, a cluster is created only if a seed cell is found. Hadrons with low momentum and an extended shower topology sometimes do not have a single cell energy deposit large enough to seed a topological cluster.

The cases where the calorimeter response is compatible with zero have been further studied. The probability that the calorimeter response is compatible with noise, $P(E=0)$, is defined as follows: the noise width (σ) is determined by finding, in the negative energy tail, the bin that contains $1/\sqrt{e}$ times fewer events than the bin at $E/P = 0$.

$P(E=0)$ is then obtained as the ratio of the number of events with $E/P < \sigma$ to the total number of events (N_{tot}) in the E/P distribution.

$$P(E=0) = \frac{N\,(E/P < \sigma)}{N_{tot}} \tag{4}$$

Fig. 3(a) shows the probability $P(E=0)$ as a function of the amount of material in nuclear interaction lengths in front of the active volume of the calorimeter in the barrel ($0.0 \leq \eta \leq 1.0$) calorimeter region. The correlation between the amount of material and the probability that no energy is associated to the track is strong. When the hadron passes more material the probability that no energy is measured close to the track impact point increases. Fig. 3(b) shows the probability (4) as a function of the track momentum in $|\eta| \leq 0.6$ region of calorimeter. This is the region where the dead material in front of the calorimeter is approximately constant. The zero response probability decreases with increasing track momentum. It is important to notice that, in general, the number of tracks with no clusters associated in the calorimeter as well as the $P(E=0)$ probability are well predicted by the MC simulations.

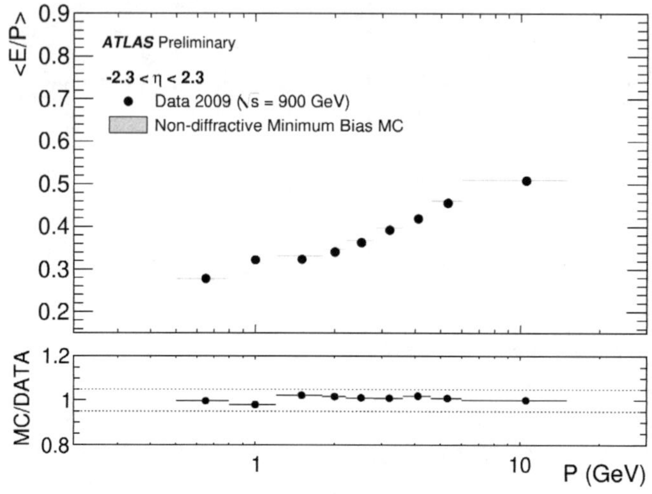

Figure 4. $\langle E/P \rangle$ as a function of the track P, in the η range $-2.3 < \eta < 2.3$.

6. Results as a Function of P and of η

The isolated tracks have been classified in bins of momentum and of pseudo–rapidity of the impact point extrapolated to the second longitudinal layer of the electromagnetic calorimeter. For each P and η bin, the estimator that was chosen to perform the comparison is the mean value, $\langle E/P \rangle$.

Fig. 4 shows $\langle E/P \rangle$ as a function of P regardless of the pseudo–rapidity of the impact point of the track. The black dots represent the data. The green shaded area corresponds to the MC prediction. The lower part of the figure shows the ratio between MC and data. The range that can be probed with the available statistics is approximately 500 MeV/c $< P <$ 10 GeV/c. In general, the agreement between MC and data is very good. The ratio between MC and data is fluctuating around one, with a maximum deviation that is well within 5%.

The dependence of $\langle E/P \rangle$ on η is shown in Fig. 5. In this case, all tracks in a given P range have been used. The region where the worst agreement between data and MC is found is the one around $\eta = 1.7$, where the MC is higher than the data. This is the transition region between the Barrel and Endcap EM calorimeters, where numerous services for the calorimeters and the inner detector are present.

7. Systematics

The following possible systematics have been evaluated:

- Track selection: the specifics of how isolated tracks are selected can have a systematic influence on the average value $\langle E/P \rangle$. The uncertainty on $\langle E/P \rangle$ due to the applied track selection was evaluated changing it within reasonable margins. The maximum variation of the $\langle E/P \rangle$ due to the track selection is in general of the same order as the statistical error , and it is below 0.01 (1.5%) for all bins. Detailed studies showed that the uncertainty of the response due to the track selection is small and well under control.
- Background contamination: the shower associated to a track can be contaminated by energy deposits from neutral particles. If the Monte Carlo modeling of the neutral component of the non–diffractive events is wrong, the contamination can be different for MC and data.

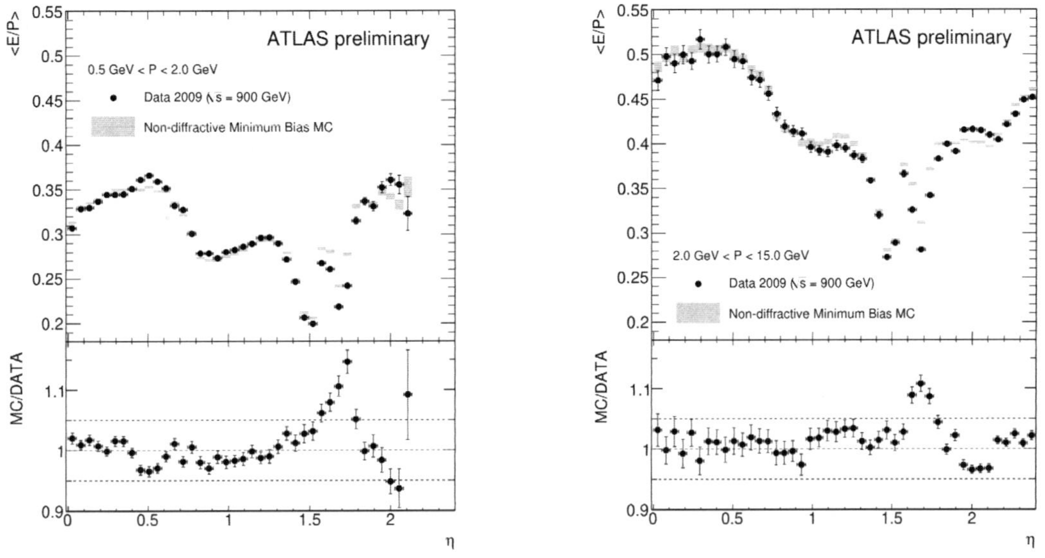

Figure 5. Average $\langle E/P\rangle$ as a function of η for tracks with momentum 0.5 GeV/c $< P <$ 2.0 GeV/c (left) and 2 GeV/c $< P <$ 15 GeV/c (right).

In this analysis, the background contamination was estimated both using the MC simulation and a data-driven background procedure [4], finding consistent results. We obtained a neutral particle contamination of $(2.5 \pm 1.5)\%$ to the energy deposit associated to a track of momentum P. This affects the absolute E/P measurement, but, since the contamination is small, even assuming a significant difference in data and MC with respect to this component, it would negligibly affect the agreement between data and MC. No additional uncertainty associated to the background contamination was therefore considered.

8. Conclusions

The data from proton–proton collisions at $\sqrt{s} = 900$ GeV collected by the ATLAS experiment have been used to compare the single track calorimeter response to the MC simulation predictions. Overall, an agreement within 5% on $\langle E/P\rangle$ has been found for particle momenta between 0.5 MeV/c and 10 GeV/c, and for a particle pseudo-rapidity range from -2.3 to 2.3. Such a level of understanding of the ATLAS detector in the early stage of the experiment is encouraging and demonstrates the accurate description of dead material before the calorimeter and of the noise in the MC.

References

[1] The ATLAS Collaboration, The ATLAS Experiment at the CERN Large Hadron Collider, *JINST (2008), 3, S08003*
[2] The ATLAS Collaboration, Charged-particle multiplicities in pp interactions at $\sqrt{s} = 900$ GeV measured with the ATLAS detector at the LHC, *Phys. Lett. B688, pp. 21–42, 2010*
[3] S. Agostinelli et al., Geant4 – a simulation toolkit, *NIM A 506 (2003) pp. 250–303*
[4] The ATLAS Collaboration, Expected Performance of the ATLAS Experiment - Detector, Trigger and Physics, *arXiv:hep-ex-0901.0512, 2009*
[5] T. Sjostrand, S. Mrenna, and P. Skands, PYTHIA 6.4 physics and manual, *JHEP05, 026, 2006*

Journal of Physics: Conference Series **293** (2011) 012056 doi:10.1088/1742-6596/293/1/012056

Signal Reconstruction of the ATLAS Hadronic Tile Calorimeter: implementation and performance

G. Usai (on behalf of the ATLAS Tile Calorimeter group)
University of Texas at Arlington
E-mail: giulio.usai@cern.ch

Abstract. TileCal, the central hadronic section of the ATLAS Calorimeter, is a sampling calorimeter made of steel and scintillating tiles. The TileCal front-end electronics reads out about 10,000 photo-multipliers at 40 Msps measuring energies ranging from $\simeq$ 30 MeV to $\simeq$ 2 TeV. The read-out system is designed to provide the ATLAS High Level Trigger with reconstructed PMT signals within the time budget allowed by the First Level Trigger (LVL1) running at a maximum trigger rate of 75 KHz. The signal amplitude, time and a reconstruction quality factor are obtained for each PMT using optimal filtering techniques implemented in the Digital Signal Processors (DSP).

1. Introduction

ATLAS [1] is a general purpose experiment designed to explore the physics landscape in proton-proton collisions at the unprecedentedly high energy regime of the Large Hadron Collider at CERN. After many years of careful preparation ATLAS received the first collisions from LHC in December 2009. After a short overview of the Tile Calorimeter read-out system we will discuss the implementation of Optimal Filtering algorithms highlighting the constraints imposed by the use of DSPs. We will report on the validation of the DSP algorithm and present the performance measured in calibration and collision events.

2. Tile Calorimeter Read Out

The Tile Calorimeter [2] is required to measure particle energies in a dynamic range corresponding to 16 bits, extending from typical muon energy deposition of a few hundreds of MeV to the highest energetic jet of particles, which in rare cases can deposit up to two TeVs in a single cell. A scheme with a double readout using two independent 10-bit ADCs was chosen to cover this range. The PMT pulse is shaped, then fanned-out and amplified in two separate branches with a nominal gain ratio of 64 [3]. The two output pulses, referred to as high gain and low gain, have a fixed width (FWHM) of about 50 ns and an amplitude that is proportional to the energy deposited in the cell. The two pulses are digitized simultaneously by two ADCs at 40 Msps; the time series (samples) for each pulse is stored in the Data Management Unit [4] that also performs some first processing. Each pulse is sampled seven times in physics mode; up to nine samples may be recorded for calibration purposes. The high gain ADC is normally used unless the time series contain measurements which saturate the ADC. In this case the low ADC readout is used. The samples are kept in digital pipelines on the detector, and if the event is accepted by the LVL1 Trigger system, are sent to the back-end electronics, the Read Out Driver boards (ROD) [11]. Each ROD (see figure 1) receives data from eight calorimeter modules through optical links and synchronizes the data so that all the event fragments are associated with the LHC bunch crossing (BCID) selected by the LVL1 trigger. Digital Signal Processors (DSP) are used for the implementation of the reconstruction algorithms, as described below. After reconstruction the event fragments are partially assembled and sent to the Read Out Buffers (ROBin) for further processing by the Second Level Trigger.

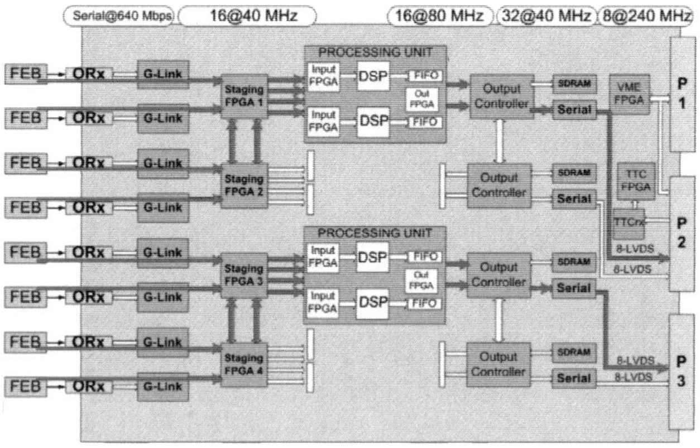

Figure 1. Scheme of the Read Out Driver data-flow.

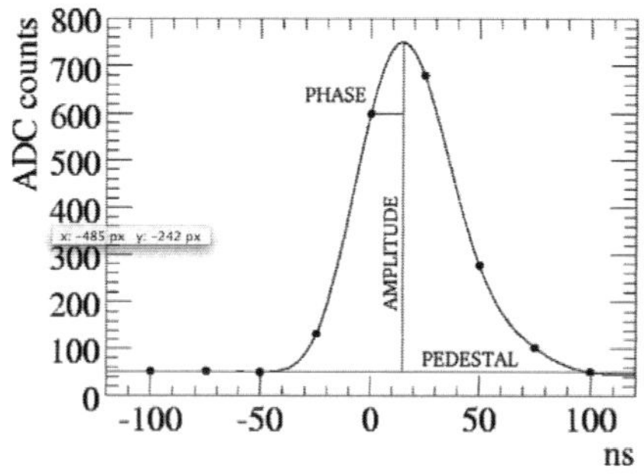

Figure 2. Pulse functional form showing the ADC samples (dots) and definition of the reconstructed quantities.

3. Optimal Filtering

Figure 2 shows an analog signal pulse and the ADC measurement samples, and illustrates the main characteristics of the pulse: amplitude, arrival phase and baseline level, or pedestal. The LHC is a synchrotron, the phase of the calorimeter signals of interaction events is expected to be synchronized with the LHC clock and constant within very small fluctuations. The residual fluctuations are mainly due to the longitudinal spread of proton bunches. The ADC measurement phase can be adjusted to compensate for delays and particle time of flight. In this condition the pulse time dependence can be linearized with a good approximation and the signal reconstructed with linear techniques. The Optimal Filtering method have been chosen for the expected performance and its simplicity [5]. The algorithm extracts the three parameters of the shaped signal: the amplitude A, the phase τ and the baseline level p using linear combinations of the samples S_i with a set of weights a_i,b_i and c_i:

$$A = \sum_{i=1}^{n} a_i S_i \ , \qquad A\tau = \sum_{i=1}^{n} b_i S_i \quad \text{and} \qquad p = \sum_{i=1}^{n} c_i S_i \ , \tag{1}$$

where n is the total number of samples. The weights are obtained minimizing the variance of the parameters against the electronic noise and Minimum Bias (MB) pileup fluctuations [6]. To give an example, the weights a_i, used in the amplitude computation are defined by the system of $n+3$ equations:

$$\begin{aligned} &\sum_{i=1}^{n} a_i g_i = 1 \ , \\ &\sum_{i=1}^{n} a_i g'_i = 0 \ , \\ &\sum_{i=1}^{n} a_i = 0 \ , \\ &\sum_{j=1}^{n} a_j R_{ij} - \lambda g_i - \kappa g'_i - \nu = 0 \quad \text{for} \quad i = 1, n \ . \end{aligned} \tag{2}$$

where $g_i = g(t_i)$ is the known signal pulse function, g'_i the pulse first derivative, λ, κ, ν are Lagrange multipliers and R_{ij} is the noise autocorrelation matrix defined as:

$$R_{ij} = \frac{\sum (n_i - \langle n_i \rangle)(n_j - \langle n_j \rangle)}{\sqrt{\sum (n_i - \langle n_i \rangle)^2 \sum (n_j - \langle n_j \rangle)^2}} \ , \tag{3}$$

here we indicate with n_i the samples recorded on the electronic noise or other fluctuation due to MB pileup. Two similar system of $n+3$ equations hold for the weights b_i and c_i used for the phase and pedestal computation.

Note that in equation 1 the baseline level is not subtracted from the samples, but the constraint $\sum_{i=1}^{n} a_i = 0$ guarantees that any constant added to the samples S_i will not contribute to the amplitude.

In order to flag possible reconstruction failures, a reconstruction quality factor (QF) is defined as the square sum of the residuals:

$$Q_{\rm F} = \sum_{i=1}^{n} \left[S_i - \left(Ag_i + A\tau g_i^{'} + ped\right)\right]^2 \quad . \tag{4}$$

The pulse shape function $g(t_i)$ used to define the weights and the QF, is measured from the data. A single pulse shape template is used to describe all the channels in the calorimeter since channel by channel differences were found to be not important [7]. Different pulse templates extracted at different amplitudes, when rescaled, show a slightly deformation of the pulse height in the tail region. The distortion is small ($\approx 1\%$) and proved to have a negligible effect on the reconstructed amplitude and time [8, 7]. The QF is more sensitive to this distortion and shows a clear increase with the amplitude.

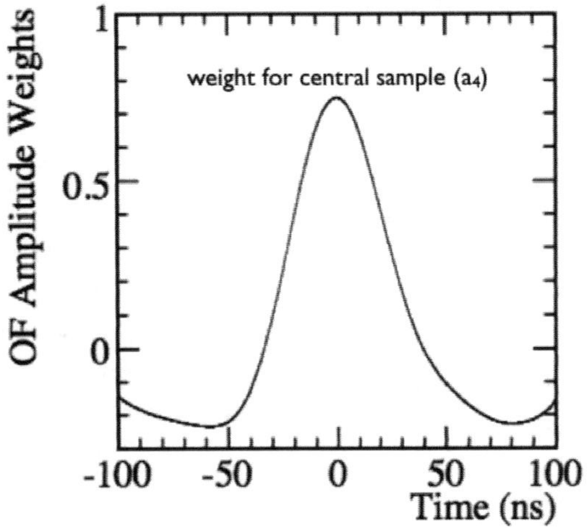

Figure 3. Behavior of the weight a_4 as a function of phase [10].

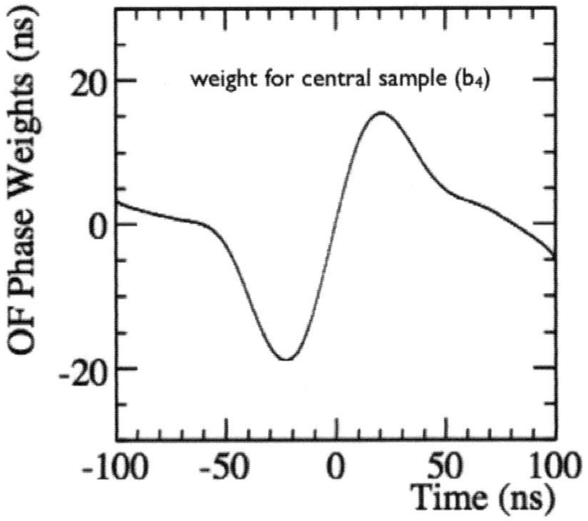

Figure 4. Behavior of the weight b_4 as a function of phase [10].

4. Implementation

The weights a_i,b_i,c_i, as defined above, depend on the phase used to sample the signals, as shown in figures 3 and 4. The timing of each detector channel was established using laser calibration and collisions events [9]. The measured offsets are stored in the database. Sets of weights are computed for all phases covering the necessary range in 0.1 ns steps.

To reconstruct asynchronous data (e.g. cosmic rays), or to avoid the use of a priori definition of phases, an iterative method can be used in the reconstruction. In this mode the equations 1 are initially evaluated with an arbitrary phase, and the successive selection of the weight is based on the result of the previous step. The iterative method is slower and more sensitive to noise fluctuations. It is worth noting that the sample acquisition window is larger than the design LHC proton bunches separation and iterative algorithm can pick up signals generated in bunch crossing different than the triggered one. For this reason the default method is the non-iterative one and we seldom use the iterative method for purposes of detector commissioning. Due to the execution time limits, the maximum LVL1 rate sustainable using iterations is about 30 KHz, the non-iterative method is well within the LVL1 trigger rate time requirement.

All the parameters needed by the reconstruction algorithm, like weights, phases and calibration constants are downloaded into the ROD/DSPs at the configuration time. The reconstructed amplitude provided to the LVL2 trigger algorithms is calibrated to the calorimeter electromagnetic energy scale by applying channel based calibration factors in the DSP.

The DSP reconstruction is necessarily limited by use of fixed point arithmetic and the internal precision available to describe the weights and calibration factors. Since the division is a time consuming operation in the DSP the phase from equation 1 is computed using a look-up table with the energy reciprocal pre-defined and stored in the DSP memory. One other limitation is in the size of output data fragments [10]. The reconstruction results are packed in a 32-bit word for each channel:15 bits for the energy, 11 bits for the time and 4 bits for the quality factor. The remaining bits encode the gain and an overal channel reconstruction good/bad status. This will be discussed more in the next section.

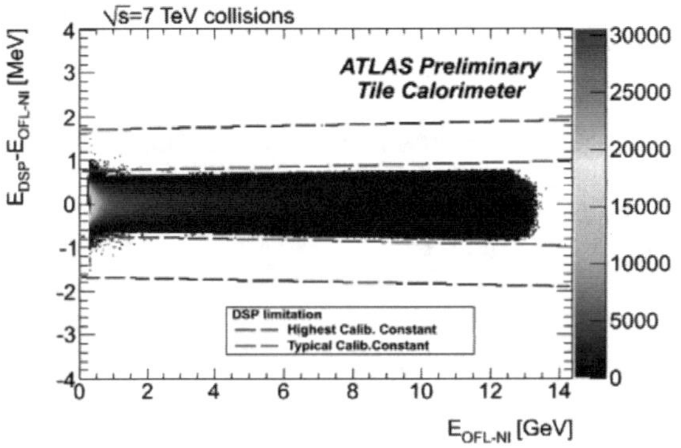

Figure 5. Numerical precision of the energy reconstructed in the ROD/DSP in the HG range evaluated using collision data.

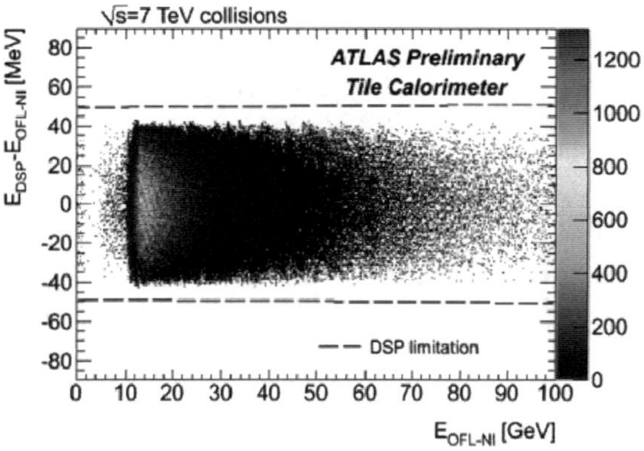

Figure 6. Numerical precision of the energy reconstructed in the ROD/DSP in the LG range evaluated using collision data.

5. Validation

The ROD can be configured to send out both the reconstructed quantities and the raw data samples. Due to the bandwidth of the output link, this configuration is sustainable only up to a LVL1 trigger rates of about 45 KHz. The raw data obtained in this way can be reconstructed using well tested offline algorithms and used to validate the DSP implementation, to study the reproducibility of the results and to fully commission the detector in the harsh environment of the LHC. When the ATLAS data taking rate increases, only a small fraction of the data will contain samples. Therefore data validation at the start is important.

The first step in the validation of the DSP results is to verify the consistency of the online and offline implementations. Figures 5 and 6 show the residuals between the energy reconstructed online by the DSP and the one reconstructed offline by an equivalent algorithm showing the loss of precision due to the DSP limitations discussed above. Note that the degradation of individual channels is slightly energy dependent. Since this error scales with the channel level calibration factors, a band is shown covering from the 99% to the worst case. The worst cases arise in few PMTs with abnormal high voltage settings. The response needs to be boosted by a large factor to compensate for the gain deficit. The maximum deviation is about 2 MeV in the high gain range. For comparison the electronic noise RMS level is about 30 MeV. The low gain scale has a precision of about 50 MeV, fully adequate in the range where signals are larger than approximately 8 GeV.

Figure 7 shows the difference between the time reconstructed in the DSP and offline as a function of the signal amplitude. Pseudo data are used for this study: data samples are generated according to the expected pulse shape, using a special module that emulates the front end electronics and covers a wide range of phases and amplitudes. The ROD is interfaced with the module and reconstructs the data in the normal fashion. The precision in the reconstructed time depends on the energy and phase granularity used in the look-up table; the oscillations apparent in figure 7 are an artifact of this granularity. Figure 8 shows the same difference as a function of the phases of the pulse. The maximum rounding error is about 0.5 ns. The larger differences up to 3 ns arise for small amplitudes and very large phases (up to 60 ns). In these conditions the numerical precision is not an issue since the reconstructed time is highly inaccurate in any case. The linear approximation discussed in section 3 does not hold. To

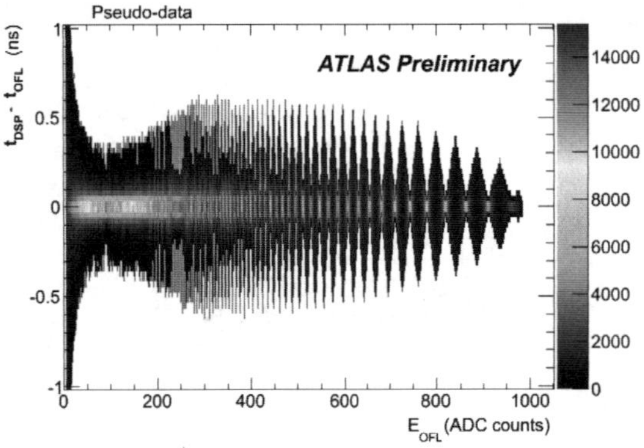

Figure 7. Numerical precision of the time reconstructed in the ROD/DSP as a function of the signal amplitude.

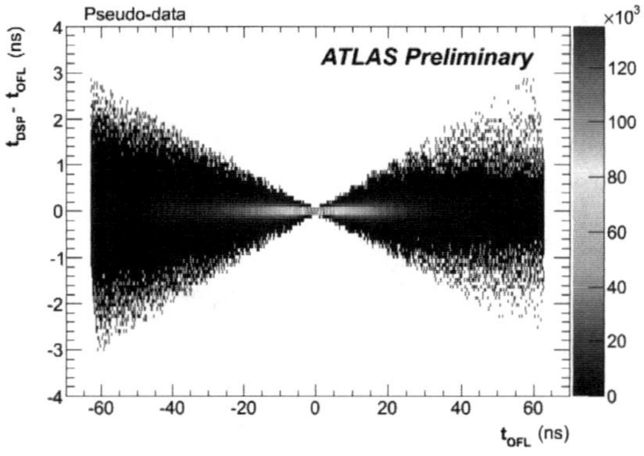

Figure 8. Numerical precision of the time reconstructed in the ROD/DSP as a function of the signal arrival time.

cross-check the DSP implementation, the fixed point arithmetic and the precision used in the DSP constants and weights can be emulated in the offline reconstruction. Identical results are obtained online and offline when the DSP emulation is used [12].

One important further step in the validation is to evaluate the linearity and the resolution of the online non-iterative algorithm with respect to the offline iterative methods. In this way we evaluate all uncertainties due to numerical precision, understanding and description of the detector response and timing and the assumptions described in section 2. These studies have just started, and we are not able yet to give fully quantitative results. Figure 9 shows the DSP reconstructed time as a function of the offline time reconstructed with iterative methods. The DSP shows a good linearity for phases within 10 ns around the expected mean time. For larger time the deviation from linearity starts becoming important. Figure 10 shows the relative error in the DSP reconstructed energy as a function of the phase. The maximum bias is about 1% for pulses arriving 3 ns from the expected time. If needed, the bias can be easily corrected online also after finalizing the understanding of the tails in the time distributions.

5.1. Handling of unexpected conditions

Failures in the reconstruction, or unexpected results, can trigger the conditional dumping of the raw data sample on a channel by channel basis in order to make the raw data available for further offline processing. Examples of these conditions are: ADC saturation, unexpected reconstructed time, bad quality of the reconstruction, etc. These conditions are defined by a programmable logic based on the comparison of the reconstructed quantities with thresholds and other external conditions.

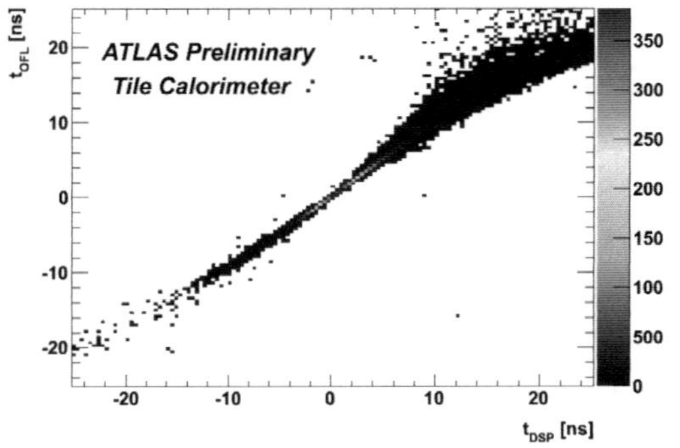

Figure 9. Time reconstructed in the DSP and time reconstructed offline with iterative method

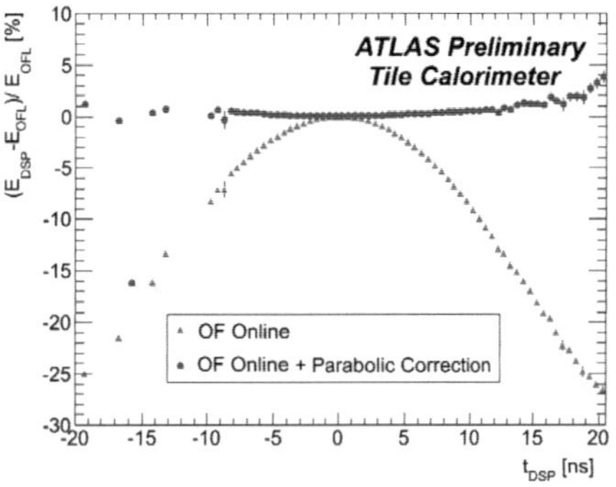

Figure 10. Bias in the DSP energy reconstruction as a function of the reconstructed time and the behavior of a second order correction.

6. Summary and Conclusions

The energy reconstruction performed online in the Read Out Drivers is an important component of the ATLAS high level trigger. The evaluation of the algorithm performance started before collisions, using calibration, cosmic ray and pseudo data. The first collision data at 900 GeV and at 7 TeV have provided an estimate of the performance of the algorithm. The DSP results can be used in the High level trigger selection. Since the LVL1 trigger rate is currently lower than the design value, it is possible to operate in a transparent mode without losing any information. These data can be recorded and made available for offline reprocessing. Future results based on increased statistics and better understanding of the data are expected.

References

[1] ATLAS Detector and Physics Performance:Technical Design Report, CERN-LHCC-99-014 (1999); CERN-LHCC-99-015 (1999).

[2] Atlas Tile Calorimeter: Technical Design Report, CERN-LHCC-98-015 (1998).

[3] K. Anderson et al. Design of the front-end analog electronics for the ATLAS Tile Calorimeter. NIM A, 551:469476, 2005.

[4] C. Bohm et al. Atlas TileCal digitizer test system and quality control. ATLAS Note, ATL- TILECAL-2004-009, 2004.

[5] W. E. Cleland and E. G. Stern. Signal processing considerations for liquid ionization calorimeters in a high rate environment. NIM A, 338:467497, 1994.

[6] E. Fullana et al. Optimal Filtering in the ATLAS Hadronic Tile Calorimeter. ATLAS Note, ATL-TILECAL-2005-001, 2005.

[7] T. Carli et al. Effect of Pulse-Shape Variations on the Energy Reconstruction in the ATLAS Tile Calorimeter. Atlas Internal Note, ATL-COM-TILECAL-2009-033, 2009.

[8] K. Anderson et al. Performance and Calibration of the TileCal Fast Readout Using the Charge Injection System. ATlas Internal Note, ATL-COM-TILECAL-2008-003. 2008.

[9] C. Clement et al. Time Calibration of the ATLAS Hadronic Tile Calorimeter using the Laser System. Atlas Note, ATL-TILECAL-PUB-2009-003, 2009.

[10] B. Salvachua et al. Online energy and phase reconstruction: commissioning phase of the ATLAS Tile Calorimeter. Atlas Note, ATL-TILECAL-2008-004, 2008.

[11] A. Valero et al., ATLAS TileCal read out driver production, JINST, vol.2, p.P05003, 2007.

[12] Compare with page 9,10,12 of the slides shown at the Conference.

Journal of Physics: Conference Series **293** (2011) 012057 doi:10.1088/1742-6596/293/1/012057

Local hadron calibration with ATLAS

Paola Giovannini, on behalf of the ATLAS Liquid Argon Calorimeter Group

Max-Planck-Institut für Physik, Werner-Heisenberg-Institut, Föhringer Ring 6, D-80805 München, Germany

E-mail: paola@mppmu.mpg.de

Abstract.
The method of Local Hadron Calibration is used in ATLAS as one of the two major calibration schemes for the reconstruction of jets and missing transverse energy. The method starts from noise suppressed clusters and corrects them for non-compensation effects and for losses due to noise threshold and dead material. Jets are reconstructed using the calibrated clusters and are then corrected for out of cone effects. The performance of the corrections applied to the calorimeter clusters is tested with detailed GEANT4 information. Results obtained with this procedure are discussed both for single pion simulations and for di-jet simulations. The calibration scheme is validated on data, by comparing the calibrated cluster energy in data with Mote Carlo simulations. Preliminary results obtained with $\sqrt{s}$ = 900 GeV collision data are presented. The agreement between data and Monte Carlo is within 5% for the final cluster scale.

1. Introduction

Precise measurement of the energies of jets is fundamental for many different physics analyses foreseen with the ATLAS detector [1]. In particular the jet energy scale uncertainty is one of the largest systematic errors on the measurement of the top quark mass and of the inclusive jet cross section, which are important steps towards new physics searches.

Jet calibration can be generally viewed as consisting of two phases. In the first phase the measured signal is corrected for detector effects and out of cone effects. This calibrates the measured jet to the particle level, which is defined as the corresponding jet obtained by running the same jet algorithm on the final state particles. In the second phase the jet energy is calibrated to the parton level.

The local hadron calibration is a complete approach to the first phase of jet calibration. In this scheme detector effects are corrected at cluster level, before any jet reconstruction. This provides a consistent calibration of all calorimeter signals, which can be used as input to missing transverse energy reconstruction as well. Clusters are obtained from calorimeter cells using a three dimensional topological algorithm, which suppresses noisy cells [2]. The energy of the input clusters is defined to be at the electromagnetic scale (EM) and includes compensation for sampling fraction losses optimized for electrons and photons. The local hadron calibration consists of four steps of corrections that are based on cluster shape variables, cluster energy and cluster pseudo-rapidity ($\eta = -\ln(\tan(\theta/2))$, θ being the angle from the beam axis) [3]:

(i) classification step: clusters are classified as stemming mainly from an electromagnetic or an

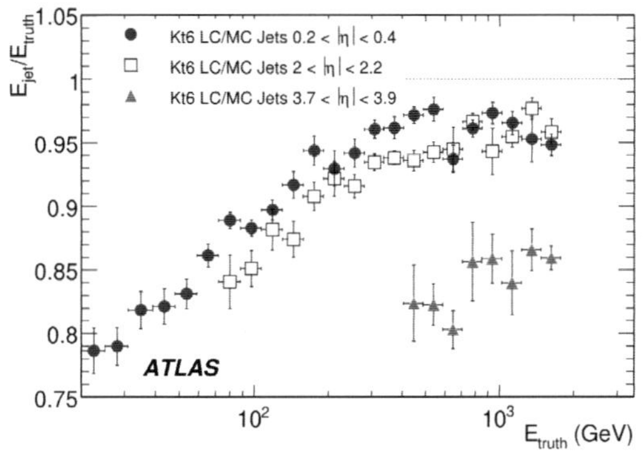

Figure 1. Energy response of K_T (R=0.6) jets, reconstructed from local hadron calibrated clusters (E_{jet}), with respect to K_T (R=0.6) jets from Monte Carlo final state particles (E_{truth}). The jet energy response is shown over the whole energy range for three different pseudo-rapidity regions. [1]

hadronic type of shower, each of the next three corrections differ for electromagnetic and for hadronic clusters;

(ii) hadronic weight step (W): clusters that are classified as hadronic receive a correction that recovers for invisible and escaped energy;

(iii) out of cluster step (OOC): clusters are corrected for the energy discarded by the clustering algorithm because of noise thresholds;

(iv) dead material step (DM): clusters are corrected for the energy lost outside the active calorimeter volumes.

The energy scale of jets reconstructed from local hadron calibrated clusters is still lower than the energy scale of jets from Monte Carlo final state particles, as shown in Figure 1. Cluster corrections cannot account for the so called out of cone effects, like energy lost by particle never reaching the calorimeter or jet-algorithm inefficiencies. In order to take into account these effects, dedicated jet level corrections, based on jet shape variables, jet energy and jet pseudo-rapidity have been developed. After this final calibration step the energy of the jets is correctly reconstructed with respect to the final state particle jets within less than 1% [4].

2. Local correction performance

A comparison to the particle jet level is not sufficient to evaluate the performance of the cluster corrections. In order to be able to disentangle effects due to the different calibration steps it is necessary to define a true cluster energy, for each of the cluster energy scales. This definition is possible using GEANT4 [5] shower simulations. In the ATLAS simulation framework GEANT4 information is organized in terms of so called calibration hits. A calibration hit can be seen as a true calorimeter cell or as true dead material deposit, depending on where the energy is released. The calibration hit energy is accessible for both active (liquid argon, scintillator) and inactive (absorber) material. For each reconstructed cluster the true cluster energy is then defined:

- at the hadronic scale as the sum of the energy of the calibration hits inside the cluster;
- at the out of cluster scale as the sum of the energy of the calibration hits near to the cluster, not included into any other cluster and deposited inside the active calorimeter regions;
- at the dead material scale as the sum of the energy of calibration hits near to the cluster, deposited in dead material regions;

The assignment of calibration hits to the out of cluster and dead material scale is performed via an algorithm that takes into account the proximity of the calibration hit to the cluster in $\eta - \phi$ space and the energy of the cluster (ϕ being the azimuthal angle measured around the

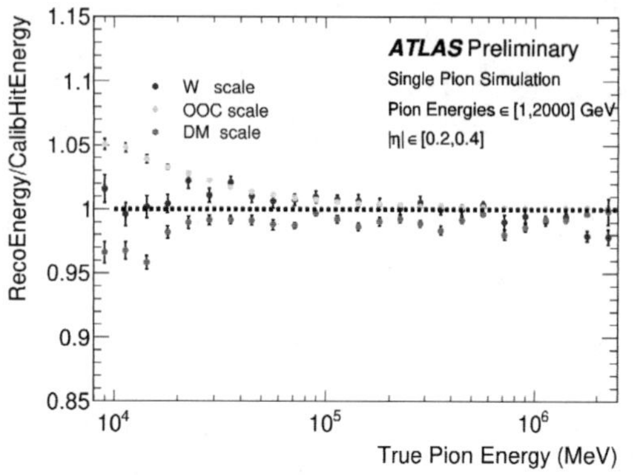

Figure 2. Linearity for the three cluster calibration correction steps with respect to true cluster energy, for single pion simulation. The hadronic weighting step L_W is shown in blue, the out of cluster step L_{OOC} in green and the dead material step L_{DM} in red. The performance is shown for the central region.

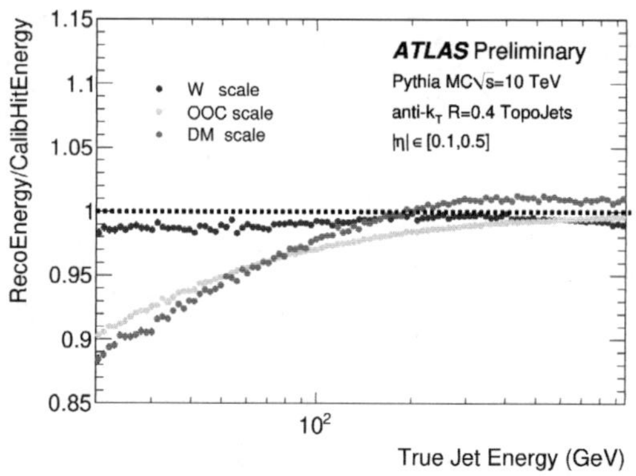

Figure 3. Linearity for the three cluster calibration correction steps with respect to true cluster energy, for di-jet simulation. The hadronic weighting step L_W is shown in blue, the out of cluster step L_{OOC} in green and the dead material step L_{DM} in red. The performance is shown for the central region.

beam axis) . The linearity performance (L) of the cluster corrections can be studied based on the following quantities:

$$L_W = \frac{R_W}{T_W} \qquad L_{OOC} = \frac{R_{OOC} + T_W}{T_{OOC} + T_W} \qquad L_{DM} = \frac{R_{DM} + T_{OOC} T_W}{T_{DM} + T_{OOC} + T_W} \tag{1}$$

where R_X is the reconstructed energy from correction X and T_X is the energy that should have been reconstructed from correction X, as assigned by the calibration hit algorithm. In case of perfect calibration ($R_X = T_X$), the linearity terms in Equation 1 should be equal to unity. Deviation from this value are dependent on the reconstructed energy contribution R_X under investigation (W or OOC or DM separately).

2.1. Single pion simulations

The local hadron calibration corrections are developed using single pion simulations [3]. In this type of events the reconstructed energy should match the initial energy of the pion, without any out of cone effects. At the same time all the clusters in the event belong to the same initial particle, without any ambiguities. For these reasons, single pion events represent the perfect benchmark for local cluster corrections.

Using the calibration hit true reference it is possible to study the linearity (L) of each correction separately, where the R_X and T_X terms of Equation 1 are calculated as a sum over all the clusters in the event. The performance for single pion simulations is shown in Figure 2 for the central region of the detector. All of the three cluster corrections perform very well, within residual non-linearities of few percent for low pion energies. This is because in the low energy regime the correlation between the lost and the deposited energy is more difficult to establish.

2.2. Di-jet simulations

The final goal of the local hadron calibration approach is the calibration of jets in high energy collisions. In order to study the performance of the cluster corrections in a typical collision event, $2 \to 2$ processes simulated with Pythia [6] have been used. In this case the terms R_X and

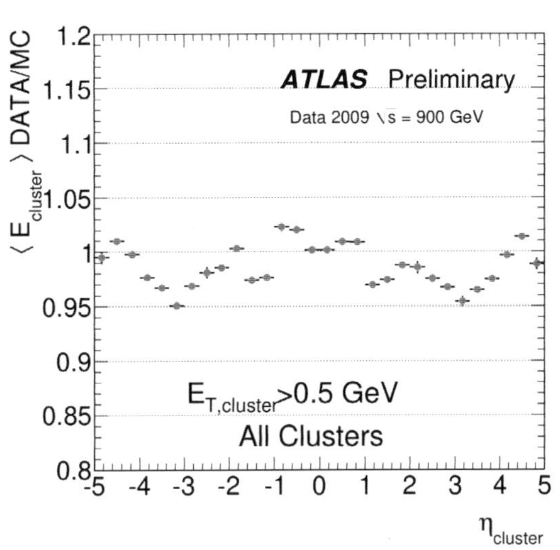

Figure 4. Ratio of the mean cluster energy in data and in Monte Carlo simulations versus η. Only clusters with transverse energy larger than 500 MeV are considered.

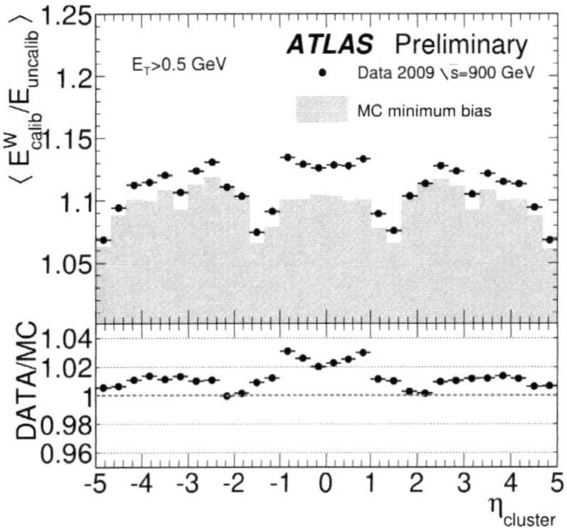

Figure 5. Mean of the ratio of the cluster energy calibrated with hadronic weights (W) over the energy before calibration in data (points) and Monte Carlo simulations (histogram), versus η. Only clusters with transverse energy larger than 500 MeV are considered. The lower plot shows the ratio of data over Monte Carlo simulations.

T_X of Equation 1 are calculated as a sum over the clusters belonging to a jet, as chosen by the Anti-K_T algorithm with R=0.4 [7]. As shown in Figure 3 for the central region, the hadronic weight corrections perform very well for all energies, while the out of cluster (in green) and the dead material (red) corrections show a non linearity of up to 10 % for low jet energies. This is mainly due to the fact that the definition of the true out of cluster and dead material energy for clusters includes calibration hit energy due to lost particles, that do not contribute to the cluster formation. In order to remove this ambiguity, a new tool that correlates each calibration hit to the parent particle has been developed and will be used to refine the true cluster energy definition.

3. Validation on data

The first step towards the validation of the local hadron calibration on data is to compare the effect of the cluster corrections in data and Monte Carlo. For this purpose ATLAS data collected in December 2009 at a center of mass energy of $\sqrt{s} = 900$ GeV have been used [4]. Events are selected using the minimum bias trigger scintillators (MBTS), installed on both sides of the ATLAS detector and covering the pseudo-rapidity region $2.09 < |\eta| < 3.84$. Offline cuts on the MBTS hits and calorimeter timing (as detailed in [8]) are used to remove contamination from events produced by the beam halo, by the interaction of beam protons with the gas in the beam pipe or by cosmic muons. Only runs in which the calorimeter and the inner detector were fully operational and the solenoid was on have been considered. These criteria select 330810 events.

Data are compared to Monte Carlo (MC) simulations of non-diffractive, single and double diffractive processes in proton-proton collisions at a center of mass energy of $\sqrt{s} = 900$ GeV. Pythia with ATLAS MC09 tune [9] is used as generator, while the detector response is simulated

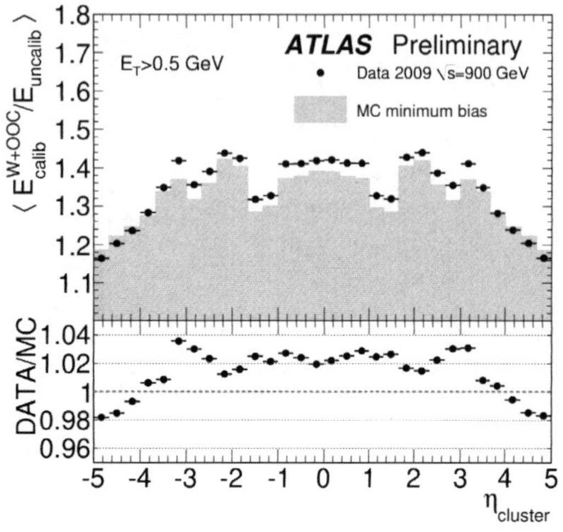

Figure 6. Mean of the ratio of the cluster energy calibrated with hadronic weights (W) and out of cluster corrections (OOC) over the energy before calibration in data (points) and Monte Carlo simulations (histogram), versus η. Only clusters with transverse energy larger than 500 MeV are considered. The lower plot shows the ratio of data over Monte Carlo simulations.

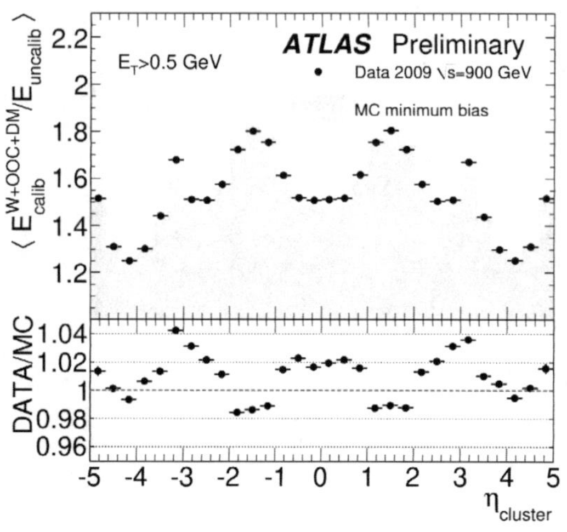

Figure 7. Mean of the ratio of the cluster energy calibrated with hadronic weights (W), out of cluster corrections (OOC) and dead material corrections (DM) over the energy before calibration in data (points) and Monte Carlo simulations (histogram), versus η. Only clusters with transverse energy larger than 500 MeV are considered. The lower plot shows the ratio of data over Monte Carlo simulation.

with GEANT4. The same trigger and event selection is used on the Monte Carlo simulations and on data.

As first step the energy of clusters at the electromagnetic scale has been compared between data and Monte Carlo. The ratio between the mean cluster energy in data and in MC for the whole pseudo-rapidity range is shown in Figure 4 for clusters with transverse energy larger than 500 MeV. The overall agreement is very good: $\approx$ 2% in the central region and $\approx$ 5% in the end-cap and forward region. Further studies are needed to understand the residual differences.

In order to validate the various calibration steps, for all clusters the ratio of the calibrated and the uncalibrated energy is considered. This procedure reduces the dependence on the absolute energy difference before calibration between data and Monte Carlo. The average of this ratio is shown versus pseudo-rapidity in Figure 5 for the hadronic weight corrections. The same study is shown in Figure 6 for clusters calibrated with hadronic and out of cluster corrections. Finally the same study is performed after hadronic weight and out of cluster and dead material corrections are applied to the clusters, as shown in Figure 7. Only clusters with a transverse energy larger than 500 MeV are considered. Since the selected minimum bias sample is dominated by low energy clusters, the corrections applied for energy lost outside clusters and in dead material are large with respect to the hadronic compensation weights. The ratio between data and Monte Carlo is shown at the bottom of Figures 5, 6 and 7. The value of the corrections is larger in data than in Monte Carlo simulations by $\approx$ 2% in the central region of the detector. This difference is likely due to incorrect noise description in the simulation, which is already under investigation. For the out of cluster and dead material corrections around $|\eta| = 3.1$ the data and the simulation show a difference of $\approx 4\%$. This region corresponds to the transition between the end-cap and the forward calorimeter, where a large amount of dead material due to the forward calorimeter support structure is present. For this reason incorrect dead material description in the simulation is the likely cause of the disagreement, which can be evaluated and corrected with more data available.

4. Conclusions

The local hadron calibration is a complete method to calibrate the energy of jets to the particle jet level. It consists of a set of cluster corrections which are applied prior to the jet reconstruction and of jet level corrections dedicated to out of cone effects.

The cluster corrections have been discussed in detail. Their performance, in terms of linearity with respect to the true cluster energy, has been shown both for the single pion simulation case and for the di-jet simulation case. In both cases the linearity is near to unity for the hadronic weight corrections, while out of cluster and dead material corrections show a discrepancy of the order of few % for single pions and of 10 % for the di-jets. In order to understand this discrepancies a refined definition of the true cluster energy is in preparation, with the use of information on the origin of each true energy deposit in the simulation.

The validation of the cluster corrections on data has started using ATLAS data collected in December 2009 at a center of mass energy of $\sqrt{s} = 900$ GeV. Data have been compared to Monte Carlo simulations of non diffractive, single and double diffractive processes. For the uncalibrated scale the overall agreement is very good: $\approx 2\%$ in the central region and $\approx 5\%$ in the end-cap and forward region. The agreement for the ratio of the calibrated scale to the uncalibrated scale is very good: $\approx 2\%$ for most η regions, and up to $\approx 4\%$ for the whole η range.

Acknowledgments

This work is partially supported by the European Commission, through the ARTEMIS Research Training Network, contract number MRTN-CT-2006-035657.

References

[1] **The ATLAS** Collaboration, *Expected Performance of the ATLAS Experiment - Detector, Trigger and Physics*, arXiv:0901.0512.
[2] W. Lampl *et. al.*, *Calorimeter clustering algorithms: Description and performance*, . ATL-LARG-PUB-2008-002.
[3] **The ATLAS** Collaboration, T. Barillari *et. al.*, *Local hadronic calibration*, . ATL-LARG-PUB-2009-001.
[4] *Inputs to jet reconstruction and calibration with the atlas detector using proton-proton collisions at* $\sqrt{s} = 900$ *gev*, Tech. Rep. ATLAS-CONF-2010-016. ATLAS-COM-CONF-2010-016, CERN, Geneva, Jul, 2010.
[5] S. Agostinelli *et. al.*, *Geant4 a simulation toolkit*, *Nucl. Instr. and Meth. A* **506** (2003), no. 3 250 – 303.
[6] T. Sjostrand, L. Lonnblad, and S. Mrenna, *PYTHIA 6.2: Physics and manual*, hep-ph/0108264.
[7] M. Cacciari, G. P. Salam, and G. Soyez, *The anti-k(t) jet clustering algorithm*, *JHEP* **04** (2008) 063, [arXiv:0802.1189].
[8] **The ATLAS** Collaboration, *Charged-particle multiplicities in pp interactions at sqrt(s) = 900 GeV measured with the ATLAS detector at the LHC*, *Phys. Lett.* **B688** (2010) 21–42, [arXiv:1003.3124].
[9] *Atlas monte carlo tunes for mc09*, Tech. Rep. ATL-PHYS-PUB-2010-002. ATL-COM-PHYS-2010-033, CERN, Geneva, Mar, 2010.

Journal of Physics: Conference Series **293** (2011) 012058 doi:10.1088/1742-6596/293/1/012058

The commissioning and first results on the performance of the CMS Preshower Detector

Chia Ming, Kuo (On behalf of the CMS ECAL Group)
Department of Physics, National Central University, Chung-Li, 32001, Taiwan
E-mail: chia-ming.kuo@cern.ch

Abstract. The Preshower is the first part of the CMS endcap electromagnetic calorimeter, as seen by the particles. It is composed of 2 layers of lead absorbers followed by silicon sensors of a total surface of 17 m^2, the biggest ever built. The use of the silicon strip sensors improves the spatial accuracy of the incidence position of electromagnetic showers. This allows an additional rejection of background (e.g. neutral pions) to new physics, such as the two-photon SM Higgs decay. In this article, the results from first commissioning, the performance with first LHC beams and first in-situ calibration with charged particles are described.

1. The CMS Preshower detector

The electromagnetic calorimeter (ECAL) of CMS comprises a barrel and two endcaps ("EE") based on lead tungstate ($PbWO_4$) scintillating crystals, with a silicon-strip preshower ("ES") in front of each EE [1].

One of the main physics goals of CMS is to discover Higgs particle. If Higgs has mass less than 150 GeV, the easiest way to detect it in CMS is via its decay to two high-energy photons. However, neutral pions can mimic high-energy photons when they decay into two closely-spaced lower energy photons that the ECAL picks up together. In the endcap regions, where the angle between two photons from the π^0 decay is likely to be small enough to cause the problem. Therefore, a highly-segmented ES is placed in front of EE to prevent such false signals.

The ES covers $1.653 < \eta < 2.6$ and has two layers: lead radiators ($2X_0$ and $1X_0$, respectively) initiate electromagnetic showers from incoming electrons/photons whilst silicon strip sensors placed after each radiator measure the energy deposited and the transverse shower profiles. Each silicon sensor measures 63×63 mm^2, with an active area of 61×61 mm^2 divided into 32 strips. The nominal thickness of the silicon is 310 μm. They are grouped into "ladders" of 7, 8, or 10 sensors. The ES front-end electronics can operate in two modes: High Gain ("HG") is used for calibration purposes and low-energy collision running, including 7 TeV; Low Gain ("LG") has a higher dynamic range more suitable for higher energy collision runs.

2. Commissioning

After the installation finished in the first half of 2009, the most important thing was to check that all connectivity at the detector side for high and low voltage power lines, optical fibres and control cables was correct and functional, prior to "closing" CMS. Data were taken using a local DAQ system based on the XDAQ framework (CMS standard) [2, 3] and analyzed by CMS Data Quality Monitoring framework. The results of the commissioning were that all cables and fibres were correctly connected and the vast majority of the detector channels were functioning perfectly. Analysis of the noise performance in HG and LG modes was also performed. The average intrinsic noise is around 5.6 (2.5) ADC counts for HG (LG) meeting with beam test performance [4].

3. Preshower Operation during LHC collisions

3.1. Timing alignment

For every ES channel, the electronic response is sampled at -5, 20 and 45 ns relative to the arrival of the trigger signal. Reconstruction of the particle arrival times is extracted by fitting the digitized amplitude samples to the known pulse shape obtained from beam test [4].

The first step of ES time alignment was the adjustment of the digitization phase of the phase locked loop ("PLL") in steps of 1.04 ns, which is common to a varying number of strips, between 192 and 320 depending on the ladder type. The PLL settings were estimated using the ES geometry and cable length. Collision data were then used to fine-tune the PLL phase adjustment for the whole ES. Figure 1 shows the synchronization of the relative phase for 7 TeV collision events, before and after fine-tuning of the PLLs.

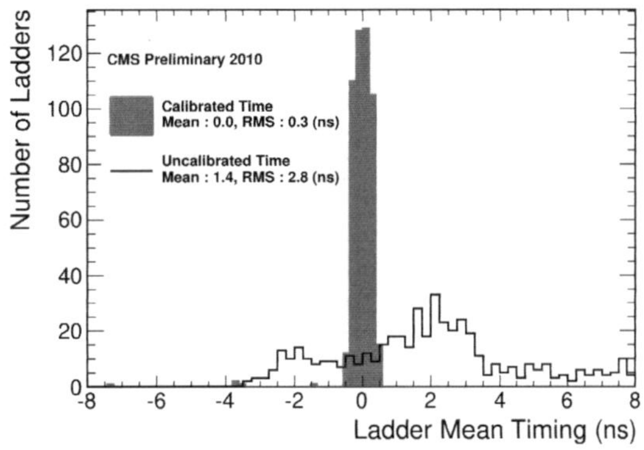

Figure 1. Synchronization of the relative phase between Preshower readout electronics (at the level of "ladders") and the passage of particles through the sensors, before and after fine-tuning.

3.2. Preshower status

Of the 4288 silicon sensors forming the ES, 9 are classed as "dead" due to hardware problems that occurred after installation. In addition, a few strips (23 out of 137216 total) are classed as noisy. The total percentage of fully-functional strips is 99.79%.

4. Data-MC comparisons of observables

4.1. Occupancy

The ES occupancy is calculated as the percentage of strips with a reconstructed energy greater than four sigma of the noise. This threshold corresponds to 32 keV of energy deposited in the strip in HG mode. Figure 2(left) shows how the occupancy varies with pseudo-rapidity (η, averaged over all ϕ), whilst fig. 2(right) shows the variation with azimuthal angle (ϕ, averaged over all η), both compared with simulation, where absolute values of the MC occupancies have been normalized to the data. The ϕ variations are due to the X-Y geometry of the ES and are well-simulated in MC.

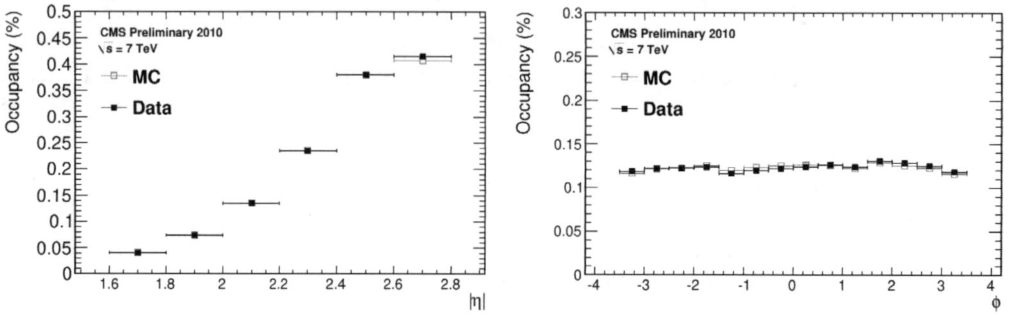

Figure 2. Preshower occupancy as functions of pseudo-rapidity (left, averaged over all ϕ) and azimuthal angle (right, averaged over all η).

4.2. Energy deposits on ES planes

Algorithm for clustering the energy deposits of incoming particles for the endcaps is done in three steps.

- Firstly, "basic clusters" (BC) of groups of crystals are formed around seed crystal having $E_T > 1$ GeV
- Then, we extrapolate their paths back to the centre of the collision and build the ES cluster around the intersected strip on each ES plane. The ES energy in MIPs is then converted to GeV and added to the EE BC energy.
- The endcap clusters are then further grouped into "superclusters" (SC), which are extended in ϕ, thus minimizing the cluster containment variations due to the strong magnetic field.

Figure 3 shows the ES cluster energies for each plane for 7 TeV minimum bias events for both data and MC. The ES clusters are associated to ECAL superclusters with raw transverse energy larger than 2 GeV and $1.7 < |\eta| < 2.5$. Figure 4 shows the energy ratio of the second to the first ES plane. These distributions show good agreement between data and simulation.

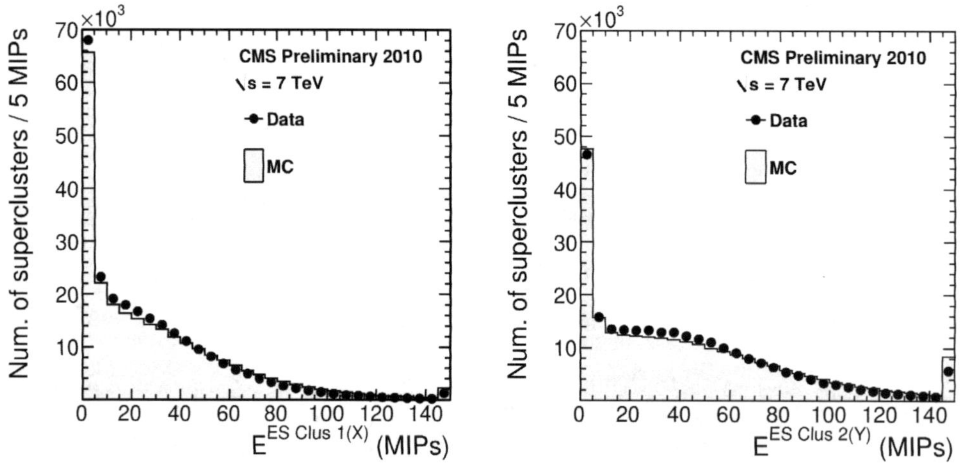

Figure 3. Energy deposited in each of the two ES planes for supercluster raw transverse energy larger than 2 GeV. Overflows are added to the last bin.

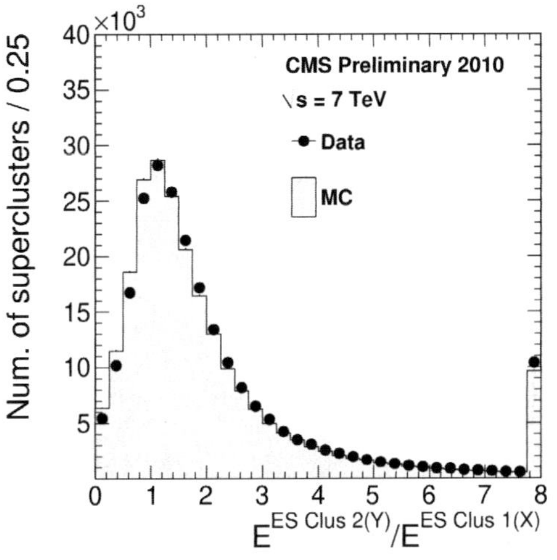

Figure 4. The ratio of the energy deposit associated with a supercluster on the second ES plane to the first one for supercluster raw transverse energy larger than 2 GeV. Overflows are added to the last bin.

5. Alignment between ES and EE

The ES disks are supported by a conical structure attached to the backplate of the endcap hadron calorimeters. This backplate also supports the EE Dees, so relative alignment between the EE and ES should be very good by design. EE clusters with $E_T > 2$ GeV, with no isolation or electron/photon identification requirements, were used as the source of this study, with position measurements being made in the EE and the ES. Note that the accuracy of these position measurements is rather poor due to the low E_T of a majority of the superclusters, but is sufficient for a first alignment. The same study was made with MC, assuming perfect alignment. The "out of the box" mis-alignment is less than half a millimetre in all planes. Figure 5 shows the alignment between the EE and ES for one of the ES planes.

6. First ES in-situ calibration with charged particles

The ES is a sampling calorimeter and essentially counts the number of charged particles passing through the silicon. The particles used to calibrate the ES are currently pions, with momentum greater than 1 GeV and an average momentum of 6.1 GeV. This means that they are nearly minimum ionizing, so for simplicity we refer to them as "MIPs". The design-goal accuracy of the channel-by-channel calibration is set to 5%. This corresponds to a contribution of about 0.25% to the overall EE+ES energy resolution for high-energy electrons since only 5% of a photon or electron energy is deposited in ES. There are two main sources of response variation (sensor-to-sensor and channel-to-channel) at the start-up: sensor thickness and gain uniformity of the electronics. Prior to their installation, all ES silicon modules were pre-calibrated with cosmic rays for at least 16 hours in the laboratory. The accuracy of the pre-calibration is estimated to be about 2.5%.

The first in-situ ES MIP calibration has been carried out by using the charged tracks with momentum greater than 1 GeV pointing to the ES from 7 TeV minimum bias events. Figure 6(left) shows the energy distribution for a silicon sensor. The distribution is fitted by a Landau function convoluted with a Gaussian function. The MIP-equivalent energy for this sensor is 47.2 ADC counts. For each sensor, the "MIP" value is defined to be equal to the fitted peak position. The distribution of the measured MIP values for all sensors having at least 1000 hits (currently

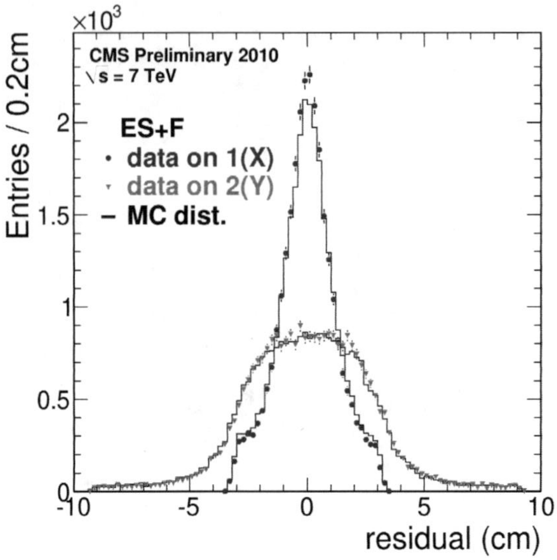

Figure 5. Alignment between ES and EE for one of the ES planes for both data and MC. The "out of the box" alignment is better than a millimetre.

3792 out of 4288 sensors) is shown in Figure 6(right).

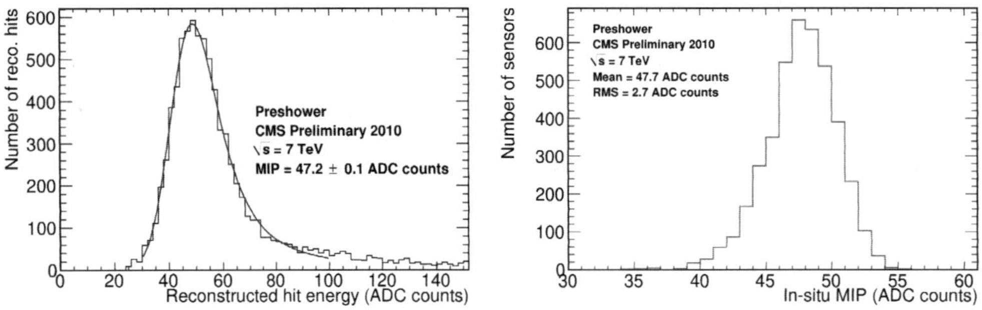

Figure 6. Energy distribution for a silicon sensor, requiring tracks with $p > 1$ GeV pointing to the ES (left). Distribution of measured MIP values for all ES sensors having at least 1000 hits. (right)

After in-situ calibration and measurement of the noise of each channel, an estimation of the signal to noise for single ionizing particles can be made. The design specification of the ES was to have an S/N ratio for single particles of about 8 for HG mode. The observed ratio, as shown in Figure 7 is between 9 and 11.

Figure 8(left) shows the correlation between sensor "MIP" values obtained from in-situ calibration and pre-calibration. The residuals plot is shown in Figure 8(right). The offset is due to the different particle momentum spectra and operating temperature in the two cases. The precision of in-situ calibration is then estimated to be about 2.2% which is better than the design-goal target.

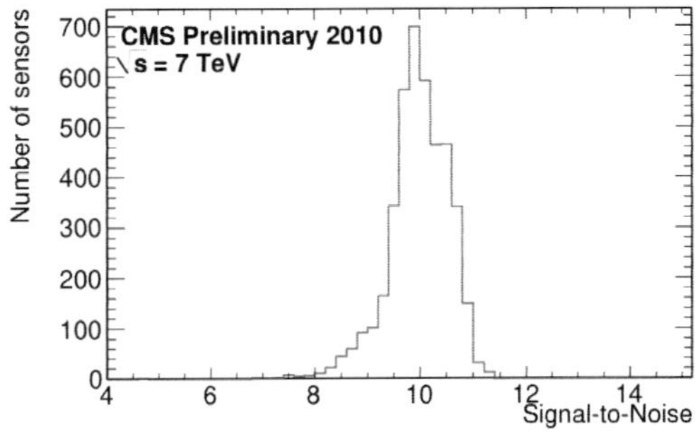

Figure 7. ES signal to noise ratio for single particles.

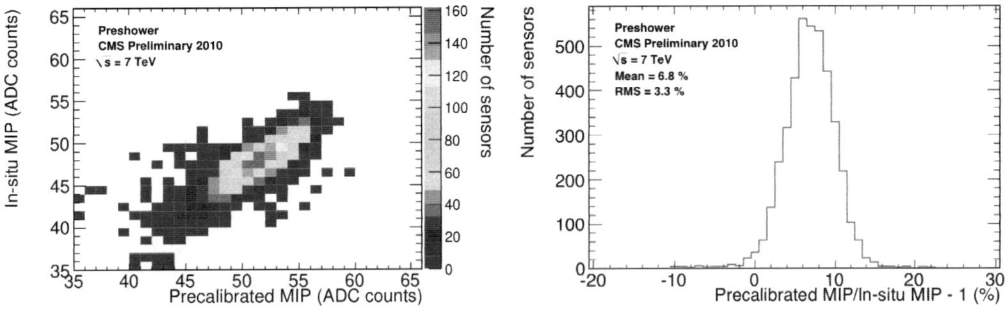

Figure 8. ES MIP values measured in CMS compared to those measured using cosmic rays in the laboratory (left). Residuals between the in-situ MIP calibration and pre-calibrations (right).

7. Summary

The CMS Preshower detector was installed and commissioned successfully in CMS in the first half of 2009. The results from first commissioning confirmed all connectivity in place and expected noise performance. Throughout the first data taking periods of the LHC, the CMS Preshower has been operational and 99.79% of channels are funcationing perfectly. The first collisions at 7 TeV have provided the opportunity to test our understanding of the basic observables, such as occupancy and energy clustering. Good agreement between data and simulation has been shown for these observables. The alignment between ES and EE is better than half a millimetre. First in-situ calibration with charged particles has been carried out, achieving required accuracy.

References

[1] CMS Collaboration, "Performance and operation of the CMS electromagnetic calorimeter", *J. of Instr.* **5** (2010) T03010.
[2] J. Gutleber, S. Murray, and L. Orsini, *Comput. Phys. Commin.* **153** (2003) 155.
[3] P. Musella, "The ECAL online software in the commissioning of the CMS detector" *Nucl. Instr. and Meth. A* **617** (2010) 303.
[4] S. W. Li, A. Go, and C. M. Kuo, "Performance of CMS ECAL Preshower in 2007 test beam", *J. Phys.: Conf. Ser.* **160** (2009) 012052.

Journal of Physics: Conference Series **293** (2011) 012059 doi:10.1088/1742-6596/293/1/012059

Commissioning of the Scintillator Pad Detector of LHCb with cosmic rays and first LHC collisions

Ricardo Vazquez Gomez on behalf of the LHCb collaboration
Universitat de Barcelona, Facultat de Fisica, Av Diagonal 647, Barcelona 08028, Spain
E-mail: rvazquez@cern.ch

Abstract. The Scintillator Pad Detector is part of the LHCb calorimeter system which includes a pre-shower with 2.5 radiation length, an electromagnetic and an hadronic calorimeter. The goal of the Scintillator Pad Detector situated upstream of the calorimeters is to tag charged particles entering the calorimeter system. This contribution discusses the requirements for the synchronization and calibration of the SPD detector. The requirements, methods and precisions obtained at the several stages are presented.

1. Introduction

The Large Hadron Collider (LHC) placed at CERN started to provide collisions in autumn 2009. In 2010 it has reached center of mass energies of 7 TeV and will start to test several extensions of the Standard Model (SM) of interactions and elementary particles. The LHCb detector [1, 2, 3] is one of the four experiments of LHC and it has been specially conceived to study B mesons physics: the study of CP violation and B rare decays [4].

The structure of this document is as follows. In Section 2, a description of the LHCb calorimeter is given. The commisioning of the SPD and its results are discused in Section 3 and 4 focusing on the time alignment and its calibration. Finally, some conclusions will be drawn in Section 5.

2. The LHCb calorimeter

The LHCb calorimeter [5] is made of four subdetectors: Scintillator Pad Detector (SPD), Pre-Shower detector (PS), Electromagnetic Calorimeter (ECAL) and Hadronic Calorimeter (HCAL). In front of the electromagnetic and hadronic calorimeters incoming particles cross two planes of scintillating cells separated by a lead layer of 2.5 radiation lengths. The first layer of cells is the SPD.

The calorimeter system is used at several stages. It provides high transverse energy hadron, electron and photon candidates for the first trigger level (L0) [6] and for the reconstruction. In particular, the SPD is used as part of the L0 which works at a frequency of 40 MHz by tagging the electromagnetic candidates as electrons or photons and by providing a fast estimation of the charged track multiplicity per event. Its multiplicity has already been efficiently used as the standalone trigger for the LHC injection test events and as part of the minimum bias trigger on any proton-proton interaction in first 2009 & 2010 collisions. It also has been used to monitor the luminosity and it can be used to veto high occupancy events when the luminosity provided by the LHC will be high.

The SPD is composed by a layer of 6016 square scintillating plastic cells. The cells have a depth of 1.5 cm, and a side length ranging from 4 to 12 cm to accommodate the fact that the occupancy decreases as the distance to the beam pipe increases. The lateral segmentation is the same as in the PS and ECAL, which is mainly used for photon identification. The cells are readout by wave-length shifting fibres, which take the light to 64-channel multi-anode phototubes. The SPD very front-end (VFE) electronics integrates the PMT signal in the 25 ns window defined by the LHC bunch crossing rate [7]. Then, it performs a binary discrimination with a threshold placed at half the peak of the deposition of minimum ionizing particles (MIP). The resulting bit is used to tag the trigger candidates as electrons or photons, and the sum of bits determines the event charged multiplicity. In order to cope with the required working frequency, the SPD VFE design includes two independent integrators and comparators for each channel, which are used alternatively.

3. Time alignment

A first requirement of the commissioning of the detector was the synchronization of its integration time to the passage of particles from LHC collisions. This needed to be done with a precision of 1 ns, in order to both maximize the signal efficiency and reduce the spill-over. The delay in the integration time is applied at the VFE level, which implies that groups of 64 channels have the same starting integration time. At the right integration time there should not be signal in the previous bunch crossing and a small signal in the following one. To do that it has been used the capability of the LHCb trigger to readout several consecutive bunch crossing around the bunch crossing where the collision takes place [8].

A first synchronization with respect to the electromagnetic and hadronic calorimeter integration time was obtained by using cosmic rays. Vertical flux in the cavern has been measured to be $0.9\,\mathrm{Hz/m^2}$ (under HCAL). In order to mimic as much as possible the particles coming from p-p collisions, a selection of horizontal cosmics ($< 30^\circ$) was performed with an estimated rate of $0.0046\,\mathrm{Hz/m^2}$. The events were triggered using the coincidence of signal at the ECAL and HCAL at a rate of 10 Hz. Although LHCb geometry is not well suited for cosmics, more than 4 million were triggered. The reconstruction of cosmic tracks is done by using only the calorimetric information. This provides information of the slope of the track and the arrival time of the cosmic muons with respect to the integration of the calorimeters. The later was obtained by comparing the energy collected by the calorimeters in consecutive integration windows [9]. As can be seen in Figure 1 the statistical precision provided by the method, using 1 million cosmic events, is around 0.5 ns with a synchronization relative to the ECAL of 2 ns. The delays obtained using this method were implemented in October 2009.

The final synchronization with respect to particles from the interaction point was obtained by scanning the starting time of integration during part of the collision data-taking of 2009 [10]. After a second scan in 2010, the analysis of the results allowed to find the optimal delay parameters with a statistical precision of about $0.3\,ns$ with 600k events and a VFE intercalibration of about $1\,ns$. Figure 2 shows the last delays applied in the electronics in April 2010.

4. Calibration

The second important aspect of the SPD commissioning is its calibration. The initial position of the discriminating thresholds were determined from a calculation based on the different gain factors that enter in the detection chain: the number of photoelectrons at the PMT, the gain of the PMT (calibrated as a function of the high voltage), and the gain of the electronics [11], this will be referred as precalibrated units. The fine calibration has to be done with charged particles from cosmic rays and LHC collisions.

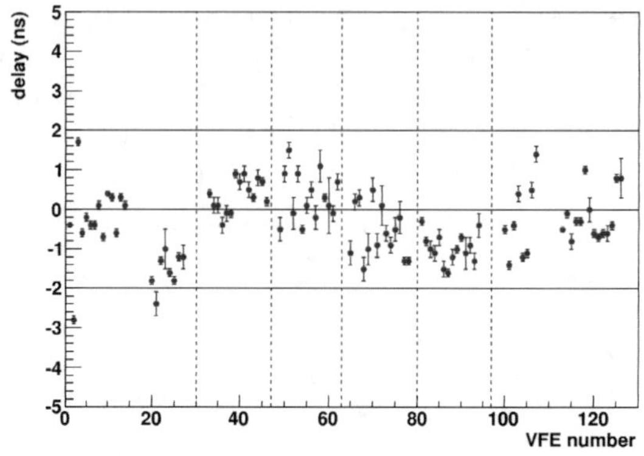

Figure 1. Delays per VFE obtained from cosmics rays.

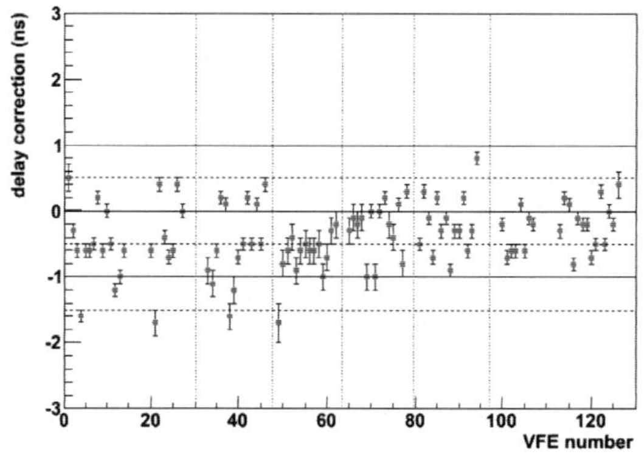

Figure 2. Delays per VFE obtained from 2010 LHC collisions.

Due to the binary output of the SPD electronics it is necessary to use dedicated methods based on the scan of the threshold position. The procedure measures the SPD efficiency to detect charged particles as a function of the discriminating threshold. The efficiency is defined as the ratio between the number of hits with a track in a given cell and the number of tracks poiting to a given cell. The result of the efficiency as a function of the threshold can be compared with the theoretical expectation given by the integral of the convolution of a Landau and a Poisson distribution. The Landau distribution describes energy lost by charged particle traversing a thin layer of material while the Poisson distribution describes de fluctuations in the number of photoelectrons generated at the photocathode. The shape of the theoretical function as a function of the SPD threshold in precalibrated MIP units can be seen in Figure 3. The resolution in the electronics to set the threshold is 5% of E_{MIP} which directly sets our objective resolution at the 5% level.

The first threshold calibration was performed using cosmic rays using a discriminant threshold of 1 precalibrated MIP. The track selection criteria was based on two main requirements. Firstly, the direction of the cosmic had to have low angle to avoid excessive ionization in the detector and secondly the arrival time of the cosmic had to be centered in the integration window to avoid any loss in the signal. As the statistics were limited, only a correction factor per VFE was

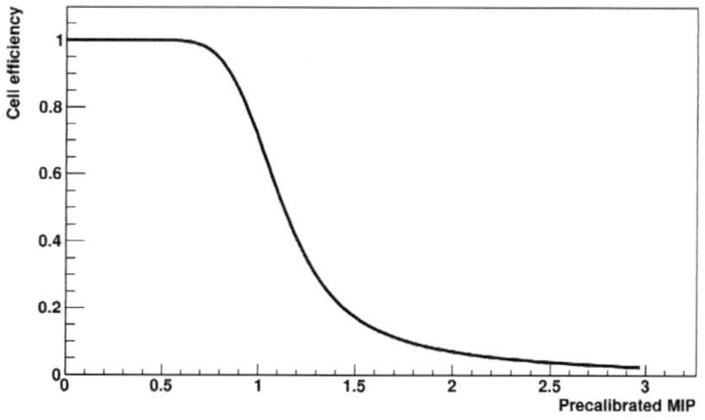

Figure 3. Theoretical efficiency as a function of the discriminant threshold in precalibrated units.

Table 1. Statistics used for calibration.

Threshold (E_{MIP})	# collisions
0.3	2.5 M
0.5	2.7 M
0.8	3.9 M
1.0	3.8 M

extracted. As cells in the same VFE have the same high voltage applied we were sensitive to possible problems in the HV settings. Figure 4 shows the correction factor per VFE obtained from cosmic data and one can see that the pre-calibration was correct up to a 20% level.

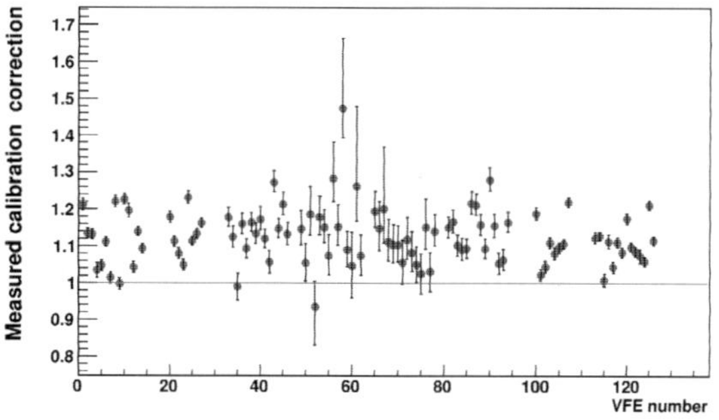

Figure 4. Threshold correction per VFE obtained from cosmic analysis. The horizontal line sets the perfect calibration where the correction is not needed.

The channel by channel calibration has been performed using the first data collected in the 2010 run. Data at four different threshold in precalibrated units has been taken with the statistics shown in Table 1.

The corrections have been obtained by fitting the theoretical function to the experimental points, using as free parameters: the calibration constant, the number of photoelectrons and

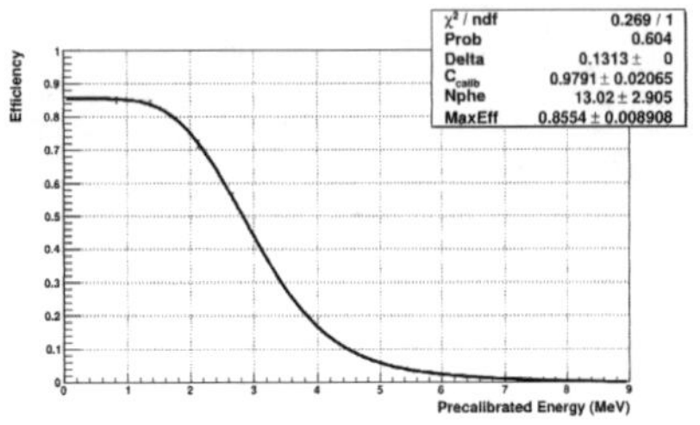

Figure 5. Example of a fit for a typical cell.

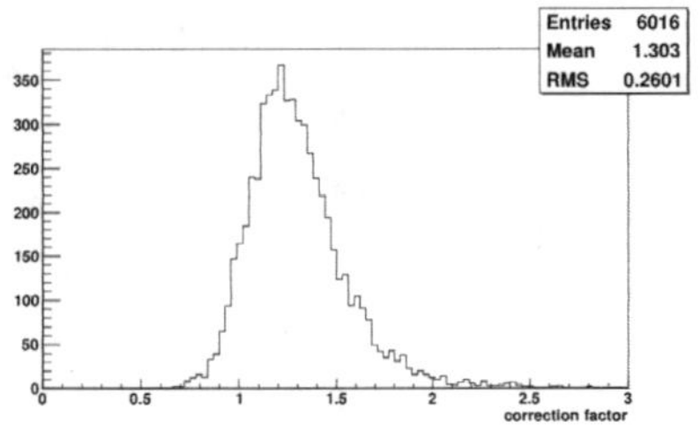

Figure 6. Distribution of calibration contants per channel obtained from 2010 collisions.

the maximum efficiency. The result of one of this fits can be seen in Figure 5. The distribution of the cell calibration constants is shown in Figure 6 where one can see that the dispersion is of the order of 20% which confirms the results obtained with cosmic rays. This corrections were applied to 97% of the cells in June 2010. The missing 3% have not abbandoned the high efficiency plateau and a dedicated study to calibrate them will be soon performed with new collision data by taking data at higher threshold values.

5. Conclusions

The SPD is a fully operational detector which is approaching its final design requirements. The commissioning with cosmic rays provided an excellent starting point for the time alignment and for the threshold calibration. The statistical precision in the relative time alignment measurement using cosmics rays was 2 ns while the threshold calibration showed that the precalibration was correct up to the 20% level. The final time alignment with collisions has provided a precision of 1 ns. The final calibration constants have confirmed the results obtained with cosmics rays and have already been applied to the 97% of the cells.

References

[1] LHCb Collaboration 1998 LHCb Technical Proposal *Preprint* CERN/LHCC/1998-004
[2] LHCb Collaboration 2008 *J. Instrumentation* **3** S08005
[3] LHCb Collaboration 2003 LHCb Reoptimized Detector Design and Performance *Preprint* CERN/LHCC/2003-0030
[4] LHCb Collaboration 2009 Roadmap for selected key measurements of LHCb *Preprint* LHCb-PUB-2009-029. %arXiv:0912.4179v2[hep-ex]
[5] LHCb Collaboration 2000 LHCb Calorimeters Technical Design Report *Preprint* CERN/LHCC/2000-0036
[6] LHCb Collaboration 2003 LHCb Trigger System Technical Design Report *Preprint* CERN/LHCC 2003-031
[7] Bota S *et al.* 2000 Scintillator Pad Detector front end electronics design *Preprint* CERN/LHCC/2000-027
[8] Calvo M *et al.* 2010 Time alignment of the SPD with cosmic rays and LHC injection events *Preprint* LHCb-2010-015
[9] Ahmis Y 2009 Time alignment of the electromagnetic and hadronic calorimeters, reconstruction of the $B \to D^-\rho(770)^+, B_s \to D_s^-\rho(770)^+$ and $B_s \to D_s^- K^{*+}(892)$ decay channels with the LHCb detector *Preprint* CERN-THESIS 2009-117
[10] Calvo M *et al.* 2010 Time alignment of the SPD with LHC proton-proton collisions *Preprint* LHCb-2010-xxx
[11] Picatoste E 2009 *J. Phys.: Conf. Series* **160** 012046

Journal of Physics: Conference Series **293** (2011) 012060 doi:10.1088/1742-6596/293/1/012060

Commissioning of the LHCb preshower detector with cosmic rays and first LHC collisions

Valentin Niess
on behalf of the LHCb Collaboration
Laboratoire de Physique Corpusculaire, Université Blaise Pascal, IN2P3/CNRS,
24 avenue des Landais, BP 80026 63177 Aubière Cedex, France

E-mail: niess@in2p3.fr

Abstract. The LHCb calorimeter system consists of a Scintillating Pad Detector (SPD), a PreShower detector (PS) followed by electromagnetic (ECAL) and hadronic (HCAL) calorimeters. The LHCb preshower detector provides the longitudinal segmentation of the electromagnetic shower detection. It is located immediately upstream from the ECAL, with one-to-one correspondence between ECAL towers, SPD and preshower cells. The PS detector is part of the first level of the LHCb trigger system (called Level-0). It is used in conjunction with the SPD, ECAL and HCAL to search for clusters of 2×2 cells and to identify e, γ, π^0, hadron of highest E_T. This contribution discusses the requirements for the synchronisation and calibration of the PS detector, as well as the methods that were used to determine the first settings before the LHC start-up. The methods used to perform the fine synchronisation and calibration once LHC collisions were available are introduced, and the precisions that were obtained are given.

1. Introduction

1.1. The LHCb detector

The LHCb detector [1], is a single arm forward spectrometer dedicated to precision studies of $b\bar{b}$ pairs and CP symmetry violation at the LHC. It covers the peculiar pseudo-rapidity region $1.9 \leq \eta \leq 4.9$, where pairs of b quarks are mostly produced in pp collisions. Furthermore, in order to maximise single pp interactions per bunch crossing it is enough to operate LHCb at a reduced luminosity of a few times 10^{32} $\text{cm}^2 \cdot \text{s}^{-1}$. The main challenge of LHCb is to perform precision measurements in a difficult hadronic environment. On the other hand, the rate of b production in LHC collisions is huge and all B species are produced yielding large samples of B_s mesons, as well as B_c and b baryons, which are poorly studied so far.

1.2. The LHCb calorimeter system

The LHCb calorimeter system is an interplay of 4 sub-systems arranged in a line: the Scintillating Pad Detector (SPD) [2], the preshower detector (PS) -which is the topic of this paper- followed by the electromagnetic (ECAL, Shashlik design, 25 X_0, 1.1 λ_I) and hadronic (HCAL, TileCal design, 5.6 λ_I) calorimeters [3, 4]. These 4 sub-systems are divided in projective cells each with their own dedicated electronics. The role of the calorimeter system is double. It is a key component of the first level trigger (L0) which is fully hardware, running at the 40 MHz LHC bunch crossing frequency. The L0 trigger is based on the fast response of the calorimeters and muon detectors in order to identify high transverse momentum hadrons, e, γ, π^0 and μ that could sign the presence of a B decay. This allows to reduce the data rate to 1 MHz with only $\sim$50% loss in efficiency. Additionally, the calorimeter system provides offline energy reconstruction for showering events, which is of particular interest for neutral particles: γ, π^0.

2. The preshower detector and its electronics

2.1. The preshower detector

Looking from the interaction point, the preshower detector is located 12 m away, after the SPD and in front of the ECAL. The SPD and PS are both made of a vertical rectangular plane (6.2 m large and 7.6 m height) of 6016 cells of 15 mm thick polystyrene scintillator pads. The cells are arranged projectively around the beam with a pass-through hole in the middle. In order to compensate for the 100 times higher density of particles close to the beam than at outer borders, there are 3 different granularity regions for cells, with cell size increasing the farther from the beam, as: 4×4 cm^2 (Inner), 6×6 cm^2 (Middle) and 12×12 cm^2 (Outer). SPD and PS cells are mapped 1 to 1 to ECAL cells, which are similarly divided in 3 granularity regions. In addition, both SPD and PS cells include an embedded LED system for monitoring purpose. Despite their identical detector designs SPD and PS however differ by their functionalities and so by their electronics. Indeed, SPD and PS are separated by a 12 mm sheet of lead absorber (2.5 X_0). Hence, while the SPD allows for the identification of charged scintillating particles the preshower provides insight on e and γ that started showering in the lead sheet as opposed to heavier leptons or hadrons.

2.2. The VFE electronics

The scintillation light in PS cells is collected by a wave length shifting fibre coiled in a groove machined in the scintillator pad. The light is then transmitted via 1 to 3 m long clear fibres to 64 channels multi-anode photomultiplier's tubes (MAPMT) mounted on Very Front End (VFE) electronic boards. The VFE boards are 100, located above and below the scintillator plane. They are shielded against the remnant magnetic field by mild steel boxes. The MAPMT's tubes are further shielded by μ-metal cylinders. Each VFE board can be mapped to a 8×8 cluster of PS cells. Since the scintillation light signals can be very fluctuating, the output signal of the MAPMT is integrated in the VFE, at 40 MHz, by two alternating chip which warrants no dead time.

2.3. The FE electronics

The shaped analogical signal from the VFE is transmitted by 27 m long RJ45 cables to PS Front End (FE) electronic boards where it is digitised by 10 bits ADCs. The PS FE electronics is located in calorimeter racks on a dedicated platform on top of the calorimeters. There are 100 FE boards running at 40 MHz and mapped 1 to 1 to VFE electronics. Hence, each FE board can handle the information from 64 PS detector channels. The FE boards incorporate 11 PGAs which allow for tunable numeric corrections per channel: pedestals, spillover and gains in order to equalise the response. The PS FE also centralises PS & SPD L0 trigger information. This requires to dial at 40 MHz with other calorimeter electronics: SPD, PS, ECAL & L0 CALO. As a result the PS FE boards are highly interconnected with 23 RJ45 or backplane links to 6 sub-systems. Synchronisation is achieved by tunable delay chip phases.

3. Calibration tasks of the preshower

There are 3 main requirements for the calibration of the preshower: the time alignment of the 6016 PS channels, their charge response calibration and the synchronisation of the L0 trigger information.

3.1. Time alignment

The scintillation signal collected in preshower cells extends to more than one cycle (25 ns). In nominal operating conditions the VFE integration start time has to be tuned to get a maximal efficiency in the triggering cycle (T0) while a fraction of the signal extends to the following cycle (Next1). The FE ADC's sampling phases and PGA's latching phases should then be adjusted consequently. There are 2 quantities of particular interest. The charge asymmetry $A = (Q_{\mathrm{T}_0} - Q_{\mathrm{Next1}})/(Q_{\mathrm{T}_0} + Q_{\mathrm{Next1}})$, for close to 0 asymmetry values, and the spillover ratio $\alpha = Q_{\mathrm{Next1}}/Q_{\mathrm{T0}}$ for nominal operating conditions ($A \simeq 1$). The charge asymmetry, close to the 0 crossing, provides an accurate measurement of the time difference between the scintillation signal start time and the VFE integration start time. It allows for a precise relative time alignment of PS channels. Once this inter-alignment is done, a global offset of $\simeq$ 14 ns is applied in order to maximise the fraction of signal in the T0 bunch. Then, and only then, we measure the resulting spillover ratio, α. Following, in nominal operating conditions, the charge in the T0 cycle is corrected in the FE from the spillover of the charge in the previous cycle, Prev1, as: $Q_{\mathrm{T0}} \to Q_{\mathrm{T0}} - \alpha\, Q_{\mathrm{Prev1}}$.

3.2. Charge calibration

The charge response of the 6016 PS channels is equalised according to the most probable energy deposition of Minimum Ionising Particles (MIP). The goal is to match 10 ADC counts in the FE to the MIP peak value. It results in a 100 MIP full dynamic. Discrimination of e and γ's at the L0 trigger is then achieved with a typical threshold of $5-10$ MIP. By tuning the high voltage (HV) of the MAPMT's one can adjust the gain of a group of 64 PS channels simultaneously. The FE numeric gain with tunable values in the range $[1;2]$ further allows a fine adjustment of the 64 channels amplitude response within a same board.

3.3. Trigger synchronisation

Because of variable cable lengths and specific connections to other calorimeters electronic boards, the synchronisation of the PS trigger data needed to be checked board per board, for the 100 FE. Therefore, a semi-automated test procedure was used. The validity of the data transmission was checked by transmitting known patterns between the calorimeters electronic boards testing all phase settings. This allowed a first tuning of PS trigger phasers. The ultimate crosscheck however came from pp collision data, by comparing the decoded trigger information to the emulated one from DAQ data.

4. Commissioning outcome

The calibration of the preshower was done in 2 main steps. First a pre-calibration was obtained from $1.5 \cdot 10^6$ cosmic events, recorded with a calorimeter only trigger. Requiring coincidences between ECAL and HCAL, the trigger rate was $\sim$ 10 Hz. The comic's tracks could be reconstructed from the clusters in the calorimeters with a pointing accuracy of $1-2°$. The fine calibration of the preshower could however only be achieved with collisions. In particular, dedicated timing runs allowed to measure the integrated charge shape as time for PS signals, by varying the global time shifts of the calorimeters.

4.1. Time alignment

4.1.1. Pre-alignment A 1st synchronisation of the PS channels was achieved by delaying the VFE integration start time from the time the light takes to travel from the scintillator cell to the MAPMT, according to measurements of the optical fibres lengths. This time alignment was crosschecked with cosmic events. However, because cosmic events are asynchronous with the experiment's clock, we could not measure directly the PS channels time alignment. Therefore, we compared the estimated time of passing of cosmic tracks in the PS, obtained from PS charge asymmetries, to the one measured by ECAL, from asymmetries also, and extrapolated to the PS. Since we can not tune the integration start time channel per channel, but only globally at the VFE level, it is relevant to average these time differences over the 64 channels of a same VFE. The resulting time difference was seen to be centred on 0 with a dispersion from VFE to VFE within ± 2 ns. Hence, we can say that following this 1st adjustment, the PS VFE were already synchronous within 2 ns, incorporating eventual biases from the ECAL.

4.1.2. Fine alignment Using *pp* collision data, dedicated runs were carried out by globally varying the time offset of the calorimeters system as respect to the LHC clock, providing the beam. These runs allowed to measure the full shape of the PS integrated charge signal as a function of the absolute VFE integration start time. A typical average shape for an Outer cell is shown on Figure 1. In particular, it was seen that the charge signal is slower in the larger cells of the Outer region than in the smaller cells of the Inner region. It results in a 2 ns systematic time shift on the signal rise time. Correcting from this effect, as well as for time of flight differences of particles originating from the interaction point, the 100 VFE could be time aligned within ± 1 ns with a 0.4 ns standard deviation.

4.1.3. Spill over ratio Once the final time alignment was done, we started measuring the spillover ratio, α, for strong deposits (≥ 7 MIP). Typical values are in the range 10−15% with a 3–5% standard deviation from event to event within a same PS channel. On average, 2% lower α values are also observed for Inner cells as compared to Middle and Outer cells, due to the signal shape difference, as can be seen on Figure 2. Within a same granularity region, the channel to channel dispersion of average α values is of 1% standard deviation. Therefore, although the spillover can be corrected channel by channel, it is not crucial to do so since intrinsic signal fluctuations, from event to event, are dominant.

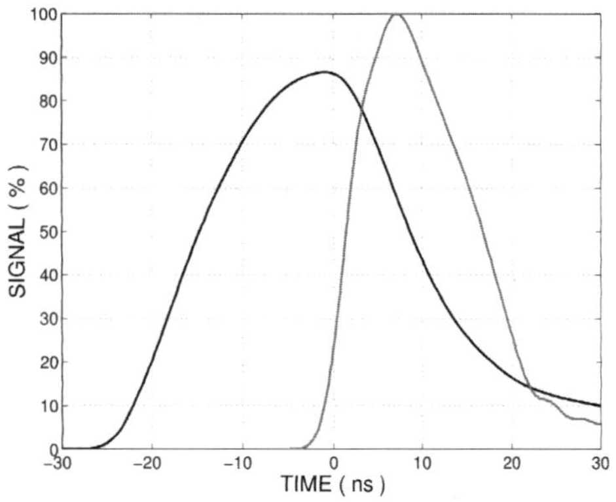

Figure 1. Average shape of the integrated PS signal (black) for an Outer cell. The red curve shows the reconstructed average shape of the signal before integration. Time $t = 0$ correspond to optimal settings for the integration start time.

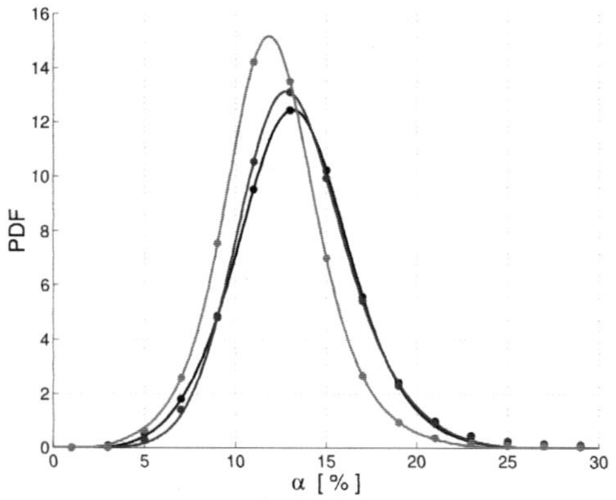

Figure 2. Probability density function of spillover ratio, α, averaged per region. Black stands for Outer cells, blue for Middle and red for Inner. Note that individual channels can have a slightly thinner distribution with typically 3–5% standard deviation.

4.2. Charge calibration

4.2.1. Pre-calibration A pre-calibration of the charge amplitude was provided by the MIP peak position for cosmic tracks. We selected well reconstructed cosmic tracks in the calorimeters that could be extrapolated to a hit in the preshower. Because cosmic tracks make a typical vertical angle of $\theta \simeq 45^\circ$ with the beam line it is important to re-scale the charge measured in PS cells by the track length in the cell, leading to a dE/dx measurement. Since we do not measure precisely the impact coordinate of the track within the PS cell, we decided to scale the charge deposit by the average track length over the cell for an homogeneous flux, $\langle \ell \rangle$. For a rectangular parallelepiped it goes as:

$$\frac{1}{\langle \ell \rangle} = \frac{\sin(\theta)\cos(\phi)}{e_x} + \frac{\sin(\theta)\sin(\phi)}{e_y} + \frac{\cos(\theta)}{e_z}, \tag{1}$$

with $e_x = e_y = 4, 6, 12$ cm, for Inner, Middle or Outer cells, and $e_z = 15$ mm. This correction is applied event by event given that θ and ϕ are measured by the calorimeters. It is worth pointing out that since cosmic tracks are triggered by requiring coincidences between ECAL and HCAL most of events are crossing PS cells from front to back sides, following the beam line or reverse, but not from sides. Consequently, the most likely track length is simply given as $\hat{\ell} = e_z / \cos(\theta)$.

For each PS channel we extracted the most likely value of dE/dx, expressed as scaled ADCs, from a fit to a Landau distribution convoluted to a Gaussian. A typical distribution for an Outer cell channel is shown on Figure 3. Note that the MIP peaks at 100 ADC instead of the nominal value of 10 ADC. This is because the MAPMT's were operated at 100 V higher HV than nominal. Making use of the embedded LED system we measured the MAPMT's gain variations with HV for each tube. The MAPMT gain, G, is well fitted by a power law as $G \propto HV^\beta$, with $\beta \in [7; 10]$. Not taking the spread of β values into account would result in a $\sim$50% miscalibration systematic down to nominal HV. With this procedure, we managed to set 99% of Outer, 95% of Middle and 90% of Inner channels within the numeric gain correction of 1 to 2. Note that the accuracy on the MIP peak value decreases with the cell size due to a lower statistic of cosmic events.

4.2.2. Calibration with collisions Using *pp* collision data, we repeated the MIP peak extraction procedure but taking benefit of the full LHCb tracking system this time. Figure 4 shows the result obtained with the PS detector pre-calibrated from cosmic data. The target was to set the MIP at 20 ADC. The non uniformity among channels, taken as σ/μ, varies from 10% in the Outer to 16% in the Inner where the pre-calibration statistics was lower. Although the typical response for Inner cells is indeed of 20 ADC one sees a systematic offset with the cell size, of 11% for Middle cells and 21% for Outer ones. This offset can be explained by our choice of normalisation of the cosmic tracks that was inadequate. Indeed, from simulations it is expected that the dE/dx peak position scales as $\hat{\ell}$ rather than $\langle \ell \rangle$. Hence, it does not depend on the cell granularity for a typical angle of $\theta = 45^\circ$. As a result, normalising the cosmic tracks by $\langle \ell \rangle$ instead of $\hat{\ell}$ results in under-rating the dE/dx peak position by 10% and 23% for Middle and Outer cells as compared to Inner cells. Consequently, the applied correction for gain equalisation is too high by a similar amount which is in good agreement with what was measured.

The charge calibration corrections obtained from MIP in *pp* collisions were compared to the one extracted from a calorimeter only energy flow calibration procedure. In the latter method one equalises the detector response by comparing the energy deposition in a cell to the deposit over its neighbour and left/right symmetric cells. By iterating this procedure over several data sets, the PS non uniformity can be reduced to an asymptotic spread of 5%. Good agreement was found between the energy flow method and the MIP scale equalisation with correlation factors of 80% for gain corrections, in the 1st iteration. The MIP methods however leads to a factor

Journal of Physics: Conference Series **293** (2011) 012060 doi:10.1088/1742-6596/293/1/012060

of ~ 2 lower spread in gain values, from which we infer that an asymptotic non uniformity of $2-3\%$ can be achieved.

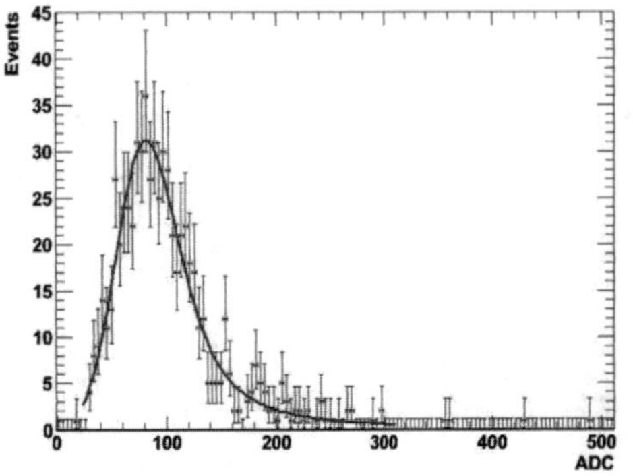

Figure 3. Typical dE/dx distribution for cosmic tracks in an Outer cell. The ADC values are scaled by the average track length $\langle\ell\rangle$.

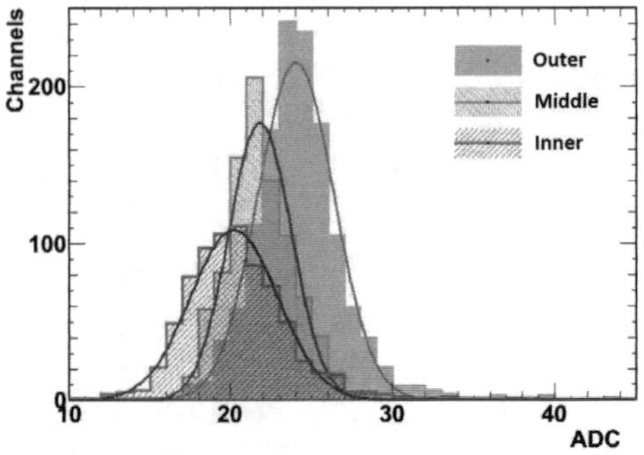

Figure 4. Distribution of MIP dE/dx peak position in pp collision data with a pre-calibrated PS detector. HV were set in order to get an expected peak value at 20 ADC.

4.3. Trigger synchronisation

The semi-automated phaser test procedure allowed to tune properly most of the PS trigger phase values. However, the comparison of the decoded trigger data to the emulated one indicated some remaining tricky issues. Those were difficult to track down because of the low statistics in the 1st days and because of the complexity of the chain involving several sub-systems. Problems could be related to cabling inversions for some peculiar PS boards which have only half of their 64 channels instrumented, as well as some out of synchronisation calorimeter boards and a few problematic cables. There is now excellent agreement between the decoded and emulated trigger data for e and γ candidates, based on SPD, PS and ECAL. Nevertheless, analysis of a larger statistics of data is ongoing in order to track down rare inconsistencies.

5. Conclusion

Starting 2008, the LHCb calorimeter was operated with a dedicated calorimeter only trigger for cosmic events. These data allowed to crosscheck the initial settings and provided a useful pre-calibration of the preshower energy scale. A non uniformity of 10% was achieved for Outer cells. The 1st collision events further refined these settings. In particular, dedicated runs allowed to adjust the PS timings within 1 ns accuracy while understanding the last systematics: signal shape differences for different cell sizes. The more accurate information from LHCb tracking systems together with the higher statistics should allow to further reduce non uniformities on the MIP scale to $2-3\%$ accuracy. Further work focuses on offline energy reconstruction, by measuring the sampling fractions in the preshower and the ECAL with π^0 events.

References

[1] Alves A A *et al.* (LHCb) 2008 *JINST* **3** S08005
[2] Gomez R V 2010 these proceedings
[3] Machikhiliyan I 2010 these proceedings
[4] Amato S *et al.* (LHCb) 2000 *LHCb Calorimeters* CERN/LHCC/2000-0036

XIV International Conference on Calorimetry in High Energy Physics (CALOR 2010) IOP Publishing
Journal of Physics: Conference Series **293** (2011) 012061 doi:10.1088/1742-6596/293/1/012061

ATLAS level-1 calorimeter trigger hardware: initial timing and energy calibration

JT Childers for ATLAS Collaboration
Universität Heidelberg, Im Neuenheimer Feld 227, D-69120 Heidelberg, Germany
E-mail: childers@kip.uni-heidelberg.de

Abstract. The ATLAS Level-1 Calorimeter Trigger identifies high-pT objects in the Liquid Argon and Tile Calorimeters with a fixed latency of up to 2.5 μs using a hardware-based, pipelined system built with custom electronics. The Preprocessor Module conditions and digitizes about 7200 pre-summed analogue signals from the calorimeters at the LHC bunch-crossing frequency of 40 MHz, and performs bunch-crossing identification (BCID) and deposited energy measurement for each input signal. This information is passed to further processors for object classification and total energy calculation, and the results are used to make the Level-1 trigger decision for the ATLAS detector. The BCID and energy measurement in the trigger depend on precise timing adjustments to achieve correct sampling of the input signal peak. Test pulses from the calorimeters were analysed to derive the initial timing and energy calibration, and first data from the LHC restart in autumn 2009 and early 2010 were used for validation and further optimization. The results from these calibration measurements are presented.

1. Introduction

The Large Hadron Collider (LHC) at CERN is designed to collide protons at a centre-of-mass energy of 14 TeV with an instantaneous luminosity of $10^{34}\,\mathrm{cm}^{-2}\,\mathrm{s}^{-1}$. The Level-1 Trigger (L1) is mainly composed of the Level-1 Calorimeter Trigger (L1Calo), the Level-1 Muon Trigger (L1Muon) and the Central Trigger Processor (CTP), but also includes the Minimum Bias, LUCID, Zero Degree Calorimeter, and Beam Pickup Triggers [1]. L1 is the first of three trigger levels that together reduce the 40 MHz input rate to approximately 200 Hz which is the typical limit on the rate for recording events. L1 itself reduces the 40 MHz input rate to a nominal rate of ~75 kHz with a current latency of 2.0 μs. The rate is then further reduced by the Level-2 Trigger to 3 kHz. The Event Filter makes the final decision before the event (of approximately 1.3 MB in size) is recorded [1].

About 200, 000 Liquid Argon and Tile Calorimeter cells are summed into 7168 analogue signals that L1Calo uses to form a trigger. These pre-summed signals, called Trigger Towers (TT), mostly have dimensions 0.1×0.1 in $\Delta\eta \times \Delta\phi$. They first arrive in the Receivers where an analogue gain can be applied to the signal. This is the primary method for calibrating the transverse energy (E_{T}) measurement in L1Calo. The PreProcessor Module (PPM) conditions and digitizes the signals and then uses the digitized signals to identify the associated bunch-crossing and measure the E_{T}. This information is sent in parallel to the Cluster Processor (CP) and Jet-Energy Processor (JEP) where high-p_{T} electrons/photons, hadrons/taus, and jets are identified. The multiplicities of these objects are counted for selected thresholds. The JEP also provides a total and missing E_{T} measurement. The multiplicities per threshold are then transmitted to the CTP where they are used in conjunction with information from other L1 systems to form the Level-1 Accept (L1A) which begins the complete readout of the detector including the L1 trigger system. For more details about the L1Calo system see [2].

The 7168 analogue signals arriving in the PPMs must be digitized to within ± 5 ns of the analogue peak in order to achieve an energy measurement of better than 2% [3]. The calibration of the timing delays applied in the PPMs aim to fulfil this requirement in all L1Calo TTs. Once the timing is optimized, the gains in the Receivers can be calibrated to ensure correct energy calculation. The process of the timing calibration is described in Section 2 and the energy calibration in Section 3.

2. Analogue Input Signal Timing Calibration

The analogue signals received by L1Calo are digitized in the PPMs at the LHC frequency of 40.08 MHz, which corresponds to one sample about every 25 ns. These digitized samples are used to assign the triggered event to the correct LHC bunch-crossing and to determine the energy deposited in the corresponding TT. It is important that the analogue signals are sampled at the peak in order to properly carry out these functions. Therefore, a nanosecond-step, or fine-timing, offset can be applied in order to adjust the sampling point [2]. As mentioned above, the analogue signals are composed of signals from multiple calorimeter cells which are summed to form a TT signal. If these calorimeter signals are not properly timed a TT signal may contain multiple peaks or other unwanted features. Extensive work has been done to ensure these timings are calibrated properly.

The fine-timing offset can be calibrated using test pulses from the charge injection and pulser systems of the Tile and Liquid Argon Calorimeters [1]. The timing of these test systems has been set to reproduce beam-like timing for the barrel regions ($|\eta| <\sim 1.5$) of both electromagnetic and hadronic calorimeters. However, variations in the timing across calorimeter partitions are difficult to determine a priori, so these and the initial per-partition TT timings need to be cross-checked and re-optimized using collision data. Only signals from real energy deposits provide the ideal input for the timing calibration.

LHC splash events from the November 2009 commissioning runs were used to cross-check and re-calibrate the initial timing of the L1Calo system derived from the calibration pulse systems. A splash event is produced by LHC beam packets interacting with the collimators located in the beam pipes at ± 145 m from the interaction point at the center of the ATLAS detector. These interactions create energetic secondary particle showers that traverse the ATLAS detector. Using the time-of-flight from the collimator to the detector and then from the detector to the collision vertex, estimates for the timing offsets can be cross-checked and corrections can be determined. Splash events were chosen for the initial timing calibrations due to the high occupancy of L1Calo channels. To reach this occupancy level with the expected collision luminosity that followed at the end of 2009 and early 2010 would have required much more time. Splash events from early 2010 were also analysed with improved calibration results, but were not available for presentation in these proceedings. However, the procedure remains the same.

2.1. Signal Fit Equation

The timing calibration begins by extracting the peak position of the analogue signal at the nanosecond level without any delay applied. To do this, a fit function is applied to the digitized signal in each TT. After investigating various fitting functions [4, 5], a Landau and Gaussian hybrid function was found to give the best fitting results for the analogue signal. The function is described by the expression,

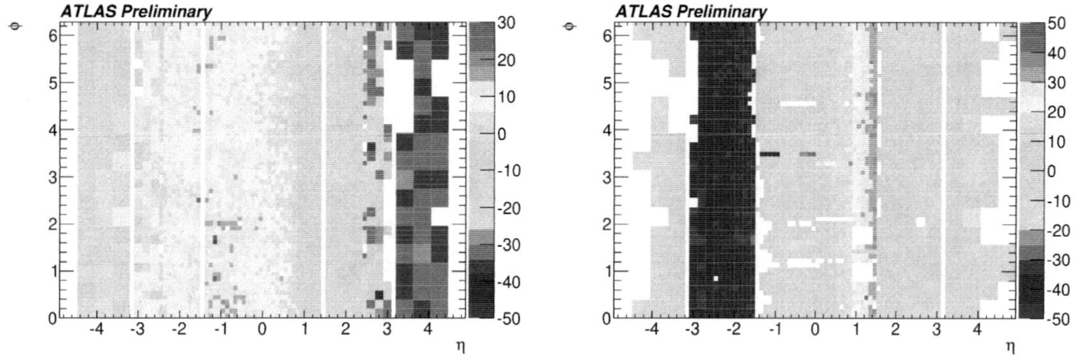

(a) Electromagnetic layer

(b) Hadronic layer

Figure 1. Here an η-ϕ plot of the signal peak time with ns precision plotted on the z-axis for a single event is presented. The peak times (t_0) are measured by fitting each trigger tower signal with a Landau/Gaussian hybrid function. The timing on the z-axis is measured relative to the center ADC sample in the case of a zero fine-timing offset. The electromagnetic layer is shown in 1(a) with beam-1 approaching in the $-\eta$ direction. The hadronic layer is shown in 1(b) with beam-2 approaching in the $+\eta$ direction.

$$
\begin{aligned}
f(t \leq t_0) &= A \cdot exp\left[-\frac{(t-t_0)^2}{2\sigma^2_{\text{gaussian}}} - \frac{1}{2}\right] + C \\
f(t > t_0) &= \left(A + D \cdot exp\left(\frac{1}{2}\right)\right) \cdot exp\left[-\frac{1}{2}\left(\frac{t-t_0}{\sigma_{\text{landau}}} + exp\left(-\frac{t-t_0}{\sigma_{\text{landau}}}\right)\right)\right] + C - D, \qquad (1)
\end{aligned}
$$

where A is the amplitude, t_0 is the peak position, C is the signal pedestal and D is the signal undershoot. With such a large parameter space and a small number of sample points, quality criteria are applied to ensure erroneous fits were excluded or corrected if possible.

Figure 1 shows the resulting peak position, t_0, after fitting all L1Calo TTs with Equation 1. Two different splash events are presented with Figure 1(a) showing the results of a splash event with beam-1 (moving in the $-\eta$ direction) in the electromagnetic layer and Figure 1(b) with beam-2 ($+\eta$ direction) in the hadronic layer. Allowing for some partition dependent offsets, the time of flight of the splash across the $\{\eta, \phi\}$ plane can easily be seen. These time-of-flight effects must be removed before calculating the fine-timing offset for collisions.

The empty bins in Figure 1 are caused by lack of suitable signals. The Tile Calorimeter in the hadronic layer ($|\eta| < 1.5$) is especially prone to pulse saturation in the splash events and this event was chosen for illustrative purposes due to the low number of saturated TT signals. Fit failures are seen in the electromagnetic and hadronic Forward Calorimeters (FCal) ($|\eta| > 3.2$) due to small signals failing the > 60 ADC units requirement. Near the outer edge of the Tile Calorimeter, signal saturation excludes extracting the peak location using the fit function.

Other empty TTs in Figure 1 are not yet fully commissioned and therefore excluded from the analysis. For example, the stripes in η in the hadronic barrel region are disabled sections of the Tile Calorimeter where L1Calo was not receiving signals. The stripes in ϕ at $|\eta| = 3.15$ are regions where the signal is suppressed by E_T corrections. The stripes at $|\eta| = 1.45$ in the electromagnetic layer are the overlap region between the Liquid Argon barrel ($|\eta| < 1.5$) and end caps ($1.5 < |\eta| < 3.2$) where TT signals will be summed from both layers. At the time, this region was not in a standard configuration and is therefore ignored in this analysis.

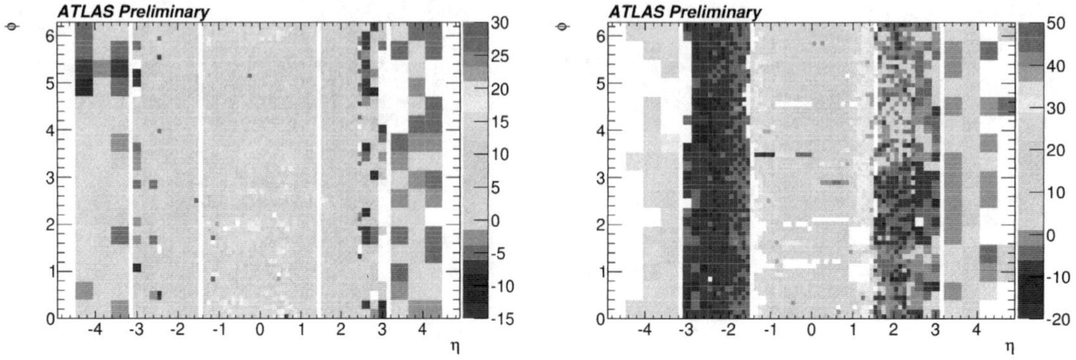

(a) Electromagnetic layer

(b) Hadronic layer

Figure 2. The corrected peak time location after correction for time-of-flight in nanoseconds is shown here for both calorimeter layers for a single event. The electromagnetic layer is shown in 2(a) with beam-1 ($-\eta$ trajectory). The hadronic layer is shown in 2(b) with beam-2 ($+\eta$ trajectory).

In the analysis presented here, the signal shapes produced by calibration pulses are assumed to be similar to those of beam collision and splash event signal shapes. However, this has not yet been verified and may be an unforeseen source of systematic errors. Also the pulse shapes vary as a function of detector partition such that Tile calibration pulses are not identical to those in the Liquid Argon hadronic end cap. Nevertheless, the Landau/Gaussian fit is adequate for pulses from both; fit parameters do, however, vary.

2.2. Time of Flight Correction
Beam splash event timing does not correspond to that of beam collision timing. Using the time-of-flight geometry of a splash event, the signal peak times from Figure 1 can be corrected to estimate the times for collisions. The positions used for the electromagnetic and hadronic layers are approximate because TTs are summed from calorimeter cells that vary in distance from the vertex. In order to minimize the effect of this approximation the halfway point between the outer and inner layers of the calorimeters was used. The correction is computed by subtracting the time-of-flight from the interaction point to the detector layer from the time-of-flight from the collimator to the same detector layer. The time-of-flight correction is then subtracted from the peak time extracted from the digitized signal using the fit function in the splash event. It is not necessary to compute this correction on a per TT basis due to the ϕ symmetry therefore it was calculated for each η-slice.

2.3. Collision Timing Offsets
The results of the time-of-flight correction described in Section 2.2 can be seen in Figure 2 for all L1Calo TTs. The electromagnetic layer, shown in Figure 2(a), is very uniform in η, ϕ with the FCal region exhibiting an overall lower offset. A perfectly timed system would have a flat response in η, ϕ. In the electromagnetic layer the timing is within ± 5 ns for the barrel and end cap regions which reflects the good timing of the calibration pulses.

The hadronic layer is less uniform than the electromagnetic layer, which can be understood as an effect of the different detector implementations involved. The border between the barrel ($|\eta| < 1.5$) and end caps ($|\eta| > 1.5$) can be recognized. These two regions were calibrated with independent signal pulser systems whose relative phase is not known. The end caps and FCal regions appear with different overall offsets for similar reasons.

A total of 55 splash events were used in this analysis, with 25 from beam-1 and 30 from beam-2. The final timing corrections applied to the L1Calo input signals were determined by averaging the results from all 55 splash events. Not all TTs include data from all 55 events due to the goodness of fit requirements. Only one splash event contributed to most TTs from the Tile Calorimeter because the signals were otherwise saturated.

The fit for some channels did not produce uniform timings for both beam-1 and beam-2 splashes, which is likely a result of the simplified geometry used in the time-of-flight correction described in Section 2.2. The variation is of the order $\sim \pm 3$ ns which is sufficient for the initial timing calibration and within the $\sim \pm 5$ ns timing goal. The effect appears mainly in the high η regions (FCal and some End Cap) where TTs are composed of larger calorimeter cells.

After considering all 55 events, the fine-timing offset could not be calculated for some TT. This included mainly Tile channels that saturated for all splash events and a few FCal channels that did not pass the minimum signal size requirement. These channels were corrected by hand using nearby channels as a guide. These calibration results will be verified using collision data, but the overall timing accuracy of L1Calo should be $\sim \pm 5$ ns using splash events.

3. Analogue Receiver Gain Energy Calibration

The energy calibration is implemented at the Receiver using an analogue gain applied to the signals arriving at L1 from the calorimeters. The energy measurement in the L1Calo PPM is based on the signal shape and height. This calibration is needed to correct for attenuation affects introduced in the long signal cables from the detector to the L1 hardware and ensures the measurement of E_{T} performed in the PPM correctly reproduces the energy seen in the corresponding calorimeter cells.

The first step is to produce input signals at various energies within the PPM readout energy range of 0 to 250 GeV with all Receiver gains set to one, which means no gain is applied. The calorimeters' calibration pulser systems are again used to provide these input signals. Using the calorimeter data one can reconstruct the transverse energy seen in a TT as measured by the calorimeter and make a linear comparison with the measured E_{T} in the PPM. The linear slope of this comparison directly corresponds to the analogue gain needed in the Receiver. This process is exhibited in Figure 3 for a single TT in the electromagnetic and hadronic layers. This calibration must be performed for all 7168 TT and the gains applied to each incoming signal in the Receivers.

About 100 $\mu\mathrm{b}^{-1}$ of 7 TeVcollision data were used to verify the results of this calibration and the result is shown in Figure 4 for the barrel region of both calorimeter layers. There is very good agreement between the calorimeter's measured transverse energy and that measured by L1Calo which indicates the calibration is effective. TTs with known problems have been excluded from these data.

4. Conclusion

The initial fine-timing offsets in the L1Calo PPM were derived using LHC splash events in preparation for collisions in late 2009. It has been shown that correcting the input signal peak time for the geometric time-of-flight in these splash events is a useful method for producing the first estimates of the timing offsets. Since these initial offset were derived, the fine-timing offsets were updated using new splash events provided by the LHC in early 2010. Additional offsets are being derived at the time of writing using beam collision data, which should represent the last system-wide change of these settings for the foreseeable future. However, some channel-by-channel adjustments may still be required.

The analogue gains applied to the input signals at the Receivers can be optimized using the calorimeters' calibration pulser systems. This calibration adjusts the gain such that the PPM accurately reproduces the energy measured by the calorimeter. The results have been

Journal of Physics: Conference Series **293** (2011) 012061 doi:10.1088/1742-6596/293/1/012061

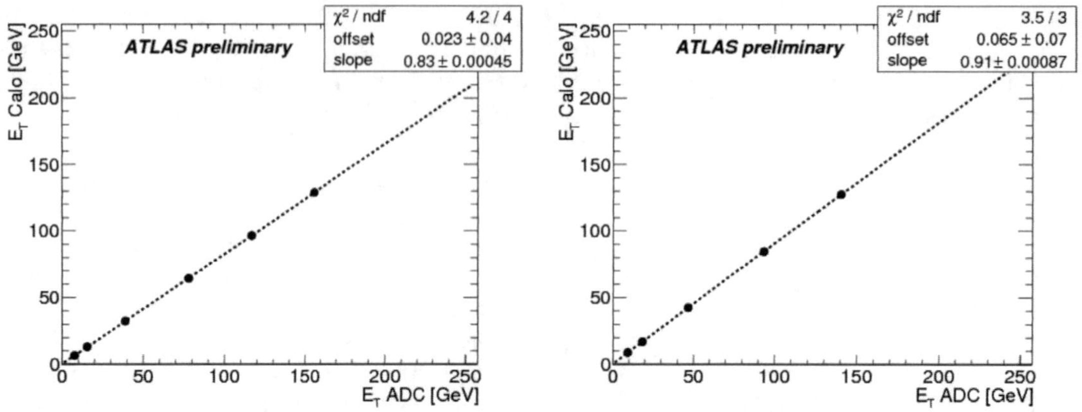

(a) Electromagnetic TT

(b) Hadronic TT

Figure 3. Here one barrel region TT per layer is shown with the reconstructed TT E_T measured by the calorimeter on the y-axis and the measured E_T from the PPM on the x-axis. The linear fit statistics are displayed on each plot.

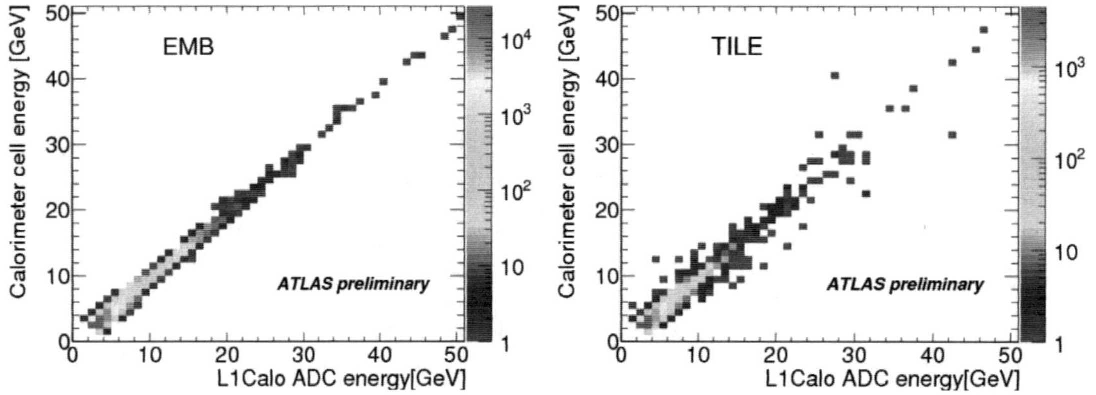

(a) Electromagnetic barrel

(b) Hadronic barrel

Figure 4. These plots show the energy correlation between the E_T measured by the PPM and the reconstructed value from the calorimeters in the barrel region of both calorimeter layers after the gain corrections have been applied. This uses about $100\,\mu b^{-1}$ of LHC collision data at 7 TeV. TTs with known problems have been excluded from these data.

verified with collision data, which showed the method is effective. As new timing offsets become available, new gain calibrations will need to be produced.

References

[1] ATLAS Collaboration 2008 *JINST* **3** 8003–8440
[2] Achenbach R *et al.* 2008 *J Inst.* **3** 3001–3041
[3] Stamen R 2008 (Naxos, Greece: TWEPP)
[4] Seidler J P 2009 *Synchronisationsstudien am ATLAS-Trigger* Diploma thesis Universität Heidelberg
[5] Gosdzik B 2007 *Offline Comparison of TileCal Signals and Trigger Signals* Diploma thesis Universität Heidelberg

Journal of Physics: Conference Series **293** (2011) 012062 doi:10.1088/1742-6596/293/1/012062

Commissioning of the ALICE-PHOS trigger

Lijiao Liu[1,2] for the ALICE Collabration

[1] Department of Physics and Technology, University of Bergen, 5007 Bergen , Norway

[2] Institute of Particle Physics, Huazhong Normal University, 430079 Wuhan , China

E-mail: Lijiao.Liu@ift.uib.no

Abstract. The Photon Spectrometer (PHOS) of the ALICE experiment at LHC is a high-resolution electromagnetic calorimeter dedicated to the precise measurement of direct photon and neutral meson yields in p+p and Pb+Pb collisions at mid-rapidity in a p_T range up to 100 GeV/c. A multi-level trigger system selects events of interest and reduces the overall data flow. PHOS generates two triggers: i) the Level 0 (L0) trigger for high luminosity p+p runs acting mainly as a spectrometer trigger and for selecting high p_T photons in Pb+Pb and in p+p collisions; ii) the Level1 (L1) trigger corresponds to a more sophisticated shower analysis which refines the p_T selection. The algorithms perform a pulse-shape analysis of the oversampled Avalanche Photo Diode (APD) signal followed by a sliding-window cluster finder. Both triggers are implemented in the firmware which, since SRAM-based FPGAs are used, can be changed on-line between runs. The trigger system was commissioned by cosmic rays and particles from interactions. The status of the trigger system and first results will be presented.

1. Introduction

ALICE (A Large Ion Collider Experiment) is an experiment designed for the study of nucleus-nucleus collisions at the Large Hadron Collider (LHC) [1][2]. It mainly focuses on the study of physics of strongly interacting matter and the deconfined state of nuclear matter which is called Quark-Gluon Plasma (QGP).

The main goal of the PHOS detector [3] is to study photon physics. One of the main physics objectives is to search for thermal photons for the determination of thermal and dynamical properties of the initial phase of the collisions. The other main physics objective is to study jet quenching through the measurements of high $p_T\,\pi^0$ and $\gamma - jet$ correlation. According to predictions [4] for heavy ion collisions, it is expected that thermal photons are in the energy range between 1 and 10 GeV. Therefore one of the two major requirements of the PHOS detector is to measure direct photons with good energy resolution in a p_T range 100 MeV/c to 100 GeV/c. The other requirement is to survive in the high multiplicity condition with up to 8000 charged particles per rapidity unit. There are 17920 PWO crystals with the size of $22 \times 22 \times 180$ mm^3 in total for the PHOS detector, which is located at a distance of 460 cm from the interaction point. It covers 100° in the azimuthal angle and $-1.2 \cdots + 1.2$ in pseudo-rapidity.

2. The PHOS trigger and readout system

Three levels of triggers, L0, L1 and L2, are used for the reduction of overall dataflow and the selection of events in the ALICE experiment. L0, L1 and L2 are required to arrive at the front end electronics at 1.2 μs, 6.5 μs, and 88 μs respectively after the collisions take place. The

PHOS detector contributes to two levels of triggers: L0 and L1. Because of the tight timing requirements of the trigger, PHOS must send the L0 trigger to the Central Trigger Processor (CTP) within 800 ns after the time of the interaction.

PHOS consists of 5 modules and each module is equipped with 64×56 PWO crystals, 112 Front End Cards (FEC) [5][6][7], 8 Trigger Region Unit (TRU)[8][9][10] and one Trigger OR unit (TOR) which is shared by all the modules. Each crystal has an Avalanche Photon Diode (APD) and a low noise Charge Sensitive Preamplifier (CSP) glued onto it. The PHOS components are divided into cold zone components at a temperature of -25 °C and warm zone components. The FECs and TRUs are located below the crystals inside the warm part of the air-tight compartments, and the TOR sits outside the air-tight compartments.

2.1. The Front End Card

The FEC is designed to fit the layout of double rows of 2×16 PWO crystals. The dynamic range between 5 MeV and 80 GeV is implemented with two shapers of gain ratio 1 : 16 and 10-bit ALTRO [11] ASICs. Each 4 output signals from CSPs in matrix 2×2 are summed by an analog adder with a fast shaper property and the summed signal goes to the TRU via differential lines.

2.2. The Trigger Region Unit

Signals from Analog sum of 14 FECs are fed to one TRU where 14 8-channel ADCs with 12-bit resolution convert the analog signals to digital signals. The onboard FPGA deserializes the ADC values, subtracts pedestals and then uses a 4×4 crystals sliding window to find clusters. The L0 decision is made by a simple threshold trigger. Both the L0 and the cluster energies are transmitted serially to the TOR for further processing. In addition, the digitized 12-bit 2×2 sums, called trigger channel signals, are buffered in the FPGA according to ALTRO format for the comparison with the energy channels in the real ALTRO on the FEC. The TRUs can be configured and programmed remotely via a Detector Control System (DCS) [12] card which is mounted on the the RCU card.

2.3. The Trigger OR

The TOR card, also with a DCS card, is connected to up to 40 TRU via differential cables. The FPGA configuration can be remotely configured and programmed with the DCS card. Its FPGA is responsible for the collection and processing of all the incoming trigger signals, including the L0 which is a simple logical OR of all the L0 trigger inputs from all the TRUs in the 5 modules. The L1 triggers are generated based on more sophisticated processing of the cluster energies only when the confirmed L0 is coming from the Local Trigger Unit (LTU)[13]. There are one L0 trigger output and three L1 trigger outputs from the TOR to the CTP[13].

2.4. The overview of the PHOS trigger and readout system

Figure 1 shows the overview of the PHOS trigger and readout system. The signals coming from 32 PWO crystals go into one FEC. There are two data paths for the signals in the FEC, one of them consists of dual gain shapers and ALTROs to digitize the energy from each PWO crystal directly. The other is Analog sum.

The L0 trigger is generated in the TRU and ORed in the TOR, whereas the L1 triggers are generated in the TOR, both L0 and L1 triggers are delivered into the CTP for the combination and synchronization of information from all the triggering detectors in the ALICE, then the trigger related sequences that consist of combined and synchronized information are sent to the LTU, where the trigger related sequences are converted into proper format to deliver to the front end electronics via an optical fiber. The RCU decodes the trigger information and sends L1 strobe (i.e. confirmed L0 for PHOS) and L2 strobe(i.e. L2a) to FEC. The ALTRO on the

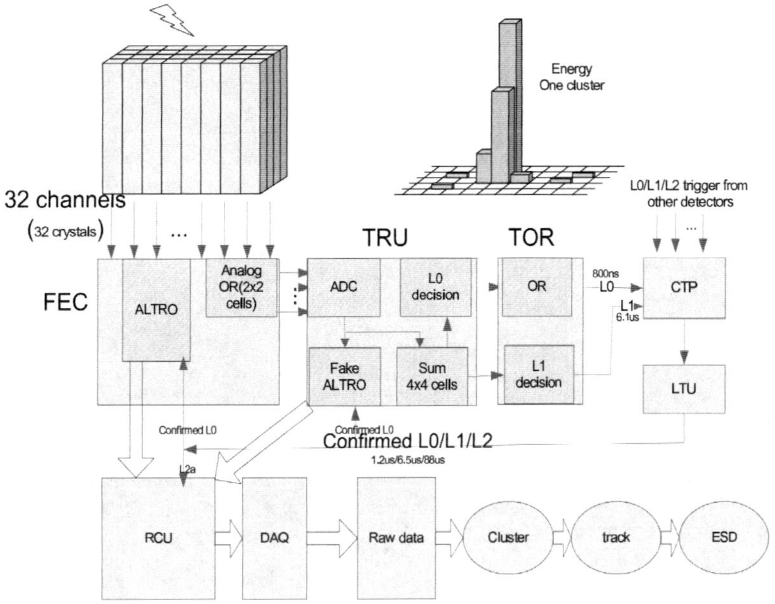

Figure 1. The dataflow of the PHOS trigger and readout system

FEC and Fake ALTRO on the TRU start to buffer the data on the L1 strobe, the RCU will ship the data to the Data Acquisition System (DAQ) when the trigger information is decoded as L2a. Moreover, there is a BusyBox that is not shown in the figure to handle busy in order to stop the CTP sending triggers whenever the buffers on the FECs are full [14].

3. Commissioning with cosmic ray and LED

The muons from cosmic rays deposit part of their energy when they traverse the PHOS detector, which can be used for trigger commissioning and calibration. Also each PHOS module is equipped with a LED Monitor System (LED MS), there is an individual LED on the top of each crystal. The brightness of the LED can be adjusted by setting different modes of amplitude: one-peak mode and grid-mode that gives a multi peak structure of the amplitude histograms. In addition, the illumination structure of the LED can be adjusted to different modes: four lines with each line covering 2×64 crystals, chess-like with four 2×16 crystals and point-like with 2×4 crystals.

3.1. The trigger channel noise test

The trigger channels were tested with both cosmic rays and LED MS at the CERN PHOS lab in August 2009. One PHOS module was used for the trigger channel test in 2009 at the lab. The one peak mode, 4 lines with 2×64 channels structure mode was set for the LED system. Figure 2 shows the pedestal of 4 trigger channels, Figure 3 shows the Root Mean Square (RMS) distribution of the pedestals from two TRU regions.

The LED calibration result is showed as Figure 4, the top part represents the matrix of energy channels on one module, the bottom part represents the matrix of trigger channels. One can see the good coincidence of energy channels and trigger channels in the figure. Figure 5 shows the cosmic rays calibration result of trigger channels, one can see that the energies of channels with hits are much higher than that of channels without hits. Therefore the trigger is expected to be issued by setting a threshold just above the noise. It is worth mentioning that the pedestal has been subtracted in the matrix of trigger channels already.

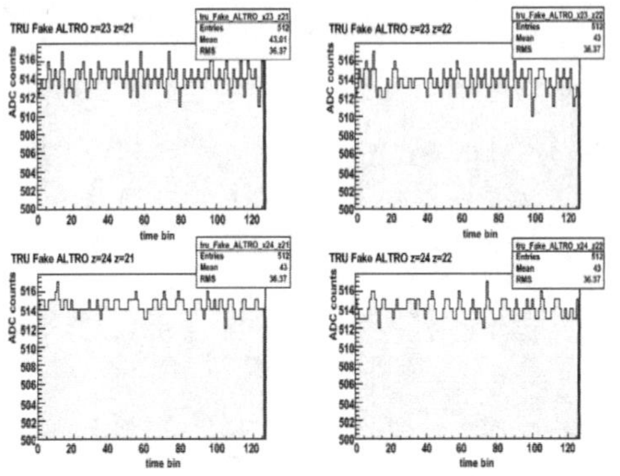

Figure 2. Pedestal of four trigger channels

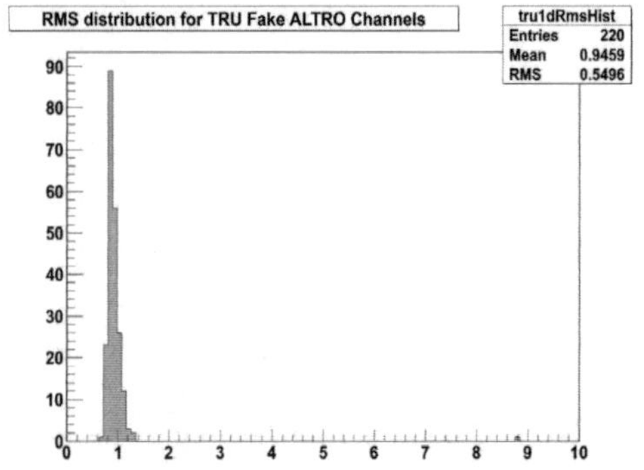

Figure 3. RMS distribution of the pedestals from two TRU regions

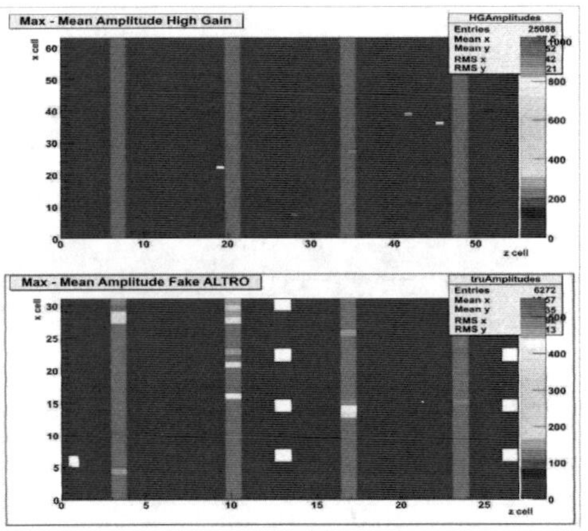

Figure 4. The energy channel matrix (top) and trigger channel matrix (bottom) with LED run

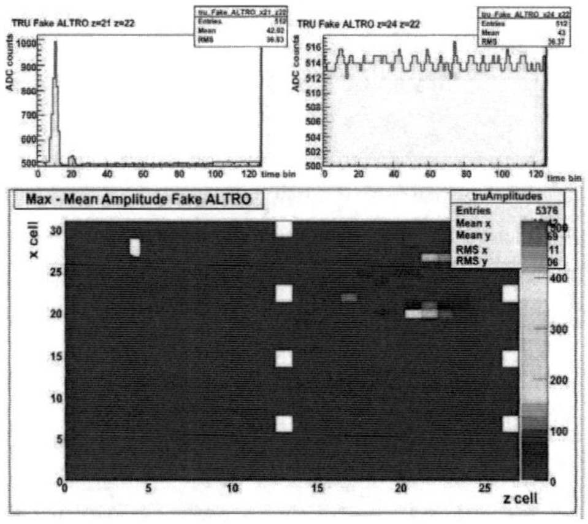

Figure 5. Trigger channels matrix with cosmic run (bottom) and two corresponding trigger channels (top)

3.2. The L0 trigger performance test

A cosmic run was taken in 2010. In this run, the PHOS is both triggering detector and readout detector. Figure 6 shows the Minimum Ionizing Particle MIP distribution of a cosmic run, a relatively low threshold was set, one can see that the MIP peak is around 210 MeV. The PHOS detector has not been fully calibrated yet till then.

3.3. The L0 trigger purity test

When the muons from cosmic rays traverse the ALICE experiment, they produce tracks in the Time Projection Chamber (TPC). There is also energy deposited in the crystals when the muons hit the PHOS detector. In principle the number of MIP tracks and the number of hits are supposed to be identical. A cosmic run was taken in 2010. In this run, both the PHOS and the TPC were read out detectors, the PHOS detector was also a triggering detector, a relatively low threshold around 180MeV was set. The trigger purity is calculated by $the\, number\, of\, triggers\, with\, MIP\, TPC\, tracks\, / the\, number\, all\, PHOS\, triggers$. In this run the purity was around 23% because the MIP energy was just above the threshold.

4. Commissioning with beam injection

The first beam injection was done in October 2009, the calibration of PHOS and the commissioning of trigger L0 is in progress during p+p collisions.

4.1. The L0 trigger timing Test

The timing test was done at P2 during p+p collisions in 2010. The Figure 7 shows the plot that has several histograms with signals from triggering detectors as a function of time in 25 ns units corresponding to bunch crossings, one can see that the PHOS L0 trigger arrives earlier than the benchmark OBPC and OBPA, which means the PHOS L0 trigger fulfills the timing requirement.

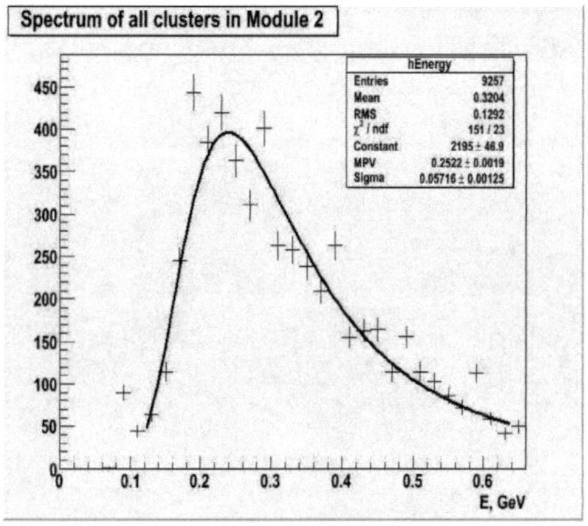

Figure 6. Clusters distribution of MIPs, solid line represents fit with Landau-Vavilov distribution

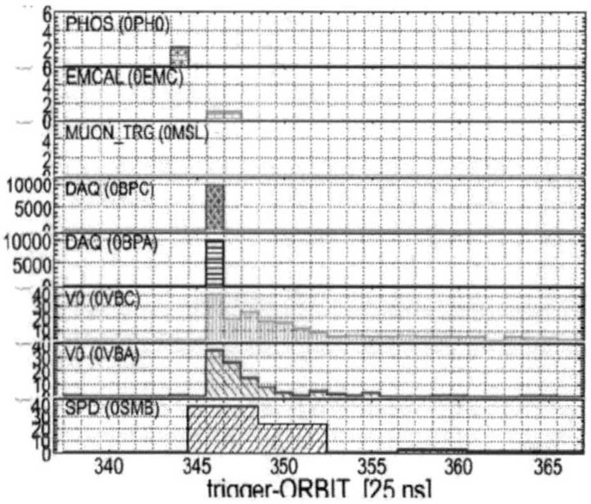

Figure 7. Orbit number versus trigger counts

4.2. L0 trigger efficiency and fake trigger rate test

Trigger channel data is supposed to be compared with energy channel data for the analysis of trigger efficiency and fake trigger rate test, therefore both trigger channel data and energy channel information have been recorded already. The analysis of trigger efficiency and fake trigger rate is ongoing.

5. Conclusions

Three of five PHOS modules have been installed at P2, the pre-commissioning of the trigger electronics has been done at the lab with cosmic rays runs and LED runs. Global timing test of the trigger system has been done during p+p collisions, also some L0 trigger performance has been analyzed during p+p collisions. The analysis of trigger efficiency and fake trigger rate is ongoing.

The update of the firmware is being developed in order to improve the L0 trigger performance. The L1 trigger development is in progress and the commissioning is supposed to be done soon.

Acknowledgments
The work is supported by the Norwegian Research Council and partly by the NSFC (10975061 and 10875051).

References
[1] The ALICE Collaboration 2004 *J, Phys.* G **30** 1517
[2] The ALICE Collaboration 2006 *J, Phys.* G **32** 1295
[3] The ALICE Collaboration 1999 ALICE Photon Spectrometer(PHOS) *ALICE Technical Design Report* (Geneva:CERN)
[4] Arleo F *et al* Photon Physics in Heavy Ion Collisions at the LHC *Preprint* hep-ph/0311131
[5] Muller H *et al* 2006 *Nucl. Instr. and Meth.* A **567** 264-67
[6] Muller H *et al* 2006 *Nucl. Instr. and Meth.* A **565** 767-83
[7] Wang Y *et al* 2010 *Nucl. Instr. and Meth.* A **617** 369-71
[8] Muller H *et al* 2004 *Nucl.Instr. and Meth.* A **518** 525-28
[9] Muller H *et al* 2005 *11th Workshop on Electronics for LHC and Future Experiments* (Heidelberg)(Geneva:CERN) pp 384-87
[10] Muller H *et al* 2010 *Nucl.Instr. and Meth.* A **617** 344-47
[11] Bosch R, DE arga A, Mota B and Musa L 2003 *IEEE Transactions on Nuclear Science* **50** 2460-69
[12] Richter M *et al* 2006 *IEEE Transactions on Nuclear Science* **53** 980-85
[13] Evans D *et al* 2004 ALICE trigger system *Proc. 10th Workshop on Electronics for LHC and Future Experiments* (Boston)(Geneva:CERN) pp 277-280
[14] Alme J *et al* 2008 *IEEE Transactions on Nuclear Science* **55** 76-83

Journal of Physics: Conference Series **293** (2011) 012063 doi:10.1088/1742-6596/293/1/012063

Electron and photon reconstruction and identification with the ATLAS detector and performance with $\pi^0 \rightarrow \gamma\gamma$ and $\eta \rightarrow \gamma\gamma$ with $\sqrt{s} = 900GeV$ data

Danilo Banfi

On behalf of the ATLAS Collaboration, ATLAS Collaboration, Università degli Studi & I.N.F.N., via Celoria 16, 20133 Milano, Italy

E-mail: danilo.banfi@cern.ch

Abstract. In this article the electron and photon reconstruction and identification with the ATLAS detector will be illustrated, both on MC simulation and on the data collected in the year 2009 at $\sqrt{s} = 900GeV$. The performance with $\pi^0 \rightarrow \gamma\gamma$ and $\eta \rightarrow \gamma\gamma$ on the same data will also be discussed.

1. Introduction

The ATLAS electromagnetic (EM) calorimeter, see Fig. 1, is a LAr-Pb sampling calorimeter with a fine segmentation in both the lateral ($\eta \times \phi$ space) and longitudinal directions of the showers [1]. At high energy, most of the EM shower energy is collected in the second layer which has a lateral granularity of 0.025×0.025 in $\eta \times \phi$ space. The first layer consists of finer-grained strips in the η-direction (with a coarser granularity in ϕ), which offer excellent $\gamma - \pi^0$ discrimination. These two layers are complemented by a presampler layer placed in front with coarse granularity to correct for energy lost in the material before the calorimeter, and by a back layer behind, which enables a correction to be made for the tail of very highly energetic EM showers. The ATLAS inner detector, see Fig. 1, provides precise track reconstruction over $\mid \eta \mid < 2.5$. It consists of three layers of pixel detectors close to the beam-pipe, four layers of silicon microstrip detectors (SCT), and a transition radiation tracker (TRT) at the outer radii, providing about 35 hits per track (in the range $\mid \eta \mid < 2.0$).

2. Data Sample

The analysis [2, 3, 4] is based on a data sample collected at $\sqrt{s} = 900GeV$. The events are triggered during LHC collisions using the Minimum-Bias-Trigger-Scintillators (MBTS), which cover the pseudo-rapidity range $2.09 < \mid \eta \mid < 3.84$ and are located in front of the end-cap EM calorimeters. Collision candidates are selected using additional timing requirements to provide further rejection against beam backgrounds: either coincident signals from the EM calorimeter end-caps or from the MBTS. The sample requires good data quality for the electromagnetic calorimeter and hadronic calorimeters. The solenoidal field is required to be at its nominal value. The data sample consists of 493,683 collision candidates which corresponds to an integrated luminosity of approximately 11.5 μb^{-1}.

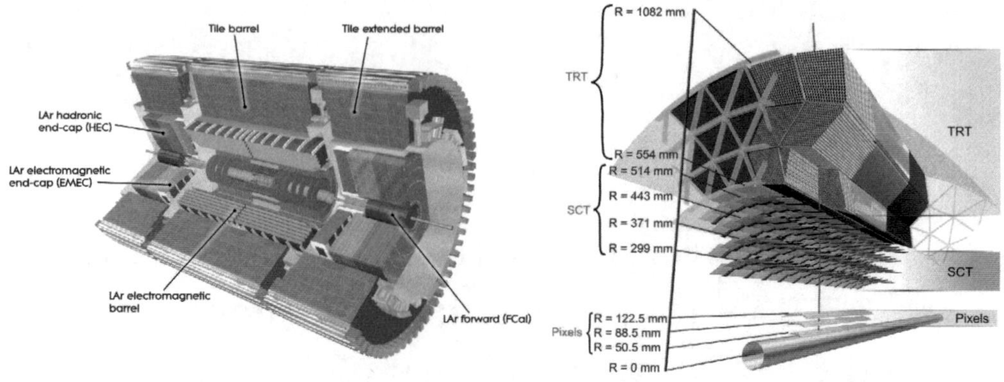

Figure 1. *Cut-away view of the ATLAS calorimeter system (left) , drawing showing the sensors and structural elements traversed by a charged track of 10GeV p_T in the barrel inner detector at η=0.3 (right).*

3. Electron and photon reconstruction

Electron and photon reconstruction [4] begins with the creation of a preliminary set of clusters in the EM calorimeter whose size corresponds to 3×5 cells in $\eta \times \phi$ in the middle layer. Electron and photon reconstruction is seeded from such clusters with $E_T > 2.5$ GeV, using a sliding window algorithm over the full acceptance of the EM calorimeter. The final cluster size is dependent on the particle hypothesis and the region of the detector (3×5 for unconverted photons in the barrel, 3×7 for converted photons and electrons in the barrel, 5×5 in all other cases). In addition to identifying efficiently EM showers, the ATLAS EM calorimeter measures their energies with high accuracy and with a linearity better than 0.5% over a large energy range, from 10 GeV to a few TeV. The cluster energy is determined precisely by computing and summing four different contributions: the energy deposited in the material in front of the EM calorimeter, that deposited in the calorimeter inside the cluster, that deposited outside the cluster (lateral leakage) and the energy deposited beyond the EM calorimeter (longitudinal leakage). The four terms are parametrised as a function of the cluster measured energies in the presampler (where it is present) and in the three accordion longitudinal layers.

4. Electron and photon identification

The baseline electron and photon identification algorithms in ATLAS rely on rectangular cuts using variables which deliver good separation between isolated electrons/photons and fake signatures from QCD jets. Three reference sets of cuts have been defined for electrons: loose, medium and tight, and two sets for photons: loose and tight. The cut values are optimised in bins of E_T and η for electrons and only in bins of η for photons, but separately for unconverted photons and converted photons. The loose selection includes shower-shape variables based on information from the middle calorimeter layer, together with hadronic leakage, the fraction of the cluster energy deposited in the hadronic calorimeter. The electron candidates will consist of clusters associated to a loosely matching track whereas the photon candidate clusters will in general lack such a track. The medium selection for electrons includes requirements on their energy deposits in the strip layer and on the track quality and track-cluster match. The tight electron selection uses the ratio between measured cluster energy and track momentum, E\p, and the fraction of high-treshold hits in the TRT, require the presence of a hit on the track in the pixel vertexing layer and rejecting candidates with a matching conversion vertex. The tight photon requirements comprise tighter cuts on the variables used for the loose cut selection and additional cuts on the middle layer and especially the strip layer with its fine granularity which provides good $\gamma - \pi^0$ separation.

Table 1. *Expected jet rejections with overall isolated and non-isolated electron efficiencies for the three sets of identification cuts and an E_T-threshold of 20 GeV. The total jet rejection includes hadron fakes and background electrons from photon conversions and Dalitz decays.*

	Efficiency (%)		Jet rejection (total)
	$Z \rightarrow ee$	$b, c \rightarrow e$	
Loose	94.30 ± 0.03	36.8 ± 0.5	1066 ± 4
Medium	89.97 ± 0.03	31.5 ± 0.5	6821 ± 69
Tight	71.52 ± 0.03	25.2 ± 0.5	$(1.38 \pm 0.06) \times 10^5$

Table 1 shows that the specific cuts optimised for the tight selection successfully remove the large backgrounds from hadrons and converted photons, which are largely dominant after the medium selection.

4.1. Electron candidates

The transverse energy spectra for all selected electron candidates are shown in Figure 2-(a). The numbers of events for different selection levels are given in Table 2. According to the MC 15% of the electrons passing the tight cuts are expected to be prompt electrons, therefore three out of the twenty electron candidates passing the tight cuts in the data would be expected to originate from heavy flavor decays.

4.2. Photon candidates

Table 2 presents the numbers of reconstructed photon candidates as a function of the selection cuts applied. Transverse energy and pseudorapidity spectra for all selected photon candidates are found to be in agreement between data and Monte Carlo, see Figs. 2-(b). From the MC approximately 71% of the candidates correspond to photons from π^0 decay, whereas $\sim$ 14% are from η, η' or ω decay into two photons, and $\sim$ 14% are from other hadrons with complex decay chains. Only $\sim$ 0.7% of all photon candidates are expected to be "prompt" at these energies, i.e. from initial or final state radiation of quarks.

Various calorimeter variables, used for candidates identification, are illustrated for the photon candidates, the equivalent distributions for electron candidates display very similar features. The lateral development of the shower, as detailed by the variables w_2 and R_η, is illustrated for the photon candidates in Figs. 3. These variables display small shifts between the shapes observed in data and those expected from simulation. The shower width w_2 is slightly larger in the data: preliminary studies show that including the cross-talk between neighbouring middle layer cells ($\sim$ 0.5%) in the simulation would explain part of the observed difference.

The variables sensitive to the shower lateral width in the strip layer, also shown in Figs. 3, F_{side} and w_{s3}, display distributions shifted slightly towards higher values in the data than in the

Table 2. *Breakdown of electron and photon candidates according to identification cuts applied. The numbers in brackets give the percentage of Monte Carlo electron candidates which are electrons from photon conversions or prompt electrons (the remainder are charged hadrons).*

Candidates	Electron 879		Photon 1694	
	Data (%)	MC (%)	Data (%)	MC (%)
Loose	46.5±1.7	50.9±0.2 (40.0±0.3)	25.4±1.0	30.5±0.1
Medium	10.6±1.0	13.1±0.2 (26.4±0.6)	-	-
Tight	2.3±0.5	2.4±0.1 (37.9±1.5)	4.1±0.5	6.6±0.1

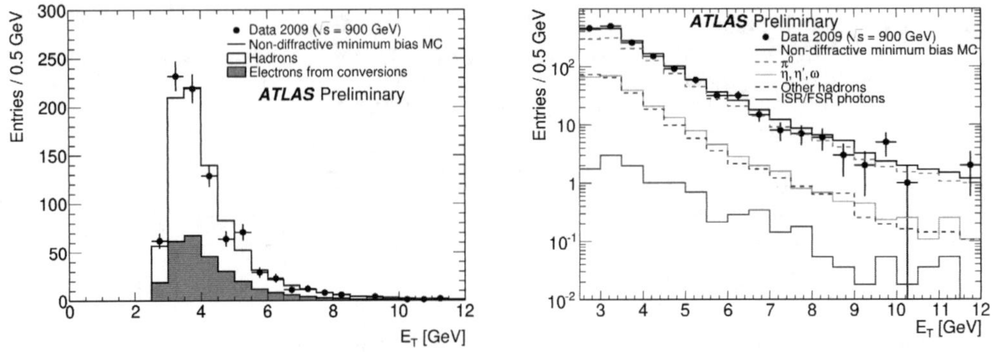

Figure 2. *Cluster E_T for all selected electron (a) and photon (b) candidates.*

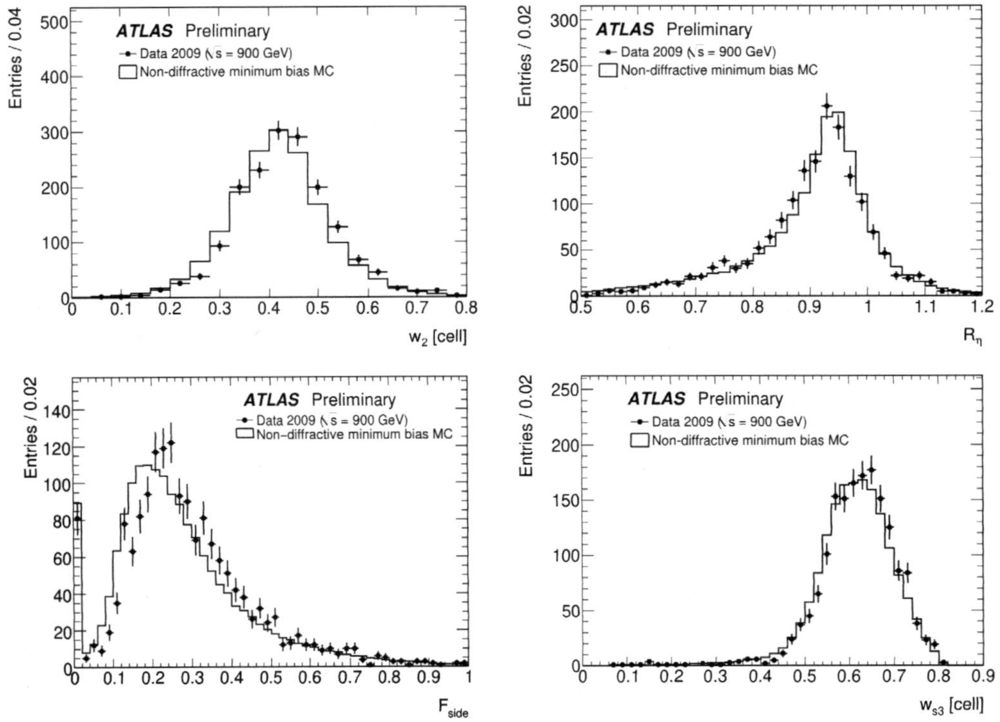

Figure 3. *Distributions of calorimeter variables compared between data and simulation for all photon candidates. Shown are the variables w_2 (lateral width of the shower) and R_η (ratio in η of cell energies in 3×7 versus 7×7 cells), F_{side} (fraction of energy outside core of three central strips but within seven strips) and w_{s3} (shower width for three strips around maximum strip)*

simulation. Neither an inaccurate cross-talk description (the effect observed is too large) nor an incorrect description of the material in front of the EM calorimeter are likely to explain entirely this effect.

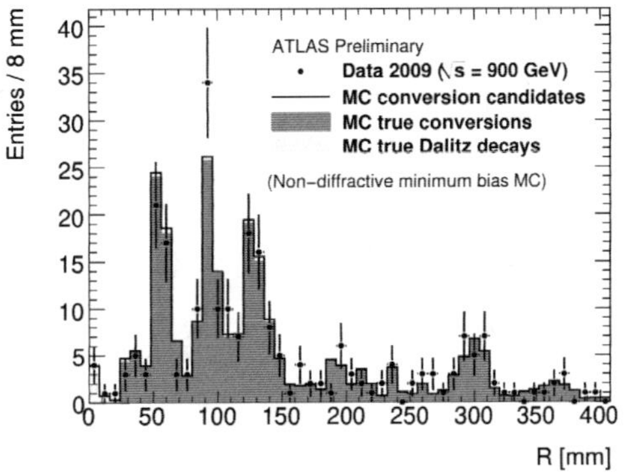

Figure 4. *Distribution of photon conversion candidate radius R integrated over all η. The distributions are normalized to the same number of photon conversion candidates in data and Monte Carlo simulation.*

5. Converted Photons and material mapping

An accurate and high-granularity map of the ID material is necessary for a precise reconstruction of high-energy photons and electrons. The ID material affects both the track trajectories (especially through bremsstrahlung effects) and the electromagnetic shower development (because of the magnetic field and the energy lost in the ID material). The radiation length of a localized amount of material is related to the fraction of photons that convert in it. The reconstruction of photon conversions [3] uses tracks found by the different track-reconstruction algorithms in ATLAS. The standard inside-out tracking is seeded with hits in the Pixel detector and in the SCT. Tracks found in the silicon detectors are extended outwards into the TRT. The photon conversion reconstruction algorithm begins by selecting single tracks with transverse momentum $p_T > 500\,\mathrm{MeV}$ and a significant fraction of high-threshold hits in the TRT, as expected from transition radiation. Photon conversion candidates are then created by pairing oppositely-charged tracks.

The results for the distributions of the photon conversion vertices in the radial direction in Fig. 4 show, albeit with limited statistics, a good match between the measured material distribution and the Monte Carlo model.

6. $\pi_0 \rightarrow \gamma\gamma$ studies

For low-energy photons [2] the sliding window algorithm developed for high-E_T objects with a seed threshold set to $E_T = 2.5$ GeV is inefficient. For the analysis of π_0 candidates, cells from the four layers are combined to form a cluster of size $\Delta\eta \times \Delta\Phi = 0.075 \times 0.125$, which corresponds to an area of 3×5 cells in the middle layer of the EM calorimeter. They are built from a seed provided by EM topological cell clusters with a seed cell threshold $|E_{cell}| = 4\sigma$ (where σ is the measured electronic noise in the cel) and with a cluster transverse energy $E_T > 300$ MeV. To account for upstream energy-loss and lateral and longitudinal leakages, the reconstructed clusters are calibrated using dedicated set of coefficients that has been extracted using low-energy photons coming only from π_0s in the minimum-bias simulation sample. In order to extract the π_0 signal from the combinatorial background, the following criteria are applied:

- A cluster-removal procedure is applied to avoid energy-sharing corrections: if two 3×5 clusters overlap, they are both rejected. Clusters in the barrel-EC transition region are rejected.
- The fraction of energy in the first layer $E_1/(E_1 + E_2 + E_3)$ is required to be larger than 0.1.

Journal of Physics: Conference Series **293** (2011) 012063 doi:10.1088/1742-6596/293/1/012063

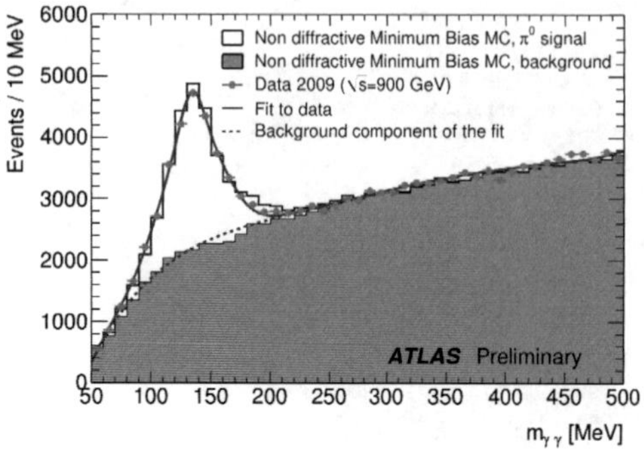

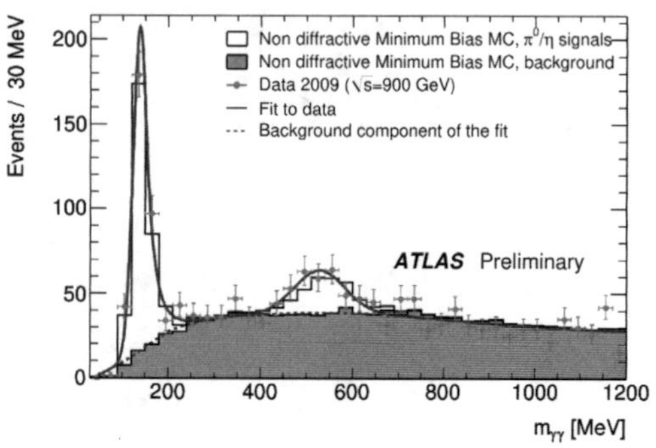

Figure 5. *Diphoton invariant mass after application of the selection criteria for π^0 (left) and for η (right) for data and Monte Carlo.*

- An additional kinematic cut requiring the 3×5 clusters to have a minimum transverse energy of $E_T = 400$ MeV is applied.

The invariant mass distribution of the photon pairs is shown in Fig. 5-left for both data and Monte Carlo. A clear excess of events can be observed near the π_0 mass.

The measured π_0 mass is 134.0±0.8(stat) MeV for the data and 132.9±0.2(stat) MeV for the Monte Carlo. The 1.5% discrepancy of the Monte Carlo mass with respect to the PDG nominal π_0 mass is consistent with the expected accuracy of the specific cluster calibration procedure for low-energy photons and the 1% uncertainty on the fitted π_0 mass arising from the background modeling and the ∼2-3% energy scale uncertainty.

7. $\eta \rightarrow \gamma\gamma$ candidates

The number of $\eta \rightarrow \gamma\gamma$ events is expected to be one order of magnitude smaller than $\pi_0 \rightarrow \gamma\gamma$ in the minimum-bias event sample. Therefore, the combinatorial background contribution in the η-mass region needs to be significantly reduced. This can be achieved by adding the following cuts to the π_0 analysis:

- Tighter kinematic cuts: $E_{cluster} > 800$ MeV, $p_{pair} > 2200$ MeV
- Track veto: only clusters which do not have any matching track are selected

The diphoton invariant mass spectrum obtained using these cuts is shown in Fig. 5-right: the number of η candidates is in agreement for the data and the Monte Carlo simulation. The η mass extracted from the fit on data, $m_\eta^{Data} = 527 \pm 11$(stat) MeV , agrees with the expected mass obtained using the same fitting function on the Monte Carlo simulation, $m_\eta^{MC} = 544 \pm 3$(stat) MeV, within the statistical and energy scale uncertainties of ∼ 2-3%. The 1% difference observed between the Monte Carlo simulation and the nominal PDG value is consistent with the expected accuracy of the specific cluster calibration.

References

[1] ATLAS Collaboration *Expected Performance of the ATLAS Experiment : Detector, Trigger and Physics* arXiv:0901.0512

[2] ATLAS Collaboration *Performance of the ATLAS electromagnetic calorimeter for $\pi^0 \rightarrow \gamma\gamma$ end $\eta \rightarrow \gamma\gamma$ events* ATLAS-CONF-2010-006

[3] ATLAS Collaboration *Photon Conversions at $\sqrt{s}$= 900GeV measured with the ATLAS Detector* ATLAS-CONF-2010-007

[4] ATLAS Collaboration *Electron and photon reconstruction and identification in ATLAS: expected performance at high energy and results at 900 GeV* ATLAS-CONF-2010-005

Journal of Physics: Conference Series **293** (2011) 012064 doi:10.1088/1742-6596/293/1/012064

First measurement of jets and missing transverse energy with the ATLAS calorimeter at $\sqrt{s} = 900$ GeV and $\sqrt{s} = 7$ TeV

David W. Miller, on behalf of the ATLAS Collaboration
SLAC National Accelerator Laboratory and Stanford University
E-mail: `David.Miller@slac.stanford.edu`

Abstract.
In December 2009, ATLAS recorded the first proton-proton collisions delivered by the LHC at a center-of-mass energy of $\sqrt{s} = 900$ GeV, and at $\sqrt{s} = 7$ TeV beginning in March 2010. We report on the first measurements of the kinematic distributions of jets and missing transverse energy at both center-of-mass regimes. These data also provide the opportunity to test the inputs to jet reconstruction and the jet calibration schemes available in ATLAS. Basic jet properties such as the number of associated charged particles, the longitudinal and radial shower topology in jets, and the transverse energy profiles of jets are compared to Monte Carlo simulation of the full detector response. We show that the basic inputs to jet reconstruction and calibration are generally well modeled by these simulations and that discrepancies will help to guide future Monte Carlo tunings and detector response descriptions.

1. Introduction
The ATLAS detector is a multi-purpose detector at the Large Hadron Collider (LHC) at CERN designed to conduct both precision Standard Model measurements as well search for signatures of physics beyond the Standard Model [1]. The collisions delivered in December 2009 at $\sqrt{s} = 900$ GeV, and at $\sqrt{s} = 7$ TeV shortly thereafter in March 2010, provide the opportunity to study the kinematics and properties of jets, as well as to conduct missing transverse energy measurements with the ATLAS calorimeters. Precise measurements of the charged particle content (measured with the Inner Detector tracking system) and energy distributions within jets at $\sqrt{s} = 900$ GeV and $\sqrt{s} = 7$ TeV lend insight into the process of jet production and internal jet structure at higher center-of-mass energies and their modeling by a Monte Carlo (MC) simulation based on phenomenological models of non-diffractive processes, including a full simulation of the detector response.

2. The ATLAS calorimeters
The ATLAS calorimeter is a non-compensating sampling calorimeter built of multiple sub-detectors with several different designs. The central calorimeter system consists of a liquid argon (LAr) electromagnetic (EM) calorimeter ($|\eta| < 1.475$)[1], and a scintillating tile (Tile) hadronic

[1] The pseudo-rapidity is defined as $\eta = -\ln(\tan(\theta/2))$, where the polar angle θ is taken with respect to the positive z direction. The anti-clockwise beam direction defines the positive z-axis, with the x-axis pointing to the

calorimeter ($|\eta| < 1.7$). The Tile calorimeter is divided into a barrel region ($|\eta| < 1.03$) and an extended barrel region, with space for services in between. The end-cap calorimeter is comprised of LAr EM and hadronic calorimeters ($1.375 < |\eta| < 3.2$ and $1.5 < |\eta| < 3.2$, respectively). A third LAr calorimeter sub-detector covers the very forward region ($3.1 < |\eta| < 4.9$), with the end-cap and forward calorimeters housed in the same cryostat. All calorimeter sub-detectors are segmented in depth to allow for measuring the longitudinal shower development.

The calibration schemes developed in ATLAS aim to correct for the non-compensation of hadronic energy deposits and to improve the jet energy resolution. A good description by the Monte Carlo simulation of the properties of jets used by the calibrations helps to reduce the systematic uncertainties in the jet energy scale. This is of key importance for many ATLAS analyses, and is a central topic of these studies.

3. Event selection and jet reconstruction

3.1. Event trigger and selection

Data collected with the tracking detectors and calorimeters fully operational and the solenoid at its full magnetic field strength are compared to a sample of fully simulated non-diffractive pp collisions ("minimum bias") and di-jet events generated with the PYTHIA [2] event generator program at $\sqrt{s} = 900$ GeV and $\sqrt{s} = 7$ TeV. For both data and MC, events are selected by requiring at least one hit on both sides of the minimum bias trigger scintillators (MBTS), located inside the inner cavity in front of the end-cap cryostats, and cover the pseudo-rapidity range $2.09 < |\eta| < 3.84$. Additional offline timing criteria are applied to reject beam background events. The simulated events are generated with a tuned set of parameters [3], and simulated using a full GEANT4 [4] simulation of the ATLAS detector. Only non-diffractive processes are considered since single and double diffractive contributions and softer collisions have been shown to be negligible for the version of PYTHIA used for these studies. The same trigger and event selection criteria are applied to the MC as to the data.

3.2. Jet reconstruction and track selection

Jets are constructed using the infrared and collinear-safe anti-k_t jet algorithm [5] with a distance parameter of $R = \sqrt{\Delta y^2 + \Delta\phi^2} = 0.6$.[2] The inputs to the jet algorithm in both data and Monte Carlo simulation are either noise suppressed$\Delta\ \eta \times \Delta\phi = 0.1 \times 0.1$ calorimeter towers or topological clusters [1, 6]. Clusters are built from a seed calorimeter cell with $|E_{\text{cell}}| > 4\sigma$ of the noise, nearest-neighbor cells with $|E_{\text{cell}}| > 2\sigma$, and an outer layer of all cells surrounding the clustered cells, where σ is defined as the RMS of the cell energy noise distribution. The noise suppression applied to towers consists in only using cells contained in clusters. Calorimeter towers, clusters and tracks are all treated as massless objects during jet finding while the four-vector recombination of the jet constituents leads to massive jets.

At $\sqrt{s} = 900$ GeV, no calibrations are applied to jets to compensate for differences in the calorimeter hadronic and electromagnetic responses [7]. In this case, only jets with transverse momenta $p_{\text{T}}^{\text{EMScale}} > 7$ GeV and pseudo-rapidity in the range $|\eta^{\text{jet}}| < 2.6$ are considered. Additional quality criteria are also applied to ensure that jets are not produced by single noisy calorimeter cells or problematic detector regions [8]. At $\sqrt{s} = 7$ TeV [9], an average jet energy scale correction is derived in the MC simulation and applied to each jet as a function of $p_{\text{T}}^{\text{EMScale}}$ and $|y^{\text{ jet}}|$, resulting in jets with a uniform response and an average jet energy scale uncertainty of about $\pm 7\%$ [10, 11].

center of the LHC ring. Rapidity is defined as $y = 0.5 \times \ln[(E + p_z)/(E - p_z)]$, where E denotes the energy and p_z is the component of the momentum along the beam direction.

[2] The effect of the jet mass is very small and pseudo-rapidity (η) is used instead of rapidity (y) in order to maintain the desired geometrical interpretation. Only in the jet finding procedure and in the detailed studies of the jet internal profile presented in Section 6 are rapidity units employed.

To measure the charged particle content of jets, tracks within the acceptance of the ATLAS Inner Detector are used with at least seven hits in the silicon detectors, transverse momenta of $p_{\mathrm{T}} > 500$ MeV, pseudo-rapidity $|\eta| < 2.5$ and a small impact parameter with respect to the primary vertex in the event [12, 13].

4. First observation of jets and measurements of missing transverse energy

The basic kinematic distributions of jets provide the first opportunity to compare the response predicted by the MC simulation to the data sample of pp collisions collected at two center-of-mass energies. Figure 1 shows the jet $p_{\mathrm{T}}^{\mathrm{EMScale}}$ distribution above 7 GeV in data and MC simulation at $\sqrt{s} = 900$ GeV. Figure 2 shows the same measurement performed at $\sqrt{s} = 7$ TeV for jets with $p_{\mathrm{T}}^{\mathrm{jet}} > 30$ GeV where a simple MC calibration scheme described in Section 3.2 is used to calibrate the jets [10, 13]. In both cases, the MC simulation is normalized to the number of jets in data. The same convention is used for the$\Delta\ \phi$ distributions for di-jet events, measuring the azimuthal separation between jets, at $\sqrt{s} = 900$ GeV (Figure 3) and $\sqrt{s} = 7$ TeV (Figure 4). These distributions indicate a good agreement between data and the MC simulation suggesting that the detector response and topology of di-jet events are well-modeled in both cases.

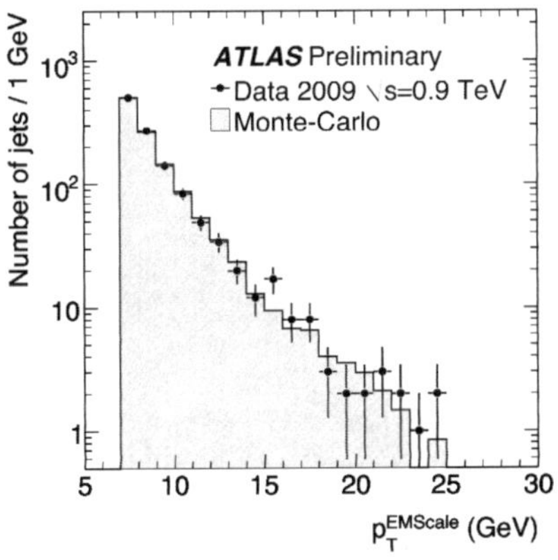

Figure 1. Jet transverse momentum distribution at $\sqrt{s} = 900$ GeV.

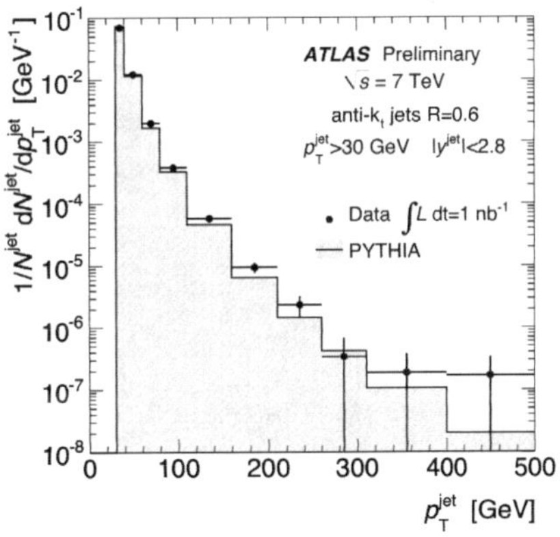

Figure 2. Jet transverse momentum distribution at $\sqrt{s} = 7$ TeV.

The missing transverse energy $(E_T^{\mathrm{miss}})^3$ depends crucially on the response of the calorimeter to both isolated objects such as jets as well as soft unclustered energy deposits, which are often very difficult to model. In soft pp collisions, such as those that dominate the sample described here, no true E_T^{miss} is expected. We observe that the shape of the E_T^{miss} spectrum and the tails of the distribution are well reflected in the MC simulation at both $\sqrt{s} = 900$ GeV (Figure 5) and $\sqrt{s} = 7$ TeV (Figure 6), indicating that the response of the calorimeter to soft diffuse energy depositions is well described.

The performance of the E_T^{miss} measurements and the agreement between data and MC simulations can be quantitatively assessed by evaluating the transverse components separately as a function of the total transverse energy of the event, $\sum E_T = \sum_{j=1}^{N_{\mathrm{cell}}} E_j \sin\theta_j$, as shown in Figure 7. In this case, the resolution is expected to increase as $\sqrt{\sum E_T}$. Fits to both the data and the MC show a reasonable agreement at $\sqrt{s} = 7$ TeV, and similarly good performance is observed at $\sqrt{s} = 900$ GeV (not shown) [14, 15]. The result of this fit as a function of $\sum E_T$

[3] The E_T^{miss} is defined using $E_x^{\mathrm{miss}} = -\sum_{j=1}^{N_{\mathrm{cell}}} E_j \sin\theta_j \cos\phi_j$, $E_y^{\mathrm{miss}} = -\sum_{j=1}^{N_{\mathrm{cell}}} E_j \sin\theta_j \sin\phi_j$ and $E_T^{\mathrm{miss}} = \sqrt{(E_x^{\mathrm{miss}})^2 + (E_y^{\mathrm{miss}})^2}$.

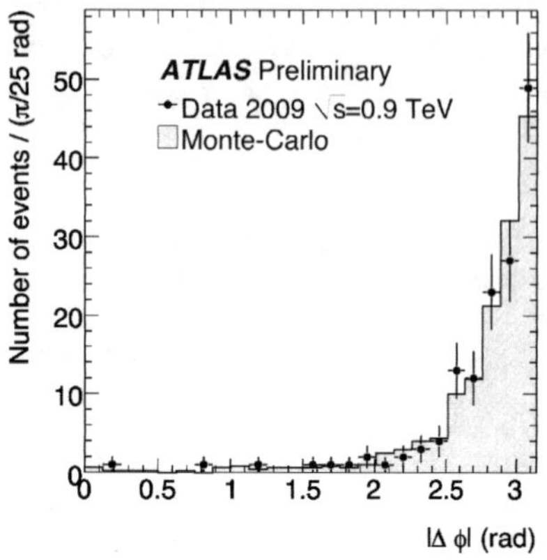

Figure 3. Di-jetΔ ϕ distribution at $\sqrt{s} = 900$ GeV

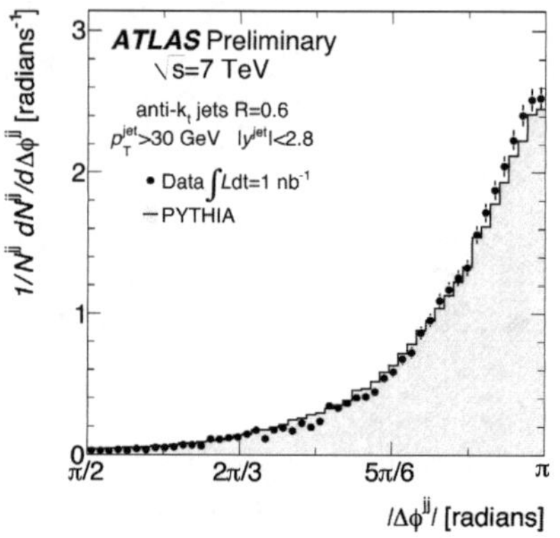

Figure 4. Di-jetΔ ϕ distribution at $\sqrt{s} = 7$ TeV

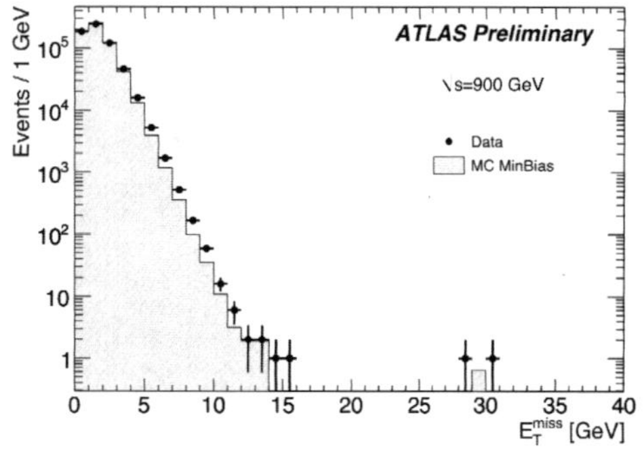

Figure 5. E_T^{miss} at $\sqrt{s} = 900$ GeV.

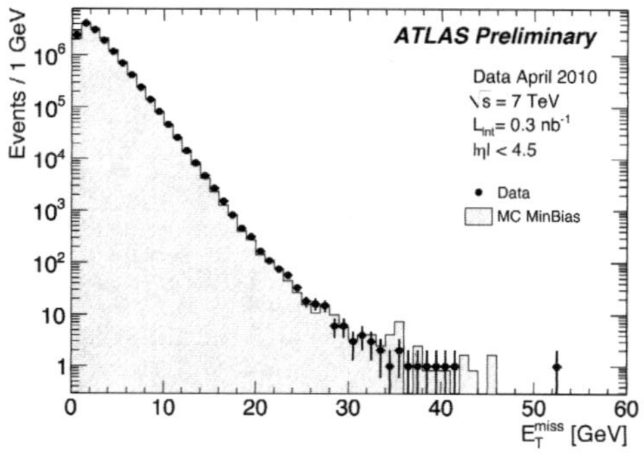

Figure 6. E_T^{miss} at $\sqrt{s} = 7$ TeV.

yields $\sigma(E_x^{\mathrm{miss}}, E_y^{\mathrm{miss}}) = 0.41 \times \sqrt{\sum E_T \text{ GeV}}$ in data and $\sigma(E_x^{\mathrm{miss}}, E_y^{\mathrm{miss}}) = 0.43 \times \sqrt{\sum E_T \text{ GeV}}$ in the MC simulation, when reconstructed at the EM scale. Both scale factors have negligible statistical uncertainty. The less than 5% difference in these resolution curves is likely due to imperfections in the PYTHIA physics modeling, and remains to be further understood.

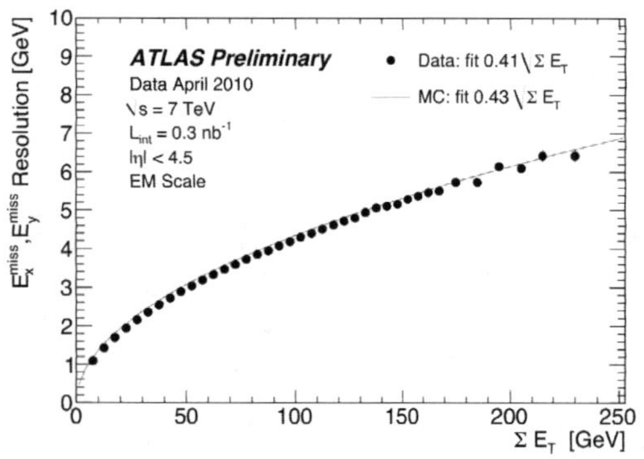

Figure 7. E_T^{miss} resolution vs. $\sum E_T$ at $\sqrt{s} = 7$ TeV, reconstructed at the EM scale.

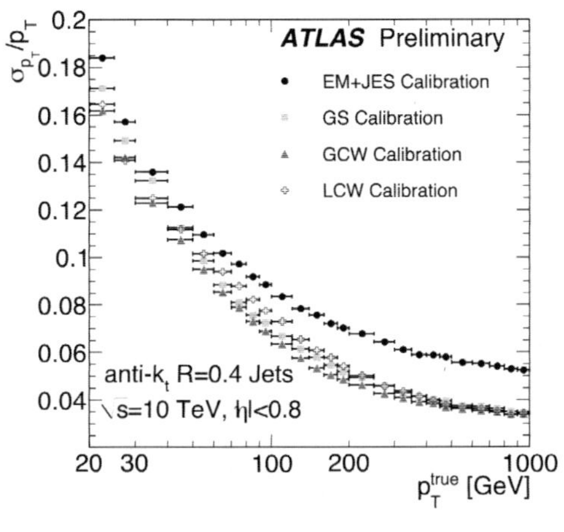

Figure 8. Fractional jet p_T resolution as a function of the p_T of the matched particle jet as calculated using a Monte Carlo simulation of proton-proton collisions at $\sqrt{s} = 10$ TeV and for jets calibrated with the four different calibrations.

5. Jet reconstruction and calibration schemes

ATLAS has adopted diverse approaches to calibrating jet energies to the hadronic scale and currently has four methods under study. These methods offer different levels of complexity and different sensitivities to systematic effects, yet all aim to establish a uniform jet energy scale and to reduce fluctuations in the measurement, thus improving the jet energy resolution. These schemes are described in detail in Refs. [12, 13], while in this section we discuss the properties of jets and clusters which form the basis of these calibrations: longitudinal energy deposition by layer, cell energy density, and topological cluster properties.

Figure 8 shows the fractional jet p_T resolution as a function of the p_T of the matched particle jet as calculated using a Monte Carlo simulation of proton-proton collisions at $\sqrt{s} = 10$ TeV and for jets calibrated with the four different calibrations. The first ("EM+JES") uses a simple p_T and η-dependent calibration scheme, and is used presently to calibrate jets in the $\sqrt{s} = 7$ TeV data. This simple approach does not obtain the ultimate resolution achievable by applying more sophisticated methods. The second method (global sequential calibration, or "GS") uses the fraction of the jet energy deposited in each of the longitudinal layers while the third method (global cell weighting, or "GCW") uses the distribution of cell energy densities inside jets to apply calibration weights. The last calibration (local cluster weighting, or "LCW") uses properties of topological clusters to classify and apply a weight to the clusters in jets themselves, and thereby bring the jet energy to the hadronic scale.

The reliance of the latter three calibration schemes on jet constituent properties to apply the calibration weights requires a detailed understanding of those properties. Several comparisons are made between data and predictions from MC simulation. At $\sqrt{s} = 900$ GeV, Figure 9 compares the layer energy fraction for the second layer of the EM barrel (the deepest layer of barrel the EM calorimeter), Figure 10 compares the cell energy densities in the same layer, and Figure 11 compares the cluster energy before and after weighting to the same ratio predicted from MC simulation. In all cases, deviations from the MC simulation predictions are much less than 5%, and are encouraging steps towards the successful commissioning of jet calibration in ATLAS and establishing a jet energy scale uncertainty for the different calibration schemes. Similar results are obtained at $\sqrt{s} = 7$ TeV [13].

6. Jet properties and internal structure

Several studies are performed of the properties of jets which can lend insight into the response of the calorimeter to the details of the hadronic shower, such as the charged particle distributions within jets and the longitudinal and transverse structure of jets.

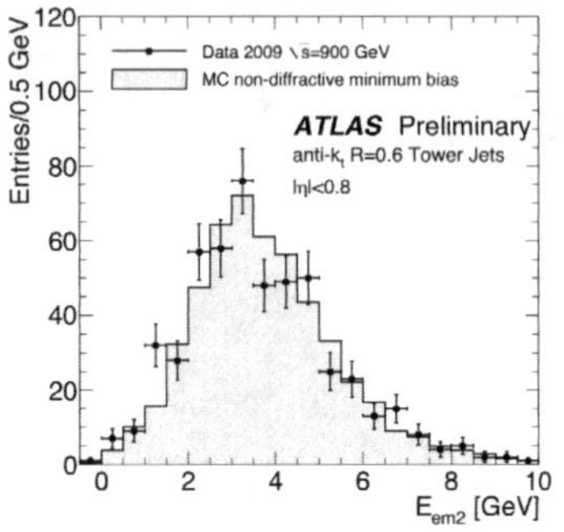

Figure 9. Energy deposited by jets in the second layer of the EM calorimeter.

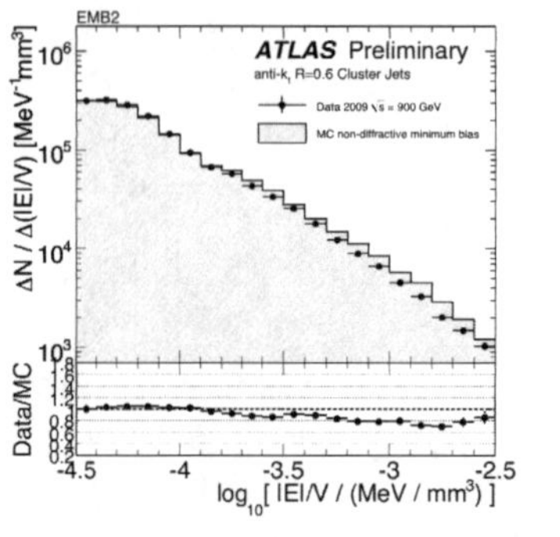

Figure 10. Cell energy densities in the second layer of the barrel EM calorimeter.

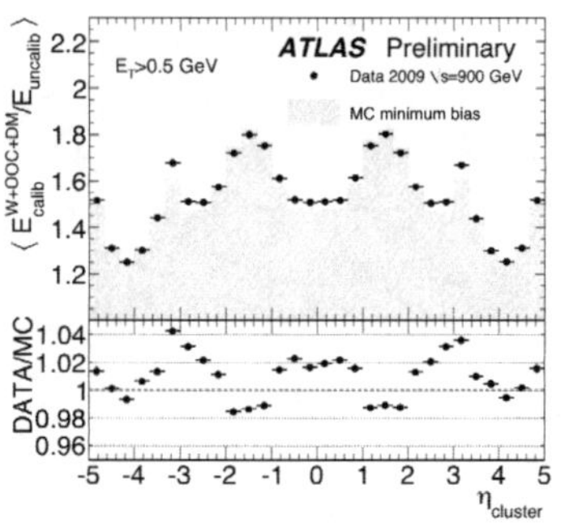

Figure 11. Mean calibrated cluster energy divided by uncalibrated cluster energy.

Figure 12 shows the measured number of tracks inside jets and its correlation with $p_{\mathrm{T}}^{\mathrm{jet}}$. As expected, the track multiplicity increases with increasing $p_{\mathrm{T}}^{\mathrm{jet}}$. The measurements are reasonably well described by the predictions from the MC simulation, suggesting that the track momentum and angular distributions in jets are well modeled, and that the response of the ATLAS detector is well described by the current GEANT4 simulation. Nevertheless, the data prefer a smaller average track multiplicity at low $p_{\mathrm{T}}^{\mathrm{jet}}$ and higher multiplicity at higher $p_{\mathrm{T}}^{\mathrm{jet}}$. At $\sqrt{s} = 7$ TeV, where the average jet momentum is much higher, we observe more low momentum tracks in the data than predicted by the MC simulation.

The transverse size of the jet can be characterized in terms of the jet width w^{jet}, defined as the weighted average distance of the jet constituents to the jet direction,

$$w^{\mathrm{jet}} = \sum (r \times E_{\mathrm{T}}^{\mathrm{constituent}}) / \sum E_{\mathrm{T}}^{\mathrm{constituent}}, \tag{1}$$

where $r = \sqrt{(\phi^{\mathrm{\ constituent}} - \phi^{\mathrm{\ jet}})^2 + (\eta^{\mathrm{\ constituent}} - \eta^{\mathrm{jet}})^2}$, $E_{\mathrm{T}}^{\mathrm{constituent}}$ is the transverse energy of the jet constituent, and the sum runs over the constituents associated to the jet by the jet algorithm. Figure 13 presents the measured jet width distribution for jets within $|\eta| < 2.6$ at $\sqrt{s} = 900$ GeV while Figure 14 shows its dependence on $p_{\mathrm{T}}^{\mathrm{jet}}$ at $\sqrt{s} = 7$ TeV. The shapes of the width distributions are well described by the predictions from the non-diffractive MC simulation, although the data at both $\sqrt{s} = 900$ GeV and $\sqrt{s} = 7$ TeV favor slightly wider jets in the central η ranges, particularly in the region $0.6 < |\eta| < 1.9$. This discrepancy persists over a large range in jet transverse momentum, suggesting that improvements in the description of both soft processes and the hard jet fragmentation are needed, which are likely to address the discrepancy together with potential improvements in the jet energy calibration.

Studies of the internal structure of jets at colliders in terms of jet shapes [16] or transverse energy profiles have demonstrated their sensitivity to a proper understanding of the jet fragmentation process, the detector response to low energy particles, and the accurate treatment of soft underlying event and pile-up contributions within the jet calibration procedure.

The differential jet profile is defined as the average jet transverse momentum that lies inside an annulus of inner radius $r - \Delta r/2$ and outer radius $r + \Delta r/2$ around the jet axis:

$$\langle \frac{1}{(2\pi r)} \frac{dp_T}{dr} \rangle_{jets} = \frac{1}{A} \frac{1}{N_{\mathrm{jet}}} \sum_{jets} p_{\mathrm{T}}(r - \Delta r/2, r + \Delta r/2), \ 0 \leq r \leq R, \tag{2}$$

where p_T is the scalar sum of the transverse momentum of jet constituents in a given annulus with area A, N_{jet} is the number of jets, $R = 0.6$, and$\Delta\ r = 0.1$ is used. Figure 15 shows

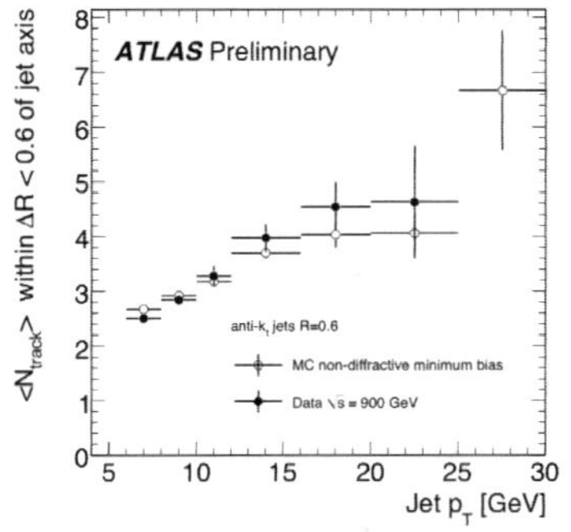

Figure 12. Number of tracks in jets with $|\eta| < 1.9$ vs. jet p_{T}at $\sqrt{s} = 900$ GeV.

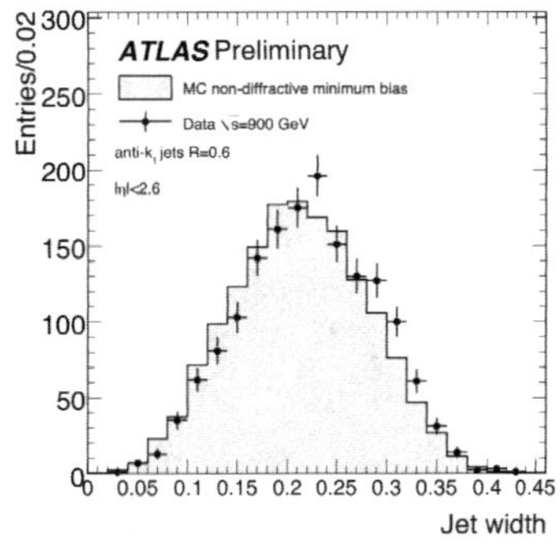

Figure 13. Jet width for jets with $|\eta| < 2.6$ at $\sqrt{s} = 900$ GeV and $p_{\mathrm{T}} > 7$ GeV.

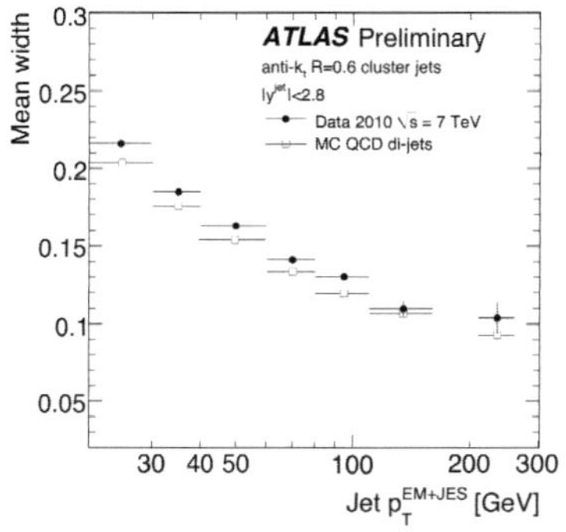

Figure 14. Mean jet width vs. jet p_{T} at $\sqrt{s} = 7$ TeV for jets with $|\eta| < 2.8$.

the measured jet profiles using calorimeter towers for jets in different jet rapidity regions up to $|y^{\text{ jet}}| = 2.6$. The jet profiles present the expected shape with a prominent peak at low r which indicates that the majority of the jet momentum is concentrated in the core of the jet. However, much like the jet width, the data favor slightly wider jets than predicted from the MC simulation.

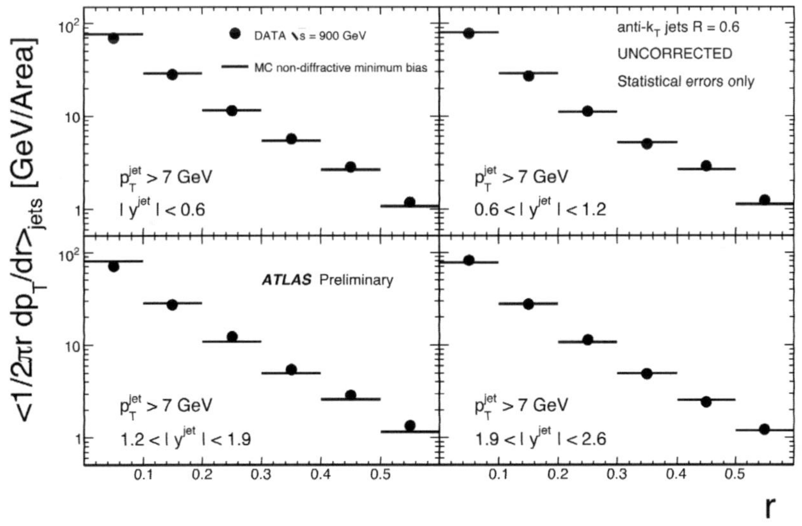

Figure 15. Measured jet profiles at $\sqrt{s} = 7$ TeV using calorimeter towers for jets with $p_{\mathrm{T}}^{\text{jet}} > 7$ GeV in different $|y^{\text{ jet}}|$ regions.

7. Conclusions

Preliminary results have been shown on the first observation and measurements of jets and missing transverse energy in the ATLAS calorimeters at both $\sqrt{s} = 900$ GeV and $\sqrt{s} = 7$ TeV. A reasonable agreement is found in basic kinematic distributions of jets. Various jet properties have been studied in the data that characterize or impact the jet reconstruction and calibration and compared to Monte Carlo simulations of the full ATLAS detector. A slightly higher soft activity is found around jets in the data, even though this effect is smaller as the p_T of the jet considered increases. This effect is also observable in the transverse profile of the jet, which points to wider jets in the data than in the Monte Carlo simulation. The measured distributions of charged tracks inside the jets also indicates a small excess in data compared to the Monte Carlo simulation. Such differences could be attributed to deficiencies on the current understanding of the detector response, the noise description, or the modeling of soft hadronic activity and fragmentation processes in the Monte Carlo simulations. A better agreement is found in the energies deposited in the different longitudinal segments of the calorimeter. These results will be considered in the future tuning of the Monte Carlo generators for ATLAS. The inputs to the different calibration schemes have also been shown to be in reasonable agreement in the data and the Monte Carlo simulation. Certain corrections that are part of the different calibration schemes have been validated directly in the data showing also good agreement. These are encouraging steps towards the successful commissioning of jet calibration in ATLAS and establishing a jet energy scale uncertainty for the different calibration schemes.

References

[1] The ATLAS Collaboration, Expected performance of the ATLAS experiment : detector, trigger and physics, **CERN-OPEN-2008-020** (2008)
[2] T. Sjöstrand *et al.*, Comp. Phys. Comm. **135**, 238 (2001)
[3] The ATLAS Collaboration, ATLAS Monte Carlo tunes for MC09, **ATL-PHYS-PUB-2010-002** (2010)
[4] S. Agostinelli *et al.*, Nucl. Instrum. Methods A **506**, 250 (2003)
[5] M. Cacciari, G. P. Salam and G. Soyez, JHEP 0804 (2008) 063. [arXiv:0802.1189]
[6] The ATLAS Collaboration, Inputs to Jet Reconstruction and Calibration with the ATLAS Detector Using Proton-Proton Collisions at $\sqrt{s} = 900$ GeV, **ATLAS-CONF-2010-016** (2010)
[7] The ATLAS Collaboration, Jet kinematic distributions in soft proton-proton collisions at $\sqrt{s} = 900$ GeV with the ATLAS detector, **ATLAS-CONF-2010-001** (2010)
[8] The ATLAS Collaboration, Data-Quality Requirements and Event Cleaning for Jets and Missing Transverse Energy Reconstruction with the ATLAS Detector in Proton-Proton Collisions at a Center-of-Mass Energy of $\sqrt{s} = 7$ TeV, **ATLAS-CONF-2010-038** (2010)
[9] The ATLAS Collaboration, Observation of Energetic Jets in pp Collisions at $\sqrt{s} = 7$ TeV using the ATLAS Experiment at the LHC, **ATLAS-CONF-2010-043** (2010)
[10] The ATLAS Collaboration, Jet energy scale and its systematic uncertainty in ATLAS for jets produced in proton-proton collisions at $\sqrt{s} = 7$ TeV, **ATLAS-CONF-2010-056**
[11] The ATLAS Collaboration, Response of the ATLAS calorimeters to single isolated hadrons produced in proton–proton collisions at a center-of-mass energy of $\sqrt{s} = 900$ GeV, **ATLAS-CONF-2010-017** (2010)
[12] The ATLAS Collaboration, Properties and internal structure of jets produced in proton-proton collisions at $\sqrt{s} = 900$ GeV, **ATLAS-CONF-2010-018** (2010)
[13] The ATLAS Collaboration, Properties of Jets and Inputs to Jet Reconstruction and Calibration with the ATLAS Detector Using Proton-Proton Collisions at $\sqrt{s} = 7$ TeV, **ATLAS-CONF-2010-053**
[14] The ATLAS Collaboration, Performance of the missing transverse energy reconstruction in minimum bias events at $\sqrt{s}$ of 900 GeV and 2.36 TeV with the ATLAS detector, **ATLAS-CONF-2010-008** (2010)
[15] The ATLAS Collaboration, Performance of the missing transverse energy reconstruction in minimum bias collisions at center-of-mass energy of $\sqrt{s} = 7$ TeV with the ATLAS detector, **ATLAS-CONF-2010-039** (2010)
[16] D. Acosta *et al* (CDF Collaboration), Phys. Rev. **D71** 112002 (2005)

Commissioning of the calorimetry in the ATLAS tau trigger system

Anna Sfyrla
for the ATLAS Collaboration

E-mail: anna.sfyrla@cern.ch

Abstract. Calorimeters are fundamental in the three levels of the ATLAS tau trigger system. The first level trigger uses the electromagnetic and hadronic calorimeters to make its decision. In the High Level Triggers, these systems are also crucial: both the second level trigger and the third level trigger (Event Filter) heavily exploit the calorimeter based information to identify tau leptons decaying hadronically. While the granularity of the first level is coarse, the second and third level triggers have the full detector read-out. This contribution focuses on the commissioning of the calorimetry in the three levels of the tau trigger with real data. Efficiency measurements with respect to tau candidates reconstructed by the offline algorithms, and distributions of calorimeter based tau information reconstructed at trigger level, are compared to the prediction of the Monte Carlo simulation and the trigger performance with first data is assessed.

1. Introduction

ATLAS is one of the two multipurpose detectors in the center of which the LHC collides proton bunches. It is built with a typical collider detector multi-layer design [1]. In its innermost part, it consists of layers of tracking devices. These devices are made of silicon pixel and strip detectors in the inner volume and transition radiation tubes in the outer part. They operate within a 2T solenoidal magnetic field. The tracker is surrounded by the calorimeters, used to identify and measure the energy of electrons, photons and jets. They consist of electromagnetic (EM) and hadronic (HAD) components. The EM calorimeter uses liquid Argon (LAr) as sampling material, while the HAD calorimeter uses scintillating tiles in the barrel region and LAr in the end-caps. The size and overall shape of the ATLAS detector is defined by the muon chambers in the outermost part, which operate within a toroidal magnetic field of up to 4T.

At the LHC design operation rate of 40 MHz, only one out of $O(10^5)$ collisions in ATLAS will be recorded for analysis. The rest of the events will be rejected by the trigger system. The ATLAS trigger system [2] consists of a hardware-based component, the Level-1 (L1), and the software parts, the Level-2 (L2) and the Event Filter (EF). The L2 and the EF are referred together as High Level Trigger (HLT). Figure 1 provides a graphical description of the ATLAS trigger system.

2. The tau trigger

2.1. Motivation

Tau leptons are the heaviest of the known leptons, with a mass of approximately 1.8 GeV/c^2. They are key signatures in searches beyond the Standard Model (SM). Due to their high mass,

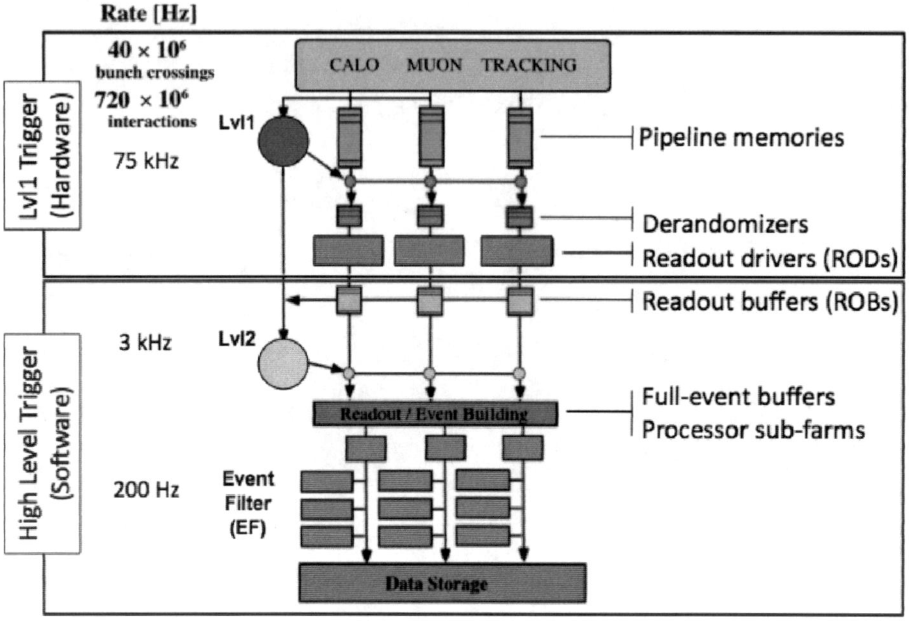

Figure 1. Schematic layout of the ATLAS trigger system. At L1, the information from the calorimeters and the muon chambers is used to evaluate each event and apply a rate reduction from 40 MHz to 75 kHz. The L1 identifies Regions of Interest (RoI) which are fed into the L2 as input. The L2 analyzes the data in these RoIs using information from all the ATLAS subsystems. The L2 selection reduces the rate down to 3 kHz. The EF has the information of the full event and runs algorithms similar to the offline reconstruction ones, to select interesting events at an output rate of 200 Hz.

they are favored in final states produced by processes such as SUSY production of tau-sleptons, charged Higgs, etc. Identifying taus not only increases the sensitivity to searches of physics beyond the SM, but it also increases the sensitivity to SM measurements where leptonic final states are involved, e.g. di-boson production.

Table 1. Tau decay modes and Branching Ratios (BR). The hadronic tau decays are distinguished in 1-prong or 3-prong decays, based on the number of charged pions (which leave tracks in the detector) involved in the decay.

Tau decay mode		**BR (%)**
Leptonic	$\tau^{\pm} \to e^{\pm} + \nu + \nu$	17.8
	$\tau^{\pm} \to \mu^{\pm} + \nu + \nu$	17.4
Hadronic 1-prong	$\tau^{\pm} \to \pi^{\pm} + \nu$	11
	$\tau^{\pm} \to \pi^{\pm} + \nu + n\pi^0$	35
Hadronic 3-prong	$\tau^{\pm} \to 3\pi^{\pm} + \nu$	9
	$\tau^{\pm} \to 3\pi^{\pm} + \nu + n\pi^0$	5
Other		5

The multiple decay modes of the tau leptons are shown in Table 1. The branching ratio into electrons or muons plus neutrinos is approximately 35%. These leptonically decaying taus are selected through the electron or muon triggers. Taus also decay in hadrons (pions, $\pi^{\pm}$ *and* π^{0}) with a branching ratio of approximately 60%. For selecting the hadronically decaying taus, a dedicated tau trigger has been developed [2].

2.2. Implementation

The tau trigger is designed to select hadronically decaying taus, as jets with special characteristics (low track multiplicity, isolation and narrowness, as described in Fig. 2). At L1, the tau trigger uses EM and HAD calorimeter trigger towers in clusters of size $\Delta\phi \times \Delta\eta = 0.2 \times 0.2$, to calculate the energy in a core region and the surrounding trigger towers between 0.2×0.2 and 0.4×0.4, to calculate the energy in an isolation region. At L2, the tau reconstruction starts from the L1 RoI, building a cluster with its initial center at the L1 RoI position. The energy weighted $\eta-\phi$ position is identified, the L2 energy is calculated in a region of $\Delta\phi\times\Delta\eta = 0.1\times0.1$, and the shape variables are obtained in a window of 0.3×0.3. Tracking algorithms are executed to identify associated tracks. Selection criteria are applied to distinguish taus from multi-jet backgrounds. The EF uses refined algorithms, similar to the offline reconstruction ones, to build tracks and clusters, and selects taus with different criteria for the 1-prong and the multi-prong decays.

Figure 2. A jet from a hadronic 3-prong tau decay, versus a jet from a QCD process. The taus are associated with low track multiplicity, narrow and isolated jets and this is how they are distinguished from the QCD jets.

The performance of the tau trigger is evaluated using Monte Carlo simulated events. Selection requirements are applied in a set of calorimeter and tracking variables, to optimize the trigger efficiency, and to acquire a steep turn-on. Because of various effects, the main being the hadronic L1 resolution, the L1 turn-on is slow and not compensated at the L2 and EF selections (Fig. 3). To select taus with a high efficiency above a certain tau p_T threshold, looser selections have to be applied at L1, L2 and EF.

3. Tau trigger menu and efficiency measurements in early data

The trigger menu is made up of a list of triggers (L1 item→ L2 chain→ EF chain), including prescales at each level. It contains the physics triggers of interest, triggers for calibrations and other physics studies (e.g. efficiency measurements and background estimations) and *pass-through* (PT) triggers for performance studies.

For the tau trigger system, various tau and tau-combined signatures are included in the menu. An indicative menu is summarized in Table 2.

In order to use a trigger in a physics analysis, its efficiency needs to be evaluated. In early data, the following three methods are being considered for efficiency measurements:

Table 2. An indicative menu for the tau trigger system. In reality the trigger menu contains many more triggers (especially for commissioning, at the initial data taking phase). Calibration triggers that use tau signatures are also listed.

Type	Motivation	Trigger
Single Tau	Higgs→ $\tau^+\tau^-$	tau50_loose
Multi Tau	Higgs→ $\tau^+\tau^-$	2tau29_loose
Combined	top, $Z \rightarrow \tau^+\tau^-$	tau12_loose_e10_loose
	top	tau16_loose_3j40
	$W \rightarrow \tau\nu$	tau12_loose_xe20
Commissioning	Trigger efficiency	tau12_loose_PT
Calibration	Hadronic	trk9_loose
	ID alignment	trk6_IDCalib

- *Tag and Probe method*: using a single-object inclusive trigger, di-object events can be studied. One object is selected by the trigger and reconstructed offline to tag the event. The second object, which was not used in the online selection, can then be used to study the trigger response. Measuring the tau trigger efficiency in $Z \rightarrow \tau^+\tau^-$ events is an example of this style of analysis.
- *QCD fake taus*: this method uses the copiously produced QCD tau-like events present in the early data, when enough statistics of real taus from W and Z bosons are not yet available. The tau trigger efficiency can be extracted using fake taus, provided the turn-on, as given by Monte Carlo simulations, is the same and the denominator well defined.
- *Bootstrap method*: the efficiency, ϵ_B, of a trigger chain B, with threshold higher than a chain A, can be indirectly measured as $\epsilon_{B|A}$ in a sample triggered by A, provided that ϵ_A is measurable: $\epsilon_B = \epsilon_{B|A} \times \epsilon_A$.

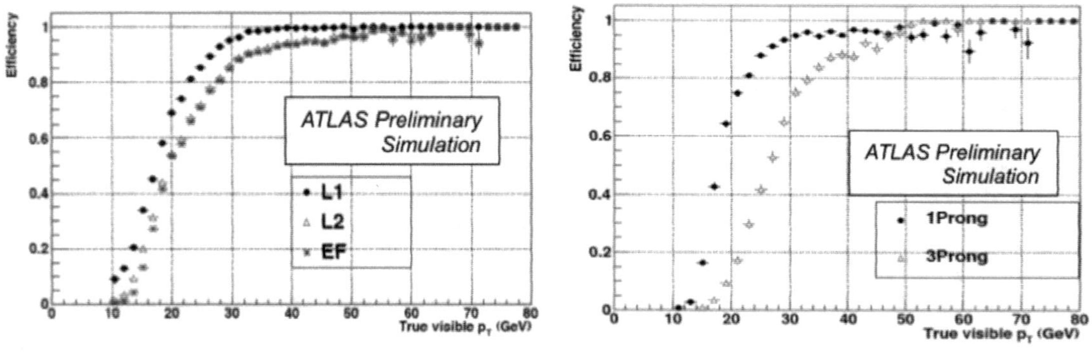

Figure 3. The trigger efficiency as a function of the simulated tau true visible p_T, for the trigger tau16_loose, as evaluated in $Z \rightarrow \tau^+\tau^-$ and $A \rightarrow \tau^+\tau^-$ events. The left plot shows the efficiency for the three trigger levels and the right plot the overall EF trigger efficiency in the 1-prong and 3-prong selections. The trigger, tau16_loose, is more than 95% efficient for 1-prong taus of $p_T >\approx 30$ GeV and 3-prong taus of $p_T >\approx 50$ GeV.

4. Rates and performance in 900 GeV proton-proton collisions

A clean sample of real hadronic tau decays is not available in the early data. It is therefore important to extract useful information from fake taus which are produced in multi-jet events, with the goal to collect a large enough data sample to check the performance of the tau trigger.

The first collision data provided by the LHC were 900 GeV $p-p$ collisions. One of the primary aspects of the tau trigger has been to check the trigger rates and whether they follow expectations from simulated events, but also extrapolations from the cosmic ray data taking, which were extensively used for the evaluation of the tau trigger algorithms [3]. In Fig. 4, the cumulative L1 tau object rate is plotted as a function of different E_T thresholds for cosmic ray and collision data. For a given threshold value, only tau objects with E_T larger than the threshold are taken into account. The black points represent all L1 tau trigger objects without imposing any requirement on the Minimum Bias Trigger Scintillators (MBTS). From this distribution two subsets have been selected, one by requiring the MBTS trigger and timing constraints consistent with collision events (blue boxes) and a second one by vetoing on the MBTS trigger decision (red triangles), selecting events dominated by cosmic ray events. A good agreement is observed between the rates from cosmic ray and non-MBTS collision data.

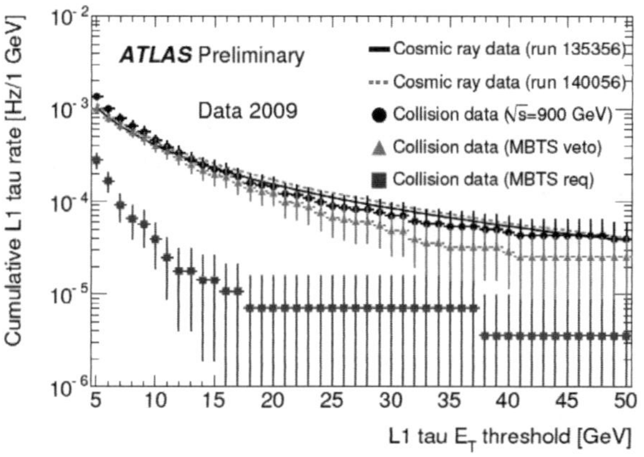

Figure 4. Cumulative L1 tau trigger rate as a function of the L1 tau object E_T threshold, normalized to one colliding bunch pair. For a given threshold value, the objects considered have a transverse energy greater than this value.

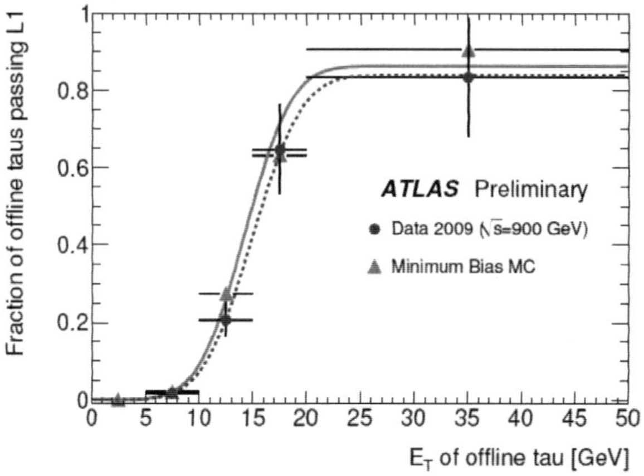

Figure 5. Fraction of the offline tau candidates matched to a L1 trigger object with $E_T > 5$ GeV as a function of the E_T of the offline tau candidate. The dashed (solid) line represents the fit to the data (Minimum Bias Monte Carlo simulated events).

In the 900 GeV collision data, the performance of the tau trigger has been evaluated by studying the events selected by the MBTS counters, requiring good runs for the calorimeter and tracking detectors. The events entering in the tau data sample were required to have at

Journal of Physics: Conference Series **293** (2011) 012065 doi:10.1088/1742-6596/293/1/012065

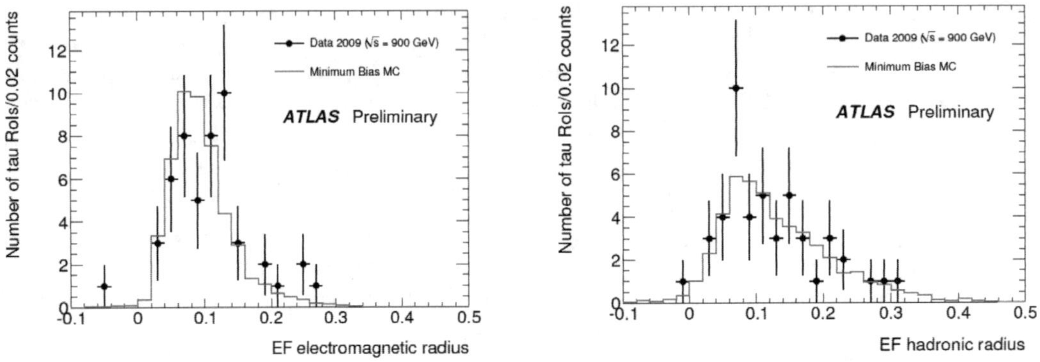

Figure 6. Comparisons between data and simulated events; the variables shown are the EF energy weighted radius, a measure of the shower size in $\eta - \phi$ around the cluster seed, at the EM calorimeter (left plot) and the HAD calorimeter (right plot).

least one offline reconstructed tau with at least one associated track. The L1 trigger efficiency is defined as the fraction of events that passed a L1 tau trigger. Figure 5 shows that the observed efficiency in data is measured to be similar to the expected one, estimated in minimum bias simulated events. The 900 GeV data variables used in the tau trigger selection were compared to the expectations from simulations. Some representative calorimeter variables are shown in Fig. 6. A reasonable agreement was found, within the limited available statistics, verifying the good detector and tau-trigger performance.

5. Conclusions

The ATLAS tau trigger system has been developed using simulated events and the evaluation of the algorithm performance had already started before collisions, using cosmic ray data. The first collision data at 900 GeV have provided an estimation of the good performance of the tau trigger. In addition to that, the calorimeters and tracking detectors, that contribute to the tau trigger reconstruction, are seen to perform well. New results with increased statistics are now expected from the 7 TeV LHC collision data. Verifying the good performance of the tau trigger system, is an important component for the commissioning of the ATLAS trigger and detector systems.

References

[1] The ATLAS Collaboration, G. Aad *et al.*, The ATLAS Experiment at the CERN Large Hadron Collider, *JINST* 3, S08003, 2008

[2] The ATLAS Collaboration, G. Aad *et al.*, Expected Performance of the ATLAS Experiment, Detector, Trigger and Physics, CERN report CERN-OPEN-2008- 020, Geneva, 2008.

[3] M.Shamim, on behalf the ATLAS Collaboration, Optimization and Performance of the ATLAS Tau Trigger with Cosmics Data, ICATPP Proceedings, http://villaolmo.mib.infn.it/ICATPP11th_2009/accepted/HEP/Shamim.pdf.

Journal of Physics: Conference Series **293** (2011) 012066 doi:10.1088/1742-6596/293/1/012066

A LYSO Calorimeter for the SuperB Factory

Claudia Cecchi[1], Valerio Bocci[3], Stefano Germani[1], Pasquale Lubrano[2], Elisa Manoni[1], Alessandro Rossi[1], Michel Lebeau[2], Marco Bizzarri[1], Giacomo Chiodi[3], Andrea Papi[2], Luigi Recchia[3].
1 Università di Perugia e INFN Sezione di Perugia
Via A. Pascoli, 06123 Perugia, Italia
2 Istituto Nazionale di Fisica Nucleare, INFN Sezione di Perugia,
Via A. Pascoli, 06123 Perugia, Italia
3 Università di Roma "La Sapienza" e INFN Sezione di Roma1
P.zzle Aldo Moro 2, 00185 Roma, Italia

E-mail: claudia.cecchi@pg.infn.it

Abstract. The SuperB project is an asymmetric e^+e^- accelerator of $10^{36}cm^{-2}s^{-1}$ luminosity, capable of collecting a data sample of 50-75 ab^{-1} in five years of running. The SuperB electromagnetic calorimeter (EMC), that will be described in this paper, provides energy and direction measurement of photons and electrons, and is used for identification of electrons versus other charged particles. In particular we will present its design, geometry study and related simulations, as well as R&D on LYSO crystals, a project for the mechanical structure and development on readout and electronics. A matrix of 6 crystals has been tested this year June 2010 at the Beam Test Facility of Frascati (BTF) at energies between 200 MeV and 500 MeV, and a beam test with the complete prototype of 25 crystals is foreseen at CERN in October 2010 to cover the energy range between 500 MeV and 7 GeV.

1. Introduction

SuperB is a proposed next generation asymmetric e^+e^- flavour factory, designed to run at a baseline luminosity of $10^{36}cm^{-2}s^{-1}$, between 50 and 80 times the peak luminosity reached by earlier B factories, KEKB and PEPII. The BaBar [1] detector has been a very effective instrument to perform precision measurements in the field of flavour physics. Therefore its main design remains a very good baseline option for an experiment at the SuperB factory [2]. The detector is composed by an Inner Vertex Detector, a Tracking System, an Electromagnetic Calorimeter, a Solenoidal Magnet and a Muon/Hadron System. A picture of the detector with the baseline (high) and the improved option (low) is shown in figure 1. The electromagnetic calorimeter is made of three parts: a central part (barrel) calorimeter, which will be reused from BaBar, a forward endcap calorimeter, which will replace the BaBar one, and the backward endcap calorimeter, not present in BaBar, a new device to improve the backward solid angle coverage. A new forward electromagnetic calorimeter has to be conceived, due to the high rate small angle Bhabhas and the radiation damage of the BaBar CsI(Tl) crystals, while the barrel of the CsI(Tl) calorimeter will be reused, with the necessary upgrades, in order to match the new requirements of an accelerator operating at the above mentioned luminosity.

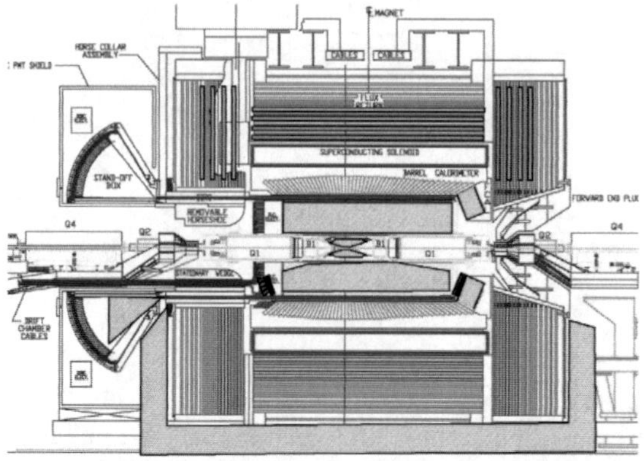

Figure 1. Detector for the SuperB factory, the high part shows the baseline while in the low part the option is reported. The green part shows new detectors.

In this paper only the new forward calorimeter will be addressed. Figure 2 shows a view of the BaBar calorimeter barrel plus endcap.

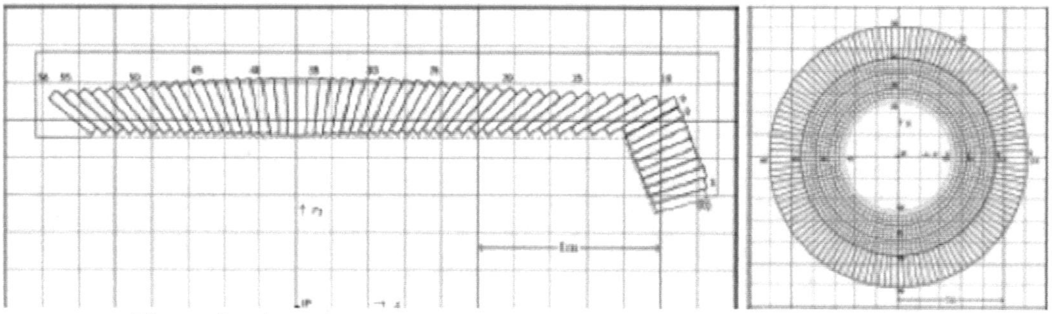

Figure 2. View of the BaBar CsI(Tl) crystal calorimeter, barrel and endcap.

The goal of the study described in this paper is to have a calorimeter with fast response, small Molière radius and high radiation hardness. A very good candidate for this purpose is the LYSO (Lutetium Yttrium Orthosilicate, with Cerium doping) crystal, which has a decay time of about 40 ns, a Molière radius of 2.07 cm, a high light yield and high density.

2. Forward Endcap Calorimeter

The forward endcap calorimeter for SuperB is a new device replacing the BaBar CsI(Tl) forward calorimeter with coverage starting at the end of the barrel and extending down to 270 mrad with respect to the beam line. It will be made of 4500 LYSO crystals arranged as described in section 2.1.2. The readout and electronics will be presented in section 2.1.3.

2.1 R&D for the LYSO crystal calorimeter

A deep R&D phase has started in 2008 for the new LYSO crystal calorimeter. In these two years many progress have been done in studying optical properties of the LYSO (under study since years in Caltech to prove the effective use in the high energy physics field, but already largely used in medical applications), a possible readout system and Very Front End (VFE) electronics to collect the signal, and a design of the mechanical structure to support the detector. In the meantime simulation packages have been developed both for the optimization of the detector in terms of geometry and performance, to study physics impact of different options, and to study the effect of the machine background in the detector. All these points will be addressed and described in the following sections.

2.1.1 LYSO crystals

R&D on LYSO crystals is ongoing mainly in Caltech and since about one year also in Italy (Perugia University and INFN). The characterization of the LYSO is being studied since years in order to have, from manufacturer, crystals with high light yield, and especially uniformity of the light along the longitudinal dimension; in particular the quality of the product in terms of Ce doping has been largely studied. A consistent uniformity has been found between PMT and APD readout, while there is a difference once reading the signal at the two ends of the crystal. This problem has been understood and solved; an extra green component was present in the radio luminescence spectrum at the tail end for some studied samples.
Figure 3 shows the light yield as a function of the distance from the end of the crystal coupled to PMT after different level of irradiation. No degradation of the performance is observed.

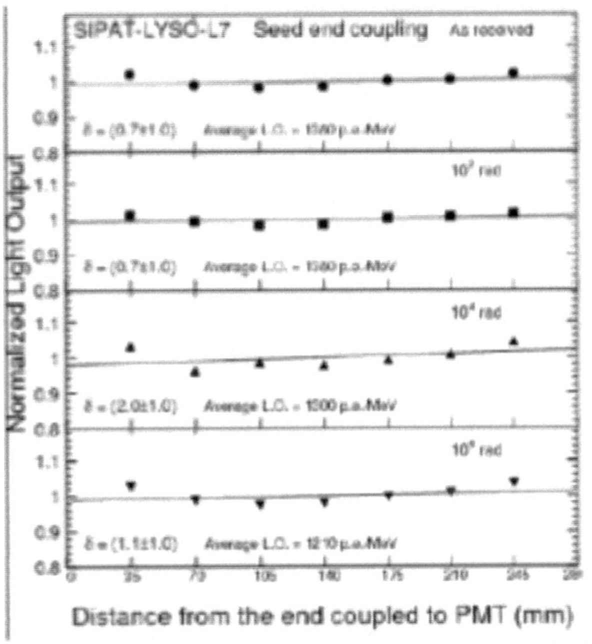

Figure 3. Light yield uniformity for a LYSO crystal after different level of irradiation. The crystal was coupled to a PMT. Consistent results have been obtained with the APD readout.

Different producers are being investigated, able to grow crystals of the requested dimensions (25mm x 25mm x 20cm) (see figure 4). The maximum transverse dimensions are dictated by the Molière radius and by the desire of obtaining two crystals from one boule. The length of 20cm corresponds to 17.5 X_0. A first production of new SIPAT (China) crystals has given very good results in terms of light yield, of the order of 86% of the best production (on this subject see dedicated talk at this conference by L. Zhang).

Figure 4. LYSO ingot growth at SIPAT. Dimensions are indicated below.

Journal of Physics: Conference Series **293** (2011) 012066 doi:10.1088/1742-6596/293/1/012066

Tests on LYSO crystals have been performed also in Italy (Perugia) with a sample of 12 crystals produced by Saint Gobain in order to measure optical properties of the crystals, like transmittance, scintillation and emission spectra, light yield and uniformity. A measurement of the light yield has been done using a Cobalt source and reading the signal with a PMT. In figure 5 (left) the spectrum of the light is shown, while the right plot shows the uniformity of the light along the crystal. The measured light yield is about 900 ph/MeV for the bare crystals, which is in agreement with the expected value; further measurements will be performed with the crystal wrapped in a tyvek foil and also putting a crystal in a cell made of the material that will be used for the mechanical structure. As can be seen the uniformity is very good, better than 1%, to be compared with our actual request of having a uniformity better than 5%.

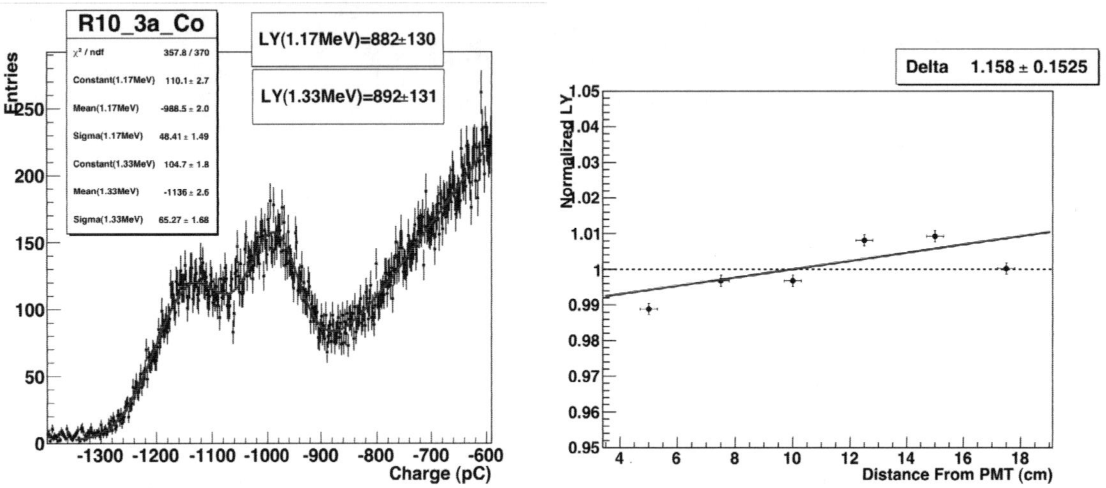

Figure 5. Distribution of the spectrum of the Cobalt source in the LYSO crystal. The LY derived from this spectrum is also indicated (left). Light yield uniformity measured on the same crystal, result is at the level of 1% (right).

An alternative choice to the LYSO would be the pure CsI even if the light output is much smaller in this case. Therefore tests on this crystal will be done in the next months.

2.1.2 Mechanics

Crystals are arranged in the calorimeter in 20 rings divided in 4 groups of 5 layers each. Each group of 5 layers is arranged in modules of 5 crystals wide, for a total of 4500 crystals. The crystals maintain the almost projective geometry of the barrel. Figure 6 shows the geometry of the calorimeter. The preferred structure for the forward LYSO calorimeter would be a continuous ring, however there is a possibility of building a structure divided in two halves if requested by installation constraints. The support structure is an alveolar, with each cell containing one crystal and a shell connecting the alveolar together, made either of carbon fibre or glass fibre. The outer diameter is 2m, the inner one is 1m. This structure is bounded by two cones at the radial extremes. The outer cone is a carbon fibre structure, 6-10 mm thick, in order to reduce the material between the barrel and the forward. The inner cone is 20-30 mm thick as there is no material issue in this part. A prototype of the mechanical structure has been built in Italy by RIBA company, to contain 25 crystals, of global overall dimensions of 130x130x250 mm3 (figure 7).

Figure 6. View of the geometry of the calorimeter. In different colours are shown the four layers of 5 rings each.

Cell walls are made by wrapping several layers of pre-impregnated glass fibre epoxy composite plus a single inner layer of aluminium reflective foil. This prototype will be used in October 2010 for a beam test at CERN. It has already been put under beam at the Beam Test facility of Frascati in June filled with only a subsample of the crystals, to test the electronics readout boards.

Figure 7. Picture of the prototype structure to be used at the Beam Test with 25 LYSO crystals.

2.1.3 Electronics

Two possible readouts are under study for this calorimeter, PiN diodes as used in the barrel and APD's (Avalanche Photodiodes). Two photodetectors per crystal are foreseen for redundancy. APD's have a low-noise gain factor of 50 and allow to calibrate crystals using sub MeV radioactive sources. The disadvantage of APD's is the strong dependence of the gain on the temperature. For more details on this see talk of R.Y. Zhu at this conference. As far as the electronics is concerned, two options are considered: one possibility is to send all data from the calorimeter over synchronous 1Gbit/s links to

Journal of Physics: Conference Series **293** (2011) 012066 doi:10.1088/1742-6596/293/1/012066

the level1 buffer in the trigger system; alternatively a “triggered” pull architecture will be used, where the trigger system receive only sums of crystals, and only triggered events are sent to the ROMs. The second option (figure 8) requires a much small number of links and is at the moment the baseline one. The signal from the APD’s or PiN’s, glued on the 4500 crystals of the forward LYSO calorimeter are sent to a charge preamplifier which translates the charge into voltage and then to a shaper with 100ns shaping time which provide a pulse of FWHM of 240ns. The shaped signal is amplified with two gains (x1 and x64) and at the end of the analog chain the chosen gain is digitized by a 12 bits ADC running at 14 MHz. With this scheme the full range from 10 MeV up to 10 GeV is covered with a resolution better than 1%. As stated above, the number of links can be reduced summing channels together at detector level and sending only the sums to the trigger. The natural granularity of the LYSO calorimeter is a module of 25 crystals. The data from the 25 crystals are summed together forming a word of 16 bits. Then 4 modules are grouped again. In this case the number of links would be only 90

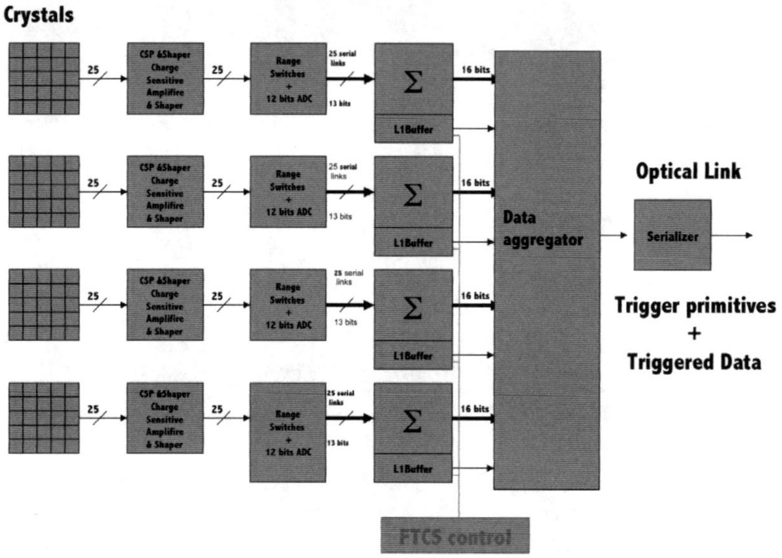

Figure 8. EMC electronics scheme.

(with respect to about 1000 in the case of non triggering architecture), but limiting the trigger granularity.

3. Simulations

Simulation studies are under way in order to optimize the detector configuration. First studies have been devoted to the geometry configuration taking into account constraints coming from the space available in the detector and from the maximum possible size of the crystals. The result of this simulation is the geometry described in paragraph 2.1.2 and shown in figure 6. In addition simulations have been developed to study the response of the calorimeter as function of the amount of dead material between crystals. Figure 9 (left) shows the resolution, for the prototype 5x5 crystal matrix, as a function of the energy for different configuration of the mechanical structure (thickness of the dead material between crystals).

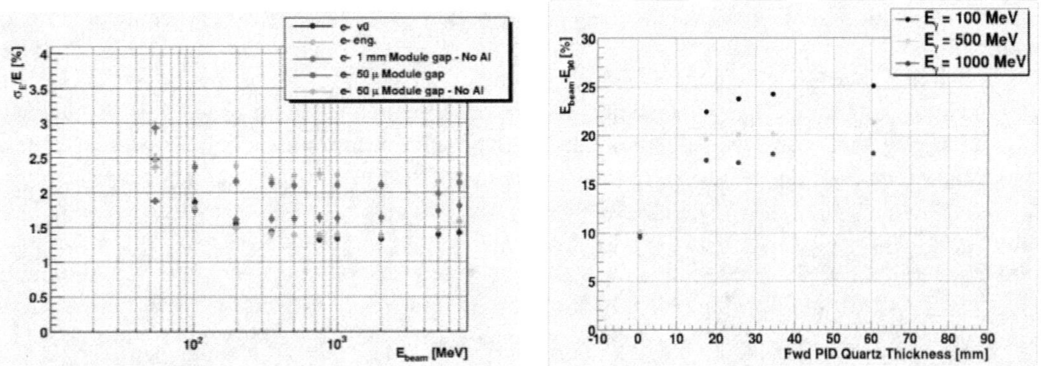

Figure 9. Left: Energy resolution as function of the energy per different types of the mechanical structure. Right: Scaled resolution (90% containment of the energy distribution) as function of the PID quartz thickness for different energies.

The simulation has gone further approaching more detailed aspects, for example the material in front of the calorimeter due to the possibility of installing a Particle Identification Detector (PID) in the forward region. A study has been performed to study the effect of the PID as a function of the thickness of the quartz bar and also as a function of the position of the PID in front of the calorimeter. Figure 9 (right) shows the resolution as a function of the thickness, showing a non negligible impact of the material on the resolution. The effect is more relevant in the case of low energy particles and slightly increase with the thickness. The evaluation of the benefits, in terms of signal over noise, in some physics channels particularly affected by the PID detector, with respect to the worsening of the resolution, is under study. Another important effect arise from the transition region between the barrel and the forward which should be ideally smooth in order to contain the electromagnetic showers and to keep pattern recognition simple. However the possible presence of the forward PID detector requires the forward calorimeter to be moved back. The effect of this on photon energy resolution has been studied and the resolution degrades in the barrel-endcap transition region as expected. In figure 10 shows the effect on the resolution due to the position of the calorimeter far from the transition region (left) and in the transition region (right).

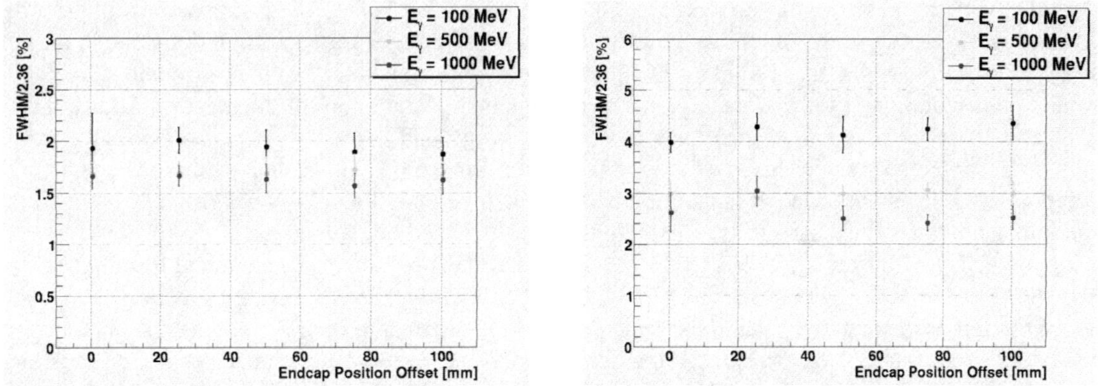

Figure 10. Effect on resolution of z-position forward calorimeter. Left: resolution as function of the calorimeter position for showers away from the transition region. Right: resolution as function of the calorimeter position for showers in the transition region.

Furthermore the impact of a realistic clustering algorithm has been studied as in the actual event environment multiple particles are present in an event and pattern recognition is then required.

Figure 11 (left) shows how the measured energy distribution changes for different reconstruction algorithms.
Finally we started to study the effect of the background coming from the machine on the calorimeter performance. In particular radiative Bhabha events have been simulated and added to the signal Monte Carlo events. This study can give valuable input on electronics requirements like the shaping time and the integration time, in order to be able to handle pile up events between signal and background. Most of the background events are coming from neutrons, while electrons and photons are very few. The physic channel used for this study is the $B^+ \rightarrow K^+ \nu\nu$. The recoil analysis approach is used where one of the two B's is fully reconstructed in the events in the semileptonic or hadronic modes (this is called the B_{tag}) and used to constraint the kinematics, while the other emisphere is used to search for the signal ($B_{signature}$). This method is used when there are undetectable particles in the signal in order to have a clean sample.

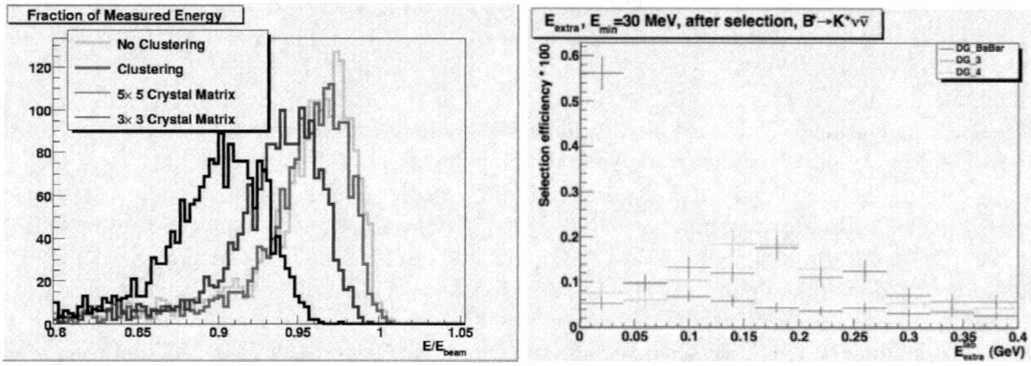

Figure 11. Left: effect on the measured energy distribution for different reconstruction algorithm. The green distribution is the ideal truth Montecarlo reconstruction. Right: E_{extra} distribution is $B^+ \rightarrow K^+ \nu\nu$ events with radiative Bhabha from background superimposed in the simulation. Blue distribution is in the case of a BaBar like geometry. Red and green are for two different configuration of the SuperB geometry.

This method has been applied in the same analysis in the BaBar experiment where it was possible to distinguish very effectively between signal and background using the E_{extra} distribution shown in figure 10 (right). A clear peak in the first bin at zero (blue crosses) is present for the simulation done in the case of a BaBar-like geometry. The other configurations, which correspond to two different possible geometry configurations for the SuperB experiment, do not show anymore this peak. This means that in the geometry configuration of SuperB, when adding background to the signal, is it not possible to distinguish it from the signal. This is due to the fact that the machine background events are superimposed to the signal ones. It has to be stressed that, a very high level of background has been taken into account for this analysis, much worse than the expected one, making this result very conservative. The electronics of the detector has to be studied in detail in order to avoid this kind of behaviour. Moreover, it has been seen that this happens only in the barrel where the electronics integration time is much larger than in the endcap and is then related to the fact that the CsI crystals have a decay time much longer than the LYSO ones. In the LYSO forward calorimeter is then possible to use a time window of only100ns (300 ns in the barrel) and the peak of the E_{extra} is clearly visible in this part of the detector. This study, strictly correlated with the development of the electronics of the detector, is still ongoing.

4. Beam Test

A beam test has been performed for one week in June 2009 at the Beam Test Facility (BTF) of LNF (Laboratori Nazionali di Frascati, INFN). One LYSO crystal has been equipped with one photodiode

PIN [Hamamatsu S2744-08 (1x2cm^2)] and with two APD's [Hamamatsu S8664-55 (0.5x0.5 cm^2 each)] as showing in figure 12.

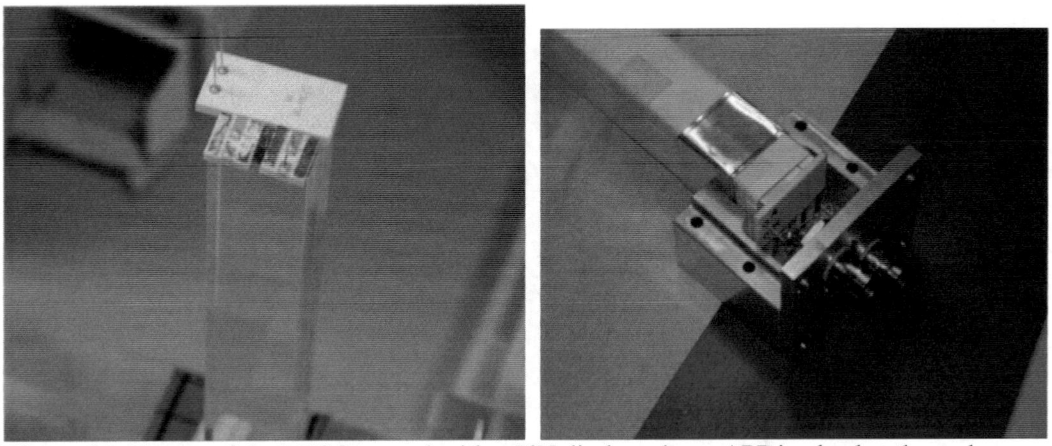

Figure 12. Picture of the LYSO crystal with a PiN diode and two APD's glued at the end.

The crystal has been studied with a beam of electrons of energies in the range between 100 MeV and 500 MeV.The BTF allows to adjust the number of electrons arriving at the detector by setting up dedicated collimators. We worked with the collimators adjusted to have 1 electron, a small contamination was present of events with 2 electrons as shown in the following. Figure 13 (left) shows the signal response in the APD and in the PiN diode and a strong correlation is observed. In figure 13 (right) the signal in ADC count for the data collected at 500 MeV is shown.

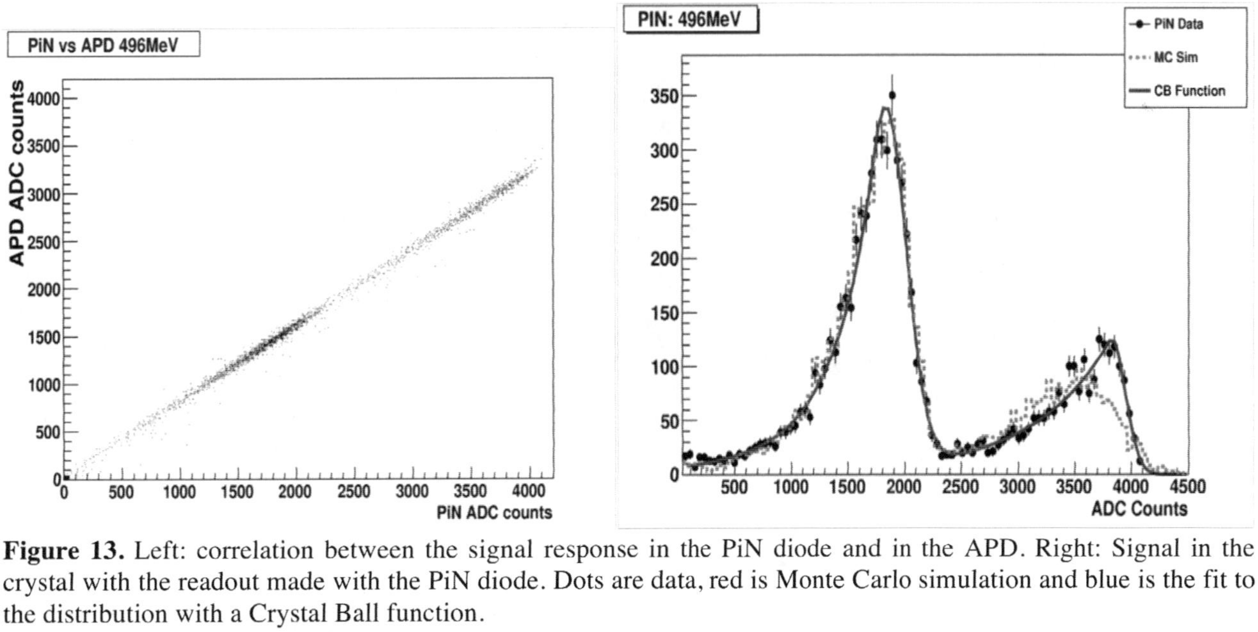

Figure 13. Left: correlation between the signal response in the PiN diode and in the APD. Right: Signal in the crystal with the readout made with the PiN diode. Dots are data, red is Monte Carlo simulation and blue is the fit to the distribution with a Crystal Ball function.

The peaks for one and two electrons are well separated and there is a good agreement between data and simulations, taking into account the poor shower containment in only one crystal. In June 2010 the matrix with 6 crystals has been put under test at the BTF to test the electronics boards. 3 crystals where equipped with PiN diodes and 3 with APD's. It was possible with 6 crystals to test the linearity and also to have an estimation of the resolution as function of the energy. Results are shown in figure 14.

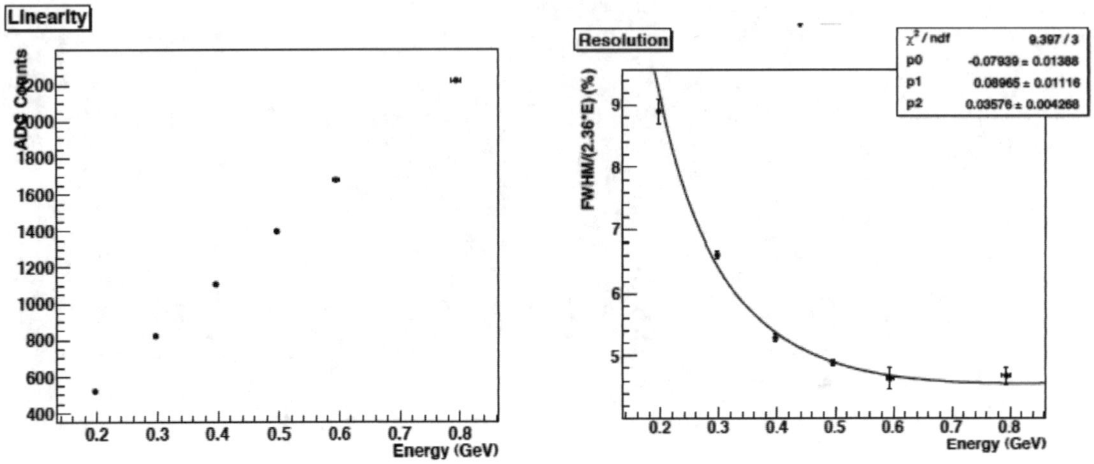

Figure 14. Left: Signal in ADC counts as a function of the energy, a good linearity is observed. Right: resolution as a function of the energy.

In October and November 2010 a matrix with 25 LYSO crystals will be put under test at CERN (1GeV-7 GeV) and then at BTF again. The LYSO will be surrounded by a ring of CsI crystals. Figure 15 shows the mechanical structure with the first column filled with crystals (left). In the right panel a CAD design of the LYSO matrix at the centre with the surrounding CsI crystals in blue and one of the five electronics board is shown.

Figure 15. Left: mechanical structure of the beam test with one column filled with LYSO crystals. Right: CAD design of the test beam matrix at the centre surrounded with CsI crystals. One of the 5 electronics boards is also shown in green.

Five electronics boards have been developed and are ready to be installed. The readout will be done with PiN diodes in one row (5 crystals) and the rest of the matrix with APD's. the DAQ system is developed with sampling ADC's to read the LYSO matrix, and peak sensing ADC's for the Cherenkov counters which will be used at CERN to separate electrons from pions, a silicon telescope which will be put in front of the matrix to have information on the beam position and to study the beam profile and finally to read a quartz bar to be put before the calorimeter to study the effect of the particle identification system.

5. Conclusions

R&D for the use of the LYSO crystals in high energy physics applications has started and is developing in order to cover many different aspects. Characterization of the LYSO is under development mainly to understand Ce doping and Light Yield uniformity response, As of today more than one manufacturer is available for the growing and production of the LYSO and quality is improving fast. The simulation for a possible LYSO calorimeter for the Superb factory has also been developed starting from the basic points of the expected resolution and going into details of needed application. A test beam has been performed with only one crystal allowing the understanding of the beam facility at LNF and showing that there is a high symmetry between the two possible readouts which are at the moment under study, PiN diodes and APD's. Two tests beam will be done in October and November one at CERN and one at the BTF.

6. Acknowledgment

This work has been supported by "Fondazione Cassa di Risparmio di Perugia". We would like to thank Prof. Ren Yuan Zhu for useful discussions and help about R&D for crystals.

References

[1] B. Aubert et al. 2002 The BaBar Detector *Nucl. Instrum. Meth. Phys. Res. Sect A* **479** 1
[2] SuperB Conceptual Design Report 2007 A High Luminosity Asymmetric e+e- Super Flavour Factory *INFN/AE-07*
[5] Ren Yuan Zhu private communications

Journal of Physics: Conference Series **293** (2011) 012067 doi:10.1088/1742-6596/293/1/012067

Electromagnetic Calorimeter Upgrade for Belle2

Isamu Nakamura (中村 勇)
KEK, 1-1 Oho, Tsukuba, Ibaraki 305-0801, Japan
E-mail: isamu.nakamura@kek.jp

Abstract. Upgrade of the Belle electromagnetic calorimeter is summarized. The Belle experiment will be upgraded to cope with higher luminosity accelerator SuperKEKB. For the electromagnetic calorimeter, readout electronics will be replaced to introduce waveform analysis. This is to reduce effect from gamma rays produced in accelerator, that will be much higher than that observed in current Belle experiment.

1. SuperKEKB and Belle2

After successful operation, the KEKB asymmetric e^+e^- collider and the Belle Detector [1] accumulated data corresponding to an integrated luminosity of 1 ab^{-1}. The experiment stopped data taking by the end of June 2010 and started to improve its performance. The SuperKEKB, successor of KEKB accelerator aims for 50 times more luminosity by squeezing the beam to about 1/20 of current KEKB and by putting twice more current. With SuperKEKB accelerator, trigger rate of Belle experiment is expected to reach as high as 20–30 kHz. Beam related background is also expected to increase. Requirement for the Belle2 detector is to have same or better performance compared to the Belle detector under 10–20 times more background, though it is very difficult to make an accurate estimation of beam background without actual accelerator operation. In the Belle2 detector upgrade, completely new tracking detector consists of two layers of DEPFET Pixel detector, four layers of Double-sided Silicon Strip detector and small cell drift chamber will be installed. Particle ID devices will also be completely new. For barrel region, Time-Of-Propagation Čerenkov detector replaces current TOF and Aerogel Čerenkov detector while proximity focusing Aerogel-RICH detector will be installed in the forward endcap. For calorimeter crystals and preamplifiers are kept but all readout electronics will be replaced. The solenoid magnet as well as the barrel muon and K_L^0 detector(KLM) will be kept. The RPC of the endcap KLM will be replaced with Scintillator strip detector. All DAQ electronics and trigger system will also be renewed. The detail of SuperKEKB/Belle2 upgrade can be found in elsewhere [2, 3].

2. The Belle Calorimeter

The main purpose of the electromagnetic calorimeter is to detect γ and π^0 from B meson decay. Energy of such γ ranges from MeV to GeV. Energy resolution is very important because for some analyses one has to separate signal and background events only with mass distribution. To separate two photons from high momentum π^0, fine granularity is required. Hermeticity is also very important to study final states with neutrino. The Belle electromagnetic calorimeter consists of 8736 CsI(Tl) crystals with typical dimension of $6\times6\times30$ cm^3. The length corresponds to about 16 X_0. Figure 1 shows the schematic view of the calorimeter. The calorimeter has pointing geometry and covers polar angle from 12° to 155°. The scintillation light of the CsI(Tl) crystal has λ = 560 nm with rather slow decay constant of 1.3 μsec but abundant light output of about 50000 p.e./MeV. The light is read out by two independent sets of silicon PIN photodiodes (Hamamatsu S2744-08) with sensitive area of 10 × 20 mm^2 connected to charge sensitive preamplifier. The signals are sent to the front-end electronics with cable of typically 10 m long. Two signals from preamplifiers are first added and then processed with two shaper with 0.2(fast) and 1μsec(slow) shaping time. The fast signal are used to generate trigger signal while the slow signal is then sent to LeCroy MQT300A charge-to-time converter. The MQT converts 3 range/12 bit charge information into 15 bit timing information which corresponds to 18 bit dynamic range. The time encoded energy information is then recorded with LeCroy 1877S Fastbus TDC. The readout system consists of about 100 TDC boards in five crates read out by three VME systems. It takes about 30 μsec to read out one physics event. The trigger rate is about 300 Hz with instantaneous luminosity of 10^{34} cm^{-2}sec^{-1}, which corresponds to about 1% readout deadtime. With 10 kHz trigger rate deadtime is 30% that is not acceptable. Although the signal is recorded with TDC, only energy is recorded. No event timing information is available.

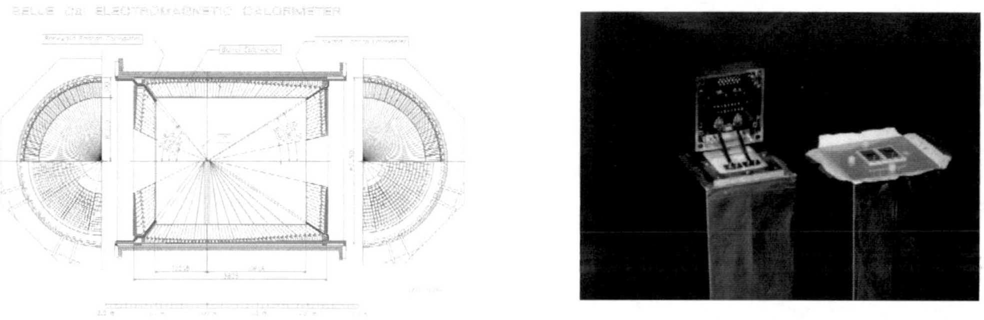

Figure 1. (Left) The Belle Electromagnetic Calorimeter. (Right) A counter.

The effect of accelerator background is estimated in the current Belle experiment using randomly triggered event. At current Belle calorimeter, each crystal receive 0.5–1 MeV energy from accelerator background as shown in Figure 2(left), where average energy deposition in one crystal is plotted as a function of polar angle. Figure 2(right) shows the energy sum in the calorimeter as functions of total beam current and average instantaneous luminosity. From the figure it seems background is linearly behaved to the beam current rather than luminosity. Therefore a factor 3–10 higher background is rough estimation for the Belle2, as beam current increase is only a factor of 2.

Radiation damage is an issue for high luminosity collider experiment and the calorimeter is no exception. Figure 3(left) shows dark current of photodiode in inner forward region as a function of time in days since the start of the calorimeter operation. Dark current increase is about 10 nA in barrel and increase toward the beam pipe upto 200 nA. The highest is inner part of forward endcap. The dark current increase is mostly due to neutron radiation. Figure 3(right) is dark current increase as a function of neutron fluence obtained by the study performed in a reactor. The 200 nA dark current corresponds to about 10^{11} neutrons/cm^2. Hence 10^{12} neutrons/cm^2 is expected in Belle2.

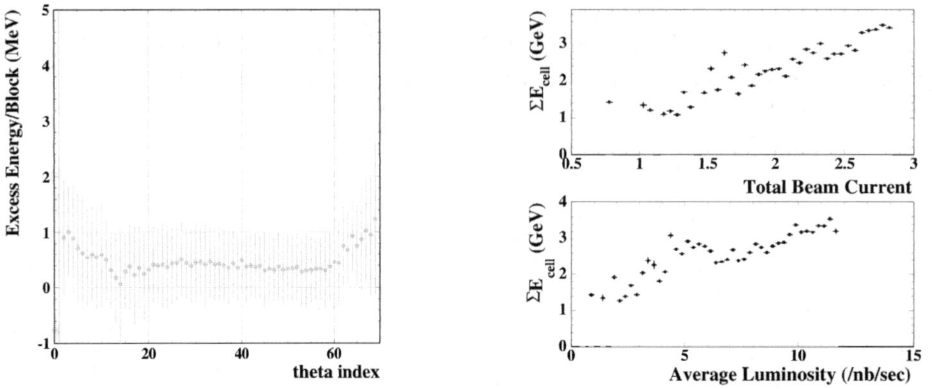

Figure 2. (Left) Energy deposit from beam background as a function of polar angle. (Right) Energy deposit as functions of total beam current(top) and instantaneous luminosity(bottom).

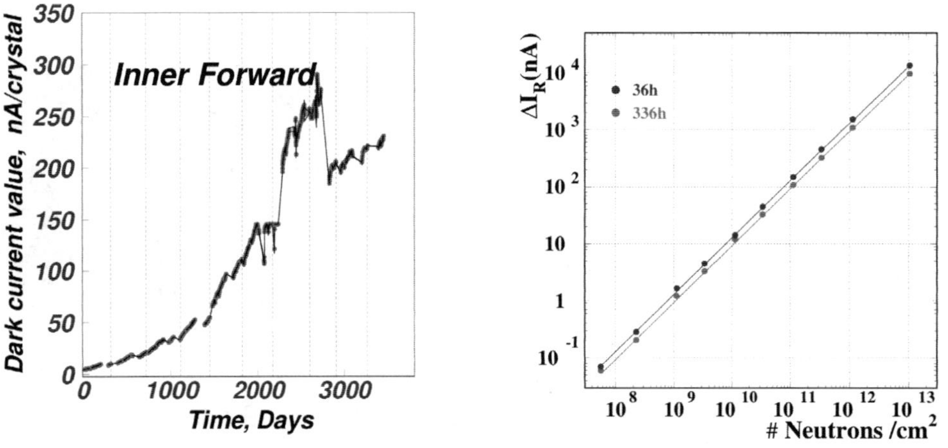

Figure 3. (Left) Dark current of Photodiode as a function of time. Large increase around 2500 is due to temperature increase. (Right) Dark current increase as a function of neutron fluence.

3. Belle2 Calorimeter Upgrade

In the Belle2 experiment trigger rate will be as high as 30 kHz and background will be 10 times as high as Belle. To cope with such situation waveform sampling together with pipelined readout will be introduced. The former enables us to use timing information to discriminate off-timing hits while the latter can eliminate effectively readout deadtime. Figure 4 shows simulated timing distribution.

To realise above readout electronics are replaced to new one. Figure 5 shows the schematic diagram of new readout system. Hart of the system is Shaper Digitizer board which receives preamplified signals from 16 crystals. Two preamplifier signals from one crystal are summed and processed with shaper with time constant of 0.5μs which is shortened from 1.0μs. Reducing shaping time further is not efficient as the time constant of scintillation is 1.3 μs. Signal after shaping is digitized by 18 bit ADC with 1.8 MHz sampling rate and sent to pipeline buffer in FPGA which is 512(290μs) deep. Once trigger is issued, waveform fitting is performed in FPGA using 16 samples to extract timing and amplitude. According to simulation study a factor of

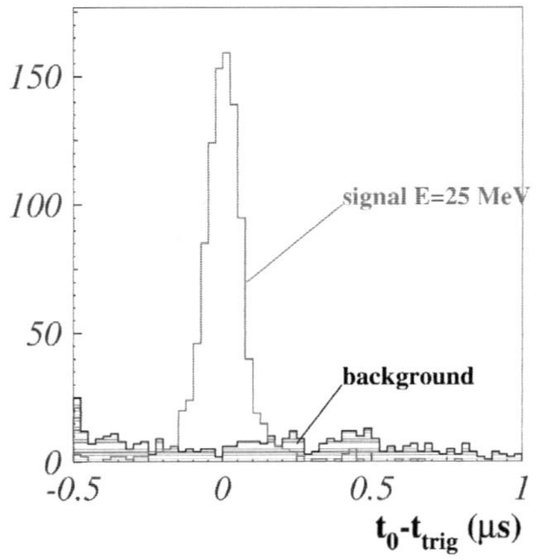

Figure 4. Expected timing distribution with new electronics.

seven reduction of fake cluster is expected with upgraded readout system.

To readout all 8736 crystals, total of 576 Shaper Digitizer boards are placed in 52 VME crates. In each VME crate a Collector board will be installed. The Collector board receives trigger signal, distribute it to Shaper Digitizer boards in the crate, receives fitted timing and amplitude and sends them to backend DAQ system via optical link. Introducing Collector board greatly reduce the number of cables from detector to electronics hut. The Collector board can also supply calibration signal.

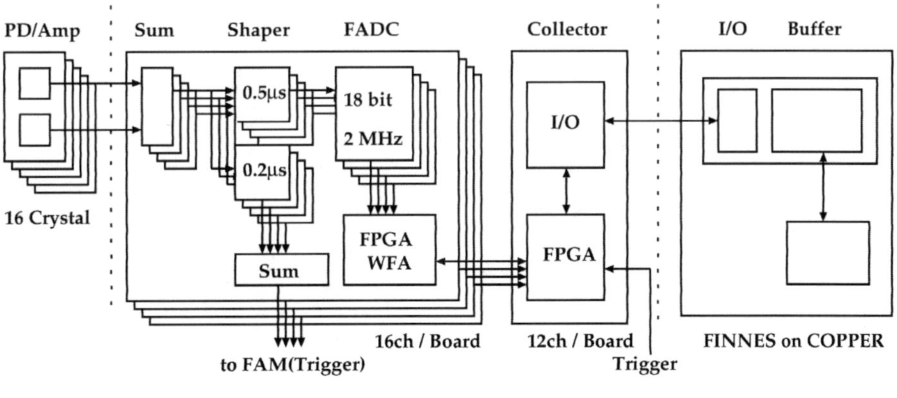

Figure 5. Schematic Diagram of new DAQ system.

4. Preparation Status

To test the scheme of waveform fitting, prototype with TKO formfactor, which is used in the Belle Calorimeter, is produced. This TKO prototype has same analog part and ADC as new Shaper Digitizer board and is compatible with current Belle DAQ system. Eight boards, corresponding to 1/8 (120 channels) of backward endcap, are installed to the Belle and the data had been taken since summer of 2009 to the end of experiment. Figure 6(left) shows energy distribution of bhabha events collected with prototype board compared with current board.

One can find that the prototype boards are working as good as current board. Figure 6(right) is difference of timing obtained from waveform fitting to the event timing reconstructed offline by other sub detectors. Timing resolution is better than 10 ns for 100 MeV gamma.

Currently R&D study with prototype VME shaper digitizer and collector boards, produced recently, are ongoing. Figure 7 shows prototype Shaper Digitizer(left) and Collector(right) boards. The plan is to finalise the board design by the end of 2010.

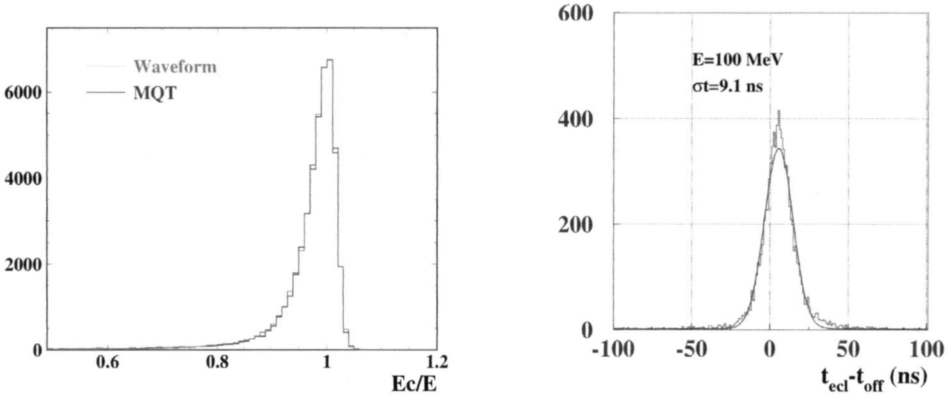

Figure 6. (left) Energy distribution for bhabha events. (right) timing resolution.

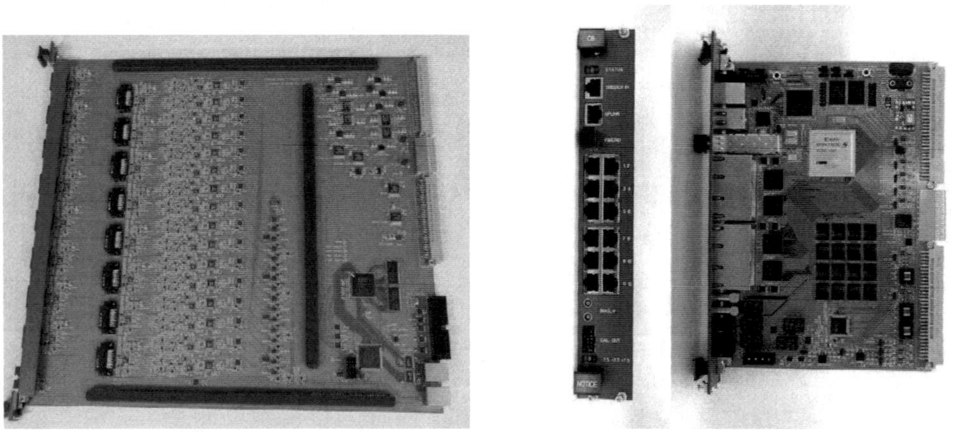

Figure 7. (Left) Shaper Digitizer prototype. (Right) Collector prototype.

References

[1] A. Abashian *et. al.*, Nuclear Instruments and Methods A479 117(2002)
[2] SuperKEKB Letter of Intent, KEK Report 04-4
[3] Belle II Technical Design Report in preperation.

Journal of Physics: Conference Series **293** (2011) 012068 doi:10.1088/1742-6596/293/1/012068

CMS Hadron Endcap Calorimeter Upgrade Studies for Super-LHC

Burak Bilki
on behalf of CMS HCAL Collaboration
University of Iowa, Iowa City, IA 52242-1479, U.S.A.
E-mail: Burak.Bilki@cern.ch

Abstract. When the Large Hadron Collider approaches Super-LHC conditions above a luminosity of $10^{34}cm^{-2}s^{-1}$, the scintillator tiles of the CMS Hadron Endcap calorimeters will lose their efficiencies. As a radiation hard solution, the scintillator tiles are planned to be replaced by quartz plates. In order to improve the efficiency of the photodetection, various methods were investigated including radiation hard wavelength shifters, p-terphenyl or 4% gallium doped zinc oxide. We constructed a 20 layer calorimeter prototype with pTp coated plates of size 20 cm x 20 cm, and tested the hadronic and the electromagnetic capabilities at the CERN H2 beam-line. The beam tests revealed a substantial light collection increase with pTp or ZnO:Ga deposited quartz plates. Here we report on the current R&D for a viable endcap calorimeter solution for CMS with beam tests and radiation damage studies.

1. Introduction

The Compact Muon Solenoid (CMS) [1] is a general-purpose detector designed to run at the highest luminosity provided by the CERN Large Hadron Collider (LHC). Current luminosity upgrade plans are divided into two 5-year phases. Phase 1 covers the period from 2013 to 2018 when the peak luminosity will increase from 10^{34} cm^{-2} s^{-1} to 4×10^{34} cm^{-2} s^{-1}. Phase 2 covers from 2018 to 2023 when the peak luminosity reaches to 10^{35} cm^{-2} s^{-1}.

Coverage between pseudorapidities of 1.3 and 3.0 is provided by the endcap hadron calorimeter (HE). The HE is composed entirely of brass absorber plates in an 18-fold ϕ-geometry matching that of the barrel calorimeter. The thickness of the plates is 78 mm while the scintillator thickness is 3.7 mm, hence reducing the sampling fraction. There are 19 active plastic scintillator layers. In the high η-region, above $|\eta| = 1.74$, the ϕ-granularity of the tiles is reduced to 10° to accommodate the bending radius of the WLS fiber readout. For the purpose of uniform segmentation in the Level-1 calorimeter trigger, the energies measured in the 10° ϕ-wedges are artificially divided into equal shares and sent separately to the trigger. The $\Delta\eta$x$\Delta\phi$ tower size matches that of the barrel in the range $1.3 < |\eta| < 1.74$. For $|\eta| > 1.74$, the η size increases [2].

2. Upgrade Requirement

Light generated in the HE scintillators (Kuraray SCSN81) is carried to hybrid photodiodes by Kuraray Y-11 double clad wavelength shifting (WLS) fibers. Both scintillators and WLS fiber have been shown to be moderately radiation hard up to 2.5 MRad [3]. The simulation studies show that the expected radiation levels have strong η dependence and they predict radiation levels up to 10 MRad in high η towers after 10 years of LHC operation at 10^{34} cm^{-2} s^{-1} luminosity [4][5]. This value reaches up to 30 MRad for the front towers of the $2.9 < |\eta| < 3.0$ region, where the Endcap Electromagnetic (EE) calorimeter does not shield the HE calorimeter.

With the increasing luminosity plan for LHC, we expect the scintillator tiles and plastic WLS fibers to face higher radiation levels than these reports. Hence, the resolution of the HE calorimeter will deteriorate faster, starting from high η towers. As a solution to this radiation damage problem, we propose to substitute the scintillators with quartz plates. Results show that quartz can withstand the radiation doses up to 1.25 GRad [6][7][8].

However, with the quartz plates, the detected photons come from Čerenkov radiation, which yields 100 times less light than the scintillation process. To improve the number of generated photons we tested different light enhancement tools, including p-Terphenyl (pTp), 4% Gallium dopped - ZincOxide (ZnO:Ga), o-Terphenyl (oTp), m-Terphenyl (mTp), and p-Quarterphenyl (pQp). The light enhancement tool on the quartz has to be radiation hard as well.

3. Selection of Light Enhancement Tool

To improve the light production inside the quartz plates, we considered various light enhancement tools: pTp, ZnO:Ga, oTp, mTp, and pQp. Other than ZnO:Ga, all of them can be evaporated in a vacuum chamber to be applied on quartz. The molecular properties of ZnO:Ga do not allow evaporation, so RF sputtering was used for this case. After the beam tests at the Fermilab Meson Test Beam Facility and at the Cern H2 beamline, the one sided coatings with 2 μm thickness of pTp and 0.2 μm of ZnO:Ga yielded the best results in pion, proton, and electron beams at various energies . Figure 1 shows that in both cases, we have increased the light yield for minimum ionizing particles at least four times compared to plain quartz plates. Since ZnO:Ga requires a more expensive and delicate deposition process, we decided to focus on pTp for the rest of our studies.

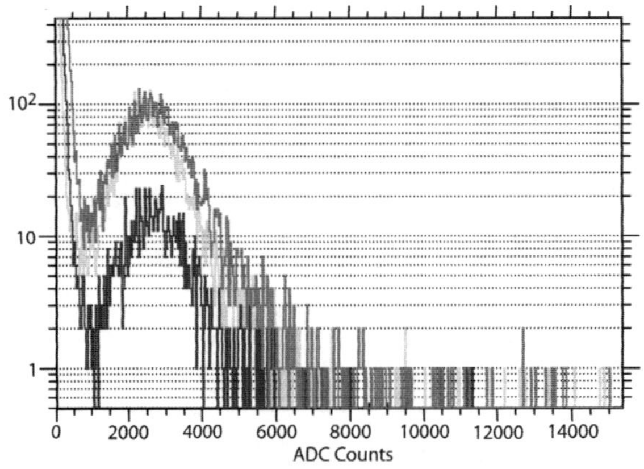

Figure 1. Light yield from 2 μm thickness of PTP (red) and 0.2 μm thickness of Ga:ZnO (green) deposited quartz plates, and clean quartz plate (blue).

Figure 2 shows the absorption and emission spectra of pTp. As the number of created Čerenkov photons increases with $1/\lambda^2$ having the pTp absorption spectra positioned in UV range helps us to enhance light production.

As the next step, we tested pTp radiation hardness with proton beams at the Indiana University Cyclotron Facility (IUCF) and CERN beam lines. The Sr^{90} activated scintillation light outputs of pTp samples both before and after irradiation were compared in the University of Mississippi CMS Laboratories. We also employed the liquid scintillation technique in which the pTp samples were mixed in a saturated toluene solution with a standard tritium beta source and scintillation light analyzed for dosed and standard samples. The toluene yields negligible scintillation light in the tests. The dosed pTp samples were also sent for chemical analysis showing slight breakdown of the tri-ring pTp molecule into simpler benzene ring forms. The results of different irradiation levels are reported in Fig.3. After 20 MRad of proton irradiation, the light output drops to 84% of the initial level. But the initial radiation damage rate slowly flattens, and after 40 MRad of radiation we still observe less than 20% loss of light production.

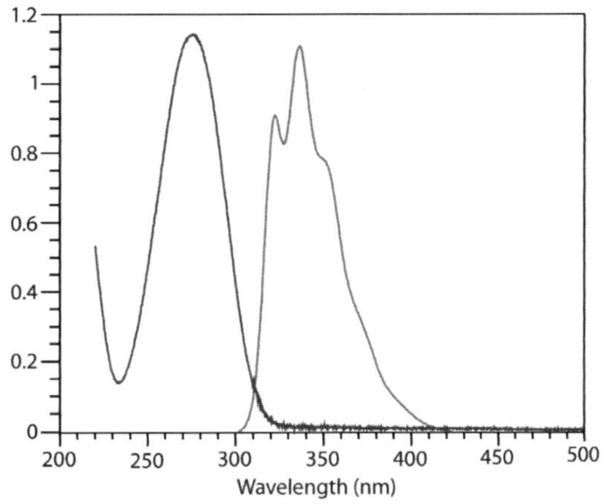

Figure 2. Absorption (blue) and emission (red) spectra of p-Terphenyl.

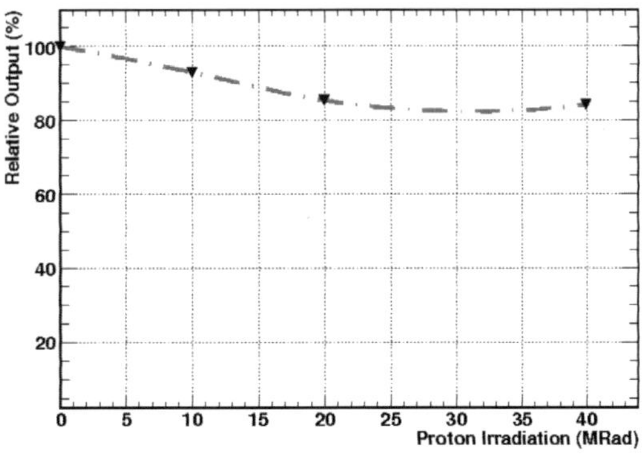

Figure 3. Light output from PTP sample after proton irradiation versus proton irradiation level with simple fitted line.

4. Construction and Tests of the Prototype

We have built a quartz plate calorimeter prototype with 0.2 μm pTp deposited on one side of the quartz tiles. The prototype consists of 20 layers of quartz plates (15 cm x 15 cm x 5 mm) with 7 cm iron absorbers between each layer making it a good representation of the small solid angle of the upgraded HE calorimeter. GE-124 quartz from GE Quartz Company was used as the material for the plates. After 2 μm pTp was deposited on every quartz plate via evaporation at Fermilab Thin Film Laboratory, one inch section on the side of each plate was polished at the University of Iowa CMS Laboratories for better PMT coupling. The light generated in the quartz plate was read out by Hamamatsu R7525-HA photomultiplier tubes from the edge of the polished side.

The quartz plates were wrapped with mylar for good reflectivity, especially in the UV range, and then with Dupont Tyvek for a robust light tight structure. Every quartz plate and PMT system was prepared to be a standalone unit. Having our prototype constructed as a collection of standalone units allowed us to change the absorber thickness between layers for a possible electromagnetic configuration.

The prototype was tested at the CERN H2 beam-line with two different configurations: a hadronic configuration with 7 cm iron absorbers between each layer and an electromagnetic configuration with 2 cm iron absorbers. Hadronic configuration was tested with 20, 30, 50, 80, 130, 200, 250, 300, and 350 GeV π^- beams and the electromagnetic configuration was tested with 50, 80, 100, 120, and 200 GeV electrons [9].

Figure 4 shows the longitudinal shape of the hadronic showers. Solid curves are the Geant4 [10] simulation results. The calibration between simulation and data was done using 80 GeV π^- data by macthing the response layer by layer. Slight variations between layers are understood to be due to alignment and optical matching differences.

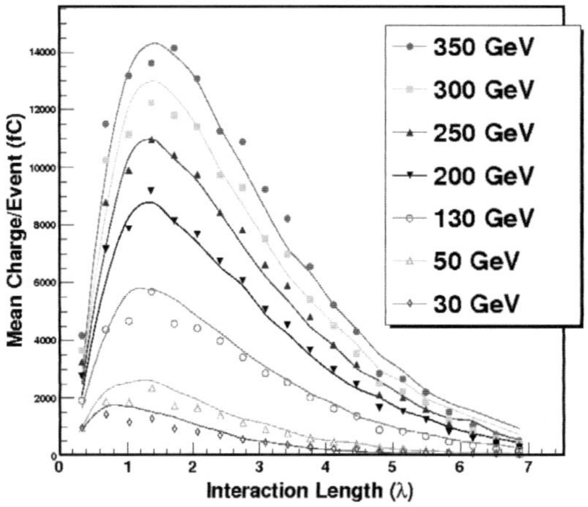

Figure 4. Longitudinal shower shape for hadronic showers. Solid curves are the Geant4 simulation results that were calibrated with 80 GeV π^- data.

Figure 5a shows the detector linearity for hadronic response. There is good agreement with the data and simulation at all energies. Figure 5b shows the energy resolution of the calorimeter. At lower energies, the hadronic resolution is better than 40% and it reaches 13% at 350 GeV. The agreement between data and simulation is better at lower energies. The main limiting factor of the prototype in terms of resolution is its small lateral size.

Although the purpose of these studies is to find a solution for radiation hardness problems of the CMS HE calorimeter, we took this opportunity to test the similar radiation hard configuration on an electromagnetic calorimeter. Figure 6 shows the longitudinal shape of the electromagnetic showers. For the calibration of the simulation results (solid curves) 80 GeV electron data was used. Figure 7a shows the detector linearity in response to electromagnetic showers and the energy resolution is shown in Fig. 7b. At energies higher than 120 GeV, both resolution curves converge to a value around 5.6%.

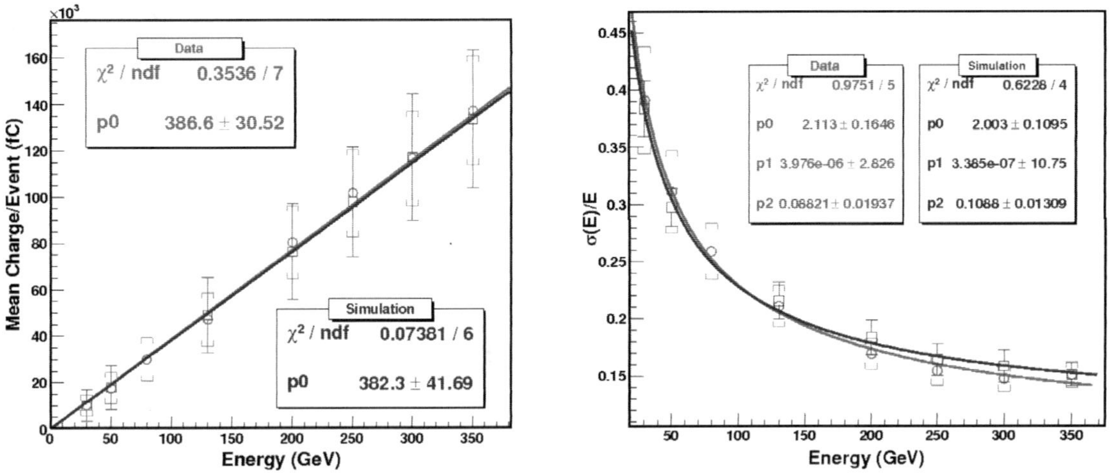

(a) (b)

Figure 5. Hadronic response linearity (a) and energy resolution (b) of the calorimeter.

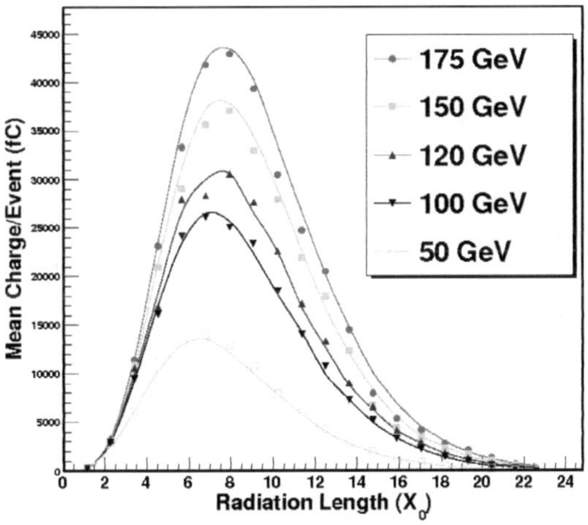

Figure 6. Longitudinal shower shape for electromagnetic showers. Solid curves are the Geant4 simulation results that were calibrated with 80 GeV e^- data.

5. Conclusions

As the peak luminosity of the LHC increases, detector upgrades are required for LHC experiments. The CMS experiment HE calorimeter consists of moderately radiation hard scintillator tiles, and they are not going to survive the high radiation environment of the SLHC. Here, we show that p-Terphenyl (pTp) deposited quartz plates are a perfect candidate to replace the scintillator tiles of the CMS HE calorimeter. Both quartz and pTp are radiation hard and cost effective options.

We have demonstrated that with one sided coatings of 2 μm pTp and 0.2 μm ZnO:Ga (4% gallium doped zinc oxide), the light production can be enhanced at least 4 times. We also report that pTp loses only 16 % of the light production after 40 MRad proton irradiation. This is well above the predicted SLHC radiation levels for most of the HE towers.

To test the calorimeter capabilities of this replacement option, we constructed a 20 layer

Journal of Physics: Conference Series **293** (2011) 012068 doi:10.1088/1742-6596/293/1/012068

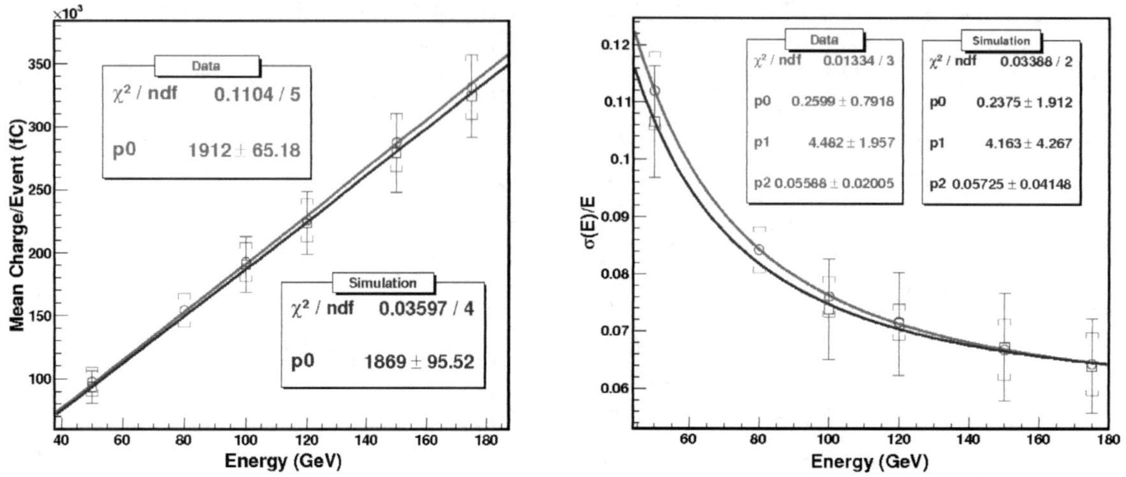

(a) (b)

Figure 7. Electromagnetic response linearity (a) and energy resolution (b) of the calorimeter.

pTp deposited quartz plate calorimeter prototype. Overall, the hadronic energy resolution of the calorimeter can be expressed as $\frac{\sigma(E)}{E} = \frac{211\%}{\sqrt{E}} \oplus 8.8\%$ with a negligible noise term, and the electromagnetic energy resolution can be expressed as $\frac{\sigma(E)}{E} = \frac{26\%}{\sqrt{E}} \oplus \frac{4.5\%}{E} \oplus 5.6\%$. Particularly hadronic resolution suffers significantly from lateral leakage but a hadronic resolution of 11% is achievable at higher energies.

pTp deposited quartz plates sandwiched between thinner absorbers can be used as a radiation hard electromagnetic or hadronic calorimeter option for future collider experiments. Considering the radiation damage issues of the CMS Endcap Electromagnetic calorimeter, we propose that pTp deposited quartz plates constitute a viable option for all CMS endcap regions.

References

[1] Chatrchyan S et al. (CMS Collabration) 2008 "The CMS Experiment at the CERN LHC" *JINST* **3** S08004 145-149

[2] CMS Collaboration 2006 "CMS Physics TDR: Volume I, Detector Performance and Software" *CERN/LHCC* 2006-001

[3] CMS Collaboration 1997 "The HCAL Technical Design Report" *CERN/LHCC* 97-31

[4] Huhtinen M 1996 "The Radiation Environment at the CMS Experiment at the LHC" Ph.D Thesis

[5] Golutvin I et al. 2002 "Simulation of Radiation Damage in HE Scintillating Tiles and Pion Energy Resolution After 10 Years of LHC Operation" CMS-Note 2002/013

[6] Dumanoglu I et al. 2002 "Radiation-hardness Studies of High-OH Content Quartz Fibres Irradiated with 500 MeV Electrons" *Nucl. Instrum. Meth.* **A490** 444-455

[7] Cankocak K et al. 2008 "Radiation-hardness Measurements of High-OH Content Fibers Irradiated with 24 GeV Protons up to 1.25 Grad" *Nucl. Instrum. Meth.* **A585** 20-27

[8] Akgun U and Onel Y 2006 "Radiation-Hard Quartz Cerenkov Calorimeters" *AIP Conf.Proc.* **867** 282-289

[9] Akgun U 2010 "CMS Hadronic Endcap Calorimeter Upgrade Studies for SLHC P-Terphenyl Deposited Quartz Plate Calorimeter Prototype" *IEEE Trans. on Nucl. Sci.* **57-2** 754-759

[10] Agostinelli S et al. 2003 *Nucl. Instrum. Meth. Phys. Res.* **A506** 250-303; Allison J et al.2006 *IEEE Trans. Nucl. Sci.* **53-1** 270-278

Journal of Physics: Conference Series **293** (2011) 012069 doi:10.1088/1742-6596/293/1/012069

R&D for a SiW Electromagnetic Calorimeter for a Future Linear Collider

Roman Pöschl on behalf of the CALICE Collaboration

LAL Orsay, B.P. 34, 91898 Orsay, France

E-mail: poeschl@lal.in2p3.fr

Abstract. This article presents an overview on the status and perspectives of a highly granular silicon tungsten electromagnetic calorimeter (SiW Ecal) conceived for the operation at a future linear electron-positron collider. Prototypes for this kind of calorimeters are developed within the R&D program of the international collaboration CALICE. This article reports on results on the "pure" calorimetric performance of a first physics prototype with an energy resolution of about $17\%/\sqrt{E}$ and an angular resolution of about 100 mrad. Beyond that, emphasis is put on the merits of the high granularity which is indispensable for the achievement of the envisaged physics goals of a linear collider. This high granularity permits the application of image processing algorithms for particle separation as well as it reveals a more detailed view into hadronic cascades than was possible hitherto.

1. Introduction

The next major worldwide project in high energy particle physics will be a linear electron positron collider at the TeV scale. This machine will complement and extend the scientific results of the LHC currently operated at CERN. The most advanced proposal for such a machine is the International Linear Collider (ILC) which will be operated at centre-of-mass energies between 90 GeV and 1 TeV. The experimentation at this machine could start around the year 2020. Introductions to the concept of the ILC and to the alternative for higher energies CLIC, short for *C*ompact *LI*near *C*ollider concept, can be found elsewhere [1, 2]. The article will focus on the status and prospects of the R&D for an electromagnetic silicon tungsten calorimeter.

2. Requirements for ILC Detectors

The well known initial state of the electron positron collisions is an ideal premise for high precision measurements at a linear collider. The extremely rich physics potential puts high requirements on the detector performance. One driving force for the detector optimisation is an optimal separation of jet final states created W and Z bosons. This measurement would become particular important in the absence of the Higgs boson which would then require a new type of interaction in order to preserve unitarity. Other important measurements would be the couplings of the Higgs boson to fermions, in particular to quarks which turn into jets.

Journal of Physics: Conference Series **293** (2011) 012069 doi:10.1088/1742-6596/293/1/012069

The reconstruction of the final state of the e^+e^- interactions will be based on particle flow algorithms (PFA) [3, 4]. The goal is to reconstruct every single particle of the final state which in turn demands a perfect association of the signals in the tracking systems with those in the calorimeters. As a consequence, this requires a perfect tracking of the particle trajectories even in the calorimeters. The jet energy resolution, a direct measurement for the detector performance, is then mostly given by residual overlap, also called *confusion*, of signals from near by particles in the final state. The concept of particle flow and the aim of resolving the jet energy E_{jet} to a value of $\sigma_{E_{jet}}/E_{jet} \approx 4\%$ lead to a challenging novel detector design.

The calorimeters have to feature an unprecedented high granularity. For an optimal particle flow, they will have to be placed inside the magnetic coil of the detector which puts tight constraints on the space available for the installation of the detectors. The design of the calorimeters have thus to follow two main guidelines

- The number of readout channels has to be driven to an unprecedented amount.
- The calorimeters have to be extremely compact and hermetic. This concerns both, the choice of the absorber material and the integration of infra-structural components such as cooling, power supply, readout lines and front end electronics.

For the electromagnetic calorimeter which surrounds the tracking system the two guidelines above lead to the choice of tungsten as absorber material having a radiation length $\mathrm{X_0} =$ 3.5 mm , Molière radius $\mathrm{R_M} = 9$ mm and interaction length $\lambda_\mathrm{I} = 96$ mm and silicon as the active material.

The CALICE collaboration [5], short for *CA*orimeter for a *LI*near *C*ollider *E*xperiment, conducts a large R&D program for the development of highly granular calorimeters as envisaged for the detectors of a linear collider. CALICE comprises 336 physicists and engineers from 17 countries and 4 continents. A physics prototype of the SiW Ecal dedicated mainly to demonstrate the physics potential of a calorimeter fulfilling the requirements above but also to validate the main mechanical concepts has been constructed and is currently examined in beam test measurements at DESY, CERN and FNAL. The results of the beam test data taking with the physics prototype as well as the steps towards the realisation of its successor, a technological prototype, will be described in the following sections.

3. The Physics Prototype

The physics prototype [6] is dedicated to demonstrate the principle of compact high granular calorimeters. Its major purpose is to verify physics observables like energy resolution and angular resolution to support the validity of the ongoing physics studies with a full detector. Another important objective of the program with the physics prototype is the verification of models of hadronic showers. These are typically less well modelled than electromagnetic showers and the uncertainty can be estimated from the fact that the simulation toolkit `GEANT4` [7] offers more than 10 different models for these hadronic showers. The Figure 1 shows a lateral view of the physics prototype. It consists of 9720 1x1 cm^2 wide calorimeter cells in 30 layers. The active zone covers 18x18 cm^2 in width and approximately 20 cm in depth. The layers are composed alternately by W absorber plates and a matrix of PIN diode sensors on Silicon Wafer basis. At normal incidence, the prototype has a total depth of $24\,X_0$ achieved using 10 layers of $0.4\,X_0$ tungsten absorber plates, followed by 10 layers of $0.8\,X_0$, and another 10 layers of another $1.2\,X_0$ thick plates. The sensitive layers are mounted onto both sides of a H-shaped tungsten plate as shown in Figure 2. Such an entity is called a slab. The readout electronics sticks out to the left of the sensitive area.

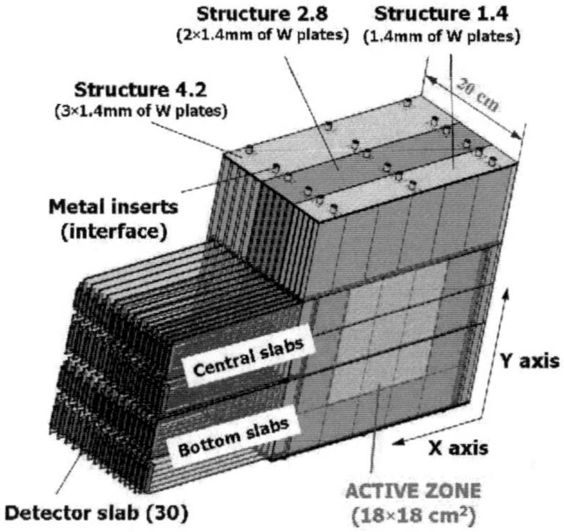

Figure 1: *Schematic 3D view of the Physics Prototype. Figure taken from [6].*

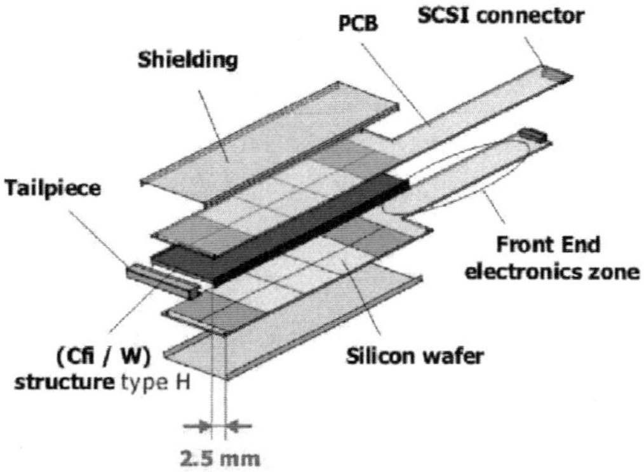

Figure 2: *Schematic diagram showing the components of a detector slab. Figure taken from [6].*

3.1. Beam test data taking

The physics prototype was subject to large scale beam test data taking in the years 2006-2008. In these beam test campaigns about 10^8 events have been recorded with electron, pion and muon beams.

The Figures 3a and 3b show the response of the prototype to electrons in terms of linearity and energy resolution [8].

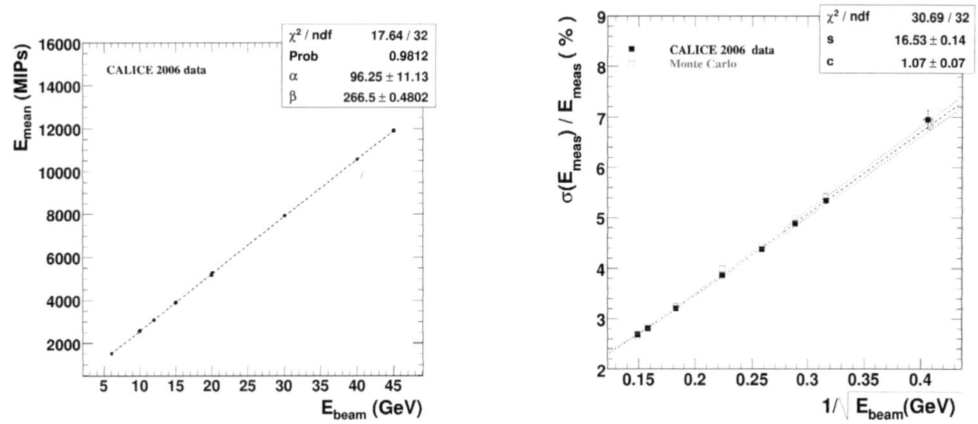

(a) (b)

Figure 3: *(a) Linearity of the energy response of the SiW Ecal as a function of the nominal energy. (b) Relative energy resolution ($\sigma(E_{\rm meas})/E_{\rm meas}$) as a function of $1/\sqrt{E_{\rm beam}}$ (solid squares), and its usual parameterisation as $s/\sqrt{E} \oplus c$. The values expected from simulation are shown (open squares).*

The results are based on an analysis of data taken in the year 2006 at CERN. Both observables are well described by a GEANT4 Monte Carlo simulation of the beam test setup. It is found that between 6 and 45 GeV the detector response is linear within 1%. The energy resolution is determined to be

$$\frac{\sigma(E_{\mathrm{meas}})}{E_{\mathrm{meas}}} = \left(\frac{16.53 \pm 0.14(\mathrm{stat}) \pm 0.4(\mathrm{syst})}{\sqrt{E(\mathrm{GeV})}} \oplus (1.07 \pm 0.07(\mathrm{stat}) \pm 0.1(\mathrm{syst}))\right)\%, \quad (1)$$

For the angular distribution the following values are reported [9]:

$$(106 \pm 2)/\sqrt{E} \oplus (4 \pm 1)\,\mathrm{mrad\ along\ the\ x\text{-}direction\ and}$$
$$(100 \pm 2)/\sqrt{E} \oplus (14 \pm 1)\,\mathrm{mrad\ along\ the\ y\text{-}direction.}$$

The differences between the two directions can be explained by the different arrangement of the detector layers in x and y direction.

3.1.1. Calibration of the SiW Ecal The details of the calibration procedure are presented in [6]. The relative calibration was performed using dedicated muon beams with an energy of up to 120 GeV, i.e. in a regime in which muons are expected to loose their energy only by ionisation and act, to a good approximation, as minimal ionising particles (MIP).

The results of the calibration reported here are valid for the data taking period 2006 in which 18 M. calibration events were recorded. The findings are however confirmed in [10]. The MIPs deposit and energy equivalent to around 45 ADC Counts in units of the CALICE Data Acquisition system. According to [6], this translates into a signal of over noise ratio of 7.5:1. The Figure 4 demonstrates the uniform response of all calorimeter cells. Residual variations can be assigned to different manufacturers (different colours) and production series of the silicon wafers. In 2006, only 1.4% of the around 6500 cells produced up to that year were found to be non-operational.

With respect to the operation in a detector at a linear collider, the stability of the calibration is an important issue. A good indicator to address this question is the comparison of calibration constants obtained in different test periods. The Figure 5 shows the good linear correlation between the calibration constants as obtained for the CERN running in October 2006 and the set of calibration constants obtained in running period two months earlier. Additionally, the constants from October 2006 are compared with those of a test bench measurement prior to the actual beam test data taking.

4. Exploiting the High Granularity

This section focuses on studies which explore the unprecedented high granularity of the SiW Ecal. As will be shown in the following, this granularity leads to a high power for the separation of close by particles [11] as well as for detailed insight into hadronic showers [12]. The results presented here are obtained with the Physics Prototype.

4.1. Particle Tracking

The final state of the e^+e^- collisions leads partially to overlapping particles in the electromagnetic calorimeter. On the other hand, the success of the Particle Flow algorithms

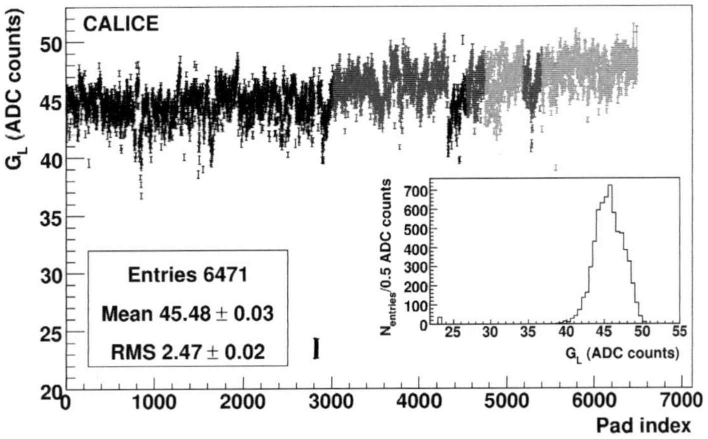

Figure 4: *Equivalent ADC value G_L for the energy deposit of a MIP in the the Ecal prototype, obtained using the October 2006 muon events sample, as a function of the pad index. The inset histogram displays the projection on the y-axis.*

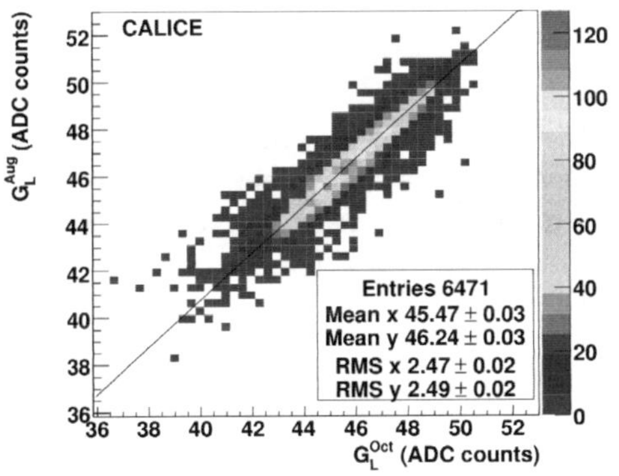

(a)

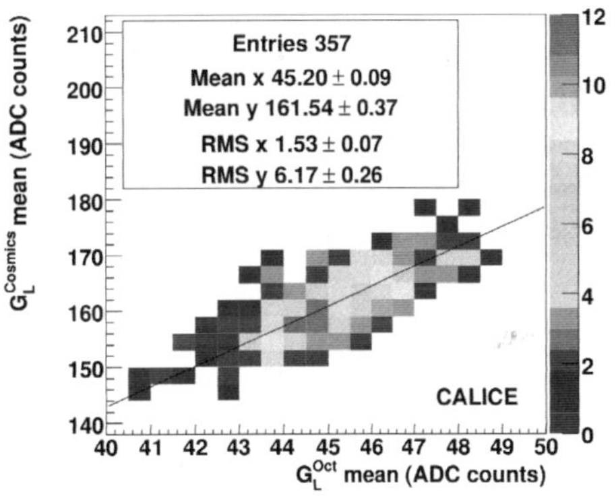

(b)

Figure 5: *(a) Calibration constants obtained with August 2006 muon runs versus those obtained with October 2006 muon runs. The colour scale represents the number of entries per bin. (b) Mean per chip of the calibration constants obtained with the cosmics test bench versus those obtained with October 2006 muon runs, discarding two outliers. The ADC counts represented in both axis are obtained with different DAQ systems.*

is based on the clean separation of signals from different particles. The high granularity of the SiW Ecal leads naturally to a rich amount of information which can be exploited using advanced e.g. imaging processing techniques. One of these techniques is the *Hough transformation* [13]. The Hough transformation provides a mapping from a n-dimensional feature space onto a m-dimensional parameter space, also called Hough space. In short words, points which are on a straight trajectory as e.g. generated by a MIP, will all result into the same parameter set in the Hough space. Thus, by subjecting the calorimeter cells

Journal of Physics: Conference Series **293** (2011) 012069 doi:10.1088/1742-6596/293/1/012069

which carry energy at or above one MIP to a Hough transformation, a MIP trajectory will lead to an accumulation at a given set of parameters. Figure 7 illustrates the merits of the Hough transformation and thus of the high granularity.

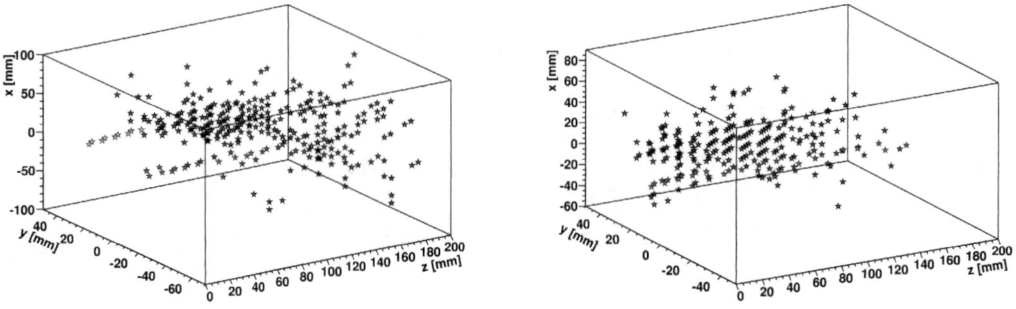

(a) Overlay of two pions. (b) Overlay of electron shower and MIP track.

Figure 6: *Examples for overlay events as observed in beam tests. The left plot shows an overlay of most probably two pions. (Selected hits are marked by red and green points). The right plot shows an electron signature and a MIP track (red points).*

Here, two examples for the separation of two close by particles are given. More quantitatively, the Figure 7a shows that a 100% detection efficiency of the MIP track is provided for fractions of shared hits of up to 50%. This result is independent of the hits actually generated by the primary MIP. At the same time, Figure 7b demonstrates that a full separation of close by particles can be achieved for distances down to 2.5 cm.

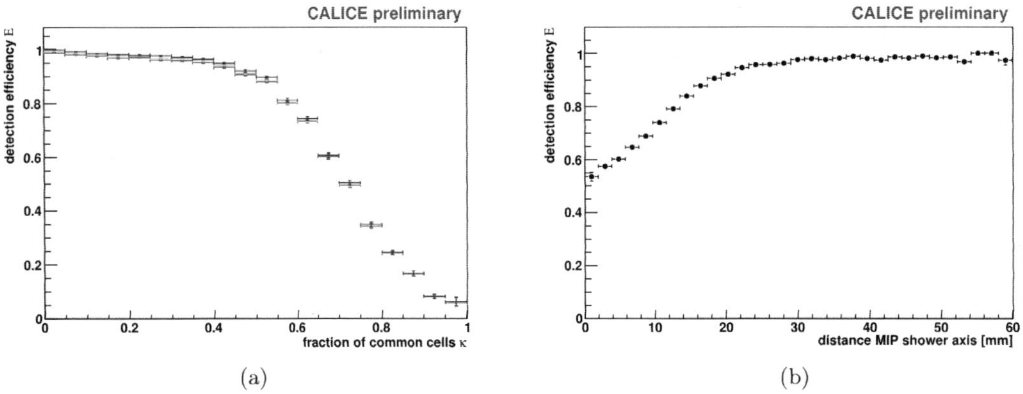

(a) (b)

Figure 7: *(a) MIP detection efficiency as a function of the MIP/shower proximity κ. The detection efficiency is broken down for different minimal numbers (red 10, green 20, blue 30) of true hits induced by the MIP. (b) MIP detection efficiency as a function of particle distance.*

4.2. Hadronic Showers in the Ecal

Given the depth of about one interaction length, about 55% of the hadrons contained in a jet will interact in the volume of the SiW Ecal. It is therefore important to understand the interactions of hadrons in the SiW Ecal.

Exploiting the longitudinal granularity of the calorimeter, the interaction point of interactions of primary hadrons can be identified by comparing energy depositions and hit densities in subsequent detector layers. The correlation between the reconstructed and the true interaction point as available from Monte Carlo simulation is shown in Figure 8a. A good correlation can be reported. Interaction layers as found in data and in Monte Carlo simulation are compared in Figure 8b. Here, the `QGSP_BERT(ini)` physics list as implemented in `GEANT4` is used. For an overview on the various physics list see [14]. The good agreement found in this case holds also when testing other `GEANT4` physics lists.

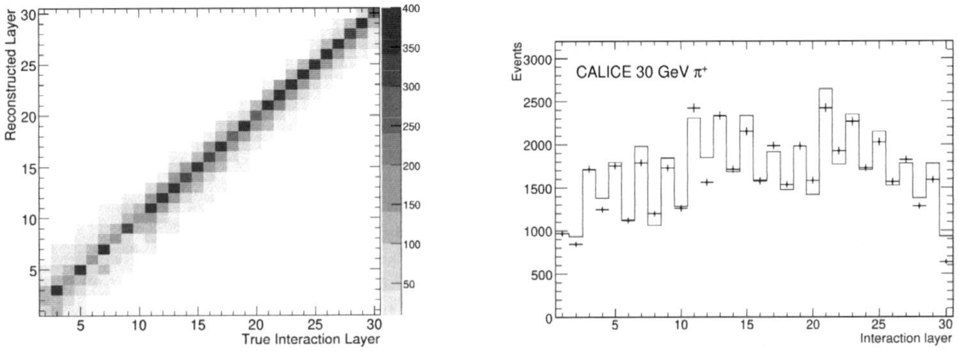

(a) (b)

Figure 8: *(a) Comparison between reconstruction and truth for the layer identified as the interaction layer. This example corresponds to a 20 GeV π^- beam simulation, using the* `QGSP_BERT` *physics list. (b) Distribution of the reconstructed interaction layer in the Ecal for 30 GeV data (points), compared with Monte Carlo predictions using the* `QGSP_BERT` *physics list (solid histogram).*

The amount of overlap of showers generated by close by particles is governed by the transversal shower radius of the particles. In turn it is easy to understand that the amount of overlap influences directly the precision which can be achieved by PFA algorithms. It is thus of major importance that the transverse properties of a hadronic shower are well modelled by Monte Carlo simulations. The Figure 9 shows the comparison of the transverse shower profile for 8 GeV and 30 GeV pions incident on the calorimeter surface. Again, the default physics list is chosen to be `QGSP_BERT`.

The mean values of the introduced transversal profiles for several physics lists and energies are given in Figure 10. From here and from the earlier result it can be concluded that those models which implement the Bertini cascade [15] give a good description at small energies. At high energies, all models however predict smaller shower radii than observed.

The longitudinal profile is composed by contributions from several components. The Figure 11 shows the longitudinal profile as a function of the shower depth after the interaction

Journal of Physics: Conference Series **293** (2011) 012069 doi:10.1088/1742-6596/293/1/012069

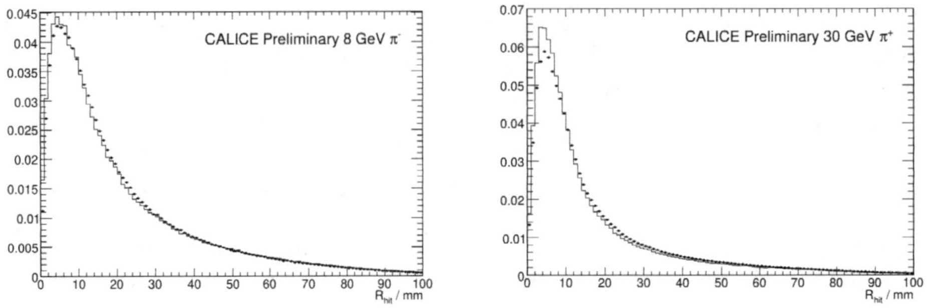

Figure 9: *Radial distribution of hits (energy weighted) for data at energies of 8 and 30 GeV (points with errors) compared with Monte Carlo (solid histograms) using the* `QGSP_BERT` *physics list. The distributions are normalised to unity.*

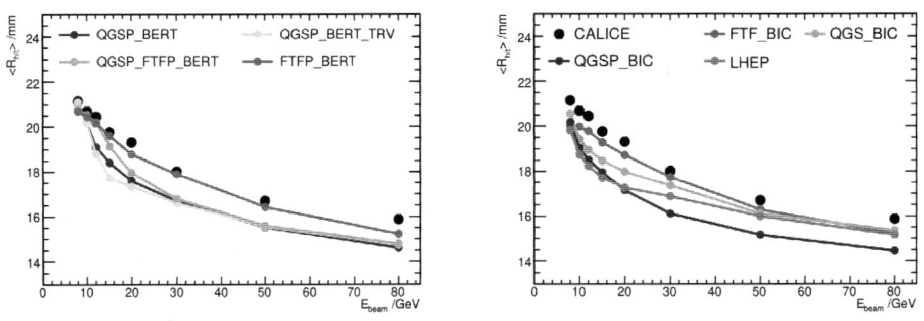

Figure 10: *Mean energy-weighted shower radius in the Ecal as a function of beam energy. The data are compared with the predictions of simulations using different* `GEANT4` *physics lists.*

point for pions with an energy of 12 GeV compared with the predictions obtained by the `QPSP_BERT` and `FTFP_BERT` models.

The profile of the Monte Carlo prediction is broken down into its various components. It is clearly visible that there are significant differences between the models and the data and between the models themselves. This is particularly true for short range components generated by heavily ionising particles such as protons. This first separation into various domains of the shower is continued in the study presented in Figure 12. Here the energy depositions in different layers after the interaction point are shown in greater detail. Again, the early stages of the shower created by nuclear break-up are not well described by either of the models. On the other hand all models, but the one dubbed LHEP, lead to a reasonable description of the energy deposition for larger shower depths. The said is particularly true for the models which feature the Bertini cascade.

5. Construction of a Technological Prototype

The next prototype, also called EUDET Module, has been conceived during the year 2008 and enters now its construction phase. It addresses, more than the first prototype, the engineering

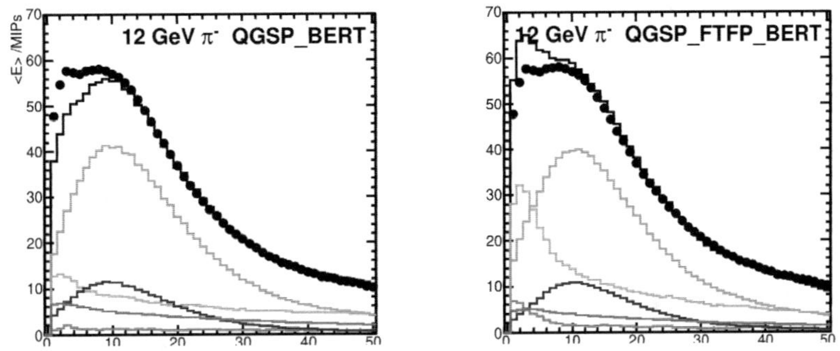

Figure 11: *Longitudinal energy profiles for 12 GeV π^- data (shown as points), compared with simulations using two physics lists. The mean energy in MIPs is plotted against the depth after the initial interaction, in units of effective 1.4 mm tungsten layers. The total depth shown corresponds to* $\sim 20\ X_0$ *or* $0.8\ \lambda_{\mathrm{int.}}$. *The breakdown of the Monte Carlo into the energy deposited by different particle categories is also indicated.*

challenges which come along with the realisation of highly granular calorimeters. The key parameters of the new prototype are

- Size of an individual cell of 5.5x5.5 mm^2.
- A depth of 24 X_0.
- Thickness of an individual layer of 3.4 mm and 4.4 mm according to the position within the calorimeter.

The Figures 13 and 14 show the mechanical housing and a cross section through two calorimeter layers which form a slab. The mechanical housing is realised by a tungsten carbon composite, which provides at the same the absorption medium and the mechanical rigidity of the detector. The silicon wafers are composed of high resistive silicon. While in principle the manufacturing of these wafers is a well known technique the challenge is to produce these wafers at small price in order to reduce the cost since a surface as large as 3000 m^2 will be needed for a linear collider detector. The final calorimeter will comprise around 10^8 channels in total. In order to reduce the non-equipped space in the detector the very front end electronics (VFE) has to be integrated into the calorimeter layers, see Figure 14, which constitutes a major challenge for the construction of the calorimeter. The room available for the readout circuits (ASICs) and the interface boards between the ASICs and the silicon wafers is about a millimetre.

Each of the ASICs of the new prototype will readout 64 calorimeter cells and realises the measuring of the analogue signal, the digitisation and the zero suppression such that only a limited number of channels are finally send to the data acquisition which is based on custom made components in order to allow for the employment of large quantities at reasonable prices for the experimentation at the linear collider.

Due to the limited space available for cooling devices the heat dissipation of the ASICs has to be minimised. In addition to the cooling, the heat dissipation will be reduced by a novel technique called "Power Pulsing". Here, the VFE will only be switched on during the millisecond of a bunch train of particles. Such a bunch train contains about 3000 particle

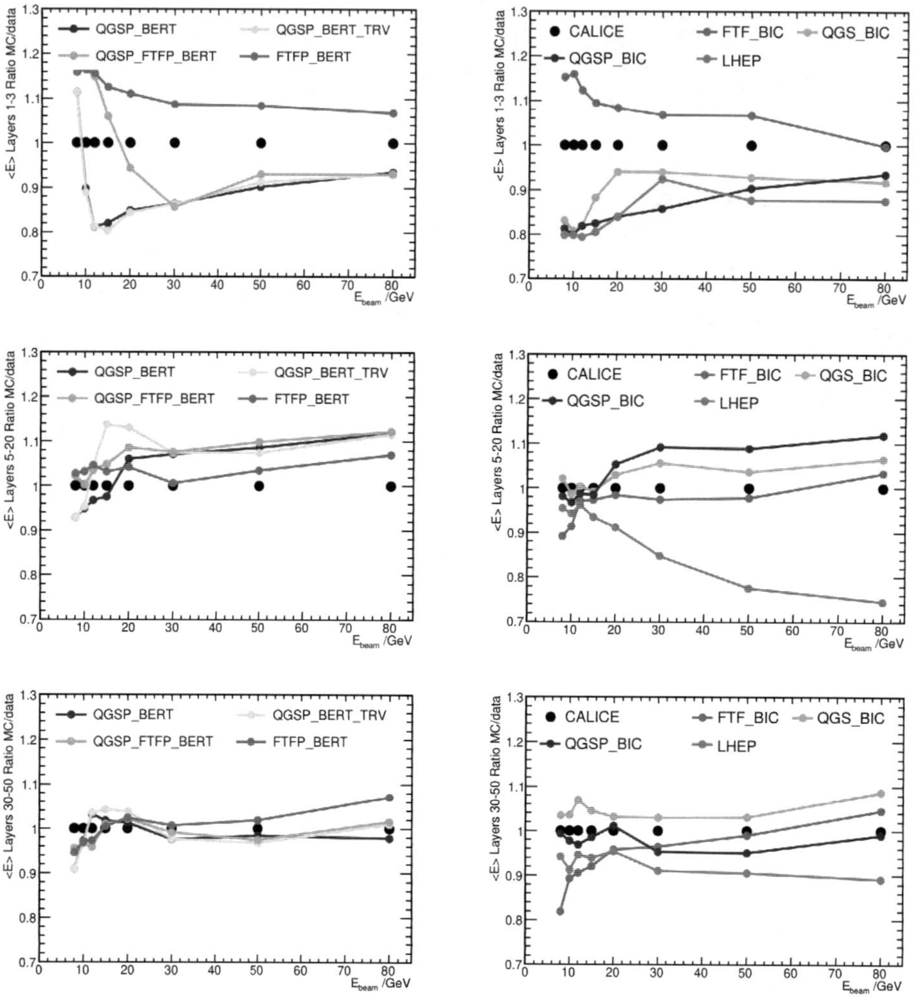

Figure 12: *Ratio of simulation to data for three different regions of the longitudinal energy profile: (top pair of plots) layers 1-3, dominated by nuclear break-up; (centre pair) layers 5-20, dominated by electromagnetic showers; and (bottom pair) layers 30-50, dominated by penetrating hadrons.*

bunches separated by around 300 ns. During the around 200 ms between these bunch trains, the electronics will be switched off. Clearly, this novel technique will require detailed studies in order to assure that the signal quality of each channel remains constant after each powering cycle. The first ASICs which incorporate the power pulsing are tested during the year 2010.

A calorimeter layer will have a length of about 1.5 m and will be composed by several units which carry silicon wafers as well as the VFE. The challenge is to integrate this fragile assembly into the alveolar structure which houses the calorimeter layers. The integration cradles are under development and a first integration test with a demonstrator has been successfully conducted in October 2009.

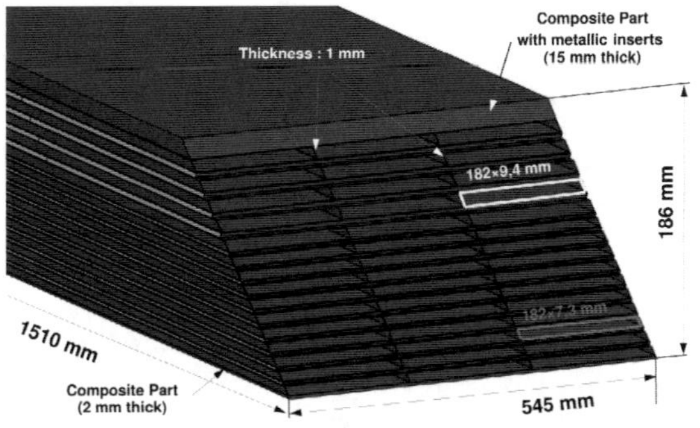

Figure 13: *Alveolar structure and its dimensions which houses the sensitive parts of the CALICE SiW electromagnetic calorimeter prototype.*

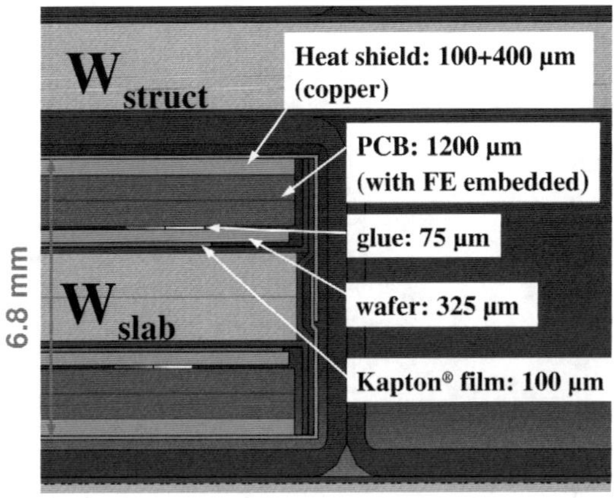

Figure 14: *Cross Section through a slab for the CALICE SiW electromagnetic calorimeter prototype.*

6. Summary and Outlook

High precision measurements with a detector at a linear electron-positron collider at the TeV scale will be based on Particle Flow. Its application requires highly granular calorimeters. The current baseline of detectors proposed for the International Linear Collider feature a highly granular silicon tungsten (SiW) electromagnetic calorimeter with a lateral segmentation of $5 \times 5\,\mathrm{mm}^2$ and a longitudinal depth of $24\,\mathrm{X}_0$ subdivided into 30 layers comprising approximately 10^8 cells in total. Between the years 2005-2008 a physics prototype with 10000 cells built to demonstrate the principle of highly granular calorimeters was operated successfully by the CALICE collaboration in several large scale test beam campaigns. The physics prototype will be succeeded by a larger technological prototype which addresses technological challenges as for example the power pulsing of the front end electronics. From the operation of the physics prototype and the first steps towards the technological prototype, it can already be concluded that it will be feasible to operate a highly granular SiW Ecal in a detector for a linear collider. This conclusion is based on the following facts:

- The signal to noise ratio is of the order of 8 which is given the early stage of the R&D already very close to the envisaged goal of 10.
- The energy resolution is approximately $17\%/\sqrt{E}$ and the angular resolution is about $(100\,\mathrm{mrad})/\sqrt{E}$. These results confirm the values used e.g. in the Letter of Intent studies for the ILD detector [16]. Ongoing studies put more emphasis on exploiting the highly granular structure of the device. In that sense, the study of details of hadronic showers as well as the optimisation of the separation of individual particles are important steps into that direction.
- By comparing data taken in different years at different locations, it could be demonstrated that the calibration is very stable over several years of operation. This is one of the key points required for the validation of the detector concepts.

- It will be possible to realise the challenging mechanics based on self-supporting alveolar structures which incorporate half of the tungsten absorbers. This was proved by the construction of a mechanical demonstrator.
- The readout electronics will be integrated into the layer structure of the calorimeter and has to integrate a multitude of tasks from first stage amplification until on-chip zero suppression and digitisation. The current status of the R&D shows that these challenging ASICs can be built. Preliminary studies indicate that the embedding of the ASICs into the layer structure will not compromise the precision measurements. One key design element for the ASICs is low power consumption. The goal is to achieve 25 μW/channel using power pulsing. Residual heat will be evacuated by a dedicated cooling system of which a first prototype was operated successfully in the context of the studies with the mechanical demonstrator.
- The technology for the silicon wafers is at hand [17]. Shortcomings observed during the operation of the physics prototype like square events could be remedied by dedicated modifications of the guard rings which surround the silicon wafers. :

References

[1] ILC Global Design Effort and World Wide Study.
Available from `http://www.linearcollider.org/cms/?pid=1000437`.
[2] M. Battaglia *et al.* .CERN-2004-005, arXiv:hep-ex/0412251.
[3] M.A. Thomson, NIM **A611** (2009) 25, arXiv:0907.3577v1 [physics.ins-det].
[4] J.C. Brient and H. Videau, eConf C010630 E3047 2001, LLR-02-001, arXiv:hep-ex/0202004v1.
[5] CALICE Collaboration, `https://twiki.cern.ch/twiki/bin/view/CALICE/WebHome`.
[6] The CALICE Collaboration, J. Repond *et al.* JINST 3:P08001 2008, arXiv:0805.4833v1 [physics.ins-det].
[7] The GEANT4 Collaboration. NIM **A506** (2003) 250, IEEE Transactions on Nuclear Science 53 No. 1 (2006) 270.
[8] The CALICE Collaboration, J. Repond *et al.* NIM **A608** (2009) 372, arXiv:0811.2354v1 [physics.ins-det].
[9] M. Faucci-Giannelli, Calice Analysis Note, CAN-017.
Available from `https://twiki.cern.ch/twiki/bin/view/CALICE/CaliceAnalysisNotes`
[10] H. Li, PhD Thesis Université Paris Sud XI and LAL Orsay, LAL 09-118.
[11] F. Fehr.Calice Analysis Note, CAN-023a.
Available from `https://twiki.cern.ch/twiki/bin/view/CALICE/CaliceAnalysisNotes`
[12] The CALICE Collaboration, C. Adloff *et al.*, 2010 JINST 5 P05007, arXiv:1003.2662v1 [physics.ins-det].
[13] P.V.C Hough, US Patent No. 3,069,654, 1962.
[14] GEANT4 Physics Reference Manual,
`http://geant4.web.cern.ch/geant4/UserDocumentation/`
`/UsersGuides/PhysicsReferenceManual/fo/PhysicsReferenceManual.pdf`
[15] M. P. Guthrie, R. G. Alsmiller, H. W. Bertini NIM **A66** (1968) 29;
A. Heikkinen, N. Stepanov and J.-P. Wellisch, Proc. Computing in High Energy and Nuclear Physics, vol. MOMT008, La Jolla, CA, 2003.
[16] The ILD Concept Group.
Available from `http://www.ilcild.org/documents/ild-letter-of-intent/LOI.pdf`.
[17] R. Cornat, Proceedings of the NSS-MIC 2008 Conference, Dresden, Germany.

Performance of the Scintillator-Strip Electromagnetic Calorimeter Prototype for the Linear Collider Experiment

Satoru Uozumi *on behalf of the CALICE Collaboration*
Kyungpook National University, Daegu 702-701, Korea
E-mail: satoru@knu.ac.kr

Abstract. The scintillator-strip electromagnetic calorimeter (ScECAL) is one of fine granular calorimeters proposed to realize Particle Flow Algorithm for the International Linear Collider experiment. The ScECAL is a sandwitch calorimeter with tungsten and scintillator layers, where the scintillator layer consists of plastic scintillator strips which size of 1 cm $\times$ 4.5 cm $\times$ 0.2 cm with a small photo-sensor (MPPC) attached at the its edge. In alternate scintillator layers, strips are orthogonally aligned to make a virtual 1 $\times$ 1 cm^2 cell with its crossing area. To establish the ScECAL technology, we have built a prototype of the ScECAL which consists of 30 layers of tungsten and scintillator layers with 2160 scintillator strips in total. In 2008 and 2009 the beam test has been performed at Fermilab meson test beam line to evaluate performance of the ScECAL prototype with various types of beams ranging 1 to 32 GeV. As a preliminary result of the beam test in 2008, we have obtained linearity of energy measurement less than 6% from the perfect linear response. Energy resolution is measured to be $\sigma/E = (15.15\pm0.03)\%/\sqrt{E}\oplus(1.44\pm0.02)\%$. Although detailed analyses are still ongoing, those results already establishes feasibility of the ScECAL as the fine granular calorimeter. However as the next step to precisely measure even higher energy jets, we will proceed to even more finely segmented calorimeter with 5 mm width scintillator strips.

1. Introduction

At the future international linear collider (ILC) experiment[1], precise jet energy measurement is crucial for many important physics studies which take advantage of the clean e^+e^- collision. Practical goal of the jet energy resolution at the ILC is set to $\sigma/\sqrt{E} = 30\%$. To achieve this high performance, Particle Flow Algorithm (PFA) [2] is considered as the most powerful jet energy reconstruction method. In the PFA, particles in a jet are categorized into charged or neutral particles. Then momenta of the charged particles are measured by tracking chamber to obtain higher resolution of the momentum than the energy resolution by calorimeter. However to realize the PFA, charged and neutral particles in a jet have to be separated with each other in the calorimeter to avoid possible over-counting. This fact results in a requirement of fine granularity for the calorimeter with very small cell size, which is 0.5-1 cm for the jet energy measurement up to E_{jet} ~200 GeV. In the CALICE collaboration[3], there are several technologies suggested to approach the fine granular calorimeter. The Scintillator-tungsten electromagnetic calorimeter (ScECAL) is one of those which adopts plastic scintillator strip technology.

Figure 1 shows structure of the ILD, which is one of the collider detector concepts proposed for the ILC, and the schematic design of the ScECAL in the ILD. The ScECAL is a sandwich calorimeter composed of 3 mm thick tungsten and 2 mm thick scintillator layers. The scintillator layer consists of small plastic scintillator strips which size of 1 cm × 4 cm × 0.2 cm. In alternate scintillator layers, strips are aligned to orthogonal directions as shown in the Figure 1. Light signal of each scintillator strip is read by Multi-Pixel Photon Counter (MPPC [4][5]) attached at a longitudinal edge of the strip. Since this type of calorimeter with the plastic scintillator strips is the world's first challenge, we have built a prototype of the ScECAL and performed beam tests to establish its feasibility. In this manuscript we will report a preliminary result of the beam test of the ScECAL prototype performed in September 2008 at Fermilab.

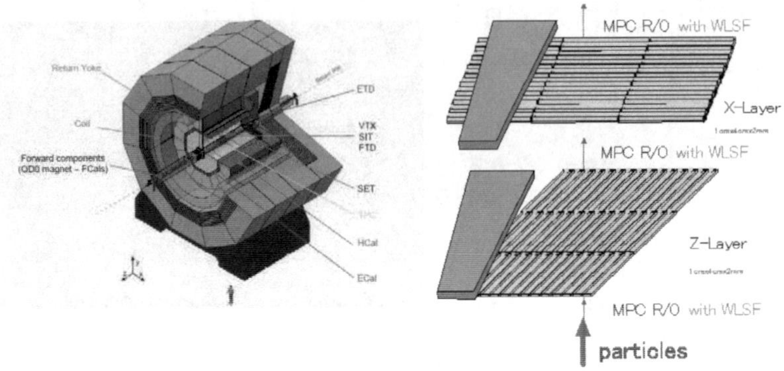

Figure 1. Picture of the ILD detector and the design of the ScECAL in the ILD.

2. The Scintillator-ECAL Second Prototype

Test of the first ScECAL prototype has been done in March 2007 at DESY with 1-6 GeV e^+ beams. Its result is reported elsewhere[6]. According to the result of the first test, the ScECAL second prototype has been built in 2008 with several improvements. In Figure 2 left side, structure and picture of the prototype are shown. It consists of 30 pairs of 3 mm thick tungsten and 2 mm thick scintillator layers. Each scintillator layer consists of 72 pieces of scintillator strips which size of 1 cm × 4 cm × 0.3 cm. In alternate scintillator layers, strips are aligned to orthogonal directions. Transverse size of the prototype is 18 cm×18 cm, and the total thickness is 26.1 cm which corresponds to about 28 radiation lengths.

Structure of each scintillator strip is also shown in right side of Figure 2. For the production of the scintillator, the plastic extrusion method is adopted [7] to reduce production cost while keeping its performance. Scintillation light signal is absorbed and re-emitted by 1 mmϕ wavelength shifting fiber (Y11) and guided by the MPPC. Response uniformity of the scintillator strip is tested both by bench test system with beta-ray and the Minimum Ionizing Particle (MIP) beams at the Fermilab beam test. The Response uniformity observed with the MIP beam is also shown in Figure 2 right side. Attenuation of the response at the far side from the MPPC is 10-15%. Effect of this non-uniformity to the calorimetry performance will be discussed in the next section. Signal from the MPPCs are read by SPIROC chip [8][9], which is developed in the CALICE collaboration for SiPM/MPPC signal readout, and converted to digital value in unit of ADC counts. Since the MPPC response is not linear to the amount of input light, response correction is done with a response function which is measured in advance.

3. Beam Test at Fermilab and the Preliminary Result

The performance of the ScECAL second prototype is evaluated at the Fermilab Meson test beam facility in 2008, using electron, pion and muon beams with its energy of 1-32 GeV. Major goals

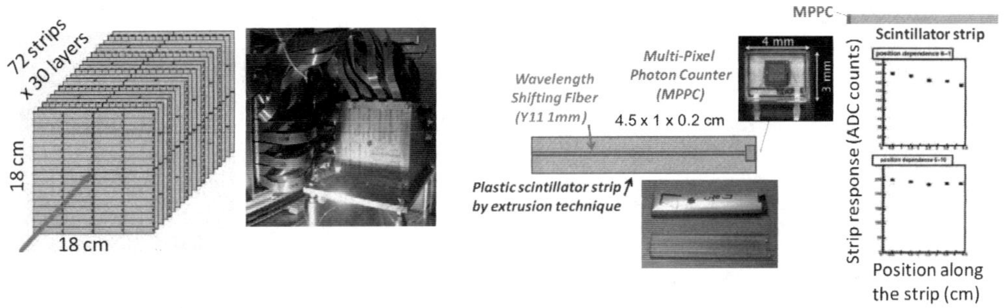

Figure 2. Schematic view and picture of the ScECAL second prototype (left), and Structure of the plastic scintillator strip combined with the MPPC (right). Response uniformity of the scintillator strip as a function of the hitting position by MIP is also shown for two typical strips.

of the beam test includes evaluation of energy resolution, linearity, and combined performance with the Analog hadron calorimeter (AHCAL) [10]. Although several detailed analyses on the data of the beam test are currently ongoing, we report some preliminary results of the beam test in this section. Setup of the beamline is shown in Figure 3. The data taking is done with trigger signal from scintillation trigger counters. Čerenkov counters located upstream of the beamline give additional information to trigger electron or pion events.

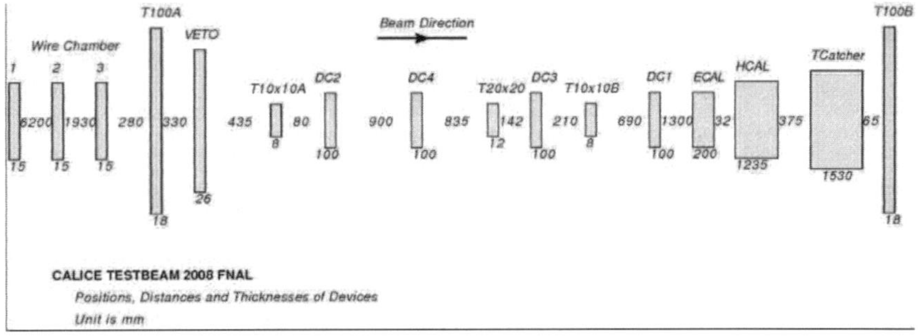

Figure 3. Setup of the beam test at the Fermilab meson test beam line. Devices starting with "T" or "DC" means scintillation trigger counters or drift chambers, respectively. T10×10A,B are used for the main beam trigger. On the upstream of those devices, Čerenkov counters are installed for electron-pion triggering.

At first response calibration among individual strips is done using 32 GeV muon beam which passes through the detectors as MIPs with constant amount of energy deposition. Shape of the typical MIP events on the ScECAL is shown in Figure 4. From the ADC distribution of the MIP events, mean response for the MIP is obtained for each individual channel as "MIP calibration constant". The MIP calibration constant is used for the response calibration of the ScECAL.

After calibrating all the channels, energies of electron beams are measured to examine the linearity and resolution of the electron energy measurement. Energy spectra for all the electron-triggered data sets at each energy points are shown in Figure 5 left side. Although electron events are taken with Čerenkov counters trigger, the electron data are still contaminated by pion and muon events. To obtain pure electron sample, offline event selection is done as follows based on various detector signal :

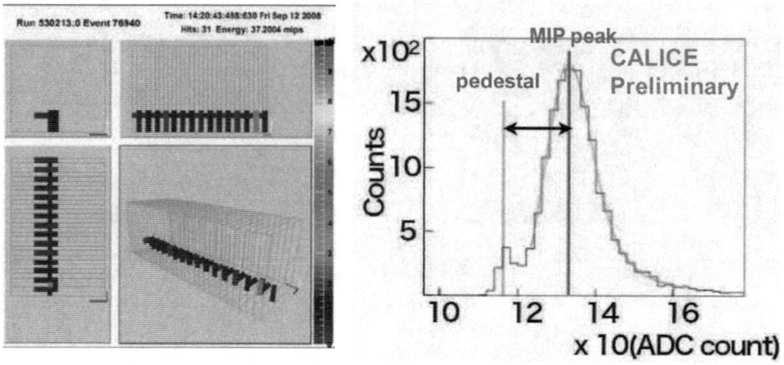

Figure 4. An event display (left) and the ADC distribution (right) of a typical MIP events after the MIP event selection.

- Signal on the AHCAL is required to be consistent with zero,
- Longitudinal and lateral shower profiles in the ScECAL are consistent with electromagnetic event.

In Figure 5 right side, measured energy spectra with different step of the event selections are shown. After all the cuts, energy spectra with clear Gaussian shape are obtained.

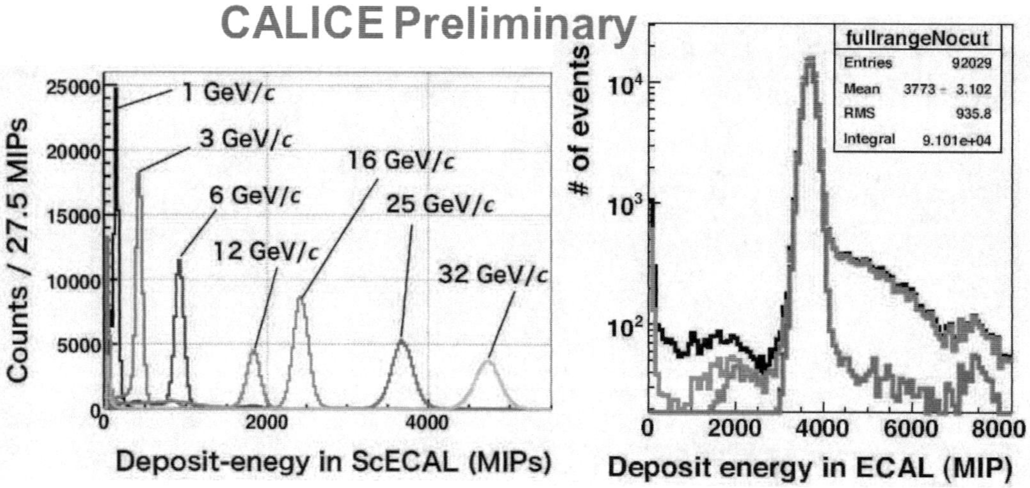

Figure 5. the energy spectra for all the electron-triggered events with various energies (right), and the measured energy spectra for 25 GeV electron events with each step of electron event selection (right).

From the measured energy spectra of the selected electron events, linearity and resolution of the electron energy measurement are evaluated. For those evaluations, the selected events are classified into two regions according to injecting position on the ScECAL surface, which are named "central" and "uniform" regions. The central region is defined as the center of the ScECAL and corresponds to the longitudinal edge of the scintillator strips, while the uniform region corresponds to ± 2.25 cm away from the ScECAL center to both vertical and horizontal directions. In the uniform region, effect of the strip response non-uniformity is expected to be small because core of the electromagnetic shower always hits around the center of the strip in every layer. Therefore if the strip response non-uniformity affects to the calorimeter performance, it should appear as larger effect in the central region, but would be smaller in the uniform region. The linearity and resolution are measured for those two regions and as a sum of those two regions. Results are shown also in Figure 6. As the preliminary result, we have obtained the linearity of the energy measurement less than 6% in terms of deviation from the perfectly linear response. This value is rather larger than expected, however ongoing detailed analyses including more precise corrections (temperature change, MPPC response curve) are expected to improve the linearity. Energy resolution is obtained to be $\sigma/E = (15.15\pm0.03)\%/\sqrt{E}\oplus(1.44\pm0.02)\%$ where E is the electron energy in GeV, and uncertainties are statistical only. This value is close to the expectation by GEANT4 simulation. Reason of non-zero constant term might be a leakage of shower from the ScECAL in addition to the same reason with the 6% of non-linearity. There is no significant difference among results in central, uniform and combined regions. It indicates the effect of strip response non-uniformity to the calorimeter performance is negligibly small.

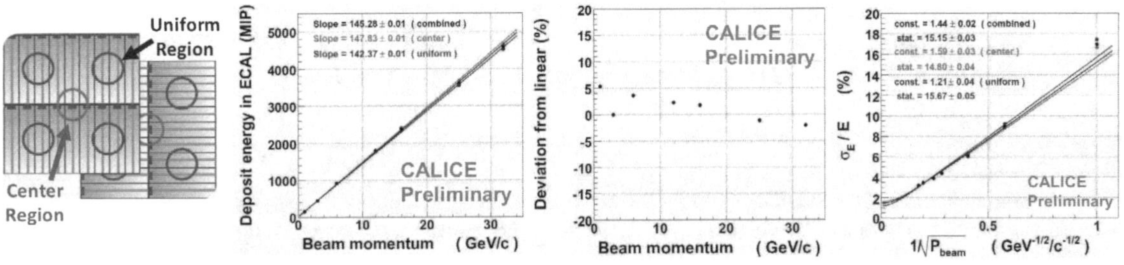

Figure 6. Definition of the center and uniform region, and the performances of the ScECAL second prototype as the preliminary result of the beam test.

4. Summary and Outlook

We have built the performance of the ScECAL second prototype. The prototype is tested by 1 to 32 GeV of electron beams at the Fermilab meson test beam line in 2008. Although detailed analyses are underway, preliminary result of the beam test shows feasibility of the ScECAL as the fine granular calorimeter with virtual cell size of 1×1 cm^2.

However recent study with Monte Carlo and PFA indicates that even smaller cell size is desirable to measure higher energy jets. Figure 7 left plot taken from reference [11] shows jet energy resolution for various energy jets, as a function of the ECAL cell size. According to this study, ECAL cell size of 1×1 cm^2 works for jets up to 100 GeV, however to measure higher energy jets, 0.5×0.5 cm^2 is desirable. To improve the granularity from 1 to 0.5 cm, we are developing new version of the scintillator strip with 5 mm width without wavelength shifting fiber for the light readout. First result of response measurement with beta-ray source is sufficiently uniform as shown in Figure 7, which encourages to proceed to the future ScECAL prototype with the 5 mm segmentation.

5. Acknowledgments

We would like to express special thanks to many people involved to this project. Especially special thanks goes to Erik Ramberg, Rick Coleman and Fermilab accelerator staffs to provide a good quality of beam. This work is partly supported by the Creative Scientific Research

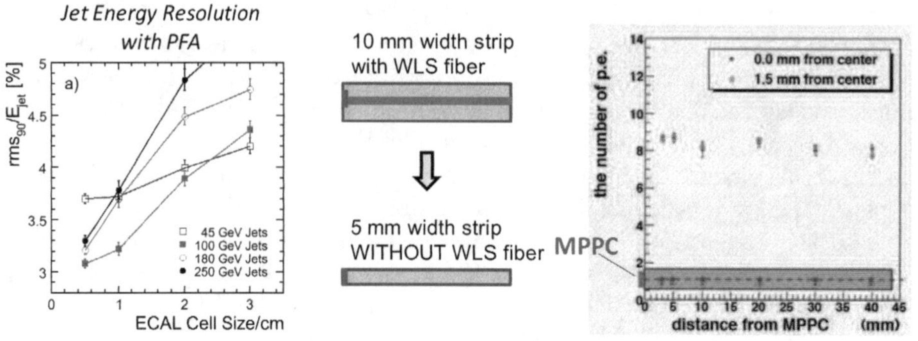

Figure 7. Jet energy resolution with various jet energies as a function of ECAL cell size, and design of the version of the scintillator strip for 5 mm granularity. Result of the response uniformity measurement is also shown.

Grants No. 18GS0202 of the Japan Society for Promotion of Science (JSPS), and the World Class University project(WCU) by Korea National Research Foundation.

6. References

[1] http://www.linearcollider.org/cms/
[2] J. C. Brient, "Improving the jet reconstruction with the particle flow method: An introduction", proceedings of the 11th International Conference on Calorimetry in High-Energy Physics (Calor 2004), Perugia, Italy, March 2004.
[3] The CALICE collaboration, https://twiki.cern.ch/twiki/bin/view/CALICE/WebHome
[4] S. Uozumi *et al.*, Nuclear Instruments and methods A 581 (2007) 427.
[5] Hamamatsu Photonics, http://sales.hamamatsu.com/en/products/solid-state-division/si-photodiode-series/mppc.php
[6] D. Jeans, "The CALICE SciW Calorimeter Prototype", CALOR08 Pavia 2008, http://agenda.infn.it/getFile.py/access?contribId=141&sessionId=14&resId=0
[7] Dong Hee Kim *et. al.*, Journal of the Korean Physical Society, Vol. 53, no. 6, pp. 3178-3181, Dec 2008.
[8] L. Raux *et al.*, Proc. 2008 IEEE Nuclear Science Symposium (NSS08), NSSMIC.2009.5401891.
[9] R. Fabbri, B. Lutz and W. Shen, arXiv:0911.1566 and EUDET-Report-2009-05, October 2009.
[10] The CALICE collaboration, arXiv:1003.2662, (2010) JINST 5 P05004, doi: 10.1088/1748-0221/5/05/P05004.
[11] M. Thomson, Progress with Particle Flow Calorimetry, arXiv:0709.1360

Journal of Physics: Conference Series **293** (2011) 012071 doi:10.1088/1742-6596/293/1/012071

What does bring space granularity to calorimetry.

Henri Videau

Laboratoire Leprince-Ringuet, École polytechnique-IN2P3/CNRS

E-mail: Henri.Videau@llr.in2p3.fr

Abstract. The development of calorimetry for the linear collider has led to a dramatic increase of granularity with respect to previous detectors. This appeared as mandatory in the framework of the so-called "particle flow approach". What does really provide granularity? Is there an optimal granularity? This will be discussed as a function of the energy reach under consideration for the electromagnetic component as well as the hadronic one.

1. Introduction

The development of calorimetry for the linear collider has led to a dramatic increase of granularity, by a factor of about 1000 shared almost equaly between directions along the showers and transverse to them. This raises few questions like, which accelerator, which physics conditions, which technological evolutions have enabled this? What is really the purpose, is there any future pursuing this way and in particular is it a key for going to higher energies?

Trying to figure out what does really provide granularity, we can examine the role it plays in electromagnetic calorimetry and then what can be expected on hadron calorimetry from the point of view of energy measurement accuracy, sensitivity to calibration or leakage.

2. Conditions of operation

Using small detecting cells read independently implies embedding the front-end electronics in the detector while integrating in this front-end functionalities like self-triggering, digital storage or huge level of multiplexing. This may also help keeping the cost reasonable. Such a constraint has as a consequence the generation inside the detector of heat which has to be drained out. An accelerator like ILC [1], with a reduced up-time may allow pulsing the power which can reduce by a factor 100 the heat generation. This is a key to the development of highly granular calorimetry, but pulsing in the presence of a 3.5 T field is another challenge. The embedding of the front-end in the detector induces also a severe constraint on the quality assurance which has to be almost at spatial level, this has a cost but is well documented.

3. The electromagnetic calorimetry

The purpose of electromagnetic calorimetry for the physics at ILC is to look for photons and electrons within jets of particles. Hence the goal is to measure their energy, position or angle once they have been properly identified, separated from neighbour photons or hadrons. Then the prime accomplishment is to make the pattern of electromagnetic showers as distinguishable, as discernible as possible. The electromagnetic showers are rather well characterised but have fluctuations, tails and halo which have to be handled in a crowded environment. Can it be expected to clean the shower to its core while keeping the energy resolution and this by playing on cell thresholds and on clustering algorithms? Trying this speaks for longitudinal as well as transverse granularity.

Restricting to sampling calorimetry it is well known that, for isolated showers, the energy resolution is directly linked to the sampling, which does not imply a longitudinal read-out grain size. The position and angle measurement precisions are in turn dictated by the energy resolution provided the cell position is properly known and the cell size well below the Moliere radius. In jets, before measuring any property, the photons have to be separated from hadrons or other photons. This statement can be qualified. Is there any point in distinguishing two close by photons, or finding π^0's? Not really because that will not modify much the energy or momentum of the system. The only specific case may be to sign the tau decay modes to be able to measure polarisation or spin correlations. Confusing photons with adjacent neutral hadron may hamper the jet signature but will enter the global energy meausrement only through the difference in response to a photon and a hadron. But confusing a photon with a charged hadron in the framework of the "particle flow analysis" (PFA [2]) induces double counting hence a first order error on energy. A mixture of photons may also be more likely to be confused with a charged hadron shower left over. The figure 1 [3] provides the probability for such confusion between a charged track and a near-by photon, this is done for 250 GeV u/d/s jets. For example 3% of the photons have a charged track hitting the calorimeter at less than 20 mm. Taking into account the number of photons suggests that a separation better than 20 mm is needed.

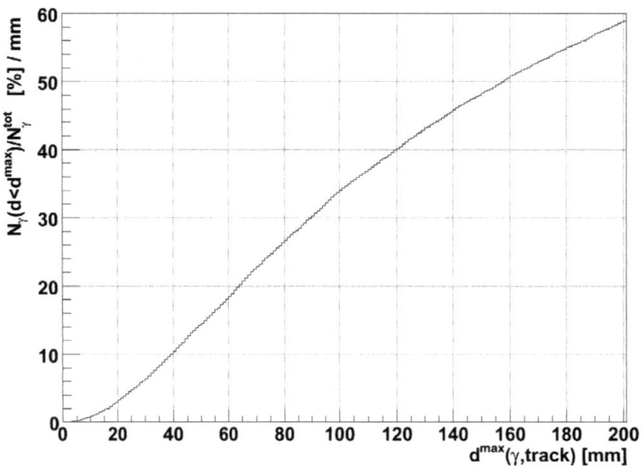

Figure 1. Average fraction of photons in a 250 GeV jet with a charged track closer than d as a function of d.

3.1. Parameters to play with for shaping the showers

The figures shown in this section come all from simulation of 10 GeV photons in the ILD [4] electromagnetic calorimeter which is a sandwich of tungsten radiators and silicon sensors. The cell size has been chosen for this study to be 1 mm^2.

The figure 2 shows that at such a level of granularity the characterisation of the shape is not totally trivial, it shows also that the density of hits outside the Moliere radius is reduced. The spectrum (figure 3) exhibits mostly few MIP peaks convoluted with Landau distributions. The radial energy distribution provides the effective Moliere radius for the given detector structure at about 17 mm. On the figure 5 the radial energy density is shown, it is basicaly the previous distribution divided by the radius; it appears that the core of the shower is very narrow, 2 to

3 mm diameter suggesting that identifying a track from a nearby photon can be done at very short distance.

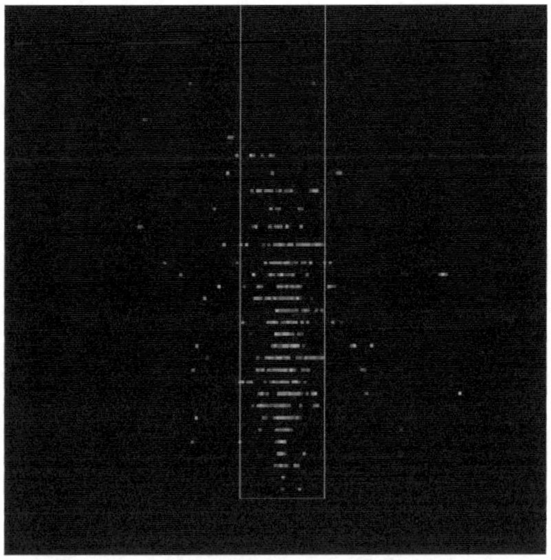

Figure 2. View of a shower, the two lines materialise the Moliere radius.

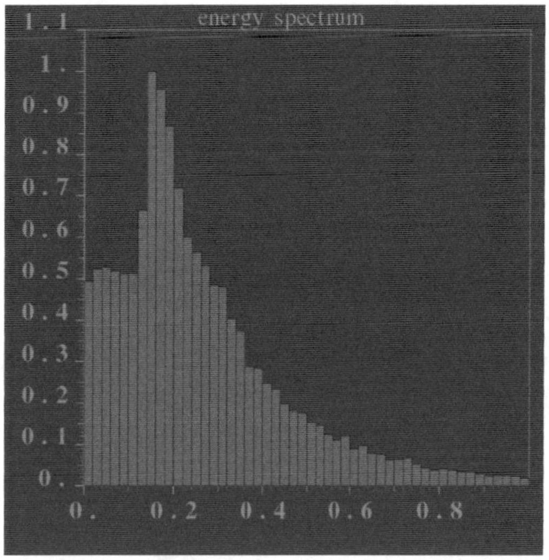

Figure 3. Energy spectrum of the shower cells.

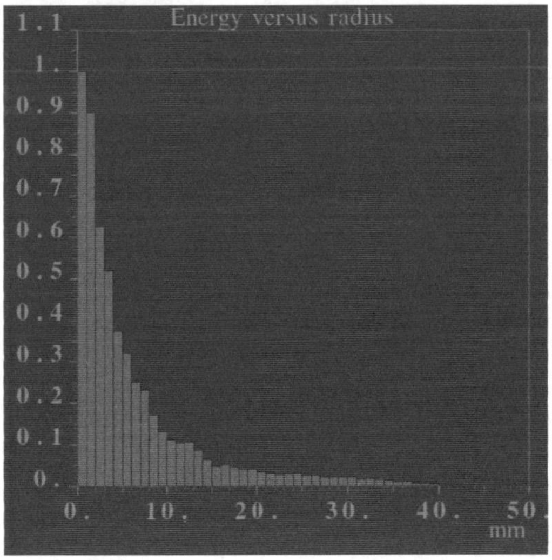

Figure 4. Radial distribution of the energy.

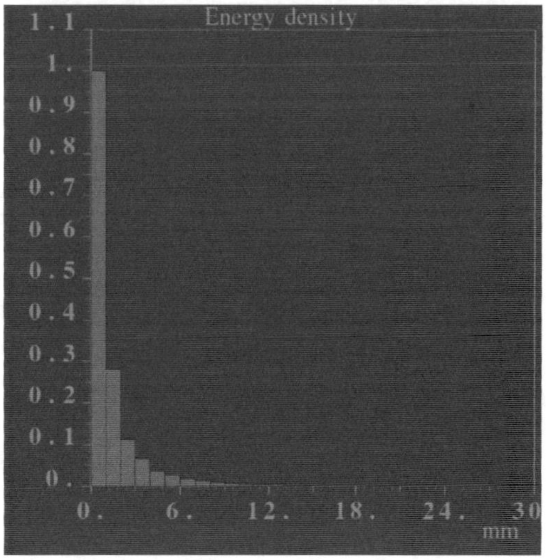

Figure 5. Radial distribution of the density of energy.

To adjust the detected shower shaping, it is possible to play with the detector structure, i.e. the Moliere radius, and on the read out or grain size. It is difficult to dream improving by going to a material denser than W at an acceptable cost. A thinner detecting medium can be envisaged but the measured Moliere radius is only a factor 1.7 above pure tungsten. The only real degree of freedom is the grain size and its impact has to be assessed. The cell threshold could also be considered but is heavily linked to the size in the cell range under consideration.

3.2. Separating nearby photons

As seen earlier, the core of the shower in this calorimeter model is about 3 mm diameter. Identifying two photons is being able to separate a real photon from a fluctuation of a nearby one. The question is then about the probability that in a given domain few mm^2 wide at a given distance of the main photon there be a significant fluctuation. This is clearly dependent on the energy of both photons but the cell size has anyway to be adapted to the core size and not to the Moliere radius, a mm level. This is illustrated in figure 6 showing that two photons of about 20 GeV each can readily be separated 12 mm apart with a millimeter grain size.

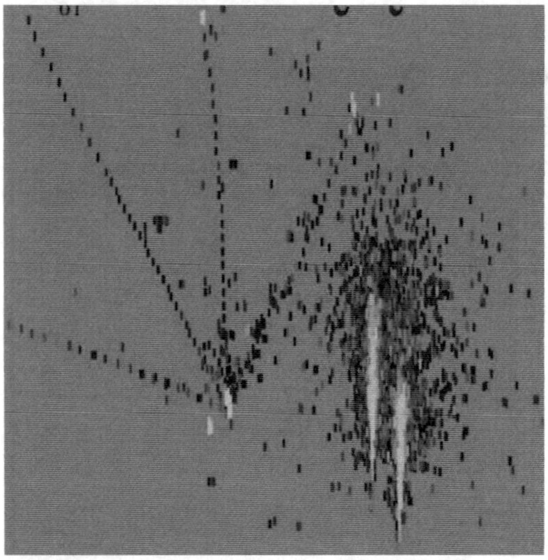

Figure 6. 100 GeV tau decaying into ρ seen with a $1\,mm^2$ grain. The cell energy is coded by a colour scale from blue to white.

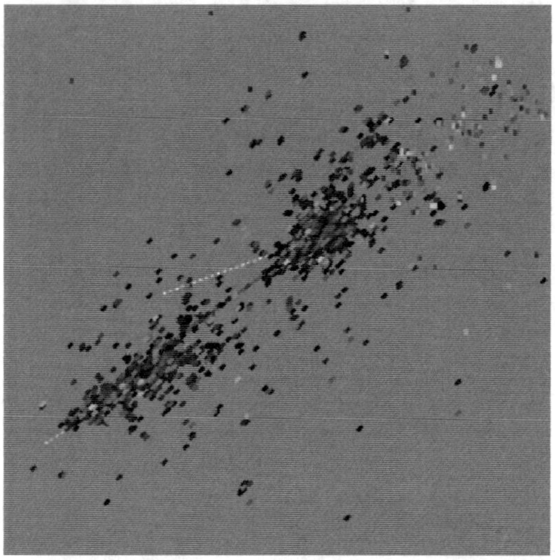

Figure 7. Display of the energy deposited by closeby photon and pion. The energy is coded by a scale from brown to white.

3.3. Separating photons from a nearby charged hadron

The problem here is to disentangle the photon from the track and its shower in order to avoid losing the photon energy or doubly counting part of the hadron energy. The strategy will depend on the interaction depth in the calorimeter. For a track long enough the charateristic shape is not anymore the shower shape but the track shape and the cell size limit is the transverse size of a track provided by the multiple scattering. This situation is illustrated on figure 7. To be more precise, the question is to know what is the probability to find n aligned fired cells at a given distance from the photon. Figure 8 provides the hit radial density of a 10 GeV photon for $1\,mm^2$ cells, the peak height being 13 cells. It can be seen that, even without looking at the longitudinal distribution of the hits, a track leaving at least 5 hits can be separated at few mm's.

3.4. Miscellaneous properties

Going to small cell sizes reduces slightly the cell response dynamics, 1 bit by going from $25\,mm^2$ to 1. At some point counting the hits will provide adequate energy measurement in most of the energy domain and anyway helps the energy measurement at low energies by killing the Landau tails. But this has a limit when the cell size reachs the length of delta rays propagating in the sensitive medium. There is an optimal cell size for counting.

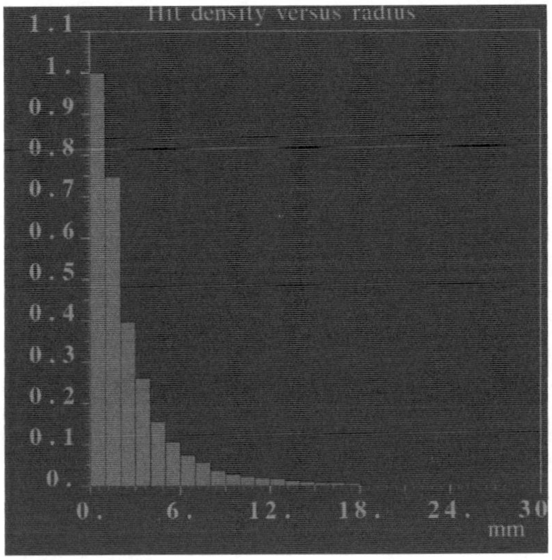

Figure 8. Distribution of the density of fired cells around the photon impact.

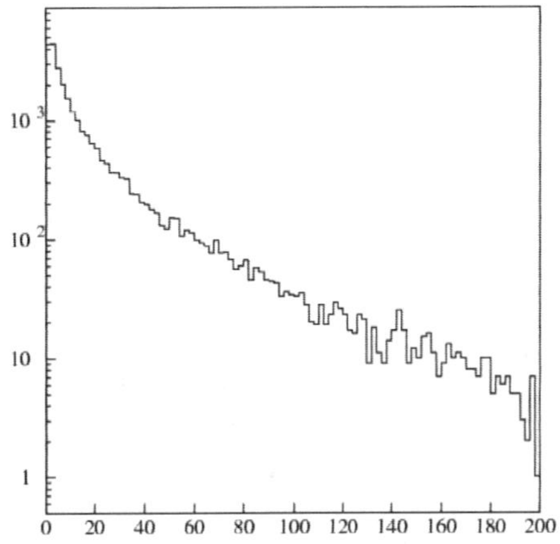

Figure 9. Energy spectrum of the neutrons and K^0s in 500 GeV jets.

4. The hadronic calorimetry

The problem here is to disentangle in jets the neutral hadrons from the majority of charged hadron showers, then to measure them at best which means master the resolution, the calibration and the leakage. Once you go below the hadronic shower size, of the order of 10 cm, the only relevant size is the track size. It is then tempting to adapt the calorimeter at looking at tracks inside the calorimeter. This may provide a lot of improvements. Identifying tracks in the showers and measuring their momenta, up to the multiple scattering, provides: a resolution improvement by weighing the electromagnetic versus hadronic components, an in situ calibration of efficiency, a way to reduce the sensitivity of the digital measurement to efficiency and multiplicity, an improvement of the shower separation by following or measuring the charged links, an improvement of the leakage estimation by measuring the momenta of tracks leaving the calorimeter.

To characterise the issue, the figure 9 (courtesy of J-C. Brient) shows the energy distribution of the K^0 and neutrons produced in 500 GeV jets of any kind except tops, produced through a Z without radiative return. 8% of these neutrals have more than 100 GeV. It can be expected that the leakage in these cases is large but the interaction probability stays the same. Does the use of the fine grain calorimeter as a true tracking calorimeter help here?

With 7 interaction lengths the sail through reaches 0.1% quite independently of the energy. The sail through is illustrated in figure 10 (be aware that the horizontal and vertical zoom factors are different). Those neutrals which interact produce second level neutrals and charged particles. The probability to have these second level neutrals escaping in turn is about 1% of their probability to be produced. The leakage is therefore linked mostly to charged particles. Why not try then to use our fine grain calorimeter truely as a tracking calorimeter profiting of the strong field and of the measurement accuracy limited though by the multiple scattering. In fact such an approach was used already in ALEPH for identifying the muons through their momentum measured in the return field.

This has been looked at with 1000 20 GeV K^0's sent into the ILD calorimeter in its digital option as simulated with MOKKA [5] for a grain size of 1 mm^2. A beautiful example is shown in figure 11 where the interaction takes place in the electromagnetic part. The charged kaon in pink has 14.5 GeV.

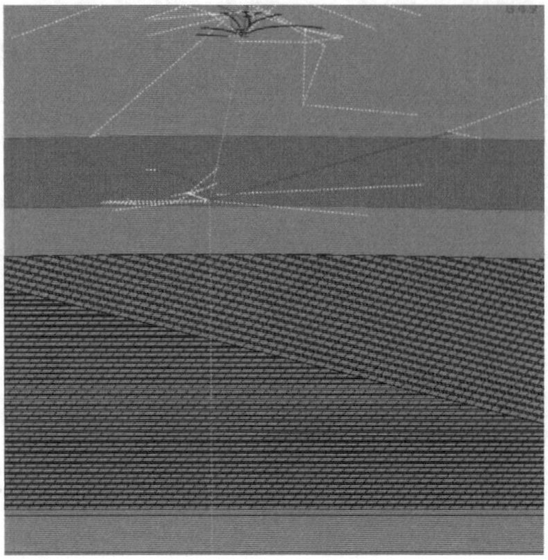

Figure 10. K0 sailing through the calorimeter.

Figure 11. Charged kaon exporting the energy outside the calorimeter.

But not only the tracks escaping can be measured, any charged secondary can be and in many cases this measurement will be better than the energy of the corresponding shower, improving the global calorimeter resolution. In our corpus of events 6.5% had a leakage larger than 2.5%. A simple procedure of fitting was able to save 2/3 of them, this is equivalent to spare one interaction length. The method was rather heuristic and needs development.

A 1 mm^2 grain may look like a dream. In view of the fact that it is used for the hadronic part in a digital mode, the main problem is probably with the connections. The data flow is not a problem, the number of cells hit does not grow by far like the inverse of the areas and the flow remains very limited at ILC. More precisely in a 10 GeV electromagnetic shower the number of cells hit grow by a factor 2 when going from 25 mm^2 to 1 mm^2 cells.

5. Conclusion

A very fine grain calorimeter helps disentangling the components of the jets and of the showers improving the resolution. Identifying properly the tracks in the showers provides a way to calibrate the detector in situ and to palliate the inefficiencies. The high field provides an adequate measurement of the charged particle momenta helping to control the leakage in the barrel. We should consider going for a true tracking calorimeter bringing PFA into the showers.

References

[1] ILC Reference Design Report 2007 *Preprint* DESY 07-046

[2] ALEPH collab.1995 *Nucl.Instrum.Meth.* **A360** 481
Videau H 1996 *Proc. Sixth Int. Conf. on Calorimetry in High Energy Phys.* (Frascati Physics Series) p 905.
Gay P 2000 *Proc. Linear Collider Workshop.* (American Institute of Physics) p 904
Videau H and Brient JC 2002 *Proc. Tenth Int. Conf. on Calorimetry in High Energy Phys.*(World Scientific) p 747.
Morgunov V 2002 *Proc. Tenth Int. Conf. on Calorimetry in High Energy Phys.*(World Scientific) p 70.
Videau H 2004 *Invited talk at the Linear Collider Workshop.* (Éditions de l'École polytechnique)
Thomson M 2009 *Nucl.Instrum.Meth.* **A611** 25

[3] Reinhard M 2010 Thesis at École polytechnique Palaiseau France

[4] ILD Letter of Intent 2009 *Preprint* DESY 2009-87

[5] MOKKA A detailed Geant4 simulation for the International Linear Collider detectors. http://polzope.in2p3.fr:8081/MOKKA/

Journal of Physics: Conference Series **293** (2011) 012072 doi:10.1088/1742-6596/293/1/012072

Engineering prototype of the CALICE analog hadron calorimeter

Erika Garutti,
on behalf of the CALICE collaboration
DESY, Notkestr. 85, 22607 Hamburg, Germany
E-mail: erika.garutti@desy.de

Abstract. A new prototype of a tile hadron calorimeter (AHCAL) for the International Linear Collider detector is currently developed within the CALICE collaboration. The aim is to improve the energy resolution by measuring details of the shower development and combining them with the data of the tracking chamber (particle flow). The prototype is based on scintillating tiles that are read out by novel Silicon-Photomultiplier (SiPM). This new prototype will take into account all design aspects that are demanded by the intended operation at the ILC It will contain about 2500 detector channels. This is the first calorimeter design which makes full use of the high integration potential of the novel photo-sensor technology.

Main focus of this contribution is the mechanical and electrical integration of the front-end electronics into the calorimeter absorber structure, with the aim of maintaining high-density calorimeter. Integration aspects and scalability to an ILC detector are discussed.

For the analog calorimeter the proposal of an integrated light-calibration system for calibration and gain monitoring are presented, addressing temperature and bias dependence of the SiPM gain.

First results from the measurements with one prototype module at the DESY test beam are presented, which demonstrate the quality of the readout system, and of the light-calibration system.

1. Introduction

The CALICE collaboration [1] is developing calorimeters with the specific aim to fulfill the hardware and physics demands of the International Linear Collider (ILC [2].) physics program [3]. The ambitious required jet energy resolution ($0.3/\sqrt{E[GeV]}$) could be achieved with extremely segmented calorimeters using the particle flow approach (PFLOW, [4]). For the Analogue Hadron Calorimeter (AHCAL) option for the ILC a physics prototype was successfully built and commissioned in test-beam campaigns at DESY, CERN and FNAL in the years 2006-09. The prototype has operated together with prototypes of an electromagnetic calorimeter, and of a tail-catcher and muon tracker. This first physics prototype [5] consisted in total of 7608 scintillator tiles with individual readout via a SiPM. The aims of this development is to establish the innovative design of highly granular sampling calorimeters; and exploit them to study hadronic showers with a new degree of accuracy to validate Monte Carlo models of hadronic physics. In addition, it is intended to show with this prototype that the requirements for particle flow applications are fulfilled, in terms of showers separability and energy resolution in a multi-particle environment.

Journal of Physics: Conference Series **293** (2011) 012072 doi:10.1088/1742-6596/293/1/012072

More recently, a new prototype for the AHCAL is being developed and commissioned to demonstrate the feasibility to build a calorimeter with fully integrated electronics meeting the constraints of a real detector. In order to cope with the expected channel number of the AHCAL barrel of about four millions, reliable assembly and production procedures have to be developed. A new engineering prototype with about 2500 detector channels is currently under development in order to test all assembly and integration issues. This prototype corresponds to one layer of the next generation analog hadronic calorimeter for the ILC.

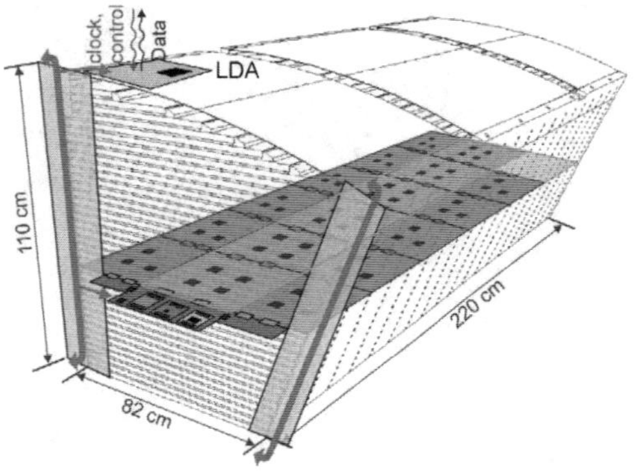

Figure 1. Drawing of an AHCAL wedge (half of a barrel octant), with integrated electronics covering one of the active layers. The detector interface electronics and the cable path are at the edge of the barrel in the gap before the end-cap detector.

Figure 2. Mechanical structure of an AHCAL wedge. Steel plates 2 cm thick are held together by a 0.5 cm panel, without spacers between the layers. The 1 cm thick cracks between wedges, resulting from two adjacent panels are pointing to the beam axis.

2. Technological prototype concept and realization

One design option for the hadronic barrel calorimeter for the ILC is to subdivide the octagonal shaped detector in 16 equivalent wedges segmented in two parts in the longitudinal direction (i.e. along the beam direction). Each resulting wedge would have the structure shown in Fig. 1. A wedge is divided in 48 layers of 2 cm-thick steel, with gaps to host the active detector layers. Each active layer is obtained connecting single detector cassettes of a size of 36×36 cm^2. Six raws and three columns of cassettes will make up one layer. The cassettes carry the active part of the calorimeter: scintillator tiles ($3 \times 3 \times 0.3$ cm^3 in size) read out via a SiPM, and the readout electronics which allows signal processing and digitization. Communication to the electronics as well as bias voltage for the photo-detectors and calibration pulses are provided by dedicated cards located at the edge of the detector, in the gap between barrel and end-cap. The control signals, for the electronics as well as the data, need to be transmitted over the 2.2 m long interconnected cassettes and back to the detector interface electronic cards.

2.1. Mechanical structure

The AHCAL mechanical structure for one wedge is a self-supporting sandwich of 48 2 cm-thick stainless-steel absorber plates. The plates are held together by a 0.5 cm steel panel and require no additional spacers between the layers. The 1 cm thick cracks between wedges, resulting from two adjacent panels are pointing to the beam axis. This satisfies the challenging requirements on minimum dead material in the wedge itself and in between adjacent wedges in the barrel.

Journal of Physics: Conference Series **293** (2011) 012072 doi:10.1088/1742-6596/293/1/012072

Studies based on particle flow reconstruction, demontrated that this amount of material does not effect jet energy resolution in a typical ILC jet energy range.

For the mechanical aspect, the flatness of the plates below 1 mm is essential for the insertion of the active layers and is obtained by a roller leveling procedure. The picture of the first AHCAL mechanical structure prototype, standing in its vertical position, is shown in Fig. 2.

2.2. *AHCAL cassette*

Each calorimeter active layer is built interconnecting several AHCAL cassettes. One cassette (see Fig. 3) can be seen as an independent piece of calorimeter hosting 144 scintillator tiles with SiPM readout. The tiles are like LEGO pieces fixed on the electronics PCB as in a puzzle. Four SPIROC chips [6] are reading out all the channels of one cassette. This chip has been designed

Figure 3. A prototype of AHCAL cassette hosting 4 SPIROC chips. To the bottom side 144 scintillator tiles with SiPM readout are connected. On the right the detector interface card is connected, which carries modules for the communication to the chips, powering the SiPM and steering the LED calibration system.

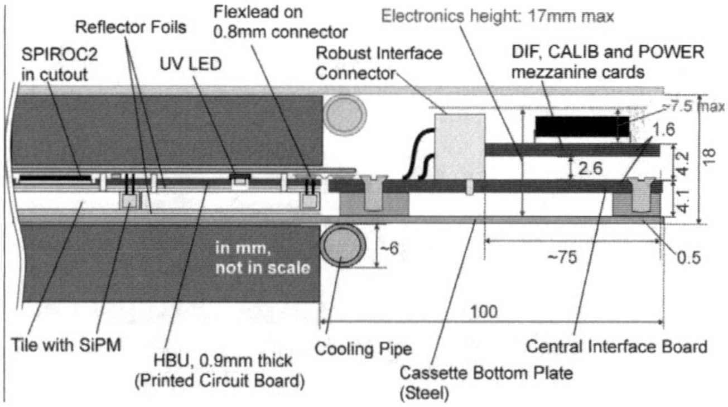

Figure 4. Vertical section of the design of the AHCAL cassette and detector interface card. The total thickness of a calorimeter active layer is 6.5 mm in the current design. The thick gray layers indicate the steel plates.

for the readout of a large number of SiPMs with minimum power consumption, for special application in the integrated electronics. Fig. 4 is a vertical section of the AHCAL cassette showing the active layer with embedded electronics sandwiched between two steel layers. The overall thickness of the cassette, including the PCB board carrying the SPIROC chips, is 6.5 mm. On the edge of the AHCAL wedge, in the gap between barrel and end-cap, sits the detector interface electronics, with a maximum thickness below 18 mm.

2.3. *SPIROC chip*

The readout chip for the AHCAL has to be embedded in the calorimeter structure. A maximum power consumption of 25μW per channel is allowed. In order to achieve this the chip is operated in power pulsed mode (according to the ILD bunch structure) with an on-time of about 1/200th. The chip offers the possibility to individually adjust the bias voltage to the SiPM. After the variable preamplifier and shaping stage, an analog memory array of 16 capacitors stores the signals to be converted by an integrated ADC. The chip features a time stamp capability, O(1 ns). Data are collected using the auto-trigger option of each channel individually. An overview of characterization studies on the SPIROC chip is given in [7].

2.4. Tile-SiPM system

Each calorimeter channel consists of a scintillators tile, $3\times3\times0.3$ cm^3 in size, with embedded wavelength shifter fiber and photo-detector (see Fig. 5). The tiles size has been optimized with PFLOW studies using the PANDORA PFA code [8]. No clear advantage is seen in choosing smaller tiles, while the option of larger calorimeter channels would lead to a deterioration of the jet energy resolution for jet energies above 100 GeV.
As photo-detector the CPTA SiPM [9] is used. The photo-detector main characteristics are:

- gain: 0.5-1×10^6,
- inter-pixel crosstalk: <10%,
- average dark noise (above 0.5 p.e. threshold): ~1.5 MHz,
- typical current: 0.2-0.3 μA,
- dark noise frequency (above 0.5 MIP threshold): <150 Hz.

This last point makes the photo-detector particularly suitable for calorimetry applications.

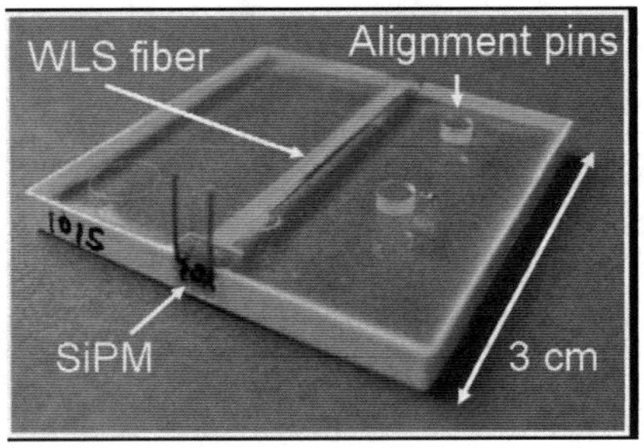

Figure 5. Single AHCAL channel: a scintillator tile readout via a wavelength shifter fiber and a SiPM (CPTA).

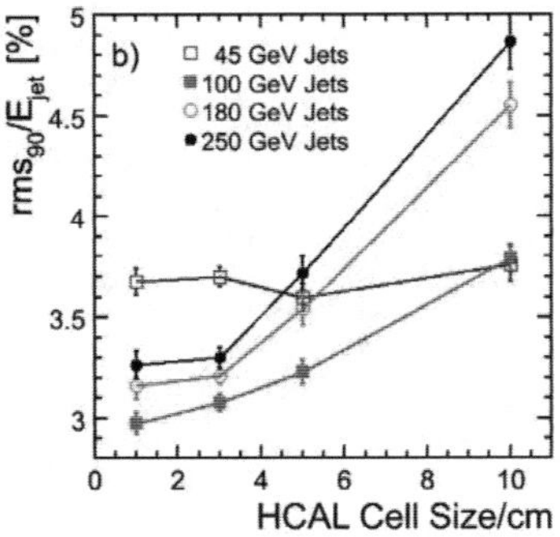

Figure 6. Tile size optimization studies performed with the PANDORA PFA code. The jet energy resolution for jets from $Z \rightarrow uds$, as a function of the AHCAL cell size.

2.5. LED monitoring system

The LED calibration and monitoring system for the technological AHCAL prototype needs to be a scalable system, addressing the needs determined during operation and analysis of the AHCAL physics prototype. The main tasks of the system are:

- calibration of the SiPM gain using single photo-electron (p.e.) peak spectra, which requires light in the order of 1-2 p.e.,
- monitoring of the long term stability of the detector using medium power light, in the order of 20-100 p.e.,
- measurement of the SiPM saturation level, with light corresponding to approximately 2000 p.e.

For each of these tasks a channel-to-channel pulse spread from the LED system as small as possible, not worse than a factor of two, is required.

Two approaches are being followed for this system, one based on a central driver and optical signal distribution (Fig. 7), one based on electrical signal distribution and an individual LED per tile (Fig. 8). The latter is already integrated in the AHCAL cassette, but not yet fully optimized. The aim is to equalize the light intensity and maximize its dynamic range.

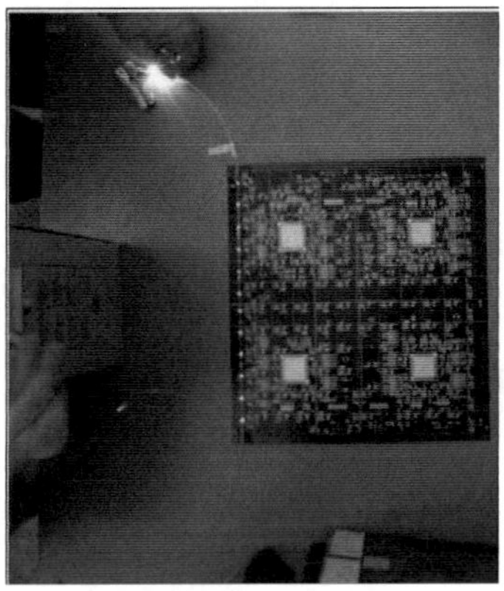

Figure 7. One option for the AHCAL LED system based on central driver and optical signal distribution [11].

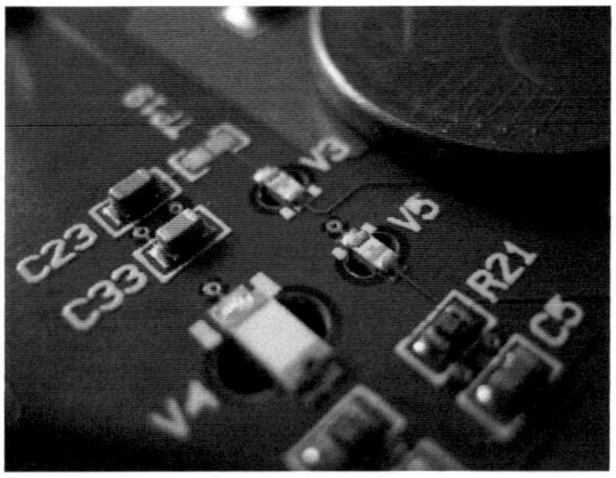

Figure 8. Another option for the AHCAL LED system based on electrical signal distribution and an individual SMD-LED per tile [10].

Results obtained with the quasi-resonant LED driver and light distribution via optical fibers are reported in [11]. A six channels UV LED driver, which meets the basic requirements for SiPM calibration, has been realized and tested. The prototype is complemented by the optical system that distributes light from the LED via a single notched fiber to a row of scintillator tiles. The uniformity of the light radiated by the notches is constant within ±20%.

Preliminary results [10] using this system show that it is possible to obtain single photo-electron peak spectra from the SiPM on tile, i.e. to extract SiPM gain, using the light injected on the tile by the SMD-type LED integrated on the PCB board. A histogram for a single SPIROC2 channel at low LED light intensity is shown in Fig. 9.

3. First test beam results

One fully equipped AHCAL cassette, described in the previous section, has been tested at the DESY test beam facility. It has been exposed to 3 GeV electron beam without absorber material in front of the cassette. The beam penetrates the thin cassette cover and leaves a MIP-like signal in the scintillator tiles. Fig. 10 shows an example of a one channel MIP calibration spectrum. A pedestal peak is visible on the left due to false trigger coincidences. The most probable value of the MIP distribution gives a light yield of 9 pixels/MIP, as expected for this system. The single photo-electron peak structure is clearly visible in the spectrum, which is a very advantageous feature of the low noise readout electronics. A signal-to-noise ratio of 5.5 is obtained for single photo-electron measurements, while the most probable value of a MIP energy distribution has a signal-to-noise ratio of 45.

4. Conclusions and outlook

A technological solution for a scintillator-based HCAL for the ILC is presented. This prototype addresses electronics integration and power pulsing issues as well as the mechanics integration and scalability.

A complex readout chip for SiPM, including signal digitization is under commissioning and the

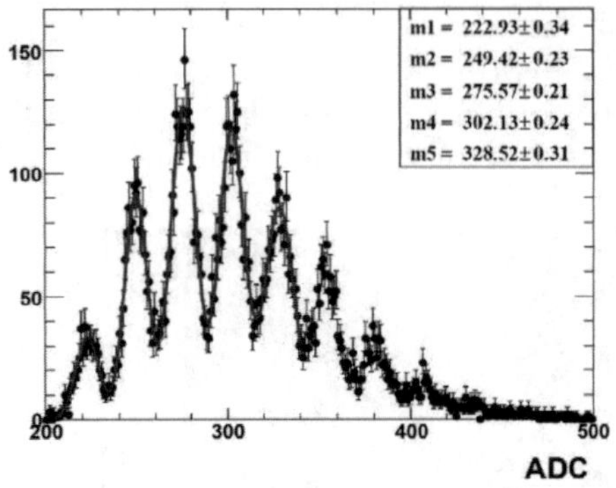

Figure 9. Single photo-electron peak spectrum obtained shining low intensity light from the LED monitoring system shown in Fig. 8 to a scintillator tile in the AHCAL cassette.

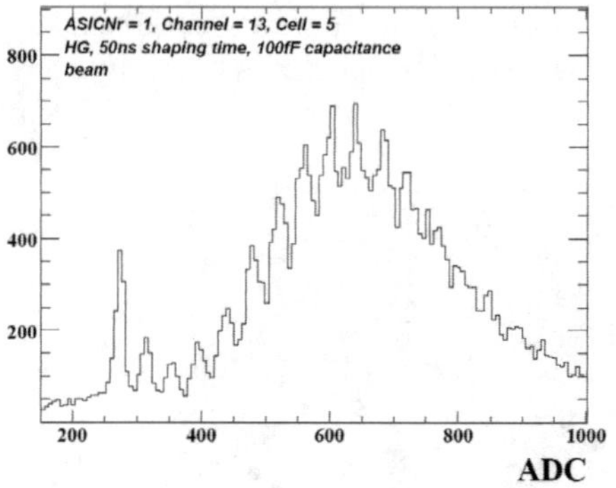

Figure 10. Energy deposition spectrum on a AHCAL tile from a 3 GeV electron beam at the DESY test beam facility.

first calibration measurements taken at the DESY test beam have been presented. Results are very encouraging.

In the future, CALICE will proceed towards the full commissioning of a calorimeter module. A 2.2 m long full calorimeter layer has to be assembled to check signal distributions and interconnectivity of cassettes. To test multi-layer operation a tower of cassettes will be equipped. In view of this test, the re-design of the detector interface is under way.

References

[1] The CALICE collaboration, see *https://twiki.cern.ch/twiki/bin/view/CALICE/WebHome*

[2] N. Phinney et al., *Reference Design Report Volume 3: Accelerator*, available at *http://lcdev.kek.jp/RDR/*

[3] T. Behnke et al., *Reference Design Report Volume 4: Detectors*, available at *http://lcdev.kek.jp/RDR/*

[4] J.-C. Brient, *Improving the jet reconstruction with the particle flow method: An introduction*, in the Proceedings of the 11th International Conference on Calorimetry in High-Energy Physics (Calor 2004), Perugia, Italy, March 2004.

[5] The CALICE collaboration, *Construction and commissioning of the CALICE Analog Hadron Calorimeter Prototype*, arXiv:1003.2662 [physics.ins-det], (2010) JINST 5 P05004, doi: 10.1088/1748-0221/5/05/P05004.

[6] L. Raux et al., *SPIROC Measurement: Silicon Photomultiplier Integrated ReadoutChips for ILC'*, paper contributed IEEE NSS (2008), NSSMIC.2009.5401891.

[7] R. Fabbri, B. Lutz and W. Shen, *Overview of Studies on the SPIROC Chip Characterisation*, arXiv:0911.1566 and EUDET-Report-2009-05, October 2009.

[8] M. Thomson, *Particle Flow Calorimetry and the PandoraPFA Algorithm*, arXiv:0907.3577 and NIMA 611 (2009) 25-40.

[9] D. McNallya, V. Golovin, *Review of solid state photomultiplier developments by CPTA and photonique SA* NIMA 610 (2009) 150-153.

[10] P.Göttlicher (for the CALICE collaboration) *First Results of the Engineering Prototype of the CALICE Tile Hadron Calorimeter*, paper contributed to IEEE NSS, 2009, Orlando, Florida.

[11] J.Cvach, M.Janata, J.Kvasnicka, *Calibration prototype for the EUDET HCAL*, EUDET-Report-07, 2008, http://www.eudet.org/e26/e26/e27/e825/EUDET_report_07.pdf.

Journal of Physics: Conference Series **293** (2011) 012073 doi:10.1088/1742-6596/293/1/012073

Particle Showers in a Highly Granular Hadron Calorimeter

Frank Simon, for the CALICE Collaboration
Max-Planck-Institut für Physik and Excellence Cluster 'Universe', Munich, Germany
E-mail: frank.simon@universe-cluster.de

Abstract. The CALICE collaboration has constructed highly granular electromagnetic and hadronic calorimeter prototypes to evaluate technologies for the use in detector systems at a future Linear Collider. The hadron calorimeter uses small scintillator cells individually read out with silicon photomultipliers. The system with 7608 channels has been successfully operated in beam tests at DESY, CERN and Fermilab since 2006, and represents the first large scale tests of these devices in high energy physics experiments. The unprecedented granularity of the detector provides detailed information of the properties of hadronic showers, which helps to constrain hadronic shower models through comparisons with model calculations. We will discuss results on longitudinal and lateral shower profiles compared to a variety of different shower models, and present studies of the energy reconstruction of hadronic showers using software compensation techniques.

1. The CALICE Calorimeters

The goal of the CALICE experimental program is to establish novel technologies for calorimetry in detectors at a future linear e^+e^- collider, and to record electromagnetic and hadronic shower data with unprecedented three dimensional spatial resolution for the validation of simulation codes and for the test and development of reconstruction algorithms. Such highly granular calorimeters are necessary to achieve an unprecedented jet energy resolution at the International Linear Collider [1] using particle flow algorithms [2, 3, 4].

The CALICE test beam setup consists of a silicon-tungsten electromagnetic calorimeter (ECAL), an analog scintillator-steel hadron calorimeter (AHCAL) and tail catcher/muon tracker (TCMT), the latter two both with individual cell readout by silicon photomultipliers (SiPMs) [5]. This setup has been tested extensively in electron, muon, and hadron beams at CERN and at the Meson Test Beam Facility at Fermilab. Figure 1 shows the schematic setup of the CALICE detectors in the CERN H6 test beam area, where data was taken in 2006 and 2007. The currently ongoing and the future program include the study of alternative technologies, such as an ECAL using scintillator strips with SiPM readout, digital hadron calorimeters with active layers based on gas detectors, and the investigation of alternative absorber materials for the hadron calorimeter for multi-TeV colliders.

The performance of the current setup for hadronic showers is driven by the analog HCAL [6], which is a 38 layer sampling calorimeter with 5 mm thick scintillator layers sandwiched by 2 cm of steel. The lateral dimensions are roughly 1×1 m^2, the total thickness amounts to 5.3 nuclear interaction lengths. The active layers are built out of scintillator tiles with sizes ranging from 30×30 mm^2 in the core of the detector to 120×120 mm^2. Each tile is read out by a built-in

Journal of Physics: Conference Series **293** (2011) 012073 doi:10.1088/1742-6596/293/1/012073

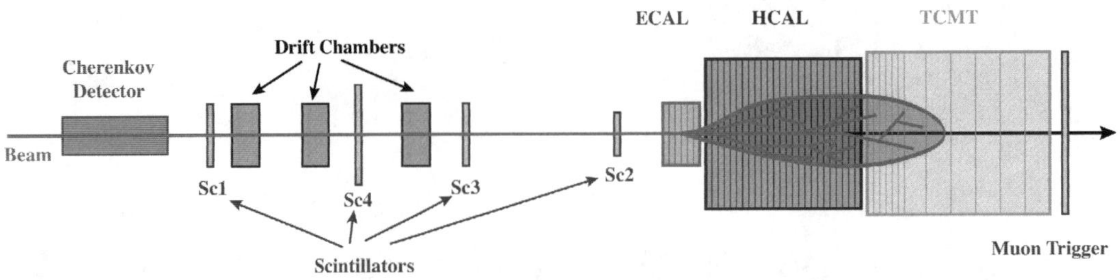

Figure 1. Schematic of the CALICE experimental setup at CERN, with electromagnetic and hadronic calorimetry as well as a tail catcher and muon tracker downstream of the calorimeters.

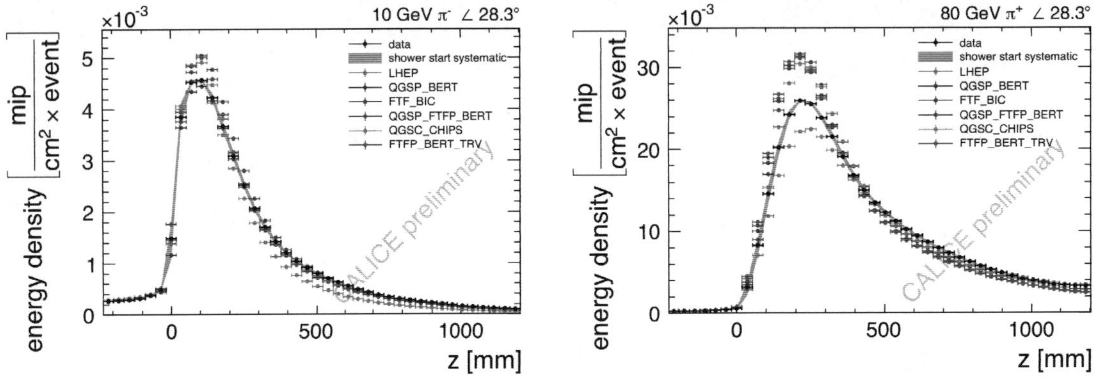

Figure 2. Longitudinal profile of hadronic showers of 10 GeV π^- (left) and 80 GeV π^+ (right) compared to various hadronic shower models.

SiPM, produced by the MEPhI/PULSAR group [5]. In total, the calorimeter has 7608 channels. This high granularity allows for detailed studies of hadronic showers, and lends itself to energy reconstruction algorithms using information of the shower substructure.

In the present paper, hadron data taken at CERN in 2007 with the complete installation of the silicon tungsten ECAL, the analog HCAL and the TCMT, are being discussed, focusing on results from the hadron calorimeter.

These experimental results are confronted with detailed simulations performed within the GEANT4 framework. For the simulations, a variety of different hadronic shower models, implemented as different "physics lists" [7] within GEANT4, are used. The simulation of the interaction and propagation of particles is coupled with a detailed simulation of the detector response, including the effects of photon statistics, saturation effects in the SiPM and in the scintillator, and the integration time of the electronics. The comparison of simulation and measurement will provide insight for the further development of GEANT4 physics lists.

2. Shower Profiles

The high granularity of the calorimeter, with layer-wise longitudinal and fine lateral segmentation, allows precise measurements of the topology of hadronic showers. Particularly powerful is the possibility to identify the starting point of the shower, by measuring the position of the first inelastic reaction of the incoming beam particle via a local increase in energy deposit and hit multiplicity over that expected from the passage of a minimum ionizing particle. This allows the measurement of shower properties with respect to the starting point, eliminating the

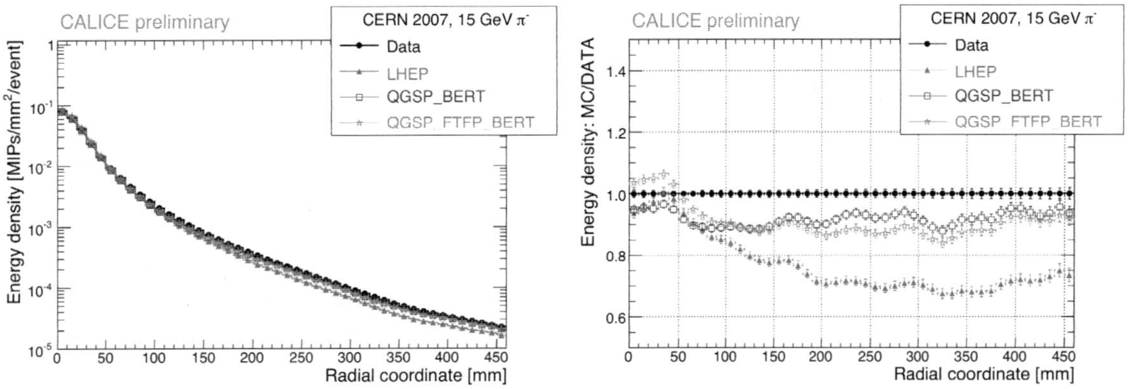

Figure 3. Lateral profile of showers of 15 GeV π^-. *Right:* Energy density as a function of radial coordinate. *Left:* Ratio of simulations based on different physics lists and data.

smearing introduced by the exponential distribution of the point of the first interaction, resulting in a significant increase in sensitivity to the longitudinal evolution of the shower.

Figure 2 shows the longitudinal shower profile measured for 10 GeV negative and for 80 GeV positive pions relative to the shower start., measured with the detector inclined by approximately 30° with respect to the beam axis. The measurement is compared to a variety of different hadronic shower models. For the 10 GeV case, in particular the theory driven model QGSP_BERT, favored by the LHC calorimeter test beams, provides a satisfactory description of the measurement, while other models show discrepancies in the region of the shower maximum and / or in the tail of the profile. At the higher energy point, all models show discrepancies with the data in particular in the shower maximum region.

Such measurements have the potential to elucidate the longitudinal dependence of the relative contributions of different components of hadronic showers, such as nuclear fragments, electromagnetic subshowers and hadrons, through detailed comparisons to simulations, as demonstrated already for the CALICE ECAL [8]. Similar studies for the analog HCAL are currently in preparation.

The performance of Particle Flow Algorithms is strongly influenced by the capability for the separation of particle showers in the calorimeter. For the evaluation of such algorithms with simulations, a realistic description of the lateral shower size is thus of importance. Figure 3 (*left*) shows the transverse shower profile, expressed by the energy density as a function of the radial coordinate. The radial distance of energy deposits to the particle axis is determined using the particle tracking information from the drift chambers in the CALICE beam setup, allowing sub-cellsize resolution.

As for the longitudinal profiles, the data are confronted with different hadronic shower models, with three such models shown in the figure. Figure 3 (*right*) shows the ratio of the simulation predictions and the measurement, showing discrepancies for all considered models. The models shown here describe the shower core within approximately 5%, with larger deviations observed in the outer regions, in particular in the case of LHEP.

3. Shower Substructure

The extremely high granularity of the CALICE calorimeters also enables measurements of the substructure of showers well beyond averaged shower profiles discussed above. Such studies test the quality of shower simulations in finer detail, by requiring a realistic modeling of the shower evolution on the particle level, not only in terms of the averaged overall energy distribution.

Journal of Physics: Conference Series **293** (2011) 012073 doi:10.1088/1742-6596/293/1/012073

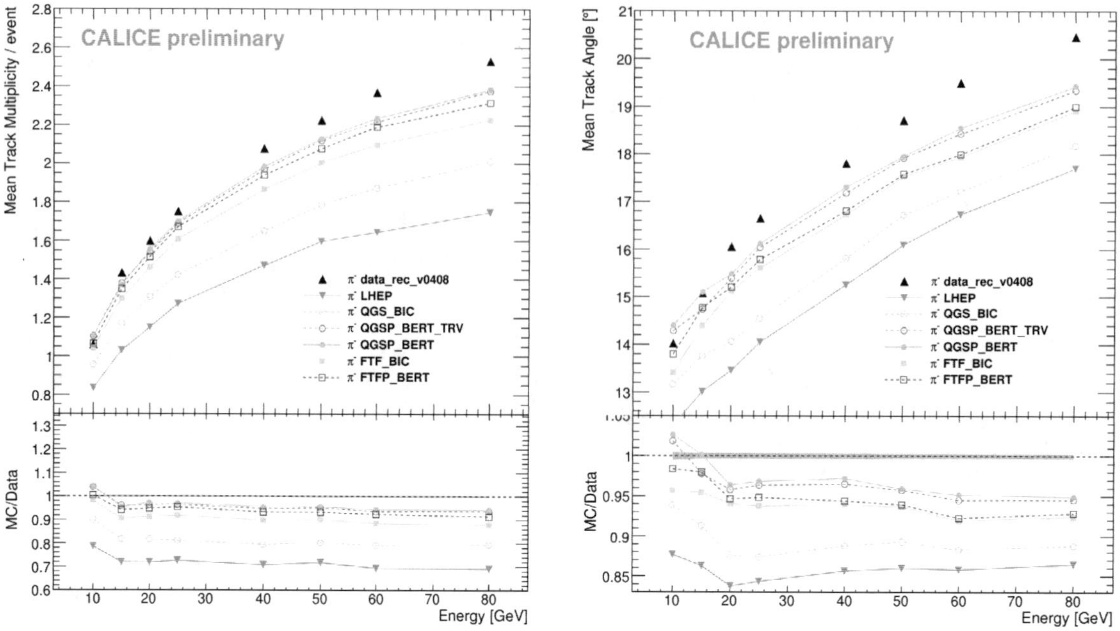

Figure 4. Average track multiplicity (left) and mean track angle (right) as a function of beam energy. Also shown are simulations based on different physics lists, and the ratio of simulations to data.

One such possibility is the detection of individual track segments within the shower, caused by single charged particles. Since such tracks behave as minimum-ionizing particles, they are a valuable calibration tool for a future linear collider calorimeter, enabling the monitoring of the calibration with regular physics data [9]. In the CALICE analog HCAL, such minimum-ionizing track segments are found from isolated hits, local energy deposits that do not have energy in next neighboring cells in the same calorimeter layer [10]. This ensures a high purity of single particle tracks, but limits the track finding to the outer, less dense regions of the hadronic showers. The minimum length of tracks was 6 layers, corresponding to 0.85 λ_I, also imposed to increase the purity of the sample.

Figure 4 shows the average multiplicity and the mean angle of the found tracks for negative pions over a wide energy range. All considered shower models reproduce the general trend of increasing track multiplicity and increasing track angle with energy. While `QGSP_BERT` and the `FTF` - based models reproduce the observations typically within 10% and 5% for the track multiplicity and angle, respectively, `LHEP` and `QGS_BIC` show larger discrepancies. These latter two models thus produce a too small particle multiplicity in the outer parts of the showers, and include not enough particle production at large angles. `QGSP_BERT`, the physics list favored by the LHC calorimeter test beams, provides the best overall description of the track segments.

4. Energy Reconstruction and Software Compensation

The detailed information of hadronic showers provided by imaging calorimeters is also beneficial for the energy reconstruction. The resolution of non-compensating hadronic calorimeters, such as the CALICE calorimeters, suffers from the fact that such detectors respond differently to electromagnetic and to purely hadronic showers. Coupled with the large event-by-event fluctuations of the electromagnetic component and with the energy dependence of the mean electromagnetic fraction, this leads to a limited energy resolution and a non-linear response of

Journal of Physics: Conference Series **293** (2011) 012073 doi:10.1088/1742-6596/293/1/012073

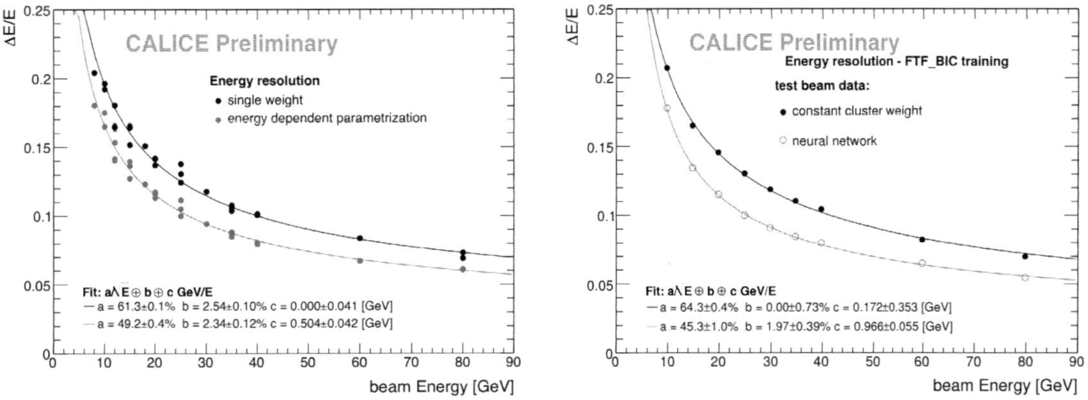

Figure 5. Energy resolution for hadrons with and without software compensation. *Left:* Local software compensation, with weights determined from data, applied to the full CALICE setup of ECAL, HCAL and TCMT. *Right:* Global software compensation on clustered data, with a neural network trained on simulations with the `FTF_BIC` physics list, applied to showers in the HCAL and TCMT.

such calorimeters.

The high granularity of the CALICE calorimeters provides excellent conditions for software compensation techniques. Such energy reconstruction algorithms treat electromagnetic and hadronic energy deposits separately to improve the energy resolution and to recover the linearity of the response. Since typical non-compensating calorimeters show a higher response to electromagnetic than to hadronic energy deposits ($e/h > 1$), electromagnetic deposits are assigned a lower weight in the overall energy sum of the shower.

For the analog HCAL, two approaches to this technique have been studied: A local and a global software compensation algorithm. Both use the fact that electromagnetic energy deposits tend to have a higher energy density than purely hadronic ones. The local software compensation technique [11] uses the energy of each cell as a measure of the local energy density at that point in the shower, and chooses a weight for the cell in question according to this. The global technique [12] uses the overall energy density of the shower, calculated from the shower volume found by a clustering algorithm to select one global weight for the shower, essentially corresponding to a determination of the overall electromagnetic content of the shower. These weights are applied relative to the naive energy calculation, which is a simple sum of the amplitudes of all cells above the noise threshold, multiplied by a constant factor to transform from the calibration scale in units of minimum ionizing particles (MIP) to the GeV scale. In both software compensation techniques considered here, the final algorithm does not require knowledge of the true beam energy, making it applicable over a wide energy range.

For a first study of local software compensation, the single-cell energy density based weights were determined from an independent real data set using a χ^2 minimization technique [11]. Due to energy-dependent factors of the shower structure, such as changing mean electromagnetic fractions and increasing core energy density for increasing particle energy, these weights are energy dependent. A parametrization of this energy dependence was derived from the first set of weights obtained from the minimization technique, making the software compensation algorithm independent from the beam energy. This algorithm was then applied to the complete CALICE setup, consisting of the ECAL, HCAL and TCMT, without any requirements on shower containment. Figure 5*(left)* shows the energy resolution obtained with the local algorithm, compared to the reconstruction with one constant conversion factor. The compensation algorithm improves the resolution by up to 20%, yielding a stochastic term of approximately $50\%/\sqrt{E}$.

The global software compensation was studied for showers in the HCAL and TCMT, by requiring an energy deposit in the ECAL consistent with a single minimum-ionizing particle. No further requirements on shower containment were made. With a simple clustering algorithm, the cells belonging to a particle shower were identified, and from this shower parameters such as the total energy and the shower volume were extracted. These parameters were used as inputs for a neural network to obtain an optimized estimate of the true particle energy. For this estimation, the energy density, given by the total shower energy divided by the shower volume, is of particular importance. The neural network was trained with a simulated data set with a quasi-continuous energy distribution to avoid artifacts from the discrete particle energies used in the test beam. Several different physics lists were tested, with `FTF_BIC` giving the best results. Figure 5 *(right)* shows the energy resolution obtained with the neural network, compared to the resolution for clustered showers with one constant conversion factor. The neural network achieves an improvement of up to 25%. The fact that the energy resolution without software compensation in the global method is worse than in the case of the local method is due to additional effects from the clustering applied here.

5. Summary

The CALICE collaboration has acquired sizeable data sets of hadronic showers with a highly granular hadronic calorimeter using a large sample almost 8 000 scintillator tiles read out with silicon photomultipliers, operated stably over several data taking periods. These data provide detailed information on the overall shape and on the substructure of hadronic showers. The comparison of longitudinal and transverse shower profiles to simulations with a variety of different hadronic shower models provides input for a further development of these models. A first study of the shower substructure using identified minimum ionizing track segments has shown qualitative agreement for some of the GEANT4 physics lists under consideration, underlining the level of detail already reached in these models. The high granularity of the detectors also proves to be a powerful tool for the improvement of the energy resolution with software compensation algorithms, both on a local cell-by-cell and on a global shower level. The remarkable improvement in performance with a neural network trained with simulations suggests that the global features of hadronic showers in the CALICE calorimeters, such as the overall energy density and its evolution with particle energy, are already well reproduced by the used physics list within GEANT4.

References

[1] Brau J *et al.* (ILC) 2007 (*Preprint* `arXiv:0712.1950 [physics.acc-ph]`)
[2] Brient J C and Videau H 2002 (*Preprint* `hep-ex/0202004`)
[3] Morgunov V L 2002 CALOR 2002, Pasadena, CA, USA
[4] Thomson M A 2009 *Nucl. Instrum. Meth.* **A611** 25–40
[5] Bondarenko G *et al.* 2000 *Nucl. Instrum. Meth.* **A442** 187–192
[6] Adloff C *et al.* 2010 *JINST* **5** P05004
[7] Apostolakis J *et al.* 2008 *2008 IEEE Nuclear Science Symposium Conference Record*
[8] Adloff C *et al.* 2010 *JINST* **5** P05007
[9] Lu S 2009 (*Preprint* `arXiv:0910.3820 [physics.ins-det]`)
[10] Weuste L 2010 (*Preprint* `arXiv:1006.3734 [physics.ins-det]`)
[11] Simon F 2009 (*Preprint* `arXiv:0911.4575 [physics.ins-det]`)
[12] Seidel K 2010 (*Preprint* `arXiv:1006.3712 [physics.ins-det]`)

Journal of Physics: Conference Series **293** (2011) 012074 doi:10.1088/1742-6596/293/1/012074

Studies of Scintillator Tiles with SiPM Readout for Imaging Calorimeters

Frank Simon, for the CALICE Collaboration
Max-Planck-Institut für Physik and Excellence Cluster 'Universe', Munich, Germany
E-mail: frank.simon@universe-cluster.de

Abstract. Imaging hadronic calorimeters with scintillator readout use small scintillator tiles individually read out by silicon photomultipliers to achieve the necessary granularity needed for sophisticated reconstruction algorithms at future collider detectors. For a second generation prototype of the CALICE analog HCAL new, 3 mm thick scintillator tiles with an embedded wavelength shifting fiber and new photon sensor from CPTA are being fabricated. The availability of blue-sensitive SiPMs also allows fiberless coupling of the photon sensor to the tile, a technique requiring modified geometries to achieve a high degree of response uniformity. We discuss results from test bench and from test beam measurements of different scintillator tile geometries as well as prospects for fiberless coupling of photon sensors.

1. Introduction

The physics program at future high energy e^+e^- colliders imposes stringent requirements on the detector performance. Of particular importance is the precise reconstruction of jets. This is crucial for the measurement of the parameters of particles decaying into hadronic final states, for the separation of W and Z bosons, and for the measurement of missing transverse energy, an important signature for new physics beyond the Standard Model.

Particle Flow Algorithms [1, 2, 3] are a promising strategy to achieve an unprecedented jet energy resolution to satisfy the requirements for precision physics at future high energy lepton colliders. These event reconstruction techniques demand extreme granularity in the calorimeters to provide the imaging capabilities necessary to separate showers of individual particles. In analog hadron sampling calorimeters with scintillator readout this is achieved by using small scintillator cells with typical sizes of 3 x 3 cm^2. The first generation prototype analog hadron calorimeter (AHCAL) [4] of the CALICE collaboration uses a total of 7608 scintillator tiles, individually read out by silicon photomultipliers (SiPMs) [5]. In each tile a wavelength shifting (WLS) fiber is embedded to match the emission spectrum of the scintillator to the range of optimal efficiency of the photon sensor.

2. Scintillator Tiles for the CALICE Technical Prototype

To address technological and integration issues for the hadronic calorimeter at a linear collider detector, a second generation "technological" prototype for the analog hadron calorimeter is currently being developed [6]. Key features of this device are fully integrated electronics and a realistic mechanical design. For an increased compactness of the detector, the thickness of the scintillator has been reduced to 3 mm, compared to the 5 mm thickness used in the physics

Journal of Physics: Conference Series **293** (2011) 012074 doi:10.1088/1742-6596/293/1/012074

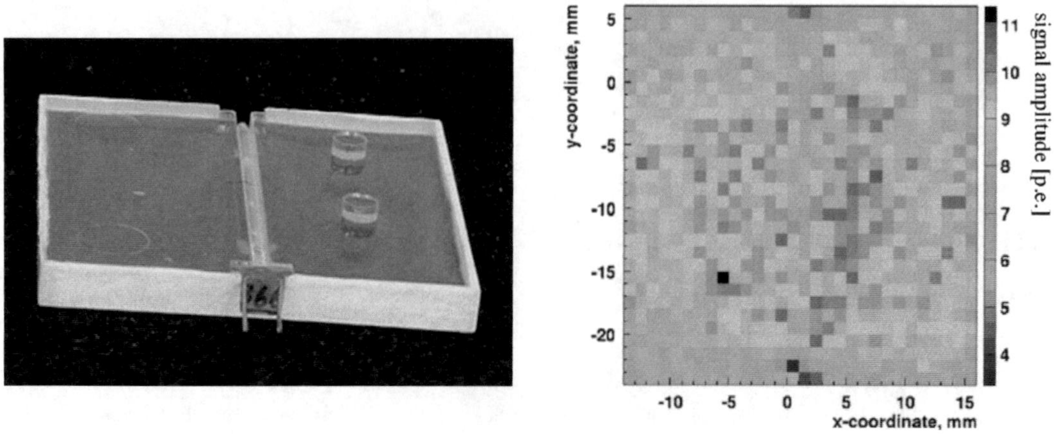

Figure 1. *Left:* Scintillator tile for the 2nd generation AHCAL prototype, with a thickness of 3 mm and a size of 30×30 mm^2, an embedded wavelength shifting fiber, and installed SiPM. *Right:* Response map measured with protons.

prototype. The size of the tiles, defining the granularity of the detector, is left unchanged at 30×30 mm^2. The scintillator tiles are fabricated by injection molding. The sides are chemically matted to prevent tile to tile light cross talk. A WLS fiber is embedded in the tile to provide wavelength matching to the SiPM and for improved light collection. The photon sensor is an improved SiPM by CPTA, with 556 pixels arranged in an active area adapted to the cross section of the WLS fiber. For future devices the number of pixels will be increased to 796 pixels for an increased dynamic range. Figure 1 *left* shows such a scintillator tile. In addition to the WLS fiber, the photon sensor and the side matting, alignment pins used for the installation of the cells on the printed circuit board housing the front-end electronics are visible.

The new photon sensors and the scintillator tile assemblies have been thoroughly tested with light sources and in particle beams. Figure 1 *right* shows a measurement of the response to protons from the ITEP synchrotron as a function of beam position on the scintillator. The SiPM operating point for the calorimeter is chosen to achieve about 10 photo-electrons for a minimum ionizing particle, since this provides a large dynamic range while ensuring an efficiency of around 96% at a threshold of 0.4 minimum ionizing particle (MIP) equivalent. As the figure shows, a high degree of uniformity is achieved, with a 20% reduction of the response in the area of the WLS fiber. The positions of the SiPM and of the opposing fiber end result in small areas of significantly reduced sensitivity, as expected.

3. Laboratory Setup for Uniformity Studies

For detailed studies of the response of small scintillator tiles with SiPM readout, a test setup that scans a ^{90}Sr source across the scintillator and measures the signal for fully penetrating electrons was constructed [7]. This setup, shown in Figure 2 *left*, uses high precision stages to position the radioactive source, which is collimated to a spot size of a few mm in diameter. The readout is triggered by a coincidence of the signal from the tile under study and of the signal in a small scintillator cube positioned underneath the tile, which moves together with the source. This ensures that only fully penetrating electrons from the high energy tail of the ^{90}Sr spectrum, originating from the secondary ^{90}Y decay, are recorded. The readout is performed with a high-speed oscilloscope, providing precision measurements and single photo-electron resolution.

The energy loss of electrons from the ^{90}Sr source in the scintillator was determined with

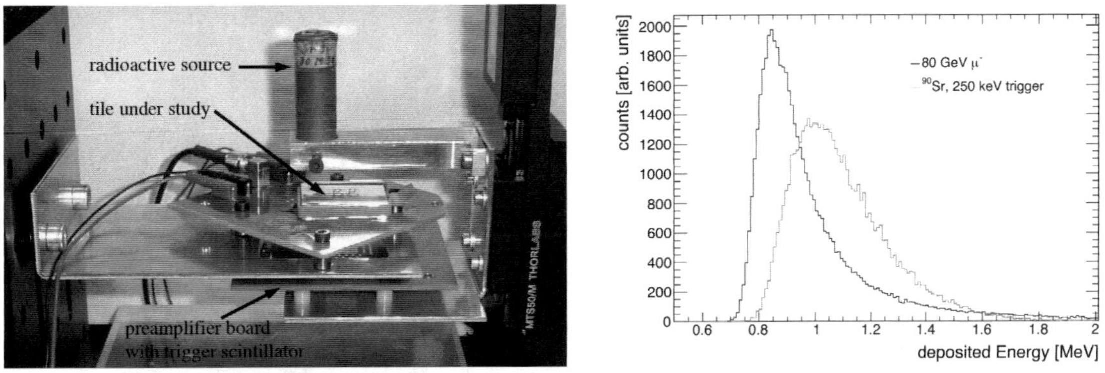

Figure 2. *Left:* Test setup for detailed response studies of scintillator tiles [7]. *Right:* Comparison of the energy loss of 80 GeV muons and of penetrating electrons from a ^{90}Sr source in 5 mm thick scintillator, simulated by GEANT4.

a GEANT4 simulation based on a realistic incident energy distribution [8] and a modeling of the trigger by requiring a minimum energy deposit of 250 keV in the trigger scintillator. This was compared to a simulation using 80 GeV muons, as used for calibration in the CALICE test beams at CERN [4]. For the muon simulation the absorber plates are included to get a realistic modeling of the energy loss distribution. Figure 2 *right* shows that electrons from the radioactive source deposit approximately 20% more energy in the scintillator than minimum-ionizing particles and also show a broader distribution with a less pronounced high energy tail. Still, the ^{90}Sr signals can serve as a good approximation for minimum-ionizing particles, once the higher mean energy deposition is taken into account.

4. Fiberless Coupling of the Photon Sensors

With the now readily available blue-sensitive SiPMs, wavelength-matching to the photon sensor with WLS fibers is no longer necessary. This allows a simplification of the construction of the scintillator tiles, and leads to relaxed mechanical tolerances for the coupling of the photon sensor since a precise alignment of the active area of the sensor with the fiber end is not necessary. The elimination of the wavelength shifter also results in a faster response of the scintillator-SiPM system since the additional time constant from the de-excitation in the fiber is no longer present. However, a fiber embedded in the scintillator serves also as a light collector, leading to a good uniformity of the response over the tile area, which is lost in the case of a simple fiberless coupling of the photon sensor.

The uniformity of the response over the full active area can be recovered by special shaping of the scintillator [7, 9]. In the case of a slit drilled in the side face of the scintillator tile, as discussed in detail in [7], the tiles are compatible with the ones used for the 2nd generation prototype discussed in Section 2. With the scanning setup, 3 mm thick tiles of that type were evaluated, using Hamamatsu Multi-Pixel Photon Counters (MPPCs) with an active area of 1 mm^2 and 1600 pixels (type S10362-11-025P), operated at their nominal operation point as given by the manufacturer. Two types of tiles, both with the same geometrical parameters for the slit at the SiPM coupling position, were studied: a tile machined from a plate of BC-420 plastic scintillator and a tile produced with the molding technique used for the CALICE technical prototype tiles. The machined tile was fully enclosed in aluminized reflective foil except for the photon sensor position, while the molded tile had chemically matted sides, with only the top and bottom face covered by reflective foil. In both tiles the slit designed to recover the response uniformity was produced by drilling.

Journal of Physics: Conference Series **293** (2011) 012074 doi:10.1088/1742-6596/293/1/012074

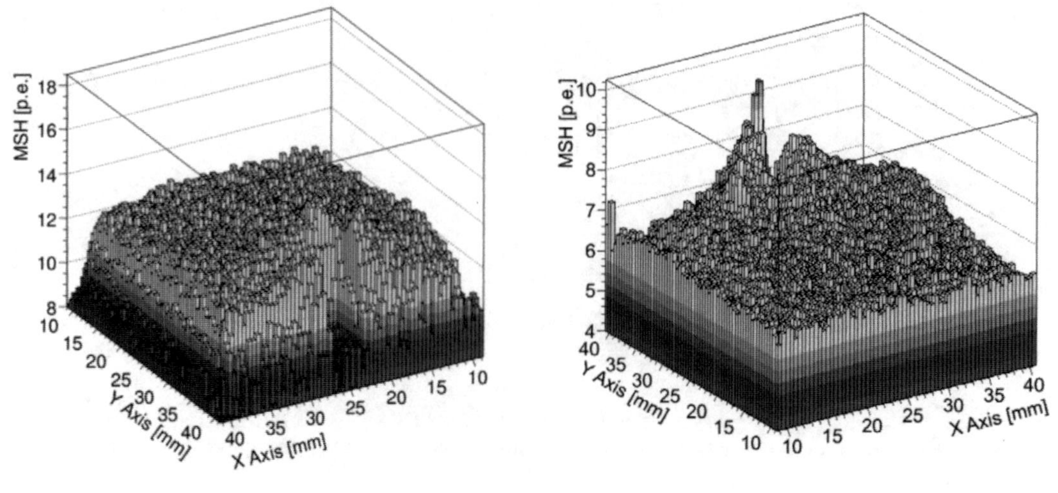

Figure 3. Response maps of 3 mm thick scintillator tiles with a coupling slit fiberlessly read out with a Hamamatsu MPPC, showing the mean signal height (MSH). *Left:* Machined scintillator fabricated from BC-420, enclosed in reflective foil. *Right:* Molded scintillator, with chemically matted sides and reflective foil on top and bottom.

Figure 3 shows the response map of these two scintillator tiles. The overall mean signal height is drastically different: While the machined tile achieves a mean amplitude of 13 p.e., only 7 p.e. are achieved in the case of the molded tile. It is known that the light yield of the scintillator used for injection molding is about 30% lower than the one of BC-420, explaining a large part of this difference. In addition, the matting of the sides very likely leads to a decreased light collection efficiency, in particular in areas further away from the photon sensor, further reducing the mean signal height. This loss in signal amplitude can be recovered by using larger photon sensors or SiPMs with fewer, larger pixels and thus higher quantum efficiency, or by increasing the overbias of the sensors.

The response uniformity of the machined tile is very good, with 82.5% (94%) of the active area within 5% (10%) of the mean signal height, excluding edge effects. The uniformity of the molded tile is slightly reduced compared to this, with 66% (87%) of the active area within 5% (10%) of the mean signal height. This reduction is in part due to the reduced reflectivity of the tile sides, which changes the light collection efficiency over the tile surface, and is likely also affected by the low overall signal. The large overshoot at the photon sensor position visible in Figure 3 is probably due to a coupling problem. Still, an overall satisfactory uniformity was achieved with this first test of fiberless coupling with an injection-molded tile.

To find tile geometries that provide a high uniformity of the response and are well suited for injection molding further studies have been performed. Figure 4 *left* shows a 5 mm thick scintillator tile with a spherical hole with a radius of 5 mm drilled to a depth of 1.8 mm into the bottom face of the tile, with the center of the hole 2 mm from the front edge. In addition to this hole, a slit to accommodate an MPPC in a compact surface-mount package was machined into the side face. The MPPC, also shown in Figure 4 *left*, was installed oriented as shown in the figure, but with the active area facing the tile. Thus, the complete active area of the photon sensor was inside the air volume of the hole in the scintillator. The reduction of the scintillation material at the SiPM position, coupled with diffuse reflection in the drilled hole leads to an excellent uniformity, as shown in Figure 4 *right*. It was observed that the performance with a polished hole surface was reduced compared to the rough, diffuse surface obtained by drilling. The overall signal amplitude is reduced by 20% compared to 5 mm thick tiles with the drilled slit,

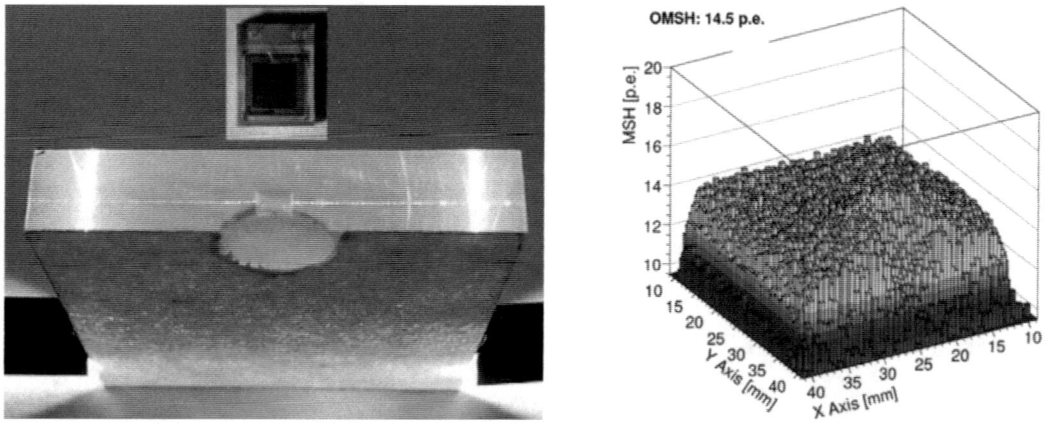

Figure 4. *Left:* 5 mm thick scintillator tile with a quarter of a spherical hole of 5 mm diameter and a slit for the photon sensor. The photon sensor, a surface-mount Hamamatsu MPPC is shown for reference.*Right:* Response map of the scintillator tile, showing the mean signal hight MSH) at each position. The overall mean (OMSH) over the full tile surface is 14.5 photoelectrons.

but an equally good uniformity is achieved with a geometry that is more favorable for injection molding. With the present size of the SiPM package, this design is not directly transferrable to 3 mm thick tiles, but there are no fundamental difficulties to extend this new geometry to thinner scintillators.

5. Conclusions

Scintillator tiles with an integrated silicon photomultiplier are key components of highly granular analog hadron calorimeters. From the original design for the CALICE AHCAL physics prototype these systems have been further developed. For the CALICE AHCAL 2nd generation technical prototype, thinner tiles with an embedded wavelength shifting fiber and newly designed SiPMs are being fabricated. These tiles show good response uniformity and the desired signal amplitude. Blue sensitive photon sensors allow the fiberless readout of the scintillator without the addition of a WLS fiber. However, specific modifications of the tile at the coupling position of the photon sensor are necessary to ensure a high degree of uniformity over the full active area. With machined tiles, high uniformity and high signal amplitudes are routinely achieved. First tests with injection molded tiles show encouraging results, suggesting that the mass production of scintillator tiles for imaging hadronic calorimeters based on a fiberless design with directly coupled photon sensors is feasible.

References

[1] Brient J C and Videau H 2002 (*Preprint* hep-ex/0202004)
[2] Morgunov V L 2002 CALOR 2002, Pasadena, CA, USA
[3] Thomson M A 2009 *Nucl. Instrum. Meth.* **A611** 25–40
[4] Adloff C *et al.* 2010 *JINST* **5** P05004
[5] Bondarenko G *et al.* 2000 *Nucl. Instrum. Meth.* **A442** 187–192
[6] Garutti E 2010 *these proceedings*
[7] Simon F and Soldner C 2010 *Nucl. Instrum. Meth.* **A620** 196–201
[8] Cross W G *et al.* 1983 *Phys. Med. Biol.* **28** 1251
[9] Blazey G *et al.* 2009 *Nucl.Instrum.Meth.* **A605** 277–281

Journal of Physics: Conference Series **293** (2011) 012075 doi:10.1088/1742-6596/293/1/012075

Tests of a Digital Hadron Calorimeter

Burak Bilki[1], John Butler[2], Ed May[3], Georgios Mavromanolakis[4,•], Edwin Norbeck[1], José Repond[3,*], David Underwood[3], Lei Xia[3], Qingmin Zhang[3,#]

1 – University of Iowa, Iowa City, IA 52242-1479, U.S.A.

2 – Boston University, 590 Commonwealth Avenue, Boston, MA 02215, U.S.A.

3 – Argonne National Laboratory, 9700 S. Cass Avenue, Argonne, IL 60439, U.S.A.

4 – Fermilab, P.O. Box 500, Batavia, IL 60510-0500, U.S.A.

Corresponding author's e-mail: repond@hep.anl.gov

Abstract. In the context of developing a hadron calorimeter with extremely fine granularity for the application of Particle Flow Algorithms to the measurement of jet energies at a future lepton collider, we report on extensive tests of a small scale prototype calorimeter. The calorimeter contained up to 10 layers of Resistive Plate Chambers (RPCs) with 2560 1×1 cm^2 readout pads, interleaved with steel absorber plates. The tests included both long-term Cosmic Ray data taking and measurements in particle beams, where the response to broadband muons and to pions and positrons with energies in the range of 1 – 16 GeV was established. Detailed measurements of the chambers efficiency as function of beam intensity have also been performed using 120 GeV protons at varying intensity. The data are compared to simulations based on GEANT4 and to analytical calculations of the rate limitations.

1. Digital Hadron Calorimetry

To fully exploit the physics potential of a future lepton collider, operating at center-of-mass energies of 0.5 TeV and above, requires an unprecedented jet energy resolution, of the order of a factor of two better than previously attained. The preferred approach to achieve this performance is through the application of Particle Flow Algorithms (PFAs) [1]. FPAs utilize the tracker to measure the momenta of charged particle, the electromagnetic calorimeter to measure the energy of photons and the combined electromagnetic and hadronic calorimeters to measure the energy of neutral hadrons, i.e. the neutrons and K_L^0's. The major challenge of PFAs is the association of energy deposits in the calorimeter to the charged or neutral particles originating from the interaction point. This challenge is met with the development of highly segmented or so-called imaging calorimeters [2].

In this context we present tests of a Digital Hadron Calorimeter (DHCAL) utilizing Resistive Plate Chambers (RPCs) as active medium. Our RPCs are based on two different designs [3] with either two or only a single glass plate. In the latter case the readout board serves as anode and closes the gas

• Also affiliated with University of Cambridge, Cavendish Laboratory, Cambridge CB3 OHE, U.K.

* Corresponding author

Also affiliated with Institute of High Energy Physics, Chinese Academy of Sciences, Beijing 100049, China and Graduate University of the Chinese Academy of Sciences, Beijing 100049, China.

volume. The chambers use 1.1 mm thick glass as resistive plates and are operated in avalanche mode. The gas gap is 1.2 mm and is maintained with the help of fishing lines spaced 5 cm apart.

The chambers are read out with a readout board containing 1×1 cm^2 pads and a binary (or digital) electronic readout system [4]. The latter is based on the DCAL chip [5], which was developed by Argonne and Fermilab and which has been optimized for the readout of large number of channels.

The project is being carried out by the DHCAL collaboration [6] including 36 people from Argonne National Laboratory, Boston University, Fermilab, Iowa University and the University of Texas at Arlington. It is part of the scientific program of the international CALICE collaboration [7], which develops finely segmented calorimetry for future lepton colliders.

2. The Vertical Slice Test

In preparation for the construction of a large size prototype calorimeter [8], the DHCAL collaboration assembled a small scale prototype calorimeter. This calorimeter utilized the entire readout chain, as envisaged for the large scale prototype, and so was called a Vertical Slice Test (VST). The VST included up to 10 RPCs, each with an area of 20×20 cm^2, and 16×16 cm^2 readout boards, each with 256 readout channels. The RPCs were based on the default design with two glass plates, except for one chamber which featured the single glass plate design.

The prototype was extensively tested with Cosmic Rays and in the Fermilab test beam [9]. In the following we shall briefly summarize the major results.

3. RPC Performance Tests

The performance tests include measurements of the noise rate, the detection efficiency for minimum ionizing particles (MIPs) and the pad multiplicity, i.e. the average number of pads firing in events where a MIP has fired at least one pad.

3.1. Measurement of the environmental dependence of the RPC performance

The performance criteria of a stack of RPCs were correlated with measurements of the ambient temperature, the atmospheric pressure and the humidity of the surrounding air [10,11]. Table 1 summarizes the results for 2-glass RPCs. The chambers were operated within the efficiency plateau, hence only small changes in efficiency were observed. As expected no evidence for any dependence on the humidity of the surrounding air was discovered.

Performance variables	Changes for $\Delta T = 1^0C$ [%]	Changes for $\Delta p = 100$ Pa [%]
Noise rate	14 ± 2	0.02 ± 0.69
Efficiency	0.26 ± 0.05	0.06 ± 0.01
Pad multiplicity	2.0 ± 0.1	0.30 ± 0.02

Table 1: Summary of environmental dependences of the performance of 2-glass RPCs

Of particular interest is the study of the performance as function of gas flow. Figure 1 shows a steep increase in noise rate and pad multiplicity for gas flow rates below 0.3 cm^3/minute, which corresponds to about eight volume changes per day. The reasons for this degradation in performance are not yet fully understood, but are subject of further investigations.

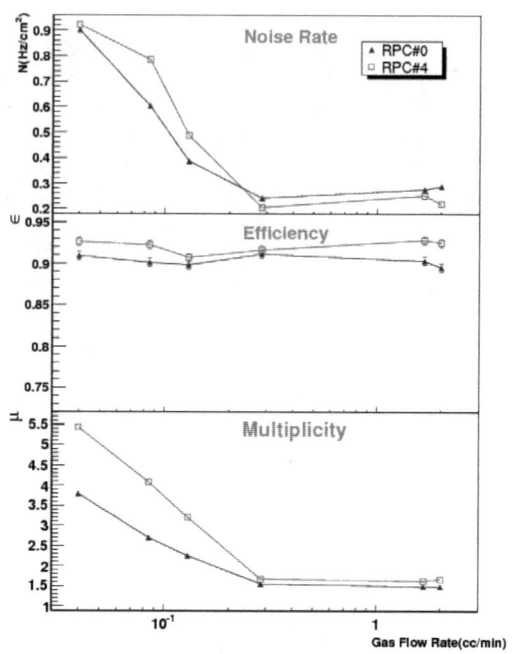

Figure 1: Performance as function of gas flow for a selected 2-glass RPC.

3.2. Measurement of the rate capability of RPCs

The MIP detection efficiency as function of beam intensity was studied [12] with 120 GeV protons, as provided by the Fermilab test beam. For rates in excess of 100 Hz/cm^2 a clear loss in efficiency was observed. As function of time after the beginning of a beam spill the loss can be parameterized by the sum of an exponential and a constant. To understand this effect, we developed an analytical calculation based on the assumption that the loss of efficiency is due to the effective voltage drop caused by the current through the chamber. The calculation is found to reproduce the observed functional form of the efficiency drop as a function of spill time. Figure 2 compares the measured efficiency at the end of the spill with the results of the calculation. Good agreement is observed.

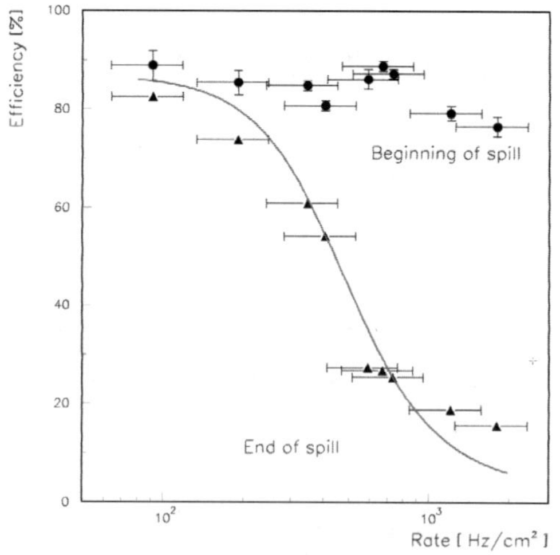

Figure 2: MIP detection efficiency at the beginning and at the end of a spill of 120 GeV protons. The red line is the result of an analytical calculation of the efficiency at the end of the spill.

4. Tests of the calorimeter prototype

The tests included six chambers interleaved with absorber plates containing a combination of a 16 mm thick stainless steel and a 4 mm thick copper plate. The stack was exposed to a broadband muon beam and to positrons and pions in the energy range of 1 – 16 GeV from the Fermilab test beam.

4.1. Simulation strategy

The set-up was simulated with a GEANT4 based program linked to a detailed simulation of the response of RPCs by a standalone program called RPC_sim [13]. The RPC_sim program generates a signal charge Q, distributes this charge over the pads, sums up all charges on a given pad, and applies a threshold T to identify the pads with hits. The response is controlled by four parameters of which three were tuned to describe the response to muons and the forth was tuned to describe the 8 GeV positron data. The pion data is subsequently simulated without further adjustment of these performance parameters.

4.2. Measurement with Muons

Broadband muons were obtained from the 120 GeV primary proton beam and a 3 m beam blocker. The muons were used to measure the average efficiency and pad multiplicity of the chambers under varying operating conditions (high voltage and threshold setting) [14]. Figure 3 shows the measured average pad multiplicity as a function of efficiency. The latter was varied applying different high voltage and threshold settings. Note, that the 1-glass design exhibits a constant value of the pad multiplicity close to unity and independent of the efficiency.

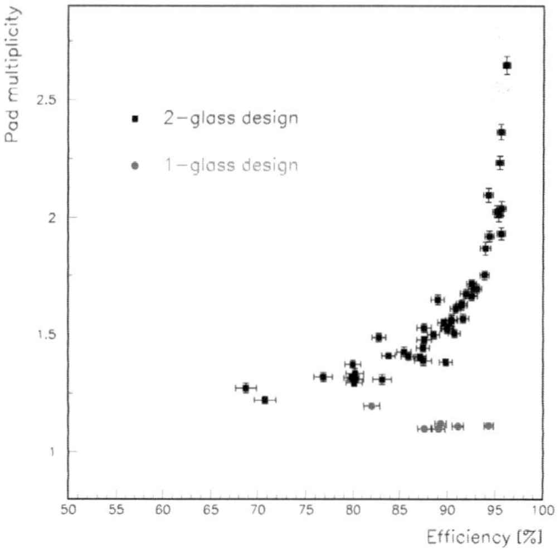

Figure 3: Average pad multiplicity versus efficiency for 2-glass (black dots) chambers and the 1-glass (red dots) chamber.

4.3. Measurement with Positrons

Positron events were obtained with the secondary beam and the requirement of a signal in one of the two upstream Cerenkov counters. Data were collected at the 1, 2, 4, 8, and 16 GeV/c momentum sedttings. Figure 4 shows the distribution of the total number of hits in the calorimeter for the different energy settings and in comparison with the GEANT4 based simulation. Reasonable agreement between data and simulation is observed [13].

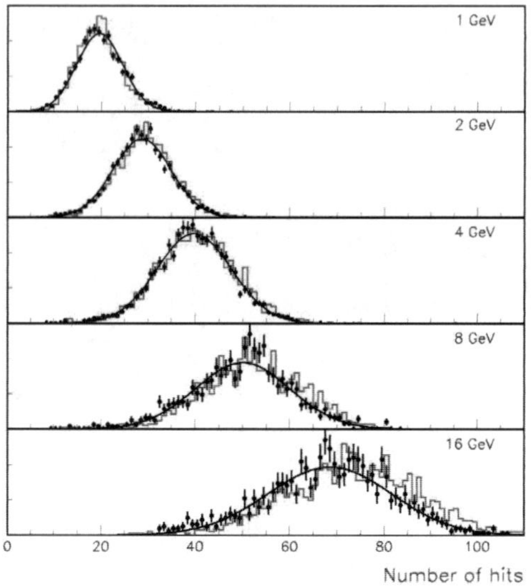

Figure 4: Distribution of the total number of hits in positron induced showers: data in black and simulation in red.

4.4. Measurements with Pions

Pion events were collected using the secondary beam and vetoing on signals from either of the upstream Cerenkov counters. Figure 5 shows the distribution of the number of hits for events selected to have initiated showering between the 1st and 2nd layer. The distributions were fit to two components: positron contamination and pion induced events as obtained from the GEANT4 based simulation. The parameters of the RPC simulation were obtained from the muon and positron shower data and no further tuning was applied to describe the pion data. Reasonable agreement between data and simulation is observed [15].

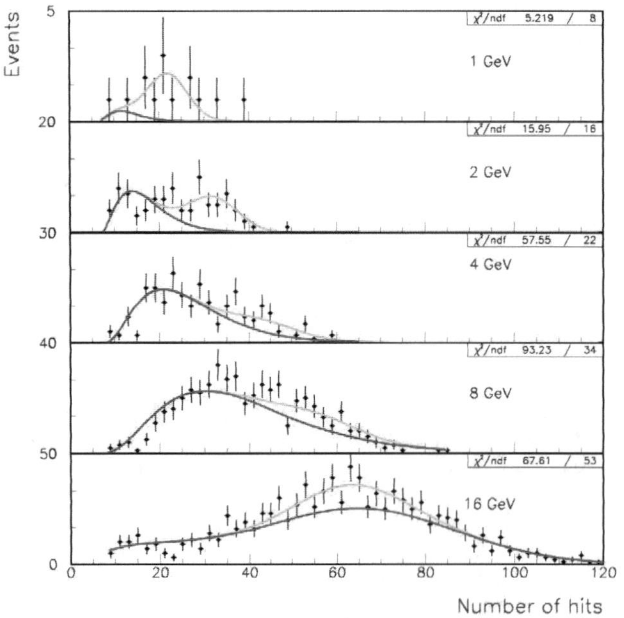

Figure 5: Distribution of the total number of hits in pion induced showers: data in black fit to the sum of two components (positron contamination in green and pion signal in red).

5. Simulation of the performance of a larger prototype calorimeter

Encouraged by the successful simulation of the pion shower data, the simulation tools were utilized to predict the performance of a larger prototype calorimeter with a front-face of 1×1 m^2 and 38 detector layers, corresponding to a depth of about four interaction lengths λ_I. The response was found to be approximately linear for pions with energies up to 25 GeV, Above this energy the response deviates from linearity due to overlapping effects in the 1×1 cm^2 pads. At 60 GeV the response is diminished by about 10 %.

Figure 6 shows the predicted resolution as function of pion energy. A fit of the points up to 28 GeV yields a stochastic term of the order of 58% with a vanishing constant term.

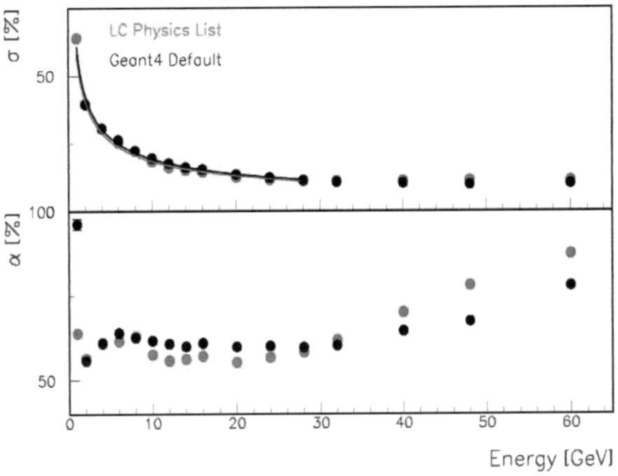

Figure 6: Top: Relative width obtained from Gaussian fits to the distribution of the number of hits versus pion energy. The lines are fits to the quadratic sum of a stochastic and a constant term. The simulations are based on the LC Physics list (magenta dots) and the Fast and Simple Physics list (black dots). Bottom: Stochastic terms (for energies up to 20 GeV) or Gaussian widths multiplied by √E for energies above 20 GeV.

6. Acknowledgements

The author would like to thank the organizers for an impeccably organized workshop and for the opportunity to present the work of the DHCAL collaboration to the calorimeter R&D community.

References

[1] M.A. Thomson, Particle Flow Calorimetry and the PandoraPFA Algorithm, Nucl. Instrum. Meth, A611:25 (2009) and references therein.
[2] J. Repond, Calorimetry at the International Linear Collider, Proceedings of the 10th Pisa Meeting on Advanced Detectors, La Biodola, Isola d'Elba, Italy, Nucl. Instrum. Meth, A572, 211 (2007).
[3] G. Drake et al., Resistive Plate Chambers for Hadron Calorimetry : Tests with Analog Readout, Nucl. Instrum. Meth, **A578**, 88-97 (2007).
[4] J. Butler et al., Prepared for the 2007 IEEE Nuclear Science Symposium, Honolulu, Hawaii, Oct. 27 – Nov. 3, 2007. Submitted to IEEE Trans. Nucl. Sci.
[5] J. Hoff et al., A custom integrated circuit for calorimetry at the International Linear Collider, IEEE Nucl. Sci. Symp., Puerto Rico, 2005, FERMILAB-CONF-05-509.
[6] http://www.hep.anl.gov/repond/DHCAL_US.html
[7] https://twiki.cern.ch/twiki/bin/view/CALICE/WebHome
[8] L. Xia, Construction of a Digital Hadron Calorimeter Prototype, these proceedings.
[9] http://www-ppd.fnal.gov/MTBF-w/
[10] Q. Zhang et al., Environmnetal Dependence of the Performance of Resistive Plate Chambers, 2010 JINST 5 P02007.
[11] Q. Zhang, R&D for a Digital Hadron Calorimeter Prototype with Glass RPCs, PhD thesis (2010), unpublished.
[12] B. Bilki et al., Measurement of the Rate Capability of Resistive Plate Chambers, 2009 JINST 4 P06003.
[13] B. Bilki et al., Measurement of Positron Showers with a Digital Hadron Calorimeter, 2009 JINST 4 P04006.
[14] B. Bilki et al., Calibration of a Digital Hadron Calorimeter with Muons, 2008 JINST 3 P05001.
[15] B. Bilki et al., Hadron Showers in a Digital Hadron Calorimeter, 2009 JINST 4 P10008.

Journal of Physics: Conference Series **293** (2011) 012076 doi:10.1088/1742-6596/293/1/012076

Test beam studies for a highly granular GRPC Semi-Digital HCAL

Vincent Boudry, for the CALICE collaboration
Laboratoire Leprince-Ringuet - École polytechnique, CNRS/IN2P3
91128 Palaiseau Cedex – France
E-mail: `Vincent.Boudry@in2p3.fr`

Abstract. The Particle Flow Analysis approach retained for the future ILC detectors requires high granularity and compact particle energy deposition. A Glass Resistive Plate Chamber based Semi-Digital calorimeter can offer both at a low price for the hadronic section. This paper presents some recent developments and results near test beam in the use of Glass Resistive Plate Chamber with embedded front-end electronics to build a prototype based on this principle. All the critical parameters such as the spatial and angular uniformity of the response as well as the noise level have been measured on small chambers and found to be appropriate. Small semi-conductive chambers allowing for high rates and a large chamber have also been tested.

1. Case for a GRPC Semi-Digital Gaseous Hadronic Calorimetry

The Particle Flow Analysis (PFA) approach [1], retained for all the detector concepts in development for the future International Linear Collider (ILC), requires calorimetric systems able to spatially separate the contributions from the individual particles constitutive of a jet. The key feature of calorimeters then shifts from the best energetic resolution in response to individual particle to the best spatial resolution. This paradigm is pushed at its limit in the case of digital calorimetry, where a single bit summarises the energy information for each cell.

A digital hadronic calorimetry has long been discussed [2, 3]. Some recent GEANT4 [4, 5] based studies on a Scintillator-Lead calorimeter [6] have shown the feasibility, using 3 thresholds ("two bits") and cell lateral sizes of $1 \times 1\,\mathrm{cm}^2$, to reconstruct the energy of single charged pions, up to 100 GeV, with an energy resolution comparable to the one obtained using an analog readout ("many bits").

The use of a gaseous, hydrogen poor, planar sensitive medium presents many advantages for a digital calorimetry. The main ones are:

- their insensitiveness to the dispersed, delayed and usually relatively badly measured neutron component of hadronic showers, resulting in narrower shower: typically a 100 GeV pion is expected to be contained at 99 % in a stainless-steel gas calorimeter of section $70 \times 70\,\mathrm{cm}^2$ [7].
- the ease to adjust the readout pad size to very small value if desired;
- more specifically for single Resistive Plate Chambers (RPC), the proven possibility to make large surface, robust and reliable sensors at very low cost.

Among the many available techniques usable for large surface gaseous sensors, 3 are now being studied in the CALICE collaboration: MicroMEsh GAseous Structure (MicroMegas) and

Gas Electron Multiplier (GEM) in which the ionisation electron signal is amplified by a mesh after a drift, and GRPC (Glass Resistive Plate Chamber), in which the amplification occurs immediately; this article focuses on the later.

The efficiency (ϵ) and cell multiplicity (μ) in response to minimum ionising particles are the two critical parameters required for the intercalibration of the channels of a (semi)digital calorimeter. They should be supplemented by the understanding of the response to a full hadronic shower, which might affect the sensor itself or the embedded electronics. Finally, although this should not be a problem at the ILC, the recovery time for various spatial scales and intensity, have to be measured, as well as the sensitivity to high magnetic fields.

Technologically, the use of RPC in a large hadronic calorimeter faces a few challenges, which should be integrated as early as possible in the conception. The first is the industrial production and handling of large surface, uniform sensors at low cost. To keep compact showers, the ratio of low density material to absorber should minimised, requiring thin sensitive and readout layer. Practically, the goal is to reach 6 mm all inclusive (RPC and electronics) for 20 mm stainless steel absorber.

Such devices have been produced in various sizes and the present paper describes the test setup in test beam and the results obtained: the aim is to validate the whole embedded electronics readout system in beam conditions (including the use of local data storage and train reconstruction), to study the GRPC behaviour as a function of the applied HV and thresholds, of incoming particle angle, position, rate and to validate the technological choices prior to the construction of a full scale 1 m^3 prototype [8].

2. Glass RPC prototypes with an embedded readout

The results shown here were obtained with various prototypes sharing the same geometry: a gap of 1.2 mm of gas between two glass plate, polarised under a nominal voltage of 7.4 kV. The gas used was a mixture 93 % of TetraFluoroEthane (TFE), 5 % of Isobutane and 2 % of Sulphur HexaFluoride (SF6) yielding 8 primary ionisations per mm of a mip track. The glass plates, respectively of 1100 and 700 μm thickness for the cathode and anode sides, are separated by Nylon fishing lines and covered on their external sides by a resistive coating.

Three types of coating were used: Graphite, Statguard® and Licron® yielding resistivity of respectively 0.4, 2 and 20 MΩ/□.

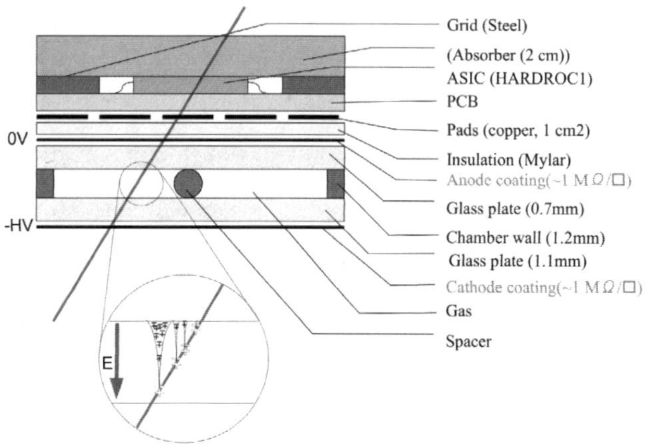

Figure 1. Transversal cut of the GRPC layout and readout electronics.

Small chambers of 8.35×33.55 cm^2 were built using either standard float glass, in combination with the 3 types of coating, or semi-conductive glass with Licron and Statguard coatings. Large chambers of 1×1 m^2 were built out of float glass coated by colloidal graphite (having resistivity of 1-2 MΩ/□).

In order to read efficiently many channels, the very front-end part of the readout electronics is integrated in the detector itself: the readout ASICs and their supporting PCB, are only separated from the anode coating by a 50 μm insulating layer of Mylar®.

With a capacity of 64 channels, the HaRDROC [9] (Hadronic RPC Detector Read Out Chip) specifically developed by the Omega micro-electronics group of Orsay, has 2 independent adjustable thresholds (above 10 fC) used for auto-triggering. It can store up to 127 events composed of an 8-bit ASIC ID (8 bits), a time stamp (24 bits by steps of 200 ns) and the hit map (2×64 bits). Each channel gain can be adjusted over a range 0–4 on 6 bits. Its internal inter-channel cross-talk has been measured to be below 2 %.

Some test boards holding 4 HaRDROC's and their control and readout logic, were produced and tested in 2007. They feature 8-layer, 800 μm thick PCB with buried and blind vias, ensuring cross-talk below 0.3 %, and the readout of 8×32 pads of $1 \times 1\,\mathrm{cm}^2$, separated by 0.5 mm gaps. First used to validate the new electronics and acquisition scheme, they then served for the beam tests equipped with GRPC sensors, as described hereafter.

The relative gain of the individual channels of each ASIC were first adjusted by levelling the threshold responses to 100 fC injected charges. A factor of 4 was gained on the built-in spread of the effective threshold, lowering it to 2.5 fC for a complete board.

For the $1 \times 1\,\mathrm{m}^2$ prototype, a new set of readout electronics have been designed: the test board functions have been split between so-called ASU (Active Sensor Unit) boards holding 24 HaRDROC ASICs each, which can be coupled together at their ends, and a readout and controller card named "DIF" (for Detector InterFace). A $1 \times 1\,\mathrm{m}^2$ chamber is readout by 6 ASU connected to 3 DIF's, for a total of 9216 channels. The acquisition is performed through USB by a software developed in the XDAQ framework.

3. Test beam setup

Up to 4 test boards were equipped with GRPC, of various types, and were used in test beam near the PS and SPS CERN facilities. They were installed perpendicularly to the beam in a structure eventually allowing for the installation of 2 cm-thick stainless steel radiators, to build a mini calorimeter, associated with a set of scintillators for triggering purposes.

Two campaigns of tests were performed in 2008 and 2009 each using low (3–12 GeV) and high (10–150 GeV) energy pions. The results presented here are restricted to the mip-like data sample (no absorbers used).

3.1. Acquisition modes

The on-board FPGA's were programmed to manage two acquisition modes relying on the auto-trigger feature of the HaRDROC's:

a train mode ("ILC like"), where the readout is started on the start-of-spill signal and stopped on the end-of-spill signal or when one of the ASIC is full;

a triggered mode used for cosmic rays and test beam data taking, where the acquisition is stopped on an external trigger. When the memory one of the ASIC's is full, a busy signal is emitted, the entire board reset and the acquisition resumed.

In both cases when the acquisition is stopped, a train of up-to 127 events is readout.

Using the coincidence of plastic scintillator plates readout by photo-multipliers as an external trigger a total of about 1,130 million triggered events were collected, at an average DAQ rate of ~ 20 Hz.

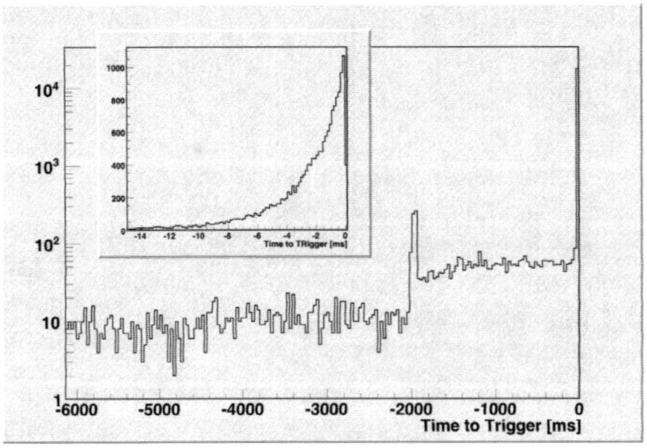

Figure 2. Time-to-trigger distribution; the events associated with the trigger are in the last 200 ns (right peak), in-spill events in the last 20 ms (detail in insert) while the previous spill is visible at -2000 ms. All events in between are considered as being noise.

3.2. Timing

At PS, 2 spills of 400 ms were received, at 2 second interval, every 48 s during the day (every 33 s during the night).

The synchronisation between the boards was ensured by a 40 MHz counter recording the time difference for each board between the last internal trigger and the external trigger. This allows to resynchronise of each board w.r.t. the trigger signal and to reconstruct the history of all boards back to the last memory reset. In most cases (low noise and particle density), the last event in memory of the ASIC in beam corresponds to the signal of the triggering particle.

By studying the correlation in time between hits of different cards, it has been observed that their clock's dispersion is small enough to allow a tight cut of 300 ns (w.r.t. their 200 ns period) to cluster hits in individual events due to a particle or noise.

This time reconstruction allows the use of events from memory which were not triggered by the scintillators in the beam. Such a time reconstruction is shown in figure 2. In particular the noise pattern can be separated from in-spill events, dominated by particles.

4. Small chambers results

4.1. Efficiency & multiplicity studies

Using the 3 other chambers as a crude tracking device one can determine the position of the particle in the 4th one and calculate the efficiency and multiplicity locally. The multiplicity μ is defined as the number of fired cells within 3 cm-radius around the impact when the chamber has responded ($\mu \geq 1$)

For the following studies one of the two following selection was applied :

- using only the last event in memory (triggering one), with $t > -400$ ns
- using the complete train statistics, in which case only the hits in time ($|\Delta t| <$ 300 ns and after the most recent board reset ($t >$ time of first event in every chamber) were used.

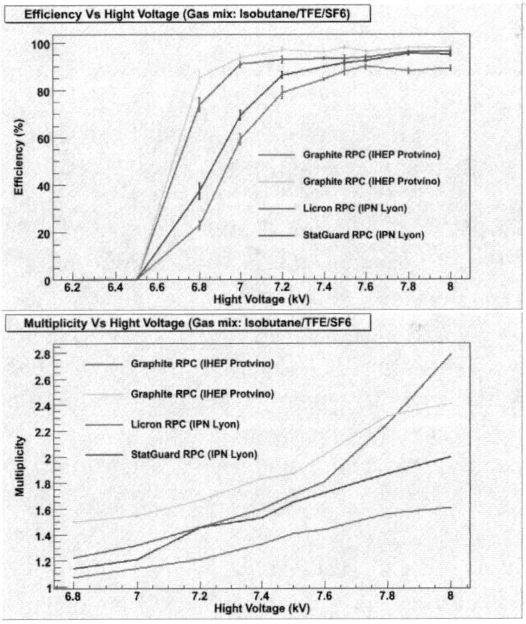

Figure 3. Efficiency and multiplicity vs HV for the 4 small float glass chambers.

A high voltage scan was performed with a threshold set to 165 fC with the different small chambers based on float glass (2 identical "Graphite", 1 "Licron" and 1 "Statguard"). Using the triggered event sample it was found that a maximum efficiency of about 98 % was reached for all chambers but for the Licron one saturating at $\sim$ 90 %. As can be seen on figure 4 the efficiency and multiplicity were found to rise faster for the graphite covered chambers w.r.t. the Licron and Statguard ones. An efficiency plateau is reached for all chambers around 7.2 kV, while the multiplicity rises continuously with the HV.

A working point of 7.4 kV (for a threshold of 165 fC) was chosen as standard for the following studies, being seen as a good compromise between high efficiency and low multiplicity. At this point the multiplicities are of 1.3 for the "Licron" chamber and $\sim$ 1.6 for the "Statguard" and "Graphite" ones.

Threshold scan: The efficiencies and multiplicities were measured as the threshold applied was varied around 165 fC, and found to behave as expected, as the inverse of the HV; the efficiency was found to be 80 % at 7.4 kV for a threshold of 1.1 pC. This will be used to model the response of the chambers.

Chamber uniformity: The response (ϵ, μ) of 4 similar small graphite chambers produced by the IHEP at Protvino were studied and found to have identical behaviour with a spread of the efficiency of 95 ± 3 % and multiplicity of 1.6 ± 0.4.

Using the train sample reconstruction individual response of the $1 \times 1\,\text{cm}^2$ cells could be studied with a good statistics. A global efficiency spread, including the statistics error for about 25k events, of 3 % was observed inside single chambers and between different chambers. The multiplicity spread in a chamber is of ~ 0.2. The main sources of variation (lower ϵ and μ) are due to the fishing lines and the chamber edges as can clearly be seen on figure 4. Some inefficiencies were also clearly observed at the centre of the beam spot for high rates run ($19\,\text{kHz/cm}^2$).

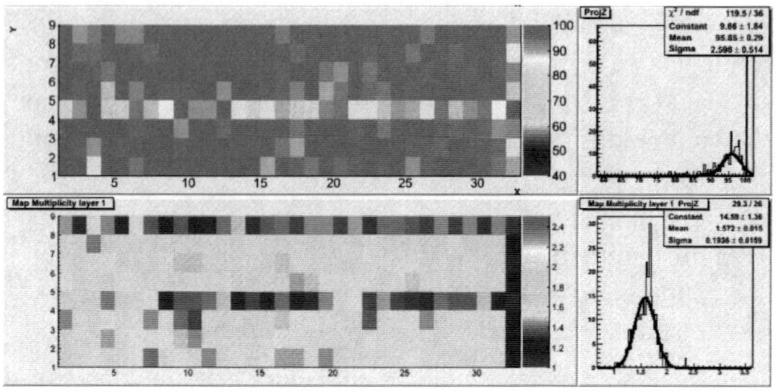

Figure 4. Efficiency (top) and multiplicity (bottom) spatial distribution on two chambers with $1\,\text{cm}^2$ resolution (left) and corresponding values distribution (right).

In the future prototypes, the fishing lines will be replaced by ceramic ball, reducing the size of the affected region by a factor of 100.

Angular response & stability: No observable difference was seen while shooting with an angle up to 60 °. This will ease the reconstruction code for large detectors.

The global behaviour was found to be constant over the time of the running (10 days), with a number of dead or inefficient cells limited to 1 %.

4.2. Noise studies

Observing only out-of-spill events and out-of-trigger cosmic data, one has access to the noise distribution. The noise rate was studied and found to be higher near the special regions (especially fishing lines and to a lesser extend borders), with rates typically of $20\,\text{Hz/cm}^2$ for our standard threshold (165 fC) going down to 10 times less when the threshold is raised at 750 fC.

In the bulk, the noise rate was found to be of $0.1\,\text{Hz/cm}^2$, with no dependence on the applied threshold (between 165 and 750 fC). It evolves slowly with HV with a slope roughly estimated $\sim 0.3\,\text{Hz/cm}^2/\text{kV}$ around our working point.

4.3. Semi-conductive GRPC's

Two small chambers provided by Tsinghua University and based on semi-conductive glass ($10^{10}\,\Omega/\text{cm}$ against $10^{13}\,\Omega/\text{cm}$ for classical float glass) have been tested at high rates. A "thin" version features glass thickness of 0.83 mm on the anode side and 1.1 mm on the cathode one, coated with Statguard. A "thick" version uses 1.1 mm glass on both sides, with a Licron coating.

As can be seen on figure 5 an efficiency curve similar to the standard chambers could be achieved, with efficiency reaching a plateau at 90 % above 7.4 kV, but here with rates above 10 to $30\,\text{kHz/cm}^2$, when the standard chambers start to become inefficient above 10–100 Hz.

5. First results on a 1 m^2 prototype

The square meter prototype was put in beam at the SPS. The acquisition worked as expected and was able to read the 9216 channels; the beam spot was observed and efficiencies of up to 93 % have been obtained. But the setup rapidly suffered from faulty mechanics on the HV connection and these results will have to be confirmed.

On a cosmic test bench, using the mini-DHCAL as a tracker device, cosmic tracks were collected. For analysis purpose, the surface was divided as a checker of 9 regions of $33 \times 33\,\mathrm{cm}^2$. One of the regions failed during this test. Between the 8 other working regions a good uniformity of response was observed with results similar to the ones of the small chambers (working HV $\sim$ 7.4 kV, $\epsilon \sim 95\,\%$ and $\mu \sim 1.6$).

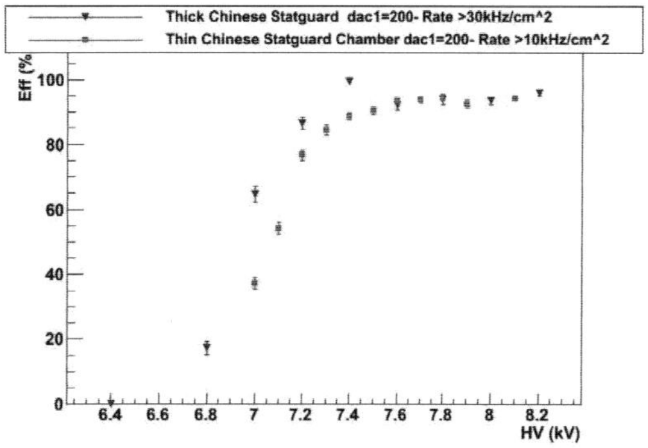

Figure 5. Efficiency of thick (blue triangles) and thin (red squares) semi-conductive glass at respectively average rates of $\gtrsim$ 30 and $\gtrsim$ 10 kHz/cm^2, a threshold of 165 fC and a voltage of 7.4 kV.

Conclusions & perspectives

A Semi-Digital Gas Hadronic Calorimeter with embedded readout is a very promising candidate for future linear collider experiments. The main critical parameters have been checked at beam tests on float glass small chambers prototypes: in response to single minimum ionising particle a high efficiency ($\epsilon \sim 95\,\%$), low multiplicity ($\mu \sim 1.6$), no angular dependency, spatial and detector uniformity ($\sigma_\epsilon \lesssim 2\,\%$, $\sigma_\mu \lesssim 3\%$ over 256 cells of $1\,\mathrm{cm}^2$) and low noise have been measured. Some prototypes using semi-conductive glass showed some promising performances for high rates ($\gtrsim$ 10 kHz). The first measures made on a square meter sensor similar to the one foreseen for a technological prototype of $1\,\mathrm{m}^3$ (including 40 of these) were found to be similarly satisfying. The full prototype is expected to be completed in 2010.

Acknowledgements

The work on the electronics and the DAQ is supported by the Commission of the European Communities under the 6th Framework Programme Structuring the European Research Area, contract number RII3-026126.

References

[1] Brient J C and Videau H 2002 *Proc. of APS / DPF / DPB Summer Study on the Future of Particle Physics, Snowmass, Co, 30 Jun - 21 Jul 2001, pp E3047., preprint* **hep-ex/0202004**
[2] Ammosov V (CALICE) 2002 *Nucl. Instrum. Meth.* **A494** 355–361
[3] Ammosov V *et al.* 2004 *Nucl. Instrum. Meth.* **A533** 130–138
[4] Agostinelli S *et al.* (GEANT4) 2003 *Nucl. Instrum. Meth.* **A506** 250–303
[5] Allison J *et al.* 2006 *IEEE Trans. Nucl. Sci.* **53** 270
[6] Matsunaga H 2007 *Pramana* **69** 1057–1061
[7] Brient J C 2004 A DAQ readout for the digital HCAL, *LC-DET-2004-029*
[8] Laktineh I 2010 Construction of a semi-digital HCAL prototype based on GRPCs *These proceedings*
[9] Callier S *et al.* 2007 *Nuclear Science Symposium Conference Record, 2007. NSS '07. IEEE* **3** 1851–1856

Construction of a technological semi-digital hadronic calorimeter using GRPC

I.Laktineh

Université de Lyon, Université de Lyon 1, CNRS/IN2P3, IPNL 4 rue E. Fermi 69622 Villeurbanne cedex

Laktineh@in2p3.fr

Abstract. A high-granularity semi-digital Hadronic calorimeter using GRPC as sensitive medium is one of the two HCAL options considered by the ILD collaboration to be proposed for the detector of the future International Linear Collider project. A prototype of $1m^3$ has been conceived within the CALICE collaboration in order to validate this option. The prototype intends to be as close as possible to the one proposed in the ILD Letter Of Intent. Few units made of $1m^2$ GRPC fully equipped with semi-digital readout electronics and new gas distribution design were produced and successfully tested. In 2010 we intend to produce 40 similar units to be inserted in a self-supporting mechanical structure. The prototype will then be exposed to TestBeams at CERN for final validation.

1. Introduction

As was explained in a previous paper [1], the use of high-granularity hadronic calorimeter is an essential ingredient for an efficient exploitation of the promising Particle Flow Algorithms technique [2]. Two kinds of such hadronic calorimeters were proposed in the ILD letter of Intent [3]. One of the two proposed options is based on using Glass Resistive Plate Chambers as the sensitive medium and stainless steel plates for absorber.

The use of gaseous detectors such as the GRPC ensures a good homogeneity of the sensitive medium. They should however, as for the scintillator-based option, satisfy strong constraints. In addition to be very efficient the GRPCs should be as thin as possible in order to reduce the construction cost of the whole detector. Indeed, the HCAL should be placed inside the magnetic field, reducing its thickness will hence reduce the radius of the magnet coil which is an important element of the detector cost. The GRPCs design should also provide negligible dead zones in order to keep the tracking capability as high as possible.

The high granularity is provided by the readout electronics system. The signal created by the passage of charged particles in GRPC detector is collected thanks to $1cm^2$ pads etched on one side of the electronics board which is put in contact with the GRPCs. A pad is read out by one of the 64 channels of the HARDROC ASICs [4] fixed on the other side of the electronics board. The HARDROC ASICs are connected with each other through the electronics board using the DAISY chain scheme. This results in limited number of connections coming out of the sensitive medium. The ASIC provides 2-bit readout. The choice of this semi-digital scheme rather than the binary one was motivated by simulation studies which show that better performance can be obtained by the semi-digital scheme at high energy as shown in Fig. 1. This result can be explained by the fact that at high energy many particles may go through one pad especially in the centre of the hadronic shower. Since the extension of the avalanche created by charged particle in GRPC is of the order of few mm^2 [5] the semi-digital readout with its three thresholds can help to distinguish among scenarios with few, many and too many particles going through one pad. This improves on the energy measurement resolution by estimating better the number of produced charged particles in the hadronic shower.

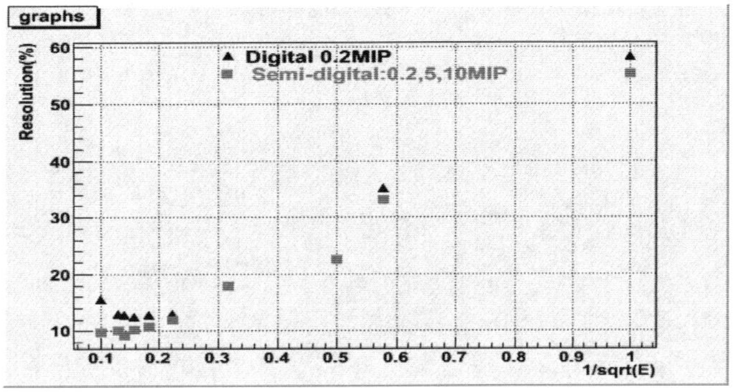

Figure1: Energy resolution of hadronic shower created by pions with binary readout (black triangles) and semi-digital readout (red squares).

In order to reduce the power consumption the HARDROC is equipped with a power pulsing scheme which reduces the consumption by a factor of 100 when applied in the envisaged ILC duty cycle.

Results obtained with small GRPCs equipped with the semi-digital readout electronics system [6] have allowed one to envisage building large GRPC chambers equipped with the same electronics.

In this paper we will describe the development realized to build $1m^2$ GRPCs satisfying the requirements mentioned above as well as the associated electronics boards. The cassette to assemble them and the mechanical structure to host the 40 envisaged ones will be detailed. The services and the acquisition system of this prototype will be described. Finally the software and analyses preparation to exploit the data of the future prototype are briefly mentioned.

2. Large GRPC development

The chambers to be used for the prototype are made of two glass plates of 0.7 mm and 1.1 mm thickness. Precision ceramic balls of diameter 1.2 mm are used as spacers to separate the glass plates. The number of balls (81) was fixed so that the deviation of the gap distance between the two plates under the glass weight and the electric force when the high voltage is applied on the two electrodes does not exceed 40 microns. The gas volume is closed by a glass fiber frame.

The resistive coating on the glass plates which is used to apply the high voltage and thus to create the electric field in the gas volume was found to play important role in the pad multiplicity associated to a mip [6]. To find the best coating for our chambers considerable efforts were made. Commercial products like Licron® and Statguard® which are used for ESD applications were used and few large chambers using those products were built and intensively tested. Both products failed to satisfy our application either for long term stability under the high voltage (Licron®) or due to the impossibility to obtain the surface uniformity needed for our application (Statguard®). Finally two products were identified, both of which are based on colloids containing graphite and both can be applied using the silk screen print method which ensures very uniform surface quality. One of these products is a single component paint with a dry surface resistivity of 1-10 MΩ/ . The second product comes as two components which must be mixed by the user. The surface resistivity may be adjusted over a wide range by changing the mix ratio. Both products require baking at around 170° C to attain a stable surface resistivity. The measured surface resistivity at various points over a 1 m^2 glass coated with the bi-component paint are shown in Fig. 2. The mean value is 1.2 MΩ/ and the ratio of the maximum to minimum values is 1.8. A study was also made of the repeatability of the surface resistivity between different mix batches. It was found that surface resistivity in the range 1-2 MΩ/ could be reliably reproduced, again with a factor of approximately two.

For large GRPCs the painting is applied on the whole glass plate except the zones situated 3 mm from the edges. This distance was optimized so the dead zone of the detector is reduced while external sparks due to the presence of absorber in the vicinity is completely eliminated.

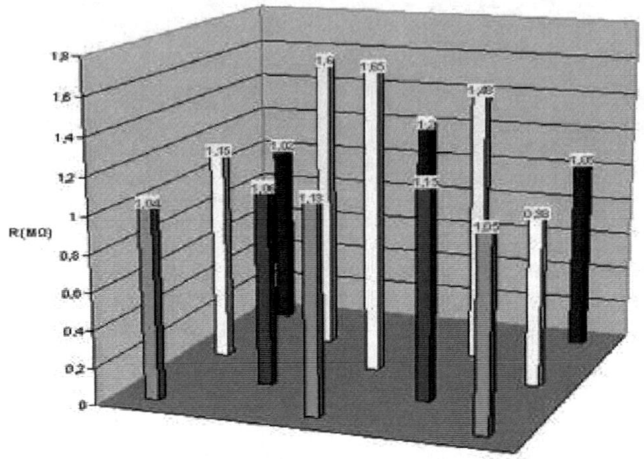

Figure 2: Surface resistivity for 1m^2 glass coated with bi-component colloidal graphite.

Another important aspect of this development concerns the gas distribution within the GRPC. In order to improve on the gas distribution in large chambers taking into account the requirement that both gas outlets should be on the same side of the detector, new schemes were studied. The one we finally adopted allows us to improve the gas distribution by channelling the gas along one side of the chamber and releasing it into the main gas volume at regular intervals. A similar system is used to collect the gas at the other side of the chamber. A finite element model has been established to verify the gas distribution [7]. The simulation confirms that the gas speed is reasonably uniform over most of the chamber area and that this design significantly improves the distribution of gas with respect to a chamber with no gas channels.

3. Readout electronics board

As described in the introduction the signal produced by charged particles crossing GRPC can be collected thanks to 1 cm^2 pads on one side of the electronics board. The signal is then funnelled out to the ASICs fixed on the other side of the board. The board is made of 1.2 mm thick, 8-layer Printed Circuit Board (PCB) which was designed to reduce the noise as well as the cross-talk among the different channels. This kind of PCB is currently limited in size so the 1m^2 area needed to read out the 1m^2 GRPC is realized using six PCBs of area 50 x 33.3 cm^2. The PCBs are linked together using special low-profile flexible connectors. Each of the six PCBs supports 24 ASICs, The PCBs are regrouped by two and connected to a small board called Detector InterFace board (DIF). The DIF which hosts FPGA device controls the transmission of configuration parameters to the 48 ASICs and the reception of data collected from them. The DIFs are connected to the electronics board through flexible Kapton® connectors.

4. Cassette description

The GRPC and its associated electronics are housed in a special cassette which protects the chamber and ensures that the readout board is in intimate contact with the anode glass (see Fig. 3). The cassette is a thin box consisting of 2 mm and 3 mm thick stainless steel plates separated by 6 mm wide stainless steel spacers which form the walls of the box. These spacers are precision machined so no space left between the two cassette plates and the sensitive medium. The dimensions of the plates are such that a space of 2 mm wide is left between the GRPC and the spacers. This space is filled with insulating product to protect against sparks. One of the two plates is 20 cm larger than the other. This allows one to fix the three DIFs as well as the gas outlets and the high voltage box.

The electronics board is assembled thanks to a polycarbonate spacer ('PCB support' in Fig. 3) which is also used to fill the gaps between the readout chips and to improve the overall rigidity of the detector. The board is then fixed on the small plate thanks to tiny screws and the new set is fixed on the other plate which hosts the detector and the spacers. The whole width of the cassette is 11 mm with only 6 of them corresponding to the sensitive medium including the GRPC detector and the readout electronics. The remaining 5 mm are those of the stainless steel plates and constitute a part of the HCAL absorber.

One fully equipped cassette was produced and tested in TestBeam at CERN recently. Efficiencies of 96±2 % were measured in many different zones of the 1m^2 including those where the electronics boards are connected. This validates the cassette concept which ensures the homogeneity of the whole sensitive medium made of the detector and the readout electronics. Other cassettes are being built and tested. These obtained results allowed us to envisage the construction of the 40 unites needed for the prototype.

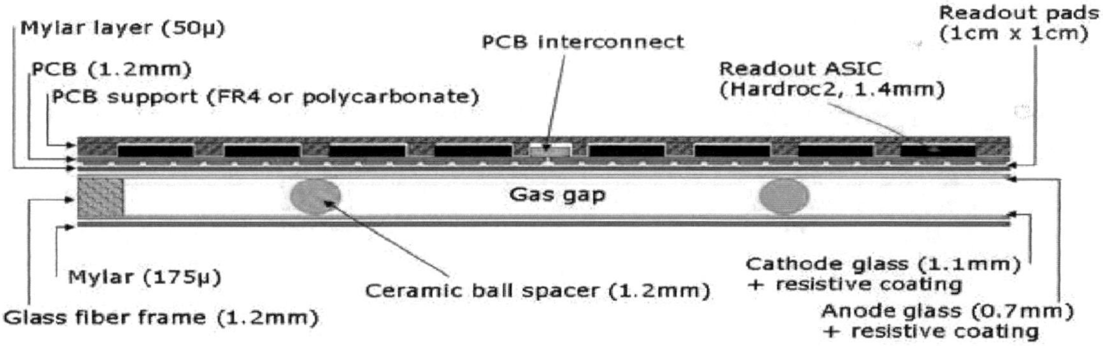

Figure 3: Cross-section through a 1 m2 chamber.

5. Mechanical structure

The prototype will be made of 40 of the previous cassettes separated by 39 stainless steel plates of thickness 1.5 cm each. With the 5 mm of each of the 40 cassettes, the total weight will exceed 6 tones. A self-supporting mechanical structure was conceived in which the absorber plates are part of the structure itself. This leads to a very robust structure. To achieve this, precision spacers of 13 X 18 mm^2 sections each will be used to fix two consecutive plates on three of the their four edges thanks to well distributed bolts. The extra 2 mm with respect to the cassette width accounts largely for the fabrication tolerance of the plates and the cassettes. This was shown from real insertion tests of the cassette in a mechanical structure made of two plates. The tests showed also that the insertion of the cassette from above is the most suitable way.

The advantage of this structure in addition to its robustness is its flexibility. Additional units can be added in the future if one wishes to increase the equivalent number of the interaction lengths.

Cassettes made of different detection technologies (MICROMEGAS, GEM…) can also be used in this structure.

6. Prototype thermal modelling

The technological prototype will feature a high number of electronics channels, the heat from which must be evacuated. Power pulsing of the electronics will in principle ensure that the power per channel is extremely low (<10 μW); nevertheless a thermal model has been developed to study the worst case scenario of no power pulsing. In this scenario, the power consumption is a factor 100 higher. Two detector planes with electronics and three absorber layers were modelled. Based on symmetry arguments, it was considered sufficient to simulate only one quarter of the full area of each layer. Cooling was assumed to be entirely passive, i.e. by convection only, for an ambient temperature of 20° C. Under these conditions, the model predicts a maximum rise in temperature of 5° C. Active cooling is therefore considered unnecessary in the present design.

7. Prototype services

The high voltage system will be based on the Cockcroft-Walton voltage multiplier. Such a system allows low voltage (0-5 V) cables to be used up to the cassettes, greatly reducing the volume of cable insulation inside the detector. This constitutes an important improvement for the cabling issues in the future ILC detector. A suitable module is being developed in collaboration with Iseg Company. The module, one for each RPC, will deliver up to 10 μA with a voltage range of 0-10 kV. In addition, it must be sufficiently low-profile (< 24 mm) to fit between cassettes. Control and monitoring of voltages will be available via Ethernet link.

A gas mixture based on tetrafluoroethane (C2H2F4) will be used, with SF6 and isobutane quenchers. A 40-channel gas distribution system has been ordered. Precision mass flow meters are used to regulate the flow of each gas component to within ±2%. The gases are mixed and then split off into 40 parallel outlets. The flow from each outlet can be independently regulated and monitored using rotameters. Each of the 40 return lines is equipped with a bubbler. The system is certified for use in flammable environments to ATEX zone 2 level.

8. Prototype acquisition system

A central acquisition system based on a common scheme adopted by the CALICE collaboration for the different technological prototypes is under construction for the Semi-Digital hadronic one. The acquisition system is based on an Off-Detector Receiver (ODR) which is a PC interface. The ODR communicates with few Link/Data Aggregator boards (LDA). Those boards play a central role. They allow to transmit the configuration parameters to the appropriate DIFs and to transmit the collected data from the previous to the ODR after sorting them out. In order to reduce the needed number of LDA a Data Concentrator Card was developed in the case of the SDHCAL. Each DCC receives the data from 9 DIFs (corresponding to 3 cassettes) and compresses them before transmitting them to the LDA. Using 14 of the DCC limits the needed number of LDA boards to two for the entire SDHCAL prototype. In order to synchronise the 120 DIFs a Clock and Control Card (CCC) was also developed. It provides a common clock to the different DIFs through the LDA-DCC. It allows also trigger distribution as well as Busy signal handling. Tests are ongoing in order to validate the whole system.

9. Software preparation

A GEANT4-based simulation of the SDHCAL prototype was elaborated. Realistic description of the sensitive medium taking into account the details of the different materials and the dead zones was adopted. To describe the GRPCs response adequately a sophisticated treatment using the avalanche description based on Polya function approach was used and its parameters deduced from a dedicated study of GRPCs realized with analogue readout electronics. In addition the pads multiplicity was correctly introduced in the simulation through a model using the distance between the incoming particle impact and the edge of the associated pad. The model parameters were fixed in order to reproduce pad multiplicity obtained from TestBeam data. This realistic description will enable us to perform efficiently comparisons between the observed data and the different hadronic shower models used in the simulation in the future.

10. Analyses preparation

In order to be able to exploit the data delivered by the future SDHCAL prototype, many analyses are ongoing. They are trying to get use of the high granularity of the SDHCAL as well as the 2-bit read-out system. Methods based on Neural Network are being developed in order to improve on the energy resolution of the hadronic shower energy. They provide important information which allows us to fix the values of the three thresholds associated to the semi-digital readout. Tracking methods using the Hough Transform to extract mips from the hadronic shower so they can be used for alignment and online control of the detector are being elaborated. Minimum Spanning Tree algorithms are also being investigated in order to separate efficiently the contribution of two particles inside the SDHCAL prototype before to measure their energy separately. This will constitute an important element to test the PFA algorithms

11. Conclusion

An important progress has been achieved toward the construction of a Semi-Digital Hadronic Calorimeter. First cassettes satisfying the requirement of efficiency, hermeticity and thickness were built and successfully tested. Services of the future prototype are under construction. A self-supporting mechanical structure is conceived and being built. Acquisition system capable of dealing with 368640 channels is well advanced and tests to validate it are ongoing. In addition, software and analyses activities are followed actively in order to be ready to exploit data from the prototype whose completion is expected by the end of 2010.

References

[1] Semi-Digital hadronic calorimeter for future high energy physics experiments , Imad Laktineh 2009 *J. Phys.: Conf. Ser.* **160** 012030.
[2] J. C. Brient, Improving the jet reconstruction with the particle flow method: An introduction, Prepared for the 11th International Conference on Calorimetry in High-Energy Physics (Calor 2004), Perugia, Italy, 28 Mar - 2 Apr 2004.PFA.
[3] ILD Concept Group, The International Large Detector - Letter of Intent, DESY-09-087, http://www.ilcild.org/documents/ild-letter-of-intent/LOI.pdf.
[4] S.Callier, et al., HARDROC1, Readout chip of the Digital HAdronic CALorimeter of ILC, Nuclear Science Symposium Conference Record, 2007, NSS 07. IEEE.
[5] Yue Qian & al., Measurement of avalanche size and position resolution of RPCs with different surfaces resistivities of the high voltage provider, CPC(HEP & NP), 2010, vol.34, pp 565-570.
[6] See V. Boudry contribution to these proceedings.
[7] Bedjidian et al., Nucl. Instr. and Meth. A, 10.1016/j.nima.2010.02.168.

Journal of Physics: Conference Series **293** (2011) 012078 doi:10.1088/1742-6596/293/1/012078

Micromegas for imaging hadronic calorimetry

C. Adloff, J. Blaha, S. Cap, M. Chefdeville, A. Dalmaz, C. Drancourt, A. Espargilière, R. Gaglione, R. Gallet, N. Geffroy, J. Jacquemier, Y. Karyotakis, F. Peltier, J. Prast and G. Vouters

Laboratoire d'Annecy-le-Vieux de Physique des Particules, Université de Savoie, CNRS/IN2P3, 9 Chemin de Bellevue 74980 Annecy-le-Vieux, France

E-mail: jan.blaha@lapp.in2p3.fr

Abstract. The recent progress in R&D of the Micromegas detectors for hadronic calorimetry including new engineering-technical solutions, electronics development, and accompanying simulation studies with emphasis on the comparison of the physics performance of the analog and digital readout is described. The developed prototypes are with 2 bit digital readout to exploit the Micromegas proportional mode and thus improve the calorimeter linearity. In addition, measurements of detection efficiency, hit multiplicity, and energy shower profiles obtained during the exposure of small size prototypes to radioactive source quanta, cosmic particles and accelerator beams are reported. Eventually, the status of a large scale chamber (1×1 m^2) are also presented with prospective towards the construction of a 1 m^3 digital calorimeter consisting of 40 such chambers.

1. Introduction

Calorimetry for experiments at a future e^+e^- collider will use Particle Flow [1] as a base technique for the physics event reconstruction. By this approach a jet energy resolution lower than 3.8% for 40 – 400 GeV jets can be reached and thus the accurate separation of W and Z hadronic decays will be possible. For optimal performance of this technique, calorimeters with very fine lateral and longitudinal segmentation must be employed. One of the proposed technology is a Micromegas (micro-mesh gaseous structure) detector with 1 cm^2 pads that are readout by embedded digital electronics processing only 1 or 2 bit information per channel. This concept, which is also called digital calorimetry, makes easier the construction of a high granular sampling hadron calorimeter with single shower imaging capability. Feasibility of the concept will be validated by a performance test of a 1 m^3 calorimeter prototype consisting of 40 layers, each equipped with 1 m^2 large Micromegas chamber, and having about 370,000 readout channels in total. Main outputs of the R&D program towards the construction of this 1 m^3 calorimeter prototype are described in what follows.

2. Micromegas basic performance

Several Micromegas chambers have been produced so far by use of the bulk Micromegas technique [2]. Their size varies from 8×8 cm^2 up to 48×32 cm^2. Each chamber consists of a PCB with top conductive layer segmented into 1 cm^2 anode pads, woven mesh supported by insulating pillars 128 μm above the anode plane, 3 mm thick plastic frame which defines the drift region, and 2 mm thick grounded steel cover with a drift copper electrode glued on its internal side. Gas distribution is provided by two holes for input and output in the plastic frame. A detailed description of the prototype design can be found elsewhere [3].

The first prototypes have been equipped with an analog readout which was intended mainly for chamber characterization and definition of necessary parameters for future prototypes with digital readout chips embedded on the external side of the chamber PCB. Three 64-channel ASICs providing digital readout were designed: HARDROC [4], DIRAC [5], and MICROROC [6]. They can operate in a power pulsing mode which is intended for matching the ILC beam time structure and therefore to reduce the power consumption down to $10\,\mu$W/channel. The first two chips have been already used for prototype readout and their performance has been evaluated during test beam campaigns. A design of the last chip, which will equip future chambers, has been completed and its production has just started.

2.1. Prototype characterization using ^{55}Fe X-rays

Fundamental characteristics of the prototype chambers, such as energy resolution, electron collection efficiency, gas gain and signal dependency on the environmental conditions, were measured at laboratory by use of an ^{55}Fe X-ray source. The measurements were performed in Ar/iC_4H_{10} (95/5) and Ar/CO_2 (80/20) gas mixtures and the signals from X-rays were processed by the analog readout.

The energy resolution of the chamber σ_E/E was measured to be 7.5 % at 5.9 keV which corresponds to a FWHM of 17.6 %. The maximum electron collection efficiency is reached for an optimal ratio between amplification and drift electric field. It was measured that the maximum of the signal is obtained for the field ratio about 150–200 for both gas mixtures. At this field ratio, the collection efficiency should be maximum. A maximum gas gain of 4×10^4 and 10^4 has been achieved in Ar/iC_4H_{10} (95/5) and Ar/CO_2 (80/20) gas mixtures, respectively. In addition, the gas gain depends strongly on the amplification gap thickness. It is estimated from test beam data analysis that gap size irregularities are smaller than 3 μm r.m.s.. Finally, the sensitivity of the gas gain to the ambient variables were measured and quantified as $(-1.37 \pm 0.01)\%/K^{-1}$ and $(-0.61 \pm 0.01)\%/mbar^{-1}$ for temperature and atmospheric pressure, respectively. These values correspond well to ones predicted from a gas gain model [7].

2.2. Prototype test in particle beams

During the last years the chambers prototypes were tested in particle beams. The major objectives of these experiments were the functional test of new prototypes in real data taking condition and evaluation of the main chamber properties such as gain uniformity, detection efficiency and hit multiplicity, and study of the prototype response in hadronic and electromagnetic showers. The test beams were carried out at CERN PS and SPS lines where chambers were exposed to low momentum electrons of 2 GeV/c and hadrons up to 10 GeV/c as well as to high momentum muons and pions of 200 GeV/c. All measurements were performed in an Ar/iC_4H_{10} (95/5) gas mixture.

2.2.1. Prototype with analog readout Three chambers with an active area of 16×6 cm^2 and one with 32×12 cm^2, with analog readout were assembled in a stack placed perpendicularly to the beam. An electronics noise of 0.3 fC corresponding to 2000 e^- was measured. The most probable value of the deposited charge is around 22 fC with a variation of 11.3 % over all 672 readout channels. Relatively large dispersion in gain among the channels is mainly due to drift or/and amplification gap non-uniformity. A detection efficiency larger than 97 % was reached at 1.5 fC readout threshold in three of four chambers with an efficiency disparity of less than 1 % per channel. The last one had efficiency around 91 % as a consequence of a higher electronic noise. The efficiency is decreasing with increasing readout threshold as is shown in Fig. 1. The hit multiplicity was measured to be between 1.06 to 1.1 at 1.5 fC readout threshold.

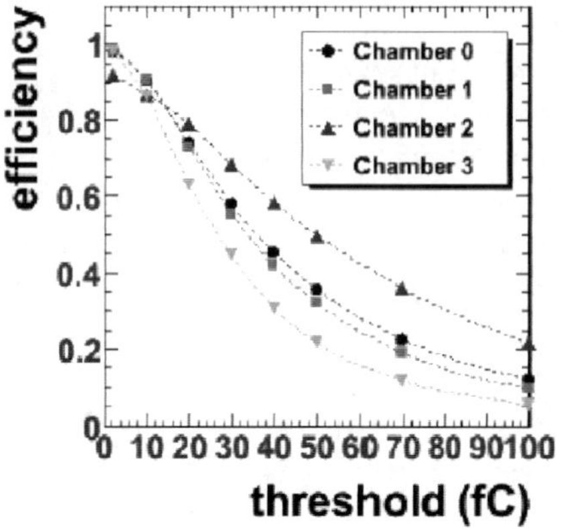

Figure 1. Detection efficiency as a function of the electronics readout threshold.

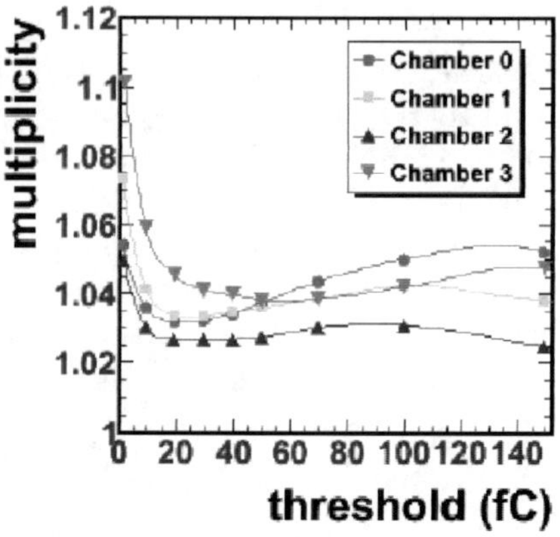

Figure 2. Hit multiplicity as a function of the electronics readout threshold.

Its dependency on the threshold is displayed in Fig. 2, where the raised values of multiplicity above 50 fC are due to the hits from long range δ-rays.

In order to verify that the prototypes also detect multi-hits events, shower profiles have been measured with the 32×12 cm^2 chamber and a variable number of 2 cm thick stainless steel absorber plates placed in front of it. This configuration allows to measure lateral and longitudinal shower profiles and also, by applying a threshold on the recorded charge signals, to compare the deposited energy and number of hits in a chamber. An example of the measured longitudinal shower profile for 2 GeV/c electrons and its comparison with Geant4 Monte Carlo data is shown in Fig. 3 and more details on this study are available in Ref. [8].

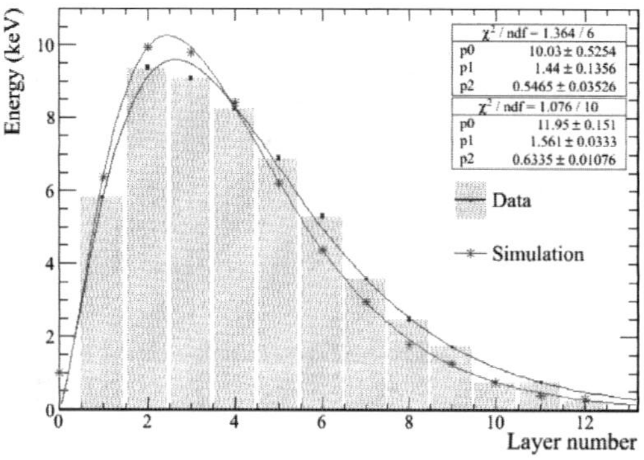

Figure 3. Shower profile for 2 GeV/c electrons measured by one Micromegas chamber and comparison with the Geant4 simulation.

2.2.2. Prototype with embedded digital readout Two stacks with embedded digital readout were assembled and exposed to 2 GeV/c electrons. The first one consists of four chambers with 32×8 cm^2 pads read out by HARDROC chips. A clear image of the beam profile has been seen in each chamber. Because of the short shaping time of the HARDROC chip, only little of the signal can be seen and, consequently, the measured efficiency is about 6 %. The second stack was assembled by four 8×8 cm^2 chambers with DIRAC chips. Unfortunately, due to insufficient spark protection, the life-time of the DIRAC chips in a beam was limited. Nevertheless, an efficiency up to 50% and a hit multiplicity between 1.06 and 1.13 have been measured. These results are affected by low statistics and therefore serve only as a first performance indication. That is why one of the main aim of the next generation readout electronics development is to obtain a reliable spark protection.

3. 1 m^2 prototype: design and sub-detector test

An essential step towards the 1 m^3 technological prototype of DHCAL, which will be made of 40 Micromegas chambers with embedded readout electronics interleaved by stainless steel absorbers, is the construction and test of a 1 m^2 prototype, the largest Micromegas chamber so far built. The chamber is formed by 6 active sensor units (ASU) of 48×32 readout pads of 1 cm^2, see Fig. 4. Each ASU is equipped with a mesh and 24 HARDROC readout chips. These ASUs are glued on a 2 mm thick stainless steel supporting plate and the chamber is closed by another 2 mm thick stainless steel plate that carries the cathode. In this arrangement, all the ASUs share the same gas volume. The drift gap is defined by a plastic frame around the chamber and spacers that are placed between the ASUs resulting in very small dead zones of around 1.6 %. The gas distribution is provided by one inlet and outlet in the chamber frame. The readout of two ASUs is chained serially and connected to the data acquisition system by a detector interface board (DIF). The DIF is a mezzanine board which allows to configure HARDROCs by setting readout thresholds and preamplifier gains, to readout the HARDORC data, and also to provide the distribution of the system clock and high voltages through another dedicated board, called inter-DIF, which stands between the DIF and the ASU. The total active area of the complete 1 m^2 chamber is 96×96 cm^2 corresponding to 9216 readout channels. The channel signals are processed by 144 HARDROC chips and three DIF boards. The effective thickness of the chamber, if one does not count the 4 mm of the supporting plates material that can be considered as a part of the calorimeter absorber, is equal to 8 mm and complies well with the ILC specification.

Figure 4. The 1 m^2 Micromegas chamber without top steel cover showing mesh side of the ASUs.

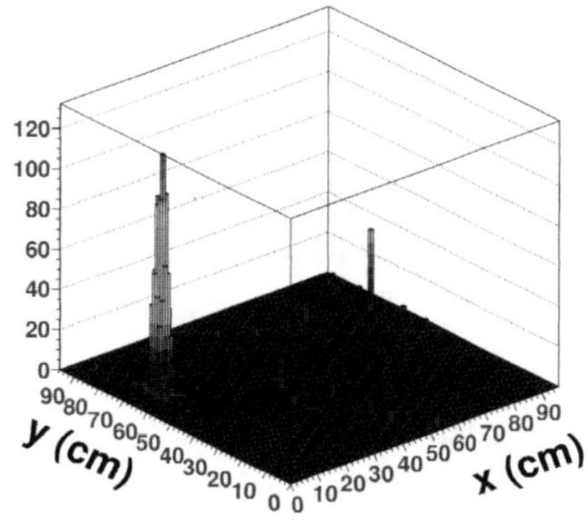

Figure 5. A beam profile of 150 GeV/c muons traversing the 1 m^2 Micromegas chamber.

3.1. ASU test and electronics calibration

Prior to the 1 m^2 assembly, the electronics characteristics of the readout chips are measured and the overall ASU functioning is verified in a test gas chamber. The electronics tests consist of the determination of the pedestals and noise level of each channel to achieve the lowest detection threshold, and the determination of the channel preamplifier gains.

The determination of the pedestals and noise level is done by measuring the trigger efficiency of each channel as a function of threshold (so-called S-curve). As the threshold is common to all chip channels, it is determined by the S-curve with the highest inflexion point. By changing the preamplifier gain of each chip channels, it is possible to squeeze the S-curves close together and lower the detection threshold to minimum (typically down to 3–4 fC).

The determination of the channel preamplifier gain is performed by injecting different test charges in the HARDROC circuitry. Once the gain distribution per chip is known, equalization constants are calculated to reduce the spread. The equalization procedure has been validated as shown in Fig. 6 where the S-curve inflexion point distributions for 100 fC test charge with and without equalized gains are plotted.

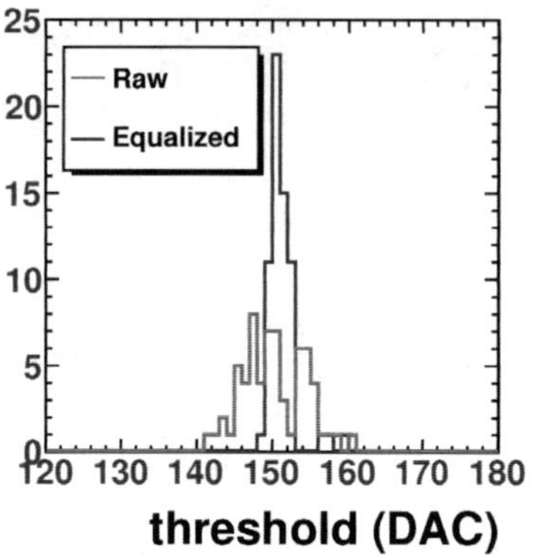

Figure 6. Distributions of the S-curve inflexion points for one HARDROC channel before and after the gain equalization for 100 fC input charge.

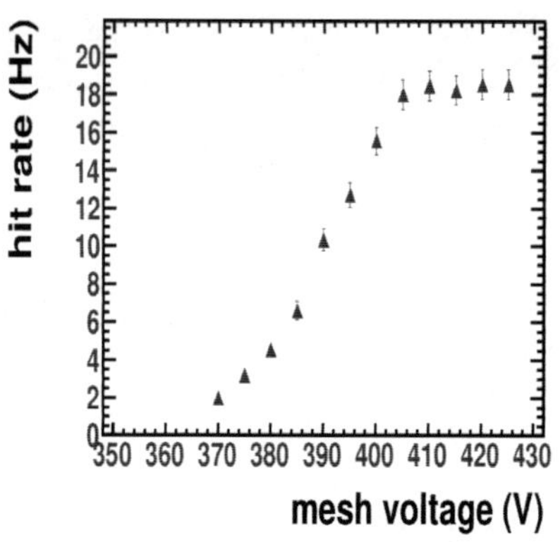

Figure 7. Hit rate from ^{55}Fe quanta as a function of the voltage on the mesh. The response is measured for one channel in a gas test box having 3 cm drift gap.

A dedicated gas chamber, which can house one ASU, has been fabricated to perform tests of the ASUs with ^{55}Fe X-rays before they are mounted into the 1 m^2 chamber. It has a perforated aluminum cover onto which is glued a cathode foil with a few nm thin conductive deposit on one side. It is therefore transparent to ^{55}Fe quanta which are almost all absorbed in the 3 cm thick drift gap. By collimating the X-ray beam and by measuring the number of hits recorded per unit time on one pad, relative variations of efficiency can be inferred. The impact of the chip settings and detector voltages can be quantified this way. As an example, the strong influence of the mesh voltage (gas gain) on the detection rate is depicted in Fig. 7. Moreover, the box allows to perform the high voltage burn-in in air, which effectively reduces ASU's sparking rate by washing out the impurities remaining in the amplification gap after the bulk production.

3.2. The of the first 1 m^2 in a beam

The first 1 m^2 prototype has been successfully built and tested in a particle beam in summer this year. This experiment has provided precious information for future mass production. During one month operation in a beam, the functionality of the DAQ system and chamber readout electronics has been verified. Gas gains up to 15.000 has been reached with very few high voltage trips and the first test in a power pulsing mode of the readout electronics has been done successfully. The data taken during this period are currently under analysis. In Fig. 5 is shown a beam profile of 150 GeV/c muons traversing 1 m^2 Micromegas prototype which is the first sign of the project feasibility and an important milestone in the large area Micromegas R&D effort.

Journal of Physics: Conference Series **293** (2011) 012078 doi:10.1088/1742-6596/293/1/012078

4. Simulation studies

In order to have better understanding of the physics performance of a digital hadron calorimeter with very fine segmentation and its capability to separate hadronic showers, detailed Monte Carlo simulations of a realistic SiD-like calorimeter module [9] have been performed [10]. These studies have been focused on the calorimeter response and linearity, energy resolution, shower profiles, and the impact of the leakage. Results obtained for digital readout with different thresholds were compared against the standard analog readout. Fig.8 depicts the energy resolution for single pions as a function of their energy for analog and digital readout with a readout threshold of 15 fC. Superior energy resolution observed for digital readout is due to the suppression of the Landau and particle path length fluctuations. On the other hand, due to the saturation in calorimeter response, the linearity of a digital calorimeter is not as good as it is in case of analog one, see Fig.9. The saturation effect can be reduced by use of semi-digital readout with two or three thresholds. The determination of the optimal thresholds and their corresponding weights is the subject of current investigation.

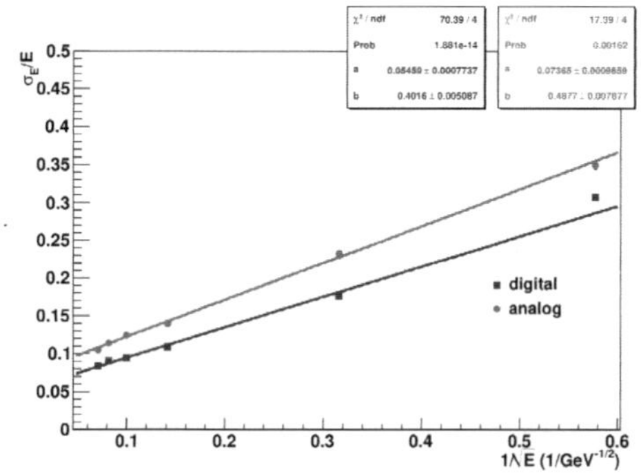

Figure 8. Energy resolution as a function of pion energy for digital and analog readout.

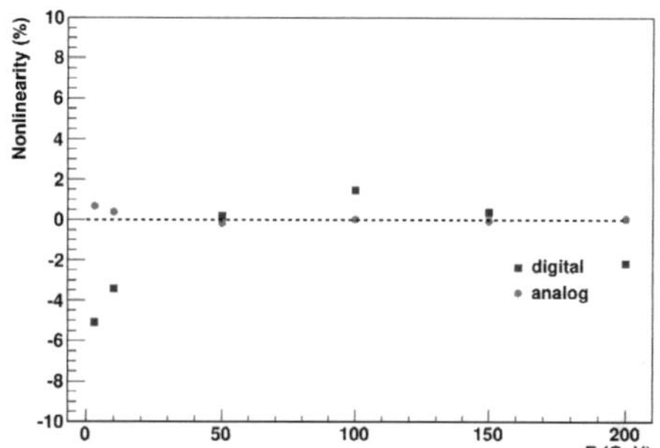

Figure 9. Non-linearity as a function of the pion energy for digital and analog readout

5. Conclusions

Several Micromegas chambers have been developed for an application in digital hadron calorimetry. The prototypes with analog readout have been characterized and extensively tested with different particle sources. The tests of these chambers have shown very good performance on detection efficiency and hit multiplicity, the parameters which are crucial for calorimetry at future linear colliders. Based on the obtained results, new generation Micromegas chambers with embedded readout electronics have been designed and produced. The feasibility of the project has been proven by construction and functionality test of the $1\,m^2$ prototype in particle beams. The simulation studies of the digital hadronic calorimeter are well advanced.

6. References

[1] M. A. Thomson, NIM **A611** (2009), pp. 25–40
[2] S. Andriamonje *et al.*, NIM **A560** (2006), pp. 405–408
[3] C. Adloff *et al.*, 2009 *JINST* 4 P11023, pp. 1–19
[4] S. Callier *et al.*, TWEPP-09 Proceeding, 2009, pp. 122–126
[5] C. Adloff *et al.*, TWEPP-09 Proceeding, 2009, pp. 117–121
[6] R. Gaglione *et al.*, to be published in TWEPP-10 Proceeding, 2010
[7] C. Adloff *et al.*, Technical Note, LAPP-TECH-2009-03, pp. 1–20
[8] C. Adloff *et al.*, 2010 *JINST* 5 P01013, pp. 1–13
[9] SiD detector, http://silicondetector.org/display/SiD/home
[10] C. Adloff *et al.*, 2009 *JINST* 4 P11009, pp. 1–6

Hadron Calorimetry -
what have we learned since CALOR 1?

Richard Wigmans
Department of Physics
Texas Tech University
Lubbock, TX 79409-1051, U.S.A.
E-mail: wigmans@ttu.edu

Abstract. In this talk, I review what has been learned about hadron calorimetry since the CALOR conference series started (1990) and I will give my view on how further progress can be made. I will also discuss some developments that may prevent further progress.

1. Introduction - the status quo in 1990

In this talk, I will review what has been learned about hadron calorimetry since the CALOR conference series started, twenty years ago. I will also give my view on how further progress can be made, and discuss developments that may prevent further progress.

Around 1990, the mysteries that had surrounded the issue of compensation in hadron calorimetry for a number of years had been solved. The reasons for the poor performance of non-compensating calorimeters were fully understood, and the same was true for the details of the compensation mechanism. One had come to realize that the use of uranium absorber was neither needed nor sufficient to achieve equal response to the em and non-em shower components, and that nuclear fission played no essential role in this matter. This was experimentally demonstrated with lead/plastic scintillator calorimeters,

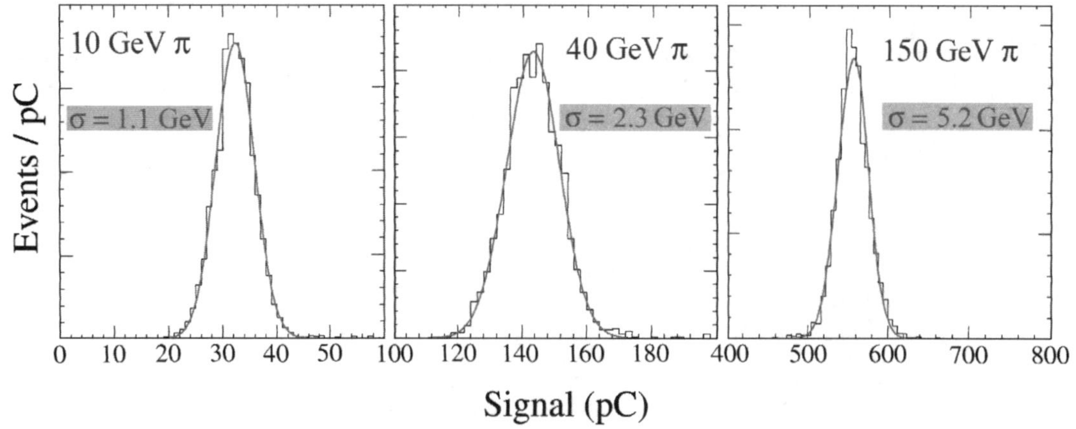

Figure 1. Signal distributions for pions of 10, 40 and 150 GeV measured with the SPACAL lead/scintillating-fiber calorimeter

independently by the ZEUS [1] and SPACAL [2] collaborations. The latter calorimeter holds until today the hadronic energy resolution record of about $30\%/\sqrt{E}$. Some of its results are shown in Figure 1. Apart from the excellent energy resolution, the figure also shows that the signal distributions were well described with Gaussian fits. In addition, the calorimeter was linear for hadron detection, with a response that was almost equal to that for the electrons with which it was calibrated. At 80 GeV, the resolution of this instrument (pictured in Figure 2) was measured to be about 3.3 GeV, sufficient to distinguish hadronically decaying Ws from Z bosons. This separation is the main goal that drives the design of calorimeters for a future high-energy electron-positron collider.

SPACAL 1989

Figure 2. The SPACAL lead/scintillating-fiber calorimeter.

The advantages of high-resolution hadron calorimetry are also clear from Figure 3, which shows the total energy measured with the compensating calorimeter of the WA80 experiment for a momentum-selected beam of heavy ions accelerated in the CERN SPS. This beam, nominally consisting of ^{32}S ions, contained at the fraction-of-1% level essentially all lower-mass ions with the same Z/A ratio. Because of the excellent energy resolution, this instrument acted in this case as a high-resolution spectrometer [3]. Note the excellent separation between the ^{14}N and ^{16}O components, where the mass resolution is comparable to that needed for separating the mentioned vector bosons.

In non-compensating calorimeters, W/Z separation has never been achieved, not even approximately. Figure 4 shows the jet-jet invariant mass distribution measured by the UA2 experiment, before (a) and after (b) subtracting the QCD background [4, 5].

Despite the obvious advantages of compensating calorimeters (signal linearity, energy resolution, response function, calibration), there are also some drawbacks. Compensation relies on amplification of the signal from neutrons produced in the shower development, such as to overcome (*i.e.*, compensate for) the effects of nuclear binding energy losses. The required amplification factor determines the sampling fraction, which is typically rather small, *e.g.*, 2.4% for Pb/plastic-scintillator. As a result, the em resolution that can be achieved with such calorimeters is limited, *e.g.*, $13\%/\sqrt{E}$ for SPACAL, $18\%/\sqrt{E}$ for ZEUS. Also, since the neutron signal results from elastic neutron-proton scattering, the signals have to be integrated over a rather large volume ($> 100\lambda_{\rm int}^3$) and a relatively long time (~ 50 ns). In practice, this is not always possible, especially in experiments at high-luminosity hadron colliders. The main aim of the DREAM project, discussed in the following, is to eliminate these drawbacks and thus improve the already impressive results shown in Figures 1 and 3.

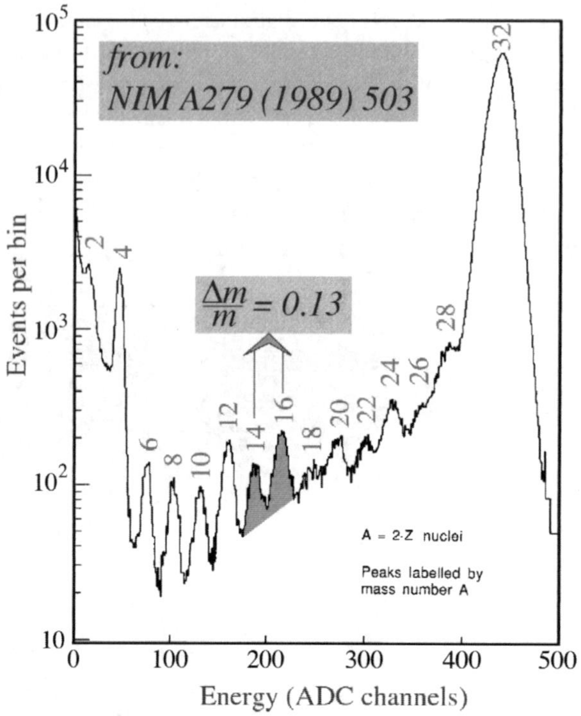

Figure 3. The WA80 calorimeter as a high-resolution spectrometer.

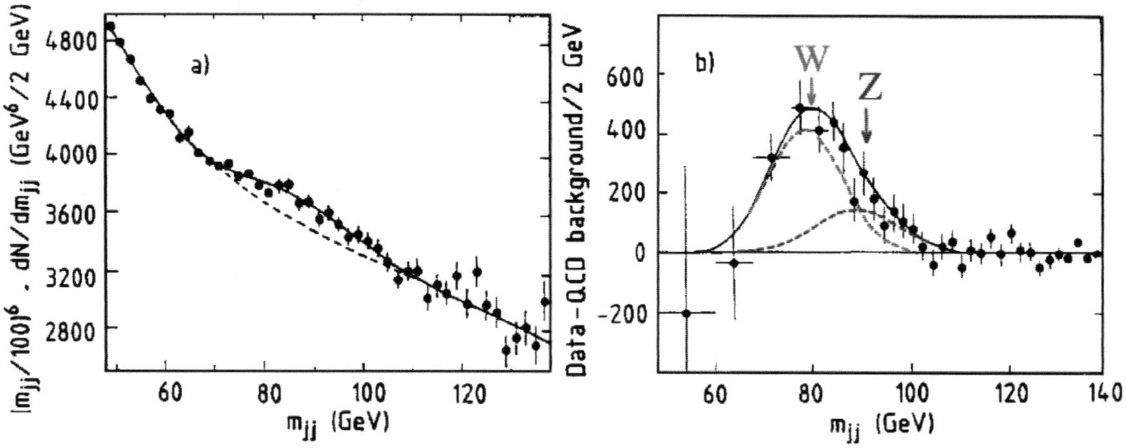

Figure 4. Two-jet invariant mass distributions from the UA2 experiment [4], before (a) and after (b) subtracting the QCD background, indicated by the dashed line in diagram a. The data in b are compatible with peaks at m_W =80 GeV and m_Z =90 GeV. The standard deviation of the mass distribution was 8 GeV, of which 5 GeV could be attributed to non-ideal calorimeter performance [5].

Even though the unraveling of the compensation puzzle was primarily achieved by building and testing generic calorimeters, some credit should also be given to Gabriel and coworkers [6], whose simulations had shown the importance of nuclear reactions in the absorption process of hadrons, and who correctly predicted the suppression of the em response of sampling calorimeters. This effect, often referred to as $e/mip < 1$, is an important ingredient for compensation in calorimeters with high-Z

absorber material. The absence of this effect makes homogeneous detectors (*e.g.,* crystals) particularly poor hadron calorimeters. It also plays an important role in the calibration problems of segmented calorimeters, which is discussed in the next section.

2. What has been learned since 1990?

In order to appreciate the developments in hadron calorimetry since 1990, it is useful to briefly summarize the relevant aspects of hadronic shower development. In the absorption process, the energy is deposited by two components, which I will call the em and non-em one (Figure 5).

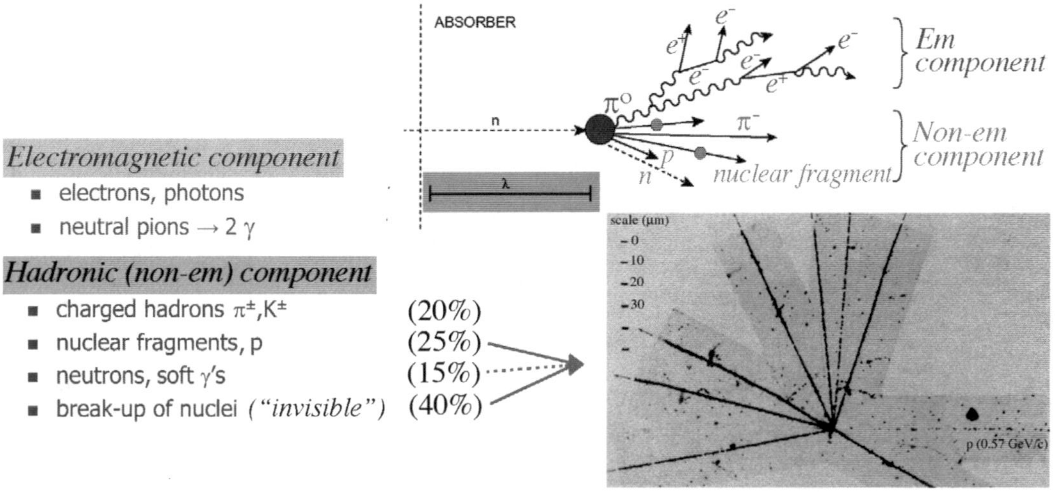

Figure 5. Schematic description of hadronic shower development, and an emulsion picture of a nuclear reaction which occur in large numbers in this process.

The calorimeter's response functions for these two components are schematically shown in Figure 6, for a generic non-compensating calorimeter. The e/h value, which represents the ratio of the average signals per unit of deposited energy for these components, has a value of 1.8 in this example. The signal from a particular hadron shower is a combination of the signals from these two components.

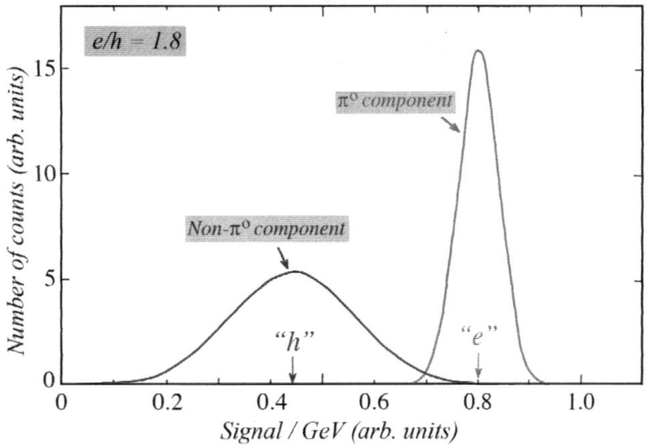

Figure 6. The response functions of a noncompensating calorimeter for the em and non-em components of hadron showers.

Event-to-event fluctuations in the energy sharing between these two components are large and non-Poissonian. These fluctuations in the em shower fraction (to be called $f_{\rm em}$ in the following) form the main contribution to the hadronic energy resolution of non-compensating calorimeters, and are also responsible for deviations from $E^{-1/2}$ scaling. In compensating calorimeters, this source of fluctuations is eliminated and the resolution is usually determined by the width of the non-em response function, which is ultimately dominated by fluctuations in nuclear binding energy losses. The latter fluctuations (in so-called *invisible energy*) can also be affected by the design of the calorimeter. Since the binding energy losses are correlated to the total kinetic energy of the numerous neutrons released in the nuclear reactions, the presence of hydrogen in the active calorimeter layers may reduce their effect significantly. The difference between the energy resolutions for 100 GeV pions measured by the ZEUS (3.5%) and D0 (7%) calorimeter systems, which both had e/h values close to 1.0, is most likely due to the absence of hydrogen in the latter.

An issue that in practice is possibly even more important than the energy resolution is the non-linearity that is typical for non-compensating hadron calorimeters. This non-linearity is a consequence of the fact that the average value of $f_{\rm em}$ increases with the shower energy. This phenomenon may lead to systematic mismeasurement of jet energies.

After this brief summary, I now discuss the developments in our understanding of hadron calorimetry that have taken place since 1990. It is a sad truth that, despite the enormous efforts that have gone into the development of GEANT-4 in the past 20 years, the contributions of Monte Carlo simulations in this domain have been negligible.

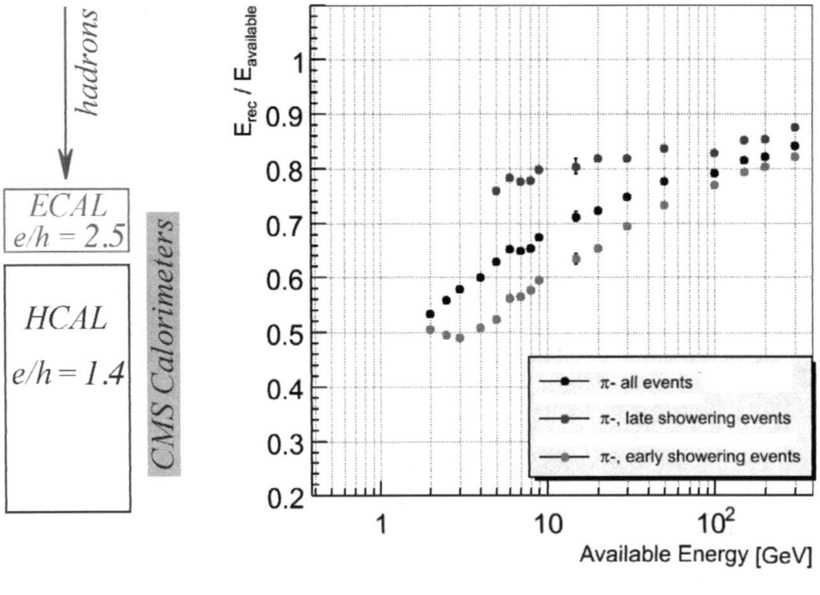

Figure 7. The response to pions as a function of energy, for the CMS barrel calorimeter. The pion events are subdivided into two samples according to the starting point of the shower, and the pion response is also shown separately for these two samples [7]. The vertical scale is normalized to the electron response.

I want to point out the contrast with em calorimetry, where Monte Carlo simulations turned out to be of crucial importance, for example for solving the complicated calibration problems of the ATLAS LAr calorimeter [8] and the calorimeter of the AMS experiment [9]. The absence of similarly reliable simulations for hadronic shower development meant, for example, that the "spikes" observed in the crystal calorimeter of the CMS experiment, which are caused by nuclear reactions close to the APDs which detect the light signals from these crystals, were not at all foreseen. Important calorimeter features such as the hadronic response and the energy resolution [10], as well as the hadronic shower profiles [11], are in general also poorly reproduced by GEANT based simulations. Such simulations are therefore most definitely not suitable for making meaningful predictions, on the basis of which calorimeter systems can be designed. The development of hadron calorimetry for the LHC experiments thus had to take place without meaningful guidance from such simulations. This has resulted in a number of unpleasant surprises, an example of which was mentioned above.

When the calorimeters were built, it turned out that the CMS experiment had to pay a high price for its complete focus on em energy resolution. Its calorimeter system consists of a crystal em section, backed up by a bronze/plastic-scintillator hadronic section. The hadronic performance of this calorimeter system was systematically studied with various types of particles ($e, \pi, K, p, \bar{p}$), covering a momentum range from 1 - 300 GeV/c. Figure 7 shows some results from this study [7]. Both sections were calibrated with 50 GeV electrons. The response to pions, represented by the black dots, indicates that the calorimeter is extremely non-linear for these particles. This non-linearity is especially evident below 10 GeV, which is important since pions in this energy range carry a large fraction of the energy of typical LHC jets, even in the TeV domain. More troublesome is the fact that the response strongly depends on the starting point of the showers. This is a direct result of the very different e/h values of the em (2.5) and hadronic (1.4) calorimeter sections. The figure shows results for two event samples, selected on that basis: showers starting in the em section or in the hadronic one. At low energies, the response is more than 50% larger for the latter (penetrating) events. In practice in an experiment, it is often hard/impossible to determine where the shower starts, especially if the pions are traveling in close proximity to other jet fragments (*e.g.*, photons from π^0 decay) which develop showers in the em section.

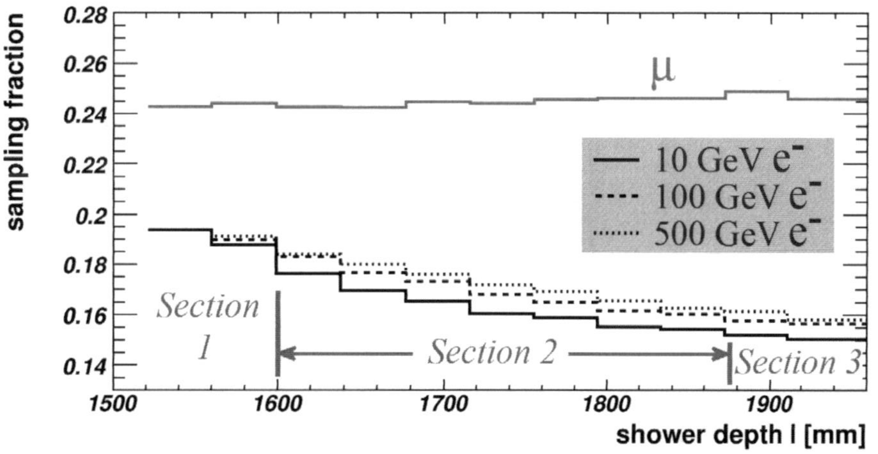

Figure 8. Evolution of the sampling fraction for electron showers of different energies in the three longitudinal segments of the ATLAS em calorimeter [8].

Another issue on which some new light has been shed since 1990 concerns the enormous complications that may arise in the process of (inter-)calibrating the different sections of a longitudinally segmented sampling calorimeter. These complications are a consequence of the fact that the sampling fraction is typically different for the different shower particles that contribute to the signals, and the fact that the shower composition changes as it develops. As a result, the relationship between measured signal and deposited energy, *i.e.*, the *calibration constant*, varies with depth and with the energy of the showering particle, and is especially for hadrons in a given longitudinal detector segment different for each event.

Unfortunately, in the absence of high-resolution longitudinally segmented hadron calorimeter systems, detailed information on the extent of these problems for hadron detection is still lacking. However, the fact that these issues already cause major problems in longitudinally segmented em calorimeters does not bode well in that regard, since the problems for hadronic showers are most likely considerably worse. Figure 8 shows how the em sampling fraction decreases with depth in the ATLAS Pb/LAr calorimeter, in an energy dependent way. The (inter)calibration of the three longitudinal segments of this calorimeter was extremely non-trivial. Signal linearity, combined with good energy resolution, was only achieved after very detailed Monte Carlo simulations, which resulted in a very complicated procedure involving a variety of energy dependent calibration parameters. Similar problems were reported by the AMS Collaboration, whose em calorimeter is longitudinally subdivided into 18 independent segments [9].

Avoiding (inter)calibration problems is one of the reasons why the DREAM project concentrates on longitudinally *unsegmented* calorimeters. The DREAM project was started to improve the already excellent hadronic performance of the compensating ZEUS and SPACAL calorimeters, and avoid the disadvantages that made these instruments less than ideal for em shower detection. In practice this implies

(i) Reducing the contribution of sampling fluctuations to the energy resolution (*the* dominating factor for ZEUS and SPACAL).
(ii) Reducing the effects of fluctuations in invisible energy. This means that the calorimeter has to be sensitive to the "nuclear" fraction of the non-em shower component. This requirement puts homogeneous detectors consisting of high-Z crystals at a clear disadvantage.
(iii) Eliminating the effects of fluctuations in $f_{\rm em}$ in such a way that the previous two points are not prevented.

The latter turned out to be possible by measuring simultaneously the signals from Čerenkov light and scintillation light produced in the shower development (hence the term dual-readout). By comparing these two signals, it turned out to be possible to measure the em shower fraction event by event, and thus eliminate the effects of the large fluctuations in this variable. This was demonstrated both for showers induced by single hadrons and for jets, as illustrated in Figure 9 [12]. Diagrams a, b show that the signal distribution of a non-compensating calorimeter is a superposition of distributions with different $f_{\rm em}$

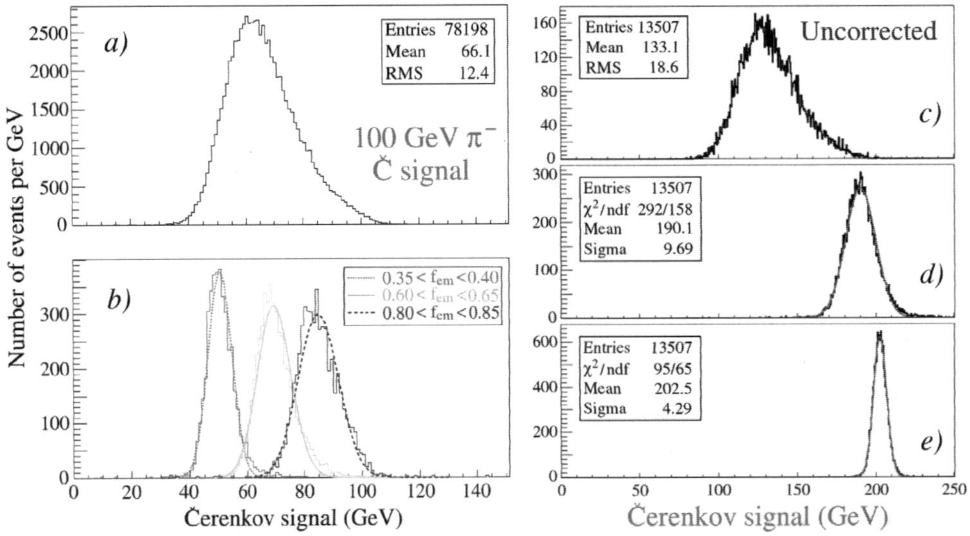

Figure 9. Čerenkov signal distribution for 100 GeV π^- (a) and distributions for subsamples of events selected on the basis of the measured $f_{\rm em}$ value (b). Signal distributions for high-multiplicity 200 GeV "jets" in the DREAM calorimeter before (c) and after (d) corrections as described in the text were applied. In diagram e, energy constraints were used, which eliminated the effects of lateral shower leakage fluctuations that dominate the resolution in d.

values (and thus different response values). Once the $f_{\rm em}$ value of a particular event has been measured, the signal can be scaled to $f_{\rm em} = 1$ (*i.e.,* the value for em showers). In doing so, the effects resulting from $f_{\rm em}$ fluctuations are eliminated. This results in improved energy resolution, a Gaussian response function and, most importantly, a correctly reproduced hadronic energy (Figure 9c, d). This was true both for single pions as well as for jets.

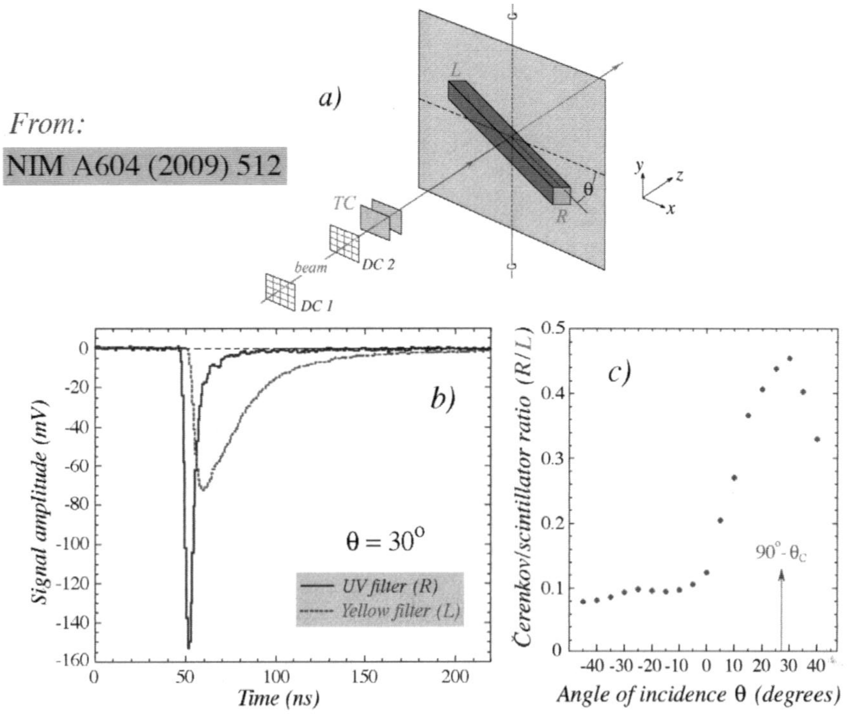

Figure 10. Unraveling of the signals from a Mo-doped $PbWO_4$ crystal into Čerenkov and scintillation components. The experimental setup is shown in diagram a. The two sides of the crystal were equipped with a UV filter (side R) and a yellow filter (side L), respectively. The signals from 50 GeV electrons traversing the crystal are shown in diagram b and the angular dependence of the ratio of these two signals is shown in diagram c [13].

We have shown that similar results can also be obtained with high-Z crystal calorimeters ($PbWO_4$, BGO), whose signals can be separated into scintillation and Čerenkov components [14, 15]. This has been achieved by making use of differences in the spectral characteristics, the time structure of the signals and/or the angular distribution of the light. Figure 10b shows that application of a suitable filter led to almost pure scintillation or Čerenkov signals from a Mo-doped $PbWO_4$ crystal. The Čerenkov nature of the UV signal is underscored by its characteristic angular distribution [13].

The DREAM Collaboration has also demonstrated the importance of measuring the "nuclear" fraction of the non-em shower component event by event. The time structure of the hadronic shower signals provided crucial information in this respect, since the contribution of neutrons was easily recognizable as a tail with a time constant of about 20 ns. This tail, caused by elastic $n-p$ scattering in the scintillating fibers, was absent in the signals from the clear fibers and also in all signals from em showers. We found that the strength of the neutron signal was rather strongly anti-correlated to $f_{\rm em}$, but did provide additional information which allowed for additional improvement of the energy resolution [16].

Further improvement of the energy resolution can only be achieved by addressing the factors that have become dominant after elimination of the effects of fluctuations in $f_{\rm em}$ was successfully accomplished. Most prominent among these are fluctuations in (lateral) shower leakage. By making the detector about five times larger, we expect to improve the resolution for 200 GeV jets to $\sim 2\%$ (see Figure 9d, e). We are currently building a new instrument that is not only sufficiently large to achieve this goal, but in which also a number of other improvements are being implemented, which should make it an excellent em calorimeter as well (em resolution better than $10\%/\sqrt{E}$). Details can be found in [17].

3. How to make further progress?

Just like in the past 30 years, further progress in the field of hadron calorimetry will entirely have to depend on *experimental verification of new ideas*, given the absence of reliable Monte Carlo simulations which otherwise might provide guidance.

Apart from the DREAM project, which will continue to explore the virtues of dual-readout techniques (see Section 2), two other ideas are currently being pursued:

(i) *Homogeneous dual-readout calorimetry.* Compared to the sampling structures developed in the framework of the DREAM project, homogeneous calorimeters have the advantage of eliminating sampling fluctuations entirely, which results in excellent energy resolution for em showers. On the other hand, such calorimeters have no handle on the "nuclear" fluctuations in the non-em component of hadron showers, a clear disadvantage compared to the fiber-based sampling detectors. Other potential disadvantages include the problems that will undoubtedly arise from attenuation of the Čerenkov light, and the complications associated with the readout of such a (longitudinally segmented) detector, not to mention the very high cost of crystal-based calorimeters.

(ii) *Particle Flow Analysis (PFA).* In this approach, the additional information provided by an upstream tracker is used to measure jet energies. The problem that limits the success of this method is of course that the calorimeter does not know or care whether the particles it absorbs are electrically charged. Therefore, one will have to correct the calorimeter signals for the contributions of the charged jet particles. Proponents of this method have advocated a fine granularity as the key to the solution of this "double-counting" problem [18]. However, it has been argued by others that this, for practical geometries, is an illusion [19]. Especially in jets with leading charged particles, the overlap between the showers from individual jet particles makes the fine granularity largely irrelevant. It should also be mentioned that calibrating a fine-grained calorimeter would also be a daunting task (see section 2). Of course, in the absence of reliable Monte Carlo simulations[1] the only way to prove or disprove the advocated merits of the proposed PFA methods is by means of dedicated experiments in realistic prototype studies.

Since progress will entirely depend on experimental tests, it is very important that new devices be tested in conditions that approach the applications for which they are designed *as closely as possible.* Since modern experiments in particle physics are strongly focussed on jet detection, this implies that one should attempt to assess the jet performance of the instruments, rather than just measuring the performance in beams of mono-energetic particles, where one has many possibilities to make results look good that are not applicable for jets (*e.g.,* weighing schemes commonly referred to as "off-line compensation").

From a calorimetric perspective, a jet is a collection of particles which are simultaneously absorbed in the same area of the calorimeter. Jet performance can be assessed experimentally with multiparticle events, in which one measures the reaction products created by a beam particle in a target placed upstream of the calorimeter. One may also generate jets by means of a suitable fragmentation function and reconstruct the jet response from a database of single-particle events with different energies measured in the calorimeter under study. This method was, for example, used to assess the jet response of the CMS calorimeter system [7].

[1] Concern about the absence of reliable Monte Carlo simulations for hadronic shower development was the main reason for a special workshop held at Fermilab in 2006 [20]. To our knowledge, the fundamental problems addressed at this workshop, *e.g.,* with regard to the (for PFA crucial) hadronic shower widths, still exist.

4. How NOT to make further progress

4.1. Avoid repeating mistakes from the past

Since progress in the field of hadron calorimetry depends entirely on experimental trial-and-error exercises, it is very important that people or groups engaged in this effort are aware of the history, if only to avoid repeating mistakes from the past. Unfortunately, this is not always the case, as I will illustrate with two well documented examples.

(i) One should avoid placing readout elements that may produce HUGE signals for one particular type of shower particle in the path of a developing shower. This may lead to a phenomenon that has become known as the "Texas Tower effect" , first observed in the forward calorimeter of the CDF experiment [21, 22, 23]. In this device, recoil protons from elastic neutron scattering deposited up to $\sim$ 1 MeV in an active (gas) layer, which because of the very small sampling fraction ($\sim 10^{-5}$) appeared as an energy deposit of 100 GeV in a single calorimeter cell. Such "hot spots" were recently also observed in the ECAL of the CMS experiment. The culprits are in that case densely ionizing particles from a nuclear reaction, such as depicted in Figure 5. Such nuclear fragments have dE/dx values that are typically 100-1000 times larger than for a mip. When traversing the APD which detects the scintillation light from the $PbWO_4$ crystals constituting the ECAL, such (MeV type) particles may generate signals that are 100 000 times larger that that from a scintillation photon, and thus fake energy deposits of tens of GeV.

(ii) "Digital calorimetry" was tried and abandoned for good reasons. Figure 11 shows the average

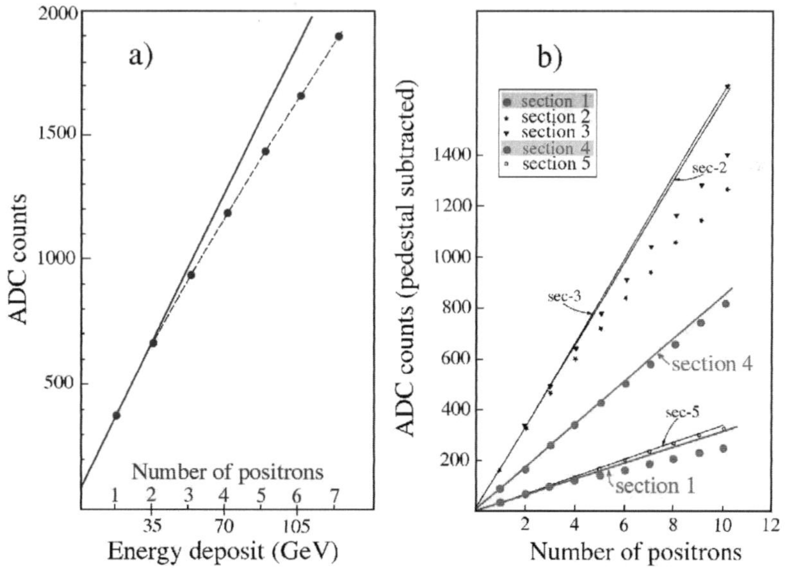

Figure 11. Average em shower signal from a calorimeter read out with wire chambers operating in the "saturated avalanche" mode, as a function of energy [24].

signals from em showers in a calorimeter that was read out with wire chambers operating in a "saturated avalanche" mode [24]. The response is strongly non-linear, as a result of the increasing *density* of shower particles, combined with the fact that in a device of this type the signal from a calorimeter cell is the same regardless of the number of particles crossing it. This calorimeter was longitudinally subdivided in five sections. In Figure 11b, the average signals recorded in these separate sections are shown as a function of the total energy deposited in the calorimeter. It illustrates that the observed saturation effect is clearly determined by the *density* of shower particles.

The degree of non-linearity, given by the deviation from the straight line, was more than a factor six larger in section 1 (where the shower starts) than in section 4 (located beyond the shower maximum), despite the fact that the energy deposit in section 1 was less than half of that in section 4.
Despite this clearly documented example, the CALICE Collaboration is apparently spending a major effort reviving this technique. Not surprisingly, test results reported at this conference exhibit very substantial response non-linearities [25].

4.2. Dishonesty in reporting results

If one wants to make real progress in this field, it is imperative that scientific integrity be the guiding principle. Of course, it is disappointing when an instrument on which a lot of effort (and money) has been spent does not perform as hoped/expected. However, by misrepresenting the measured performance, one does a serious disservice to our discipline. I want to illustrate these statements with two recent, well-documented examples.

(i) The term *linearity* has a precisely defined meaning. A calorimeter is said to be linear if the measured signal is proportional to the deposited energy. The best way to present experimental test results of signal linearity is by means of the residuals: $(E_{\text{meas}} - E_{\text{dep}})/E_{\text{dep}}$ as a function of the deposited energy E_{dep}. Figure 12 shows linearity results recently published by the CALICE

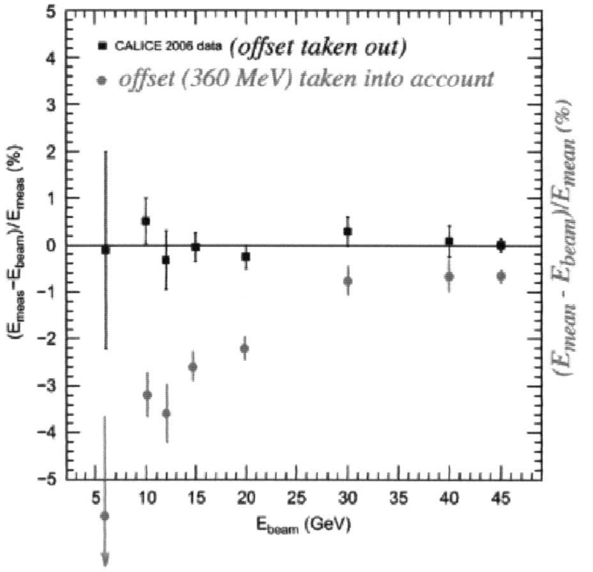

Figure 12. The response of the CALICE W/Si em calorimeter to electrons, as a function of energy. Shown is the relative difference between the average measured signal and the beam energy, as a function of energy. Results are given with and without correcting for a 360 MeV offset observed in the measured signals [26].

Collaboration for their W/Si em calorimeter [26]. The black squares around the line at zero represent the experimental data, and the authors conclude that "the response of their calorimeter is linear to within approximately 1%". However, upon careful reading of the paper one sees that the authors have fitted the measured mean values of their signal distributions to an expression of the type: $E_{\text{mean}} = \beta \cdot E_{\text{beam}} - 360$ MeV, and subsequently redefined the *measured* energy as $E_{\text{meas}} = E_{\text{mean}} + 360$ MeV. If one just takes the measured signals at face value, then the result, represented by the red dots in Figure 12 looks all of a sudden very different. The correct conclusion from the measured data is that this calorimeter exhibited a signal non-linearity for em showers of about 5% in the energy range of 6 - 45 GeV, manifesting itself in the form of a 360 MeV offset .

(ii) The term *energy resolution* has a very precisely defined meaning in calorimetry. It represents the precision with which the energy deposited by a particle absorbed in the calorimeter can be measured. It is usually determined from the signal distribution for a beam of mono-energetic particles sent into the calorimeter. In general, the energy resolution is defined as the *rms* value of this distribution. Only when the signal distribution is well described by a Gaussian function is this *rms* value equal to the sigma of the Gaussian fit, in all other cases $\sigma_{\rm rms} > \sigma_{\rm Gauss}$.
The CALICE Collaboration has found that the signal distributions resulting from their Monte Carlo simulations of PFA performance have substantial tails. They don't like these tails, since they spoil the energy resolution. Therefore, they have invented a new variable, called *rms90*, defined as the $\sigma_{\rm rms}$ of the events in the smallest reconstructed energy interval that contains 90% of the events [27]. As an aside, we mention that even for a perfectly Gaussian signal distribution, *rms90* is more than 20% smaller than $\sigma_{\rm Gauss}$; for non-Gaussian response functions, the discrepancy is correspondingly larger. Of course, everyone is free to use whatever phony statistics suits their purpose. However, the claims subsequently made by the CALICE Collaboration about the energy resolution achievable with their PFA methods are grossly misleading, since *rms90* has now taken on the meaning of *energy resolution* in their vocabulary.

My criticism of CALICE may sound very harsh. However, it is only the academic dishonesty of which I have given some examples above that bothers me. I believe that the CALICE project offers wonderful opportunities to investigate the experimental possibilities and impossibilities of the PFA approach. This should be done on the basis of *experimental data*, and it is very straightforward how to do that. In calorimetric terms, a jet is simply a collection of particles, mainly pions and photons (from π^0 decay). An event generator, *e.g.*, PYTHIA, provides all needed information about these particles, *i.e.*, their energies and the p_T with respect to the jet axis (the direction of the fragmenting quark). One can then place the calorimeter at a large distance from the vertex, 10 meters or so. This distance defines the impact points of the different jet fragments. Next, one can generate a jet event in the calorimeter, using the extensive database built up in several CALICE test beam campaigns over the years. Of course, one does not have data at the precise energies of all the jet fragments, but things change rather slowly with energy, so it is OK to interpolate. So if the jet calls for an 11.6 GeV π^+, one takes an event from a 10 GeV pion run, and multiplies the signals in all the cells that contributed in that event by 1.16. For a photon with energy 4.2 GeV, one takes an event from a 5 GeV electron run and multiplies the signals in all contributing cells by 0.84. Etcetera. In doing so, a hit pattern for this particular jet event in the calorimeter system is generated. Of course, this procedure can be repeated for the same jet event as many times as desired, using other events from the database to reconstruct event patterns. All for the same jet. For each event pattern, the measured energy can be determined as the sum of the exactly known energies of the charged jet fragments plus the calorimeter energy measured after removing hits from the event pattern using a PFA algorithm. In this way, a total measured energy distribution for the particular jet event in question is obtained. This entire procedure can be repeated for other jet events. One can study differences between jet events with leading charged hadrons and leading π^0s. One can study differences between events with a fragmenting u or d quark and fragmenting heavier quarks. In the latter case, leading K^0s may be expected as well.

At a distance of 10 meters, the whole PFA scheme will probably work pretty well. Subsequently, the calorimeter can be moved to 9 meters, 8 meters, etcetera. At some point, the confusion will kick in. One can study where that happens, for what type of events, and one could try out the effectiveness of various remedies in practice. The effects of a magnetic field could also be implemented in a rather straightforward way.

A procedure as described above would be a great contribution to the fundamental study of hadron calorimetry, it would offer countless opportunities for students to make their mark with individual contributions. In CMS, a similar approach was followed to investigate the effects of the very different e/h values of the ECAL and HCAL on the jet response of the combined calorimeter system [7], and in that process many things were learned that no MC simulation would have been able to tell us.

Journal of Physics: Conference Series **293** (2011) 012079 doi:10.1088/1742-6596/293/1/012079

Unfortunately, from what I have heard at this conference [28] it seems that the CALICE Collaboration has chosen *not* to follow such a path. Instead, they prefer to resort to rhetoric ("PFA gives better jet energy resolution than any other calorimeter"), based on Monte Carlo simulations that use phony statistics (see example above) and hadronic shower simulations that are notoriously unreliable. Statements such as "there is no need for a test beam based demonstration of PFlow calorimetry" and "it is widely believed that the most promising strategy for the ILC jet energy is the Particle Flow approach" [27] just illustrate to me that the scientific community has little to expect in terms of contributions to further understanding of hadron calorimetry from this Collaboration. Which is a real pity.

5. Summary

- It is, and has been for 20 years, possible to build calorimeters that can separate hadronically decaying W's from Z's.
- Calibrating a longitudinally segmented calorimeter continues to be a very complicated and usually grossly underestimated job.
- The DREAM approach combines the advantages of compensating calorimetry with a reasonable amount of design flexibility.
- The dominating factors that limited the hadronic energy resolution of compensating calorimeters such as ZEUS and SPACAL to $30 - 35\%/\sqrt{E}$ can be eliminated, and the theoretical resolution limit for hadron calorimeters ($15\%/\sqrt{E}$) seems to be within reach.
- The DREAM project holds the promise of high quality calorimetry for **all** types of particles, with an instrument that can be calibrated with electrons.
- Rhetoric, combined with academic dishonesty is *not* helpful for further progress in this field.

References

[1] Bernardi E *et al.* 1987, Nucl. Instr. and Meth. **A262**, 229.
[2] Acosta D *et al.* 1991, Nucl. Instr. and Meth. **A308**, 481.
[3] Young GR *et al.* 1989, Nucl. Instr. and Meth. **A279**, 503.
[4] Alitti J *et al.* 1991, *Z. Phys.* **C49**, 17.
[5] Jenni P 1988, *Nucl. Phys. B (Proc. Suppl.)* **3**, 341.
[6] Gabriel TA *et al.* 1994, Nucl. Instr. and Meth. **A338**, 336.
[7] Akchurin N *et al.* 2007, *The Response of the CMS Combined Calorimeters to Single Hadrons, Electrons and Muons*, CMS Note 2007/012, CERN, Genève, Switzerland.
[8] Aharrouche M *et al.* 2006, Nucl. Instr. and Meth. **A568**, 601.
[9] Cervelli F *et al.* 2002, Nucl. Instr. and Meth. **A490**, 132.
[10] *et al.* 2010, Nucl. Instr. and Meth. **A607**, 372.
[11] *et al.* 2010, Nucl. Instr. and Meth. **A615**, 158.
[12] Akchurin N *et al.* 2005, Nucl. Instr. and Meth. **A537**, 537.
[13] Akchurin N *et al.* 2009, Nucl. Instr. and Meth. **A604**, 512.
[14] Akchurin N *et al.* 2007, Nucl. Instr. and Meth. **A582**, 474.
[15] Akchurin N *et al.* 2007, Nucl. Instr. and Meth. **A595** 359.
[16] Akchurin N *et al.* 2009, Nucl. Instr. and Meth. **A598**, 422.
[17] Wigmans R 2010, *Dual-Readout Calorimetry for High-Quality Energy Measurements*, CERN-SPSC 2010-012.
[18] *TESLA Technical Design Report* 2001, report DESY 2001-011, DESY, Hamburg, Germany.
[19] Lobban O, Sriharan A and Wigmans R 2002, Nucl. Instr. and Meth. **A495**, 107.
[20] Raja R and Albrow M 2006, *Hadronic Shower Simulations* (HSS06, Fermilab); AIP Conference Proceedings # 896.
[21] Brandenburg G *et al.* 1988, Nucl. Instr. and Meth. **A267**, 257.
[22] Cihangir S *et al.* 1988, Nucl. Instr. and Meth. **A267**, 249.
[23] Fukui Y *et al.* 1988, Nucl. Instr. and Meth. **A267**, 280.
[24] Atač M *et al.* 1983, Nucl. Instr. and Meth. **205**, 113.
[25] Repond J 2010, Contribution to these Proceedings.
[26] Adloff C *et al.* 2009, Nucl. Instr. and Meth. **A608**, 372.
[27] Thomson M 2009, Nucl. Instr. and Meth. **A611**, 25.
[28] Sefkow F 2010, Contribution to these Proceedings.

Journal of Physics: Conference Series **293** (2011) 012080 doi:10.1088/1742-6596/293/1/012080

Measurement of the neutron fraction event-by-event in DREAM

John Hauptman, for the DREAM Collaboration[1]

Department of Physics and Astronomy, Iowa State University, Ames, IA 50014, USA

E-mail: hauptman@iastate.edu

Abstract. We have measured the neutron fraction event-by-event in beam test data taken at CERN by the DREAM collaboration. I will review these measurements in the context of the importance of neutrons to future high-precision calorimetry, and bring together the data from SPACAL , the GLD compensating calorimeter, and DREAM to estimate the impact neutron fraction measurements will make on hadronic energy resolution in dual-readout calorimeters.

Introduction: compensation and dual-readout ("dynamic compensation")

After reviewing the history of compensating calorimetry in SPACAL and the GLD concept calorimeter, I will describe the benefits of neutron measurements in dual-readout calorimetry, both integral to the DREAM module and separately in "leakage counters".

The measurement of the slow neutrons from nuclear break-up in hadronic showers is one means of achieving compensation, that is, the equality of the mean electromagnetic (e) and hadronic (h) response of a hadron-induced shower, often referred to as "$e/h = 1$". This was first demonstrated in the SPACAL test module [1], and later in the ZEUS calorimeter tests, and again in the GLD test module [2]. However, this compensation required a specific ratio of absorber (Pb) to sensor (plastic scintillator) of approximately 4-to-1 by volume, which in turn limited the electromagnetic energy resolution due to the small scintillator sampling fraction. Dual-readout calorimetry was invented [3] to solve this problem: achieve compensation with almost *any ratio* of absorber to sensor, and for almost *any absorber* and *any dual sensors*. This flexibility in achieving compensation is made possible with dual-readout calorimeters.

1. SPACAL compensating calorimeter

Neutrons liberated from broken-up nuclei will have kinetic energies, T, in the MeV region from their nuclear Fermi motion, corresponding to velocities of $v \sim \sqrt{2T/M_n} \sim 0.04c \sim 1$ cm/ns, and will

[1] DREAM collaboration (2010): N. Akchurin[a], F. Bedeschi[b], A. Cardini[c], R. Carosi[b], G. Ciapetti[d], R. Ferrari[e], S. Franchino[f], M. Fraternali[f], G. Gaudio[e], J. Hauptman[g], M. Incagli[b], F. Lacava[d], L. La Rotonda[h], S. Lee[g], M. Livan[f], E. Meoni[h], A. Negri[f], D. Pinci[d], A. Policicchio[h], S. Popescu[a], F. Scuri[b], A. Sill[a], G. Susinno[h], W. Vandelli[i], T. Venturelli[h], C. Voena[d], I. Volobouev[a], and R. Wigmans[a],
[a]Texas Tech University, Lubbock (TX), USA; [b]Dipartimento di Fisica, Università di Pisa and INFN Sezione di Pisa, Italy; [c]Dipartimento di Fisica, Università di Cagliari and INFN Sezione di Cagliari, Italy; [d]Dipartimento di Fisica, Università di Roma"La Sapienz" and INFN Sezione di Roma, Italy; [e]INFN Sezione di Pavia, Italy; [f]INFN Sezione di Pavia and Dipartimento di Fisica Nucleare e Teorica, Università di Pavia, Italy; [g]Iowa State University, Ames (IA), USA; [h]Dipartimento di Fisica, Università della Calabria and INFN Cosenza, Italy; [i]CERN, Genève, Switzerland.

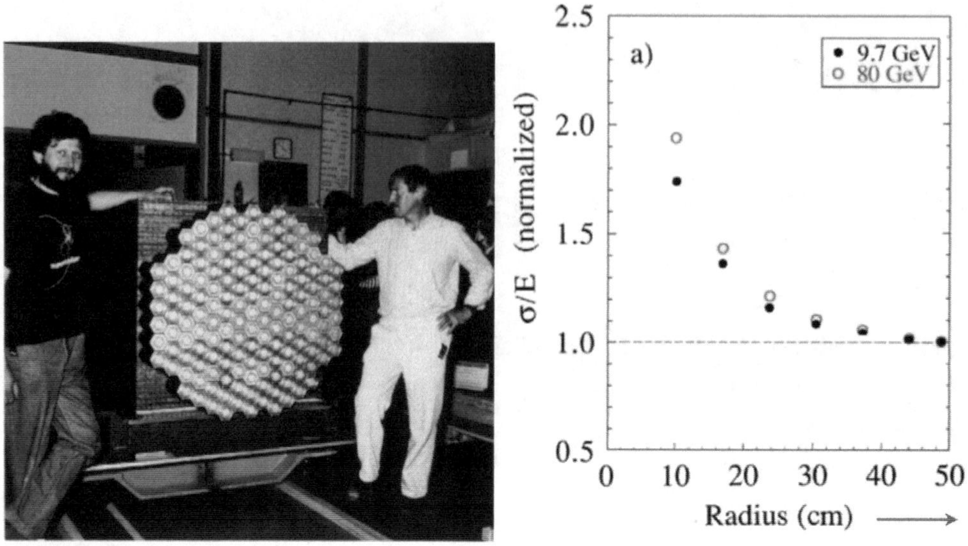

Figure 1. (a) Photograph of the compensating SPACAL module that achieved a hadronic energy resolution of nearly $30\%/\sqrt{E}$ in 9.5 $\lambda_{\rm int}$ of Pb, and with 155 channels; and, (b) the energy resolutions for hadronic showers at 9.7 and 80 GeV as a function of the volume of the calorimeter over which signals are collected. The resolution improvement is nearly a factor of 2 going from a radius of 10cm to 50cm.

therefore persist in the calorimeter volume for tens of nanoseconds as the neutrons elastically scatter from the absorber nuclei.

The protons in the H-C molecules of the scintillating fibers are the only free protons in the calorimeter. A neutron signal is generated when a neutron elastically scatters from a proton in a scintillating fiber, $np \rightarrow np$, and the recoil proton ionizes in the scintillator generating scintillation light.[2]

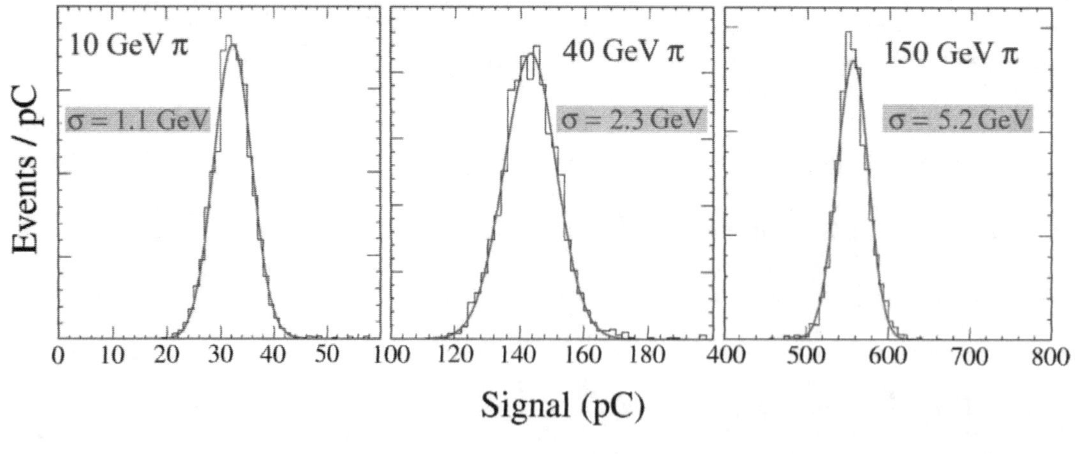

Figure 2. The direct pulse height distributions from the SPACAL module at 10, 40 and 150 GeV π^- beam energy. Distributions such as these the only honest way to correctly assess the quality of a calorimeter: the mean and width, the low side tails and the high-side tails, the linearity of the response in energy, and the degree to which the distributions are non-Gaussian. From *Nucl. Instr. Meths.* **A308** (1991) 481.

The overall energy resolution of the SPACAL module (1989, Fig. 1) is still the best ever achieved, and the pulse height response at three energies is shown in Fig. 2. These data also have a constant term similar to that found in the GLD test (Sec. 2), about 1% or less.

We emphasize that showing the full pulse height response distributions, such as those shown in Fig. 2 in units of pC, is the only honest way to correctly assess the quality of a calorimeter. Everything about these distributions is important: the mean and width, the low-side tails and the high-side tails, the linearity of the response in energy, and the degree to which the distribution is non-Gaussian.

2. GLD concept detector compensating calorimeter

The GLD ("Global Large Detector") concept detector [2] was designed for the ILC with a compensating Pb-scintillator sampling calorimeter in the ratio 8mm:2mm. This sampling is a factor of 2 coarser than the ZEUS prototype calorimeter tests and therefore one expects an electromagnetic energy resolution of $\sqrt{2} \times (\sigma_E/E)_{\text{zeus}} \approx \sqrt{2} \times 18\%/\sqrt{E} \approx 25\%/\sqrt{E}$. This design was thoroughly simulated and the corresponding calorimeter module was tested from 1 to 200 GeV with e^- and π^- beams.

The time-dependence of the energy resolution for π^--induced showers simulated at 4 GeV is shown in Fig. 3(a) out to 1000 ns. The hadronic energy resolution improves from about 32% to 25% essentially due to the accumulation of the neutrons as part of the total signal, thereby reducing the overall energy fluctuations. Further plots from this simulation in Ref. [2] show a 30% increase in the total signal due to the neutrons, and the distribution of the neutron arrivals in both space and time.

Both the electromagnetic and the hadronic energy resolutions are shown in Fig. 3(b) and both demonstrate a dominant $1/\sqrt{E}$ dependence with stochastic terms of 23.9% for e^- and 46.7% for π^-. Also impressive are the small constant terms of 0.8% and 0.9% for e^- and π^-, respectively.

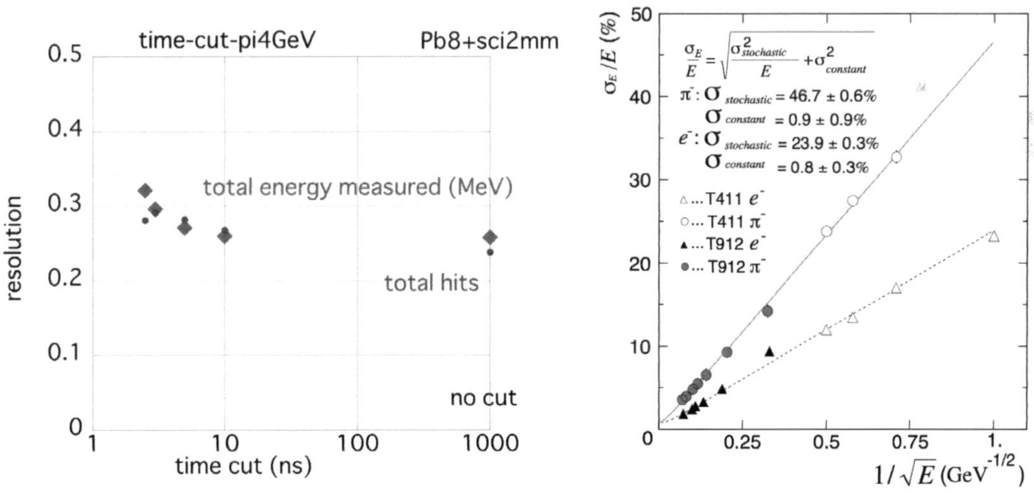

(a) (b)

Figure 3. GLD concept detector: (a) simulation showing the energy resolution as a function of an upper time cut (in ns) on the accumulated signal improving from about 32% for $t < 3$ns to 25% for $t < 1000$ns. (b) the corresponding data for both e^- and π^- beams demonstrating excellent energy resolutions and, more importantly, small constant terms of 0.9% and 0.8% for π^- and e^-, respectively.

[2] This scintillation light luminosity is suppressed by Birks recombination for densely ionizing slow protons by a factor of approximately $\sim (1 + k_B \cdot dE/dx)^{-1}$ for dE/dx in MeV/g-cm^{-2} and $k_B \approx 0.008$ g-cm^{-2}/MeV.

Journal of Physics: Conference Series **293** (2011) 012080 doi:10.1088/1742-6596/293/1/012080

3. DREAM dual-readout calorimeter

The DREAM geometry used in these neutron measurements is shown in Figs. 4(a,b). Four signals are collected and recorded in the digital oscilloscope: the scintillation signals summed for the central channel (chan-1), the inner ring of 6 channels (chan-2), the outer ring of 12 channels (chan-3), and the Čerenkov signal (chan-4) summed for 16 of the 19 channels[3].

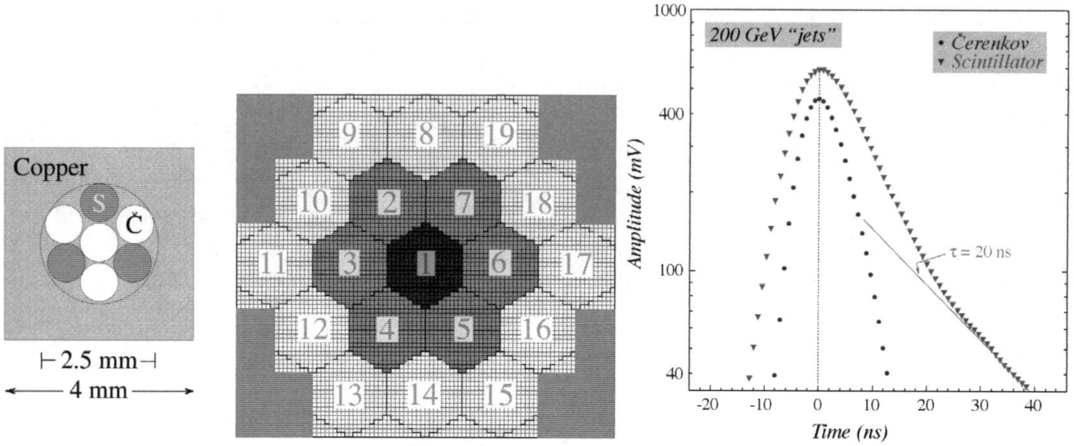

(a) (b) (c)

Figure 4. (a) The basic building block of DREAM showing the Cu absorber and the seven fibers (3 scintillating and 4 clear); (b) the channel geometry of the module. The filling material on the periphery is Al; and, (c) the average time distribution of the Čerenkov and scintillation light for the signal summed over the whole module showing the fast Čerenkov light and the slower scintillation light with a clear neutron component with a 20ns effective lifetime inside the module.

The mean Čerenkov pulse and the mean scintillation pulse are shown in Fig. 4(c), and the neutron signal is defined as the integral under the scintillation pulse from 20 to 40 ns for each event. The distribution of this neutron signal for 200 GeV "jets"[4] divided by the total energy is called the "neutron fraction", f_n, and is shown in Fig. 5(a) plotted against the usual dual-readout electromagnetic fraction, $f_{\rm EM}$, defined in Ref. [4]. The anti-correlation between f_n and $f_{\rm EM}$ is purely physical and represents fluctuations between π^0 and $\pi^\pm$ production in hadronic shower development.[5]

The Čerenkov pulse height distribution is shown in Fig. 5(b) as broad, asymmetric and non-Gaussian. Selections on the neutron fraction, f_n, reveal that this Čerenkov distribution is a sum of narrower Gaussians also shown in Fig. 5(b). Correcting the whole distribution linearly by f_n, the original raw Čerenkov distribution results in a narrower distribution with an effective resolution (σ/μ) improvement of about 20% at all three jet energies 100, 200, and 300 GeV. This is shown in Fig. 6(a) as a function of $1/\sqrt{E}$. The f_n-corrected response has a negligible constant term as indicated by the line drawn through the origin (not a fit). In addition, the effect on the resolution due to neutrons in the leakage counters is shown in Fig. 6(b).

All of this argues strongly for the benefits of event-by-event neutron measurements in high-precision hadronic calorimetry.

The energy resolution of a hadronic calorimeter is limited by fluctuations, primarily fluctuations in $\pi^\pm$ *vs.* π^0 production, secondarily by fluctuations in the binding energy losses from broken-up

[3] The three channels left out were on the periphery of the module and chosen uniformly

[4] These are "interaction" jets generated by placing a 0.1-$\lambda_{\rm int}$ lucite block in front of DREAM and accepting events with a charged particle multiplicity above 12 exiting the lucite.

[5] There are two physical effects which suppress the mean value of f_n: the small 1 kt volume of the DREAM module and Birks suppression.

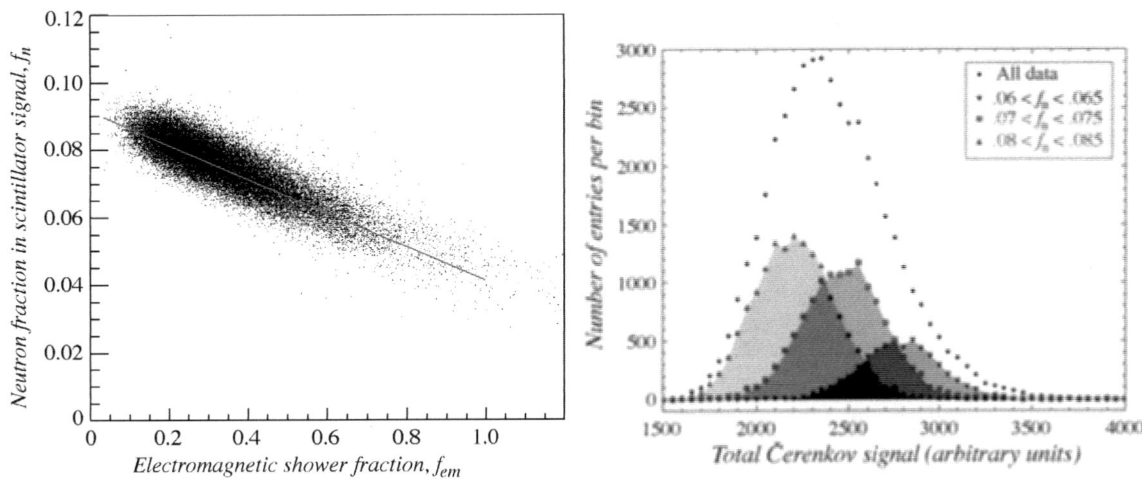

(a) (b)

Figure 5. A plot of the neutron fraction, f_n, against the electromagnetic fraction, $f_{\rm EM}$, shows their approximately linear anti-correlation and demonstrates the fundamental fluctuation between π^0 and $\pi^{\pm}$ production in hadronic interactions; and, (b) the decomposition of the full Čerenkov pulse height distribution into its Gaussian components as a function of f_n.

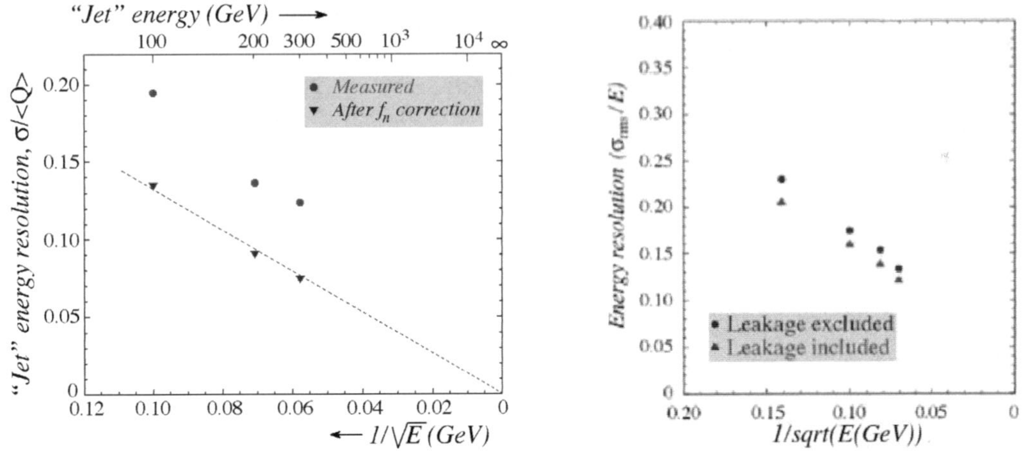

(a) (b)

Figure 6. (a) The resolution (σ/μ) of the overall raw measured Čerenkov response (upper red circles), and after linearly correcting the Čerenkov response by f_n (lower blue triangles); and, (b) the resolution with and without the addition of the neutron signals from the leakage counters.

nuclei, and thirdly, by leakage. In the DREAM module, the leakage is lateral and primarily consists of neutrons from the small module. In Fig. 7(a) we select small ranges in the electromagnetic fraction, $f_{\rm EM}$ (*e.g.*, $0.40 < Q/S < 0.42$), and look at the remaining fluctuations in the neutron fraction, f_n, plotted against the total Čerenkov signal. The f_n fluctuations are about 5%. For a slice at larger $f_{\rm EM}$ ($0.070 < Q/S < 0.75$) and therefore smaller neutron fraction, the fluctuations in f_n are about 10%. Quantitatively, for this $f_{\rm EM}$ selection and an additional selection on f_n, $0.045 < f_n < 0.065$, the distribution in total Čerenkov signal is shown in Fig. 7(b) and fitted to a Gaussian resolution of 4.7%. For a tighter selection of $0.050 < f_n < 0.055$, the fitted resolution narrows to 4.4%. This is close to the expected fluctuations due to leakage in the DREAM module.

It is perfectly clear that the measurement of neutron energy, both internal to DREAM (Fig. 6(a)) and external (Fig. 6(b)) in the leakage counters, improves the estimate of hadronic shower energy. The theoretical lower limit of $\sigma/E \sim 15\%/\sqrt{E}$ on hadronic energy resolution [3] will be sought

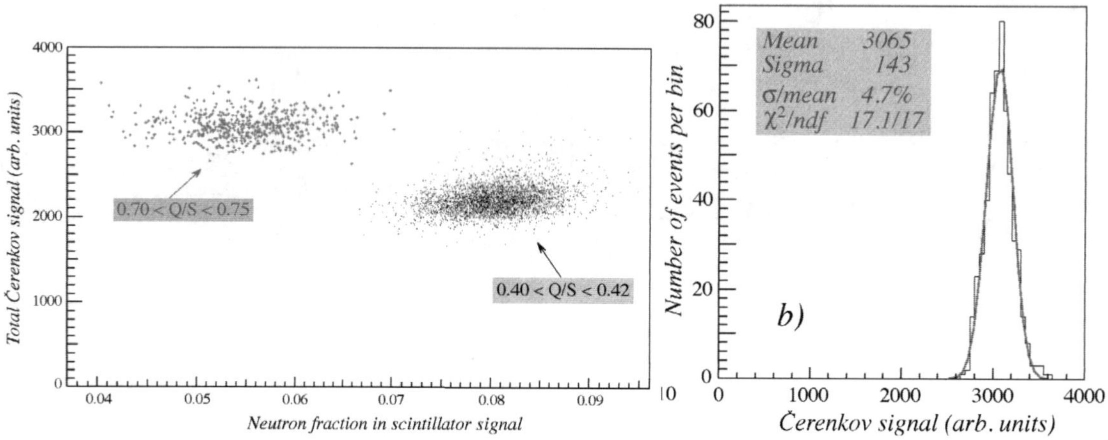

(a) (b)

Figure 7. For 200 GeV jets, (a) a scatter plot of the Čerenkov signal and the neutron fraction for two selections on the electromagnetic fraction, $f_{\rm EM}$; and, (b) the resulting resolution (σ/μ) for fixed $f_{\rm EM}$ given by $0.70 < Q/S < 0.75$ and restricted neutron content, $0.045 < f_n < 0.065$, is fitted to be 4.7%. Narrowing the f_n content to $0.050 < f_n < 0.055$ narrows the resolution distribution to 4.4%.

experimentally with a larger calorimeter of the DREAM type with expected mean leakage of about 1%. The measurements and calculations in Figs. 1(b) and 3(a) demonstrate that a large volume and long times are both required for good hadronic energy resolution that depends on the measurement of the neutrons liberated in nuclear break-up.

3.1. Acknowledgments

We acknowledge the US Department of Energy through grant DE-PS02-09ER09-02 for this work, and for its support of the original DREAM module and for continued support of these and other measurements by the DREAM collaboration. We also acknowledge the support of INFN in our plans for improving the neutron measurements. Finally, we acknowledge the excellent quality and variety of beams at CERN.

References

[1] Acosta D *et al.* 1990, *Nucl. Instr. and Meth.* **A294**, 193.
[2] Uozumi S, *et al.*, 2002, *Nucl. Instr. and Meth.* **A487**, 291.
[3] Wigmans R 2000, *Calorimetry: Energy Measurement in Particle Physics*, International Series of Monographs on Physics, Vol. 107, Oxford University Press (2000).
[4] Akchurin N *et al.* 2005, *Nucl. Instr. and Meth.* **537**, 537: "Hadron and Jet Detection in a Dual-Readout Calorimeter".

Journal of Physics: Conference Series **293** (2011) 012081 doi:10.1088/1742-6596/293/1/012081

Performance of a dual readout calorimeter with a BGO electromagnetic section

Gabriella Gaudio

on behalf of the Dream Collaboration

E-mail: `gabriella.gaudio@pv.infn.it`

Abstract. The dual readout technique has been tested on a hybrid calorimeter. The electromagnetic section of this instrument consists of 100 BGO crystals and the hadronic section is made out scintillating and Cherenkov fibers embedded in a copper matrix (DREAM). The electromagnetic fraction of hadronic showers is evaluated on an event-by-event basis from the relative amounts of Cherenkov and scintillation lights produced in the shower development. The performance of such a calorimeter in terms of energy resolution is presented. Effects of side leakage on detector performance are also studied.

1. Introduction

The response of an hadronic calorimeter can be expressed in term of the em shower fraction (f) and of the e/h ratio as

$$R = [ef + h(1 - f)]E_0$$

where e and h are the detector response to em and non-em shower components, respectively. The event by event fluctuations of f represent one of the main limitation to the hadronic resolution of non compensating calorimeter, even if the h/e value is measured.

With the dual readout technique, one uses two different active media, one producing Cherenkov light (C) and one scintillation light (S), therefore with very different h/e ratios (let's say c and s, respectively). Their responses will be:

$$C = [f + c(1 - f)]E_0 \qquad S = [f + s(1 - f)]E_0$$

If c and s are known, it is possible to measure the electromagnetic fraction:

$$f = \frac{c - s(C/S)}{(C/S)(1 - s) - (1 - c)}$$

and thus the released energy, corrected for the fluctuations of f

$$E_0 = \frac{C - \lambda S}{1 - \lambda}$$

where $\lambda = (1 - c)/(1 - s)$ is a constant of the calorimeter depending only on its mechanical structure [1]

In the following the application of this technique to an hybrid system made out a BGO matrix

Journal of Physics: Conference Series **293** (2011) 012081 doi:10.1088/1742-6596/293/1/012081

and the DREAM module is described. Before combining the energy measurements of the two sections, it is necessary to calibrate them and to determine the suitable corrections to the electromagnetic fraction fluctuation on an event-by-event basis. In particular, for the fiber detector the leakage contribution is evaluated, while for the BGO matrix we have focused on the resolution measurements.

Figure 1. The hybrid calorimeter system: the 100-crystal BGO matrix located upstream of the fiber module. Scintillator paddles for the measurement of the leakage are visible.

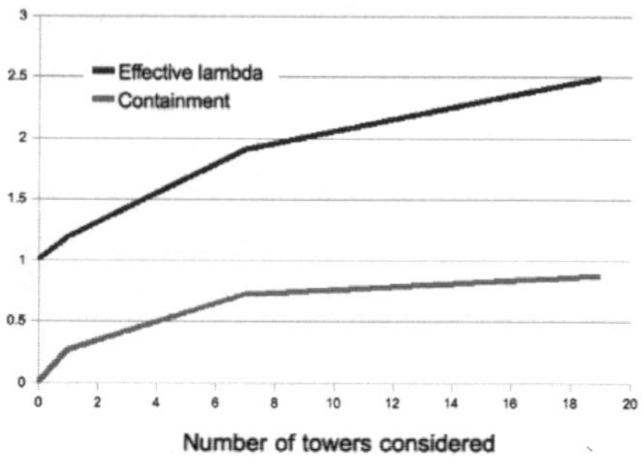

Figure 2. Effective lambda constant and the containment parameter as function of the number of the towers of the DREAM module summed in the measurement

2. Lateral leakage

Once the fluctuations of the e.m. fraction were successfully eliminated using the method described above, the resolution of the DREAM fiber calorimeter turned out to be limited mainly by the uncertainties on the lateral leakage. In fact, in present setup the fiber module was surrounded by 8 plastic scintillator paddles, as shown in Fig. 1, and the leakage was measured to be 5% - 10% for an 100 GeV pion beam steered in center of the detector. Measurements also show a good anti-correlation between the electromagnetic fraction measured in the fiber module and the leakage detector. As expected, the higher the electromagnetic content of the shower the better it is contained; moreover the em fraction increases with the beam energy as expected. From this information one can write the response for a not fully-containing detector for the two sensitive media, separately, as

$$C = [f + c\alpha(1-f)]E_0 \qquad S = [f + s\alpha(1-f)]E_0$$

where α is the contained fraction of the non-em part. Therefore the measured energy will be:

$$E_0 = \frac{C - \hat{\lambda}S}{1 - \hat{\lambda}}$$

defining $\hat{\lambda} = (1-\hat{c})/(1-\hat{s})$, $\hat{c} = \alpha c$ and $\hat{s} = \alpha s$. In Fig. 2 the effective constant $\hat{\lambda}$ and the containment parameter α are shown as a function of the number of towers of the DREAM module summed in the measurement. From the $\hat{\lambda}$ value extracted from the data, it is possible to determine the parameter α and therefore to correct the energy measurement for both the em fraction fluctuation and the leakage.

3. The BGO electromagnetic section

The electromagnetic section is a matrix made of 100 BGO crystal (Fig. 1). The matrix is readout by 12 PMTs (Photonis 3392B) arranged in 3 rows of 4 PMTs each. In order to extract the Cherenkov component from the light produced, we used UV filters [3]. For the data presented in this paper, the UV filters were used on the central PMT row only. The waveforms of the PMT signals were recorded using Domino Ring Samplers IV. A typical filtered signal is shown in Fig. 3. In order to extract the C and S contribution for filtered PMTs we fitted the acquired signals event-by-event.

The performance of the BGO matrix in terms of resolution and linearity has been evaluated

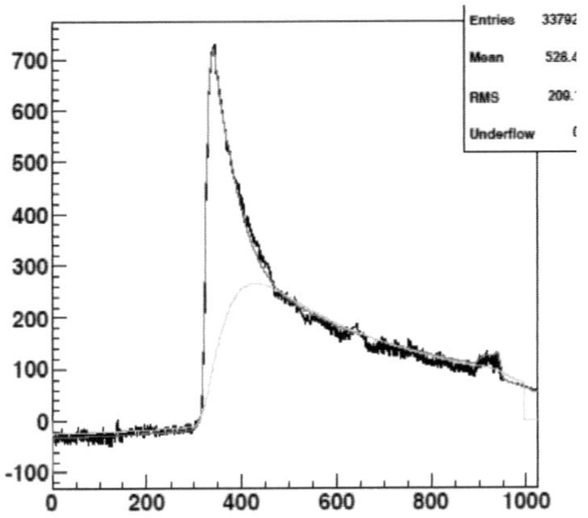

Figure 3. Fits used to disantangle Cherenkov (red line) and scintillation (green line) contributions in the signal from filtered PMT's

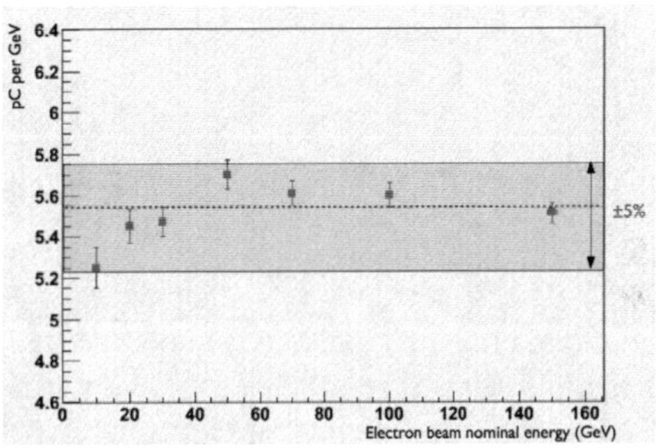

Figure 4. Linearity of the response for the Cherenkov signals as function of the electron beam nominal energy.

using the sum of the scintillation contribution from all the PMTs and, for the Cherenkov light, the sum of the signals of the 4 filtered PMTs. The energy resolution measured with electrons in the range 10 - 150 GeV is

$$\frac{\sigma_{E_C}}{E} = \frac{36\%}{\sqrt{E}} \oplus 3\% \qquad \frac{\sigma_{E_S}}{E} = \frac{33\%}{\sqrt{E}} \oplus 1\%$$

for Cherenkov and Scintillation respectively. Linearity for the Cherenkov signal as a function of the nominal beam energy is shown in Fig. 4 and measured to be within ±5%. The resolution obtained is strongly limited as a consequence of the non optimized readout. An improved scheme has been realized for the 2010 test beam.

4. A complete Dual Readout calorimetry system

After a suitable calibration with electrons of the ECAL section, we started to study the performance of the whole calorimeter with pion beams. The information collected with the two sections are combined to obtain an overall evaluation of the electromagnetic fraction of the shower. In order to perform such a measurement we selected only those events in which pions start showering in the crystal section. For showers developing late in the BGO matrix, one could expect some correlation between the e.m. fractions measured in the two sections. In fact, the C/S ratios in the two calorimeters shows a good correlation (Fig. 5).

In both sections, the resolution depends on the single resolutions on C and S, respectively, and on the correlation between these two quantities. The poor energy resolution resulting from both

Journal of Physics: Conference Series **293** (2011) 012081 doi:10.1088/1742-6596/293/1/012081

the Cherenkov and scintillation contributions in the BGO matrix, as described above, makes the correlation in the ECAL section as low as 0.45. In DREAM the correlation was measured to be 0.8 0.9, despite the presence of lateral leakage.
The total energy measured is obtained by summing the energy measurements of each section, after corrections for both em fractions and leakage. The distribution of the total energy has a symmetric and amost gaussian shape (Fig. 6), with respect to the total Cherenkov and Scintillation spectra respectively, that show the typical response of a non compensating calorimeter.

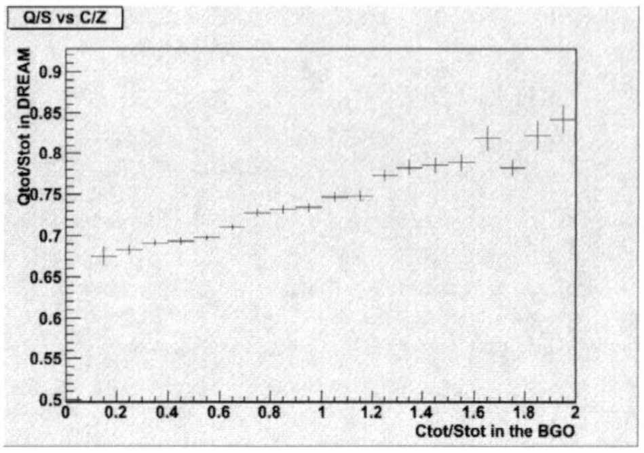

Figure 5. Correlation between the ratios of the scintillation and the Cherenkov signals in the two calorimeters

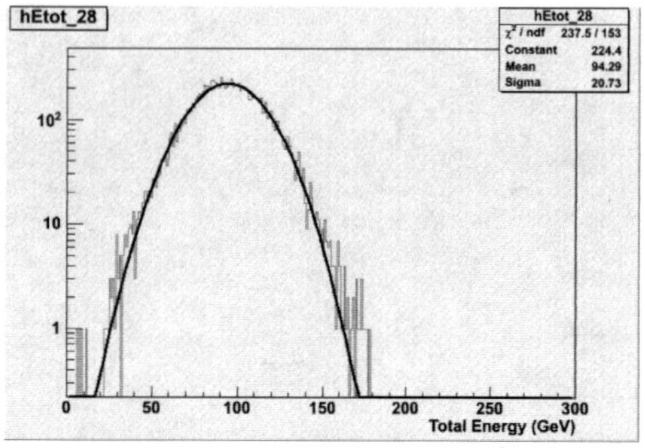

Figure 6. Total energy distribution obtained using both BGO and fiber detectors, corrected for both leakage and em fraction.

5. Conclusions

Dual readout calorimetry has been successful in measuring event by event the electromagnetic fraction of an hadronic shower. The knowledge of f allows to correct the response of a non compensating calorimeter. To realize a calorimeter with very good both em and hadronic resolution one could use an homogeneous em section. Scintillating crystals produce Cherenkov light in addition to scintillation. Therefore one can apply the dual readout technique also to this kind of detector. For the first time a complete dual readout calorimeter, with a BGO e.m. section, was built and tested. Although the non-perfect readout system adopted (only 4 PMTs with filters) the Dual Readout principle worked quite well. A clear correlation between the e.m. fraction measured in the two sections was found and the energy measured by the complete system showed a gaussian shape and didn't show the tail typical of a non-compensating calorimeter system. A better readout for the BGO matrix is being presently used for test beam at cern.

References

[1] N. Akchurin et al, NIM A537 (2005) 537-561
[2] N. Akchurin et al, NIM A610 (2009) 488-501
[3] N. Akchurin et al, NIM A595 (2008) 359-374

Calorimetry with meta-crystals

Etiennette Auffray, Paul Lecoq, Georgios Mavromanolakis

European Organization for Nuclear Research, CERN, CH-1211 Geneva, Switzerland

E-mail: gmavroma@mail.cern.ch

Abstract. We present the meta-crystals concept, an approach that consists of using both undoped and properly doped heavy crystal fibers of identical material as the active medium of a calorimeter. The undoped fibers behave as Cherenkov radiators while the doped ones behave as scintillators. A dual readout calorimeter can be built with its sensitive volume composed of a mixture of both types of crystals. In addition if the calorimeter is adequately finely segmented it can also function as a particle flow calorimeter at the same time. In this way one could possibly combine the advantages of both the particle flow concept and the dual-readout scheme. We discuss the approach of dual readout calorimetry with meta-crystals made of Lutetium Aluminium Garnet (LuAG) and present studies on the material development, first testbeam activities and results based on simulation for understanding the performance trends. We close with a brief outlook on open issues and further R&D needed to proceed from an ideal conceptual case to the design of a realistic detector.

1. Introduction

An experiment at a future lepton collider sets strict requirements for vertex, tracking and calorimetric detectors. The R&D effort on the design of the calorimeters is directed mainly into two fronts, the particle flow approach [1] and the dual-readout scheme [2]. Within the former both the electromagnetic and the hadronic calorimeters are very finely segmented in order to allow very efficient pattern recognition and thus identify and track all particles in a jet. In the dual readout approach one would like to have a calorimeter that is capable of measuring both the ionisation/scintillation and the Cherenkov signals generated by a hadronic shower. By doing so it is possible to determine on an event by event basis the electromagnetic fraction of the shower in order to cancel/correct for this source of fluctuation that degrades the energy resolution of the calorimeter. In the following we discuss a dual readout approach that is based on meta-crystals [3].

2. Calorimetry with meta-crystals

The meta-crystals concept consists of using both undoped and properly doped heavy crystal fibers of identical material for the generation of the signal of the detector. The undoped fibers behave as Cherenkov radiators while the doped ones behave as scintillators. A dual readout calorimeter can be built with its sensitive volume composed of a mixture of both types of crystals. In addition if the readout volume is adequately finely segmented longitudinally and transversely it can also function as a particle flow calorimeter at the same time. The conceptual design of a readout unit of such a calorimeter is illustrated in figure 1. The unit consists of a structured distribution of different types of fibers. Typical dimensions of a unit are of the

Journal of Physics: Conference Series **293** (2011) 012082 doi:10.1088/1742-6596/293/1/012082

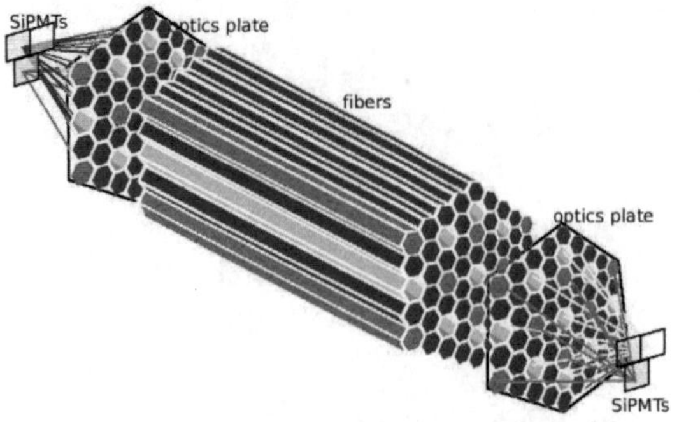

Figure 1. Conceptual design of a readout unit composed of a structured distribution of three different types of crystal fibers, Cherenkov, scintillating and neutron sensitive. The light from the different types of fibers is directed to different silicon photomultipliers by micro-lenses plates at both ends of the unit. (Not to scale)

order of $1 - 1.5\ R_M$ and $20 - 25\ X_0$ in the transverse and longitudinal direction respectively. In addition to Cherenkov and scintillating fibers one could also have fibers that would be sensitive to neutrons and so to have a triple readout calorimeter. Candidate materials are those based on lithium or boron which have high neutron capture cross-section. As shown in the figure the light from the different types of fibers is directed to different silicon photomultipliers or other solid state photodetectors by using diffractive optics light concentrators (micro-lenses). The diffractive optics plate is patterned so to match the structure of fibers and can be attached to both ends of the unit. The development of such a concept requires a comprehensive program of R&D studies in various fronts, from material development and crystal fiber production to simulation studies, testbeam activities and prototyping. Recent progress on these topics is discussed below.

2.1. Material development and related studies

A candidate material under study to be used for this concept is the Lutetium Aluminium Garnet $Lu_3Al_5O_{12}$ (LuAG) crystal. Its properties are summarised in table 1. This crystal material combines a high density of 6.73 gr/cm^3 and relatively short radiation and interaction lengths of 1.41 cm and 23.3 cm respectively sufficient to build a dense and compact detector. Its high refractive index of 1.842 at 633 nm and low energy threshold for Cherenkov radiation at 97 keV for electrons makes it a good Cherenkov radiator (to be compared with 1.55 and 190 keV respectively for quartz). Furthermore when LuAG is doped with cerium (Ce) it shows excellent scintillating performance with high light yield of more than 25000 photons per deposited MeV, emission peaked at 520 nm and decay time of about 60 nsec. A faster scintillation signal can be achieved if LuAG is doped with praseodymium (Pr), then the scintillation signal is characterized by UV emissions peaked at 320 nm and 370 nm and decay time of 20 nsec (see figure 4 and [5]).

Table 1. Physical and optical properties of Lutetium Aluminium Garnet (LuAG) crystal.

Physical properties		Optical properties	
Density	6.73 gr/cm^3	Light yield (Ce doped)	> 25000 ph/MeV
Zeff	62.9	Emission wavelength	520 nm (Ce doped)
Radiation length X_0	1.41 cm	Decay time	60 nsec (Ce doped)
Interaction length λ_I	23.3 cm	Refractive index	1.842 at 633 nm
Melting point	2260 °C	Cherenkov threshold	97 keV
Thermal expansion	8.8 10^{-6}/°C	Max Cherenkov angle	57 °
Thermal conductivity	31 W/m°C	Total reflection angle	33 °

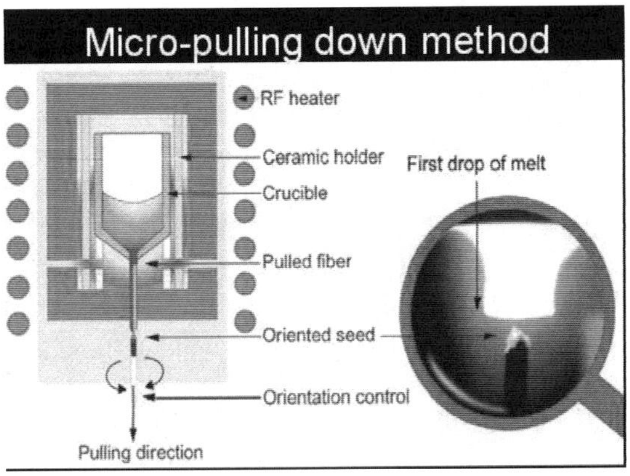

Figure 2. Schematic of the micro-pulling down method used in the production of crystal fibers.

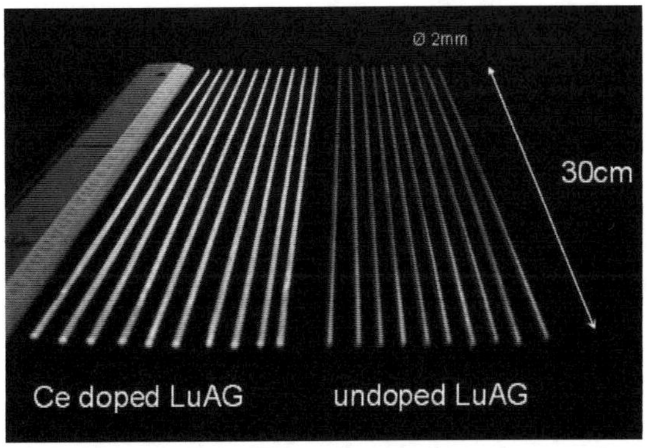

Figure 3. Samples of Ce doped and undoped LuAG crystal fibers produced by Fibercryst-Lyon with the micro-pulling down method.

Bulk LuAG crystals can be produced with traditional growth methods like the Czochralski or Bridgman-Stockbarger technique. The ingot produced then undergoes cutting and polishing procedures to the desired shape and dimensions. With these methods it is very difficult to produce fibers or in general elongated crystals with large aspect ratio. A technology that can be used for this purpose is the micro-pulling down method, a schematic of which is shown in figure 2. The University of Lyon1 in collaboration with Fibercryst-Lyon and Cyberstar-Grenoble have improved the method and applied it successfully to the production of fibers made of LuAG and other crystal materials [4]. With this method crystal fibers with diameter between 0.3 and 3 mm and length up to 2 m can be produced. The pulling rate is ranging from 0.1 to 0.5 mm/min and is considerably faster than the traditional methods, about a factor 10 and 50 times faster than Czochralski and Bridgman-Stockbarger method respectively. Another advantage of this technique is the fact that the capillary die can be non-cylindrical e.g. square, hexagonal etc, and so fibers with these shapes in cross-section can be produced as well. The method is yet to be optimised further for industrial application and it is expected that its overall cost per unit volume of production will be comparable to that of standard crystal growth methods. Samples of Ce doped and undoped LuAG crystal fibers, 2 mm in diameter and 30 cm in length, produced with this method are shown in figure 3. Figure 4 shows a typical example of the series of measurements we perform on a sample to study its properties with respect to light transmission, excitation-emission, attenuation, diffusion etc for optical characterisation and further development of the crystal material [5].

2.2. Testbeam activities and simulation studies

Bundles of undoped and Ce doped LuAG crystal fibers have been exposed to electron beam at CERN SPS during short periods in 2009 and 2010. The fibers were read-out from both ends with photomultiplier tubes. The setup was temperature controlled and could be rotated so that the beam particles can transverse the fibers at given angle. A bundle of 20 fibers, with dimensions of 2 mm in diameter and 80 mm in length each, wrapped in teflon box can be seen in figure 5. The corresponding signal pulses recorded are also shown. They are typical Cherenkov and scintillation light pulses as expected respectively. Some results of the data analysis are presented in figure 6. Figure 6(a) shows the left-right asymmetry of the recorded signal for different angles of orientation of the fibers with respect to the axis of the incident beam and for different setup configurations and samples. The undoped LuAG fibers show strong asymmetry

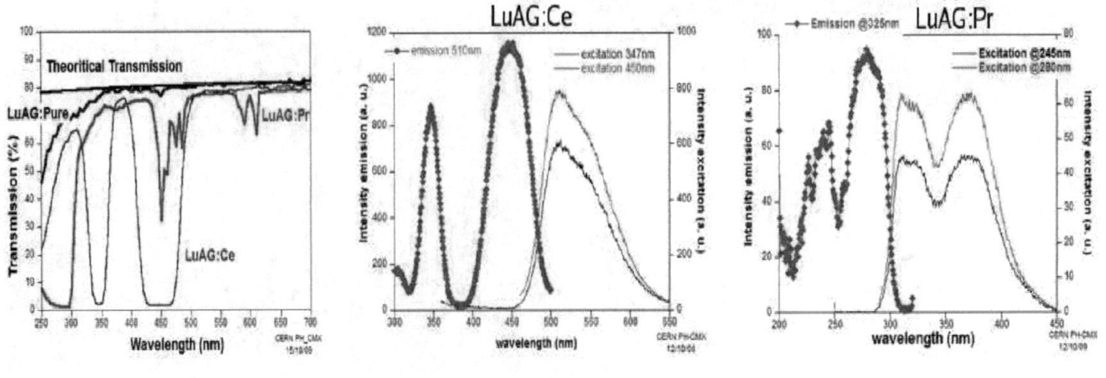

Figure 4. Illustration of series of measurements for optical characterisation of crystal fibers.

as expected from a Cherenkov based detector due to the directionality of the signal generation mechanism. While the Ce doped fibers show less or no asymmetry since the scintillation light is isotropically emitted. The full-width at half maximum of the average pulses, of the order of 100 nsec for scintillating fibers and 7 nsec for the Cherenkov ones, and the rise time, about 20 nsec and 5 nsec respectively, can be seen in figure 6(b) and (c). Further studies included the temperature dependence of the behavior of the fibers [6]. Similar studies are also done at a testbench using cosmic muons. So far we have performed small scale tests equivalent to the level of a single calorimetric channel, our plans in the near future include the construction and study of a multichannel module.

In the simulation front we are performing systematic scanning of the parametric space with design parameters under study such as transverse and longitudinal granularity, sampling frequency/readout fraction, total length of calorimeter, mixture of conventional and dual readout components, corresponding composition etc. Figure 7(a) summarises the results from a set of case studies with a $4.3\times4.3\times8.6\ \lambda_I^3$ ($\approx 1\times1\times2$ m^3 LuAG) calorimetric volume and for different readout/sampling fractions of the scintillation/ionisation and Cherenkov signals. The studies are with single pions in the energy range up to 50 GeV. The figure shows the performance in terms of the stochastic term of the energy resolution after the dual readout technique is applied. The left most series of points represent the resolution performance of a conventional detector i.e. without any readout of the Cherenkov signal and so where the dual readout correction scheme is not applied. Various sampling configurations of scintillation and Cherenkov signal readout are studied, covering the range of readout fraction from 100% down to 6.25%. We observe an improvement in the energy resolution achieved with the dual readout technique compared to conventional single signal readout and with a trend of improvement being more significant as we approach cases with smaller sampling readout fractions. The influence of the total length of the calorimeter is shown in figure 7(b) and (c) for two cases of scintillation sampling readout 100% and 12.5% respectively. Similar systematic studies combined with very preliminary cost considerations and technical challenges to be addressed help us to evaluate the performance trends, to understand any showstoppers and to proceed from an ideal case to a realistic detector.

3. Summary and outlook

We discussed the meta-crystals concept an approach that consists of using undoped and Ce doped LuAG crystal fibers as the active medium to build a dual readout calorimeter. We presented some aspects of the comprehensive R&D effort that has been established in our group with respect to material development and characterisation, fiber crystal production, first small scale testbeam activities and simulation studies for systematic scanning of the design parameters

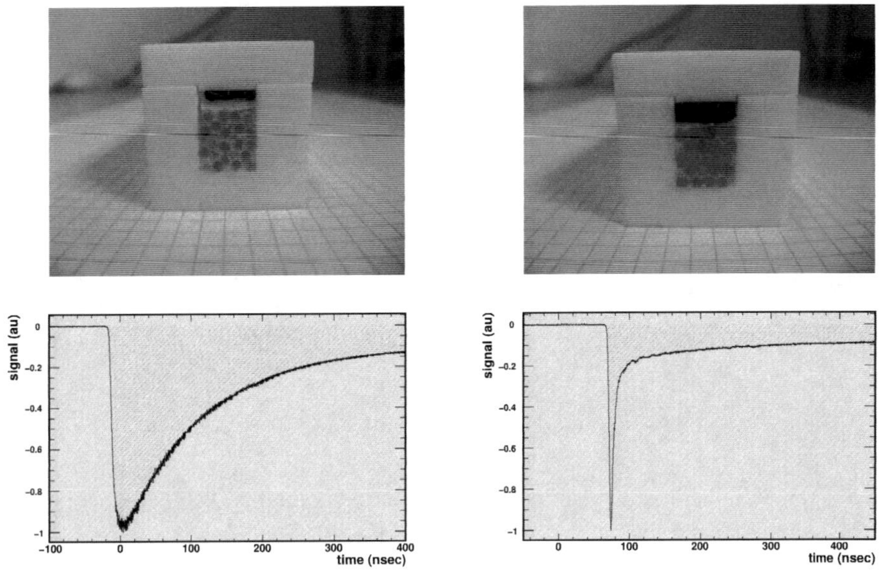

Figure 5. Bundles of Ce doped (top left) and undoped (top right) LuAG fibers and corresponding typical signal pulses recorded (bottom row).

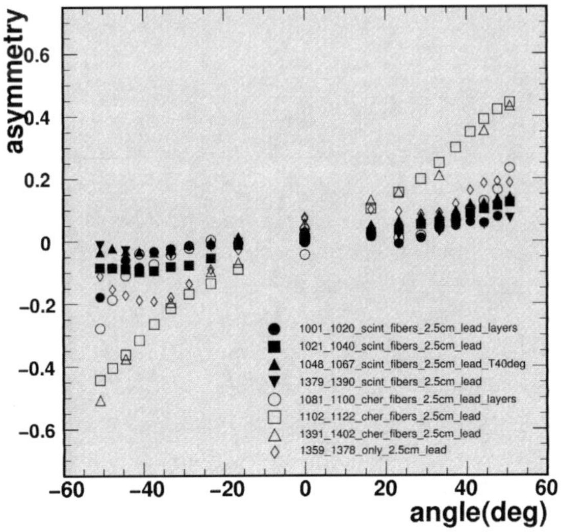

(a)

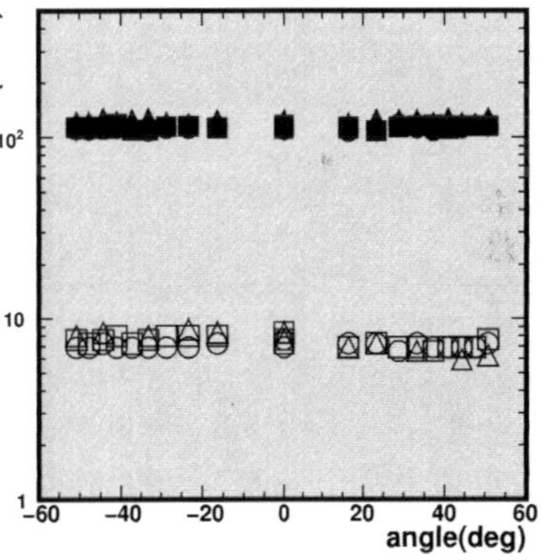

(b)

Figure 6. Normalised left-right asymmetry of recorded signal with bundles of Ce doped (solid symbols) and undoped (open symbols) LuAG fibers for different orientation angle with respect to the incoming electron beam (a). Full-width at half-maximum (b) and risetime (c) of average signal pulses recorded.

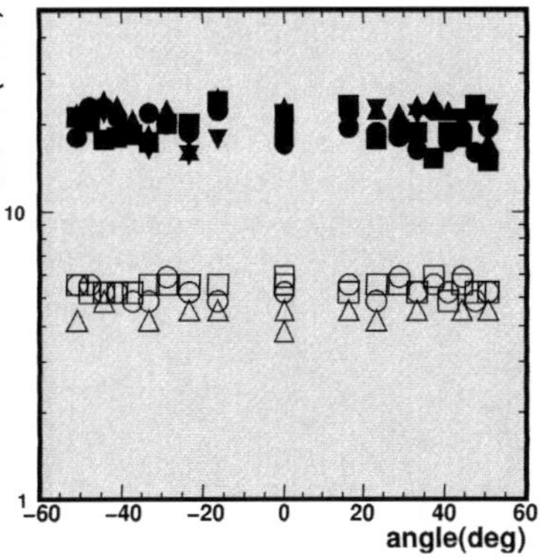

(c)

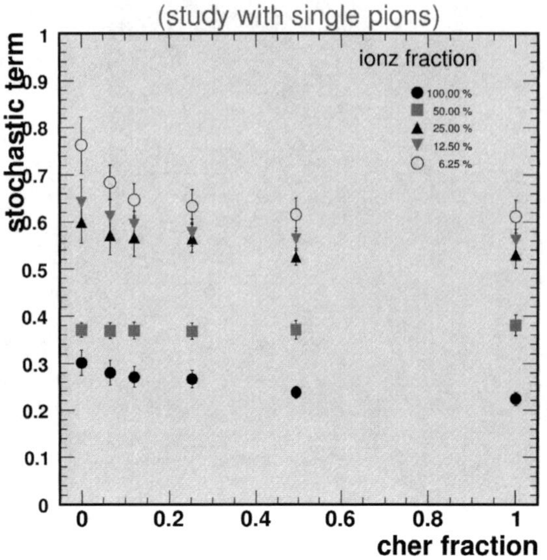

(a)

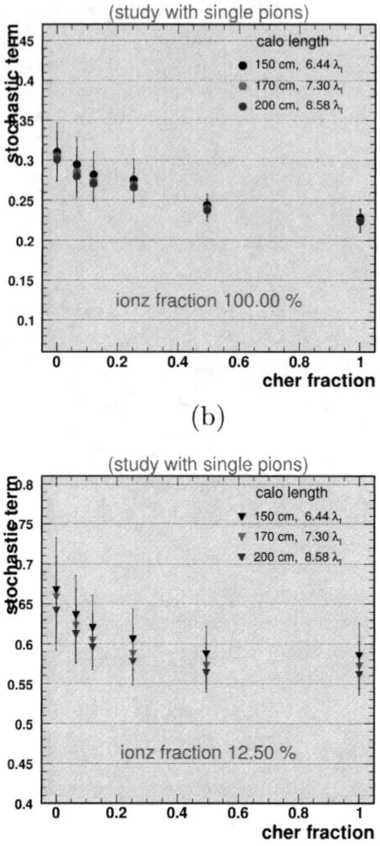

(b)

(c)

Figure 7. Simulated performance, in terms of the energy resolution's stochastic term, of $4.3 \times 4.3 \times 8.6\ \lambda_I^3$ dual readout calorimeters with various sampling configurations of scintillation/ionisation and Cherenkov signal readout (a). Results for various calorimeter lengths and with scintillation/ionisation readout fraction of 100% (b) and 12.5% (c).

of a calorimeter. We are at an early stage of development of such a concept with many design issues that should be studied further and which need rigorous R&D and prototyping. Open questions and challenges include among others the optimisation of crystal fiber production in large quantities and related cost drivers, development of a readout scheme based on micro lenses, construction challenges and scale-up problems. These can only be answered through the usual phase of prototype development, test and study of 1 permille to 1 percent and 10 percent modules of the final detector. We continue our effort and plan to study further some of the open issues in the near future.

References

[1] see articles of the CALICE Collaboration, *in these proceedings*
[2] N. Akchurin *et al, Nucl. Instr. and Meth. A 537(2005)537*
[3] P. Lecoq, *Jour. Phys. Conf. Series 160(2009)012016, Proc. of CALOR2008*
[4] B. Hautefeuille *et al, Jour. Cryst. Growth 289(2006)172*
[5] E. Auffray *et al, Proc. of the 2009 IEEE Nuclear Science Symp. and Medical Imaging Conf., p.2245*
[6] E. Auffray *et al, IEEE Trans. on Nucl. Sci. 57(2010)1454, Proc. of SCINT2009*

Journal of Physics: Conference Series **293** (2011) 012083 doi:10.1088/1742-6596/293/1/012083

Total Measurement Calorimeter TMC

Tohru Takeshita

Shinshu University

tohru@shinshu-u.ac.jp

Abstract. A new type of total absorbing calorimeter with active absorber is investigated . Some initial results are presented and further activities are also discussed to archive TMC.

1. Introduction

The total absorption calorimeter had been constructed and played relevant role in many high energy experiments for the electro-magnetic calorimeter (ECA). However, at higher energy experiments, the crucial physics will be derived by measuring the energy of jets, where those jets are the final states of heavy relevant particles, such as Higgs, W/Z, and top quarks. The jet energy measurement is dominated by the energy measurement of hadrons in jets.

There are two ideas to resolve the jet measurement. One is called the PFA approach, which is divided into sum of three measurements, say, track momentum, photon energy, and neutral hadron energy[1]. The other one is so called dual read out solution. There are several approaches to achieve the dual read out system to have energy measurement in two ways [2]. This work is classified into the dual read out approach.

In the context of the dual read out scheme, two kinds of non-independent measurement are extracted from the calorimeter. They are charged hadron interactions and neutral pion EM shower. Moreover we add here additional factor of the total measurement in the absorber in this work.

In the sandwich calorimeter detector, usually we lose information from the absorber, However, we introduce an active absorber system which leads to a total measurement calorimeter with longitudinal shower profile, since the calorimeter composed of sandwich type. A sort of information from the absorber energy measurement reflects a specific part the hadron interactions. The hadrons drop energy in the absorber material through hadronic interaction, which mostly create charged and neutral pions. The charged hadrons lose their energies by electro-magnetic (EM) interaction with the electrons though their tracks. However, the neutral pions decay in short time and generate two photons, which emerge as the EM showers. This shower is easily detected and measured in the heavy transparent absorber such as lead glass, by producing Cherenkov lights.

2. Hadron interaction and the calorimeter

The hadron interaction consists of mainly the strong interaction as well as the electromagnetic interaction. The strong interaction, furthermore, include the nuclear breakup or nuclear fusion. This makes the hadron interactions to be complicated phenomena. The calorimeter construction is difficult due to those facts, as well. Furthermore, those ejected neutrons from the nuclear interactions, will be detected as a scattered protons after some time in widely spread out areas.

2.1. Hadron energy measurement

Jet final states should be identified and measured to determine the physics interaction process, therefore measurement of the jet energy plays a crucial role on the physics analysis. The jets are composed of mostly hadrons at the fragmentation level, then the neutral pions decay into two photons immediately. Thus a jet is detected as a group of charged hadrons and photons. A common idea to measure the jet energy effectively is to equip two calorimeters to fit the photon and hadron energy measurement. They are called the electromagnetic calorimeter (ECAL) and the hadronic calorimeter (HCAL) , being optimize the absorber and active material for both interactions.

In some ECAL, total absorption electromagnetic calorimeter which composed of the lead glass and photo-sensor only has been constructed and employed widely. The electromagnetic (EM) shower is detected and measured by the Cherenkov lights. The total absorption calorimeter, generally speaking, has a good energy resolution, since total EM-shower track length is affecting for the measurement of the EM shower energy [3].

On the other hand the hadron energy measurement has a relatively poor resolution, because of shorter track length of charged hadrons passing through the HCAL Furthermore the production of neutral pions generates EM showers in the hadronic shower. The contribution in the hadronic energy measurement of the neutral pions is sometimes dominating, because the energy deposit of the EM shower is much bigger than that of charged hadron passing. The fluctuation of the production of the neutral pions makes the hadron energy resolution poor.

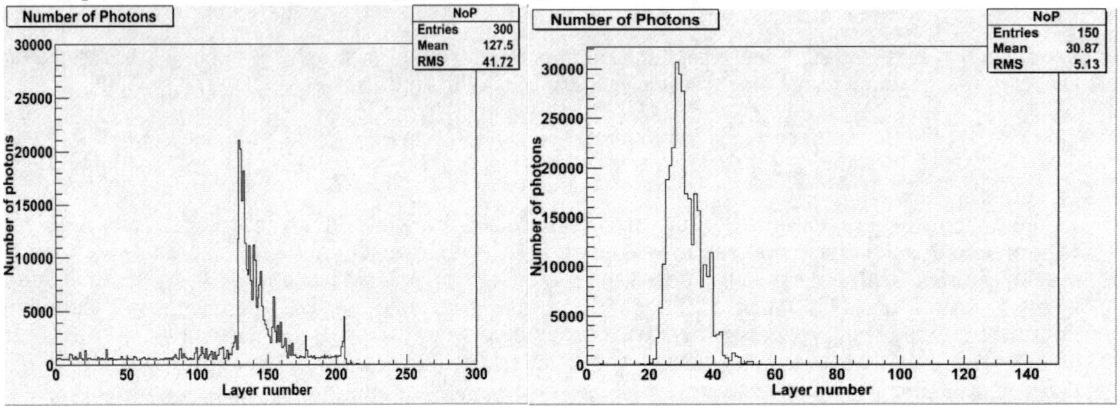

Figure1. (left) A simulated 20GeV char4ged pion interactions in the fine segmented HCAL in the longitudinal direction. A energy peak around 130th layer is thought to be a production of a neutral pion.

Figure2. (right) A simulated 5GeV neutral pion at 20th layer. The pulse hight profile in the longitudinal direction looks similar to the figure1.

By introducing the fine segmentation in the longitudinal direction of the HCAL, we are able to find the neutral pion production event by event basis. An example event, which contains a clear production of neutral pion is shown in figure 1 in a simulation of a 20 GeV pion shower. In figure 1 right, we add and a 5GeV neutral pion generated at 20th layer. One can find the sequential large pulse height in successive layers in the longitudinal segmentation.

2-2 Total energy measurement

In order to improve the hadron energy measurement, avoiding the big fluctuation due to the EM shower production is apparent. Thus we need to identification and precise measurement of the neutral pions. To achieve this, we introduce active absorber which is made of such as lead glass or other transparent , high Z , heavy material by detecting the Cherenkov light. Those heavy material is used as an absorber of the charged hadrons, as well.

The figure 3 shows an idea of such detector with fine longitudinal segmentation. In this scheme, lead glass block is employed for the absorber and active Cherenkov radiator with a small photo-sensor. For the charged energy measurement, as usual hadron calorimeter, we set scintillator layers. This scintillator will work as a neutron radiator as well, so as to detect really the total energy measurement in the hadronic interaction.

Since this detector works as a combination of ECAL and HCAL, we need to test the performance in electrons and pions. In figure 4, a scatter plot of simulated 1GeV electron are shown. The total number of photons generated in the lead glass absorbers and that of scintillators are scatter-plotted. A clear anti-correlation is visible, in addition to showing better energy resolution in the Cherenkov light detection. The end point of the anti-correlation plot at lowest energy corresponds to the initial energy ; 1 GeV, in this case.

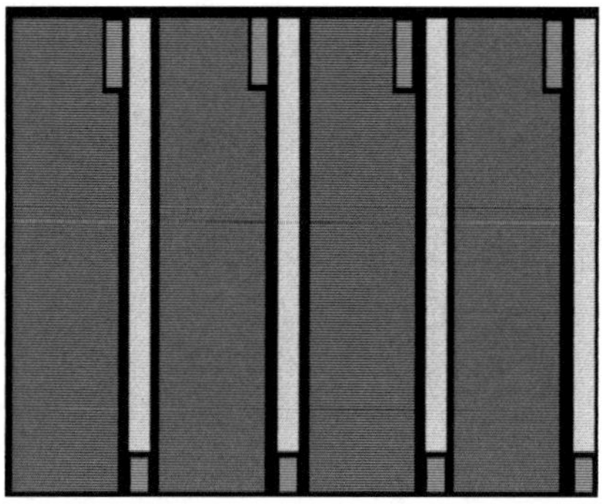

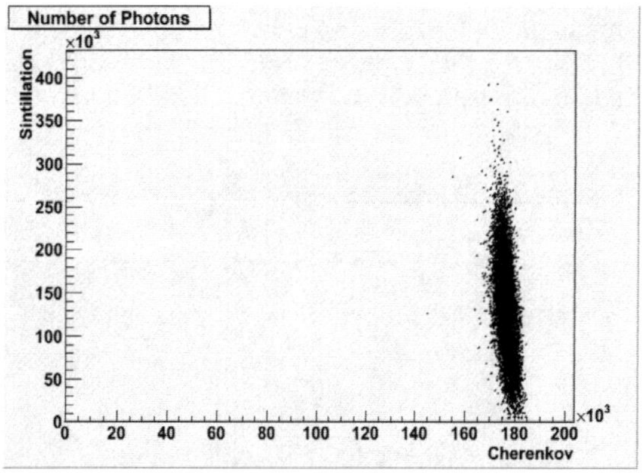

Figure 3 (left) An example TMC detector, which composed of a sandwich calorimeter with the lead glass (in green) and the scintillator (in light blue). The small blocks (in pink) indicate the photo-sensors for Cherenkov and scintillation lights.
Figure 4 (right) A simulated scatter plot with number of Cherenkov photon vs number of scintillator photons.

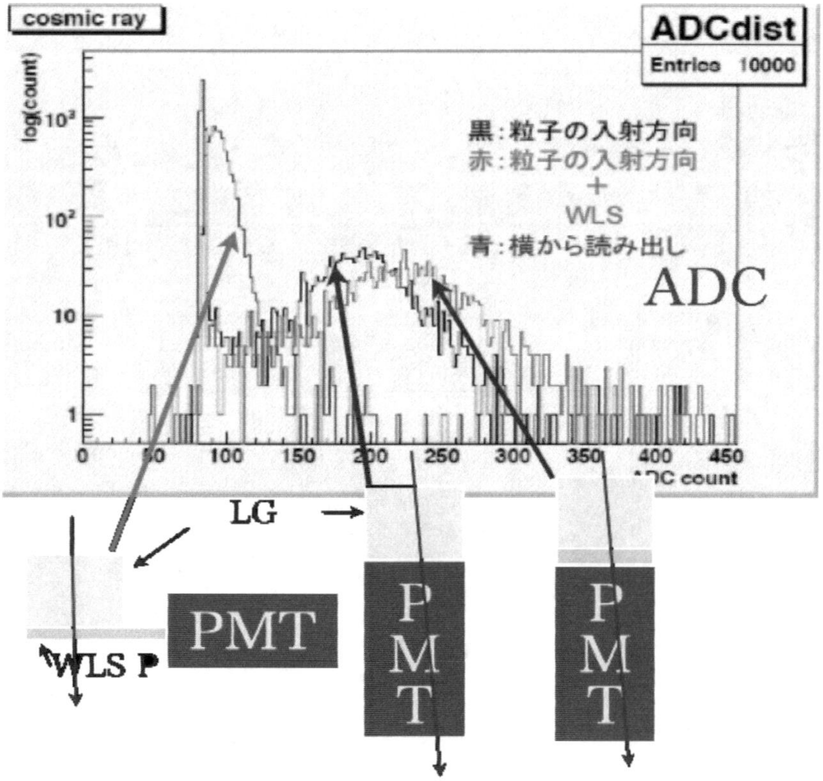

Figure 5; ADC distribution of cosmic muons, the blue line indicates with using a wave length shifting plate (WLSFP) , the black with direction connection and the red with a WLSP on the PMT between the lead glass.

3 . Cherenkov measurement

With using a lead glass block of 60mm long, we detected the cosmic ray muons under different photo-coupling to a photo-multiplier. It is shown in figure 5.

Although the accumulation of Cherekov light is not sufficient, further trial will be carried out, soon. However, if we give up to detect MIP particles, such as muons, the EM shower detection is quite easier. We have carried out a test beam experiment with 3GeV electrons. There we found a much

Journal of Physics: Conference Series **293** (2011) 012083 doi:10.1088/1742-6596/293/1/012083

bigger number of Cherenkov light by passing the electron shower, even in a slice of a EM shower. At this test experiment, we used the same thickness of 5cm for the lead glass blocks. On the other hand the calibration scheme must be resolved by other than high energy muons.

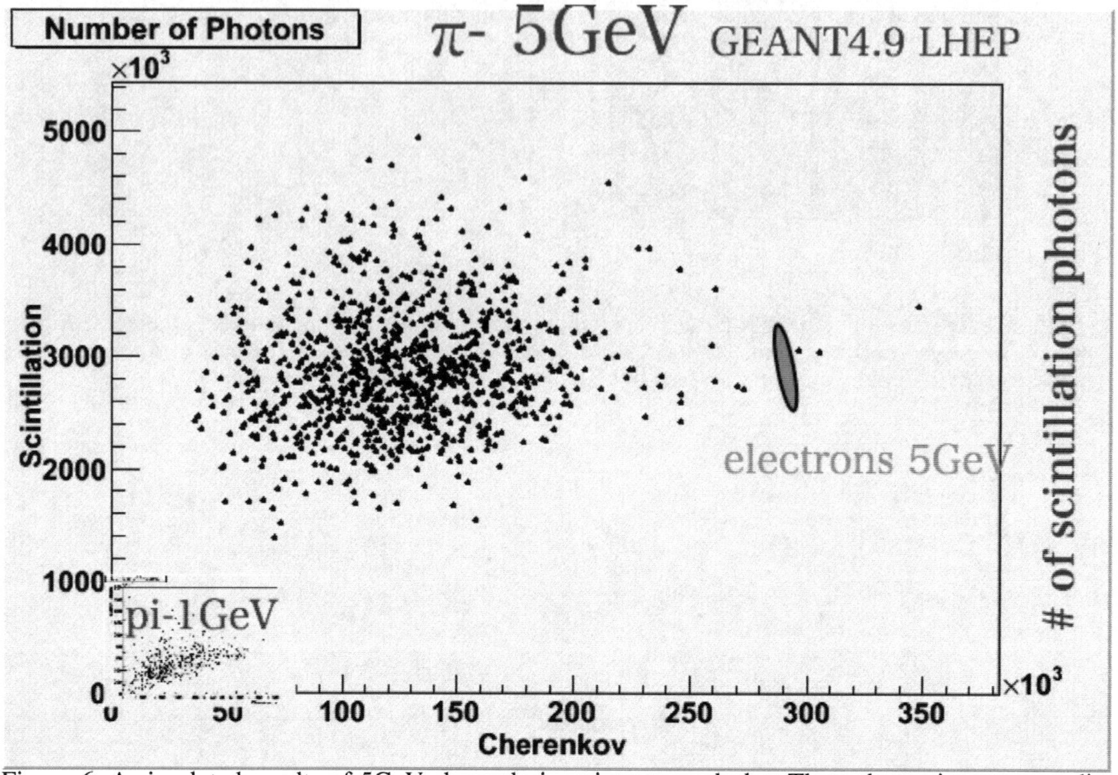

Figure 6; A simulated results of 5GeV charged pions in scattered plot. The red area is corresponding area for the 5 GeV electrons. A clear separation is seen. More over, 1GeV pions are added with the same scale in it.

4 Hadron measurement in the TMC

Hadron response is simulated with 5 GeV charged pions. The electron signal area is added by hand in the same energy with red. The same energy electrons and charged pions are clearly separated in the plot, once we measure Chrenkov and scintillator signal simultaneously in a event. The figure 6 shows also contribution from the charged pions of 1 GeV at the bottom end of the smallest region in the plot. This indicates separation between 5 GeV and 1 GeV pions is easily done in the TMC.

5 Summary

An idea of Total Measurement Calorimeter (TMC) is described. Fine segmented HCAL will give us active neutral pion generation in the hadronic interactions. To detect neutral pions effectively, active absorber with lead glass system is introduced, tested and simulated. The initial trials look fine and promising for the bigger and deeper calorimeter test.

References

[1] Performance of the ALEPH detector at LEP. ALEPH collaboration, Nucl. Inst. and Meth. A360 (1995) 481.

[2] Hadron and jet detection with a dual-readout calorimeter. N. Akchurin et al. Nucl. Inst. and Meth. A 537 (2005) 537

[3] The OPAL detector at LEP. OPAL collaboration, Nucl. Inst. and Meth. A305 (1991) 275-319.